Human Embryology and Developmental Biology

Human Embryology and Developmental Biology

Bruce M. Carlson, MD, PhD
Professor and Chairman
Department of Anatomy and Cell Biology
University of Michigan School of Medicine
Ann Arbor, Michigan

Original artwork by Margaret Croup Brudon

with 465 *illustrations*

St. Louis Baltimore Boston Chicago London Madrid Philadelphia Sydney Toronto

Editor: Robert Farrell
Developmental Editor: Emma D. Underdown
Project Manager: Carol Sullivan Wiseman
Senior Production Editor: Linda McKinley
Designer: Betty Schulz
Cover Design: Rokusek Design

Printed in the United States of America
Composition by The Clarinda Company
Printing/binding by Von Hoffmann Press, Inc.

Mosby–Year Book, Inc.
11830 Westline Industrial Drive
St. Louis, Missouri 63146

Library of Congress Cataloging in Publication Data

Carlson, Bruce M.
Human embryology and developmental biology/ Bruce M. Carlson;
original artwork by Margaret Croup Brudon.
p. cm.
Includes bibliographical references and index.
ISBN 0-8016-6415-2
1. Embryology, Human. 2.Developmental biology. 3. Human growth.
I. Title.
[DNLM: 1. Abnormalities—embryology. 2. Developmental Biology.
3. Fetal Development—physiology. QS 604 C284h 1994]
QM601.C29 1994
612.6'4—dc20
DNLM/DLC
for Library of Congress

93-34393
CIP

94 95 96 97 98 / 9 8 7 6 5 4 3 2 1

To my parents, who got me started,

and to Jean and the boys, who keep me going

Reviewers

Jo Ann Cameron, PhD
Associate Professor, Department of Cell and Structural Biology
University of Illinois College of Medicine
Urbana, Illinois

Allen C. Enders, PhD
Professor, Department of Cell Biology and Human Anatomy
University of California-Davis School of Medicine
Davis, California

Raymond F. Gasser, PhD
Professor, Department of Anatomy
Louisiana State University Medical School
New Orleans, Louisiana

Roger R. Markwald, PhD
Professor and Chairman, Department of Cell Biology and Anatomy
Medical University of South Carolina
Charleston, South Carolina

Preface

Like the entire field of medicine, human embryology has entered a period of revolutionary changes. New tools and concepts in molecular biology and genetics have already opened the door to a deeper understanding of how the genetic information inherent in the fertilized egg is expressed and how it controls fundamental developmental processes. Techniques such as ultrasound and other imaging modalities have provided new ways of visualizing living embryos. Still other techniques allow the control of reproduction, whether by preventing or permitting pregnancy.

One of the real challenges in teaching embryology to medical and other health care students is to present adequate coverage of the structural changes within the embryo while introducing sufficient molecular, experimental, and technological information to understand the mechanisms underlying both normal and abnormal development. Even more challenging is to do this in the small number of hours that are typically allocated to embryology courses. Similarly, the textbook author must package the essential components of human development into a book that will not overwhelm the student with its length or complexity.

The chapters are organized into two parts. Part One covers early development and the relationship between the mother and the fetus. The function of the structures and tissues is explained, both in normal development and in the development of anomalies. It is only by understanding the molecular basis for both normal and abnormal development that the newer modalities for manipulating embryonic development will realize a clinical benefit (Chapter 7). Part Two discusses the development of the body systems. The neural crest is given its own chapter (Chapter 13) as a model system that affects most of the other organ systems and that illustrates many of the fundamental concepts of developmental biology. The last chapter, Chapter 19, brings together the entire process of embryonic development to result in the birth of the fetus and the immediate adaptations necessary for life outside the womb.

The intent of this text, then, is to provide a coherent story of human embryonic development that takes into account (1) normal morphology and function, (2) the developmental basis for a number of the more important congenital anomalies, (3) the new technology that allows the manipulation of embryonic development, and (4) the links between the data generated from molecular and experimental studies on a variety of developmental systems and the descriptive database on human embryology.

This text includes numerous pedagogic aids to reinforce the important points as the story unfolds. Key terms are introduced in boldface type and defined for emphasis and for easy retrieval during review. Also useful in this regard are the tables and boxes that summarize and compare information. The four-color line art explicitly indicates the changes in structure and tissues over time. As much as possible, these figures present the spatial relationships in a multidimensional manner. The striking scanning electron micrographs go one step further to show these relationships in vivo. The summaries at the end of each chapter highlight the important concepts.

Clinically oriented questions guide the student to apply the basic science of embryology to the clinical setting. Brief answers to these questions are supplied at the back of the book for self-assessment and to indicate areas for further review. Within each section the common anomalies of the organ or system are described and discussed from their developmental basis to their clinical presentation to emphasize further the practical implications. Finally, for an overview of the development from fertilized oocyte to newborn, p. xiii includes a table on the Carnegie stages of early human embryonic development, and pp. xiv-xv show the major developmental events during the fetal period.

The approach to this book stems from my having taught a course that combined medical embryology and undergraduate developmental biology in the Inteflex (combined B.S.-M.D.) Program at the University of Michigan. The basis is the conviction that it is most valuable for the student to approach human embryonic development with a multidimensional view. Such an approach, within the constraints mentioned, necessitates the selection of the core information that will allow the student to acquire a good understanding of the main themes of embryonic development without getting lost in a sea of details. Of equal importance is exposing the student to the areas of contemporary research that are likely to have a major impact on our understanding of the mechanisms that control normal and abnormal development. In my view, preparing the student for the clinical applications of research advances is as important as presenting the current base of knowledge.

Acknowledgments

Thanks are due to many people for their role in bringing this book to fruition.

To Emma D. Underdown, Developmental Editor, who did a spectacular job in overseeing the gestation of this book;

To Linda McKinley, Production Editor, for a smooth passage of the manuscript through the production process;

To Mason Barr, MD, Department of Pediatrics; Alphonse Burdi, PhD, Department of Anatomy and Cell Biology; and Kathy Tosney, PhD, University of Michigan, for their many original slides of congenital malformations;

To Jan E. Jirásek, MD, DSc, Institute of the Care of the Mother and Child, Prague, for sending me prints of scanning electron micrographs of human embryos;

To William Brudon, who displayed his usual darkroom magic in printing and retouching most of the photographs used in the book;

To Margaret Croup Brudon, who once again was able to convert my almost undecipherable sketches into artwork;

To my secretary, Sharon Moskwiak, whose efficiency and organizational skills kept things running smoothly at the Michigan end; and

To the manuscript reviewers, who were unusually thorough and helpful in their comments.

Bruce M. Carlson

Contents

PART TWO DEVELOPMENT OF THE BODY SYSTEMS

Carnegie Stages of Early Human Embryonic Development (Weeks 1-8)

CARNEGIE STAGE	AGE (DAYS)	CROWN-RUMP LENGTH (mm)	PAIRS OF SOMITES	EXTERNAL FEATURES
1	1	0.1		Fertilized oocyte
2	2-3	0.1		Morula (4-16 cells)
3	4	0.1		Free blastocyst
4	5-6	0.1		Attachment of blastocyst to endometrium
5	7-12	0.1-0.2		Implantation, bilaminar embryo with primary yolk sac
6	13-15	0.2-0.3		Trilaminar embryo with primitive streak, chorionic villi
7	16	0.4		Gastrulation, formation of notochordal process
8	18	1.0-1.5		Hensen's node and primitive pit, notochord and neurenteric canal, appearance of neural plate, neural folds, and blood islands
9	20	1.5-2.5	1-3	Appearance of first somites, deep neural groove, elevation of cranial neural folds, early heart tubes
10	22	2.0-3.5	4-12	Beginning of fusion of neural folds, formation of optic sulci, presence of first two pharyngeal arches, beginning of heart beat, curving of embryo
11	24	2.5-4.5	13-20	Closure of cranial neuropore, formation of optic vesicles, rupture of oropharyngeal membrane
12	26	3-5	21-29	Closure of caudal neuropore, formation of pharyngeal arches 3 and 4, appearance of upper limb buds and tail bud, formation of otic vesicle
13	28	4-6	30+	Appearance of lower limb buds, lens placode, separation of otic vesicle from surface ectoderm
14	32	5-7		Formation of lens vesicle, optic cup, and nasal pits
15	33	7-9		Development of hand plates, primary urogenital sinus, prominent nasal pits, evidence of cerebral hemispheres
16	37	8-11		Development of foot plates, visible retinal pigment, development of auricular hillocks, formation of upper lip
17	41	11-14		Appearance of finger rays, rapid head enlargement, six auricular hillocks, formation of nasolacrimal groove
18	44	13-17		Appearance of toe rays and elbow regions, beginning of formation of eyelids, tip of nose distinct, presence of nipples
19	48	16-18		Elongation and straightening of trunk, beginning of herniation of midgut into umbilical cord
20	51	18-22		Bending of arms at elbows, distinct but webbed fingers, appearance of scalp vascular plexus, degeneration of anal and urogenital membranes
21	52	22-24		Longer and free fingers, distinct but webbed toes, indifferent external genitalia
22	54	23-28		Longer and free toes, better development of eyelids and external ear
23	57	27-31		More rounded head, fusion of eyelids

Data taken largely from O'Rahilly R, Müller F: *Developmental stages in human embryos,* Washington, DC, 1987, Carnegie Institution of Washington, Pub 637.

Major Developmental Events During the Fetal Period

EXTERNAL FEATURES	INTERNAL FEATURES
8 WEEKS	
Head is almost half the total length of fetus Cervical flexure is about 30° Indifferent external genitalia are present Eyes are converging Eyelids are unfused Tail disappears Nostrils are closed by epithelial plugs	Midgut herniation into umbilical cord occurs Extraembryonic portion of allantois has degenerated Ducts and alveoli of lacrimal glands form Paramesonephric ducts begin to regress in males Recanalization of lumen of gut tube occurs Lungs are becoming glandlike Diaphragm is completed First ossification begins in skeleton Definitive aortic arch system takes shape
9 WEEKS	
Neck develops and chin rises from thorax Cranial flexure is about 22° Chorion is divided into chorion laeve and chorion frondosum Eyelids meet and fuse External genitalia begin to become gender specific	Intestines are herniated into umbilical cord Early muscular movements occur ACTH and gonadotropins are produced by pituitary Corticosteroids are produced by adrenal cortex Semilunar valves in heart are completed Fused paramesonephric ducts join vaginal plate Urethral folds begin to fuse in males
10 WEEKS	
Cervical flexure is about 15° Gender differences are apparent in external genitalia Fingernails appear Eyelids are fused	Intestines return into body cavity from umbilical cord Bile is secreted Blood islands are established in spleen Thymus is infiltrated by lymphoid stem cells Prolactin production by pituitary occurs First permanent tooth buds form Deciduous teeth are in early bell stage Epidermis is three layered
11 WEEKS	
Cervical flexure is about 8° Nose begins to develop bridge	Urine is excreted into aminotic fluid Stomach musculature can contract T lymphocytes emigrate into bloodstream Colloid appears in thyroid follicles
12 WEEKS	
Head is erect Neck is almost straight and well defined External ear is taking form and has moved close to its definitive position in the head Yolk sac has shrunk Fetus swallows amniotic fluid Fetus can respond to skin stimulation	Ovaries descend below pelvic rim Parathyroid hormone is produced Blood can coagulate

Major Developmental Events During the Fetal Period—cont'd

EXTERNAL FEATURES	INTERNAL FEATURES
4 MONTHS	
Skin is thin—blood vessels can easily be seen through it Nostrils are almost formed Fetus may begin to suck its thumb Eyes have moved to front of face Legs are longer than arms Fine lanugo hairs appear on head Fingernails are well formed, toenails are forming Epidermal ridges appear on fingers and palms of hand Enough amniotic fluid is present to permit amniocentesis Mother can feel fetal movements	Seminal vesicle forms Transverse grooves appear on dorsal surface of cerebellum Bile is produced by liver and stains meconium green Gastric glands bud off from gastric pits Brown fat begins to form Pyramidal tracts begin to form in brain Hematopoiesis begins in bone marrow Ovaries contain primordial follicles
5 MONTHS	
Epidermal ridges form on toes and soles of feet Vernix caseosa begins to be deposited on skin Abdomen begins to fill out Eyelids and eyebrows develop Lanugo hairs cover most of body	Myelination of spinal cord begins Sebaceous glands begin to function Thyroid-stimulating hormone is released by pituitary Testes begin to descend
6 MONTHS	
Skin is wrinkled and red Decidua capsularis degenerates because of reduced blood supply Lanugo hairs darken	Surfactant begins to be secreted Tip of spinal cord at S1 level
7 MONTHS	
Eyelids begin to open Eyelashes are well developed Scalp hairs are lengthening (longer than lanugo) Skin is slightly wrinkled	Sulci and gyri begin to appear on brain Subcutaneous fat storage begins Testes are descending into scrotum Termination of splenic erythropoiesis occurs
8 MONTHS	
Skin is pink and smooth Eyes are capable of pupillary light reflex Fingernails have reached tip of fingers	Regression of hyaloid vessels from lens occurs Testes enter scrotum
9 MONTHS	
Toenails have reached tip of toes Most lanugo hairs are shed Skin is covered with vernix caseosa Attachment of umbilical cord becomes central in abdomen About 1 L of amniotic fluid is present Placenta weighs about 500 g Fingernails extend beyond fingertips Breasts protrude and secrete "witches' milk"	Larger amounts of pulmonary surfactant are secreted Ovaries are still above brim of pelvis Testes are descended into scrotum Tip of spinal cord is at L3 Myelination of brain begins

PART ONE

Early Development and the Fetal-Maternal Relationship

CHAPTER 1

Getting Ready for Pregnancy

Human pregnancy begins with the fusion of an egg and a sperm, but a great deal of preparation precedes this event. First, both male and female sex cells must pass through a long series of changes **(gametogenesis)** that converts them genetically and phenotypically into mature **gametes,** which are capable of participating in the process of fertilization. Next, the gametes must be released from the gonads and make their way to the upper part of the uterine tube, where fertilization normally takes place. Finally, the fertilized egg, now properly called an **embryo,** must make its way into the uterus, where it sinks into the uterine lining **(implantation)** to be nourished by the mother. All these events involve interactions between the gametes or embryo and the adult body in which they are housed, and most of them are mediated or influenced by parental hormones. This chapter focuses on gametogenesis and the hormonal modifications of the body that enable reproduction to occur.

GAMETOGENESIS

Gametogenesis is typically divided into four phases: (1) the extraembryonic origin of the germ cells and their migration into the gonads, (2) an increase in the number of germ cells by mitosis, (3) a reduction in chromosomal material by meiosis, and (4) structural and functional maturation of the eggs and sperms. The first phase is identical in both males and females, whereas distinct differences exist between the male and female patterns in the last three phases.

Origin and Migration of Germ Cells

Primordial germ cells, the earliest recognizable precursors of gametes, arise outside the gonads and migrate to the gonads during early embryonic development. Primordial germ cells first become recognizable at 24 days postfertilization in the endodermal layer of the yolk sac due to their large size and high content of the enzyme alkaline phosphatase. By mechanisms that are still not well understood, these cells soon exit from the yolk sac into the hindgut epithelium and then migrate through the dorsal mesentery until they reach the primordia of the gonads (Fig. 1-1). In the mouse, an estimated 100 cells leave the yolk sac, and through mitotic multiplication (six to seven rounds of cell division), about 5000 primordial germ cells enter the primitive gonads.

The way the primordial germ cells get from their site of origin to the gonads is still poorly understood, but it has been long suspected that the developing gonads give off a chemotactic attractant for the germ cell. An alternative hypothesis is that the germ cells are guided toward the gonads by the nature of the cellular and noncellular microenvironment that surrounds them. Although the primordial germ cells may initially be carried toward the gonads by some of the major tissue rearrangements that occur in the early embryos, by the time they reach the dorsal mesentery, they are capable of independent migration by ameboid movement.

Misdirected primordial germ cells that lodge in extragonadal sites usually die, but if such cells survive, they may develop into **teratomas.** Teratomas are bizarre growths that contain scrambled mixtures of highly differentiated tissues, such as skin, hair, cartilage, and even teeth (Fig. 1-2). They are found in the mediastinum, the sacrococcygeal region, and the oral region.

Increase in the Number of Germ Cells by Mitosis

Once they arrive in the gonads, the primordial germ cells begin a phase of rapid mitotic proliferation. In a mitotic division, each germ cell produces two **diploid** progeny that are genetically equal. Through several series of mitotic di-

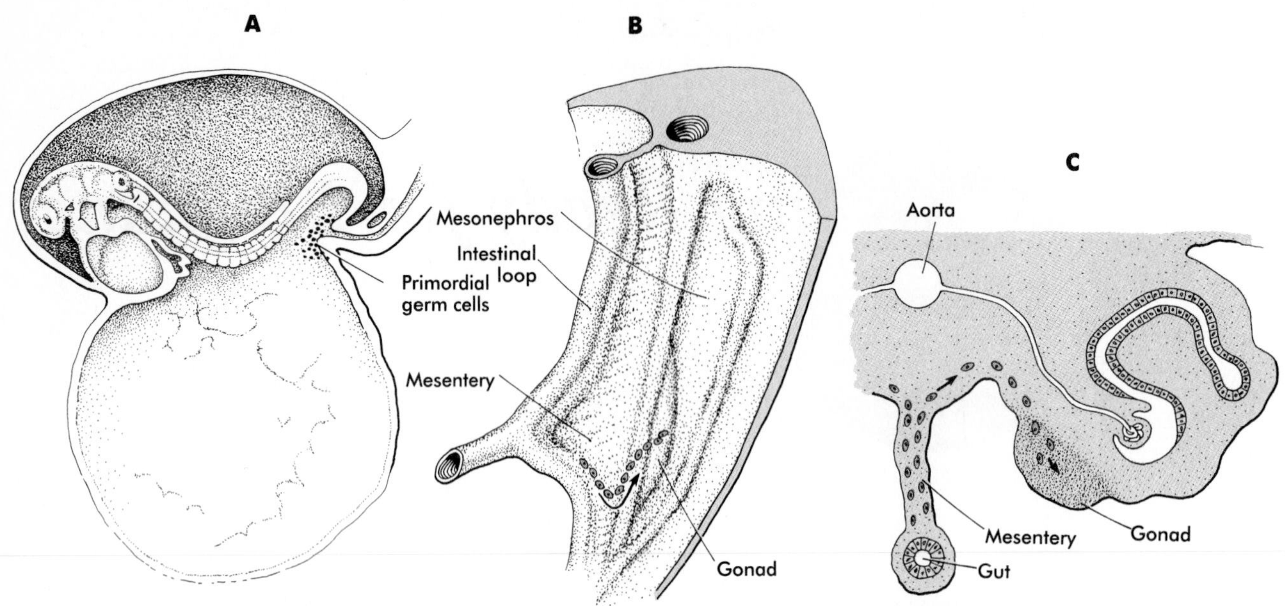

FIG. 1-1 Origin and migration of primordial germ cells in the human embryo. **A,** Location in the 16-somite human embryo. **B,** Pathway of migration through the dorsal mesentery. **C,** Cross section showing the pathway of migration through the dorsal mesentery and into the gonad.

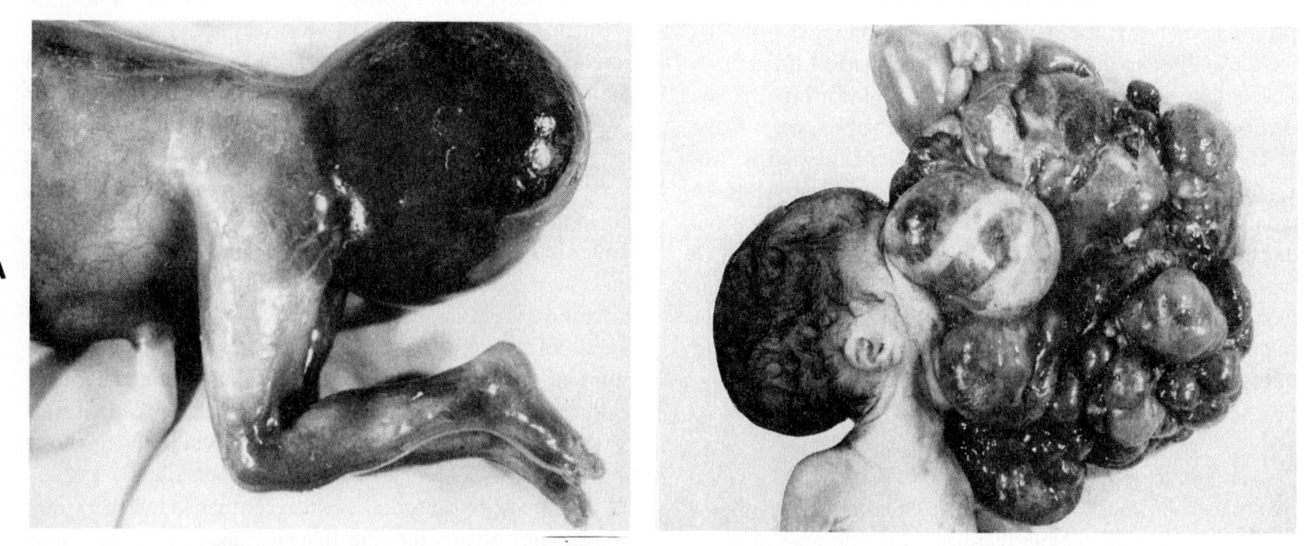

FIG. 1-2 **A,** Sacrococcygeal teratoma in a fetus. **B,** Massive oropharyngeal teratoma. (Courtesy Mason Barr, Ann Arbor, Mich.)

visions, the number of primordial germ cells increases exponentially from hundreds to millions. The pattern of mitotic proliferation differs markedly between male and female germ cells. **Oogonia,** as mitotically active germ cells in the female are called, go through a period of intense mitotic activity in the embryonic ovary from the second through the fifth month of pregnancy in the human. During this period the population of germ cells increases from only a few thousand to nearly 7 million (Fig. 1-3). This number represents the maximum number of germ cells that is ever found in the ovaries. Shortly thereafter, large numbers of oogonia undergo a natural degeneration called **atresia.** Atresia of germ cells is a continuing feature of the histological landscape of the human ovary until menopause.

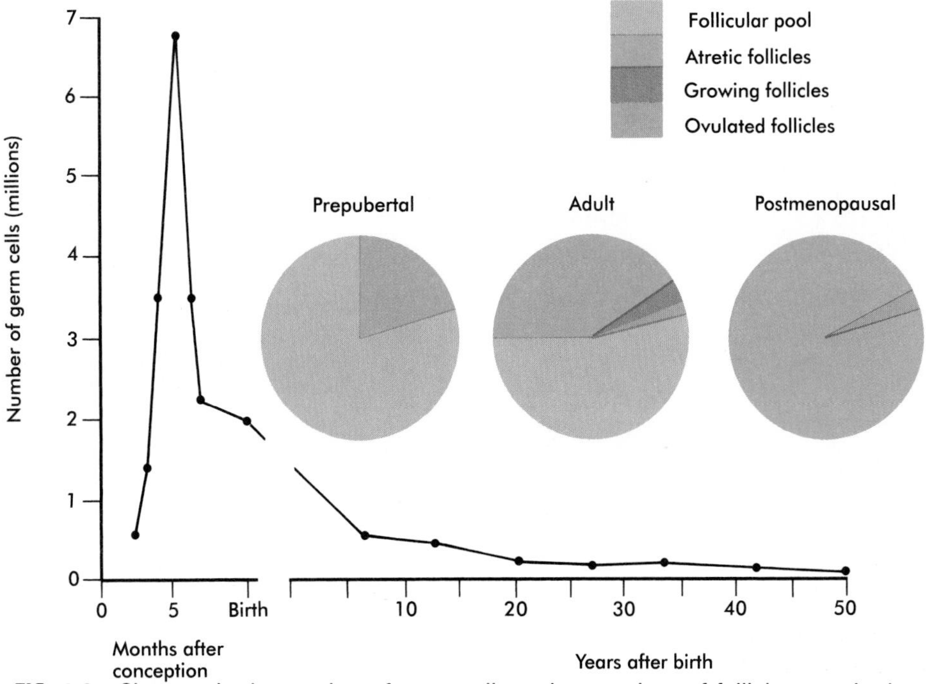

FIG. 1-3 Changes in the number of germ cells and proportions of follicle types in the human ovary with increasing age.
(Based on studies by Baker in Austin and Short [1970] and Goodman and Hodgen [1983].)

Spermatogonia in the male, which are the counterparts of oogonia, follow a pattern of mitotic proliferation that differs greatly from that in the female. Mitosis also begins early in the embryonic testes, but in contrast to the female, male germ cells maintain the ability to divide throughout life. The seminiferous tubules of the testes are lined with a germinative population of spermatogonia. Beginning at puberty, subpopulations of spermatogonia undergo periodic waves of mitosis. The progeny of these divisions enter meiosis as synchronous groups. This pattern of spermatogonial mitosis continues throughout the life of the male.

Meiosis

The biological significance of meiosis in the human is similar to that in other species. Of primary importance are (1) reduction of the number of chromosomes from the diploid (2n) to **haploid** (1n) number so that the species number of chromosomes can be maintained from generation to generation, (2) independent reassortment of maternal and paternal chromosomes for better mixing of genetic characteristics, and (3) further redistribution of lesser amounts of maternal and paternal genetic information through the process of crossing-over during the first meiotic division.

Meiosis involves two sets of divisions (Fig. 1-4). Before the first meiotic division, DNA replication has already occurred, so at the beginning of meiosis, the cell is 2n, 4c. (In this designation, *n* is the species number of chromosomes and *c* is the amount of DNA in a single set [n] of chromosomes before DNA replication has occurred.) The cell contains the normal number (2n) of chromosomes, but as the result of replication, its DNA content (4c) is double the normal amount (2c).

In the first meiotic division, often called the **reductional division,** a prolonged prophase (see Fig. 1-4) results in the pairing of homologous chromosomes and frequent **crossing-over,** resulting in the exchange of segments between members of the paired chromosomes. During metaphase of the first meiotic division, the chromosome pairs (tetrads) line up at the metaphase (equatorial) plate in such a way that at anaphase I, one chromosome of a homologous pair moves toward one pole of the spindle and the other chromosome moves toward the opposite pole. This represents one of the principal differences between a meiotic and mitotic division. In a mitotic anaphase, the centromere between the sister chromatids of each chromosome splits after the chromosomes have lined up at the metaphase plate, and one chromatid from each chromosome migrates to each pole of the mitotic spindle. This results in genetically equal daughter cells after a mitotic division, whereas the daughter cells are genetically unequal after the first meiotic division. Each daughter cell of the first meiotic division contains the haploid (1n) number of chromosomes, but each chromosome still consists of two chromatids (2c) connected by a centromere. No new duplication of chromosomal DNA is required between the first and second mei-

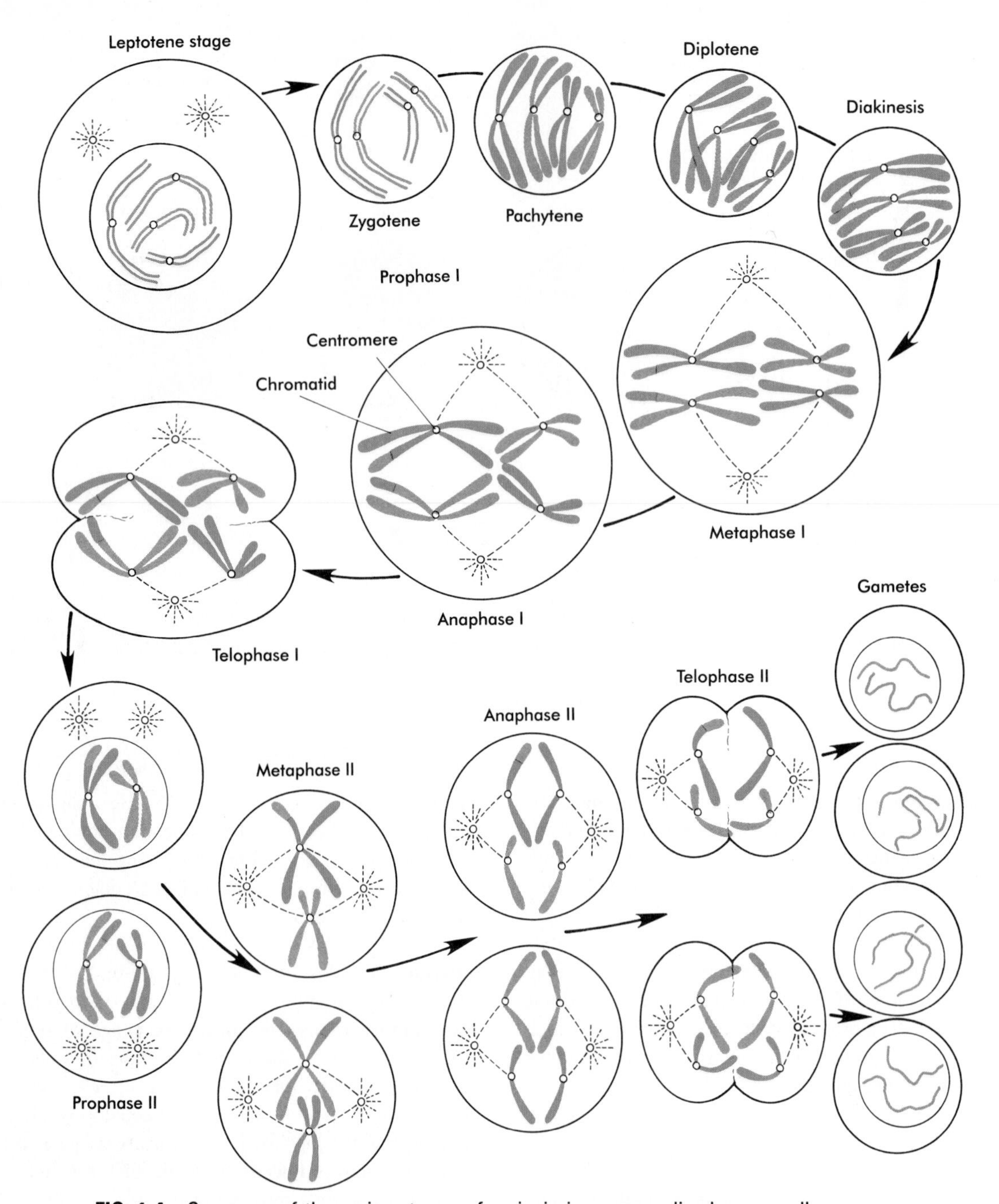

FIG. 1-4 Summary of the major stages of meiosis in a generalized germ cell.

otic divisions because each haploid daughter cell resulting from the first meiotic division already contains chromosomes in the replicated state.

The second meiotic division is similar to an ordinary meiotic division, except that the cell is haploid (1n, 2c). When the chromosomes line up along the equatorial plate at metaphase II, the centromeres between sister chromatids divide, allowing the sister chromatids of each chromosome to migrate to opposite poles of the spindle apparatus during anaphase II. Each daughter cell of the second meiotic division is truly haploid (1n, 1c).

Meiosis in Females. In addition to the purely genetic aspects of meiosis are important features that characterize the development of male and female sex cells. As the oogonia enter the first meiotic division late in the fetal period, they are called **primary oocytes.**

Meiosis in the human female is a very leisurely process. As the primary oocytes enter the diplotene stage of the first meiotic division in the early months after birth, the first of two blocks in the meiotic process occurs (Fig. 1-5). The suspended diplotene phase (sometimes called the **dyctotene** or **dictyate** stage) of meiosis is the period when the pri-

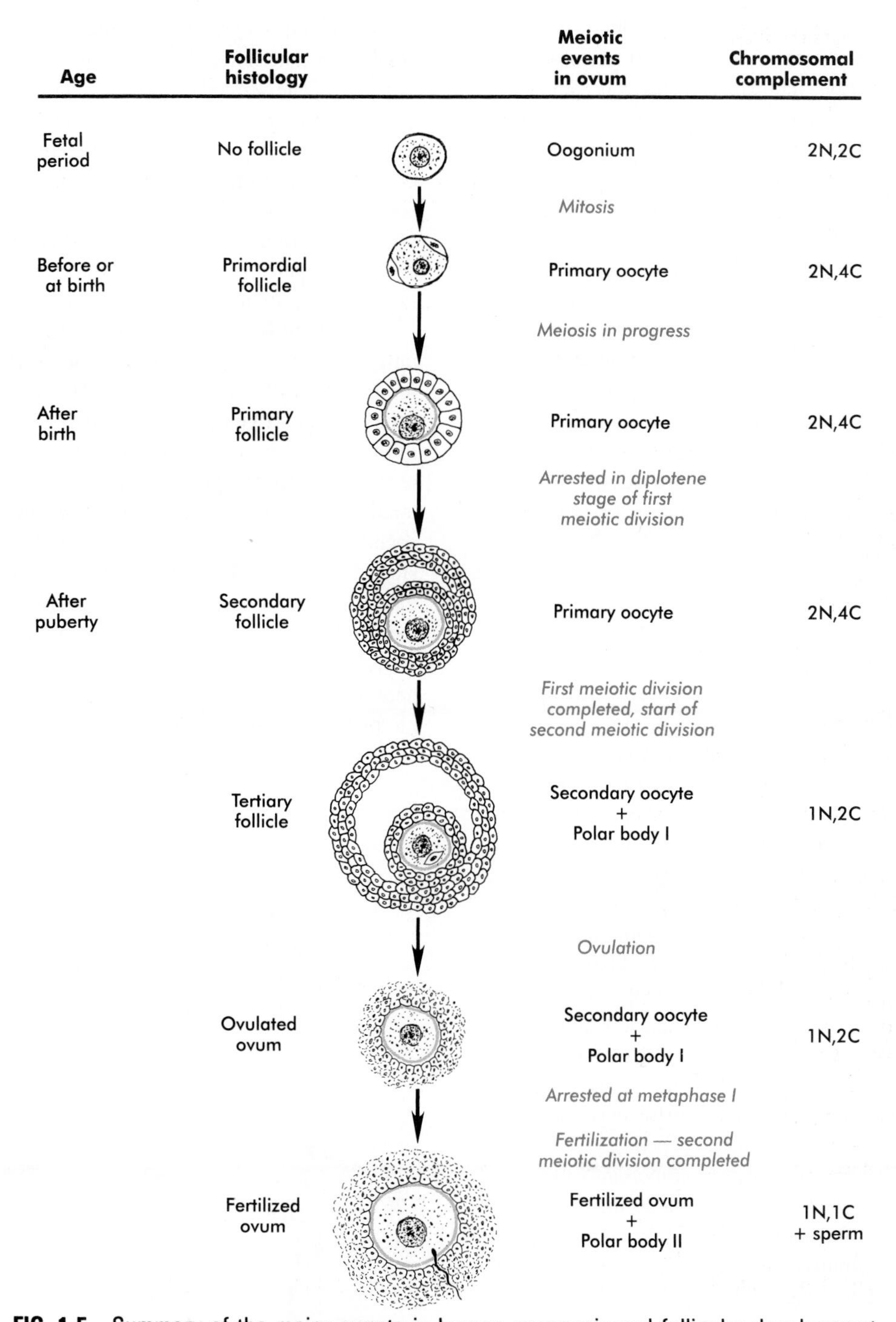

FIG. 1-5 Summary of the major events in human oogenesis and follicular development.

mary oocyte prepares for the needs of the embryo. In oocytes of amphibians and other lower vertebrates, which must develop outside the mother's body and often in a hostile environment, it is highly advantageous for the early stages of development to occur very rapidly so that the stage of independent locomotion and feeding is attained as soon as possible. These conditions necessitate a strategy of storing up the materials needed for early development well in advance of ovulation and fertilization because normal synthetic processes would not be rapid enough to produce the materials required for the rapidly cleaving embryo. In such species, yolk is accumulated, the genes for producing ribosomal RNA are amplified, and many types of RNA molecules are synthesized and stored in an inactive form.

The morphological substrate for RNA synthesis in the amphibian oocyte is represented by the lampbrush chromosomes, which are characterized by many prominent loops of spread-out DNA on which many mRNA molecules are synthesized. The amplified genes for producing rRNA are manifested by the presence of 600 to 1000 nucleoli within the nucleus. In addition, the primary oocytes anticipate the needs of the cleaving embryo by producing several thousand cortical granules, which are of great importance during the fertilization process (see Chapter 2).

The mammalian oocyte prepares for an early embryonic period that is much more prolonged than that of amphibians and takes place in the nutritive environment of the maternal reproductive tract. Therefore it is not faced with the need to store as great a quantity of materials as are the eggs of lower vertebrates. As a consequence, the buildup of yolk is negligible. However, some evidence indicates a low level of ribosomal DNA amplification (two to three times) in diplotene human oocytes; the presence of 2 to 40 small (2 μm) RNA-containing micronuclei (miniature nucleoli) per oocyte nucleus correlates with the molecular data. Because the need for rDNA amplification in the developing eggs of primates is unlikely, the presence of micronuclei may be an evolutionary vestige.

Analysis of RNA accumulation in mammalian oocytes is based principally on results obtained in the mouse, in which some RNA accumulation begins during the diplotene stage. Human diplotene chromosomes do not appear to be arranged in a true lampbrush configuration, and massive amounts of RNA synthesis seem unlikely. It is not yet possible to present an overall picture of RNA synthesis in the developing oocytes of primates. The developing mammalian (mouse) oocyte produces 10,000 times less rRNA and 1000 times less mRNA than does its amphibian counterpart. Nevertheless, there is a steady accumulation of mRNA and a proportional accumulation of rRNA. Mammals accumulate relatively greater amounts of stable mRNAs as a higher proportion of the total RNA than do amphibians. These amounts of maternally derived RNA seem to be enough to take the fertilized egg through the first couple of cleavage divisions, after which the embryonic genome takes control of macromolecular synthetic processes.

Because cortical granules play an important role in preventing the entry of excess spermatozoa during fertilization in human eggs, the formation of cortical granules (mainly from the Golgi apparatus) continues to be one of the functions of the diplotene stage that is preserved in humans. Roughly 4500 cortical granules are produced in the mouse oocyte. A somewhat higher number is likely in the human oocyte.

Unless they degenerate, all primary oocytes remain arrested in the diplotene stage of meiosis until puberty. During the reproductive years, small numbers of primary oocytes (10 to 30) complete the first meiotic division with each menstrual cycle and begin to develop further. The other primary oocytes remain arrested in the diplotene stage, some for as long as 50 years.

With the completion of the first meiotic division shortly before ovulation, two unequal cellular progeny result. One is a large cell, called the **secondary oocyte.** The other is a small, nonfunctional cell called the **first polar body** (see Fig. 1-5). The secondary oocytes begin the second meiotic division, but again the meiotic process is arrested, this time at metaphase. The stimulus for the release from this meiotic block is fertilization by a spermatozoon. Unfertilized secondary oocytes fail to complete the second meiotic division. The second meiotic division is also unequal for the cellular progeny; one of the daughter cells is again relegated to becoming a small, nonfunctional second polar body.

Meiosis in Males. Meiosis in the male does not begin until after puberty. In contrast to the primary oocytes in the female, not all spermatogonia enter meiosis at the same time. In fact, large numbers of spermatogonia remain in the mitotic cycle throughout much of the reproductive lifetime of males. Once the progeny of a spermatogonium have entered the meiotic cycle as **primary spermatocytes,** they spend several weeks passing through the first meiotic division (Fig. 1-6). The result of the first meiotic division is the formation of two **secondary spermatocytes,** which immediately enter the second meiotic division. About 8 hours later, the second meiotic division is completed and four haploid (1n, 1c) **spermatids** remain as progeny of the single primary spermatocyte.

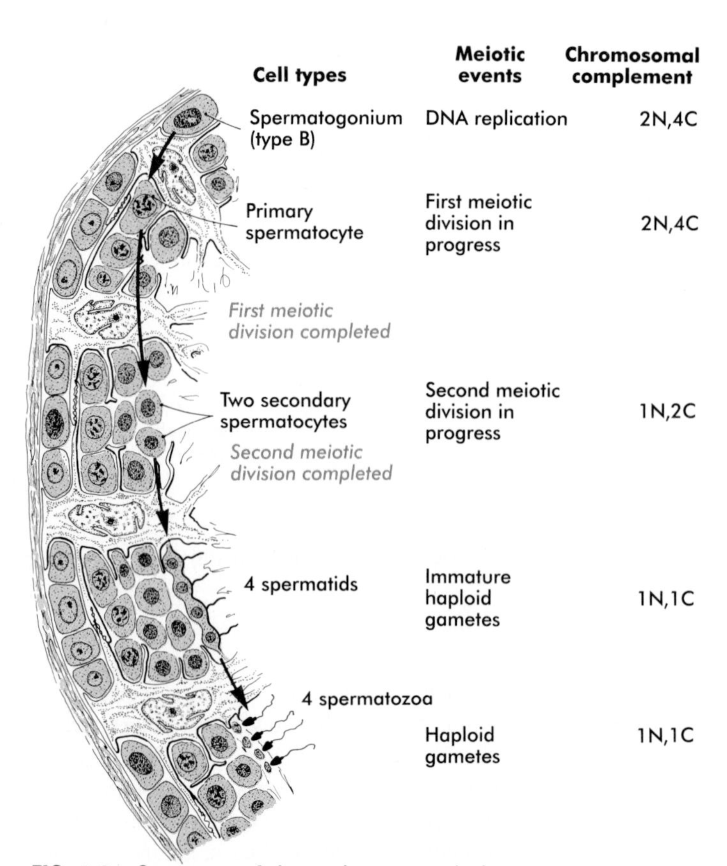

FIG. 1-6 Summary of the major events in human spermatogenesis.

Meiotic Disturbances Resulting in Chromosomal Aberrations. Chromosomes sometimes fail to separate during meiosis, a phenomenon known as **nondisjunction.** As a result, one haploid daughter gamete contains both members of a chromosomal pair for a total of 24 chromosomes, whereas the other haploid gamete contains only 22 chromosomes (Fig. 1-7). When such gametes combine with normal gametes of the opposite sex (with 23 chromosomes), the resulting embryos contain 47 chromosomes (with a **trisomy** of one chromosome) or 45 chromosomes (**monosomy** of one chromosome). (Specific syndromes associated with the nondisjunction of chromosomes are summarized in Chapter 8.) The generic term given to a condition characterized by an abnormal number of chromosomes is **aneuploidy.**

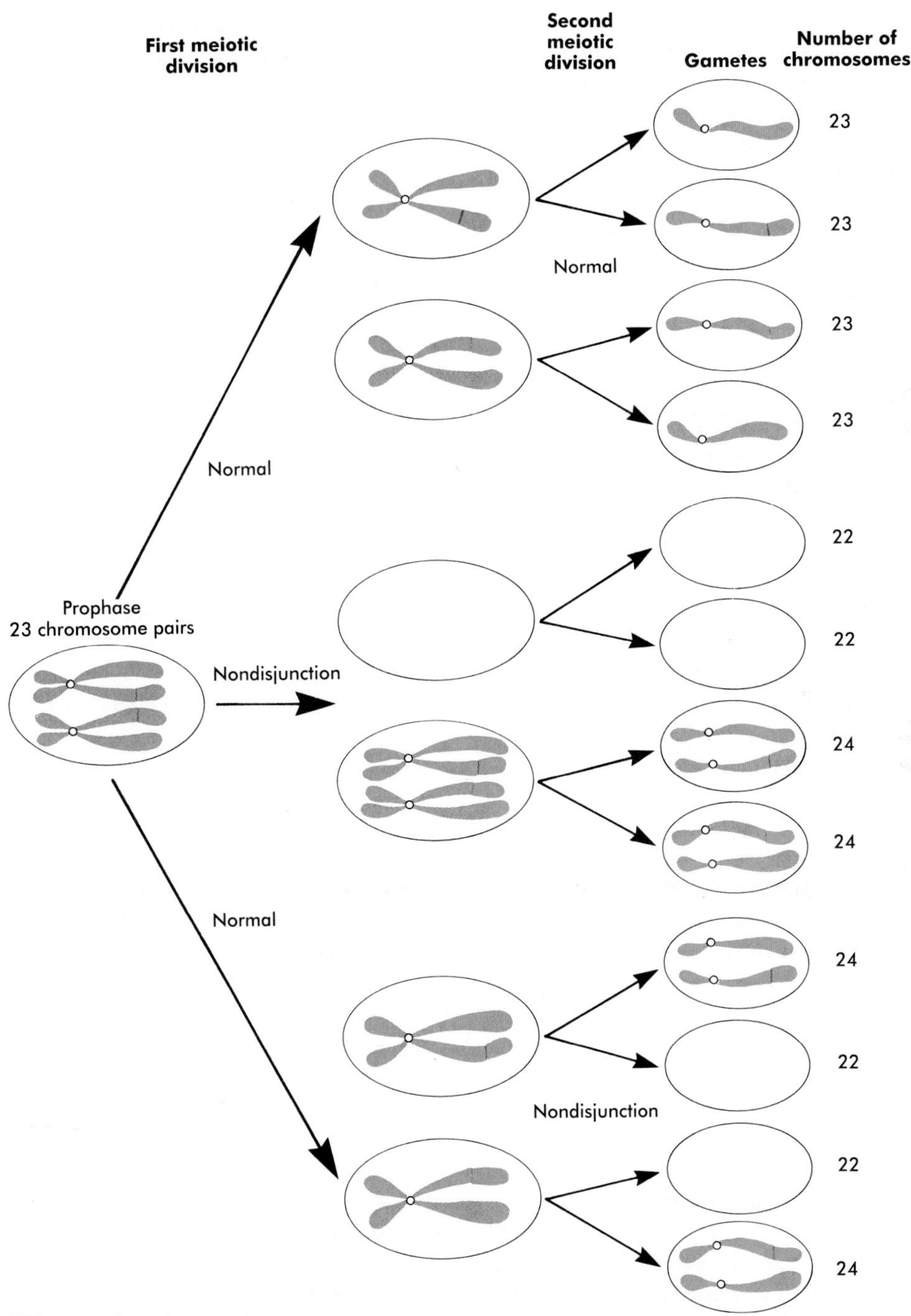

FIG. 1-7 Possibilities for nondisjunction. *Top arrow*: Normal meiotic divisions. *Middle arrow*: Nondisjunction during the first meiotic division. *Bottom arrow*: Nondisjunction during the second meiotic division.

In other cases, part of a chromosome can be **translocated** to another chromosome during meiosis, or part of a chromosome can be **deleted.** Similarly, duplications or inversions of parts of chromosomes occasionally occur during meiosis. These conditions may result in syndromes similar to those seen after the nondisjunction of entire chromosomes.

Under some circumstances (e.g., simultaneous fertilization by two spermatozoa, failure of the second polar body to separate from the oocyte during the second meiotic division), the cells of the embryo contain more than two multiples of the haploid number of chromosomes **(polyploidy).**

Chromosomal abnormalities are the underlying cause of a high percentage of spontaneous abortions during the early weeks of pregnancy. Over 75% of the spontaneous abortions occurring before the second week and over 60% of those occurring during the first half of pregnancy contain chromosomal abnormalities ranging from trisomies of individual chromosomes to overall polyploidy. Although the incidence of chromosomal anomalies declines with stillbirths occurring after the fifth month of pregnancy, it is close to 6%, a tenfold higher incidence over the 0.5% of living infants who are born with chromosomal anomalies. In counseling patients who have had a stillbirth or a spontaneous abortion, it can be useful to mention that this is often nature's way of handling an embryo destined to be highly abnormal.

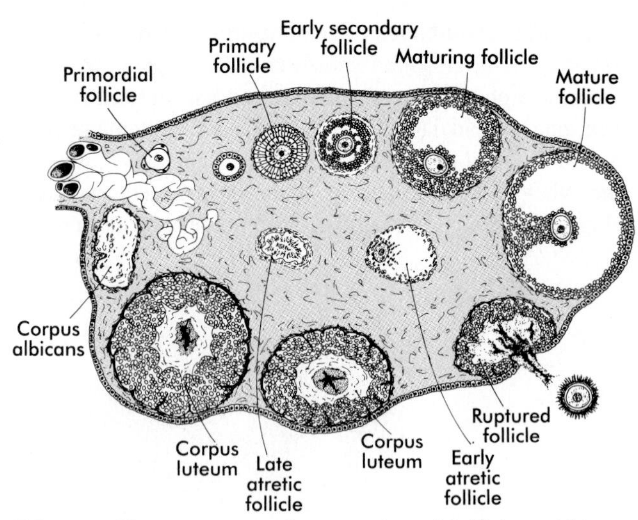

FIG. 1-8 The sequence of maturation of follicles within the ovary.

Final Maturation of Eggs and Sperm

Oogenesis. Of the roughly 2 million primary oocytes that are present in the ovaries at birth, only about 40,000—all of which are arrested in the diplotene stage of the first meiotic division—survive until puberty. From this number, approximately 400 (1 per menstrual cycle) are actually ovulated. The remainder of the primary oocytes degenerates without leaving the ovary, but many of them undergo some further development before becoming atretic.

The egg along with its surrounding cells is called a **follicle.** Maturation of the egg is intimately bound with the development of its cellular covering. Because of this, considering the development of the egg and its surrounding follicular cells as an integrated unit is a useful approach in the study of oogenesis.

In the embryo, oogonia are naked, but after meiosis begins, cells from the ovary partially surround the primary oocytes to form **primordial follicles** (see Fig. 1-5). By birth, the primary oocytes are invested with one or two complete layers of follicular cells, and the complex of primary oocyte and the follicular cells is called a **primary follicle** (Fig. 1-8). Both the oocyte and the follicular cells develop prominent microvilli and gap junctions that connect the two cell types (Fig. 1-9).

The gap junctions permit the exchange of amino acids and glucose metabolites that are required for growth of the oocyte. Considerable evidence indicates that the follicular cells secrete a **meiotic inhibitory factor** responsible for maintaining the first arrest of meiosis in the diplotene stage. The inhibitory factor, which appears to involve cAMP (cyclic adenosine monophosphate) and other purines, is transferred from the follicular cells to the oocyte via the gap junctions that connect them. As will be described more fully in Chapter 2, the release of meiotic inhibition shortly before ovulation is associated with the disruption of the gap junction connections. The oocyte also appears to influence the development of the surrounding follicular cells, especially those of the cumulus oophorous. Follicular functions that depend on interactions between the oocyte and surrounding follicular cells are summarized in the box.

As the primary follicle takes shape, a prominent, translucent, noncellular membrane called the **zona pellucida** forms between the primary oocyte and its enveloping follicular cells (see Fig. 1-9). The microvillous connections between the oocyte and follicular cells do persist through the substance of the zona pellucida. In rodents, the components of the zona pellucida (three glycoproteins and glycosaminoglycans) are synthesized almost entirely by the egg, but in some mammals, follicular cells may also contribute materials to the zona. The zona pellucida contains sperm receptors and other components that are important in fertilization and early postfertilization development. (Their functions are more fully discussed in Chapter 2.)

In the prepubertal years, many of the primary follicles enlarge, mainly because of an increase in the size of the oocyte and the number of follicular cells. A basement membrane called the **membrana granulosa** surrounds the epithelial **granulosa cells** of the primary follicle. The membrana granulosa forms a barrier to capillaries, and as a result, both the oocyte and the granulosa cells depend on diffusion of oxygen and nutrients for their survival.

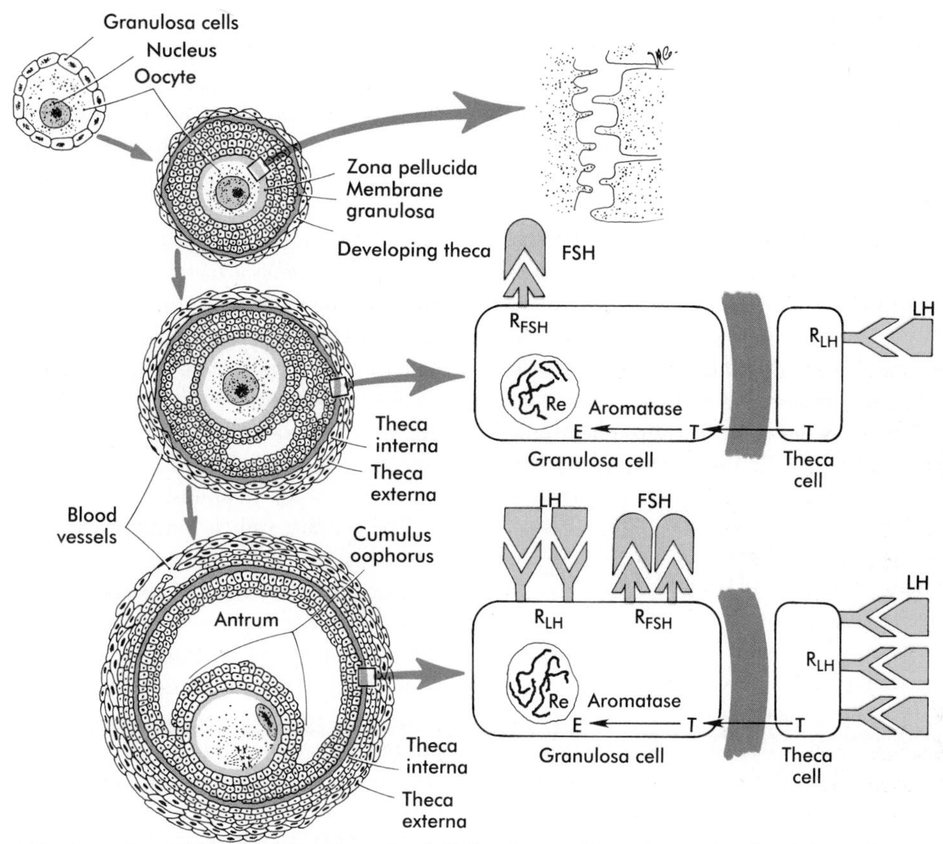

FIG. 1-9 Growth and maturation of a follicle along with major endocrine interactions in the theca cells and granulosa cells. *E,* Estrogen; *R,* receptor; *T,* testosterone.

An additional set of cellular coverings, derived from the ovarian connective tissue **(stroma),** begins to form around the developing follicle after it has become two to three cell layers thick. Known initially as the **theca folliculi,** this covering ultimately differentiates into two layers, a highly vascularized and glandular **theca interna** and a more connective, tissuelike, outer capsule called the **theca externa.** The early thecal cells appear to secrete an **angiogenesis factor,** which stimulates the growth of blood vessels in the thecal layer. This nutritive support facilitates growth of the follicle.

Early development of the follicle occurs without the significant influence of hormones, but as puberty approaches, continued follicular maturation requires the action of the pituitary gonadotrophic hormone **FSH (follicle-stimulating hormone)** on the follicular cells, which have by this time developed FSH receptors on their surfaces (see Fig. 1-9). After blood-borne FSH is bound to the FSH receptors, the stimulated granulosa cells produce small amounts of **estrogens.** The most obvious indication of the further develop-

Functional Processes in Ovarian Follicles that Depend on Factors Exchanged between Oocytes and Follicular Cells

Oocyte Functions Dependent on Factors from Granulosa Cells

Metabolism
Growth
Meiotic arrest
Maturation

Granulosa Cell Functions Dependent on Factors from Oocytes

Proliferation
Differentiation
Follicle organization
Cumulus expansion

From Eppig JJ: *Bioessays* 13:569-574, 1991.

ment of some of the follicles is the formation of an **antrum,** a cavity filled with a fluid called **liquor folliculi.** Initially formed by secretions of the follicular cells, the antral fluid is later formed mostly as a transudate from the capillaries on the other side of the membrana granulosa. With the appearance of the antrum, the follicle is a **secondary follicle.**

Responding to the stimulus of pituitary hormones, secondary follicles produce significant amounts of steroid hormones. By this time, the cells of the theca interna possess receptors for **LH (luteinizing hormone),** also secreted by the anterior pituitary (see Fig. 1-14). The theca interna cells produce **androgens,** which pass through the membrana granulosa to the granulosa cells. The influence of FSH induces the granulosa cells to form the enzymes **(aromatase)** that convert the theca-derived androgens into estrogens (mainly 17β-estradiol). Not only does the estradiol leave the follicle to exert important effects on other parts of the body, but it stimulates the formation of LH receptors on the granulosa cells. Through this mechanism, the follicular cells are able to respond to the large LH surge that immediately precedes ovulation.

Under multiple hormonal influences, the follicle increases rapidly (Figs. 1-9 and 1-10) and presses against the surface of the ovary. At this point, it is called a **tertiary (graafian) follicle.** About 10 to 12 hours before ovulation, meiosis resumes. The resumption of meiosis is closely correlated with the disappearance of the gap junctions between the oocyte and granulosa cells. Disruption of the gap junctions presumably stops the infusion of the meiosis inhibitory factor from the granulosa cells into the oocyte.

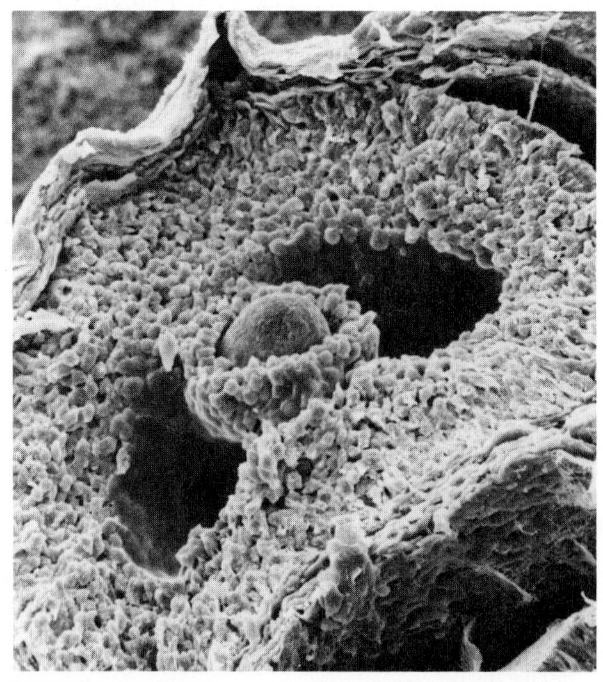

FIG. 1-10 Scanning electron micrograph of a mature follicle in the rat ovary. The spherical oocyte *(center)* is surrounded by smaller cells of the corona radiata, which projects into the antrum. (×840.)
(Courtesy P. Bagavandoss, Ann Arbor, Mich.)

The egg, now a secondary oocyte, is located in a small mound of cells known as the **cumulus oophorus,** which lies on one side of the greatly enlarged antrum. The oocyte appears to secrete a factor or factors that pass through the gap junctions into the surrounding cumulus cells. The factor or factors enable the cumulus cells to respond to gonadotropic hormones and to secrete hyaluronic acid, which results in an expansion of the cumulus oophorus. In keeping with the hormonally induced internal changes, the diameter of the follicle increases from about 6 mm early in the second week to almost 2 cm at ovulation.

The tertiary follicle protrudes from the surface of the ovary like a blister. The granulosa cells contain large numbers of both FSH and LH receptors, and LH receptors are abundant in the cells of the theca interna. The follicular cells secrete large amounts of estradiol (see Fig. 1-15), which prepares many other components of the female reproductive tract for gamete transport. Within the antrum, the follicular fluid contains (1) a complement of proteins similar to that seen in serum but in a lower concentration; (2) as many as 20 enzymes; (3) dissolved hormones, including FSH, LH, and steroids; and (4) proteoglycans. The strong negative charge of the proteoglycans attracts water molecules, and with greater amounts of secreted proteoglycans, the volume of antral fluid increases correspondingly. The follicle is now poised for ovulation and awaits the stimulus of the preovulatory surge of FSH and LH released by the anterior pituitary gland.

The reason only one follicle normally matures to the point of ovulation is still not completely understood, but some aspects of the selection process are becoming clearer. Early in the cycle, as many as 50 follicles begin to develop, but only about 3 attain a diameter as great as 8 mm. Initial follicular growth is gonadotropin independent, but continued growth depends on a minimum "tonic" level of gonadotropins, principally FSH. With gonadotropin-induced growth occurring, one dominant enlarging follicle becomes independent of FSH and secretes large amounts of inhibin (see p. 19). Inhibin suppresses the secretion of FSH by the pituitary, and when the FSH levels fall below the tonic threshold, the other developing follicles become atretic. The dominant follicle acquires its status about 7 days before ovulation. It may also secrete an inhibiting substance that acts directly on the other growing follicles.

Spermatogenesis. After the onset of puberty, **spermatogenesis** begins in the seminiferous tubules of the testis. In the broadest sense, the process begins with mitotic proliferation of the spermatogonia. At the base of the **seminiferous epithelium** are several populations of spermatogonia. **Type A spermatogonia** represent the stem-cell population that mitotically maintains proper numbers of spermatogonia throughout life. Dark type A spermatogonia are noncycling cells that may represent a long-term reserve population. Some of these cells enter the mitotic cycle as pale type A cells. Type A cells give rise to **type B spermatogonia,** which are destined to leave the mitotic cycle and enter meiosis. Many spermatogonia and their cellular descendants are connected by intercellular cytoplasmic bridges, which may be instrumental in maintaining the synchronous development of large clusters of sperm cells.

All spermatogonia are sequestered at the base of the seminiferous epithelium by interlocking processes of **Sertoli cells,** which are very complex cells that are regularly distributed throughout the periphery of the seminiferous epithelium and that occupy about 30% of its volume (see Fig. 1-6). As the progeny of the type B spermatogonia (now primary spermatocytes) complete the leptotene stage of the first meiotic division, they pass through the Sertoli cell barrier to the interior of the seminiferous tubule. This is accomplished by the formation of a new layer of Sertoli cell processes beneath these cells and, slightly later, the dissolution of the original layer that was above them. The Sertoli cell processes are very tightly joined and form an immunological barrier **(blood-testis barrier)** between the forming sperm cells and the rest of the body, including the spermatogonia. Once they have begun meiosis, developing sperm cells are immunologically different from the rest of the body. Autoimmune infertility could arise if the blood-testis barrier were broken down.

The progeny of the type B spermatogonia, which have entered the first meiotic division, are the primary spermatocytes (see Fig. 1-6). Located in a characteristic position just inside the layer of spermatogonia and still deeply embedded in Sertoli cell cytoplasm, primary spermatocytes spend 24 days passing through the first meiotic division. During this time, the developing sperm cells use a strategy similar to that of the egg, namely, producing in advance molecules that are needed at later periods when changes occur very rapidly. This involves the production of mRNA molecules and their storage in an inactive form until they are needed to produce the necessary proteins.

A well-known example of preparatory mRNA synthesis involves the formation of **protamines,** which are small, arginine-rich proteins that displace the nuclear histones and allow the high degree of compaction of nuclear chromatin required during the final stages of sperm formation. Protamine mRNAs are first synthesized in primary spermatocytes but are not translated into proteins until the spermatid stage. In the meantime, the protamine mRNAs are complexed with proteins and are not active in translation.

After completion of the first meiotic division, the primary spermatocyte gives rise to two secondary spermatocytes, which remain connected by a cytoplasmic bridge. The secondary spermatocytes enter the second meiotic division without delay. This phase of meiosis is very rapid, typically completed in approximately 8 hours. Each secondary spermatocyte produces two immature haploid gametes, the spermatids. The four spermatids produced from a primary spermatocyte progenitor are still connected to one another and typically to as many as 100 other spermatids as well. In mice, some genes are transcribed as late as the spermatid stage.

Spermatids do not divide further, but they undergo a series of profound changes that transforms them from relatively ordinary looking cells to highly specialized **spermatozoa** (singular, **spermatozoon**). The process of transformation from spermatids to spermatozoa is called **spermiogenesis** or **spermatid metamorphosis.**

Several major categories of change occur during spermiogenesis (Fig. 1-11). One is the progressive reduction in the size of the nucleus and tremendous condensation of the chromosomal material, which is associated with the replacement of lysine-rich histones by arginine- and cysteine-rich protamines. Along with the changes in the nucleus, a

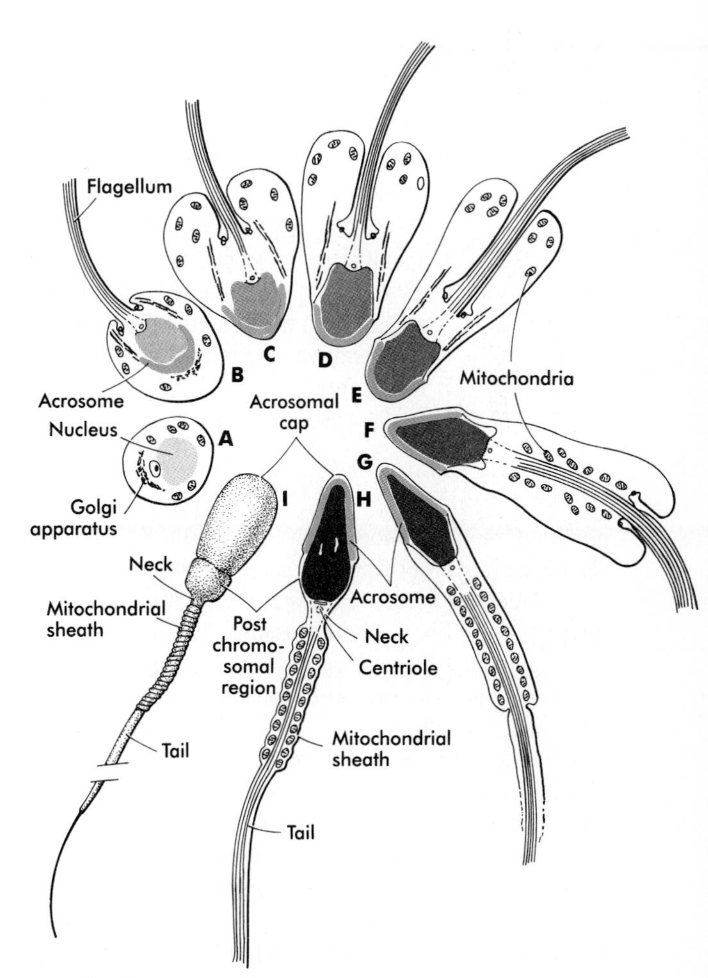

FIG. 1-11 Summary of the major stages in spermiogenesis, starting with a spermatid (**A**) and ending with a mature spermatozoon (**I**).

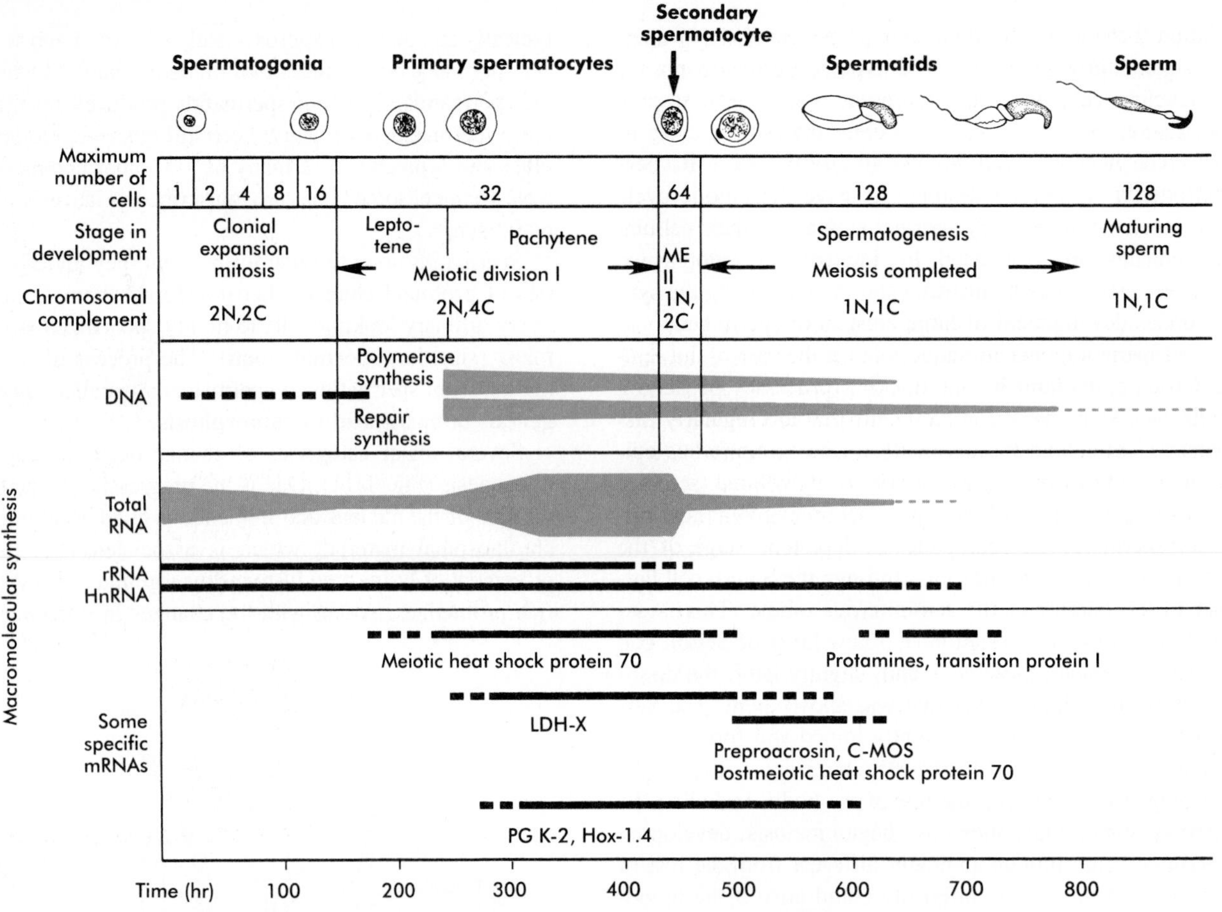

FIG. 1-12 Summary of the major events in spermatogenesis in the mouse, with specific examples of macromolecular synthesis superimposed. *ME II,* Second meiotic division. The solid lines represent qualitative periods of known synthesis. Broken segments in the lines represent uncertainty about the onset or duration of the activity. Bars with varying thicknesses represent relative amounts of the gene product that is represented.

(Modified from Erickson RP: *Trends Genet* 6:264-269, 1990.)

profound reorganization of the cytoplasm occurs. Cytoplasm streams away from the nucleus, but a condensation of the Golgi apparatus at the apical end of the nucleus ultimately gives rise to the **acrosome.** The acrosome is an enzyme-filled structure that plays a very important role in the fertilization process. At the other end of the nucleus, a prominent **flagellum** grows out of the centriolar region. **Mitochondria** are arranged around the proximal part of the flagellum in a spiral. During spermiogenesis, the plasma membrane of the head of the sperm is partitioned into a number of antigenically distinct molecular domains. These domains undergo numerous changes as the sperm cells mature in the male and at a later point when the spermatozoa are traveling through the female reproductive tract. As spermiogenesis continues, the remainder of the cytoplasm (**residual body**) moves away from the nucleus and is shed along the developing tail of the sperm cell. The residual bodies are phagocytosed by Sertoli cells.

For many years, gene expression in postmeiotic (haploid) spermatids was considered to be impossible. Recent molecular biological research on mice has shown that postmeiotic gene expression in postmeiotic spermatids is not only possible but common. Nearly 100 proteins that are produced only after the completion of the second meiotic division have been identified, and many additional proteins are synthesized both during and after meiosis. Some of these proteins are of obvious importance to the function of spermatozoa, but the relevance of other proteins resulting from gene expression in the spermatid stage is not understood. (Fig. 1-12 summarizes certain patterns of synthesis in developing mouse sperm cells.) In many mammalian species, postmeiotic gene expression in the egg is a moot point because the final meiotic division is not completed until after penetration of the egg by a sperm cell.

After spermiogenesis (approximately 64 days after the start of spermatogenesis), the spermatozoon is a highly specialized cell well adapted for motion and the delivery of its packet of DNA to the egg. The sperm cell consists of a head (2 to 3 μm wide and 4 to 5 μm long) containing the nucleus and acrosome; a midpiece containing the centrioles, the proximal part of the flagellum, and the mitochondrial helix; and the tail (about 50 μm long), which consists of a highly specialized flagellum (see Fig. 1-11). (Specific functional properties of these components of the sperm cell are discussed in Chapter 2.)

Although spermatozoa in the seminiferous tubule appear to be mature by morphological criteria, they are nonmotile and incapable of fertilizing an egg. From the testis, they undergo a leisurely transit to the **epididymis** via fluid currents that originate in the seminiferous tubule. During this time, they are exposed to a variety of different environments and secretions of the male reproductive tract.

While in transit, the spermatozoa undergo biochemical maturation, becoming covered with a glycoprotein coating and experiencing other surface modifications. The glycoprotein coating is removed in the female reproductive tract through the **capacitation** reaction, thus rendering the spermatozoa capable of fertilizing an egg. Final biochemical maturation and provision of the spermatozoa with an external energy source occur when the ejaculated sperm are mixed with secretions of the **prostate gland** and **seminal vesicles** during ejaculation. At this point, the spermatozoa are provided with an environment conducive to independent motion.

Abnormal spermatozoa. Substantial numbers (up to 10%) of mature spermatozoa are grossly abnormal. The spectrum of anomalies ranges from double heads or tails to defective flagella or variability in head size. Such defective sperm cells are highly unlikely to fertilize an egg. If the percentage of defective spermatozoa rises above 20% of the total, reduced fertility may result.

PREPARATION OF THE FEMALE REPRODUCTIVE TRACT FOR PREGNANCY

Structure of the Female Reproductive Tract

The structure and function of the female reproductive tract are well adapted for the transport of gametes and maintenance of the embryo. Many of the subtler features of this adaptation are under hormonal control and are cyclic. This section briefly reviews the aspects of female reproductive structure that are of greatest importance in understanding gamete transport and embryonic development.

Ovaries and Uterine Tubes. The **ovaries** and **uterine** (or **fallopian**) **tubes** form a functional complex devoted to the production and transport of eggs. The almond-shaped ovaries, located on either side of the uterus, are positioned very near the open, funnel-shaped ends of the uterine tubes (Fig. 1-13). Numerous, small, fingerlike projections called **fimbriae** project from the open **infundibulum** of the uterine tube toward the ovary and are involved in directing the ovulated egg into the tube. The uterine tube is character-

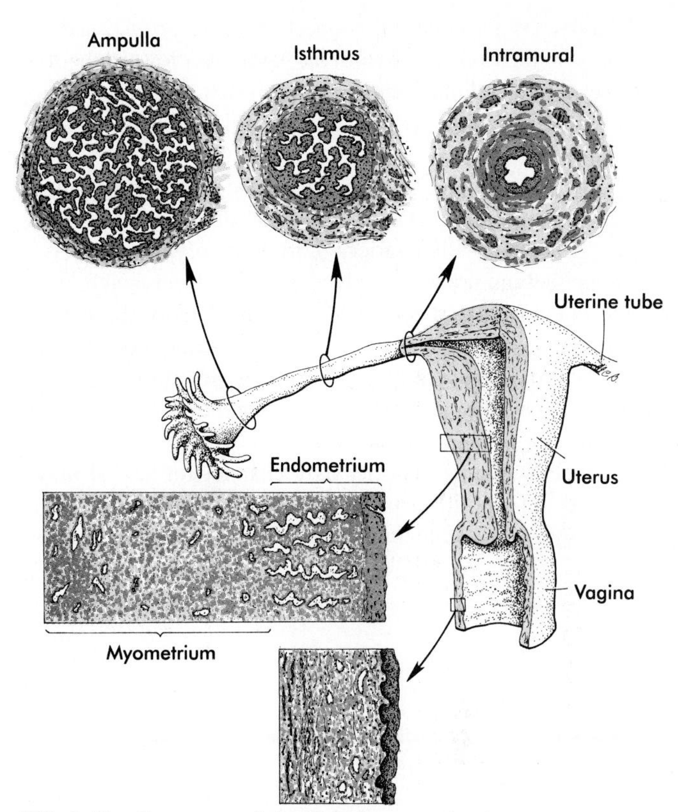

FIG. 1-13 Structure of the female reproductive tract.

ized by a very complex internal lining with a high density of prominent longitudinal folds in the upper **ampulla** and a simpler lining nearer to the uterus. The lining epithelium of the uterine tubes contains a mixture of ciliated cells that assists in gamete transport and secretory cells that produce a fluid supporting early development of the embryo. Layers of smooth muscle throughout the uterine tubes provide the basis for peristaltic contractions. The amount and function of many of these components are under cyclic hormonal control, and the overall effect of these changes is to facilitate the transport of gametes and the fertilized egg.

Uterus. The principal functions of the uterus are to receive and maintain the embryo during pregnancy and to expel the fetus at the termination of pregnancy. The first function is carried out by the uterine mucosa (endometrium) and the second by the muscular wall (myometrium). Under the cyclic effect of hormones, the uterus undergoes a series of prominent changes throughout the course of each menstrual cycle.

The **uterus** is a pear-shaped organ with thick walls of smooth muscle **(myometrium)** and a complex mucosal lining (see Fig. 1-13). The mucosal lining, called the **endometrium,** has a structure that changes daily throughout the menstrual cycle. The endometrium can be subdivided into two layers, a **functional layer** that is shed with each menstrual period or after parturition and a **basal layer** that remains intact. The general structure of the endometrium consists of (1) a columnar **surface epithelium,** (2) **uterine**

glands, (3) a specialized connective tissue stroma, and (4) **spiral arteries** that coil from the basal layer toward the surface of the endometrium (see Fig. 1-13). All these structures participate in the implantation and nourishment of the embryo.

The lower outlet of the uterus is the **cervix.** The mucosal surface of the cervix is not typical uterine endometrium but is studded with a variety of irregular crypts. The cervical epithelium produces a glycoprotein-rich cervical mucus, the composition of which varies considerably throughout the menstrual cycle. The differing physical properties of cervical mucus make it easier or more difficult for spermatozoa to penetrate through the cervix and find their way into the uterus.

Vagina. In its capacity as a channel for sexual intercourse and the birth canal, the vagina is lined with a stratified squamous epithelium, but the epithelial cells contain deposits of **glycogen,** which vary in amount throughout the menstrual cycle. Glycogen breakdown products contribute to the low acidity (pH 4.3) of the vaginal fluids. The low pH of the upper vagina appears to serve a bacteriostatic function and prevents infectious agents from entering the upper genital tract through the cervix and ultimately spreading to the peritoneal cavity through the open ends of the uterine tubes.

Hormonal Control of the Female Reproductive Cycle

Reproduction in the human female is governed by a complex series of interactions between hormones and the tissues that they influence. The hierarchy of cyclic control begins with input to the **hypothalamus** of the brain (Fig. 1-14). The hypothalamus influences hormone production by the anterior lobe of the pituitary gland **(hypophysis).** The

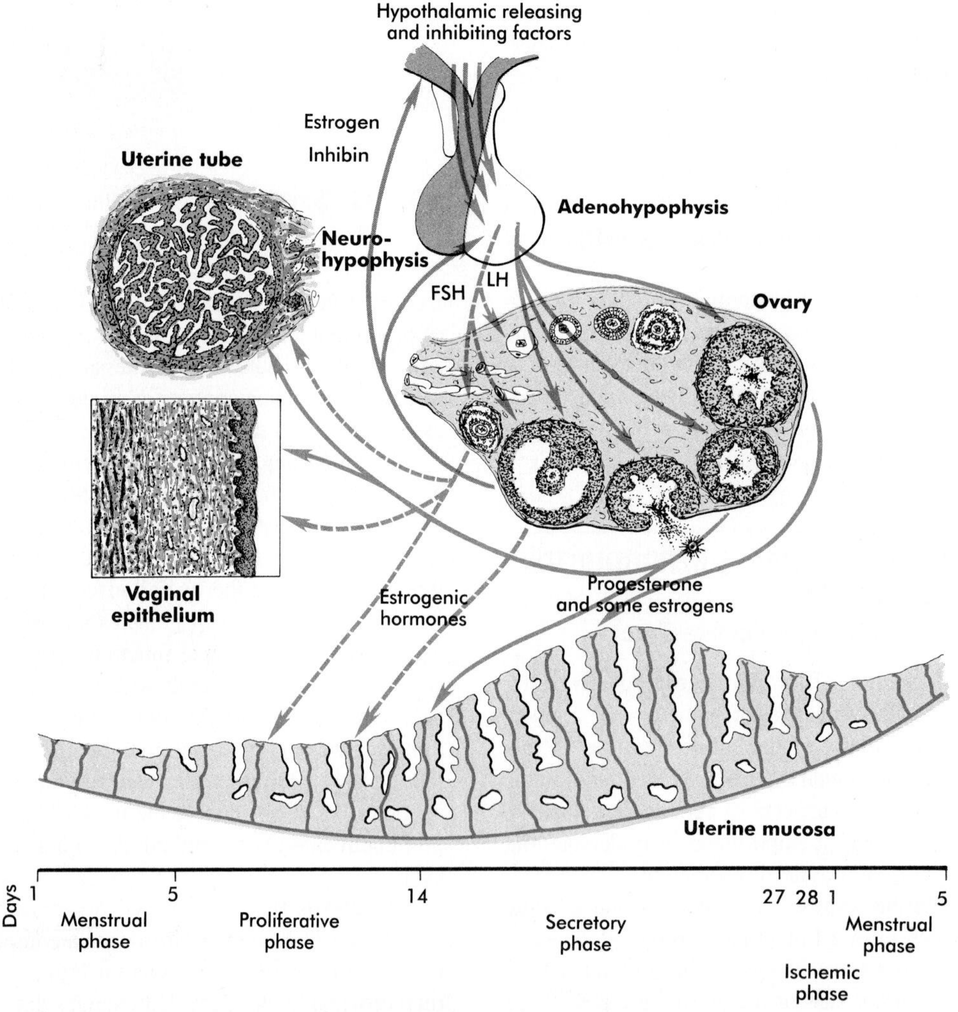

FIG. 1-14 General scheme of hormonal control of reproduction in the female. Inhibitory factors are represented by purple arrows. Stimulatory factors are represented by red arrows. Hormones involved principally in the proliferative phase of the menstrual cycle are represented by dashed arrows; those involved principally in the secretory phase are solid.

pituitary hormones are spread via the blood throughout the entire body and act on the ovaries, which are in turn stimulated to produce their own sex steroid hormones. During pregnancy, the placenta exerts a powerful effect on the mother by producing a number of hormones. The final level of hormonal control of female reproduction is that exerted by the ovarian or placental hormones on other reproductive target organs (e.g., the uterus, uterine tubes, vagina, breasts).

Hypothalamic Control. A variety of inputs, many of which are still poorly defined, stimulates neurosecretory cells in the hypothalamus to produce **gonadotropin-releasing hormone (GnRH),** along with releasing factors for other pituitary hormones. Releasing factors as well as an inhibiting factor are carried to the anterior lobe of the pituitary gland by blood vessels of the **hypothalamohypophyseal portal system,** where they stimulate the secretion of pituitary hormones (Table 1-1).

TABLE 1-1 Major Hormones Involved in Mammalian Reproduction

HORMONE	CHEMICAL NATURE	FUNCTION
HYPOTHALAMUS		
Gonadotropin-releasing hormone (GnRH, LHRH)	Decapeptide	Stimulates release of LH and FSH by anterior pituitary
Prolactin-inhibiting factor	Dopamine	Inhibits release of prolactin by anterior pituitary
ANTERIOR PITUITARY		
Follicle-stimulating hormone (FSH)	Glycoprotein (α and β subunits) (MW ~35,000)	Male: Stimulates Sertoli cells to produce androgen-binding protein Female: Stimulates follicle cells to produce estrogen
Luteinizing hormone (LH)	Glycoprotein (α and β subunits) (MW ~28,000)	Male: Stimulates Leydig cells to secrete testosterone Female: Stimulates follicle cells and corpus luteum to produce progesterone
Prolactin	Single-chain polypeptide (198 amino acids)	Promotes lactation
POSTERIOR PITUITARY		
Oxytocin	Oligopeptide (MW ~1100)	Stimulates ejection of milk by mammary gland
OVARY		
Estrogens	Steroid	Has multiple effects on reproductive tract, breasts, body fat, and bone growth
Progesterone	Steroid	Has multiple effects on reproductive tract and breast development
Testosterone	Steroid	Is precursor for estrogen biosynthesis, induces follicular atresia
Inhibin	Protein (MW ~56,000), several forms	Inhibits FSH secretion
TESTIS		
Testosterone	Steroid	Has multiple effects on male reproductive tract, hair growth, and other secondary sexual characteristics
Inhibin	Protein	Inhibits FSH secretion, has local effects on testis
PLACENTA		
Estrogens	Steroid	Has same functions as ovarian estrogens
Progesterone	Steroid	Has same functions as ovarian progesterone
Human chorionic gonadotropin (HCG)	Glycoprotein (MW ~30,000)	Maintains activity of corpus luteum during pregnancy
Human placental lactogen (somatomammotropin)	Polypeptide (MW ~20,000)	Promotes development of breasts during pregnancy

LHRH, Luteinizing hormone–releasing hormone; *MW,* molecular weight.

Pituitary Gland (Hypophysis). The pituitary gland consists of two components, the **anterior pituitary (adenohypophysis),** an epithelial glandular structure that produces a variety of hormones in response to factors carried to it by the hypothalamohypophyseal portal system, and the **posterior pituitary (neurohypophysis),** a neural structure that releases hormones by a neurosecretory mechanism.

Under the influence of GnRH and direct feedback by steroid hormone levels in the blood, the anterior pituitary secretes two polypeptide **gonadotrophic hormones,** FSH and LH, from the same cell type (see Table 1-1). In the absence of an inhibiting factor **(dopamine)** from the hypothalamus, the anterior pituitary also produces **prolactin,** which acts on the mammary glands.

The only hormone from the posterior pituitary that is directly involved in reproduction is **oxytocin,** an oligopeptide involved in childbirth and the stimulus for milk letdown from the mammary glands in lactating women.

Ovaries and Placenta. The ovaries and, during pregnancy, the placenta constitute a third level of hormonal control. Responding to blood levels of the anterior pituitary hormones, granulosa cells of the ovarian follicles convert androgens (**androstenedione** and **testosterone**) synthesized by the theca interna into estrogens (mainly **estrone** and the tenfold more powerful **17β-estradiol**), which then pass into the bloodstream. After ovulation, **progesterone** is the principal secretory product of the follicle after its conversion into the corpus luteum (see Chapter 2). During later pregnancy, the placenta supplements ovarian steroid hormone production by synthesizing its own estrogens and progesterone. It also produces two polypeptide hormones (see Table 1-1). **Human chorionic gonadotropin (HCG)** acts on the ovary to maintain the activity of the corpus luteum during pregnancy. **Human placental lactogen (somatomammotropin)** acts on the corpus luteum; it also promotes breast development by enhancing the effects of estrogens and progesterone and stimulates the synthesis of milk constituents.

Reproductive Target Tissues. The last level in the hierarchy of reproductive hormonal control constitutes the target tissues, which ready themselves both structurally and functionally for gamete transport or pregnancy in response to ovarian and placental hormones binding to specific cellular receptors. Changes in the number of ciliated cells and in smooth muscle activity in the uterine tubes, the profound changes in the endometrial lining of the uterus, and the cyclic changes in the glandular tissues of the breasts are some of the more prominent examples of hormonal effects on target tissues. These changes are described more fully later in the text.

A general principle recognized some time ago is the efficacy of first priming reproductive target tissues with estrogen so that progesterone can exert its full effects. Estrogen induces the target cells to produce large quantities of progesterone receptors, which must be in place for progesterone to act on these same cells.

Hormonal Interactions with Tissues during Female Reproductive Cycles

Knowledge of the changes the ovaries undergo is necessary to understand hormonal interactions and tissue responses during the female reproductive cycle. Responding to both FSH and LH secreted by the pituitary just before and during a menstrual period, a set of secondary ovarian follicles begins to mature and secrete 17β-estradiol. By ovulation, all of these follicles except one have undergone atresia, their main contribution having been to produce part of the supply of estrogens needed to prepare the body for ovulation and gamete transport.

During the preovulatory, or **proliferative, phase** (days 5 to 14) of the menstrual cycle, estrogens produced by the ovary act on the female reproductive tissues (see Fig. 1-14). The uterine lining becomes reepithelialized from the just-completed menstrual period. Then, under the influence of estrogens, the endometrial stroma progressively thickens, the uterine glands elongate, and the spiral arteries begin to grow toward the surface of the endometrium. The mucous glands of the cervix secrete a glycoprotein-rich but relatively watery mucus, which facilitates passage of spermatozoa through the cervical canal. As the proliferative phase progresses, a higher percentage of the epithelial cells lining the uterine tubes becomes ciliated, and smooth muscle activity in the tubes increases. In the days preceding ovulation, the fimbriated ends of the uterine tubes move closer to the ovaries.

Toward the end of the proliferative period, a pronounced increase in estradiol secreted by the developing ovarian follicle acts on the hypothalamohypophyseal system, causing increased responsiveness of the anterior pituitary to GnRH and a surge in the hypothalamic secretion of GnRH. Approximately 24 hours after the level of 17β-estradiol reaches its peak in the blood, a preovulatory surge of LH and FSH is sent into the bloodstream by the pituitary gland (Fig. 1-15). The LH surge is not a steady increase in gonadotropin secretion, but rather constitutes a series of sharp pulses of secretion that appears to be responding to a hypothalamic timing mechanism. (For further details, the reader can consult texts or reviews of reproductive endocrinology.)

The LH surge leads to ovulation, and the Graafian follicle becomes transformed into a corpus luteum (yellow body). The basal lamina surrounding the granulosa of the follicle breaks down and allows blood vessels to grow into the layer of granulosa cells. Through proliferation and hypertrophy, the granulosa cells undergo major structural and biochemical changes and now produce progesterone as their primary secretory product. This is done directly, rather than converting androgens to estrogens. Some estrogen is still secreted by the corpus luteum. After ovulation, the menstrual cycle, which is now dominated by the secretion of progesterone, is said to be in the **secretory phase** (days 14 to 28 of the menstrual cycle).

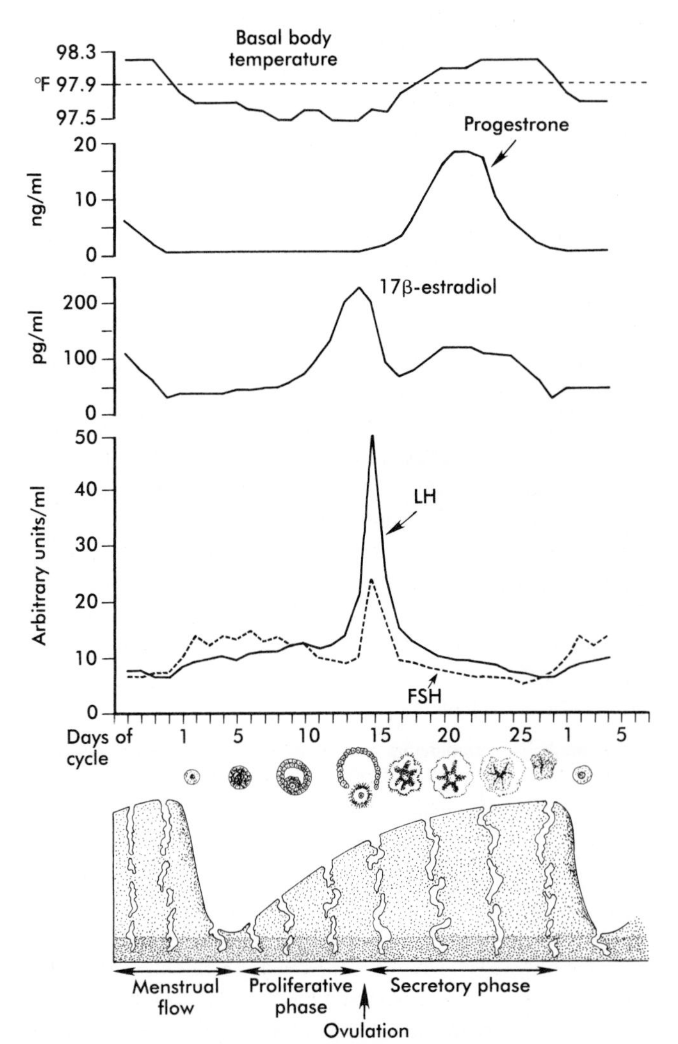

FIG. 1-15 Comparison of curves representing daily serum concentrations of gonadotropins and sex steroids and basal body temperature in relation to events in the human menstrual cycle.

(Redrawn from Midgley AR et al. In Hafez, Evans, eds: *Human reproduction,* New York, 1973, Harper & Row.)

Following the LH surge and increasing concentration of progesterone in the blood, the basal body temperature rises (see Fig. 1-15). Because of the link between a rise in basal body temperature and the time of ovulation, accurate temperature records are the basis of the **rhythm method** of birth control.

Around the time of ovulation, the combined presence of estrogen and progesterone in the blood causes the uterine tube to engage in a rhythmic series of muscular contractions designed to promote transport of the ovulated egg. Progesterone prompts epithelial cells of the uterine tube to secrete fluids that provide nutrition for the cleaving embryo. Later during the secretory phase, high levels of progesterone induce regression of some of the ciliated cells in the oviductal epithelium.

In the uterus, progesterone prepares the estrogen-primed endometrium for implantation of the embryo. The endometrium, which has thickened under the influence of estrogen during the proliferative phase, undergoes further changes. The straight uterine glands begin to coil and accumulate glycogen and other secretory products in the epithelium. The spiral arteries grow farther toward the endometrial surface, but mitosis in the endometrial epithelial cells decreases. Through the action of progesterone, the cervical mucus becomes highly viscous and acts as a protective block, inhibiting the passage of materials into or out of the uterus. During the secretory period, the vaginal epithelium becomes thinner.

In the mammary glands, progesterone furthers the estrogen-primed development of the secretory components and causes water retention in the tissues. However, more extensive development of the lactational apparatus awaits its stimulation by placental hormones.

Midway through the secretory phase of the menstrual cycle, the epithelium of the uterine tubes has already undergone considerable regression from its midcycle peak, whereas the uterine endometrium is at full readiness to receive a cleaving embryo. If pregnancy does not occur, a series of hormonal interactions brings the menstrual cycle to a close. One of the early feedback mechanisms is the production of the protein **inhibin** by the granulosa cells. Inhibin is carried by the bloodstream to the anterior pituitary, where it directly inhibits the secretion of gonadotropins, especially FSH. Through mechanisms that are still unclear, the secretion of LH is also reduced. This inhibition results in the regression of the corpus luteum and a marked reduction in the secretion of progesterone by the ovary.

Some of the main consequences of the regression of the corpus luteum are the infiltration of the endometrial stroma with leukocytes, the loss of interstitial fluid, and the spasmodic constriction and breakdown of the spiral arteries, causing local ischemia. The ischemia results in local hemorrhage and the loss of integrity of areas of the endometrium. These changes initiate menstruation (by convention, constituting days 1 to 5 of the ideal menstrual cycle). Over the next few days, the entire functional layer of the endometrium is shed in small bits, along with the attendant loss of about 30 ml of blood. By the time the menstrual period is over, only a raw endometrial base interspersed with the basal epithelium of the uterine glands remains as the basis for the healing and reconstitution of the endometrium during the next proliferative period.

HORMONAL INTERACTIONS INVOLVED WITH REPRODUCTION IN MALES

Along with the homologies of certain structures between the testis and ovary, some strong parallels between the hormonal interactions involved in reproduction in males and females exist. The most important homologies are between

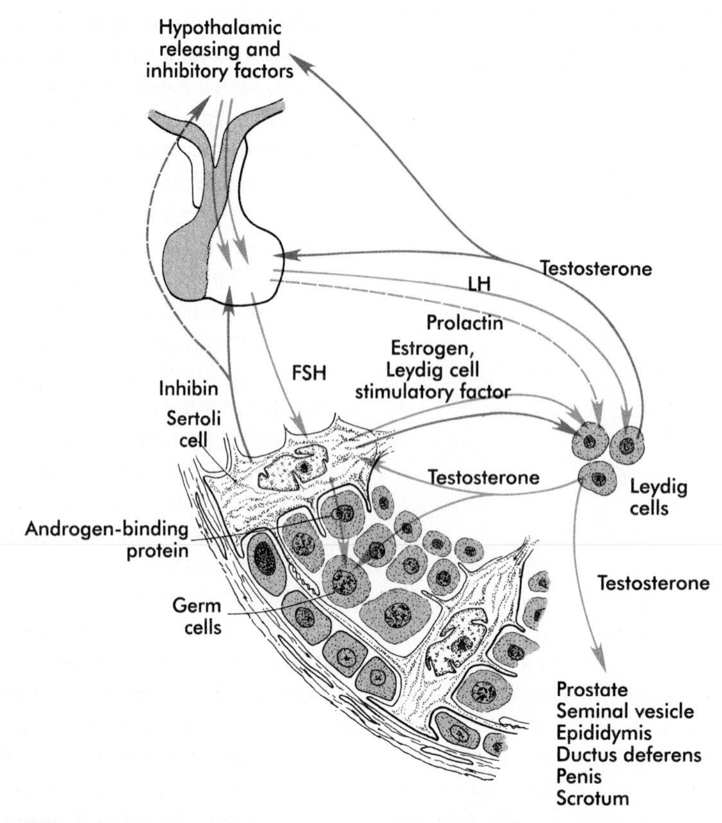

FIG. 1-16 General scheme of hormonal control in the male reproductive system. Red arrows represent stimulatory influences. Purple arrows represent inhibitory influences. Suspected interactions are represented by dashed arrows.
(Based on studies by Weiss [1988].)

TABLE 1-2 Homologies between Hormone-Producing Cells in Male and Female Gonads

PARAMETER	GRANULOSA CELLS (FEMALE)	SERTOLI CELLS (MALE)	THECA CELLS (FEMALE)	LEYDIG CELLS (MALE)
Origin	Rete ovarii	Rete testis	Stromal mesenchyme	Stromal mesenchyme
Major receptors	FSH	FSH	LH	LH
Major secretory products	Estrogens, progesterone, inhibin	Estrogen, inhibin, androgen-binding protein	Androgens	Testosterone

granulosa cells in the ovarian follicle and Sertoli cells in the seminiferous tubule of the testis and between theca cells of the ovary and Leydig cells in the testis (Table 1-2).

The hypothalamic secretion of GnRH stimulates the anterior pituitary to secrete FSH and LH. The LH binds to the nearly 20,000 LH receptors on the surface of each Leydig (interstitial) cell, and through a cascade of second messengers involving cAMP and protein phosphorylation, LH stimulates the synthesis of testosterone from cholesterol. Testosterone is released into the blood and is taken to the Sertoli cells and throughout the body, where it affects a variety of secondary sexual tissues, often after it has been locally converted to dihydrotestosterone.

Sertoli cells are stimulated by pituitary FSH via surface FSH receptors and by testosterone from the Leydig cells via cytoplasmic receptors. After FSH stimulation, the Ser-

Major Functions of Sertoli Cells

- Maintenance of the blood-testis barrier
- Secretion of tubular fluid (10 to 20 μl/gm of testis/hr)
- Secretion of androgen-binding protein
- Secretion of inhibin
- Secretion of a wide variety of other proteins (e.g., growth factors, transferrin, retinal-binding protein, metal-binding proteins)
- Maintenance and coordination of spermatogenesis
- Phagocytosis of residual bodies of sperm cells

toli cells convert some of the testosterone to estrogens (like the granulosa cells in the ovary do). Some of the estrogen diffuses back to the Leydig cells along with a **Leydig cell stimulatory factor,** which is produced by the Sertoli cells and reaches the Leydig cells by a **paracrine** (non–blood-borne) mode of secretion (Fig. 1-16). The FSH-stimulated Sertoli cell produces **androgen-binding protein,** which binds testosterone and is carried into the fluid compartment of the seminiferous tubule where it exerts a strong influence on the course of spermatogenesis. Like their granulosa cell counterparts in the ovary, the hormone-stimulated Sertoli cells produce inhibin, which is carried by the blood to the anterior pituitary and possibly the hypothalamus. There it acts by negative feedback to inhibit the secretion of FSH. In addition to inhibin and androgen-binding protein, the Sertoli cells have a wide variety of other functions, the most important of which are summarized in the box.

SUMMARY

1. Gametogenesis is divided into four phases:
   a. Extraembryonic origin of germ cells and their migration into the gonads
   b. An increase in the number of germ cells by mitosis
   c. A reduction in chromosomal material by meiosis
   d. Structural and functional maturation
2. Primordial germ cells are first readily recognizable in the yolk sac endoderm and migrate through the dorsal mesentery to the primordia of the gonads.
3. In the female, oogonia undergo intense mitotic activity in the embryo only. In the male, spermatogonia are capable of mitosis throughout life.
4. Meiosis involves reduction in chromosome number from diploid to haploid, independent reassortment of paternal and maternal chromosomes, and further redistribution of genetic material through the process of crossing-over.
5. In the oocyte, there are two meiotic blocks—in diplotene of prophase I and metaphase II. In the female, meiosis begins in the 5-month embryo; in the male, meiosis begins at puberty.
6. Failure of chromosomes to separate properly during meiosis results in nondisjunction, which is associated with multiple anomalies depending on which chromosome is affected.
7. Developing oocytes are surrounded by layers of follicular cells and interact with them through gap junctions. When stimulated by pituitary hormones (e.g., FSH, LH), the follicular cells produce steroid hormones (estrogens and progesterone). The combination of oocyte and follicular (granulosa) cells is called a *follicle*. Under hormonal stimulation, certain follicles greatly increase in size, and each month, one of these follicles undergoes ovulation.
8. Spermatogenesis occurs in the testis and involves successive waves of mitosis of spermatogonia, meiosis of primary and secondary spermatocytes, and final maturation (spermiogenesis) of postmeiotic spermatids into spermatozoa. Functional maturation of spermatozoa occurs in the epididymis.
9. Female reproductive tissues undergo cyclic, hormonally induced preparatory changes for pregnancy. In the uterine tubes, this involves the degree of ciliation of the epithelium and smooth muscle activity of the wall. Under the influence of estrogens and then progesterone, the endometrium of the uterus builds up in preparation to receive the embryo. In the absence of fertilization and with the subsequent withdrawal of hormonal support, the endometrium breaks down and is shed (menstruation). Cyclic changes in the cervix involve thinning of the cervical mucus at the time of ovulation.
10. Hormonal control of the female reproductive cycle is hierarchical, with releasing or inhibiting factors from the hypothalamus acting on the adenohypophysis, causing the release of pituitary hormones (e.g., FSH, LH). The pituitary hormones sequentially stimulate the ovarian follicles to produce estrogens and progesterone, which act on the female reproductive tissues. In pregnancy, the remains of the follicle (corpus luteum) continue to produce progesterone, which maintains the early embryo until the placenta begins to produce sufficient hormones to maintain pregnancy.
11. In the male, LH stimulates the Leydig cells to produce testosterone, and FSH acts on the Sertoli cells, which support spermatogenesis. In both the male and female, feedback inhibition decreases the production of pituitary hormones.

REVIEW QUESTIONS

1. In a routine chest x-ray examination, the radiologist sees what appears to be teeth in a mediastinal mass. What is the likely diagnosis and what is a probable embryological explanation for its appearance?
2. When does meiosis begin in the female and the male?
3. At what stages is meiosis arrested in the female?
4. What is the underlying cause of most spontaneous abortions during the early weeks of pregnancy?
5. What is the difference between spermatogenesis and spermiogenesis?
6. The actions of what hormones are responsible for the changes in the endometrium during the menstrual cycle?
7. Sertoli cells in the testis are stimulated by what two major reproductive hormones?

REFERENCES

Browder LW, ed: *Developmental biology. Vol 1. Oogenesis,* New York, 1985, Plenum.

Clermont Y: The cycle of the seminiferous epithelium in man, *Am J Anat* 112:35-51, 1963.

Eddy EM and others: Origin and migration of primordial germ cells in mammals, *Gamete Res* 4:333-362, 1981.

Eppig JJ: Intercommunication between mammalian oocytes and companion somatic cells, *Bioessays* 13:569-574, 1991.

Erickson RP: Post-meiotic gene expression, *Trends Genet* 6(8):264-269, 1990.

Fawcett DW: *A textbook of histology,* Philadelphia, 1986, Saunders.

Gilbert EF, Opitz JM: Developmental and other pathological changes in syndromes caused by chromosome anomalies, *Perspect Pediatr Pathol* 7:1-63, 1982.

Kerr JB: Functional cytology of the human testis, *Bailleres Clin Endocrinol Metab* 6(2):235-250, 1992.

Larson WJ and others: *Expansion of the cumulus-oocyte complex during the preovulatory period: possible roles in oocyte maturation, ovulation, and fertilization.* In Familiari G, Makabe S, Motta PM, eds: *Ultrastructure of the ovary,* Boston, 1991, Kluwer Academic, pp 45-61.

Moodbidri SB, Garde SV, Sheth AR: Inhibin: unity in diversity, *Arch Androl* 28:149-157, 1992.

Verhoeven G: Local control systems within the testis, *Bailleres Clin Endocrinol Metab* 6(2):313-333, 1992.

Willard HF: Centromeres of mammalian chromosomes, *Trends Genet* 6(12):410-416, 1990.

Wynn RM: *Biology of the uterus,* New York, 1977, Plenum.

Zamboni L: Physiology and pathophysiology of the human spermatozoon: the role of electron microscopy, *J Electron Microsc Tech* 17:412-436, 1991.

CHAPTER 2

Transport of Gametes and Fertilization

Chapter 1 described the origins and maturation of male and female gametes and the hormonal conditions that make such maturation possible. It also described the cyclic, hormonally controlled changes in the female reproductive tract that ready it for fertilization and the support of embryonic development. This chapter first explains the way the egg and sperm cells come together in the female reproductive tract so that fertilization can occur. It also outlines the complex set of interactions involved in fertilization of the egg by a sperm.

OVULATION AND EGG AND SPERM TRANSPORT

Ovulation

Toward the midpoint of the menstrual cycle, the mature Graafian follicle, containing the egg that has been arrested in prophase of the first meiotic division, has moved to the surface of the ovary. Under the influence of follicle-stimulating hormone (FSH) and luteinizing hormone (LH), the follicle expands dramatically. The first meiotic division is completed and the second meiotic division proceeds until the metaphase stage, at which the second meiotic arrest occurs. After the first meiotic division, the first polar body is expelled. By this point, the follicle bulges from the surface of the ovary. The apex of the protrusion is the **stigma.**

The stimulus for ovulation is the surge of LH that is secreted by the anterior pituitary at the midpoint of the menstrual cycle. Within minutes after the sharp rise in LH concentration in the blood, local blood flow increases in the outer layers of the follicular wall and throughout the ovary as well. Along with increased blood flow, plasma proteins leak into the tissues through the postcapillary venules, resulting in local edema. The edema and the release of a number of pharmacologically active compounds such as prostaglandins, histamine, and vasopressin provide the starting point for a series of reactions that results in the local production of collagenase (Fig. 2-1). At the same time, secretion of hyaluronic acid by the granulosa cells results in loosening of the granulosa layers. The collagen degradation and ischemia and the death of some of the overlying cells cause a weakness of the outer follicular wall. The combination of the weakened follicular wall, pressure of the antral fluid (about 5 to 6 mm of water), and possibly the contraction of smooth muscle elements leads to rupture of the outer follicular wall about 28 to 36 hours after the LH surge.

Ovulation results in the expulsion of both antral fluid and the ovum from the ovary into the peritoneal cavity. The ovum, however, is not ovulated as a single naked cell but as a complex consisting of (1) the ovum, (2) the zona pellucida, (3) the two- to three-cell-thick corona radiata, and (4) a sticky matrix containing surrounding cells of the cumulus oophorus. By convention, the last cells are designated as **corona radiata** after ovulation has occurred. Normally, one egg is released at ovulation. The release and fertilization of two eggs can result in fraternal twinning.

Some women experience mild to pronounced pain at the time of ovulation. Often called **Mittelschmerz** (German for "middle pain"), such pain may accompany slight bleeding from the ruptured follicle. Another sign of ovulation is a slight rise in basal body temperature. This has often been used as a reference point in rhythm methods of contraception (see Chapter 7).

Egg Transport

The first step in egg transport is capture of the ovulated egg by the uterine tube. Shortly before ovulation, the epithelial cells of the uterine tube become more highly ciliated, and smooth muscle activity in the tube and its suspensory liga-

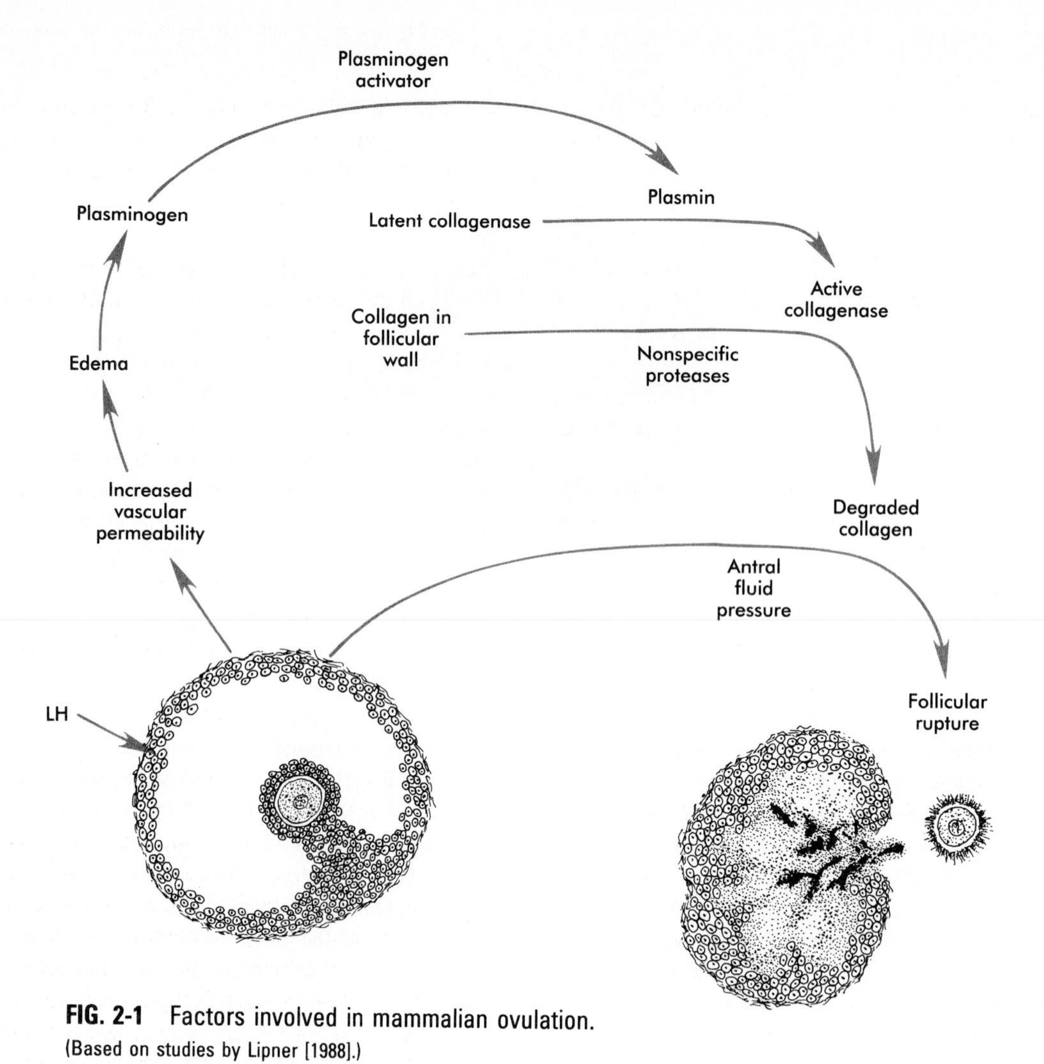

FIG. 2-1 Factors involved in mammalian ovulation.
(Based on studies by Lipner [1988].)

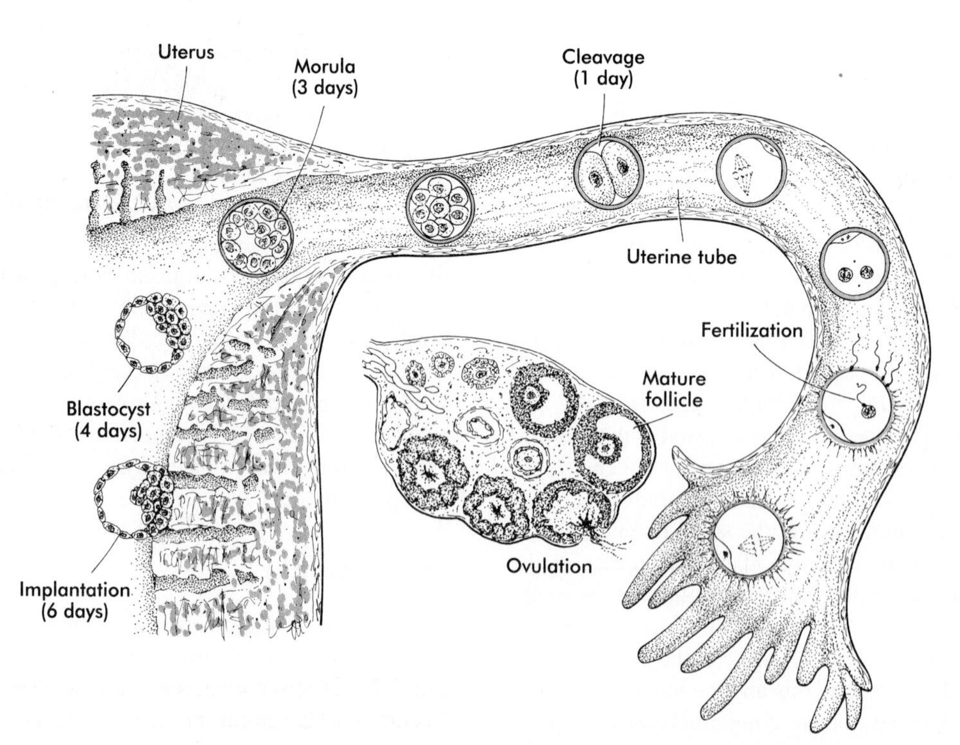

FIG. 2-2 Follicular development in the ovary, ovulation, fertilization, and transport of the early embryo down the uterine tube and into the uterus.

ment increases as the result of hormonal influences. By the time of ovulation, the fimbriae of the uterine tube move closer to the ovary and actually seem to rhythmically sweep over its surface. This action, plus the currents set up by the cilia, efficiently captures the ovulated egg complex. Experimental studies on rabbits have shown that the bulk provided by the cellular coverings of the ovulated egg is important in facilitating the egg's capture by the uterine tube. Denuded ova or inert objects of that size are not so readily transported. Some evidence suggests that the human egg may rely less on its cellular coverings for transport than does that of the rabbit.

Even without these types of natural adaptations, the ability of the uterine tubes to capture eggs is remarkable. If the fimbriated end of the tube has been removed, egg capture occurs remarkably often, and pregnancies have occurred in women who have had one ovary and the contralateral uterine tube removed. In such cases, the ovulated egg would have to travel free in the pelvic cavity for a considerable distance before entering the ostium of the uterine tube on the other side, or an extra long tube could possibly swing to reach the contralateral ovary.

Once inside the uterine tube, the egg appears to be transported toward the uterus mainly as the result of contractions of the smooth musculature of the tubal wall. Although the cilia lining the tubal mucosa may also play a role in egg transport, their action is not obligatory because women with **immotile cilia syndrome** are normally fertile.

Tubal transport of the egg usually takes 3 to 4 days, regardless of whether fertilization occurs (Fig. 2-2). Egg transport typically occurs in two phases—slow transport in the ampulla (approximately 72 hours) and a more rapid phase (8 hours) during which the egg or embryo passes through the isthmus and into the uterus (see p. 33). By a poorly understood mechanism, possibly local edema or reduced muscular activity, the egg is temporarily prevented from entering the isthmic portion of the tube, but under the influence of progesterone, the uterotubal junction relaxes and permits entry of the ovum.

While in the uterine tube, the egg is bathed in **tubal fluid,** which is a combination of secretion by the tubal epithelial cells and transudate from capillaries just below the epithelium. In some mammals, exposure to oviductal secretions is important to survival of the ovum, but the role of the tubal fluid in humans is less clear.

By roughly 80 hours, the ovulated egg or embryo has passed from the uterine tube into the uterus. If fertilization has not occurred, the egg degenerates and is phagocytized. (Implantation of the embryo is discussed in Chapter 3.)

Sperm Transport

Sperm transport occurs in both the male and the female. In the male, sperm transport is closely connected with structural and functional maturation, whereas in the female reproductive tract, it is important for as many spermatozoa as possible to pass quickly through the tract to the upper uterine tube where they can meet the ovulated egg.

After spermiogenesis in the seminiferous tubules, the spermatozoa are morphologically mature but nonmotile and incapable of fertilizing an egg (Fig. 2-3). Spermatozoa are passively transported via testicular fluid from the seminif-

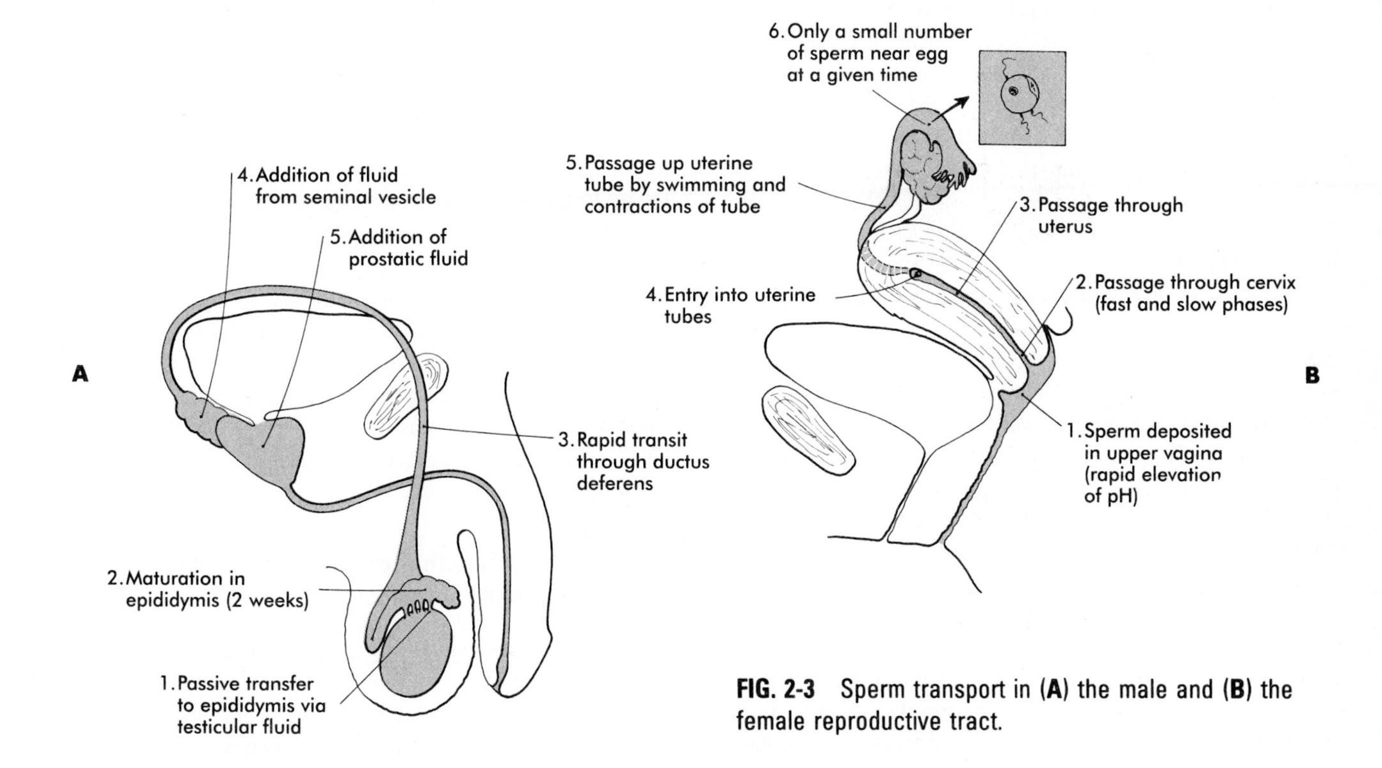

FIG. 2-3 Sperm transport in **(A)** the male and **(B)** the female reproductive tract.

erous tubules to the caput (head) of the epididymis through the rete testis and the efferent ductules. They are propelled by fluid pressure generated in the seminiferous tubules and are assisted by smooth muscle contractions and ciliary currents in the efferent ductules. Spermatozoa spend from 4 to 12 days in the highly convoluted duct of the epididymis, during which time they undergo biochemical maturation. This period of maturation is associated with changes in the glycoproteins in the plasma membrane of the sperm head. By the time the spermatozoa have reached the cauda (tail) of the epididymis, they are capable of fertilizing an egg.

On ejaculation, the spermatozoa rapidly pass through the **ductus deferens** and become mixed with fluid secretions from the **seminal vesicles** and **prostate gland.** Prostatic fluid is rich in citric acid, acid phosphatase, zinc, and magnesium ions, whereas fluid of the seminal vesicle is rich in fructose (the principal energy source of spermatozoa) and prostaglandins. The 2 to 6 ml of ejaculate **(semen,** or **seminal fluid)** typically consists of 40 to 250 million spermatozoa mixed with alkaline fluid from the seminal vesicles (60% of the total) and acid secretion (pH 6.5) from the prostate (30% of the total). The pH of normal semen is typically between 7.0 and 8.3. Despite the large number of spermatozoa normally present in an ejaculate, a number as small as 25 million per ejaculate may be compatible with normal fertility.

During copulation, the seminal fluid is normally deposited in the upper vagina (see Fig. 2-3), where its composition and buffering capacity immediately protect the spermatozoa from the acid fluid found in the upper vagina. The acidic vaginal fluid normally serves a bacteriocidal function in protecting the cervical canal from pathogenic organisms. Within about 10 seconds, the pH of the upper vagina is raised from 4.3 to as much as 7.2. The buffering effect lasts only a few minutes in humans but provides enough time for the spermatozoa to approach the cervix in an environment (pH 6.0 to 6.5) optimal for sperm motility.

The next barrier that the sperm cells must overcome is the cervical canal and the cervical mucus that blocks it. Changes in intravaginal pressure may suck spermatozoa into the cervical os, but swimming movements of spermatozoa appear to be important for penetrating the mucus.

The composition and viscosity of cervical mucus vary considerably throughout the menstrual cycle. Composed of **cervical mucin** (a glycoprotein with a high carbohydrate composition) and soluble components, cervical mucus is not readily penetrable. However, between days 9 and 16 of the cycle, its water content increases, which facilitates the passage of sperm through the cervix around the time of ovulation.

There are two main modes of sperm transport through the cervix. One is a phase of initial rapid transport, by which some spermatozoa can reach the uterine tubes within an hour of ejaculation. Such rapid transport relies more on muscular movements of the female reproductive tract than on the motility of the spermatozoa themselves. The second, slow phase of sperm transport involves the swimming of spermatozoa through the cervical mucus (traveling at a rate of 2 to 3 mm/hr), their storage in cervical crypts, and their final passage through the cervical canal as much as 2 to 3 days later.

Relatively little is known about the passage of sperm through the uterine cavity, but contraction of the uterine smooth muscle, rather than sperm motility, seems to be the main intrauterine transport mechanism. At this point the spermatozoa enter one of the uterine tubes. For years, many reproductive biologists have believed that pure chance dictates which uterine tube the spermatozoa enter, but it has been suggested that the egg gives off some type of chemoattractant partially directing the sperm toward the egg. The existence of sperm chemoattractants has been well demonstrated in certain aquatic animals, but evidence about such a phenomenon in mammals has been equivocal. Recently, however, sperm-attracting activity has been detected in mature human follicular fluid. Further research is required to determine whether follicular fluid in vivo attracts spermatozoa into the uterine tube containing the ovum or whether there is a short-range attraction of spermatozoa by the ovulated egg and its membranes.

A surprisingly small number of spermatozoa (only about 200) are found in a uterine tube at a given time. With muscular movements of the tube and some swimming movements, sperm cells work their way up the tube and even out through the ostium and into the peritoneal cavity. The simultaneous transport of an egg down the tube and sperm up the tube is currently explained on the basis of peristaltic contractions of the musculature of the uterine tube. These contractions subdivide the tube into compartments. Within a given compartment, the egg and/or sperm are caught up in churning movements that over a day or two result in a mixing of the gamete contents of the ampullary portion of the tube.

The most recent estimates suggest that spermatozoa can retain their function in the female reproductive tract for about 80 hours. During their passage through the reproductive tract, the spermatozoa undergo the capacitation reaction. Capacitation, which is the alteration of the glycoprotein surface of spermatozoa under the influence of secretions of the tissues of the female reproductive tract, is required for spermatozoa to be able to fertilize an egg in many mammals. The extent to which capacitation is involved during the transport of human sperm is less clearly defined. Fertilization of the egg typically occurs in the ampullary portion of the uterine tube.

Formation and Function of the Corpus Luteum of Ovulation and Pregnancy

While the ovulated egg is passing through the uterine tubes, the ruptured follicle from which it arose undergoes a series of striking changes that are essential for the progression of events leading to and supporting pregnancy (see Fig. 1-8).

Soon after ovulation, the basement membrane that separates the granulosa cells from the theca interna breaks down, allowing thecal blood vessels to grow into the cavity of the ruptured follicle. The granulosa cells simultaneously undergo a series of major changes in form and function **(luteinization).** Within 30 to 40 hours of the LH surge, these cells, now called **granulosa lutein cells,** begin secreting increasing amounts of progesterone along with some estrogen. This pattern of secretion provides the hormonal basis for the changes in the female reproductive tissues during the last half of the menstrual cycle. During this period, the follicle continues to increase in size. Because of its yellow color, it is known as the **corpus luteum.** The granulosa lutein cells are terminally differentiated. They have stopped dividing and secrete progesterone for 10 days. These cells are unresponsive to human chorionic gonadotropin.

In the absence of fertilization and a hormonal stimulus provided by the early embryo, the corpus luteum begins to deteriorate late in the menstrual cycle. In some mammals, action by a local **uterine luteolytic factor** may facilitate regression of the corpus luteum. Deterioration of the corpus luteum and the accompanying reduction in progesterone production cause the hormonal withdrawal that results in the degenerative changes of the endometrial tissue during the last days of the menstrual cycle.

During the deterioration of the corpus luteum, the granulosa lutein cells degenerate and are replaced with collagenous scar tissue. Because of its white color, the former corpus luteum now becomes known as the **corpus albicans** ("white body").

If fertilization occurs, the production of the protein hormone **chorionic gonadotropin** by the future placental tissues maintains the corpus luteum in a functional condition and even causes an increase in its size and hormone production. Chorionic gonadotropin binds to the same receptors on the follicular cells as does LH. According to Jones (1990), the thecal cells (**theca lutein cells**) respond to the chorionic gonadotropin secreted by the extraembryonic tissues by dividing and producing large amounts of progesterone. Because the granulosa lutein cells are unable to divide and also cease producing progesterone after 10 days, the large **corpus luteum of pregnancy** is composed principally of theca lutein cells. The corpus luteum of pregnancy remains functional for the first few months of pregnancy. After the second month, the placenta produces enough estrogens and progesterone to maintain pregnancy on its own. At this point the ovaries can be removed and pregnancy will continue.

FERTILIZATION

Fertilization is a series of processes rather than a single event. Viewed in the broadest sense, these processes begin when spermatozoa start to penetrate the corona radiata that surrounds the egg and end with the intermingling of the maternal and paternal chromosomes after the spermatozoon has entered the egg.

Penetration of the Corona Radiata

When the spermatozoa first encounter the ovulated egg in the ampullary part of the uterine tube, they are confronted by the corona radiata and possibly some remnants of the cumulus oophorus, which represents the outer layer of the egg complex (Fig. 2-4). The corona radiata is a highly cellular layer with an intercellular matrix consisting of proteins and a high concentration of carbohydrates, especially hyaluronic acid. In contrast to earlier beliefs, only a few dozen spermatozoa are found in the vicinity of the egg in the uterine tube.

Although it is widely believed that hyaluronidase emanating from the acrosome of the sperm head plays a major role in penetration of the corona radiata, the evidence is not unequivocal. Rooster spermatozoa, which lack hyaluronidase, can penetrate the mammalian corona radiata. The active swimming movements of the spermatozoa appear to play a significant role in penetration of the corona radiata.

Attachment to and Penetration of the Zona Pellucida

The **zona pellucida,** which is 13 microns thick in humans, is composed principally of sulfated glycoproteins. Although surprisingly simple in composition, it plays a number of important roles:

1. It serves as a barrier that normally allows only sperm of the same species access to the egg.
2. After fertilization, the modified zona prevents any additional spermatozoa from reaching the egg.
3. During the early stages of cleavage, it acts as a porous filter through which certain substances secreted by the uterine tube can reach the egg.
4. It normally prevents premature implantation of the cleaving embryo into the wall of the uterine tube.

In the mouse, by far the best-studied mammal, the zona consists of three glycoproteins—ZP1, ZP2, and ZP3—with molecular weights of 200,000, 120,000, and 83,000 daltons, respectively. ZP2 and ZP3 combine to form basic units that polymerize into long filaments. These filaments are periodically linked by cross-bridges of ZP1 molecules (Fig. 2-5). The zona pellucida of the unfertilized mouse egg is estimated to contain over a billion copies of the ZP3 protein.

After they have penetrated the cumulus oophorus, spermatozoa bind tightly to the zona pellucida by means of the plasma membrane of the head of the sperm. The ZP3 molecule, specifically the O-linked oligosaccharides attached to the polypeptide core, acts as the sperm receptor in the zona in the mouse. Molecules on the surface of the sperm head act as specific binding sites for the ZP3 sperm receptors on

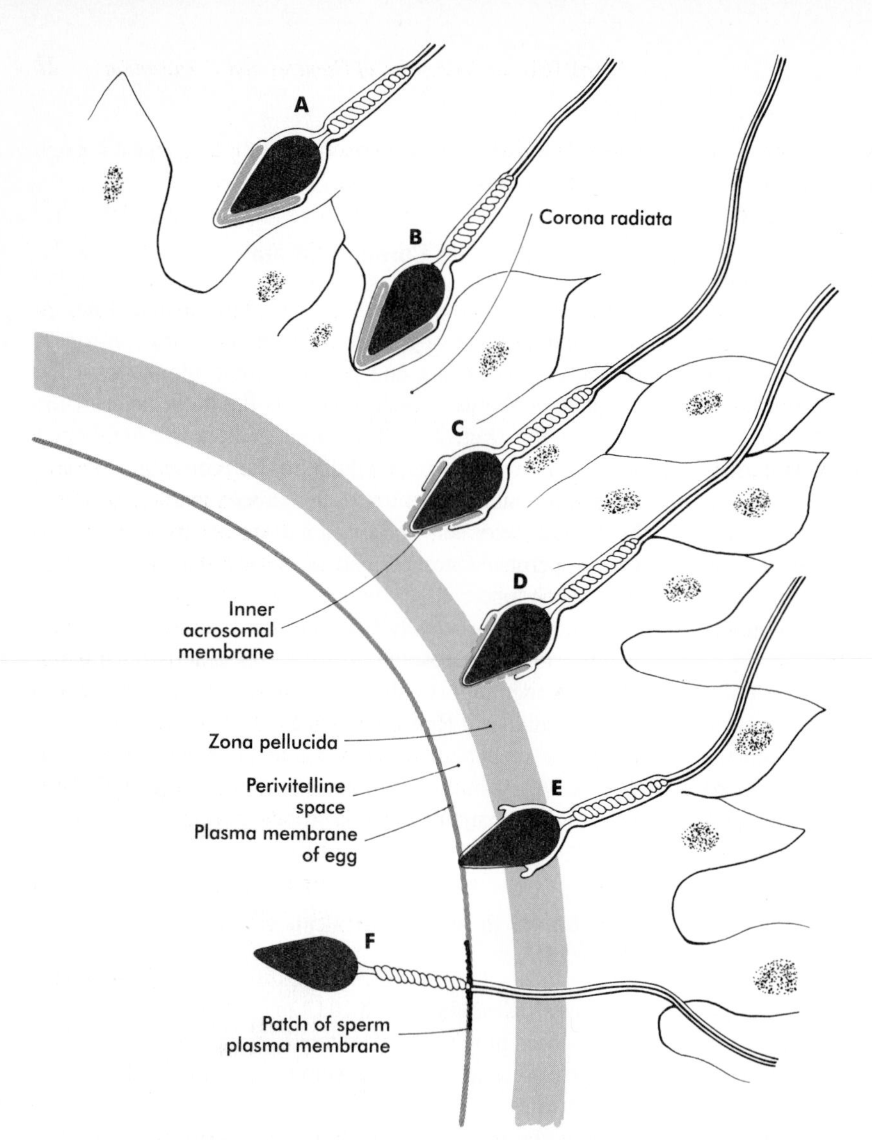

FIG. 2-4 The sequence of events in penetration of the coverings and plasma membrane of the egg. **A** and **B,** Penetration of the corona radiata. **C** and **D,** Attachment to the zona pellucida and acrosomal reaction. **E** and **F,** Binding to plasma membrane and entry into the egg.

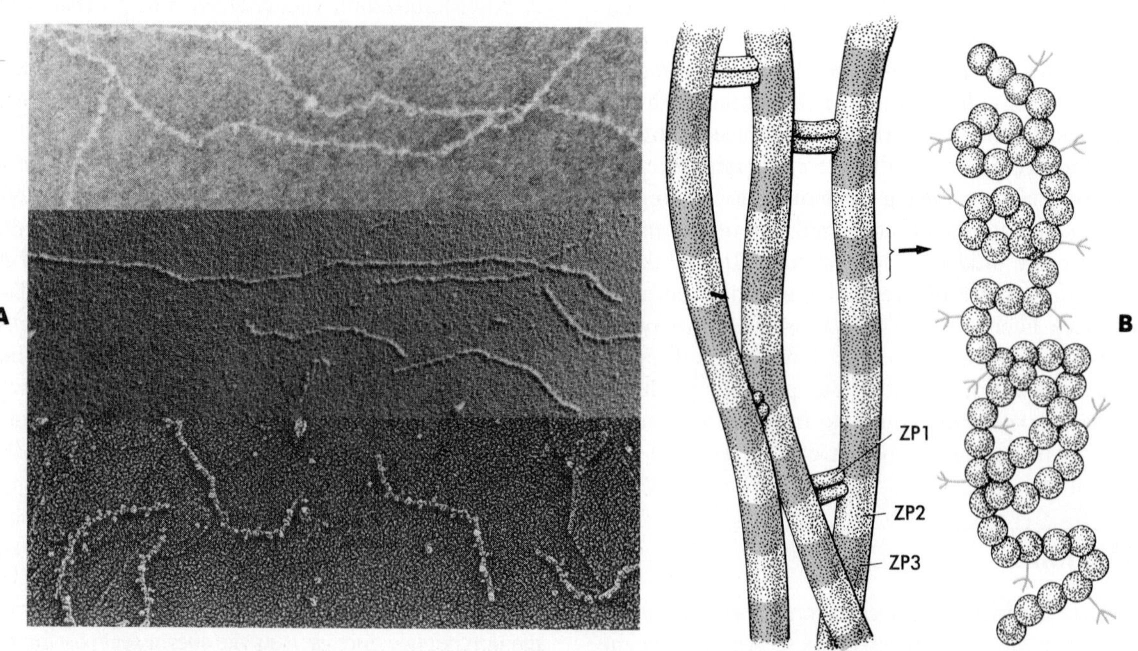

FIG. 2-5 **A,** Filamentous components of the mammalian (mouse) zona pellucida.
(From P. Wasserman, *Sci Am.*)
B, Molecular organization of the filaments in the zona pellucida. *Far right*: Structure of the ZP3 glycoprotein.
(Modified from P. Wassarman, *Sci Am.*)

the zona pellucida. The specific molecules have not been definitively determined, but one of the likely candidates is a galactosyl transferase.

Only after binding to the zona pellucida do mammalian spermatozoa undergo the **acrosomal reaction.** The essence of the acrosomal reaction is the fusion of parts of the outer acrosomal membrane with the overlying plasma membrane and the pinching off of fused parts as small vesicles. This results in the liberation of the multitude of enzymes that are stored in the acrosome (box).

The mechanism of the acrosomal reaction in mammals is not completely understood, but it appears to be stimulated by the ZP3 molecule (see Fig. 2-5). In contrast to the sperm receptor function of ZP3, a large segment of the polypeptide chain of the ZP3 molecule must be present to induce the acrosomal reaction. One of the initiating events of the acrosomal reaction is a massive influx of Ca^{++} through the plasma membrane of the sperm head. This process, accompanied by an influx of Na^+ and an efflux of H^+, raises the intracellular pH. Fusion of the outer acrosomal membrane with the overlying plasma membrane soon follows. As the vesicles of the fused membranes are shed, the enzymatic contents of the acrosome are freed and can assist the sperm in making its way through the zona pellucida.

After the acrosomal reaction, the inner acrosomal membrane forms the outer surface covering of most of the sperm head (see Fig. 2-4, *D*). Toward the base of the head of the sperm (in the equatorial region), the inner acrosomal membrane fuses with the remaining **postacrosomal plasma membrane** to maintain membrane continuity around the head of the sperm.

Only after completing the acrosomal reaction can the spermatozoon successfully begin to penetrate the zona pellucida. Penetration of the zona appears to be accomplished by a combination of mechanical propulsion by movements of the sperm's tail and by digestion of a pathway through the action of acrosomal enzymes. Although the action of several acrosomal enzymes, including hyaluronidase, may be involved in initiating penetration through the zona, the most important enzyme is **acrosin,** a serine proteinase that is bound to the inner acrosomal membrane. Many investigators feel that the well-defined tunnel that marks the pathway of the sperm through the zona can be attributed to the fact that acrosin is membrane bound rather than diffusible. Once the sperm has made its way through the zona and into the **perivitelline space** (the space between the egg's plasma membrane and the zona pellucida), it can make direct contact with the plasma membrane of the egg.

Some Major Mammalian Acrosomal Enzymes

Acid proteinase	β-galactosidase
Acrosin	β-glucuronidase
Arylaminidase	Hyaluronidase
Arylsulfatase	Neuraminidase
Collagenase	Phospholipase C
Esterase	Proacrosin

Fusion of Sperm and Egg

After a brief transit period through the perivitelline space, the sperm makes contact with the plasma membrane of the egg. Fusion between the sperm and egg occurs at the **equatorial region** of the sperm head, where the inner acrosomal membrane has previously fused with the remaining plasma membrane of the sperm. The acrosomal reaction seems to cause some change in the membrane properties of the sperm, because if the acrosomal reaction has not occurred, the spermatozoon is unable to fuse with the egg. After initial fusion, the contents of the sperm (the head, midpiece, and usually the tail) sink into the egg (Fig. 2-6), whereas the plasma membrane of the sperm, which is antigenically distinct from that of the egg, becomes incorporated into the egg's plasma membrane and remains recognizable at least until the start of cleavage. Although mitochondria located in the neck of the sperm enter the egg, it is not clear that they contribute to the functional mitochondrial complement of the zygote.

Prevention of Polyspermy

Once a sperm has fused with an egg, entry of other spermatozoa into the egg **(polyspermy)** must be prevented or abnormal development will likely result. Two blocks to polyspermy, fast and slow, are typically present in vertebrate fertilization.

The **fast block to polyspermy,** which has been best studied in sea urchins, consists of a rapid electrical depolarization of the plasma membrane of the egg. The resting membrane potential of the egg changes from about −70 mV to +10 mV within 2 to 3 seconds after fusion of the sperm with the egg. This change in membrane potential prevents other spermatozoa from adhering to the egg's plasma membrane. Little is known about the fast block to polyspermy in the human egg, but because at the moment of fertilization there are far fewer spermatozoa in the vicinity of the mammalian egg than in the sea urchin, the need for an effective fast block may be less. The fast block is short lived, usually lasting only about a minute. This time is sufficient for the egg to mount the permanent slow block.

The **slow block to polyspermy** begins with the propagation of a wave of Ca^{++} from the site of sperm-egg fusion. Within a couple of minutes, the Ca^{++} wave has passed through the egg, sequentially acting on the cortical granules as it passes by them. Exposure to Ca^{++} causes

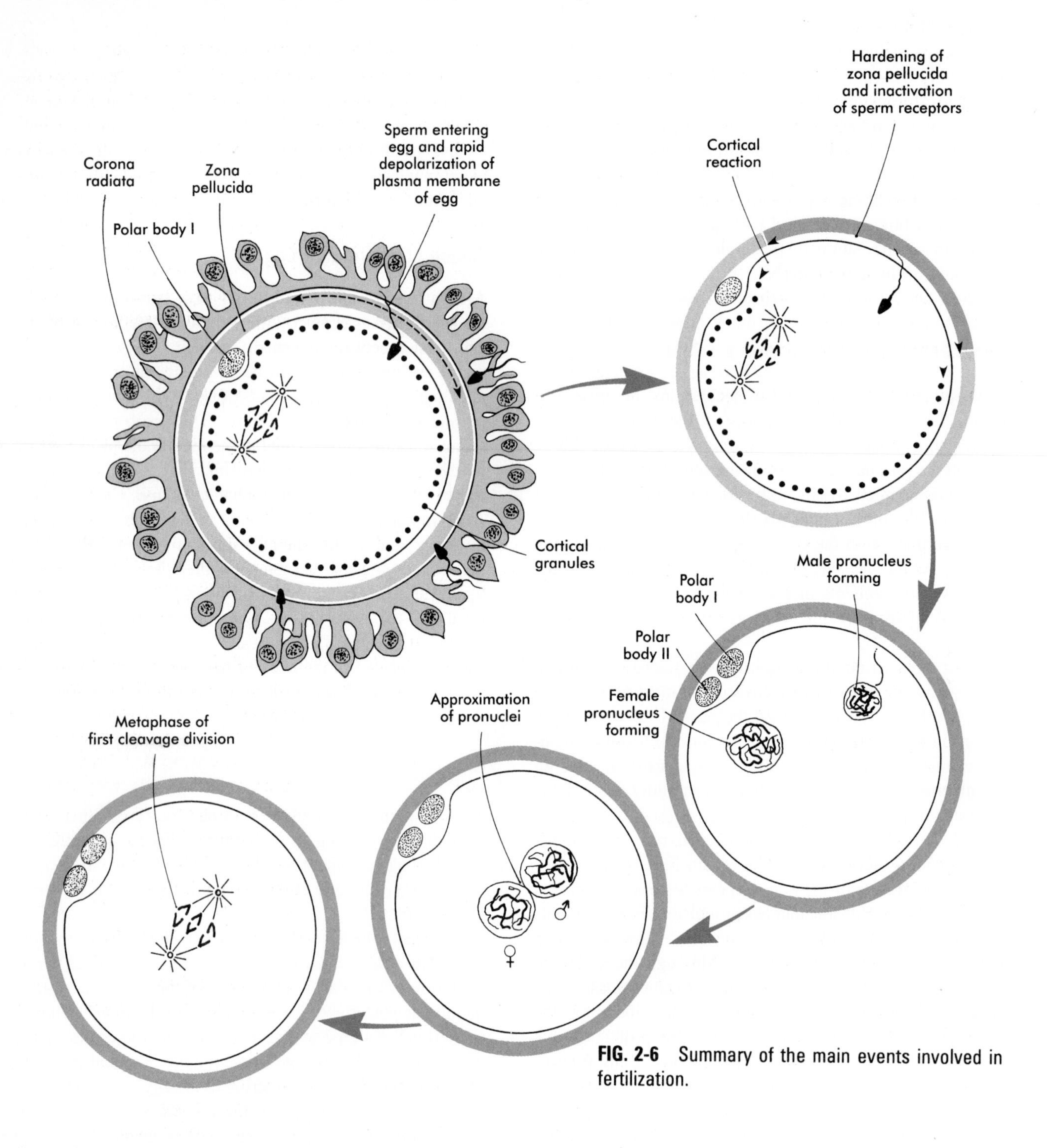

FIG. 2-6 Summary of the main events involved in fertilization.

the cortical granules to fuse with the plasma membrane and to release their contents (hydrolytic enzymes and polysaccharides) into the perivitelline space.

The secretory products of the cortical granules diffuse into the porous zona pellucida and hydrolyze the sperm receptor molecules (ZP3 in the mouse) in the zona. This reaction, called the **zona reaction,** essentially eliminates the ability of spermatozoa to adhere to and penetrate the zona. The zona reaction has been visually observed in human eggs that have undergone in vitro fertilization.

Interspecies molecular differences in the sperm-binding regions of the ZP3 molecule may serve as the basis for the inability of spermatozoa of one species to fertilize an egg of another species. In mammals, there is less species variation in the composition of ZP3; this may explain the reason penetration of the zona pellucida by spermatozoa of closely related mammalian species is sometimes possible, whereas it is rare among lower animals. The specificity between the ZP3 receptor site and binding sites on the sperm could be used as the basis for an immunologically based mode of contraception.

Metabolic Activation of the Egg

One of the significant changes brought about by the penetration of a sperm cell is a rapid intensification of the egg's respiration and metabolism. The mechanisms underlying these changes are not fully understood even in the best-studied systems, but the early release of Ca^{++} from internal stores is believed to be the initiating event. In some species, Ca^{++} release is shortly followed by an exchange of extracellular Na^{+} for intracellular H^{+} through the plasma membrane. This results in a rise in intracellular pH, which precedes an increase in oxidative metabolism.

Decondensation of the Sperm Nucleus

In the mature spermatozoon, the nuclear chromatin is very tightly packed, due in large part to the −SS− (disulfide) cross-linking that occurs among the protamine molecules complexed with the DNA during spermatogenesis. Shortly after the head of the sperm enters the cytoplasm of the egg, the permeability of its nuclear membrane begins to increase, allowing cytoplasmic factors within the egg to affect the nuclear contents of the sperm. Protamines are rapidly lost from the chromatin of the sperm, and the chromatin begins to spread out within the nucleus (now called a **pronucleus**) as it moves closer to the nuclear material of the egg. After a short period during which the male chromosomes are naked, histones begin to associate with the chromosomes.

Completion of Meiosis and the Development of Pronuclei in the Egg

After penetration of the egg by the sperm, the nucleus of the egg, which had been arrested in metaphase of the second meiotic division, completes the last division, releasing a **second polar body** into the perivitelline space (Fig. 2-6). A pronuclear membrane, probably derived from the endoplasmic reticulum of the egg, forms around the female chromosomal material. Cytoplasmic factors appear to control the growth of both the female and male pronuclei. DNA replication occurs in the developing pronuclei (each becomes 1n [haploid], 2c [two chromatids]) as they approach one another. When the male and female pronuclei come into contact, their membranes break down and the chromosomes intermingle. The maternal and paternal chromosomes quickly become organized around a mitotic spindle in preparation for an ordinary mitotic division. At this point, the process of fertilization can be said to be complete and the fertilized egg is called a **zygote.**

What Is Accomplished by Fertilization?

The process of fertilization ties together a number of biological loose ends. First, it stimulates the egg to complete the second meiotic division. Second, it restores the normal diploid number of chromosomes (46 in the normal human) to the zygote. Third, the sex of the future embryo is determined by the chromosomal complement of the spermatozoon. (If the sperm contains 22 autosomes and 2 X chromosomes, the embryo will be a genetic female, and if it contains 22 autosomes and an X and a Y chromosome, the embryo will be a male. See Chapter 16 for further details.) Fourth, through the mingling of maternal and paternal chromosomes, the zygote is a genetically unique product of chromosomal reassortment, which is important for the viability of any species. Finally, the process of fertilization causes metabolic activation of the egg, which is necessary for cleavage and subsequent embryonic development to occur.

SUMMARY

1. Ovulation is stimulated by a surge of LH and FSH in the blood. Expulsion of the ovum from the graafian follicle involves local edema, ischemia, collagen breakdown, fluid pressure, and smooth muscle activity in rupturing the follicular wall.
2. The ovulated egg is swept into the uterine tube and transported through it by ciliary action and smooth muscle contractions as it awaits fertilization by a sperm cell.
3. Sperm transport in the male involves a slow exit from the seminiferous tubules, maturation in the epididymis, and rapid expulsion at ejaculation, where the spermatozoa are joined by secretions from the prostate and seminal vesicles to form semen.
4. In the female, sperm transport involves entry into the cervical canal from the vagina, passage through the cervical mucus, and transport through the uterus into the uterine tubes, where capacitation occurs. The meeting of egg and sperm typically occurs in the upper third of the uterine tube.
5. The fertilization process consists of several sequential events:
   a. Penetration of the corona radiata
   b. Attachment to and penetration of the zona pellucida
   c. Fusion of sperm and egg
   d. Prevention of polyspermy
   e. Metabolic activation of the egg
   f. Decondensation of the sperm nucleus
   g. Completion of meiosis in the egg
   h. The development and fusion of male and female pronuclei
6. Attachment of the sperm cell to the zona pellucida is mediated by the ZP3 protein, which also stimulates the acrosomal reaction of the spermatozoon.
7. The acrosomal reaction involves fusion of the outer acrosomal membrane with the plasma membrane of the sperm call and the fragmentation of the fused membranes, leading to the release of the acrosomal

enzymes. One of the acrosomal enzymes, acrosin, is a serine proteinase, which digests components of the zona pellucida and assists the penetration of the swimming sperm through the zona.

8. After fusion of the sperm to the egg membrane, a rapid electrical depolarization produces the first block to polyspermy in the egg. This is almost immediately followed by a wave of calcium ions that causes the cortical granules to release their contents into the perivitelline space and ultimately inactivate the sperm receptors in the zona pellucida.
9. After sperm penetration, a rapid intensification of respiration and metabolism of the egg occurs.
10. After entry of the sperm into the egg, the nuclear material of the sperm decondenses and forms the male pronucleus. At the same time, the egg completes the second meiotic division and the remaining nuclear material becomes surrounded by a membrane, thereby forming the female pronucleus.
11. When the male and female pronuclei join and their chromosomes become organized for a mitotic division, fertilization is complete, and the fertilized egg is properly called a *zygote*.

REVIEW QUESTIONS

1. What is the principal hormonal stimulus for ovulation?
2. What is capacitation?
3. Where does fertilization take place?
4. Name two functions of the ZP3 protein of the zona pellucida.
5. What is polyspermy, and how is it prevented after a sperm enters the egg?

REFERENCES

Austin CR, Short RV, eds: *Reproduction in mammals, book 1,* ed 2, Cambridge, England, 1982, Cambridge University.

Blandau RJ, Hayashi R: Ovulation and egg transport in mammals, film, Seattle, University of Washington Press.

Chang MC: Experimental studies of mammalian fertilization, *Zool Sci* 1:349-364, 1984.

Dean J: The zona pellucida genes encode essential proteins for mammalian fertilization and early embryogenesis, *Proc Soc Exp Biol Med* 196:141-146, 1991.

Dietl J, ed: *The mammalian egg coat: structure and function,* Berlin, 1989, Springer-Verlag.

Dietl JA, Rauth G: Molecular aspects of mammalian fertilization, *Hum Reprod* 4:869-875, 1989.

Eisenbach M, Ralt D: Precontact mammalian sperm-egg communication and role in fertilization, *Am J Physiol* 262(3):C1095-C1101, 1992.

Familiari G, Makabe S, Motta PM: *The ovary and ovulation: a three-dimensional study*. In Van Blerkom J, Motta PM, eds: *Ultrastructure of human gametogenesis and early embryogenesis,* Boston, 1989, Kluwer Academic, pp 85-124.

Garbers DL: Molecular basis of fertilization, *Annu Rev Biochem* 58:719-742, 1989.

Gaunt SJ: Spreading of the sperm surface antigen within the plasma membrane of the egg after fertilization in the rat, *J Embryol Exp Morph* 75:257-270, 1983.

Gwatkin RBL: *Fertilization mechanisms in man and mammals,* New York, 1977, Plenum.

Harper MJK: *Gamete and zygote transport*. In Knobil E, Neill JD, eds: *The physiology of reproduction,* vol 1, New York, 1988, Raven, pp 103-134.

Hartmann JF, ed: *Mechanism and control of animal fertilization,* New York, 1983, Academic.

Jones GS: Corpus luteum: composition and function, *Fertil Steril* 54:21-26, 1990.

Jones R: Identification and functions of mammalian sperm-egg recognition molecules during fertilization, *J Reprod Fertil* 42(suppl):89-105, 1990.

Klemm U, Muller-Esterl W, Engel W: Acrosin, the peculiar sperm-specific serine protease, *Hum Genet* 87:635-641, 1991.

Kopf GS: Zona pellucida-mediated signal transduction in mammalian spermatozoa, *J Reprod Fertil* 42(suppl):33-49, 1990.

Lipner H: *Mechanism of mammalian ovulation*. In Knobil E, Neill J, eds: *The physiology of reproduction,* vol 1, New York, 1988, Raven, pp 447-488.

Metz CB, Monroy A, eds: *Biology of fertilization,* vols 1-3, New York, 1985, Academic.

Myles DG, Koppel DE, Primakoff P: *Defining sperm surface domains*. In Alexander NJ and others, eds: *Gamete interaction: prospects for immunocontraception,* New York, 1990, Wiley-Liss, pp 1-11.

Niswender GD, Nett TM: *The corpus luteum and its control*. In Knobil E, Neill JD, eds: *The physiology of reproduction,* vol 1, New York, 1988, Raven, pp 489-526.

O'Rand MG and others: *Receptors for zona pellucida on human spermatozoa*. In Alexander NJ and others, eds: *Gamete interactions: prospects for immunocontraception,* New York, 1990, Wiley-Liss, pp 213-224.

Phillips DM: *Structure and function of the zona pellucida*. In Familiari G, Makabe S, Motta PM, eds: *Ultrastructure of the ovary,* Boston, 1991, Kluwer Academic, pp 63-72.

Saling PM: How the egg regulates sperm function during gamete interaction: facts and fantasies, *Biol Reprod* 44:246-251, 1991.

Topfer-Peterson E and others: *Sperm acrosin and binding to the zona pellucida*. In *Gamete interaction: prospects for immunocontraception,* New York, 1990, Wiley-Liss, pp 197-212.

Wassarman PM: The biology and chemistry of fertilization, *Science* 235:553-560, 1987.

Wassarman PM: Profile of a mammalian sperm receptor, *Development* 108:1-17, 1990.

Yanagimachi R: *Mammalian fertilization*. In Knobil E, Neill J, eds: *The physiology of reproduction,* vol 1, New York, 1988, Raven, pp 135-185.

Zaneveld LJD, DeJonge CJ: *Mammalian sperm acrosomal enzymes and the acrosome reaction*. In Dunbar BS, O'Rand MB, eds: *A comparative overview of mammalian fertilization,* New York, 1991, Plenum, pp 63-79.

CHAPTER 3

Cleavage and Implantation

The act of fertilization releases the ovulated egg from a depressed metabolism and prevents its ultimate disintegration within the female reproductive tract. Immediately after fertilization, the zygote undergoes a pronounced shift in metabolism and begins several days of **cleavage.** During this time the embryo, still encased in its zona pellucida, is transported down the uterine tube and into the uterus. Roughly 6 days later, the embryo sheds its zona pellucida and attaches to the uterine lining.

With intrauterine development and a placental connection between the embryo and mother, higher mammals, including humans, have evolved greatly different modes of early development from those found in most invertebrates and lower vertebrates. The eggs of lower animals, which are typically laid outside the body, must contain all the materials required for the embryo to attain the stage of independent feeding. Two main strategies have evolved. One is to complete early development as rapidly as possible, a strategy that has been adopted by *Drosophila,* the sea urchin, and many amphibians. This involves storing a moderate amount of yolk in the oocyte and also preproducing much of the molecular machinery necessary for the embryo to move rapidly through cleavage to the start of gastrulation. The oocytes of such species typically produce and store huge amounts of ribosomes and mRNA and tRNA. These represent maternal gene products, and it means that early development in these species is predominantly controlled by the maternal genome. The other strategy of independent development, adopted by birds and reptiles, consists of producing a very large egg containing enough yolk that early development can proceed at a slower pace. This eliminates the need for the oocyte to synthesize and store large amounts of RNAs and ribosomes before fertilization.

Mammalian embryogenesis employs some fundamentally different strategies from those used by the lower vertebrates. Because the placental connection to the mother obviates the need for the developing oocyte to store large amounts of yolk, the eggs of mammals are very small. However, because mammalian development is internal and the embryo receives nourishment from the mother, cleavage is a prolonged process that typically coincides with the time required to transport the early embryo from the site of fertilization in the uterine tube to the place of implantation in the uterus. One of the prominent innovations in early mammalian embryogenesis is the formation of the **trophoblast,** the specialized tissue that forms the trophic interface between the embryo and the mother, during the cleavage period. The placenta represents the ultimate manifestation of the trophoblastic tissues.

CLEAVAGE

Morphology

In comparison with most other species, mammalian cleavage is a leisurely process measured in days rather than hours. Development proceeds at the rate of roughly one cleavage division per day for the first 3 or 4 days (Fig. 3-1). After the 2-cell stage, mammalian cleavage is asynchronous, with one of the two cells **(blastomeres)** dividing to form a 3-cell embryo. When the embryo consists of approximately 16 cells, it is sometimes called a **morula** (derived from Greek and Latin words meaning "mulberry").

After several cleavage divisions, the embryos of placental mammals enter into a phase called **compaction,** during which the individual outer blastomeres tightly adhere through gap and tight junctions. These blastomeres tend to lose their individual identity when viewed from the surface (Fig. 3-2). With these changes the outer blastomeres develop polarity, with apical surfaces facing the outside of the embryo and basal surfaces facing the inside.

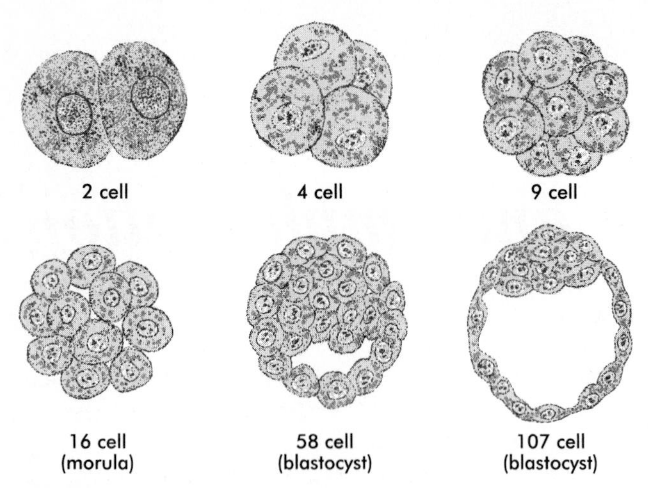

FIG. 3-1 Drawings of early cleavage stages in human embryos. The drawings of the 58- and 107-cell stages represent sections made through the embryos.

About 4 days after fertilization, a fluid-filled space begins to form inside the embryo. The space is known as the **blastocoele,** and the embryo as a whole is called a **blastocyst** (see Fig. 3-1). The transition from morula to blastocyst and the formation of a fluid-filled blastocoele depend first on the maintenance of intercommunications between superficial blastomeres via gap junctions. In the absence of gap junctions, embryos fail to **cavitate** (form a blastocoele). Cavitation involves the buildup of fluid within the blastocoele. Fluid accumulation is a function of a sodium transport system based on Na^+, K^+-ATPase that develops in the outer blastomeres. The net effect of this enzyme is the movement of Na^+ and H_2O across the blastomeres and the buildup of fluid in the spaces forming among the inner blastomeres.

At the blastocyst stage, the embryo consists of two types of cells: an outer superficial layer (the trophoblast) that surrounds a small inner group of cells called the **inner cell**

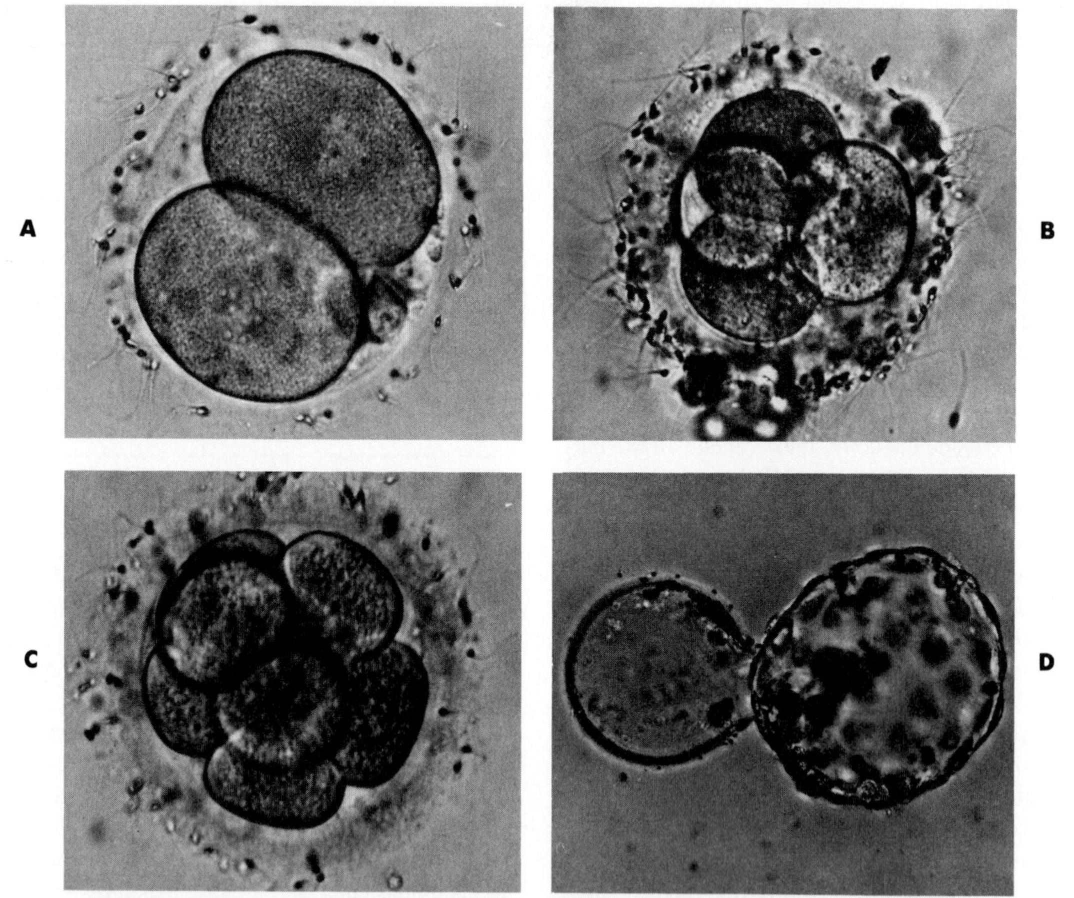

FIG. 3-2 Photomicrographs of cleavage stages of human eggs fertilized in vitro. **A,** Two blastomeres 39 hours after fertilization. A polar body is seen to the right of the boundary between the blastomeres. **B,** Four blastomeres 42 hours after fertilization. **C,** Eight blastomeres 49 hours after fertilization. **D,** Hatching blastocyst 123 hours after fertilization. The empty zona pellucida is on the left. In **A** to **C,** numerous spermatozoa can be seen clinging to the zona pellucida.

(From Veeck LL: *Atlas of the human oocyte and early conceptus,* vol 2, Baltimore, 1991, Williams & Wilkins.)

mass. The appearance of these two cell types reflects major organizational changes that have occurred within the embryo and represents the specialization of the blastomeres into two distinct cell lineages. Cells of the inner cell mass give rise to the body of the embryo itself plus a number of extraembryonic structures, whereas cells of the trophoblast form only extraembryonic structures, including the outer layers of the placenta.

Control of the Cell Cycle during Cleavage

The hallmark of the cleavage period is the successive waves of mitosis that sweep through the embryo. Each mitotic division is under the control of proteins so basic that their structures and functions have been preserved for almost a billion years. Such an interpretation is based on their presence in organisms as diverse as humans and yeasts.

Early research on frog eggs suggested the presence of a **maturation-promoting factor (MPF)** that induced meiosis in early oocytes. MPF was found to be a regulator of both meiosis and mitosis. Further research showed that active MPF is a complex of two proteins called **cdc2** (cell division cycle) and **cyclin,** which guide a cell through its mitotic cycle.

The **mitotic cycle** in any cell is divided into four phases (Fig. 3-3). The **interphase** (often called the **G_1 phase**) is the period during which the cell typically carries out its assigned functions. As it prepares for mitosis, it moves into the **S phase,** during which its nuclear DNA is replicated. DNA synthesis is followed by a usually brief **G_2** (gap 2) **phase,** which precedes actual mitosis **(M phase).**

The cdc2 protein is present throughout the mitotic cycle. Cyclin is synthesized and accumulates in the cell during interphase, but it combines with cdc2 to form a prematuration-promoting factor (pre-MPF) before mitosis (see Fig. 3-3). Enzymatic modification converts this complex to an active form of MPF that initiates mitosis. Among the specific actions of MPF are the initiation of the breakdown of the nuclear envelope and stimulation of assembly of the mitotic spindle. Many of the actions of MPF involve the phosphorylation of proteins. For example, breakdown of the nuclear envelope is due to the phosphorylation of **nuclear lamins,** or envelope proteins. Phosphorylation causes the lamins to dissociate, leading to disintegration of the nuclear envelope. Active MPF also activates enzymes that abruptly break down cyclin.

When cyclin levels fall below a certain threshold, the cdc protein of MPF loses it activity, thus ending mitosis. The loss of MPF activity allows cellular phosphatase enzymes to remove phosphate groups that were added to proteins under the influence of MPF. One effect of this is reformation of the nuclear membrane as the nuclear lamins become dephosphorylated. The phosphatases also inactivate the enzymes that break down cyclin, allowing the cyclin to again accumulate in the cell during interphase. This sets the stage for a repetition of the mitotic cycle.

Molecular Biology and Genetics

Most studies on the molecular biology and genetics of early mammalian development have been performed on the mouse. Until more information on early primate embryogenesis becomes available, results obtained from experimentation on mice must be used as a guide.

The consequence of the lack of advance storage of ribosomes and RNAs during mammalian oogenesis is that

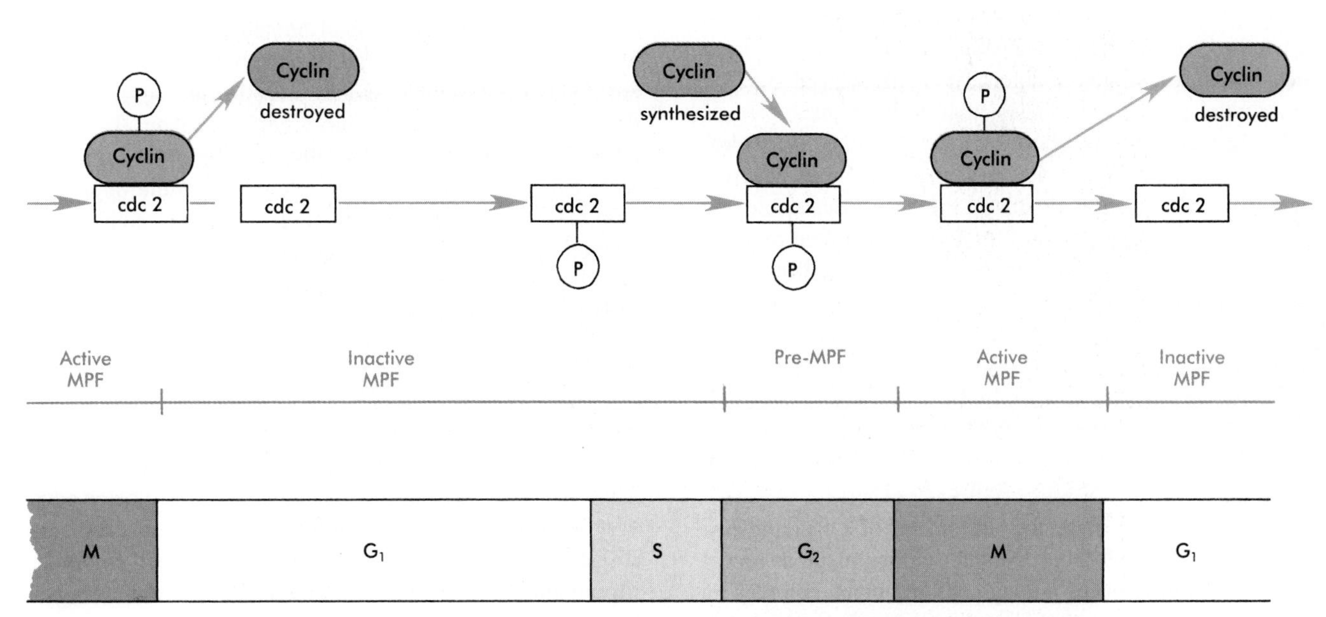

FIG. 3-3 The cell cycle and its control. *MPF,* Maturation-promoting factor. G_1, S, G_2, and M are stages of the cell cycle (see text).

the zygote must rely on embryonic gene products very early during cleavage, typically by the 2- or 4-cell stage. There does not appear, however, to be a sharp transition between the cessation of reliance on purely maternal gene products and the initiation of transcription from the embryonic genome. For example, paternal gene products (e.g., isoforms of β-glucuronidase and β_2-microglobulin) appear in the embryo very early, whereas maternal actin and histone mRNAs are still being used for the production of corresponding proteins. As an indication of the extent to which the early embryo relies on its own gene products, development past the 2-cell stage does not take place in the mouse if mRNA transcription is inhibited. In contrast, similar treatment of amphibian embryos does not disrupt development until late cleavage, at which time the embryos begin to synthesize the mRNAs that are required to control morphogenetic movements and gastrulation.

A very important gene in early development is **oct-3,** which is detected as a DNA-binding protein. Oct-3 protein is a specific transcription factor that binds the octamer ATTTGCAT on DNA. There is a close relationship between the expression of the *oct-3* gene and the highly undifferentiated state of cells. In the mouse, maternally derived oct-3 protein is found in developing oocytes and is active in the zygote. After the experimentally induced loss of oct-3 protein, development is arrested at the 1-cell stage. Such studies indicate that maternally derived oct-3 protein is required to permit development to proceed to the 2-cell stage, when the transcription of embryonic genes begins.

Oct-3 gene is expressed in all blastomeres up to the morula stage. As various differentiated cell types begin to emerge in the embryo, the level of *oct-3* gene expression decreases until it is no longer detectable. Such a decrease is first noted in cells that become committed to forming extraembryonic structures and finally in cells of the specific germ layers as they emerge from the primitive streak (see Chapter 4). Even after virtually all cells of the embryo have ceased to express the *oct-3* gene, it is still detectable in the primordial germ cells as they migrate from the region of the allantois to the genital ridges. In the adult, oct-3 protein is detectable in oocytes, especially maturing oocytes, and in some unidentified cells in the testis. Because of its patterns of distribution, oct-3 protein is suspected of playing a regulatory role in early determination or differentiation decisions by cells.

Along with a small amount of preformed maternal mRNAs in mammalian embryos is a correspondingly low capacity for translation of mRNAs. A variety of injection experiments suggest that the factor limiting translational efficiency may be the small number of ribosomes stored in the egg.

Even at an early stage the blastomeres of a cleaving embryo are not homogeneous. Simple staining methods reveal pronounced differences among cells in human embryos as early as the 7-cell stage (Fig. 3-4). Autoradiographic studies have shown that all blastomeres of 4-cell human embryos have low levels of extranucleolar and no nucleolar RNA synthesis. By the 8-cell stage, some blastomeres have very high levels of RNA synthesis, but other blastomeres still show the pattern seen in blastomeres of 4-cell embryos. Morphological studies show corresponding differences between transcriptionally active and inactive blastomeres.

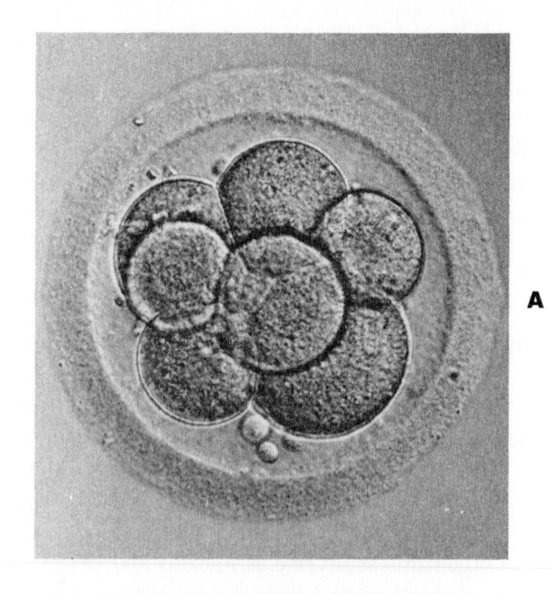
A

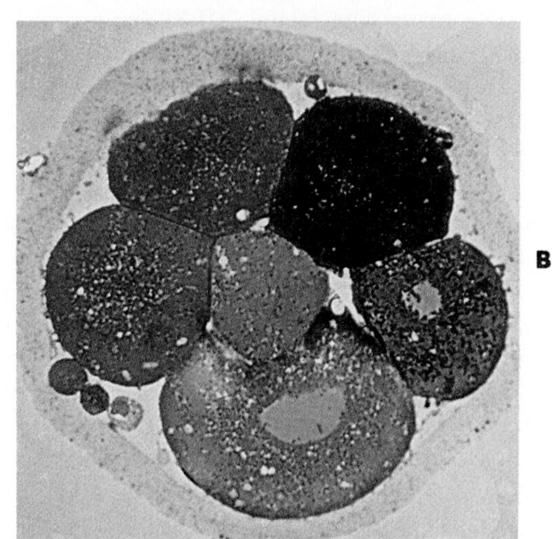
B

FIG. 3-4 A 7-cell human embryo recovered from the uterine tube. Three small polar bodies are seen in the lower part of each photograph. The zona pellucida surrounds the blastomeres and polar bodies. **A,** Photograph of the intact embryo before fixation. **B,** Photomicrograph of a section through the embryo stained with toluidine blue. Two nuclei are visible in the upper right blastomeres. Two blastomeres stain differently (metachromatically) from the others.

(From Avendaño S and others: *Fertil Steril* 26:1167-1172, 1975. Photographs courtesy H.B. Croxatto, Santiago, Chile.)

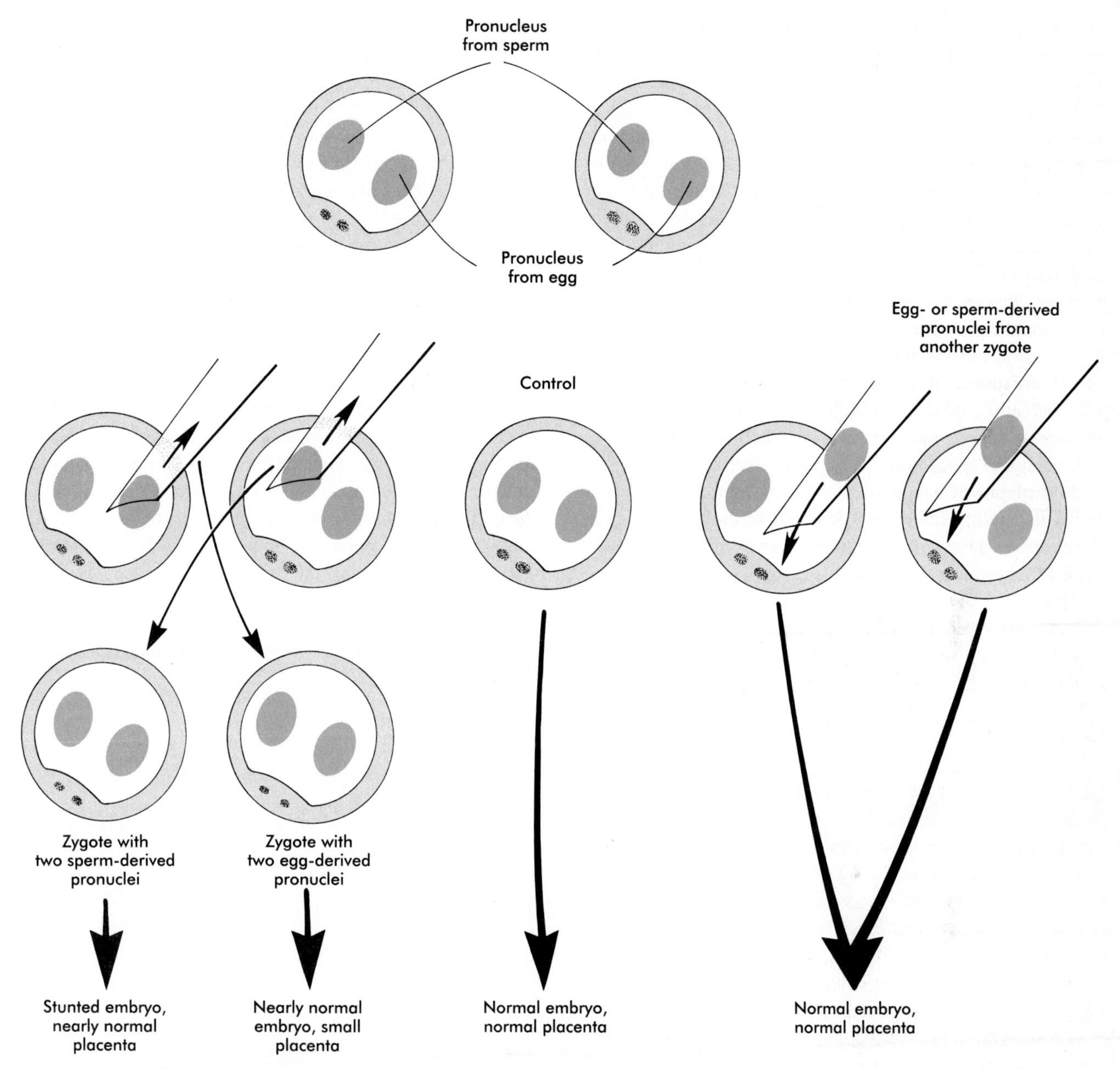

FIG. 3-5 Parental imprinting by the use of pronuclear transplants.

Once cleavage begins, transcription products from both maternally and paternally derived chromosomes are active in guiding development. Haploid embryos commonly die during cleavage or just after implantation. However, there is increasing evidence that the control of early development involves more than simply having a diploid set of chromosomes in each cell.

Experimentation, coupled with observations on some unusual developmental disturbances in mice and humans, has suggested that chromosomes derived from the egg possess different qualities from those derived from the sperm. Called **parental imprinting,** the effects are manifest in different ways. It is possible to remove a pronucleus from a newly inseminated mouse egg and replace it with a pronucleus taken from another inseminated egg at a similar stage of development (Fig. 3-5). If a male or female pronucleus is removed and replaced with a corresponding male or female pronucleus, development is normal. However, if a male pronucleus is removed and replaced with a female pronucleus (resulting in a zygote with two female pronuclei), the embryo itself develops fairly normally, but the placenta and yolk sac are poorly developed. Conversely, a zygote with two male pronuclei produces a severely stunted embryo, whereas the placenta and yolk sac are nearly normal.

Such experiments suggest that the different environments to which chromosomes are exposed in the parental gonads imprint information that affects certain aspects of development in the embryo. Paternal imprinting seems to selectively turn off certain genes involved in development of the embryo itself, whereas maternal imprinting seems to turn off some genes involved in the formation of extraembryonic structures such as the placenta. In the human, some chromosomes appear to be affected by maternal imprinting, others by paternal imprinting, and still others do not seem to be imprinted by either parent. A striking example of paternal imprinting in the human is a **hydatidiform mole** (see Fig. 6-17), which is characterized by the overdevelopment of trophoblastic tissues and the extreme underdevelopment of the embryo (see Chapter 6). This condition results from the fertilization of an egg by two spermatozoa and the consequent failure of the maternal genome of the egg to participate in development or from the duplication of a sperm pronucleus in an "empty" egg. This form of highly abnormal development is consistent with the hypothesis that paternal imprinting favors development of the trophoblast at the expense of the embryo. Parental genetic imprinting lasts only for the duration of the life of the individual. When that individual produces gametes, the original imprinting is erased and a new imprint, based on the sex of that individual, is imposed on the chromosomes of the gametes.

Another example of the inequality of genetic expression during early development is the pattern of X-chromosome inactivation in female embryos. It is well known from cytogenetic studies that one of the two X chromosomes in the cells of females is inactivated by extreme condensation. This is the basis for the **sex chromatin** or **Barr body** that can be demonstrated in cells of females but not normal males. The purpose of X-chromosome inactivation is dosage compensation, or preserving the cells from an excess of X-chromosomal gene products.

Studies on mice using levels of enzymes encoded in the DNA of the X chromosome have shown that both X chromosomes are transcriptionally active during early cleavage of female embryos. After differentiation of the blastomeres into trophoblast or inner cell mass, the pattern changes (Fig. 3-6). Both X chromosomes continue to be active in cells of the inner cell mass, whereas in all cells of the trophoblast **(trophectoderm),** the paternally derived X chromo-

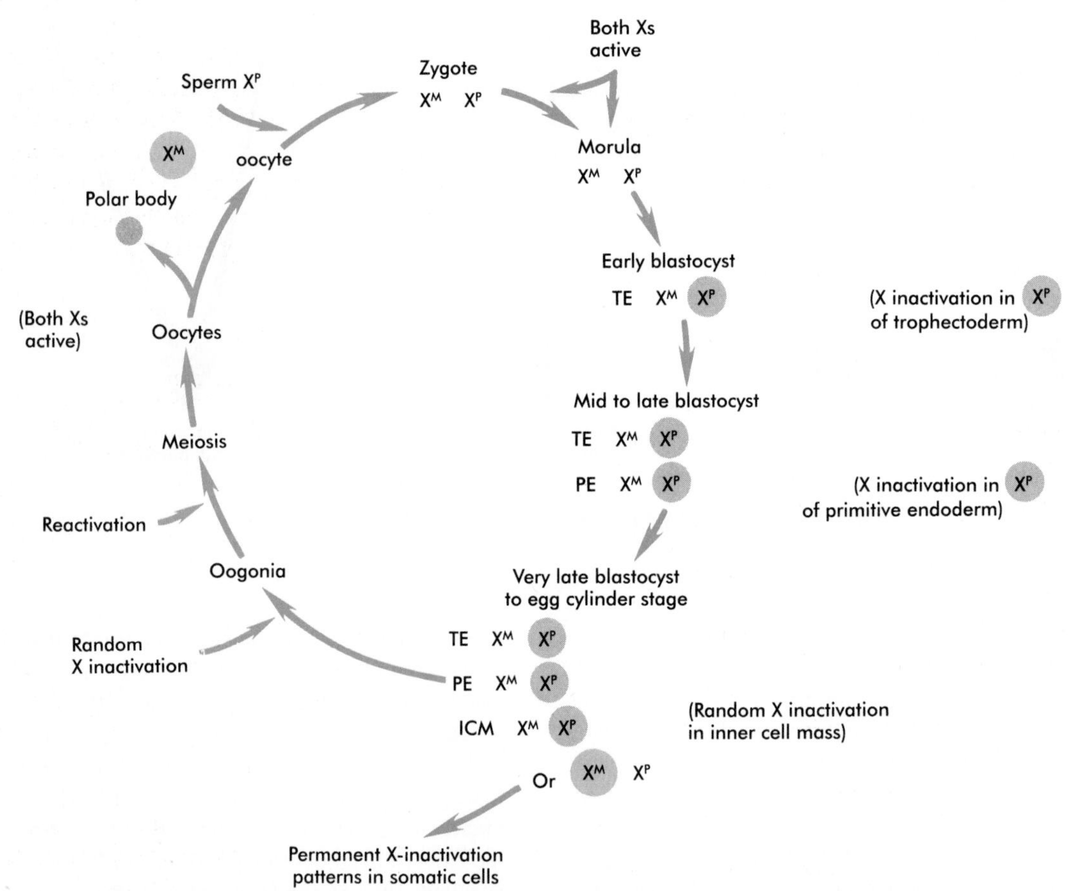

FIG. 3-6 X-chromosomal inactivation and reactivation during the mammalian life cycle. *ICM,* Inner cell mass; *PE,* primitive ectoderm; *TE,* trophectoderm; X^M, maternal X chromosome; X^P, paternal X chromosome.
(Based on studies by Gartler and Riggs [1983].)

somes are selectively inactivated. As the cells of the inner cell mass become subdivided into other lineages, differential X-chromosomal inactivation also occurs. X inactivation ultimately occurs in all cells, and only during oogenesis do both X chromosomes of the oocytes become active again. The selective inactivation of the paternal X chromosome in the trophoblast is another example of parental genetic imprinting.

Certain genetic defects inherited in a mendelian fashion are known to affect cleavage and early development in the mouse. When homozygous, these types of defects are almost invariably lethal. Nonetheless, the analysis of lethal genetic conditions in a number of species has proved extremely valuable in uncovering mechanisms of normal development. Virtually nothing is known about early lethal mutants in human embryos; however, considering the high percentage of human embryos that are spontaneously aborted during the early weeks of pregnancy, it would be surprising if similar genetic defects did not exist.

Developmental Properties of Cleaving Embryos

Early mammalian embryogenesis is considered to be a highly regulative process. **Regulation** is the ability of an embryo or an organ primordium to produce a normal structure if parts have been removed or added.* At the cellular level, it means that the fates of cells in a regulative system are not irretrievably fixed and that the cells can still respond to environmental cues. Because the assignment of blastomeres into different cell lineages is one of the principal features of mammalian development, identifying the environmental factors that are involved is important.

Of the experimental techniques used to demonstrate regulative properties of early embryos, the simplest is to separate the blastomeres of early cleavage–stage embryos and determine whether each one can give rise to an entire embryo. This method has been used to demonstrate that single blastomeres from 2- and sometimes 4-celled embryos can form normal embryos. In mammalian studies, a single cell is more commonly taken from an early cleavage–stage embryo and injected into the blastocoele of a genetically different host. Such injected cells become incorporated into the host embryo, forming cellular **chimeras** or **mosaics.** When genetically different donor blastomeres are injected into host embryos, the ultimate fate of the donor blastomeres can be determined by histochemical or cytogenetic analysis. Such experiments, which involve different isozymes of the enzyme **glucose phosphate isomerase,** have shown that all blastomeres of an 8-cell mouse embryo remain **totipotent;** that is, they retain the ability to form any cell type in the body. Even at the 16-cell stage of cleavage, some blastomeres are capable of producing progeny that are found in both the inner cell mass and the trophoblastic lineage.

Another means of demonstrating the regulative properties of early mammalian embryos is to dissociate mouse embryos into separate blastomeres and then to combine the blastomeres of two or three embryos (Fig. 3-7). The combined blastomeres soon aggregate and reorganize to become a single large embryo, which then goes on to become a normal-appearing **tetraparental** or **hexaparental mouse.** By various techniques of making chimeric embryos, it is even possible to combine blastomeres to produce interspecies chimeras (e.g., a sheep-goat).

One of the most important steps in early mammalian development is the decision that results in the appearance of two separate lines of cells—the trophoblast and the inner cell mass—from the early blastomeres. Up to the 8-cell stage, all blastomeres are virtually identical in their developmental potential. Shortly thereafter, however, differences are noted between cells that have at least one surface situated on the outer border of the embryo and those that are completely enclosed by other blastomeres.

The relationship between the position of the blastomeres and their ultimate developmental fate was incorporated into the **inside-outside hypothesis.** The outer blastomeres ultimately differentiate into the trophoblast, whereas the inner blastomeres form the inner cell mass, from which the body of the embryo arises. Although this hypothesis has been supported by a variety of experiments, the mechanisms by

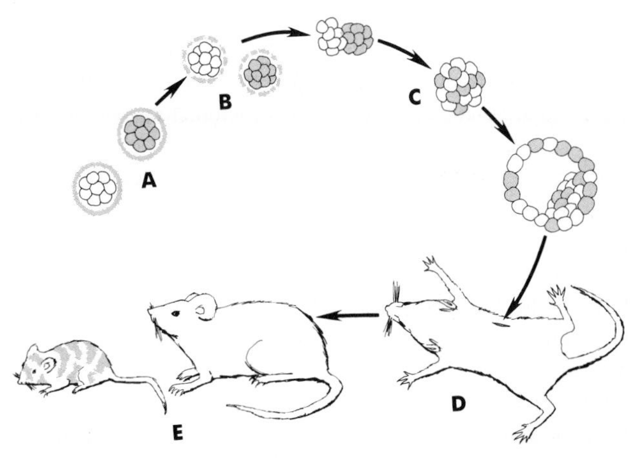

FIG. 3-7 The procedure for producing tetraparental embryos. **A,** Cleavage stages of two different strains of mice. **B,** Removal of the zona pellucida. **C,** Fusion of the two embryos. **D,** Implantation of embryos into a foster mother. **E,** Chimeric offspring obtained from the implanted embryos.

(Based on studies by Mintz [1962].)

*Opposed to regulative development is **mosaic development,** which is characterized by the inability to compensate for defects or to integrate extra cells into a unified whole. In a mosaic system, the fates of cells are rigidly determined, and removal of cells results in an embryo or a structure that is missing the components that the removed cells were destined to form. Most regulative systems have an increasing tendency to exhibit mosaic properties as development progresses.

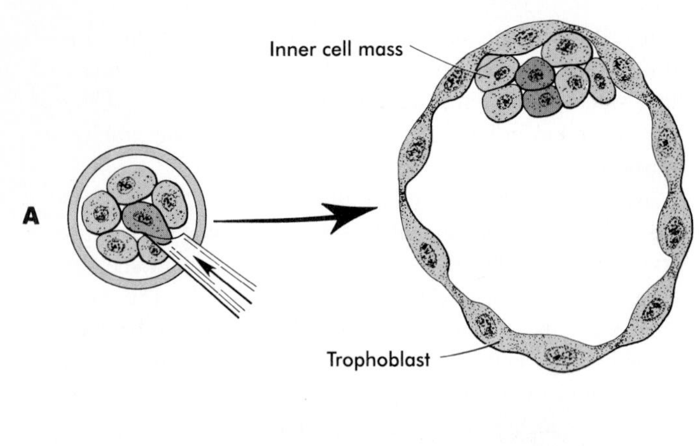

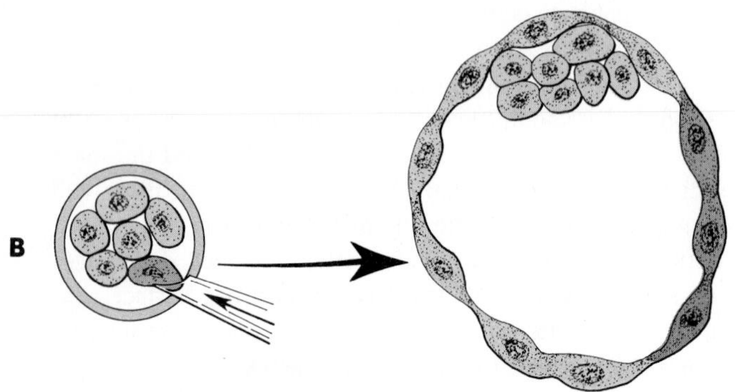

FIG. 3-8 Experiments illustrating the inside-outside hypothesis of cell determination in early mammalian embryos. **A,** If a marked blastomere is inserted into the interior of a morula, it and its progeny become part of the inner cell mass. **B,** If a marked blastomere is placed on the outside of a host morula, it and its descendants contribute to the trophoblast.

which the blastomeres recognize their positions and then differentiate accordingly have remained elusive and are still little understood. If marked blastomeres from disaggregated embryos are placed on the outside of another early embryo, they typically contribute to the formation of the trophoblast. Conversely, if the same marked cells are introduced into the interior of the host embryo, they participate in formation of the inner cell mass (Fig. 3-8). Outer cells in the early mammalian embryo are linked by tight and gap junctions, but whether this morphological characteristic is a cause or an effect of their differentiation into trophoblast is not known.

Another characteristic of cleavage-stage mammalian embryos is the absence of rigidly fixed body axes. In many lower vertebrates the dorsoventral axis is fixed in the egg before fertilization, and the anteroposterior axis is determined at fertilization by the site of penetration of the egg by the sperm. Avian and mammalian embryos, on the other hand, do not show evidence of axial fixation until considerably later during cleavage. In the 4- to 5-day-old human blastocyst with a well-defined inner cell mass, the dorsal surface is the part of the inner cell mass that abuts the outer trophoblast, and the ventral surface is the part that faces the blastocyst cavity. Not until almost 2 weeks postfertilization can the longitudinal axis be identified. Recognizing the cranial from the caudal end of the embryo (longitudinal axis) is only possible after the appearance of the primitive streak early in the third week.

Experimental Manipulations of Cleaving Embryos

Much of the knowledge about the developmental properties of early mammalian embryos is the result of recently devised techniques for experimentally manipulating them. Typically, the use of these techniques must be combined with other techniques that have been designed for in vitro fertilization, embryo culture, and embryo transfer. (The two last techniques, as applied to humans, are described in Chapter 7.)

Classic strategies for investigating developmental properties of embryos are (1) removing a part and determining the way the remainder of the embryo compensates for the loss (such experiments are called **deletion experiments**) and (2) adding a part and determining the way the embryo integrates the added material into its overall body plan (such experiments are called **addition experiments**). Although some deletion experiments have been done, the strategy of addition experiments has proved to be most fruitful in elucidating mechanisms controlling mammalian embryogenesis.

Blastomere removal and addition experiments (Fig. 3-9) have convincingly demonstrated the regulative nature (i.e., the strong tendency for the system to be restored to wholeness) of early mammalian embryos. Such knowledge is important in understanding the reason exposure of early human embryos to unfavorable environmental influences typically results in either death or a normal embryo.

One of the most powerful experimental techniques of the last two decades has been the injection of genetically or artificially labelled cells into the blastocyst cavity of a host embryo (see Fig. 3-9). This technique has been used to show that the added cells become normally integrated into the body of the host embryo, additional evidence of embryonic regulation. An equally powerful use of this technique has been in the study of cell lineages in the early embryo. By identifying the progeny of the injected marked cells, investigators have been able to determine the **potency** (the range of cell and tissue types that an embryonic cell or group of cells is capable of producing) of the donor cells.

A technique that is providing new insight into genetic control mechanisms of mammalian development is the production of **transgenic embryos.** Transgenic embryos (commonly mice) are produced by directly injecting foreign DNA into the pronuclei of zygotes. The DNA, typically recombinant DNA for a specific gene, can be fused with a different regulatory element that can be controlled by the

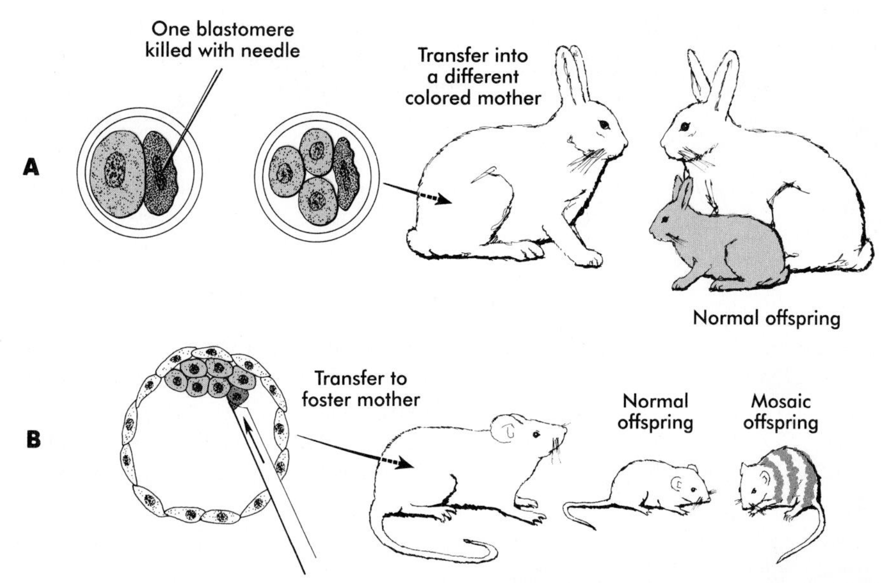

FIG. 3-9 Blastomere addition and deletion experiments. **A,** If one blastomere is killed with a needle and the embryo is transferred into a different colored mother, a normal offspring of the color of the experimentally damaged embryo is produced. **B,** If a blastomere of a different strain is introduced into a blastocyst, a mosaic offspring with color markings characteristic of the strain of the introduced blastomere is produced.

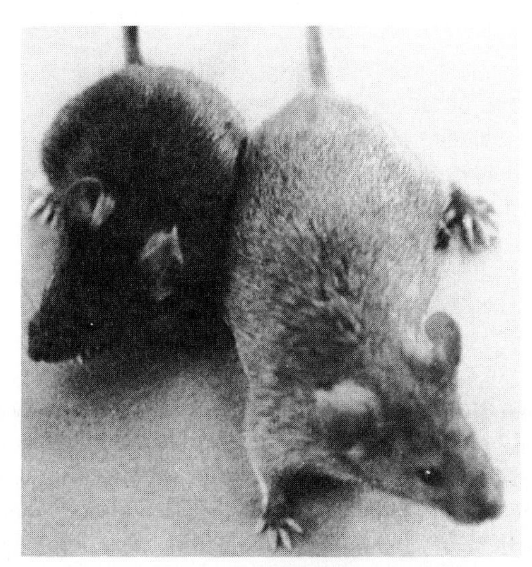

FIG. 3-10 Photograph of two 10-week-old mice. The one on the left (normal mouse) weighs 21.2 grams. The one on the right (a transgenic littermate of the normal mouse) carries a rat gene coding for growth hormone. It weighs 41.2 grams.

(From Palmiter RD and others: *Nature* 300:611-615, 1982.)

investigator. For example, transgenic mice have been created by injecting the rat growth–hormone gene coupled with a metallothionein promoter region *(MT-I)* into the pronuclei of mouse zygotes. The injected zygotes are transplanted into the uteri of foster mothers, which give birth to normal-looking transgenic mice. Later in life, when these transgenic mice are fed a diet rich in zinc, which stimulates the MT-I promoter region, the rat growth–hormone gene is activated, causing the liver to produce large amounts of the polypeptide hormone. The function of the transplanted gene is obvious; under the influence of the rat growth hormone that they were producing, the transgenic mice grew to a much larger size than their normal littermates (Fig. 3-10). The technique of producing transgenic embryos is being increasingly used both to examine factors regulating the expression of specific genes in embryos and to disrupt genes in the host embryos. In addition, the efficacy of this technique to correct known genetic defects is being increasingly explored in mice.

Twinning

Some types of twinning represent a natural experiment that demonstrates the highly regulative nature of early human embryos. In the United States, roughly 1 pregnancy in 90 results in twins, and 1 per 8000 results in triplets. Of the total number of twins born, approximately two thirds are **fraternal,** or **dizygotic,** twins and one third are **identical,** or **monozygotic,** twins. Dizygotic twins are the product of the fertilization of two ovulated eggs, and the mechanism of their formation involves the endocrine control of ovulation. Monozygotic twins and some triplets, on the other hand, are the product of one fertilized egg. They arise by the subdivision and splitting of a single embryo. Although

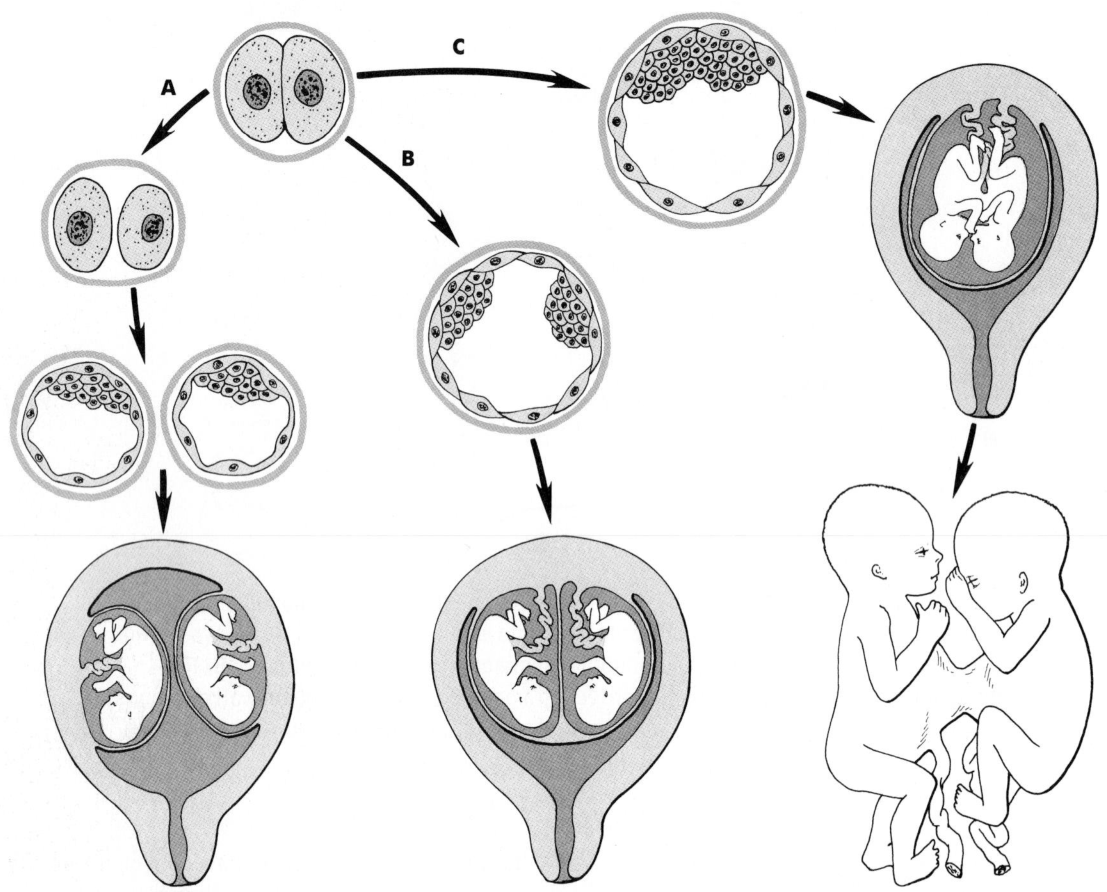

FIG. 3-11 Modes of monozygotic twinning. **A,** Cleavage of an early embryo, with each half developing as a completely separate embryo. **B,** Splitting of the inner cell mass of a blastocyst and the formation of two embryos enclosed in a common trophoblast. This is the most common mode of twinning. **C,** If the inner cell mass does not completely separate, conjoined twins may result.

monozygotic twins could theoretically arise by the splitting of a 2-cell embryo, it is commonly accepted that most arise by the subdivision of the inner cell mass in a blastocyst (Fig. 3-11). Because the vast majority of monozygotic twins are perfectly normal, the early human embryo can obviously be subdivided and each component part regulated to form a normal embryo. Inferences on the origin and relations of multiple births can be made from the arrangement of the extraembryonic membranes at the time of birth (see Chapter 6).

Quadruplets or higher orders of multiple births occur very rarely. In previous years, these could be combinations of multiple ovulations and splitting of single blastocysts. In the modern era of reproductive technology, the majority of multiple births, sometimes up to septuplets, can be related to the side effects of antiinfertility drugs taken by the mother.

The separation of portions of the inner cell mass in an embryo is sometimes incomplete, and although two embryos take shape, they are joined by a tissue bridge of varying proportions. When this occurs, the twins are called **conjoined twins** (sometimes colloquially called "Siamese twins"). The extent of bridging between the twins varies from a relatively thin connection in the chest or back to massive fusions along much of the body axis. Examples of the wide variety of types of conjoined twins are illustrated in Figs. 3-12 and 3-13. With increasing sophistication of surgical techniques, more complex degrees of fusion can be separated. A much less common variety of conjoined twin is a **parasitic twin,** in which a much smaller but often remarkably complete portion of a body protrudes from the body of an otherwise normal host twin (Fig. 3-14). Common attachment sites of parasitic twins are the oral region, the mediastinum, and the pelvis. The mechanism of conjoined twinning has not been directly demonstrated experimentally, but a likely theoretical explanation is the partial secondary fusion of originally separated portions of the inner cell mass.

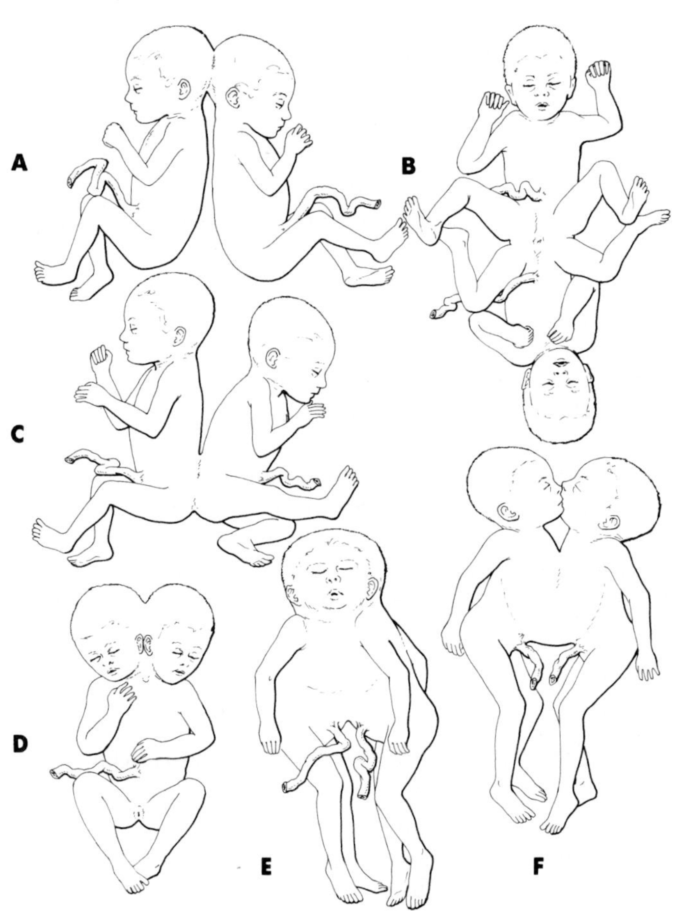

FIG. 3-12 Types of conjoined twins. **A,** Head-to-head fusion (cephalopagus). **B** and **C,** Rump-to-rump fusion (pyopagus). **D,** Massive fusion of head and trunk, resulting in a reduction in the number of appendages and a single umbilical cord. **E,** Fusion involving both head and thorax (cephalothoracopagus). **F,** Chest-to-chest fusion (thoracopagus).

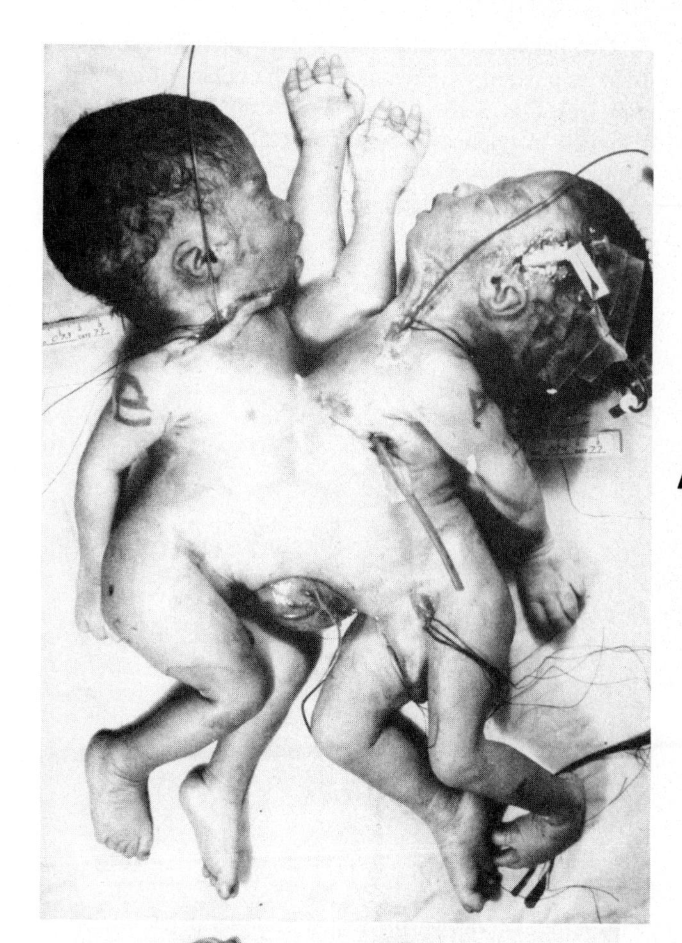

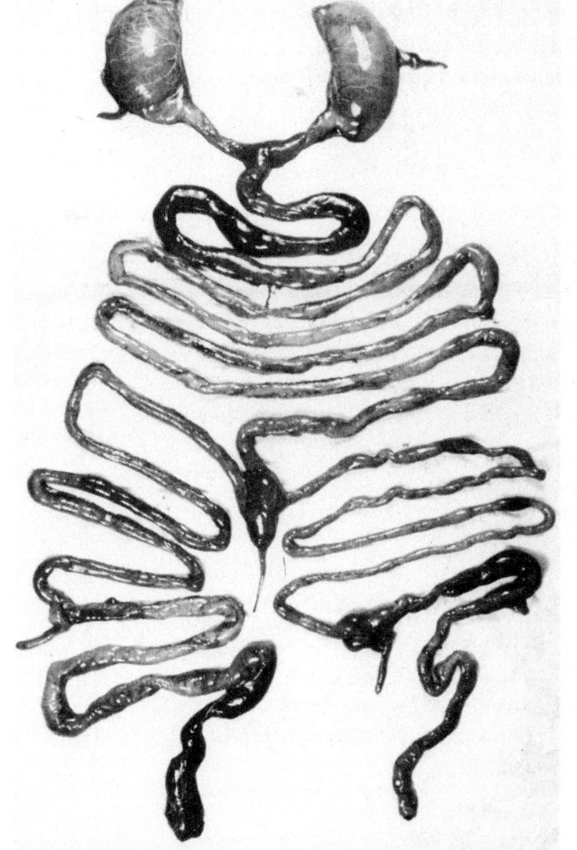

FIG. 3-13 **A,** Conjoined twins with broad truncal attachment. **B,** Dissected intestinal tracts from the same twins showing partial fusion of the small intestine and mirror image symmetry of the stomachs.
(Courtesy Mason Barr, Ann Arbor, Mich.)

One phenomenon that can be encountered in conjoined twins is a reversal of symmetry of the organs of one of the pair (see Fig. 3-13, *B*). Such reversals of symmetry are common in duplicated organs or entire embryos. Over a century ago, this phenomenon was recorded in a large variety of biological situations and was incorporated into what is now called **Bateson's rule.** Bateson's rule states that when duplicated structures are joined during critical developmental stages, one structure is the mirror image of the other. Despite the long recognition of this phenomenon, only in recent years has there been theoretical understanding of the mechanism behind the reversal of symmetry.

Reversal of symmetry, called **situs inversus,** is also seen in roughly 1 in 10,000 superficially normal individuals born from single births (Fig. 3-15). Such an individual is often not recognized until examined relatively late in life by an astute diagnostician. The cause of situs inversus in **singletons** (individuals born as single births) is not known. One possibility is a genetic mutant (populations of genetically based individuals with reversed symmetry are known in a number of invertebrate species); another is that the individual is a sole surviving member (the one with reversed symmetry) of a set of unrecognized intrauterine conjoined twins. A recessive mutation that causes situs inversus in mice has recently been discovered.

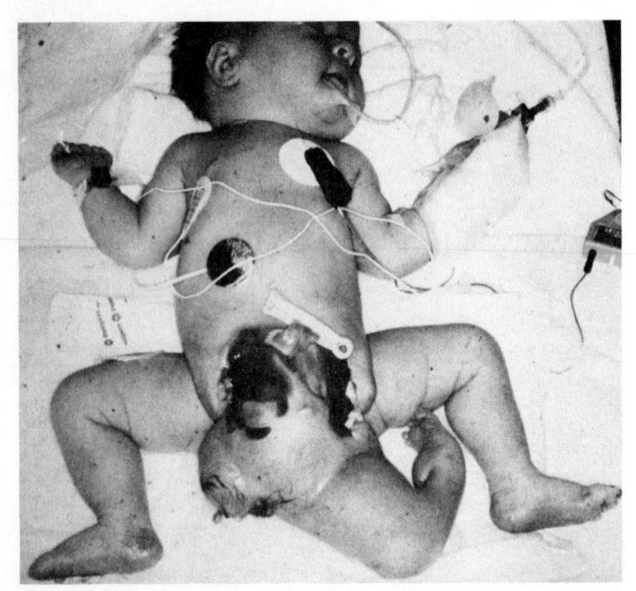

FIG. 3-14 Parasitic twin arising from the pelvic region of the host twin. One well-defined leg and some hair can be seen on the parasitic twin.
(Courtesy Mason Barr, Ann Arbor, Mich.)

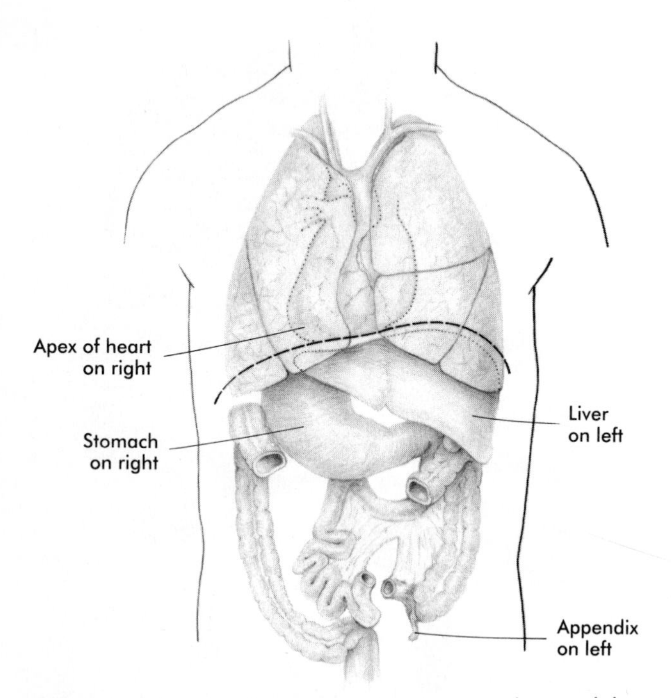

FIG. 3-15 A case of complete situs inversus in an adult.

EMBRYO TRANSPORT AND IMPLANTATION

Transport Mechanisms by the Uterine Tube

The entire period of early cleavage takes place while the embryo is being transported from the place of fertilization to its implantation site in the uterus (see Fig. 2-2). It is increasingly apparent that the early embryo and the female reproductive tract influence one another during this period of transport, but knowledge is still fragmentary.

At the beginning of cleavage, the zygote is still encased in the zona pellucida and the cells of the corona radiata. The corona radiata is lost within 2 days of the start of cleavage. The zona pellucida, however, remains intact until the embryo reaches the uterus.

For reasons that are not well understood, the embryo remains in the ampullary portion of the uterine tube for approximately 3 days. It then traverses the isthmic portion of the tube in as little as 8 hours. Some evidence indicates that under the influence of progesterone, the uterotubal junction relaxes, allowing the embryo to enter the uterine cavity. A couple of days later (7 to 8 days after fertilization), the embryo implants into the midportion of the posterior wall of the uterus.

Zona Pellucida

During the entire period from ovulation until its entry into the uterine cavity, the ovum and then the embryo are surrounded by the zona pellucida. From the time of ovulation, the zona pellucida fulfills a variety of functions. Just after ovulation, the zona and the cells of the corona radiata facilitate the transport of the ovum into the ampullary part of the uterine tube. At fertilization, the zona first promotes the acrosome reaction of the spermatozoa and then, after the cortical reaction has taken place, acts as a barrier to the entry of further sperm cells into the egg (see Chapter 2). After fertilization, the zona helps to keep the blastomeres of the cleaving embryo together. The zona pellucida, which lacks HLA (histocompatibility) antigens, also serves as an immunological barrier between the mother and the antigenically different embryo. A very important function is to prevent the premature implantation of the embryo into the wall of the uterine tube (see p. 47).

After the embryo has reached the uterine cavity, it sheds the zona pellucida in preparation for implantation. In rodents, the blastocyst "hatches" from the zona by digesting a hole through it by means of a trypsinlike enzyme that is secreted by a few trophoblastic cells. The blastocyst then extrudes itself through the hole. Very few specimens of human embryos have been taken in vivo from the period just preceding implantation, but in vitro studies on human embryos suggest a similar mechanism, which probably occurs 1 to 2 days before implantation (see Fig. 3-2, *D*).

Implantation into the Uterine Lining

Approximately 6 to 7 days after fertilization, the embryo begins to make a firm attachment to the epithelial lining of the endometrium. Soon thereafter it sinks into the endometrial stroma, and its original site of penetration into the endometrium becomes closed over by the epithelium, much like a healing skin wound.

Successful implantation (Table 3-1) requires a high degree of preparation and coordination by both the embryo and the endometrium. The complex hormonal preparations of the endometrium that began at the close of the previous menstrual period are all aimed at providing a suitable cellular and nutritional environment for the embryo. Dissolution of the zona pellucida signals the readiness of the embryo to begin implantation.

The first stage in implantation consists of apposition of the expanded blastocyst to the endometrial epithelium. According to one hypothesis, progesterone and especially estrogen produced by the blastocyst cause enough edema in the endometrium to almost obliterate the already flattened uterine cavity. This presses the blastocyst against the epithelium. Adherence between the blastocyst and endometrium may be mediated by cell surface glycoproteins, but the specific nature of the attachment is still poorly understood. The uterine epithelium must be conditioned, probably hormonally, to become receptive to attachment of the trophoblast, and the time period of receptivity is relatively short. Both in vivo and in vitro studies have shown that actual attachment of the blastocyst occurs at the area above the inner cell mass **(embryonic pole),** strongly suggesting that all surfaces of the trophoblast are not the same. Molecules as diverse as epidermal growth factor, platelet-activating factor, and fibronectin have been implicated in the attachment phase.

The next stage of implantation is penetration of the uterine epithelium.* In primates, the cellular trophoblast un-

*Several major types of attachments occur between embryo and uterus, and several implantation mechanisms exist among mammals. Because of the difficulty in extrapolating from other species, this section concentrates on human implantation, even though considerably less is known about implantation in primates than in certain other mammals.

TABLE 3-1 Stages in Human Implantation

AGE (DAYS)	DEVELOPMENTAL EVENT IN EMBRYOS	CHANGES LEADING TO NEXT STAGE
5	Maturation of blastocyst	
5	Loss of zona pellucida from blastocyst	Blastocyst develops syncytial trophoblast that adheres to uterine surface
6?	Attachment of blastocyst to uterine epithelium	Trophoblast intrudes between uterine epithelial cells
6-7	Epithelial penetration	Trophoblast spreads along basal lamina
7½-8	Trophoblastic plate formation	Trophoblast penetrates basal lamina, invades stroma, and taps maternal vessels
9-11	Lacuna formation	Trophoblast differentiation forms microvillus-lined clefts and lacunae
12-13	Primary villus formation	Focal cytotrophoblast proliferations extend from chorionic plate into syncytium
13-15	Secondary placental villi, secondary yolk sac formation	Mesenchymal cells proliferate on fetal side of primary villi
16-18	Branching and anchoring villi formation	Cytotrophoblast extends through syncytial trophoblast and spreads along basal plate
18-22	Tertiary villi formation	Vessels differentiate in situ and are filled with blood when heart beat links yolk sac, embryo, and chorioallantoic placenta

Modified from Enders AC: *Implantation.* In *Encyclopedia of human biology,* vol 4, New York, 1991, Academic.

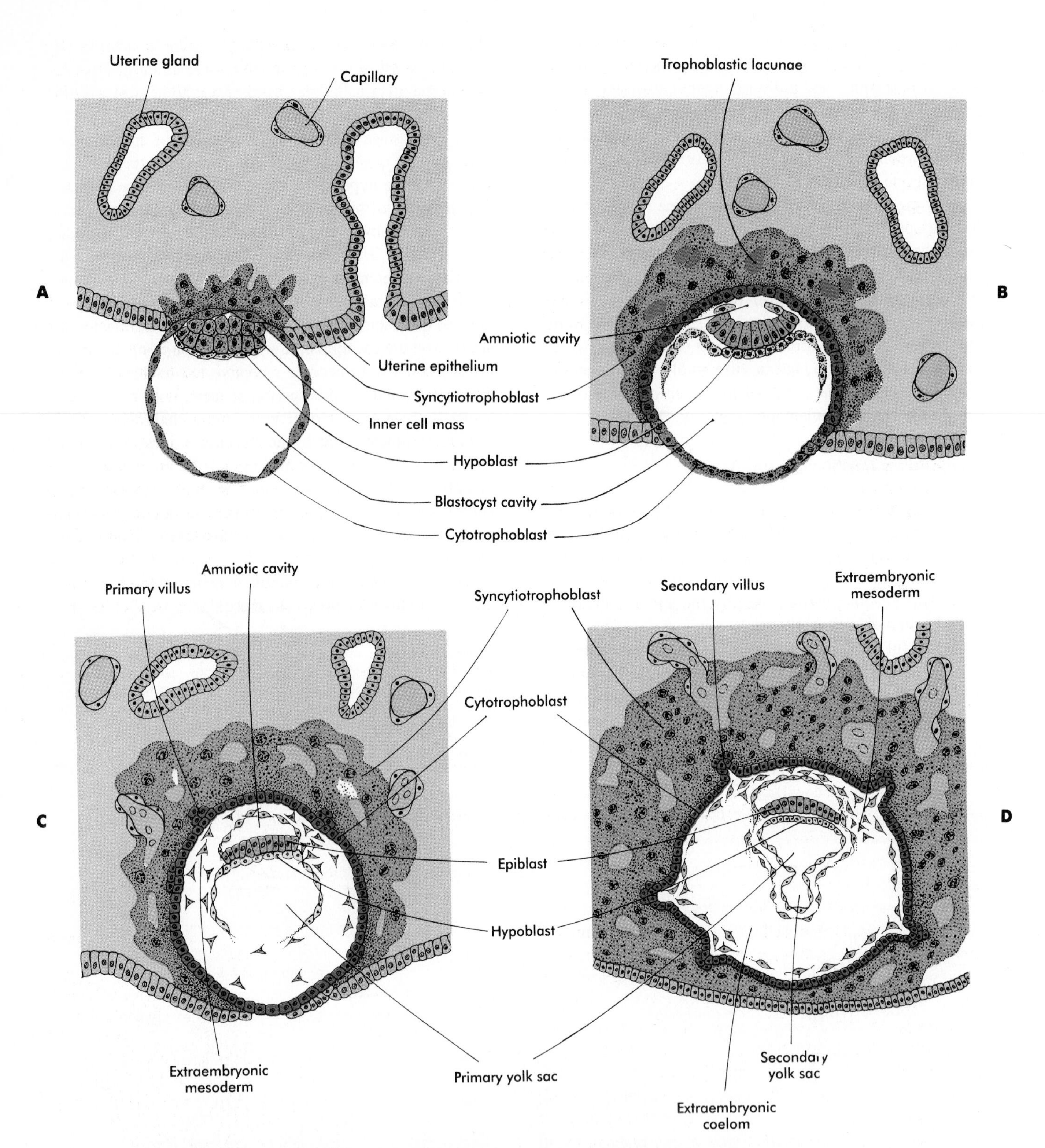

FIG. 3-16 Major stages in implantation of a human embryo. **A,** The syncytiotrophoblast is just beginning to invade the endometrial stroma. **B,** Most of the embryo is embedded in the endometrium; there is early formation of the trophoblastic lacunae. The amniotic cavity and yolk sac are beginning to form. **C,** Implantation is almost complete, primary villi are forming, and the extraembryonic mesoderm is appearing. **D,** Implantation is complete; secondary villi are forming.

dergoes a further stage in its differentiation just before it contacts the endometrium. In the area around the inner cell mass, cells derived from the cellular trophoblast **(cytotrophoblast)** fuse to form a multinucleated **syncytiotrophoblast.** Although only a small area of syncytiotrophoblast is evident at the start of implantation, this structure (sometimes called the **syntrophoblast**) soon surrounds the entire embryo. Small projections of syncytiotrophoblast insert themselves between uterine epithelial cells. They then spread along the epithelial surface of the basal lamina that underlies the endometrial epithelium to form a somewhat flattened **trophoblastic plate.** Within a day or so, syncytiotrophoblastic projections from the small trophoblastic plate begin to penetrate the basal lamina (Fig. 3-16, *A, B*). The early syncytiotrophoblast is a highly invasive tissue, and it quickly expands and erodes its way into the endometrial stroma. Although the invasion of the syncytiotrophoblast into the endometrium is obviously enzymatically mediated, the biochemical basis in humans is not known. By 10 to 12 days after fertilization, the embryo is completely embedded in the endometrium. The site of initial penetration is first marked by a bare area or a noncellular plug and is later sealed by migrating uterine epithelial cells (Fig. 3-16, *C, D*).

As the embryo burrows into the endometrium and some cytotrophoblastic cells fuse into syncytiotrophoblast, the stromal cells of the uterus begin to undergo a profound transformation of their own. This transformation, called the **decidual reaction,** had been poorly understood. It now appears that a primary function of the decidual reaction is to provide an immunologically privileged site for the developing embryo. An embryo is antigenically different from the mother and consequently should be rejected by a cellular immune reaction similar to the type that rejects an incompatible heart or kidney transplant. One of the major mysteries of mammalian development is why the embryo is not rejected like a foreign body.

In response to a poorly defined signal by the embryo, the fibroblastlike stromal cells of the somewhat edematous endometrium swell, with glycogen and lipid droplets accumulating. The decidual cells are tightly adherent and form a massive cellular matrix that first surrounds the implanting embryo and later occupies most of the endometrium. Concurrent with decidualization of the endometrial stroma, the leukocytes that have infiltrated the endometrial stroma during the late progestational phase of the menstrual cycle secrete **interleukin-2,** which prevents maternal recognition of the embryo as a foreign body during the early stages of implantation.

As early implantation continues, projections from the invading syncytiotrophoblast envelop portions of maternal endometrial blood vessels. They erode into the vessel walls, and maternal blood begins to fill the isolated lacunae that have been forming in the trophoblast (see Fig. 3-16, *C, D*). Trophoblastic processes enter the blood vessels and even share junctional complexes with the endothelial cells. By the time blood-filled lacunae have formed, the trophoblast changes character, and it is not as invasive as it was during the first few days of implantation.

Ectopic Pregnancy

The blastocyst typically implants into the posterior wall of the uterine cavity, but in a low percentage (0.25% to 1.0%) of cases, implantation occurs in an abnormal site. Such a condition is known as an **ectopic pregnancy.**

Tubal pregnancies are by far the most common type of ectopic pregnancy. Although most tubal pregnancies are found in the ampullary portion of the tube, they can be located anywhere from the fimbriated end to the uterotubal junction (Fig. 3-17). Tubal pregnancies (Fig. 3-18) are most commonly seen in women who have had **endometriosis** (a condition characterized by the presence of endometriumlike tissue in abnormal locations), prior surgery, or **pelvic inflammatory disease** (e.g., gonorrhea, tuberculosis). Scarring from inflammation or sometimes anatomical abnormalities result in blind pockets among the mucosal folds of the uterine tube that can trap a blastocyst. Typically, the woman shows the normal signs of early pregnancy, but at about 2 to 2½ months, the implanted embryo and its asso-

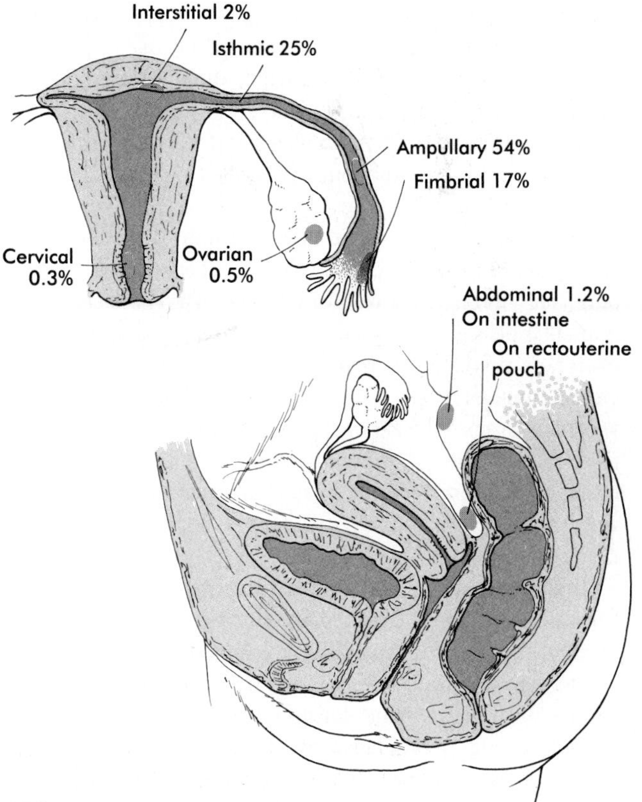

FIG. 3-17 Sites of ectopic pregnancy and the frequency of their occurrence.

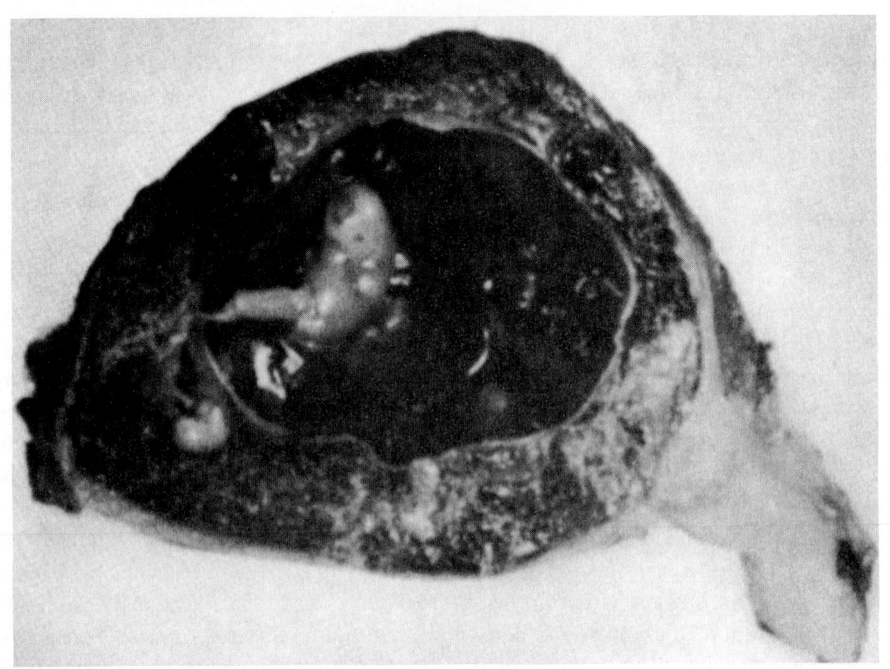

FIG. 3-18 Tubal pregnancy containing two embryos, one much smaller than the other. The trophoblast has penetrated the attenuated muscular wall of the tube. Rupture of the tube was imminent.
(From Kraus F: *Gynecologic pathology*, St Louis, 1967, Mosby.)

ciated trophoblastic derivatives have grown to the point where the stretching of the tube causes acute abdominal pain. If untreated, a tubal pregnancy typically ends with rupture of the tube and hemorrhage, often severe enough to be life threatening to the mother.

Very rarely an embryo implants in the ovary **(ovarian pregnancy)** or in the abdominal cavity **(abdominal pregnancy).** Such instances can be the result of fertilization of an ovum before it enters the tube, the reflux of a fertilized egg from the tube, or very rarely, the penetration of a tubal pregnancy through the wall of the tube. The most common implantation site for an abdominal pregnancy is in the **rectouterine pouch (pouch of Douglas),** located behind the uterus. Implantation on the intestinal wall or mesentery is very dangerous because of the likelihood of severe hemorrhage as the embryo grows. In some instances, an embryo has developed to full term in an abdominal location. If not delivered, such an embryo can calcify, forming a **lithopedion.**

Within the uterus, an embryo can implant close to the cervix. Although embryonic development is likely to be normal, the placenta typically forms a partial covering over the cervical canal. This condition, called **placenta previa,** can result in hemorrhage during late pregnancy and, if untreated, would likely cause the death of the fetus and/or mother because of premature placental detachment with accompanying hemorrhage. Implantation directly within the cervical canal is extremely rare.

Embryo Failure and Spontaneous Abortion

A high percentage of fertilized eggs (over 50%) do not develop to maturity and are spontaneously aborted. Most spontaneous abortions **(miscarriages)** occur during the first three weeks of pregnancy. Because of the small size of the embryo at that time, spontaneous abortions are often not recognized by the mother, who may equate the abortion and attendant hemorrhage with a late and unusually heavy menstrual period.

Examinations of early embryos obtained after spontaneous abortion or from uteri removed by hysterectomy during the early weeks of pregnancy have shown that many of the aborted embryos are highly abnormal. Chromosomal abnormalities represent the most common category of abnormality in abortuses (about 50% of the cases). When viewed in the light of the accompanying pathology, spontaneous abortion can be viewed as a natural mechanism for reducing the incidence of severely malformed infants.

SUMMARY

1. Early human cleavage is slow, with roughly one cleavage division occurring per day for the first 3 to 4 days. The cleaving embryo passes through the morula stage (16 cells) and enters a stage of compaction. By 4 days, a fluid-filled blastocoele forms within the embryo, and the embryo becomes a blastocyst, with an inner cell mass surrounded by a trophoblast.

2. Cleavage divisions (and all cell divisions) are regulated by a maturation-promoting factor, which is a complex of cdc2 and cyclin proteins. After mitosis, cyclin is broken down. Before the next cell division, cyclin is built up and complexes with cdc2; this complex becomes activated to become maturation-promoting factor, which initiates mitosis.
3. The zygote relies on maternal mRNAs, but by the 2-cell stage, the embryonic genome becomes activated. The *oct-3* gene is important in very early development, and its expression is associated with the undifferentiated state of cells.
4. Through parental imprinting, specific homologous chromosomes derived from the mother and father exert different effects on embryonic development. In female embryos, one X chromosome per cell becomes inactivated, forming the sex chromatin body. The early embryo has distinct patterns of X-chromosome inactivation.
5. The early mammalian embryo is highly regulative. It can compensate for the loss or addition of cells to the inner cell mass and still form a normal embryo. According to the inside-outside hypothesis, the position of a blastomere determines its developmental fate (i.e., whether it will become part of the inner cell mass or trophoblast).
6. Transgenic embryos are produced by injecting recombinant DNA into the pronuclei of zygotes. Such embryos are used to study the effects of specific genes on development.
7. Monozygotic twinning, usually caused by the complete separation of the inner cell mass, is possible because of the regulative properties of the early embryo. Incomplete splitting of the inner cell mass can lead to the formation of conjoined twins.
8. After fertilization, the embryo spends several days in the uterine tube before entering the uterus. During this time, it is still surrounded by the zona pellucida, which prevents premature implantation.
9. Implantation of the embryo into the uterine lining involves several stages: apposition of the expanded (hatched) blastocyst to the endometrial epithelium, penetration of the uterine epithelium, invasion into the tissues underlying the epithelium, and erosion of the maternal vascular supply. Connective tissue cells of the endometrium undergo the decidual reaction in response to the presence of the implanting embryo. Implantation is accomplished through the invasive activities of the syncytiotrophoblast, which is derived from the cytotrophoblast.
10. Implantation of the embryo into a site other than the upper uterine cavity results in an ectopic pregnancy. Ectopic pregnancy is most often encountered in the uterine tube.
11. A high percentage of fertilized eggs and early embryos do not develop and are spontaneously aborted. Many of these embryos contain major chromosomal abnormalities.

REVIEW QUESTIONS

1. What is the importance of the inner cell mass of the cleaving embryo?
2. What intracellular event stimulates mitosis?
3. Parental imprinting is a phenomenon demonstrating that certain homologous maternal and paternal chromosomes have somewhat different influences on the development of the embryo. Excess paternal influences result in abnormal development of what type of tissue at the expense of development of the embryo itself?
4. What property of cleaving mammalian embryos is responsible for the production of normal monozygotic twins and the repair of damage to the embryo?
5. What is the cellular origin of the syncytiotrophoblast of the implanting embryo?
6. A woman who is 2 to 3 months pregnant suddenly develops severe lower abdominal pain. In the differential diagnosis, the physician must include the possibility of what condition?

REFERENCES

Avendano S and others: A seven-cell human egg recovered from the oviduct, *Fertil Steril* 26:1167-1172, 1975.

Bateson W: *Materials for the study of variation*, London, 1894, MacMillan.

Cruz YP: Role of ultrastructural studies in the analysis of cell lineage in the mammalian pre-implantation embryo, *Microsc Res Tech* 22:103-125, 1992.

Denker HW: *Trophoblast-endometrial interactions at embryo implantation: a cell biological paradox*. In Denker HW, Aplin JD, eds: *Trophoblast research,* vol 4, New York, 1990, Plenum, pp 3-29.

Enders AC: Trophoblast differentiation during the transition from trophoblastic plate to lacunar stage of implantation in the rhesus monkey and human, *Am J Anat* 186:85-98, 1989.

Enders AC: *Implantation*. In *Encyclopedia of human biology,* vol 4, New York, 1991, Academic.

Gardner RL: Clonal analysis of early mammalian development, *Philos Trans R Soc Lond [Biol]* 312:163-178, 1985.

Gartler SM, Riggs HD: Mammalian X-chromosome inactivation, *Annu Rev Genet* 17:155-190, 1983.

Gualtieri R, Santella L, Dale B: Tight junctions and cavitation in the human pre-embryo, *Mol Reprod Dev* 32:81-87, 1992.

Kidder GM, Watson AJ: *Gene expression required for blastocoel formation in the mouse*, In Heyner S, Wiley LM, eds: *Early embryo development and paracrine relationships,* New York, 1990, Liss, pp 97-107.

Leese HJ: *The energy metabolism of the preimplantation embryo*. In *Early embryo development and paracrine relationships,* New York, 1990, Liss, pp 67-78.

Lindenberg S: Ultrastructure in human implantation: transmission and scanning electron microscopy, *Bailleres Clin Obstet Gynecol* 5:1-14, 1991.

McConnell J: Molecular basis of cell cycle control in early mouse embryos, *Int Rev Cytol* 129:75-90, 1991.

Monk M, Surani A, eds: Genomic imprinting, *Development* 1990(suppl):1-155, 1990.

Moore T, Haig D: Genomic imprinting in mammalian development: a parental tug-of-war, *Trends Genet* 7:45-49, 1991.

Murray AW, Kirschner MW: What controls the cell cycle? *Sci Am* 264:56-63, 1991.

O'Rahilly R: The manifestation of the axes of the human embryo, *Z Anat Entwickl Gesch* 132:50-57, 1970.

Palmiter RD and others: Dramatic growth of mice that develop from eggs microinjected with metallothionein-growth hormone fusion genes, *Nature* 300:611-615, 1982.

Pederson RA: *Early mammalian embryogenesis*. In Knobil E, Neill J, eds: *The physiology of reproduction*, New York, 1988, Raven, pp 187-230.

Rosner MH and others: Oct-3 and the beginning of mammalian development, *Science* 253:144-145, 1991.

Sapienza C: Parental imprinting of genes, *Sci Am* 263:52-60, 1990.

Solter D: Differential imprinting and expression of maternal and paternal genomes, *Annu Rev Genet* 22:127, 1988.

Spencer R: Conjoined twins: theoretical embryologic basis, *Teratology*, 45:591-602, 1992.

Strange C: Cell cycle advances, *Bio Science* 4:252-256, 1992.

Tarkowski AK, Wroblewska J: Development of blastomeres of mouse eggs isolated at the 4- and 8-cell stage, *J Embryol Exp Morph* 18:155-180, 1967.

Tesarik J and others: Early morphological signs of embryonic genome expression in human preimplantation development as revealed by quantitative electron microscopy, *Dev Biol* 128:15-20, 1988.

Uchida IA: Twinning in spontaneous abortions and developmental abnormalities, *Issues Rev Teratol* 5:155-180, 1990.

Weitlauf HM: *Biology of implantation*. In Knobil E, Neill J, eds: *The physiology of reproduction*, New York, 1988, Raven, pp 231-262.

Yokoyama T and others: Reversal of left right symmetry: a situs inversus mutation, *Science* 260:679-682, 1993.

CHAPTER 4

The Formation of Germ Layers and Early Derivatives

As it is implanting into the uterine wall, the embryo undergoes profound changes in its organization. Up to the time of implantation, the blastocyst consists of the inner cell mass, from which the body of the embryo proper arises, and the outer trophoblast, which represents the future tissue interface between the embryo and mother. Both components of the blastocyst serve as the precursors of other tissues that appear in subsequent stages of development. Chapter 3 discussed the way the cytotrophoblast begins to give rise to an outer syncytial layer, the syncytiotrophoblast, shortly before attaching to uterine tissue (see Fig. 3-16). Not long thereafter the inner cell mass begins to give rise to other tissue derivatives as well. The subdivision of the inner cell mass ultimately results in an embryonic body that contains the three primary embryonic germ layers: the **ectoderm** (outer layer), **mesoderm** (middle layer), and **endoderm** (inner layer). The process by which the germ layers are formed through cell movements is called **gastrulation.**

After the germ layers have been laid down, the continued progression of embryonic development depends on a series of signals called **embryonic inductions** that are exchanged between the germ layers or other tissue precursors. In an inductive interaction, one tissue (the **inductor**) acts on another **(responding tissue)** so that the developmental course of the latter is different from what it would have been in the absence of the inductor. Understanding the nature of embryonic induction, especially the inductive signal, had been a time-consuming and frustrating obsession of embryologists for many decades. Only in recent years have some breakthroughs in the understanding of inductive processes occurred.

The developments that can be seen with a microscope during this period are tangible reflections of profound changes in gene expression and cellular properties of implanting embryos. In recent years a virtual explosion of molecular information on early development in *Drosophila* has begun to be translated to studies on early development in amphibia and mammals. Some of the major findings are introduced in this chapter and Chapter 5. In a few years, this information will play a central role in the understanding of many fundamental aspects of both normal and abnormal human development.

THE TWO GERM-LAYER STAGE

Just before the embryo implants into the endometrium early in the second week, significant changes begin to occur in the inner cell mass as well as the trophoblast. As the cells of the inner cell mass rearrange into an epithelial configuration, a thin layer of cells appears ventral to the main mass of cells (see Fig. 3-16). The main upper layer of cells is known as the **epiblast,** and the lower layer is called the **hypoblast,** or **primitive endoderm** (Fig. 4-1).

How the hypoblast forms in human embryos is not understood, but comparative embryological data suggest that cells of this layer arise by **delamination** (a separation or dropping down) from the inner cell mass. The hypoblast is considered an **extraembryonic endoderm,** and it ultimately gives rise to the endodermal lining of the **yolk sac** (see Fig. 3-16). After the hypoblast has become a well-defined layer and the epiblast has taken on an epithelial configuration, the former inner cell mass is transformed into a **bilaminar disk,** with the epiblast on the dorsal surface and the hypoblast on the ventral surface.

The epiblast contains the cells that will compose the embryo itself, but extraembryonic tissues also arise from this layer (Fig. 4-2). The next layer to appear after the hypoblast is the **amnion,** a layer of extraembryonic ectoderm that ultimately encloses the entire embryo in a fluid-filled

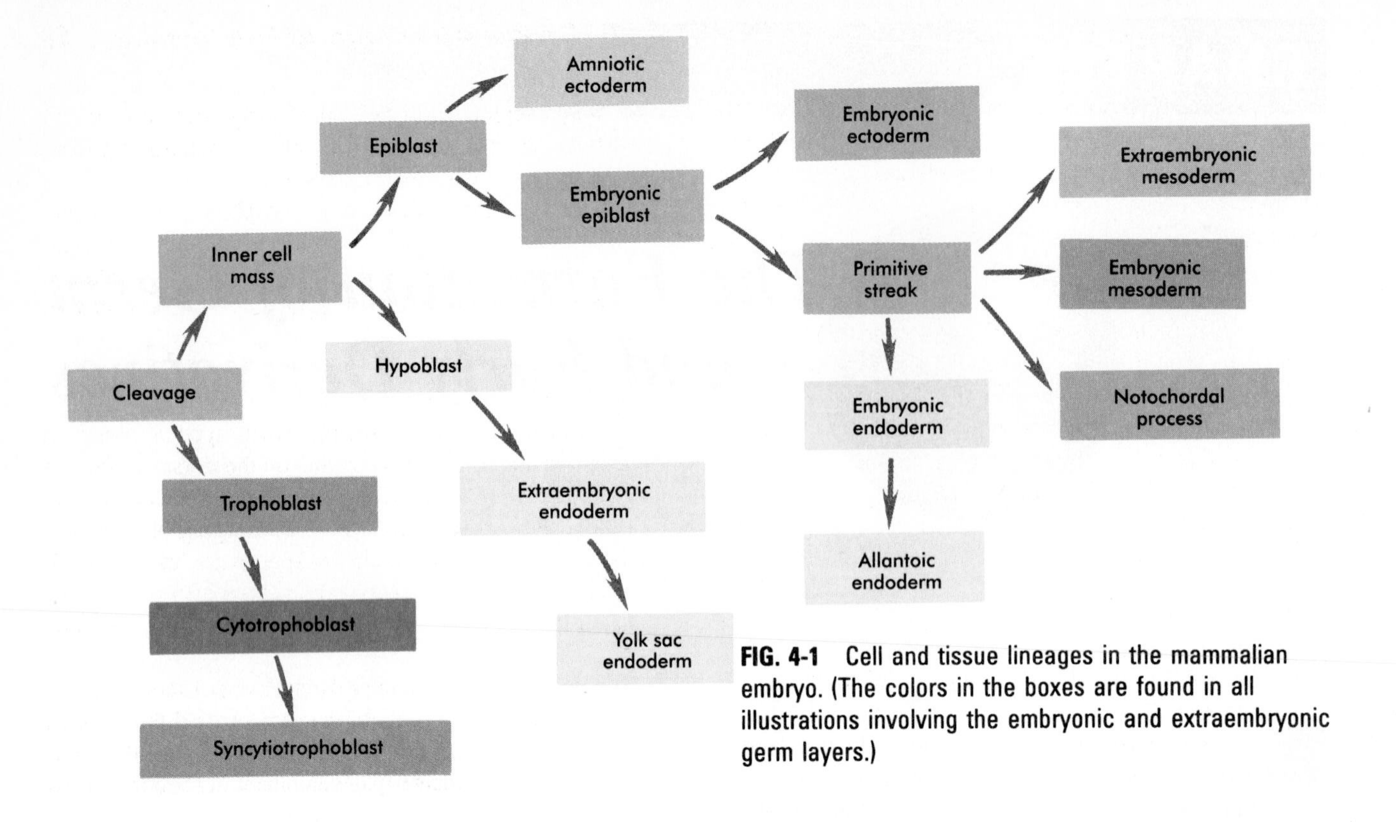

FIG. 4-1 Cell and tissue lineages in the mammalian embryo. (The colors in the boxes are found in all illustrations involving the embryonic and extraembryonic germ layers.)

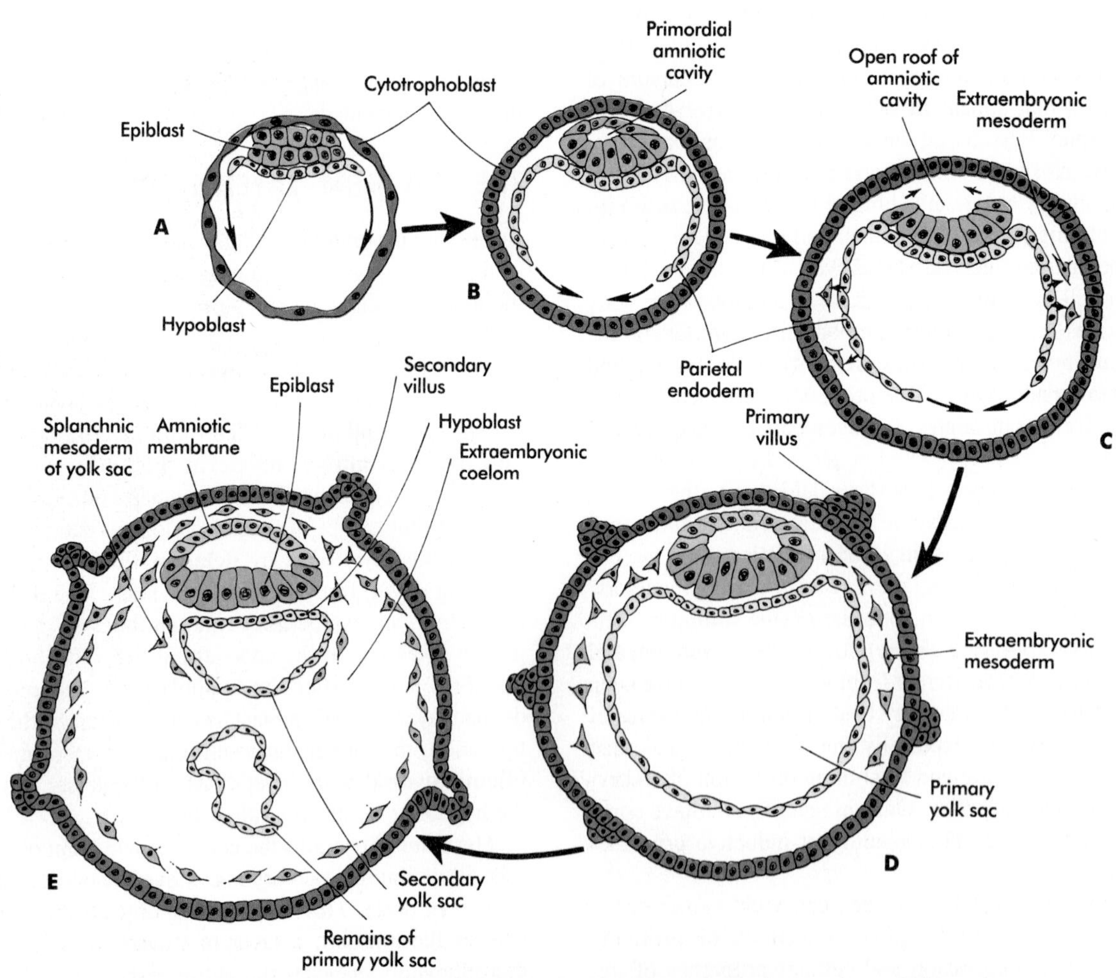

FIG. 4-2 The origins of the major extraembryonic tissues. The syncytiotrophoblast is not shown. **A,** Beginning of implantation. **B,** Implanted blastocyst at 7½ days. **C,** Implanted blastocyst at 8 days. **D,** Embryo at 9 days. **E,** Beginning of third week.

chamber called the **amniotic cavity** (see Chapter 6). Due to a paucity of specimens, the earliest stages in the formation of the amnion and amniotic cavity are not completely understood. Studies on primate embryos indicate that a primordial amniotic cavity first arises by **cavitation** (formation of an internal space) within the preepithelial epiblast; it is covered by cells derived from the inner cell mass. According to some investigators, the roof of the amnion then opens, exposing the primordial amniotic cavity to the overlying cytotrophoblast. Soon thereafter (by about 8 days after fertilization), the original amniotic epithelium re-forms a solid roof over the amniotic cavity.

While the early embryo is still sinking into the endometrium (about 9 days after fertilization), cells of the hypoblast begin to spread to line the inner surface of the cytotrophoblast with a continuous layer of extraembryonic endoderm called **parietal endoderm** (see Fig. 4-2). When the endodermal spreading is completed, a vesicle called the **primary yolk sac** has taken shape (see Fig. 3-16). At this point (roughly 10 days after fertilization), the embryo complex consists of the bilaminar germ disk, which is located between the primary yolk sac on its ventral surface and the amniotic cavity on its dorsal surface.

Starting at about 12 days after fertilization, another extraembryonic tissue, the **extraembryonic mesoderm,** begins to appear (see Fig. 4-2). It now appears that the first extraembryonic mesodermal cells arise from a transformation of parietal endodermal cells. These cells are later joined by extraembryonic mesodermal cells that have originated from the primitive streak. Contemporary morphological evidence does not support the classic viewpoint that extraembryonic mesoderm arises from the trophoblast. The extraembryonic mesoderm becomes the tissue that supports the epithelium of the amnion and yolk sac as well as the **chorionic villi,** which arise from the trophoblastic tissues (see Chapter 6). The support supplied by the extraembryonic mesoderm is not only mechanical, but also trophic, since the mesoderm serves as the substrate through which the blood vessels supply oxygen and nutrients to the various epithelia.

GASTRULATION AND THE THREE EMBRYONIC GERM LAYERS

At the end of the second week the embryo consists of two flat layers of cells, the epiblast and hypoblast. As the third week of pregnancy begins, the embryo enters the period of gastrulation, during which the three embryonic germ layers become clearly established. The morphology of human gastrulation follows the pattern in birds. Because of the large amount of yolk in birds' eggs, the avian embryo forms the primary germ layers as three overlapping flat disks that rest on the yolk much like a stack of pancakes. Only later do the germ layers fold to form a cylindrical body. Although the mammalian egg is essentially devoid of yolk, the morphological conservatism of early development still constrains the human embryo to follow a pattern of gastrulation similar to that seen in reptiles and birds. Because of the scarcity of material, even the morphology of gastrulation in human embryos is not known in detail. Nevertheless, extrapolation from avian and mammalian gastrulation can provide a reasonable working model of human gastrulation.

All embryonic germ layers originate from the epiblast (see Fig. 4-1). The first evidence of gastrulation is the formation of the **primitive streak,** which appears first as a thickening and then as a short line on the dorsal surface of the epiblast (Fig. 4-3). The early primitive streak is actually a condensation caused by the converging of epiblastic cells toward that area. With the appearance of the primitive streak, the anteroposterior (craniocaudal) and right-left axes of the embryo can be readily identified (Figs. 4-4 and 4-5).

Recognition of the cellular dynamics associated with the primitive streak makes it easier to understand the detailed structure of this area. As cells of the epiblast reach the primitive streak, they change shape and pass through it on their

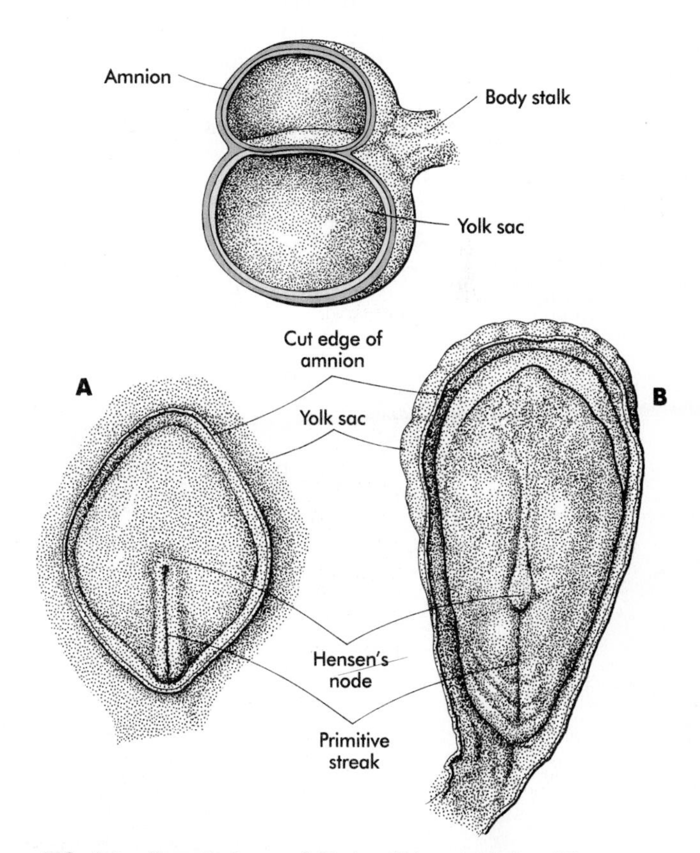

FIG. 4-3 Dorsal views of 16-day **(A)** and 18-day **(B)** human embryos. *Top:* Sagittal section through an embryo and its extraembryonic membranes during early gastrulation.

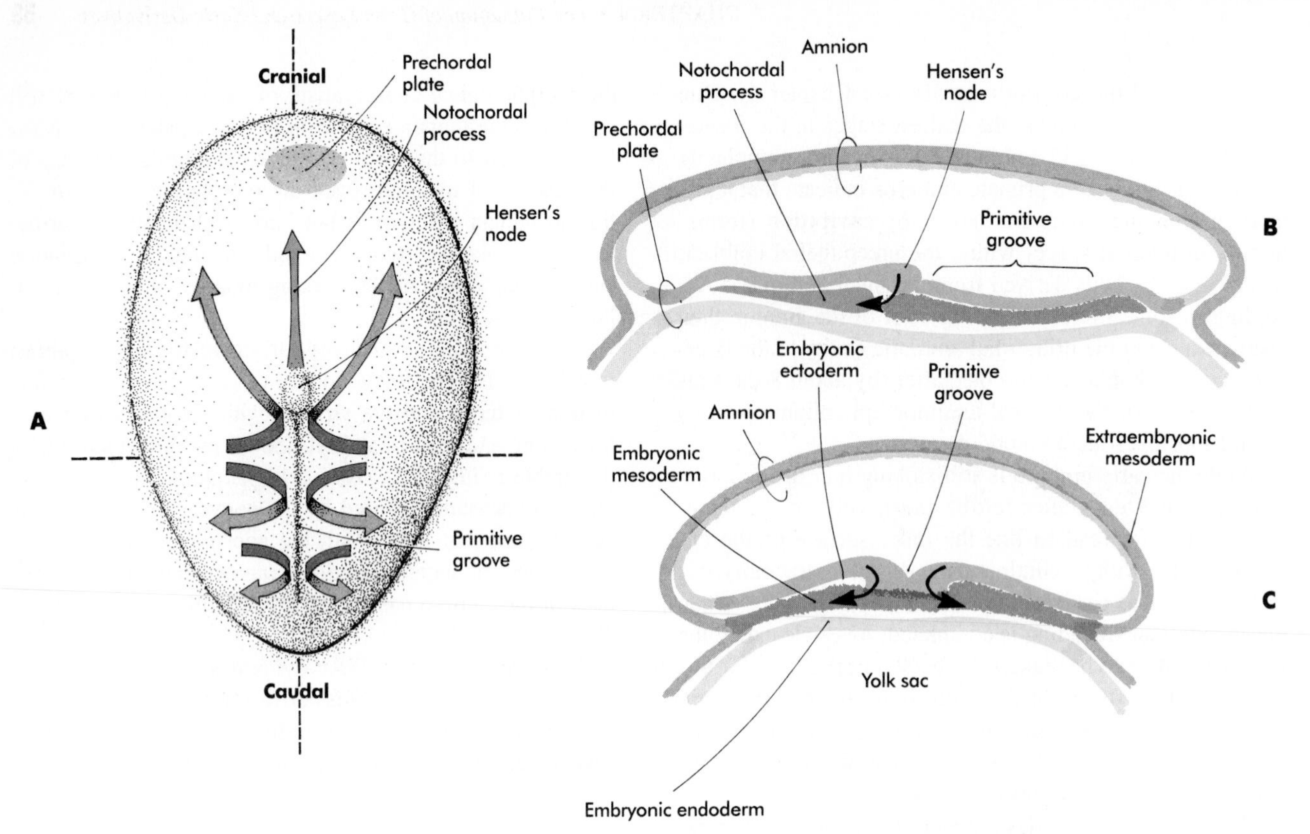

FIG. 4-4 **A,** Dorsal view through a human embryo during gastrulation. The arrows show the directions of cellular movements across the epiblast toward the primitive streak, through the primitive streak, and away from the primitive streak as newly formed mesoderm. **B,** Sagittal section through the craniocaudal axis of the same embryo. **C,** Cross section through the level of the primitive streak in **A** *(dashed lines).*

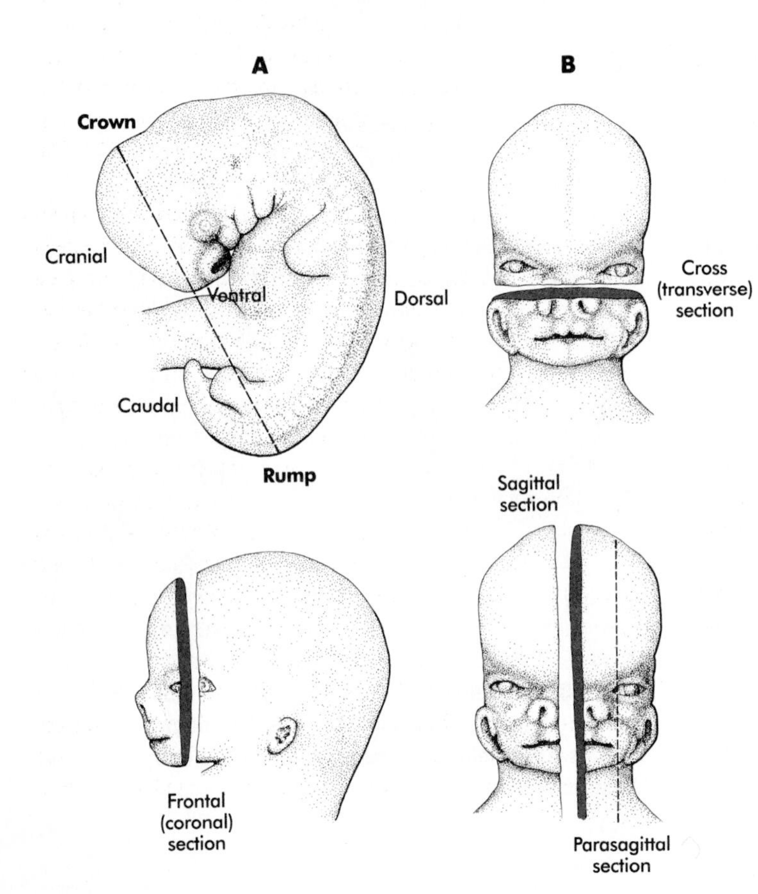

FIG. 4-5 Various planes, sections, and other descriptive terms used in descriptions of embryonic material. **A,** Lateral view of 6-week-old embryo. The dashed line indicates the crown-rump length, one of the standard ways of measuring human embryos. The crown-rump length is the greatest straight-line distance between the apex of the head and the caudal end of the trunk. **B, C,** and **D,** Heads of 8-week-old embryos, illustrating common planes of section.

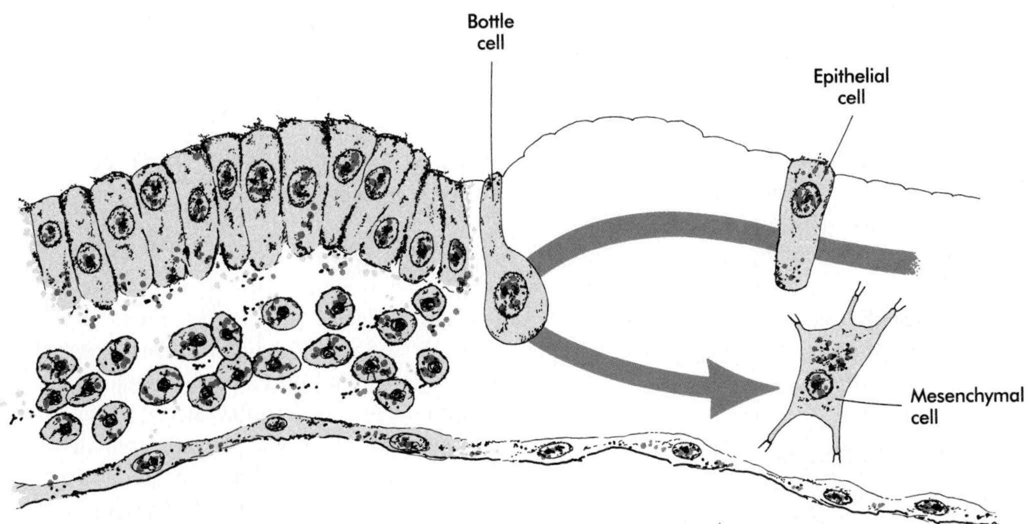

FIG. 4-6 Changes in the shape of a cell as it migrates along the epiblast (epithelium), through the primitive streak (bottle cell), and away from the groove as mesoderm (mesenchyme). The same cell can later assume an epithelial configuration as part of a somite.

way to forming new cell layers beneath (ventral to) the epiblast (see Fig. 4-4). The movement of cells through the primitive streak results in the formation of a groove (the **primitive groove**) along the midline of the primitive streak. At the end of the primitive streak is a small but well-defined accumulation of cells called the **primitive knot,** or **Hensen's node.** This structure is of great developmental significance because it is the area through which migrating epiblastic cells are channeled into a rodlike mass of mesenchymal cells called the **notochord.** (The notochord and its functions in the early embryo are discussed on p. 56.)

The movements of the cells passing through the primitive streak are accompanied by major changes in their structure and organization (Fig. 4-6). While in the epiblast, the cells have the properties of typical epithelial cells, with well-defined apical and basal surfaces, and they are associated with a basal lamina that underlies the epiblast.* As they enter the primitive streak, these cells elongate and take on a characteristic morphology that has led to their being called **bottle cells.** When they become free of the epiblastic layer in the primitive groove, the bottle cells assume the morphology and characteristics of mesenchymal cells, which are able to migrate as individual cells if provided with the proper environment (see Fig. 4-6).

In birds, the first cells to leave the primitive streak enter the hypoblast, displacing some of the extraembryonic endodermal cells to form the definitive **embryonic endoderm.** This process has not been confirmed in the human embryo, but the morphology of human gastrulation is compatible with the occurrence of a similar process.

The most prominent feature of human gastrulation is the formation of mesoderm. Some cells migrate through the primitive streak at its earliest stage to form the extraembryonic mesoderm (see Fig. 4-2). In addition to spreading beneath the trophoblast and around the yolk sac, a mass of extraembryonic mesodermal cells called the **body stalk** forms a connection between the embryo proper and the **chorion,** which is the combination of trophoblast and underlying extraembryonic mesoderm (see Fig. 6-1). The body stalk later becomes the umbilical cord.

After the primitive streak is well established, the majority of cells passing through it spread out between the epiblast and hypoblast to form the **embryonic mesoderm** (see Fig. 4-4). The transformations of morphology and behavior of these cells passing through the primitive streak are associated with profound changes not only in their surface properties and internal organization but also in the way that they relate to their external environment.

Starting in early gastrulation, cells of the epiblast produce **hyaluronic acid,** which enters the space between the epiblast and hypoblast. Hyaluronic acid, a polymer consisting of repeating subunits of ***D*-glucuronic acid** and ***N*-acetylglucosamine,** is frequently associated with cell migration in developing systems. The molecule has a tremendous capacity to bind water (up to 1000 times its own volume), and it seems to function to keep mesenchymal cells from aggregating during cell migrations. After leaving the primitive streak, the mesenchymal cells of the embryonic mesoderm find themselves in a hyaluronic acid–rich envi-

*During germ layer formation, the basal lamina that lines the inner surface of the trophoblast continues into the embryo proper and separates the epiblast from the hypoblast.

ronment, but that alone is apparently not enough to support their migration from the primitive streak. In all vertebrate embryos that have been investigated to date, the spread of mesodermal cells away from the primitive streak or the equivalent structure is found to depend on the presence of **fibronectin** associated with the basal lamina beneath the epiblast. The embryonic mesoderm ultimately spreads laterally as a thin sheet of mesenchymal cells between the epiblast and hypoblast (see Fig. 4-4). By the time the mesoderm has formed a discrete layer in the human embryo, the epiblast and hypoblast are conventionally the ectoderm and endoderm, respectively. This terminology is used for the remainder of the text.

Regression of the Primitive Streak

After its initial appearance at the extreme caudal end of the embryo, the primitive streak extends cranially until about 18 days after fertilization (see Fig. 4-3). Thereafter, the primitive streak is said to regress caudally (see Fig. 4-11), although overall there is probably less actual regression than growth of the caudal part of the embryo. Nevertheless, a relative shortening of the primitive streak occurs in comparison with its overall length, and it strings out the notochord in its wake. Vestiges of the primitive streak remain into the fourth week. During that time the formation of mesoderm continues by means of cells migrating from the epiblast through the primitive groove.

The primitive streak normally disappears without a trace, but in rare instances, large tumors called **teratomas** appear in the sacrococcygeal region (see Fig. 1-2, *A*). Teratomas often contain bizarre mixtures of many different types of tissue, such as cartilage, muscle, fat, hair, and glandular tissue. Because of this, sacrococcygeal teratomas are thought to arise from remains of the primitive streak (which can form all germ layers). Teratomas are also found in the gonads and the mediastinum. These tumors are thought to originate from germ cells.

The Notochord

Starting at Hensen's node, a population of epiblastic cells drops to the level of the mesodermal layer and migrates directly cranially to form a loose cellular rod called the **chordamesoderm,** or the **notochordal process** (see Fig. 4-4). The notochordal process extends cranially until it encounters the **prochordal plate,** an area where ectoderm and endoderm are tightly apposed. The prochordal plate represents the site of the future oral cavity.

In what is sometimes called the second phase of gastrulation, the primitive streak regresses toward the caudal end of the embryo. As Hensen's node follows the regressing primitive streak, it lays the remainder of the notochordal process in its wake. The morphology of formation of the notochordal process in the human embryo is complex, and depending on which embryo is studied, various scenarios have been proposed. Details of these hypotheses are beyond the scope of this text, but one commonly noted structural feature is the **neurenteric canal** (Fig. 4-7), a channel of unknown function that provides temporary continuity between the amniotic cavity and the yolk sac.

The notochordal process soon consolidates to form the

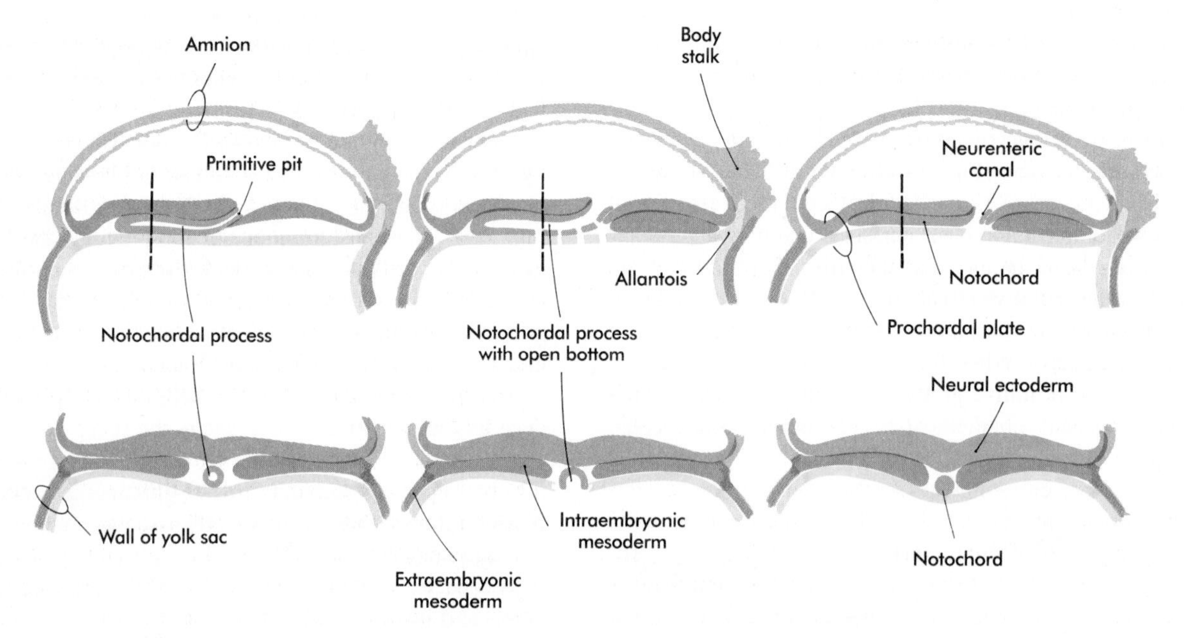

FIG. 4-7 Later stages in formation of the notochord. *Top:* Sagittal sections. *Bottom:* Cross sections at the level of the vertical line in the upper figure. In the upper row, the cranial end is on the left. The function of the neurenteric canal remains obscure.

notochord, a cellular rod that is the basis for naming the phylum to which all vertebrates belong Chordata. From the standpoint of comparative anatomy, the notochord serves both phylogenetically and ontogenetically as the original longitudinal support mechanism for the body, a function that certainly operates in the human embryo. However, its most important role in higher vertebrates has little to do with mechanical support. The notochord (chordamesoderm) functions as the primary inductor in the early embryo; in other words, it is a prime mover in a series of signal-calling episodes that ultimately transform unspecialized embryonic cells into definitive adult tissues and organs. Of particular relevance is the role of the notochord in the transformation of portions of the embryonic ectoderm into the nervous system.

Comparative embryological research has demonstrated quite convincingly that Hensen's node is the equivalent of the dorsal lip of the blastopore in amphibian embryos. Furthermore, like the chordamesoderm that passes through the dorsal lip and migrates beneath the ectoderm in the amphibian embryo, the mammalian notochordal process functions as the primary inductor of the nervous system in the early embryo. The cells in Hensen's node synthesize retinoic acid, an important morphogenetic signaling molecule (see Chapter 11).

PRIMARY INDUCTION OF THE NERVOUS SYSTEM

The inductive relationship between the chordamesoderm and the overlying ectoderm in the genesis of the nervous system was recognized in the early 1900s. Although original experiments were done on amphibians, similar experiments in higher vertebrates have shown that the essential elements of **primary,** or **neural, induction** are the same in all vertebrates.

Deletion and transplantation experiments in amphibians set the stage for present understanding of neural induction. (See Chapters 5 and 12 for further details on formation of the nervous system.) In the absence of chordamesoderm moving from the dorsal lip of the blastopore, the nervous system, which is first represented by a thickened plate of transformed ectodermal cells situated along the dorsal midline of the embryo, does not form from the dorsal ectoderm. On the other hand, if the dorsal lip of the blastopore is grafted beneath the belly ectoderm of another host, a secondary nervous system and body axis form in the area of the graft (Fig. 4-8). The dorsal lip has been called the **organizer** because of its ability to stimulate the formation of a secondary body axis. Subsequent research has shown that the interactions occurring in the region of the dorsal lip in amphibians are far more complex than a single induction between chordamesoderm and ectoderm.

Experiments of the type described have also been conducted on embryos of birds and mammals (see Fig. 4-8); clearly Hensen's node and the notochordal process are homologous in function to the dorsal lip and chordamesoderm. This means that Hensen's node and the notochordal process act as the primary inductor, and the overlying ectoderm is the responding tissue in higher vertebrates. This fundamental relationship was established more than a half century ago. Since that time, embryologists have devoted an enormous amount of research to identifying the nature of the inductive signal that passes from the chordamesoderm to the ectoderm.

Early attempts to uncover the nature of the inductive stimulus were marked by a great deal of optimism. As early as the 1930s, various laboratories had proposed that molecules as diverse as proteins and steroids were the inductive stimulus. Soon thereafter came the discovery that an even wider variety of stimuli, including inorganic ions, killed tissues and that even slight damage to the cells of the responding tissue could elicit neural induction. With such a plethora of possible inductors, attention turned to properties of the responding tissue (the dorsal ectoderm) and the way it might react through a final common pathway to the inductive stimulus. The quest for the neural inductive molecules and their mode of action has been arduous and frustrating, with many blind alleys and wrong turns encountered along the way.

A number of laboratories found that isolated ectoderm could respond in vitro to inductive stimuli and become transformed into neural tissue. A very useful technique for studying induction in vitro involved separating the responding tissue from the inducing tissue by a filter with pores that permitted the passage of molecules but not cells. This technique has been used in the analysis of a number of mammalian inductive systems.

Various experimental manipulations have shown clearly that neural induction is not a simple all-or-nothing process. Rather, considerable regional specificity exists (e.g., certain artificial inductors stimulate the formation of more anterior neural structures, and others more posterior ones). In amphibian embryos, anterior chordamesoderm has different inducing properties than posterior chordamesoderm. The recent availability of molecular probes, or monoclonal antibodies, against segment-specific proteins that were initially discovered in *Drosophila* has greatly facilitated the analysis of regional induction of the early nervous system. (These molecules and their roles in normal development are discussed in Chapter 5.)

When initially described, primary (neural) induction was considered to be the first inductive process that takes place in the embryo. Subsequent experimentation, conducted principally on amphibians, has shown that other important inductions occur before neural induction. The best understood of these is the induction of mesoderm in the amphibian blastula. The mesoderm normally arises from a ring of cells around the equatorial region of the blastula (Fig. 4-9). If the ectoderm that is located at the

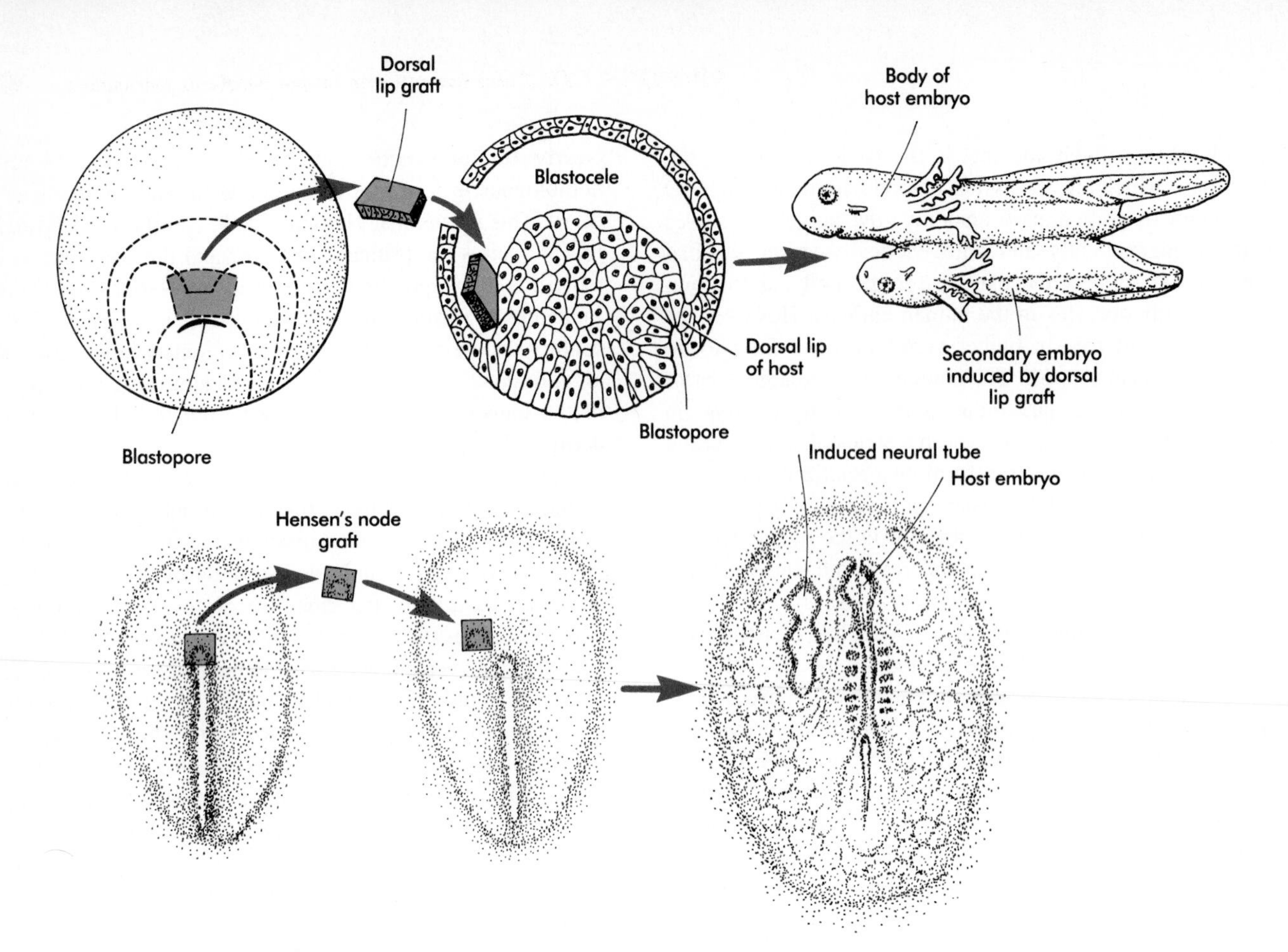

FIG. 4-8 Early experiments demonstrating primary induction. *Top:* A graft of the dorsal lip of the blastopore in a salamander embryo induces a secondary embryo to form.
(Based on studies by Spemann [1938].)
Bottom: A graft of Hensen's node from one chick embryo to another induces the formation of a secondary neural tube.
(Based on studies by Waddington [1933].)

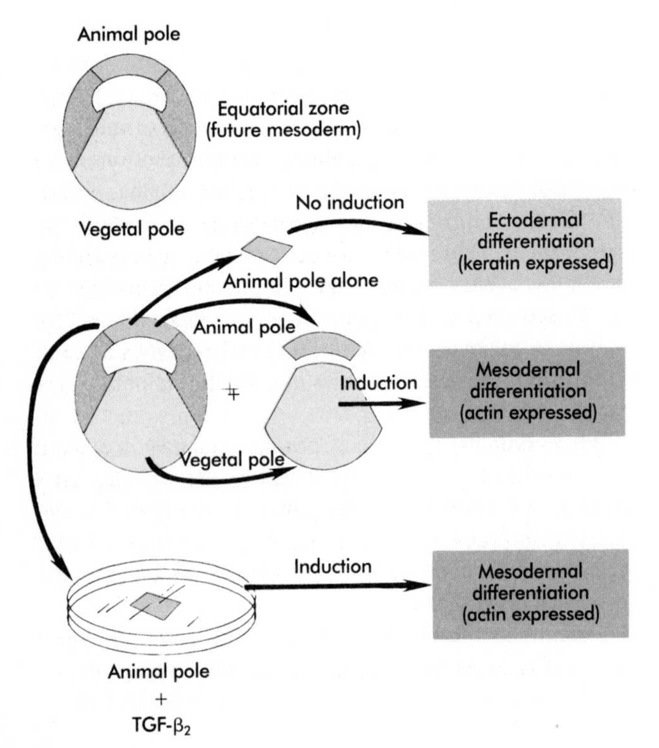

FIG. 4-9 Experiments illustrating the induction of mesoderm in the amphibian blastula. Animal pole is the region of yolk-poor cells corresponding to the future rostral end of the amphibian embryo. Vegetal pole is the region of yolk-rich cells corresponding to the future caudal region of the amphibian embryo.

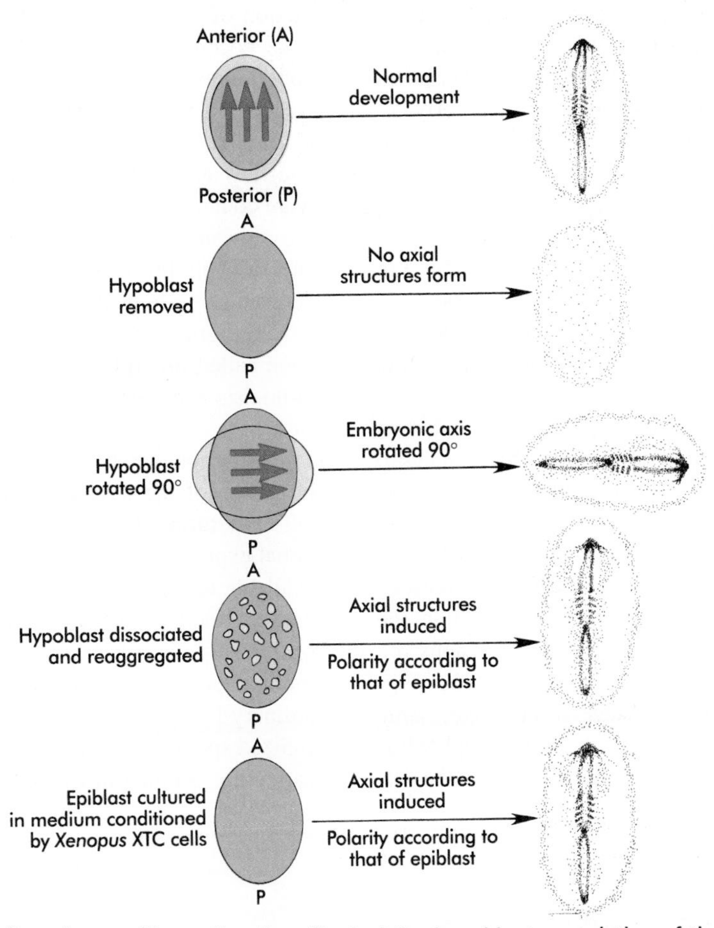

FIG. 4-10 Experiments illustrating the effect of the hypoblast on axiation of the epiblast in the avian embryo.
(Based on studies by Mitrani and Shimoni [1990].)

roof of the blastocoele is isolated, it remains general ectoderm and produces normal levels of **keratin** proteins, which are ectoderm-specific molecules. If the same piece of ectoderm is apposed to endoderm, it differentiates into mesoderm, as indicated by its producing **α-actin,** a molecule characteristic of muscle.

In recent years, understanding of the nature of mesodermal induction has been greatly enhanced by the demonstration that certain growth factors, especially **transforming growth factor-β_2 (TGF-β_2)** and a peptide growth factor called **activin,** also a member of the TGF gene family, can effect mesodermal induction acting alone. Despite many more years of being analyzed, neural induction has not been as clearly defined molecularly as has been the induction of mesoderm.

Although the vast majority of current research on induction in early embryos is being conducted on amphibian embryos, there is hope that the lessons learned from amphibians can be transferred to mammalian and human embryos. Experimental studies on avian embryos have shown that the primary hypoblast exerts an inductive and morphogenetic influence on the overlying epiblast. Rotating the hypoblast 90 degrees with respect to the orientation of the epiblast demonstrates that the hypoblast determines the origin and orientation of the primitive streak. After this manipulation, the primitive streak forms according to the orientation of the rotated hypoblast, 90 degrees from normal (Fig. 4-10). If an early avian epiblast is isolated and grown in culture, a notochord and axial mesoderm do not form. If an intact hypoblast or even dissociated and then reaggregated hypoblastic cells are juxtaposed to the cultured epiblast, axial structures form. Another experiment has shown that adding tissue culture medium exposed to a mesodermal inductor (XTC cells) to an isolated avian epiblast results in the induction of the epiblast to form axial mesodermal structures.

Induction studies of this type have rarely been conducted on mammalian embryos because of the difficulty in obtaining sufficient materials at the appropriate stages. However, with the increasing availability of sensitive probes and molecular techniques that can amplify extremely small amounts of RNAs, the analysis of inductions in early rodent and even human embryos is not too distant.

Early Formation of the Neural Plate

The first obvious morphological response of the embryo to primary induction is the transformation of the dorsal ectoderm overlying the notochordal process into an elongated patch of thickened epithelial cells called the **neural plate** (Fig. 4-11).

With the formation of the neural plate, the ectodermal germ layer becomes subdivided into two developmental lineages, neural and nonneural. This example illustrates several fundamental developmental concepts—restriction, determination, and differentiation. The zygote and blastomeres resulting from the first couple of cleavage divisions are **totipotent** (i.e., capable of forming any cell in the body). As development progresses, certain decisions are made that narrow the developmental options of cells (Fig. 4-12). For example, at an early stage in cleavage, some cells become committed to the extraembryonic trophoblastic line and are no longer capable of participating in the formation of the embryo itself. At the decision point where cells become committed to becoming trophoblast, a **restriction** event has occurred. When a cell or group of cells has passed its last decision point (e.g., the transition from cytotrophoblast to syncytiotrophoblast), their fate is fixed, and they are said to be **determined.** These terms, which were coined in the early days of experimental embryology, are now understood to reflect limitations in gene expression as cell lineages follow their normal developmental course. The rare instances in which cells or tissues strongly deviate from their normal developmental course, a phenomenon called **metaplasia,** are of considerable interest to pathologists and those who study the control of gene expression.

Restriction and determination signify the progressive limitation of the developmental capacities in the embryo. **Differentiation** describes the actual morphological or functional expression of the portion of the genome that remains available to a particular cell or group of cells. Differentiation commonly connotes the course of phenotypic specialization of cells. One example of differentiation occurs in spermatogenesis, during which spermatogonia, relatively ordinary-looking cells, become transformed into highly specialized spermatozoa. (Other examples of cellular differentiation are detailed in Chapter 9.)

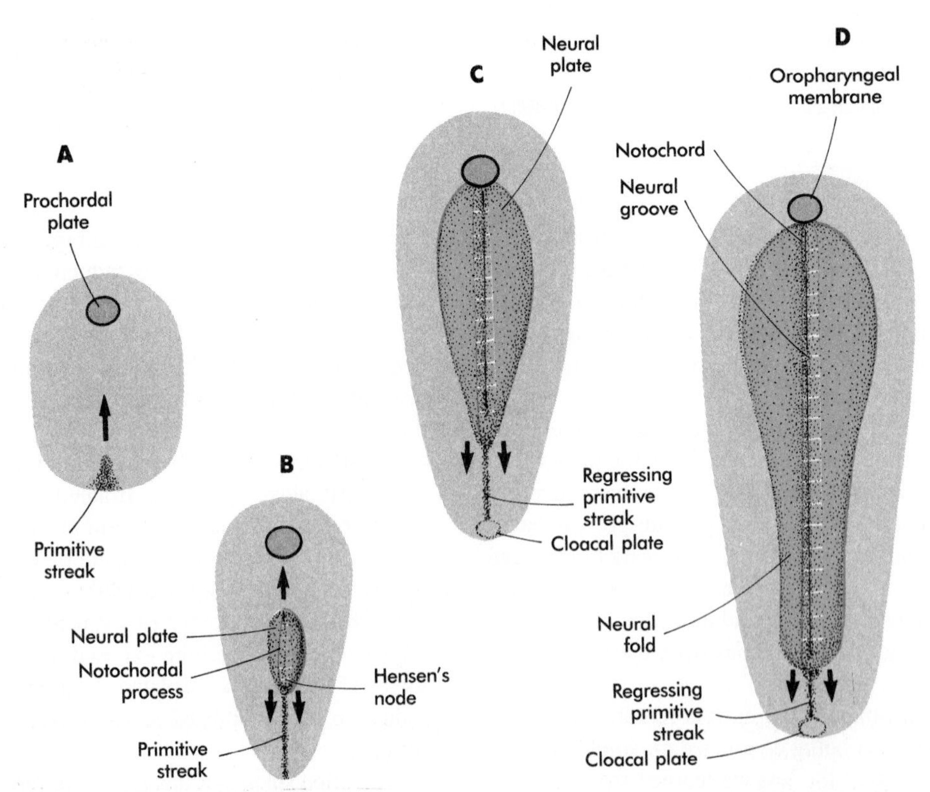

FIG. 4-11 Relationships between the neural plate and primitive streak. **A,** Day 15. **B,** Day 18. **C,** Day 19. **D,** Days 20 to 21.

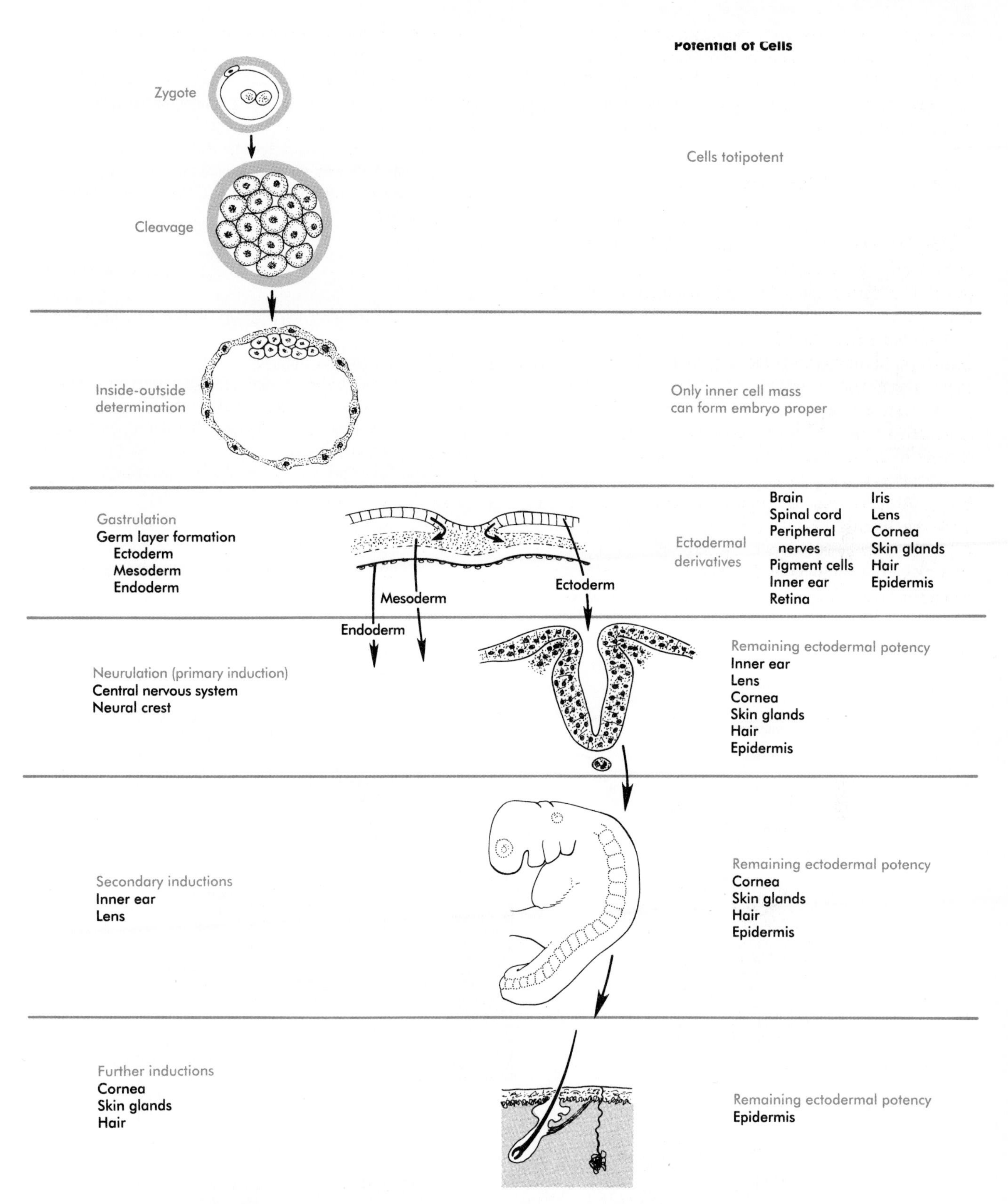

FIG. 4-12 Restriction during embryonic development. The labels on the right illustrate the progressive restriction of the developmental potential of cells that are in the line leading to formation of the epidermis. On the left are developmental events that remove groups of cells from the epidermal track.

TABLE 4-1 Cell Adhesion Molecules

MOLECULE	BINDING MECHANISM	ION DEPENDENCE	TISSUE DISTRIBUTION
N-CAM (D_2, BSP-2)	Homophilic*	Ca^{++} independent	*Early:* Epiblast, neural plate and tube, placodes, mesoderm *Late:* Nervous system, skeletal muscle (motor end plate), cardiac muscle, smooth muscle, adrenal cortex, gonads, renal epithelia, spleen, gut, mesenteries
Ng-CAM (L1, NILE)	Heterophilic†	Ca^{++} independent	*Late:* Neurons and glial cells
L-CAM (E-cadherin, uvomorulin, Cell CAM 120/80, Arc-1)	Homophilic	Ca^{++} dependent	*Early:* Inner cell mass, trophoblast *Late:* Nonneural ectoderm, urogenital epithelium (mesonephric duct, paramesonephric duct), epithelia of gut (liver, pancreas), respiratory epithelia, pharyngeal glands
N-cadherin (A-CAM, C-Cal-CAM)	Homophilic	Ca^{++} dependent	*Early:* Mesoderm, notochord *Late:* Nervous system, lens, striated muscle, primordial germ cells, renal primordia
P-cadherin	Homophilic	Ca^{++} dependent	*Early:* Extraembryonic ectoderm, endoderm, notochord, lateral plate mesoderm *Late:* Epidermis, pigment layer of retina, placenta

Data from Edelman GM: *Annu Rev Cell Biol* 2:81-116, 1986; and Takeichi M: *Development* 102:639-655, 1988.
*Homophilic: binding to cells of the same type.
†Heterophilic: binding to cells of a different type.

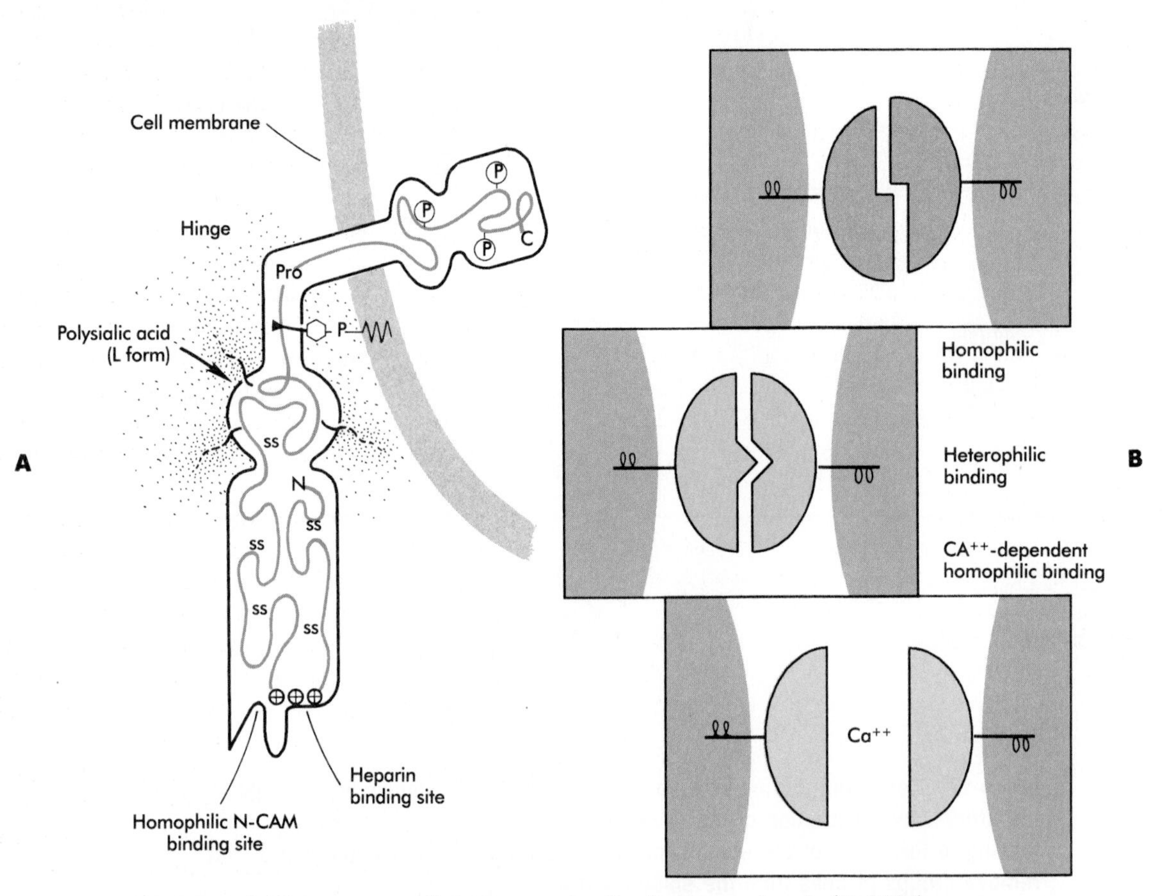

FIG. 4-13 **A,** The structure of the neuronal cell adhesion molecule (N-CAM). (Based on studies by Rutishauser and Jessell [1988].)
B, Types of intercellular adhesion.

CELL ADHESION MOLECULES

In the early 1900s, researchers determined that suspended cells of a similar type have a strong tendency to aggregate. If different types of embryonic cells are mixed together, they typically sort according to tissue type. Their patterns of sorting even give clues to their properties and behavior in the mature organism. For example, if embryonic ectodermal and mesodermal cells are mixed, they come together into an aggregate with a superficial layer of ectodermal cells surrounding a central aggregate of mesodermal cells. The organization of certain abnormal remnants of embryonic cells (such as cysts) and certain wound-healing phenomena can be related to differential patterns of cell adhesion.

Research conducted in recent years has begun to provide a molecular basis for many of the cell aggregation and sorting phenomena described by the earlier embryologists. Several families of **cell adhesion molecules (CAMs)** have been characterized (Table 4-1).

CAMs have been extensively studied in relation to early developmental events. In pregastrulation avian embryos, cells of both epiblast and hypoblast contain two CAMs (N-CAM and L-CAM) on their surfaces. A significant change occurs when cells of the epiblast migrate through the primitive streak. They lose the expression of both CAMs during the migratory phase and while they are beginning to form an organized mesoderm. Later, certain mesodermal cell derivatives reexpress CAMs. In general, when an epithelial cell type becomes transformed into a mesenchymal cell, its surface CAMs are lost.

The expression of CAMs is a sensitive indicator of primary induction in the early embryo. Before induction, the epiblast expresses both N-CAM and L-CAM. After primary induction of the nervous system, the cells of the neural plate retain N-CAM but lose the expression of L-CAM. Conversely, in nonneural ectoderm, N-CAM expression is lost but L-CAM expression is retained.

Some CAMs require the presence of Ca^{++} to function, whereas others are independent of Ca^{++}. N-CAM, a Ca^{++}-independent CAM, binds directly to other N-CAM molecules on neighboring cells of the same type (Fig. 4-13). N-CAM is unusual in having a high concentration of **negatively charged sialic acid** groups in the carbohydrate component of the molecule, and embryonic forms of N-CAM have three times as much sialic acid as the adult form of the molecule.

Undoubtedly many more types of CAMs will be discovered. A knowledge of the way this class of molecules functions during normal development will likely provide important clues to the genesis of a number of genetic and environmentally induced birth defects.

SUMMARY

1. Just before implantation, the inner cell mass becomes reorganized as an epithelium (epiblast), and a second layer (hypoblast) begins to form beneath it. Within the epiblast, the amniotic cavity forms by cavitation; outgrowing cells of the hypoblast give rise to the endodermal lining of the yolk sac. Extraembryonic mesoderm appears to form by an early transformation of parietal endodermal cells and cells migrating through the primitive streak.
2. During gastrulation, a primitive streak forms in the epiblast at the caudal end of the bilaminar embryo. Cells migrating through the primitive streak form the three embryonic germ layers—ectoderm, mesoderm, and endoderm.
3. Hensen's node, located at the cranial end of the primitive streak, is the source of the cells that become the notochord. It also functions as the organizer or primary inductor of the future nervous system.
4. As they pass through the primitive streak, future mesodermal cells change in morphology from epithelial epiblastic cells, to bottle cells, to mesenchymal cells. Extraembryonic mesodermal cells form the body stalk, connecting the embryo to the chorion. The migration of mesenchymal cells during gastrulation is facilitated by extracellular matrix molecules such as hyaluronic acid and fibronectin.
5. Late in the third week after fertilization, the primitive streak begins to regress caudally. Normally the primitive streak disappears, but sacrococcygeal teratomas occasionally form in the area of regression.
6. The essential elements of primary (neural) induction are the same in all vertebrates. In mammals, Hensen's node and the notochordal process act as the primary inductor of the nervous system. Mesodermal induction occurs even earlier than neural induction. Growth factors such as TGF-β_2 and activin are the effective agents in mesodermal induction. The hypoblast determines the origin and orientation of the primitive streak.
7. Early blastomeres are totipotent. As development progresses, cells pass through restriction points that limit their differentiation. When the fate of a cell is fixed, the cell is said to be *determined*. Differentiation refers to the actual expression of the portion of the genome that remains available to a determined cell, and the term connotes the course of phenotypic specialization of a cell.
8. Embryonic cells of the same type adhere to one another and will reaggregate if separated. The molecular basis for cell aggregation and adherence is the presence of cell adhesion molecules (CAMs) on their surfaces. There are several families of adhesion molecules, some of which are Ca^{++} dependent and some of which are Ca^{++} independent.

REVIEW QUESTIONS

1. Which layer of the bilaminar (two-layered) embryo gives rise to all of the embryonic tissue proper?
2. Of what importance is Hensen's node in embryonic development?
3. The migration of mesodermal cells from the primitive streak is facilitated by the presence of what molecules of the extracellular matrix?
4. What molecules can bring about mesodermal induction in an early embryo?
5. At what stage in the life history of many cells are cell adhesion molecules lost?

REFERENCES

Edelman GM: Cell adhesion molecules in the regulation of animal form and tissue pattern, *Annu Rev Cell Biol* 2:81-116, 1986.

Edelman GM and others: Early epochal maps of two different cell adhesion molecules, *Proc Natl Acad Sci USA* 80:4384-4388, 1983.

Enders AC: Trophoblastic differentiation during the transition from trophoblastic plate to lacunar stage of implantation in the rhesus monkey and human, *Am J Anat* 186:85-98, 1989.

Enders AC, Hendricks AG, Schlafke S: Implantation in the rhesus monkey: initial penetration of endometrium, *Am J Anat* 167:275-298, 1983.

Enders AC, King BF: Formation and differentiation of extraembryonic mesoderm in the rhesus monkey, *Am J Anat* 181:327-340, 1988.

Hemmati-Brivanlou A, Stewart RM, Harland RM: Region-specific neural induction of an *engrailed* protein by anterior notochord in *Xenopus, Science* 250:800-802, 1990.

Hogan BLM, Thaller C, Eichele G: Evidence that Hensen's node is a site of retinoic acid synthesis, *Nature* 359:237-241, 1992.

Luckett WP: The development of primordial and definitive amniotic cavities in early rhesus monkey and human embryos, *Am J Anat* 144:149-168, 1975.

Luckett WP: Origin and differentiation of the yolk sac and extraembryonic mesoderm in presomite human and rhesus monkey embryos, *Am J Anat* 152:59-98, 1978.

Mitrani E, Shimoni Y: Induction by soluble factors of organized axial structures in chick epiblasts, *Science* 247:1092-1094, 1990.

Mitrani E and others: Activin can induce the formation of axial structures and is expressed in the hypoblast of the chick, *Cell* 63:495-501, 1990.

Riou JF and others: Tenascin: a potential modulation of cell-extracellular matrix interactions during vertebrate embryogenesis, *Biol Cell* 75:1-9, 1992.

Rutishauser U, Jessell TM: Cell adhesion molecules in vertebrate neural development, *Physiol Rev* 68:819-857, 1988.

Smith JC, Dale L, Slack JMW: Cell lineage labels and region-specific markers in the analysis of inductive interactions, *J Embryol Exp Morph* 89(suppl):317-331, 1985.

Spemann H: *Embryonic development and induction,* New York, 1938, Hafner.

Spemann H, Mangold H: Ueber Induktion von Embryonenanlagen durch Implantation artfremder Organisatoren, *Arch Microskop Anat Entw-Mech* 100:599-638, 1924.

Takeichi M: The cadherins: cell-cell adhesion molecules controlling animal morphogenesis, *Development* 102:639-655, 1988.

Tiedemann H: Cellular and molecular aspects of embryonic induction, *Zool Sci* 7:171-186, 1990.

Townes PL, Holtfreter J: Directed movements and selective adhesion of embryonic amphibian cells, *J Exp Zool* 128:53-120, 1955.

Yamada T: Regulations in the induction of the organized neural system in amphibian embryos, *Development* 110:653-659, 1990.

CHAPTER 5

Establishment of the Basic Embryonic Body Plan

After gastrulation is complete, the embryo proper consists of a flat, three-layered disk containing the ectodermal, mesodermal, and endodermal germ layers. Its cephalocaudal axis is defined by the location of the primitive streak. Because of the pattern of cellular migration through the primitive streak and the regression of the streak toward the caudal end of the embryo, a strong **cephalocaudal gradient** of maturity is established. This gradient is marked initially by the formation of the notochord and later by the appearance of the neural plate, which results from the primary induction of the dorsal ectoderm by the notochord.

Despite the relatively featureless morphological appearance of the early postgastrulation embryo during the third week, evidence is increasing that during this period, if not earlier, the pattern of the basic body plan is being established. One of the earliest manifestations of this pattern is the regular segmentation that becomes evident along the craniocaudal axis of the embryo. Such a segmental plan, which is a dominant characteristic of early embryos, becomes less obvious as development progresses. Nonetheless, even in the adult, the regular arrangement of the vertebrae, the ribs, and the spinal nerves persists as a reminder of humans' highly segmented phylogenetic and ontogenetic past. Only in recent years have embryologists begun to understand the molecular and cellular controls that underlie the process of segmentation.

Another major change critical in understanding the fundamental basis of the body plan is the lateral folding of the early embryo from three essentially flat, stacked, pancake-like disks of cells (the primary embryonic germ layers) to a cylindrical configuration, with the ectoderm on the outside, the endoderm on the inside, and the mesoderm in between. The basis for the lateral folding still remains better described than understood.

This chapter concentrates on the establishment of the basic overall body plan. In addition, it charts the appearance of the primordia of the major organ systems of the body from the undifferentiated primary germ layers (Table 5-1).

THE MOLECULAR BASIS FOR THE ORGANIZATION OF THE VERTEBRATE BODY PLAN

For many years, vertebrate embryologists viewed *Drosophila* as an arcane organism that was marvellous for the study of genetics but of questionable relevance for any real understanding of developmental mechanisms in vertebrates, especially mammals. This viewpoint has dramatically changed as the result of an increasingly large number of molecular embryological studies that have been conducted since the mid-1980s. Presently, considerable evidence suggests that very similar molecular and developmental mechanisms underlie some fundamental aspects of early embryological development in both *Drosophila* and humans as well as all other vertebrates and an unknown number of invertebrate species. These similarities are most striking when the setup of the basic body plan in the early embryo is examined. A brief introduction to *Drosophila* development serves as a reference point when developmentally important genes homologous to those in *Drosophila* are discussed in this chapter and later in the text.

Embryonic development of *Drosophila* is under tight genetic control. In the earliest stages, the dorsoventral and anteroposterior axes are fixed by the actions of batteries of **maternal effect genes** (Fig. 5-1). Once these broad parameters have been established, the oval-shaped embryo undergoes a series of three sequential steps that result in the segmentation of the entire embryo along its anteroposterior axis. The first step in segmentation, under the control of

TABLE 5-1 Germ Layer Origins of Major Organs in the Body

EMBRYONIC PRECURSOR STRUCTURE	ADULT STRUCTURE
ECTODERM	
Neural tube	Brain: neurohypophysis, cranial motor nerves, epiphysis, optic nerve and retina
	Spinal cord: spinal motor nerves
Neural crest	Cranial crest derivatives: sensory ganglia, parasympathetic ganglia, glial and Schwann cells, leptomeninges, melanocytes, carotid body and parafollicular cells, many bones of face and cranium, visceral cartilages (throat), connective tissue, minor muscles
	Trunk crest derivatives: spinal ganglia, parasympathetic ganglia, satellite and Schwann cells, melanocytes, adrenal medulla
Outer epithelium of body	Epidermis
	Skin glands, hair, nails
	Nasal epithelium
	Oral epithelium and tooth enamel
	Adenohypophysis
	Lens of eye, cornea
	Inner ear
MESODERM	
Paraxial (somites)	Connective tissue of skin
	Skeletal muscles
	Axial skeleton
Intermediate	Kidneys
	Genital structures
	Renal and genital ducts
Lateral	Somatic: connective tissue of ventral body wall, parietal peritoneum, blood vessels, limbs
	Splanchnic: adrenal cortex, visceral peritoneum, heart, blood vessels
ENDODERM	
Primitive gut	Digestive tube
	Respiratory epithelium
	Digestive glands
	Pharyngeal glands
	Eustachean tube and lining of middle ear
	Urinary bladder

what are called **gap genes,** subdivides the embryo into broad regional domains. In the second step, a group of **pair-rule genes** is involved in the formation of individual body segments. The third level in the segmentation process is controlled by the **segment-polarity genes,** which work at the level of individual segments and are involved in their anteroposterior organization.*

The segmentation process results in a regular set of subdivisions along the entire anteroposterior axis of the early *Drosophila* embryo, but none of the previously mentioned developmental controls impart specific or regional characteristics to the newly formed segments. This function is relegated to two large families of **homeotic genes** called the **antennapedia** complex and the **bithorax** complex. The specific genes in these two complexes determine the morphogenetic character of each body segment (Fig. 5-2). Mutations of homeotic genes have long been known to produce bizarre malformations in insects, such as extra sets of wings or limbs instead of antennae (hence the term "antennapedia"). Close parallels exist between homeobox genes in the mouse and human (Fig. 5-3).*

Recent molecular biological research has shown that the

*In *Drosophila,* each segment is subdivided into anterior and posterior halves. The posterior half of one segment and the anterior half of the next are collectively known as a **parasegment.** The genetics and developmental aspects of insect parasegments are beyond the scope of this text, but later in this chapter when formation of the vertebral column is discussed, a similar set of divisions of the basic body segments in vertebrate embryos is introduced.

*With the discovery of homeobox-containing genes in many vertebrate species, several complex sets of terminology of the genes in the main homeobox complexes have arisen. Just before this book went to press, a unifying set of nomenclature for the four main homeobox complexes had been proposed (Scott, 1992). For ease of transition from old to new terminology, the original terminology is listed first, with the new terminology following in parentheses, whenever mammalian homeobox-containing genes of the clusters illustrated in Figure 5-3 are mentioned.

Genetic hierarchy	Functions	Representative genes	Effects of mutation
Maternal effect genes	Establish gradients from anterior and posterior poles of the egg	Bicoid Swallow Oskar Caudal Torso Trunk	Major disturbances in anteroposterior organization
Segmentation genes			
Gap genes	Define broad regions in the egg	Hunchback Krüppel Knirps Tailless	Adjacent segments missing in a major region of the body
Pair-rule genes	Define 7 segments	Hairy Even skipped Runt Fushi tarazu Odd paired Odd skipped Paired	Part of pattern deleted in every other segment
Segment polarity genes	Define 14 segments	Engrailed Gooseberry Hedgehog Wingless	Segments replaced by their mirror images
Homeotic genes	Determine regional characteristics	Antennapedia complex Bithorax complex	Inappropriate structures form for a given segmental level

FIG. 5-1 The sequence of genetic control of early development in *Drosophila.* Within each level of genetic control are listed representative genes.
(Modified from Dressler GR, Gruss P: *Trends Genet* 4:214-219, 1988.)

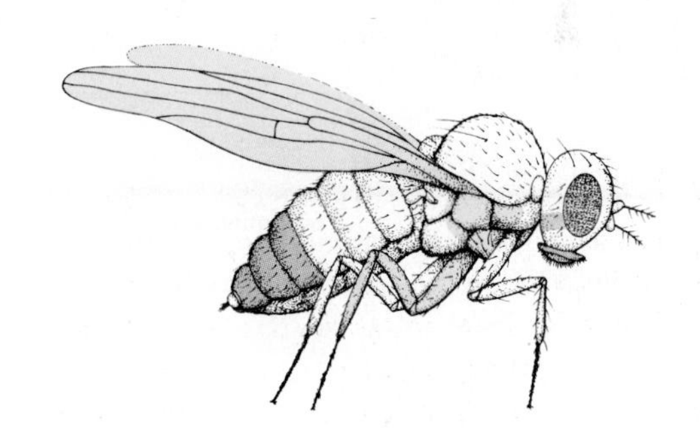

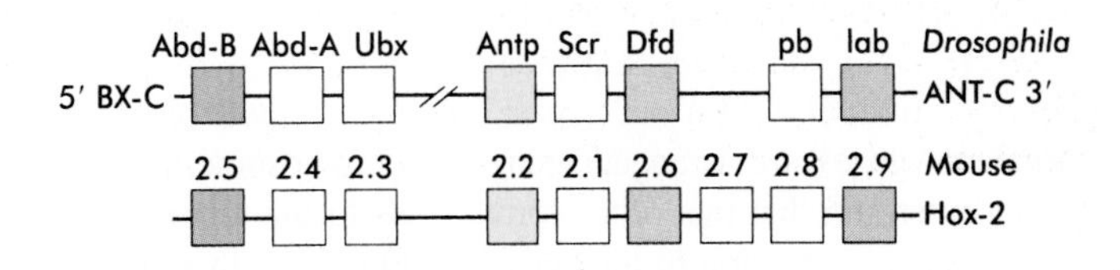

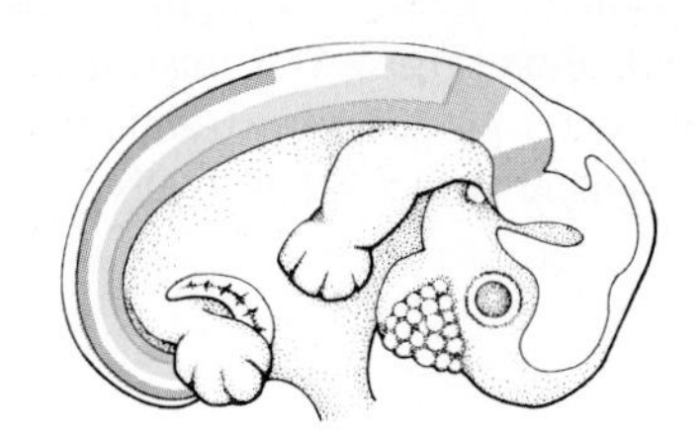

FIG. 5-2 The organization of certain homeobox-containing genes of *Drosophila* and the mouse and their segmental expression in the body. (For current *Hox* terminology, see Fig. 5-3.)
(Based on studies by DeRobertis et al [1990] and Schindler [1990].)

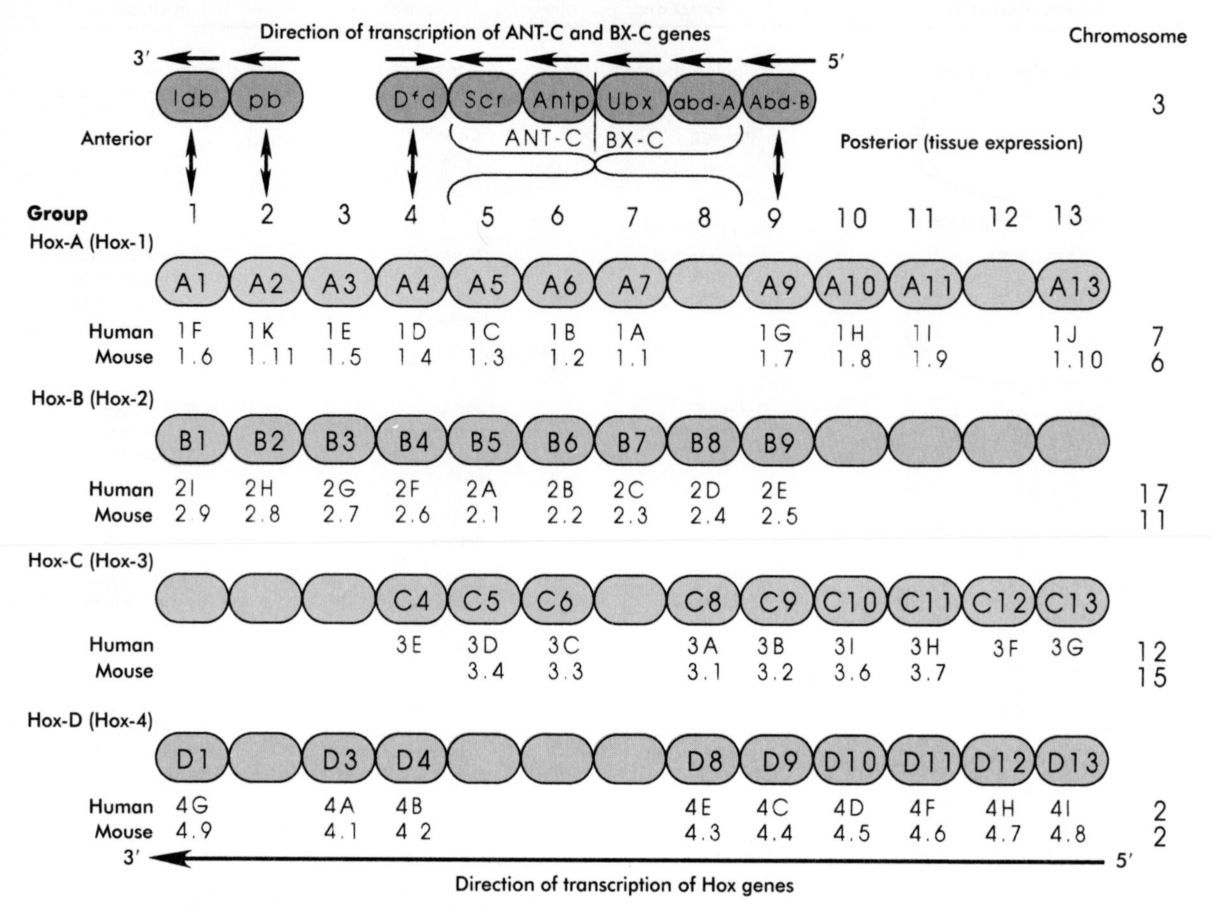

FIG. 5-3 Comparison of mouse and human *Hox* families. The new terminology is represented by a letter followed by a number.
(Based on review by Scott [1992].)

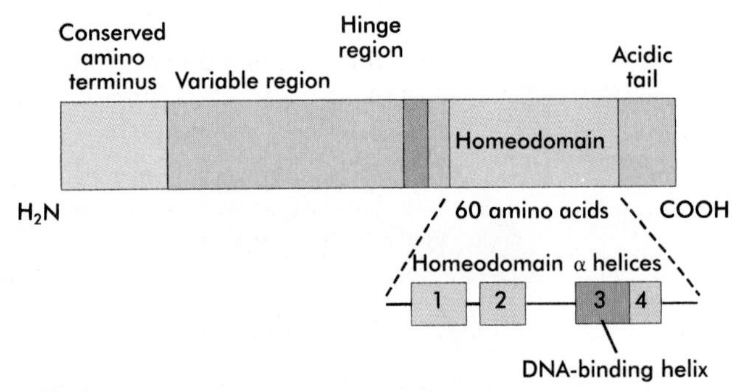

FIG. 5-4 The structure of a typical homeodomain protein.

homeotic genes are arranged along the chromosome in a strict order and that this order corresponds to the topography of their expression in the body (see Fig. 5-2). In addition, within the homeotic genes is a highly conserved region called the **homeobox** (Fig. 5-4). Consisting of 183 nucleotides, the homeobox produces a 61-amino acid homeodomain protein segment, which binds to DNA (sometimes within its own gene) and acts as a **transcription factor** (see p. 137). *Drosophila* has 8 homeobox genes located in two clusters on one chromosome, whereas mice and humans are known to possess at least 38 homeobox genes, which are found in four clusters on four different chromosomes (see Fig. 5-3). Homeoboxes have been found in a wide spectrum of invertebrates and vertebrates, including humans. Their specific functions are still little understood, but because of their high degree of preservation throughout phylogeny, they are assumed to have very important and fundamental functions in both development and evolution.

Because of its highly regulative nature, early mammalian development appears to operate under a less strict degree of genetic control than that of *Drosophila*. However, increasing numbers of homologues of *Drosophila* genes are found to function in the early mammalian embryo. Of particular importance are the homeotic and certain segmentation genes.

Embryologists are employing the technique of in situ hybridization to localize the products (**messenger RNAs [mRNAs]**) of the mammalian homologues of *Drosophila*

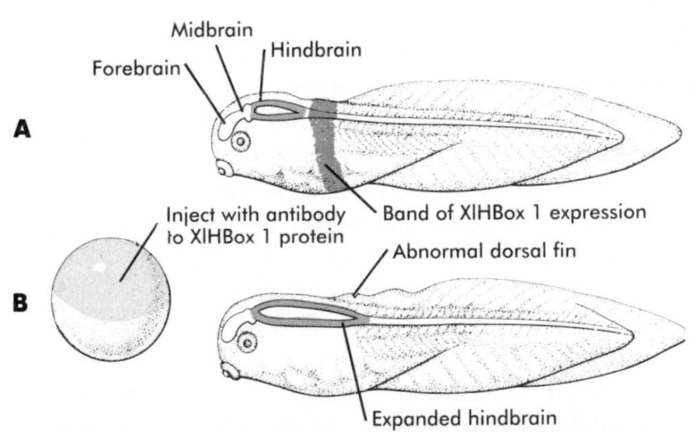

FIG. 5-5 Effect of interference of *XlHbox 1* function on development in *Xenopus.* **A,** Normal larva, showing a discrete band (green) of *XlHbox 1* expression. **B,** Caudal expansion of the hindbrain after antibodies to XlHbox 1, protein are injected into the early embryo.
(Based on studies by Wright et al [1989].)

genes in the mammalian embryo. Interestingly, the segmental expression of many of these gene products is seen most strongly in the central nervous system (see Fig. 5-2). Some of these genes become reexpressed later in development as other structures such as the limbs begin to take form. As of this writing, mapping the expression of segmentation and homeotic gene products in mammals is still in the early stages, and generating complete maps of the location of specific gene products during the various stages of mammalian embryonic development is not possible. Nevertheless, information about the location of such gene products is presented throughout the text as the development of specific organs is described. Current understanding of the functions of these gene products is still very fragmentary.

Studies on the activation of homeobox genes show a connection between certain agents that are known to affect morphogenesis (e.g., retinoic acid, growth factors) and the activation of homeobox genes. For example, **fibroblast growth factor** selectively activates posterior homeobox genes, whereas **transforming growth factor-β (TGF-β)** selectively activates anterior homeobox genes. These agents are expressed in regions of the body (e.g., Hensen's node) that are important centers of morphogenetic control.

Interference of the function of homeotic genes can cause significant alterations in morphology. For example, in *Xenopus* tadpoles, the **XlHbox 1** protein is expressed in a discrete band of both ectodermal and mesodermal cells at a level just behind the hindbrain (Fig. 5-5). If antibody to this protein is injected into a one-cell embryo, the area of anterior spinal cord that normally expresses the XlHbox 1 protein does not form. Instead the hindbrain is abnormally elongated posteriorly into that area.

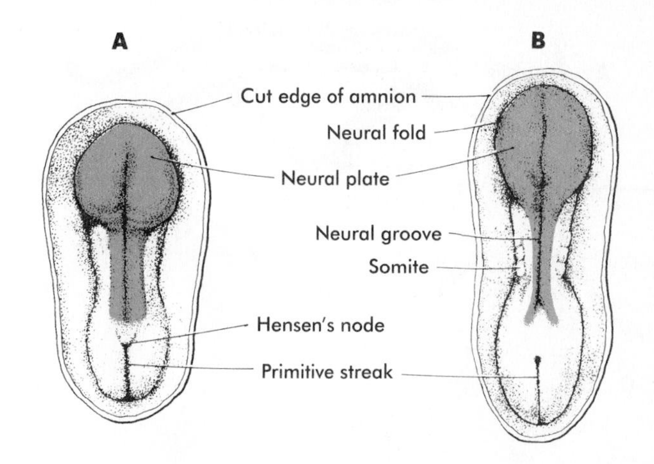

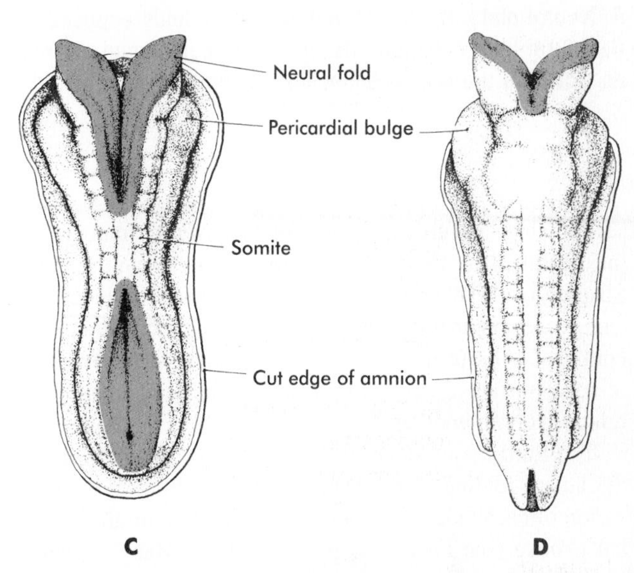

FIG. 5-6 Early stages in the formation of the human central nervous system. **A,** At 18 days. **B,** At 20 days. **C,** At 22 days, **D,** At 23 days.

DEVELOPMENT OF THE ECTODERMAL GERM LAYER

Neurulation: Formation of the Neural Tube

The principal early morphological response of the embryonic ectoderm to primary induction is an increase in the height of the cells that are destined to become components of the nervous system. These transformed cells are evident as a thickened **neural plate,** visible on the dorsal surface of the early embryo (Figs. 5-6, *A* and 5-7). Unseen but also important is the restriction in expression of cell adhesion molecules (CAMs) from N-CAM and L-CAM in the preinduced ectoderm to N-CAM in the neural plate.

Transformation of the general embryonic ectoderm into a thickened neural plate is the first of four major stages in formation of the neural tube. The principal activity of the

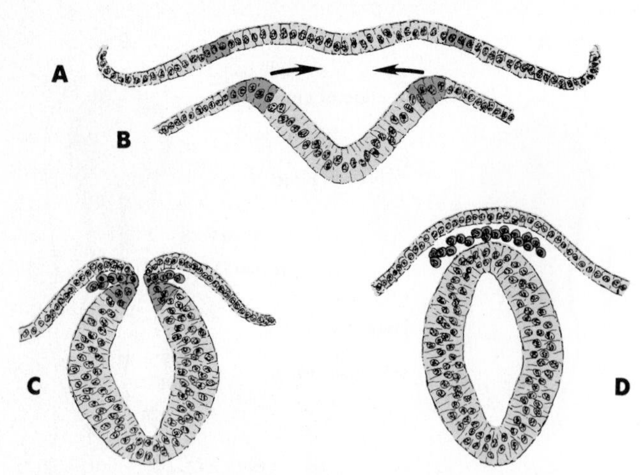

FIG. 5-7 Cross sections through the forming neural tube. **A,** Neural plate. **B,** Neural fold. **C,** Neural folds apposed. **D,** Neural tube complete. (Neural crest before and after its exit from the neural epithelium is shown in green.)

second stage is further shaping of the overall contours of the neural plate so that it becomes narrower and longer. Shaping of the neural plate is accomplished to a great extent by region-specific changes in the shape of the neuroepithelial cells (e.g., an increase in height at the expense of basal surface area) and by rearrangements of these cells relative to one another.

The third major stage in this process of **neurulation** is the lateral folding of the neural plate, resulting in the elevation of each side of the neural plate along a midline **neural groove** (see Figs. 5-6, *B* and 5-7, *B*). Many explanations have been proposed for lateral folding of the neural plate and ultimate closure of the neural tube. Most have invoked a single or dominant mechanism. It is now becoming apparent that lateral folding is the result of a number of region-specific mechanisms, both intrinsic and extrinsic to the neural plate.

The ventral midline of the neural plate, sometimes called the **median hinge point,** appears to act like an anchoring point about which the two sides become elevated at a sharp angle from the horizontal. At the median angle, bending can be accounted for to a great extent by changes in the shape of the neuroepithelial cells of the neural plate. These cells become narrower at their apex and broader at their base (see Fig. 5-7, *C*) through a combination of a basal position of the nuclei (causing a lateral expansion of the cell in that area) and a purse string–like contraction of a ring of actin-containing microfilaments in the apical cytoplasm. Throughout lateral folding of the neural plate in the region of the spinal cord, much of the structure remains flat, and elevation of the neural folds appears to be accomplished largely by factors extrinsic to the neural epithelium, in particular forces generated by the surface epithelium lateral to the neural plate.

The fourth stage in formation of the neural tube consists of apposition of the two most lateral apical surfaces of the neural folds, their fusion, and the separation of the completed segment of the neural tube from the overlying ectodermal sheet (see Fig. 5-7, *C, D*). At the same time, cells of the neural crest separate from the neural tube.

Closure of the neural tube begins almost midway along the craniocaudal extent of the nervous system of the 21- to 22-day-old embryo (see Fig. 5-6, *C*). Over the next couple of days, closure extends both cephalically and caudally in a manner superficially resembling the closing of a zipper. The unclosed cephalic and caudal parts of the neural tube are called the **anterior** (cranial) and **posterior** (caudal) **neuropores.** The neuropores also ultimately close off, and the entire future central nervous system is organized similar to an irregular cylinder sealed at both ends. Occasionally one or both neuropores remain open, resulting in serious birth defects (see p. 234).

Caudal to the posterior neuropore, the remaining neural tube (more prominent in animals with large tails) is formed by the process of **secondary neurulation.** Secondary neurulation in mammals appears to begin with the formation of a rodlike condensation of mesenchymal cells beneath the dorsal ectoderm of the tail bud. Within the mesenchymal rod, a central canal forms directly by **cavitation** (the formation of a space within a mass of cells). This central canal is continuous with the one that was formed during primary neurulation by the lateral folding of the neural plate. Because of the poor development of the tail bud, secondary neurulation in humans is not a prominent process.

Segmentation in the Neural Tube

Soon after the neural tube has taken shape, the region of the future brain can be distinguished from the remaining spinal cord. The brain-forming region undergoes a series of subdivisions that constitute the basis for the fundamental organization of the adult brain. The first set of subdivisions results in a three-part brain, consisting of a forebrain **(prosencephalon),** midbrain **(mesencephalon),** and hindbrain **(rhombencephalon).** (The further subdivision and development of the brain are covered in Chapter 12 [see Fig. 12-2].) Superimposed on this traditional organization of the developing brain is another level of segmentation, the significance of which is just beginning to be understood.

Starting with the earliest embryological investigations over 150 years ago, investigators have described transiently visible regular segments in the posterior brain-forming regions of the neural tube in almost all vertebrate embryos that have been studied. These segments were called **neuromeres** (Fig. 5-8), but their significance has been debated since their discovery. Some investigators considered them evidence that vertebrates evolved from a segmented ancestral form, but others dismissed neuromeres as biologically inconsequential. Neuromeres are only transiently demon-

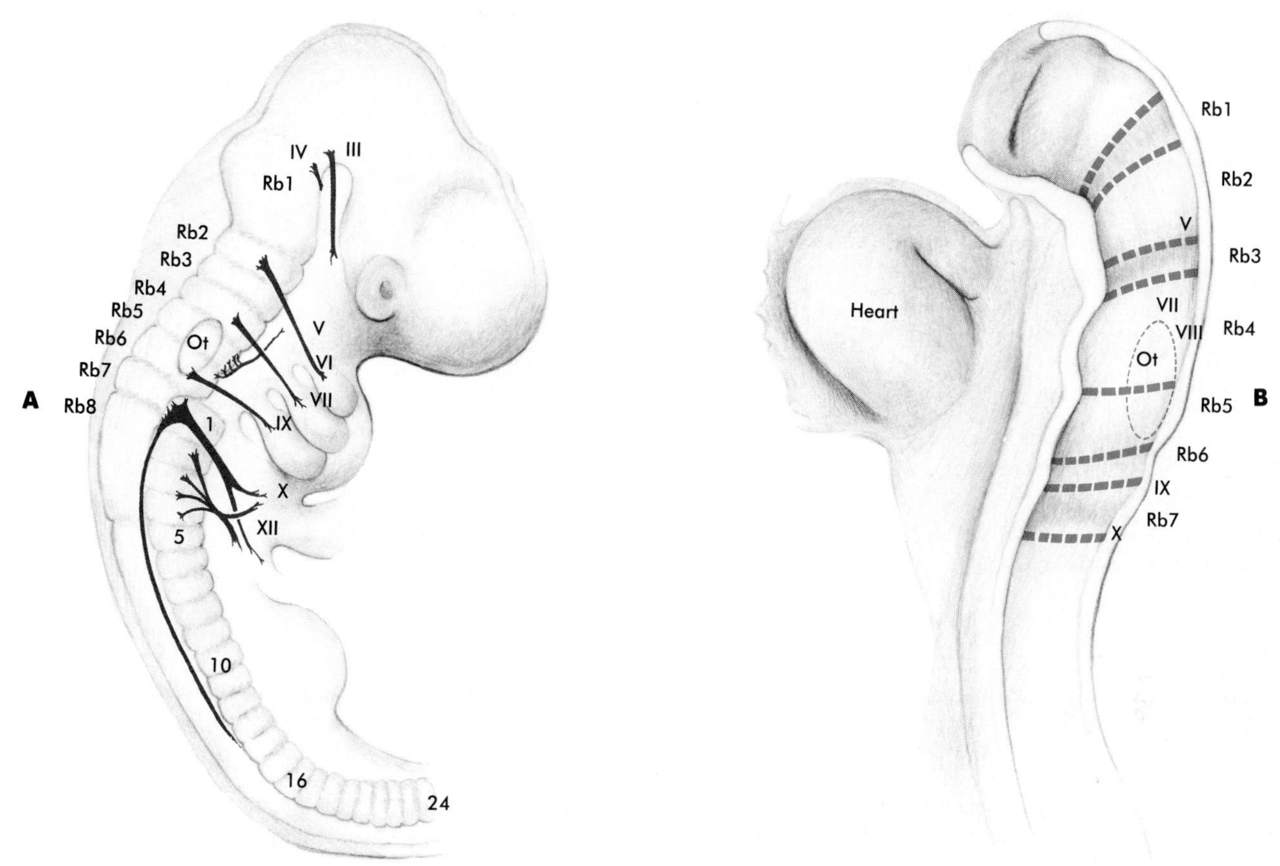

FIG. 5-8 Neuromeres in **(A)** 3-day-old chick and **(B)** 24-day-old human brain. *Ot,* Otic vesicle; *Rh,* rhombomeres (neuromeres in rhombencephalon); Roman numerals, cranial nerves; numbers, somites.

strable. In the human embryo, they can be seen from early in the fourth to late in the fifth week (Fig. 5-8, *B*). However, during their brief existence, they provide the basis for the fundamental organization of the nervous system. For instance, in the chick, the cell bodies of many of the cranial nerves are known to arise from specific neuromeres.

Individual proliferation centers in the neural tube give rise to the neuromeres. Their slightly bulging appearance may be due to local areas of rapid cell proliferation. Once established, neuromeres act like isolated compartments in insect embryos, and cells in adjacent neuromeres do not intermingle. The controls and significance of this cellular behavior in vertebrates are still poorly understood.

Studies on the expression of homeobox genes in vertebrates have shown a correspondence between the localization of specific gene products and individual neuromeres (see Fig. 15-3). Some of genes (related to the families of segmentation genes of *Drosophila*) are members of the **Hox-2** *(Hox-B)* cluster, **Krox-20, Wnt,** and **engrailed.** These homeotic genes may specify the identity of the corresponding neuromeres and their later derivatives.

Although neuromeres are not seen in the region of the neural tube from which the spinal cord arises, the regular arrangement of the exiting motor and sensory nerve roots is evidence of a fundamental segmental organization in this region of the body as well. (The relationship between the spinal nerves and the mesodermal segments of the body [the somites and their derivatives] is discussed on p. 211.)

The Neural Crest

When the neural tube has just been completed and is separating from the general cutaneous ectoderm, a population of cells called the **neural crest** leaves the dorsal part of the neural tube and begins to spread throughout the body of the embryo (see Fig. 5-7). The neural crest produces an astonishing array of structures in the embryo (see Table 13-1), and its importance is such that the neural crest is sometimes called the fourth germ layer of the body. (Further consideration of the neural crest is presented in Chapter 13.)

Sensory Placodes and Secondary Inductions in the Cranial Region

As the cranial region begins to take shape, several series of ectodermal **placodes** (ectodermal thickenings) appear lat-

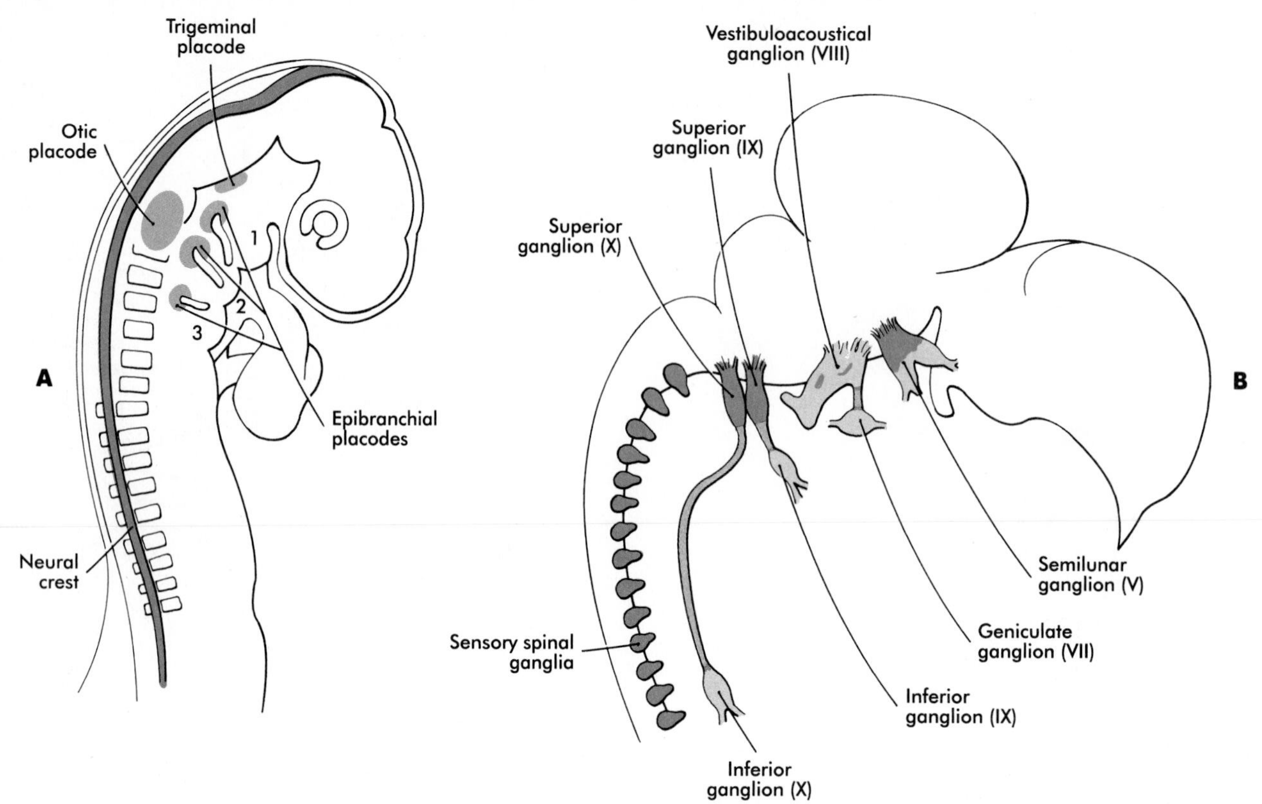

FIG. 5-9 Ectodermal placodes and neural crest in the formation of sensory ganglia of cranial and spinal nerves in the chick embryo. **A,** At 2 days. **B,** At 8 days. Neural crest (green); placodes (blue).

(Modified from LeDouarin and others: *Trends Neurosci* 9[4]:175, 1986.)

eral to the neural tube. Most, if not all, result from secondary inductive processes between other tissues (in most cases, the neural tube or neural crest) and the overlying ectoderm. Among the prominent early placodes are the paired lens placodes, which ultimately form the lens of the eye (see Fig. 14-1) and the otic placodes, which form the inner ear (see Fig. 14-21). In the most rostral regions of the head, ectodermal placodes give rise to the olfactory sensory epithelium of the nose, and a similar invagination from the roof of the stomodeum gives rise to the anterior lobe of the pituitary gland. In the region of the hindbrain, several sets of placodes, developing in concert with local neural crest, share in the formation of the sensory ganglia of the cranial nerves (Fig. 5-9).

DEVELOPMENT OF THE MESODERMAL GERM LAYER

Basic Plan of the Mesodermal Layer

After passing through the primitive streak, the mesodermal cells spread laterally between the ectoderm and endoderm as a continuous layer of mesenchymal cells (see Fig. 4-6). Subsequently, three regions can be recognized in the mesoderm of cross-sectioned embryos (Fig. 5-10). Nearest the neural tube is a thickened column of mesenchymal cells known as the **paraxial mesoderm,** or **segmental plate.** This tissue is soon organized into somites. Continuous with the paraxial mesoderm is a compact region of **intermediate mesoderm,** which ultimately gives rise to the urogenital system. Beyond that, the **lateral mesoderm** ultimately splits into two layers and forms the bulk of the tissues of the body wall, the wall of the digestive tract, and the limbs (see Table 5-1).

The Paraxial Mesoderm

As Hensen's node and the primitive streak regress toward the caudal end of the embryo, they leave the notochord and the induced neural plate. Lateral to the neural plate, the paraxial mesoderm appears to be a homogeneous strip of closely packed mesenchymal cells. However, if scanning electron micrographs of this mesoderm are examined with stereoscopic techniques, a series of regular pairs of segments can be discerned. These segments, called **somitomeres,** have been most studied in avian embryos, but they are also found in mammals. New pairs of somitomeres form along Hensen's node as it regresses toward the caudal end of the embryo (Fig. 5-11). Not until almost 20 somitomeres

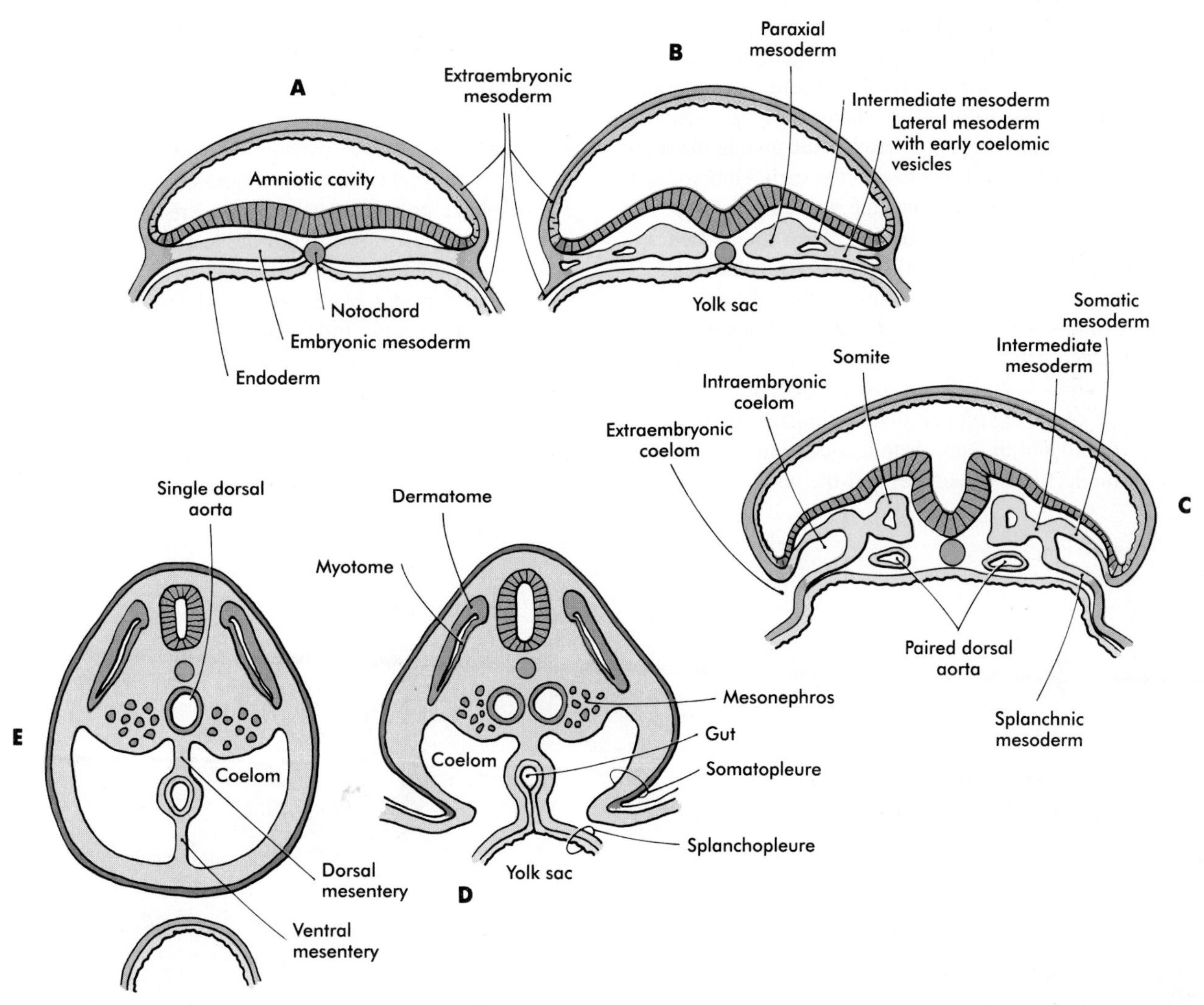

FIG. 5-10 The development of intraembryonic and extraembryonic mesoderm in cross-sections of human embryos.

FIG. 5-11 Relationship between somitomeres and somites in the early chick embryo. Cranial somitomeres *(open circles)* take shape along Hensen's node until seven pairs have formed. Caudal to the seventh somitomere, somites *(rectangles)* form from caudal somitomeres *(ovals)*. As the most anterior of the caudal somitomeres transform into somites, additional caudal somitomeres take shape posteriorly. For a while, the equilibrium between transformation into somites anteriorly and new formation posteriorly keeps the number of caudal somitomeres at 11.

(Modified from Jacobson AG: *Development* 104:209-220, 1988.)

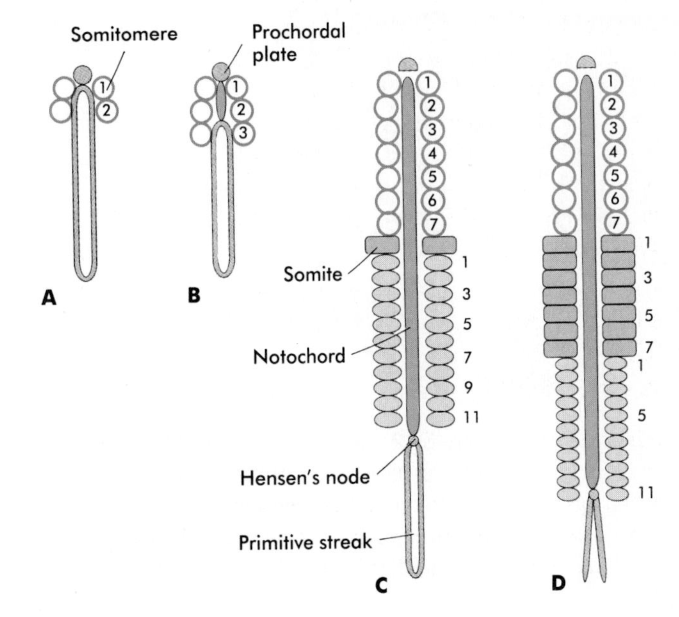

have formed and Hensen's node has regressed quite far caudally does the first pair of **somites** (brick-shaped masses of paraxial mesoderm) form behind the seventh pair of somitomeres. In contrast to the somitomeres, whose existence was not established until 1979, somites are among the most prominent topographical features of the early embryo, and they have been recognized since the sixteenth century.

After the first pair of somites has been established approximately 20 days after fertilization, a regular relationship develops between the regression of the primitive streak and the formation of additional somites and somitomeres. The first seven pairs of somitomeres in the cranial region remain as such, with the first pair of somites forming at the expense of the eighth pair of somitomeres. In the types of embryos studied to date, there is a constant relationship between the last pair of somites and the number of somitomeres (usually 10 to 11) that can be demonstrated behind them. Every few hours the pair of somitomeres located caudal to the last somites becomes transformed into a new pair of somites, and a new pair of somitomeres is laid down at the caudal end of the paraxial mesoderm near Hensen's node (see Fig. 5-11). Once the regression of Hansen's node is complete, no more somitomeres are formed, but new somites continue to form until the last of the caudal somitomeres are obliterated.

Despite considerable experimental research over recent decades, surprisingly little is understood about the mechanism and overall control of somite formation. Early models of somite formation focused on segmentation of the somites themselves, but with the discovery of somitomeres, attention is turning toward pattern formation in these structures. The way somites and somitomeres stop forming is

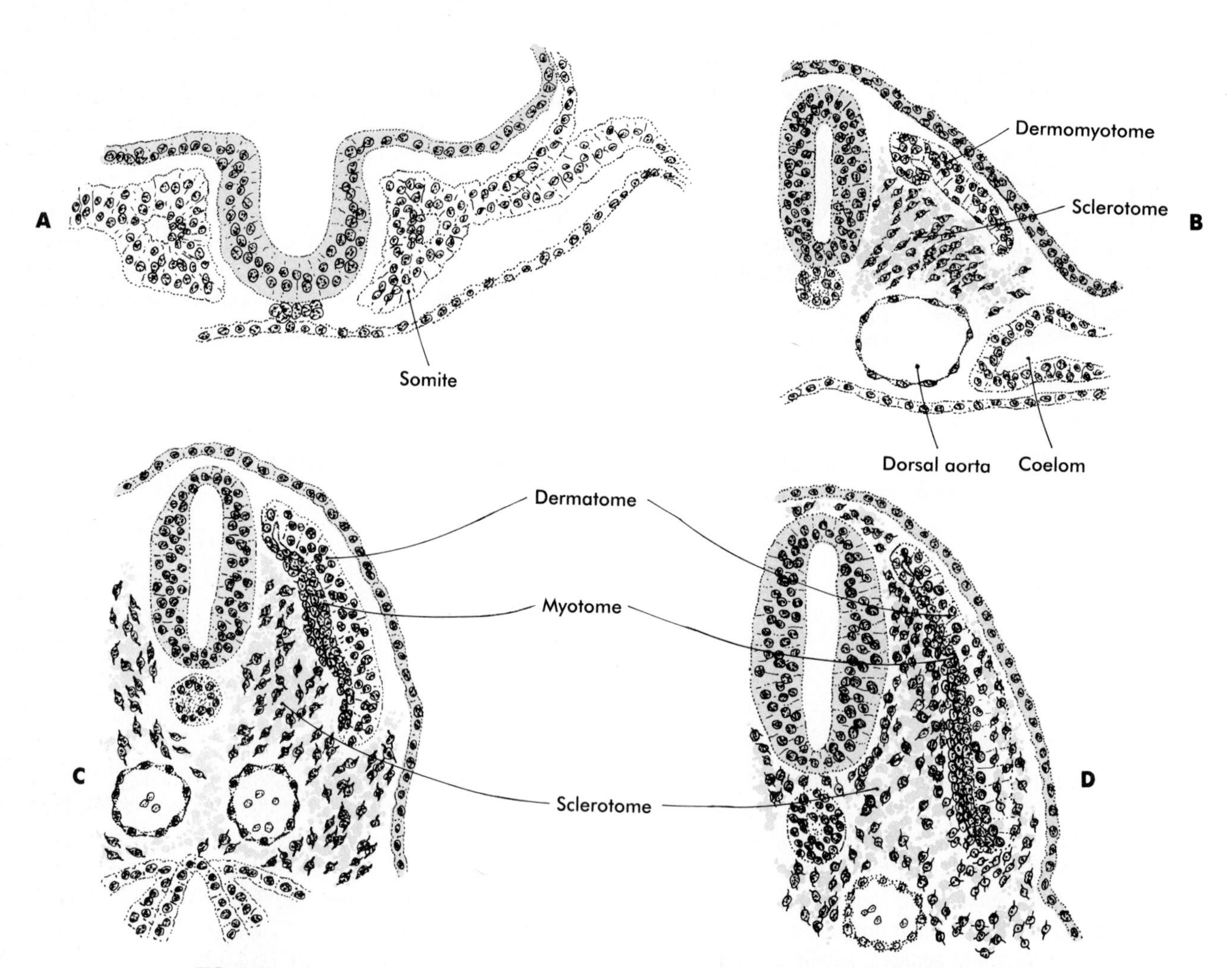

FIG. 5-12 Stages in the life history of a somite in a human embryo. **A,** Epithelial stage of a somite in the preneural tube stage. **B,** Epitheliomesenchymal transformation of the ventromedial portion into the sclerotome. **C,** Appearance of a separate myotome from the original dermomyotome. **D,** Early stage of breakup of the epithelial dermatome into dermal fibroblasts.

no better understood than the way they are generated, but one possibility is that genetically programmed cell death in the tail disrupts the paraxial mesoderm or late-forming somitomeres and acts as a stop signal.

The Formation of Individual Somites. The formation of an individual somite involves the transformation of cells with a mesenchymal morphology to a sphere of epithelial cells within the paraxial mesoderm (Fig. 5-12, *A*). This transformation is preceded by an increase in the intercellular adhesive properties of the presomitic cells. The cells of the epithelial somite are arranged with their apical surfaces around a small central lumen (which contains a small number of core cells) and their outer basal surfaces surrounded by a basal lamina (containing laminin, fibronectin, and other components of the extracellular matrix). Cell-marking studies have shown that not all cells within a somite are descended from a single cellular clone.

Shortly after the formation of the purely epithelial somite, the cells of the ventromedial wall **(sclerotome)** of the somite are subjected to an inductive stimulus from the notochord and ventral wall of the neural tube. Their response is a burst of mitosis, the loss of intercellular adhesion molecules **(N-cadherin),** the dissolution of the basal lamina in that region, and the transformation of the epithelial cells (and core cells) in that region to a mesenchymal morphology that is called **secondary mesenchyme.** These cells migrate or are otherwise displaced medially from the remainder of the somite (Fig. 5-12, *B*) and begin to produce **chondroitin sulfate proteoglycans** and other molecules characteristic of cartilage matrix. The preinduced cells of the future sclerotome produce small amounts of chondroitin sulfate, but the induction by the notochord and spinal cord causes a great increase in the synthesis of this substance. Ultimately, the cells of the sclerotome surround the notochord and neural tube to form the vertebrae, but they also form the ribs (see Chapter 10).

After the cells of the sclerotome have broken away, the remainder of the epithelial somite is known as the **dermomyotome** (see Fig. 5-12, *B*). Mesenchymal cells arising from the dorsomedial border of the dermomyotome form a separate layer, the **myotome,** beneath the remaining somitic epithelium, which is now called the **dermatome** (Fig. 5-12, *C*). The cells of the dermatome lose their epithelial configuration (Fig. 5-12, *D*) and migrate as secondary mesenchymal cells toward the ectoderm to form the **dermis** of the skin, whereas the cells of the myotome form much of the musculature of the body. In the area of the future limbs, some myotomal cells migrate into the limb buds to form the limb musculature (see Chapter 11).

Organization of the Somite and the Basic Segmental Body Plan. By the time the somite has reached the stage of sclerotome formation, it has become functionally divided into anterior and posterior half-segments, even though no morphological evidence of such a subdivision initially exists (Fig. 5-13). A major difference between the two halves is the ability of neural crest cells to penetrate the anterior half but not the posterior half. One possible reason for this may be the production by the posterior half-segment of

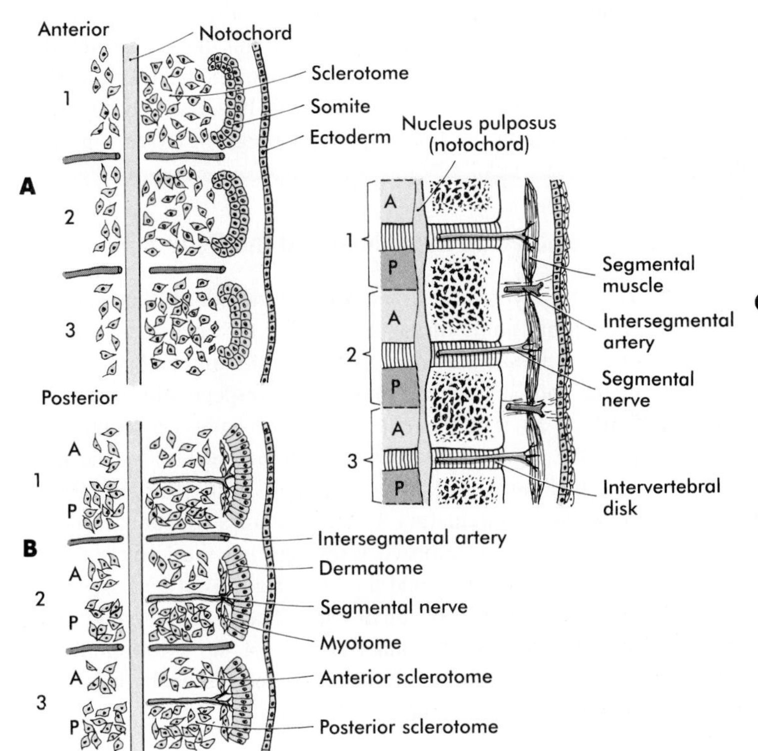

FIG. 5-13 A, Early movement of seemingly homogeneous schlerotome from the somite. **B,** The breakup of the sclerotomal portions of the somites into anterior *(A)* and posterior *(P)* halves and the coalescence of the posterior portion of one somite with the anterior portion of the one caudal to it to form the body of a vertebra. **C,** With this rearrangement, the segmental muscles (derived from the myotomes) extend across intervertebral joints and are supplied by spinal nerves that grow out between the anterior and posterior halves of the somites.

chondroitin sulfate, known to be a poor substrate for migrating neural crest cells.

There is also evidence for an intrinsic medial and lateral organization within the somite. Tracing experiments have shown that medial and lateral halves of somites are derived from different lineages during gastrulation. Other experiments have shown that the lateral halves of somites contain the myogenic (muscle-forming) cells that populate the limb buds, whereas the medial half of the somite contains the precursor cells of the axial musculature (see Chapter 10). In a third subdivision of the somite, cells of the dorsomedial quadrant form the axial muscles (myotome), and those of the ventromedial quadrant form the axial skeleton (sclerotome). The fate of the cells within the early somites is not irreversibly fixed; if segments of somites are transplanted to other locations within the somite, the fate of the transplanted cells often corresponds to the new site rather than their original locations within the somite.

As the cells of the sclerotome disperse around the notochord, cells of the anterior half of one somite aggregate with cells of the posterior half of the more cranial somite. Ultimately, this aggregate forms a single vertebra. Such an arrangement places the bony vertebrae out of phase with the myotomally derived segmental muscles of the trunk (Fig. 5-13, *C*). This allows the contracting segmental muscles to move the vertebral column laterally. The relationship between the anterior half of one somite and the adjoining posterior half of its neighboring somite is reminiscent of the parasegments of *Drosophila* (similarly arranged subdivisions of the segments into two parts), but whether they are functionally similar in terms of genetic control is undetermined.

The Intermediate Mesoderm

Connecting the paraxial mesoderm and the lateral plate mesoderm in the early embryo is a small cord of cells called the **intermediate mesoderm** that runs along the entire length of the trunk (see Fig. 5-10, *C*). The intermediate mesoderm is the precursor of the urogenital system. The earliest signs of differentiation of the intermediate mesoderm are in the most cranial regions, where vestiges of the earliest form of the kidney, the **pronephros,** briefly appear. In the lateral region of the intermediate mesoderm, a longitudinal **pronephric duct** appears on each side of the embryo. The pronephric duct is important in organizing the development of much of the adult urogenital system, which forms largely from cells of the caudal portions of the intermediate mesoderm (see Chapter 17).

The Lateral Mesoderm

The **lateral plate mesoderm** soon divides into two layers as the result of the formation and coalescence of coelomic (body cavity) spaces within it (see Fig. 5-10, *B, C*). The dorsal layer, which is closely associated with the ectoderm, is called **somatic mesoderm,** and the combination of somatic mesoderm and ectoderm is called the **somatopleure** (see Fig. 5-10, *D*). The ventral layer, called **splanchnic mesoderm,** is closely associated with the endoderm, and the combined endoderm and splanchnic mesoderm is called the **splanchnopleure.** The intraembryonic somatic and splanchnic mesodermal layers are continuous with the layers of extraembryonic mesoderm that line the amnion and yolk sac.

While the layers of somatic and splanchnic mesoderm are taking shape, the entire body of the embryo undergoes a lateral folding process that effectively transforms its shape from three flat germ layers to a cylinder, with a tube of endoderm (gut) in the center, an outer tubular covering of ectoderm (epidermis), and an intermediate layer of mesoderm. This transformation occurs before the appearance of the limbs.

Formation of the Coelom. As the embryo undergoes lateral folding, the small coelomic vesicles that formed within the lateral mesoderm coalesce into the **coelomic cavity** (see Fig. 5-10). Initially, the **intraembryonic coelom** is continuous with the **extraembryonic coelom,** but as folding is completed in a given segment of the embryo, the two coelomic spaces are separated. The last region of the embryo to undergo complete lateral folding is the area occupied by the yolk sac. In this area, small channels connecting the intraembryonic and extraembryonic coeloms persist until the ventral body wall is completely sealed.

In the cylindrical embryo, the somatic mesoderm constitutes the lateral and ventral body wall, and the splanchnic mesoderm forms the mesentery and the wall of the digestive tract. The somatic mesoderm of the lateral plate also forms the mesenchyme of the limb buds, which begin to appear late in the fourth week of pregnancy (see Fig. 11-1).

Extraembryonic Mesoderm and the Body Stalk

The thin layers of extraembryonic mesoderm that line the ectodermal lining of the amnion and the endodermal lining of the yolk sac are continuous with the intraembryonic somatic and splanchnic mesoderm, respectively (see Fig. 5-10, *A, B*). The posterior end of the embryo is connected with the trophoblastic tissues (future placenta) by the mesodermal **body stalk** (see Fig. 3-16, *D*). As the embryo grows and a circulatory system becomes functional, blood vessels from the embryo grow through the body stalk to supply the placenta, and the body stalk itself becomes better defined as the **umbilical cord.** The extraembryonic mesoderm that lines the inner surface of the cytotrophoblast (see Fig. 6-4) ultimately becomes the mesenchymal component of the placenta.

Early Stages in the Formation of the Circulatory System

As the embryo grows during the third week, it attains a size that does not permit diffusion to distribute oxygen and nutrients to all of its cells or the efficient removal of waste products. The early development of the heart and circulatory system is an embryonic adaptation that permits the rapid growth of the entire embryo complex. The circulatory system faces the daunting task of having to grow and become continuously remodeled to keep pace with the embryo's overall growth while remaining fully functional in supplying the needs of the embryo's cells.

The earliest aspects of development of the circulatory system consist of the formation of the heart and great vessels from bilaterally paired vascular tubes derived from the splanchnic mesoderm in the region of the head. The blood has a different origin, arising from **blood islands** in the mesodermal lining of the yolk sac.

The Heart and Great Vessels. The heart is derived from splanchnic mesoderm as bilateral tubular primordia located ventrolateral to the early pharynx (Fig. 5-14). Experimental studies on lower vertebrates have traced the precardiac mesoderm to earlier stages in development, where the heart-forming region constitutes a horseshoe-shaped region of mesoderm extending back on either side from the region of the prechordal plate (Fig. 5-15). An inductive influence from the neighboring endoderm has been postulated to stimulate the early formation of the heart.

In human embryos, the earliest recognizable precardiac mesoderm is a crescent-shaped zone of thickened mesoderm

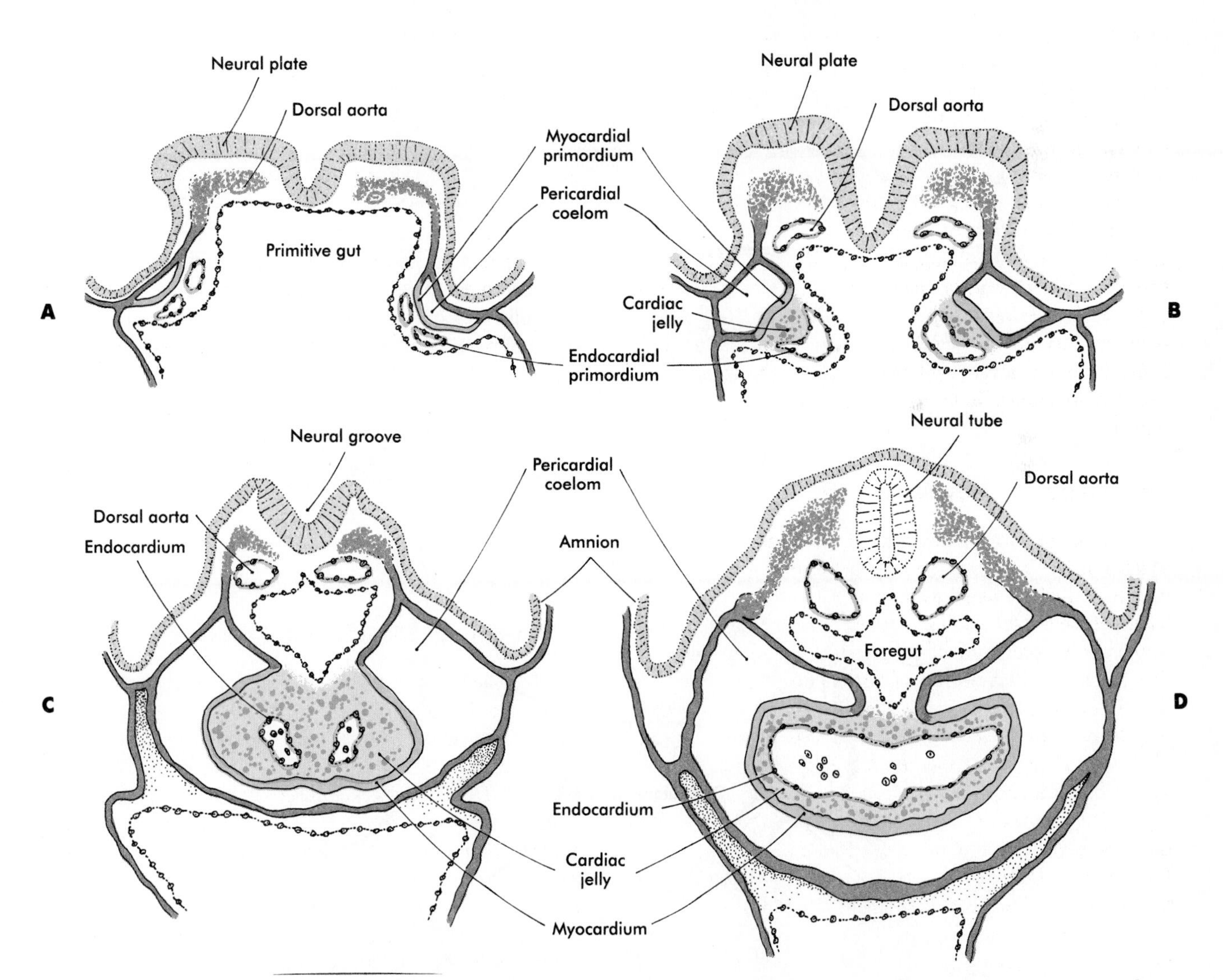

FIG. 5-14 Cross sections through the level of the developing heart from 20 to 22 days. **A,** A 2-somite embryo. **B,** A 4-somite embryo. **C,** A 7-somite embryo. **D,** A 10-somite embryo.

rostral to the embryonic disk of the gastrulating embryo early in the third week (Fig. 5-16). As the mesoderm begins to split into the splanchnic and somatic layers, a **cardiogenic plate** is recognizable in the splanchnic mesoderm underlying the anterior margins of the early neural plate (Fig. 5-16, *B*). In this area, the space between the two layers of mesoderm is the forerunner of the **pericardial cavity.** The main layer of splanchnic mesoderm in the precardiac region thickens to become the **myocardial primordium.** Between this structure and the endoderm of the primitive gut appear isolated mesodermal vesicles, which soon fuse to form the tubular **endocardial primordia** (see Fig. 5-14, *A, B*). The endocardial primordia ultimately fuse and become the inner lining of the heart.

As the head of the embryo takes shape by both lateral and ventral folding, the bilateral cardiac primordia come together in the midline ventral to the gut and fuse to form a primitive single tubular heart. This structure consists of an

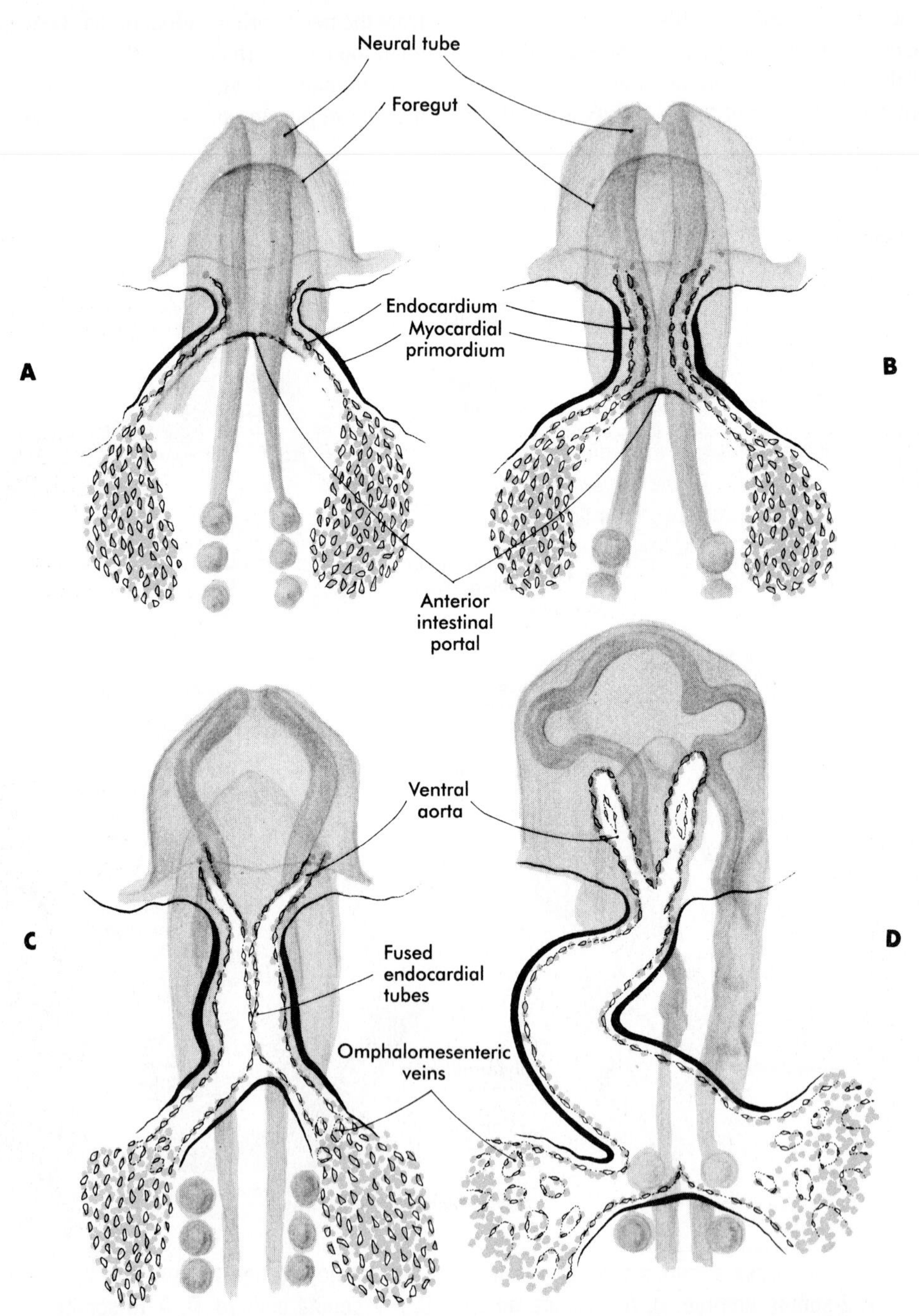

FIG. 5-15 Formation of the tubular heart in the chick embryo from paired primordia. The embryo is viewed from the ventral side. **A,** Stage 8. **B,** Stage 9−. **C,** Stage 10. **D,** Stage 11.

inner **endocardial lining** surrounded by a loose layer of specialized extracellular matrix that has historically been called **cardiac jelly** (see Fig. 5-14, *C*). Outside the cardiac jelly is the **myocardium,** which will ultimately form the muscular part of the heart. The outer lining of the heart, called the **epicardium,** is derived from the dorsal mesentery, which migrates to cover the surface of the heart. The entire tubular heart is located in the space known as the **pericardial coelom.** Shortly after the single tubular heart is formed, it begins to form a characteristic S-shaped loop that presages its eventual organization into the configuration of the adult heart (Fig. 5-17). (The cellular and molecular aspects of early cardiogenesis are covered in Chapter 18.)

The early heart does not form in isolation. At its caudal end, the endocardial tubes do not fuse but rather extend toward the posterior part of the body as the venous inflow tract of the heart (see Fig. 5-17). Similarly, the paired endothelial tubes leading out from the heart at its cranial end loop around the pharynx as the aortic outflow tract. By 21 or 22 days after fertilization, differentiation of cardiac muscle cells in the myocardium is sufficiently advanced to allow the heart to begin beating (see Chapter 12).

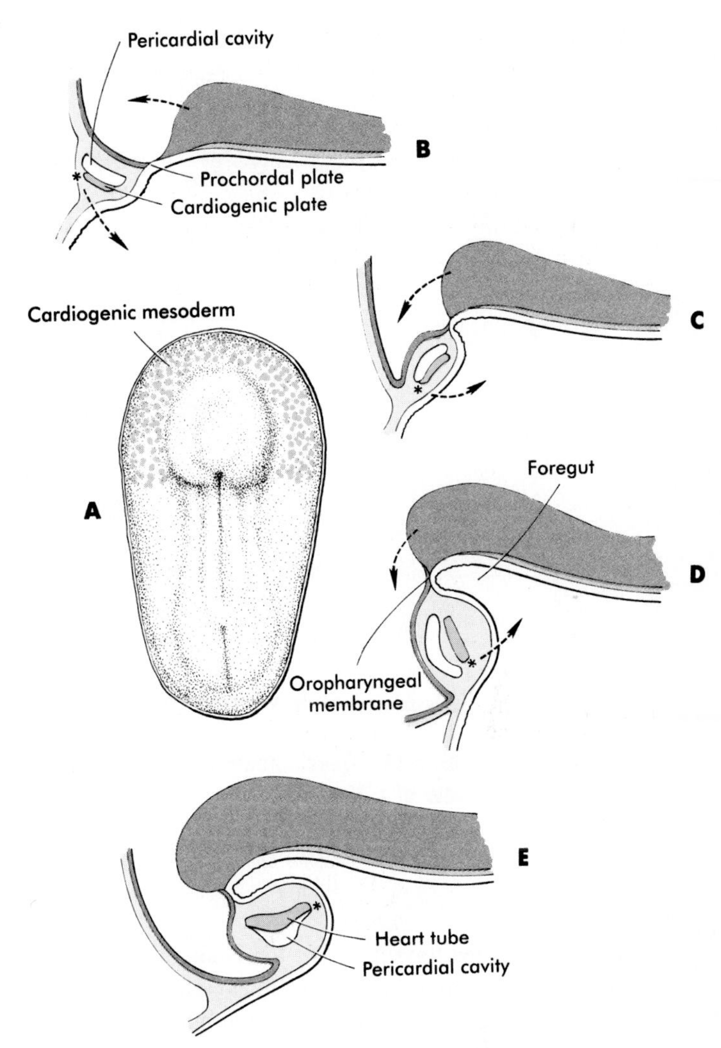

FIG. 5-16 Formation of the heart from precardiac mesoderm in the human embryo. **A,** Dorsal view of an 18-day-old embryo. **B, C, D, E,** Sagittal sections through the cranial ends of 18- to 22-day-old embryos showing the roughly 180 degree rotation of the primitive heart tube and pericardium with the expansion of the cranial end of the embryo.

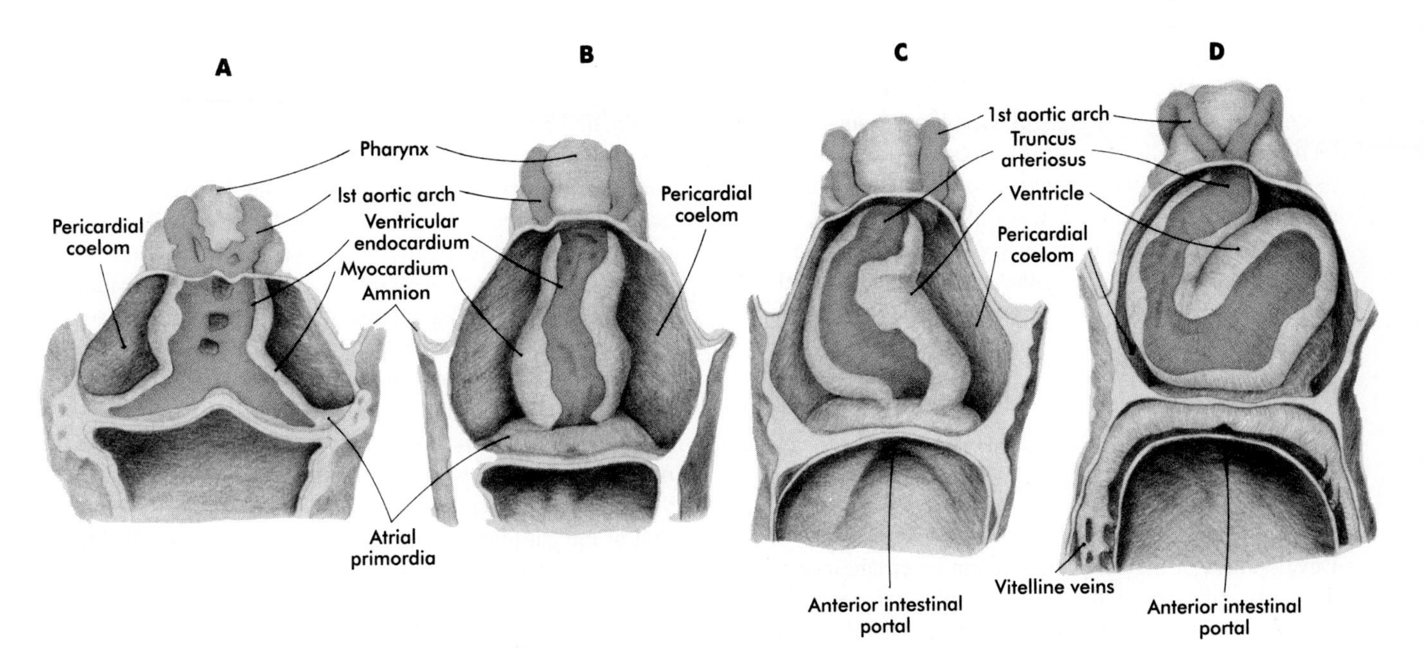

FIG. 5-17 Formation of the S-shaped heart from fused cardiac tubes in the human embryo at about 21 to 23 days. **A,** An 4-somite embryo. **B,** An 8-somite embryo. **C,** A 10- to 11-somite embryo. **D,** A 12-somite embryo.

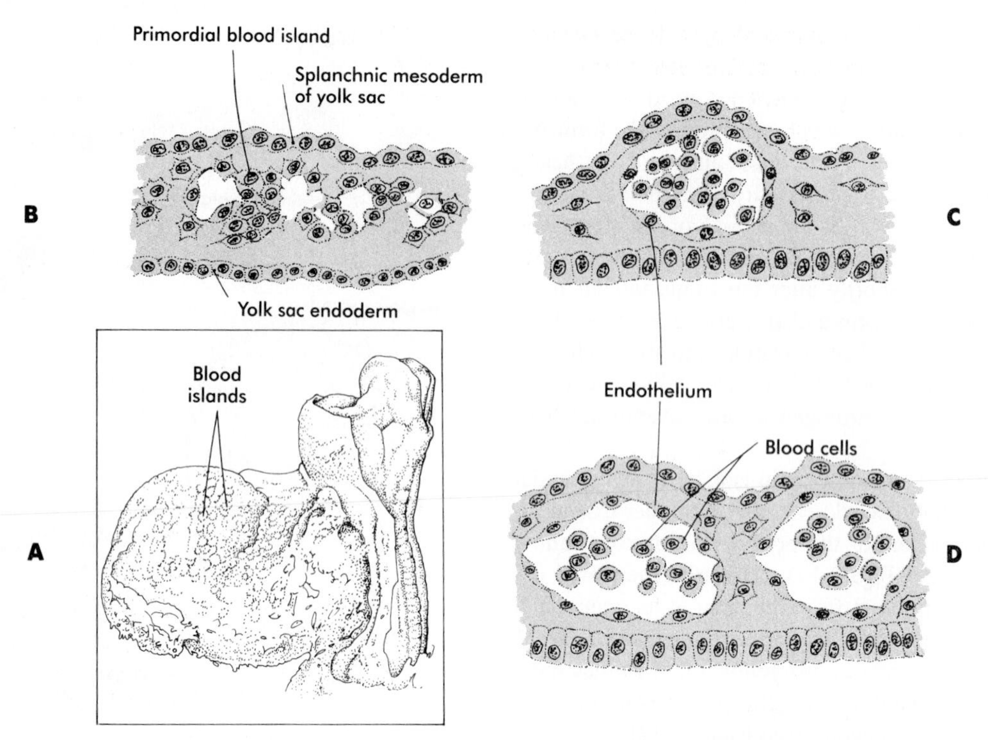

FIG. 5-18 Development of blood islands in the yolk sac of human embryos. **A,** Gross view of a 10-somite human embryo showing the location of blood islands on the yolk sac. **B, C, D,** Successive stages in the formation of blood islands.
From Corner: *Carnegie Contr Embryol* 20:81-102, 1929.)

Blood and Blood Vessels. The formation of blood and blood vessels begins in the mesodermal wall of the yolk sac as well as in the wall of the chorion outside the embryo proper. Stimulated by an inductive interaction with the endoderm of the yolk sac, many small blood islands appear in the extraembryonic splanchnic mesoderm of the yolk sac (Fig. 5-18). As the result of a poorly understood mechanism, the central cells of the blood islands become blood-forming cells **(hemocytoblasts),** whereas those on the outside acquire the characteristics of **endothelial lining cells,** which form the inner walls of blood vessels. As the vesicular blood islands in the wall of the yolk sac fuse, they form primitive vascular channels that extend toward the body of the embryo. Connections are made with the endothelial tubes associated with the tubular heart and major vessels, and the primitive plan of the circulatory system begins to take shape.

DEVELOPMENT OF THE ENDODERMAL GERM LAYER

Development of the endodermal germ layer continues with the transformation of the flat intraembryonic endodermal sheet into a tubular gut as a result of the lateral folding of the embryonic body and the ventral bending of the cranial and caudal ends of the embryo into a roughly C-shaped structure (Figs. 5-10 and 5-19). A major morphological consequence of these folding processes is the sharp delineation of the **yolk sac** from the digestive tube.

Early in the third week, when the three embryonic germ layers are first laid down, the intraembryonic endoderm constitutes the roof of the roughly spherical yolk sac (see Fig. 5-19). Expansion of either end of the neural plate, particularly the tremendous growth of the future brain region, results in the formation of the **head fold** and **tail fold** along the sagittal plane of the embryo. This process, along with concomitant lateral folding, results in the formation of the beginnings of the tubular **foregut** and **hindgut.** This process also begins to delineate the yolk sac from the gut proper. The sequence of steps in the formation of the tubular gut can be likened to a purse string constricting the ventral region of the embryo, although the actual mechanism is more related to overall growth of the embryo than a real constriction. The region of the imaginary purse string becomes the **yolk stalk** (also called the **omphalomesenteric** or **vitelline duct**), with the embryonic gut above and the yolk sac below (see Figs. 5-10, *D* and 5-19, *D*). The portion of the gut that still opens into the yolk sac is called the **midgut,** and the points of transition between the open-floored midgut and the tubular anterior and posterior regions of the gut are called the **anterior** and **posterior intestinal portals.**

At the anterior end of the foregut, the original prochordal plate temporarily remains an ectodermal-endodermal bi-

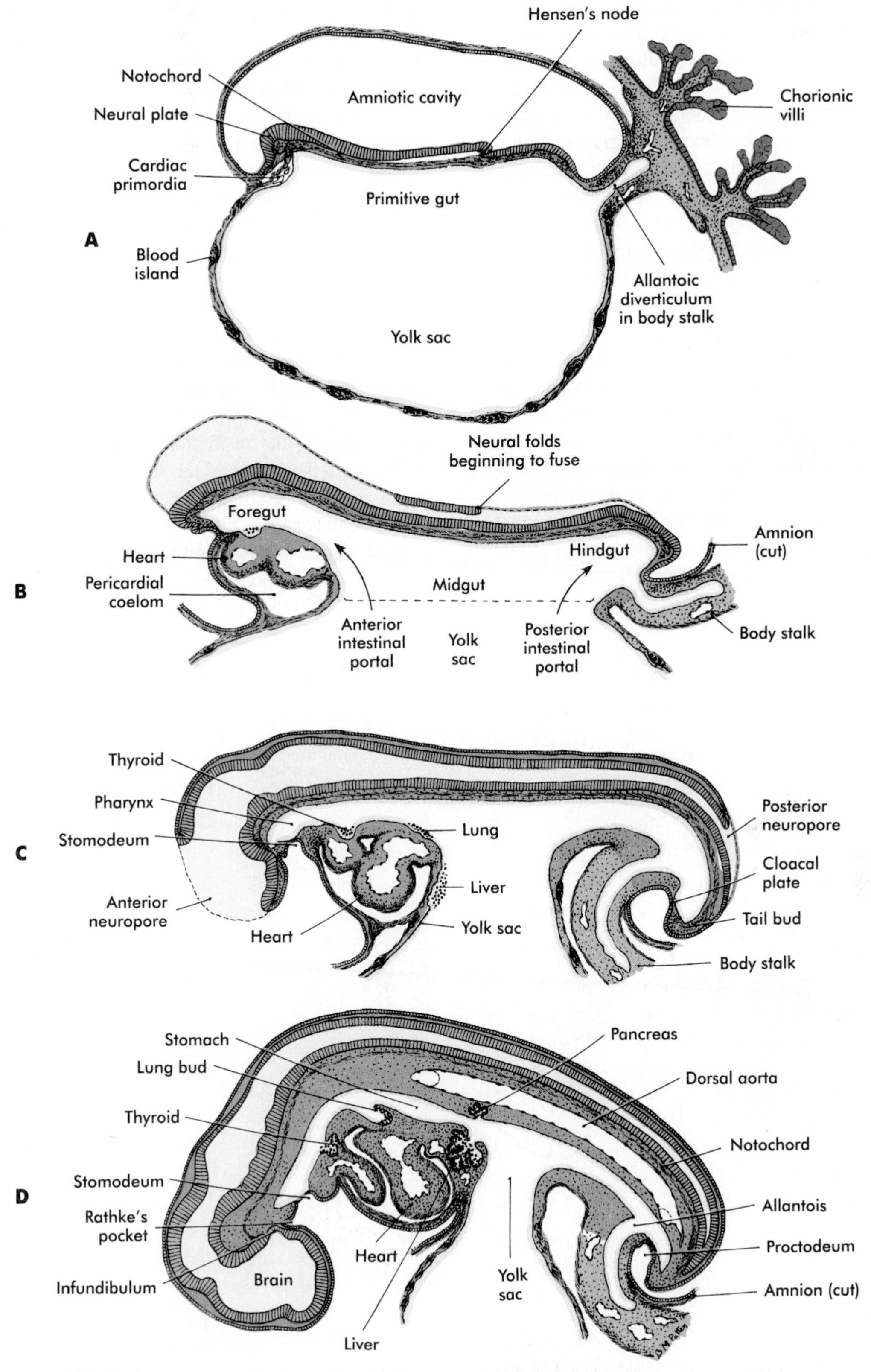

FIG. 5-19 Sagittal sections through human embryos showing the early establishment of the digestive system. **A,** At 16 days. **B,** At 18 days. **C,** At 22 days. **D,** At the end of the first month.
(After Patten.)

layer called the **oropharyngeal membrane** (see Fig. 5-16). This membrane separates the future mouth **(stomodeum),** which is lined by ectoderm, from the **pharynx,** the endodermally lined anterior part of the foregut. The rapid bulging of the cephalic region, in conjunction with the constriction of the ventral region, has a major topographical effect on the rapidly developing cardiac region. In the early embryo, the cardiac primordia are located cephalic to the primitive gut. The forces that shape the tubular foregut, however, cause the bilateral cardiac primordia to turn 180 degrees in the craniocaudal direction while the paired cardiac tubes are moving toward one another in the ventral midline.

In the region of the hindgut, the expansion of the embryo's body is not as prominent as it is in the cranial end, but nevertheless a less exaggerated ventral folding also occurs in that region (see Fig. 5-19). Even as the earliest signs of the tail fold are taking shape, a tubular evagination of the hindgut extends into the mesoderm of the body stalk. This evagination is called the **allantois.** In most mammals and birds the allantois represents a major structural adaptation for the exchange of gases and the removal of urinary wastes. Because of the efficiency of the placenta, the allantois never becomes a prominent structure in the human embryo. Nevertheless, because of the blood vessels that become associated with it, it remains a vital part of the link between the embryo and mother (see Chapter 6).

Caudal to the allantois is another ectodermal-endodermal bilayer called the **cloacal plate,** or **proctodeal membrane** (see Fig. 5-19, *C*). This membrane, which will ultimately break down, separates the **proctodeum,** or future cloacal region, from the posterior part of the endodermally lined hindgut and the exit point of the urogenital system.

As the gut becomes increasingly tubular, a series of local inductive interactions between the epithelium of the digestive tract and the surrounding mesenchyme initiate the formation of most of the major digestive glands (e.g., the thyroid gland), the respiratory system, the liver, and the pancreas. In the region of the stomodeum, an induction between forebrain and stomodeal ectoderm initiates the formation of the anterior pituitary gland. (Further development of these organs is discussed in Chapters 15 and 16.)

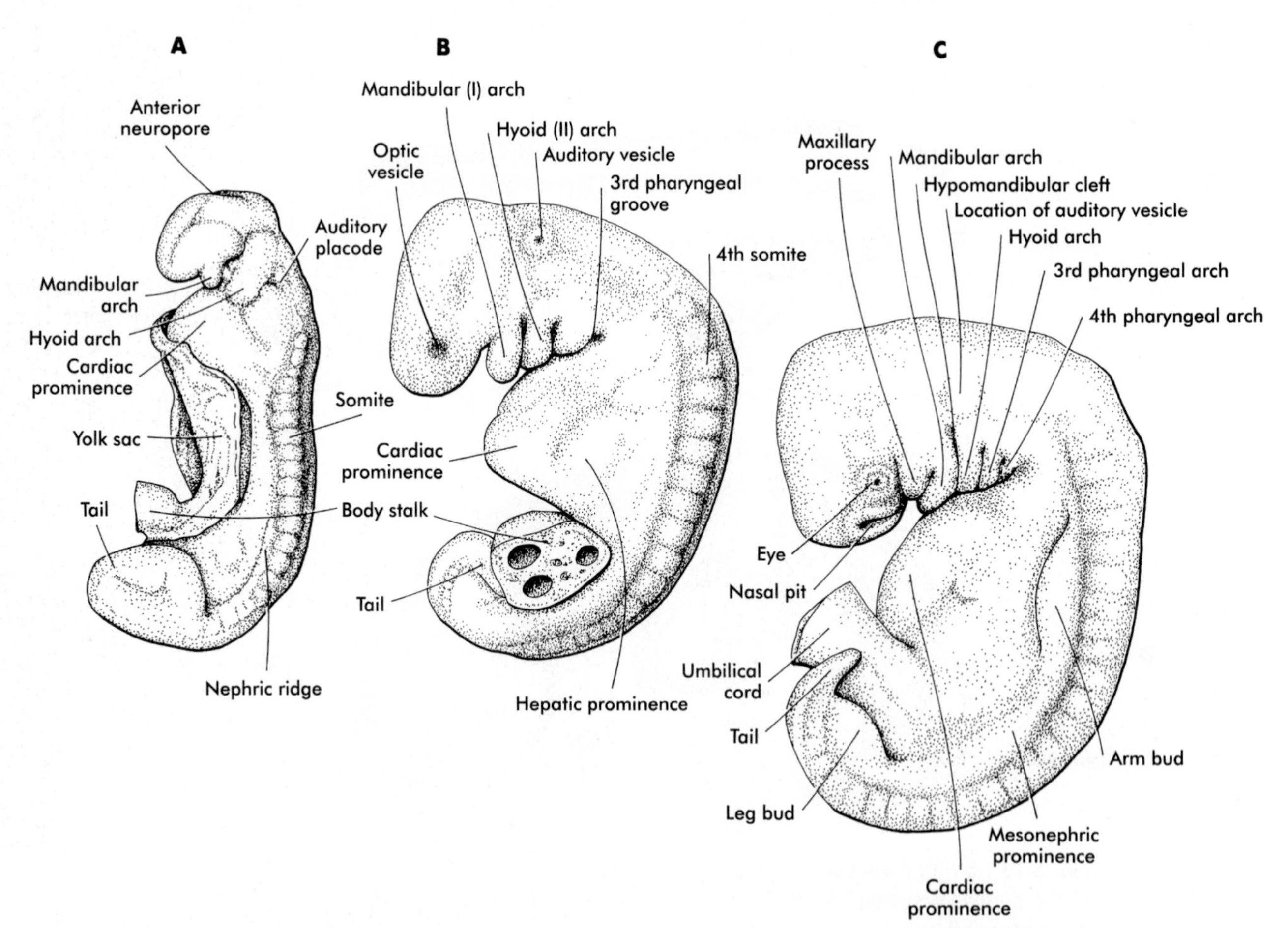

FIG. 5-20 The gross development of human embryos during the period of early organogenesis. **A,** Early in the fourth week. **B,** The middle of the fourth week. **C,** The end of the fourth week.

BASIC STRUCTURE OF THE 4-WEEK-OLD EMBRYO

Gross Appearance

By the end of the fourth week of pregnancy the embryo, which is still only about 4 mm long, has established the rudiments of most of the major organ systems except for the limbs (which are still absent) and the urogenital system (which has developed only the earliest traces of the embryonic kidneys). Externally, the embryo is C-shaped, with a prominent row of somites situated along either side of the neural tube (Figs. 5-20 and 5-21). Except for the rudiments of the eyes and ears and the stomodeal plate, which is beginning to break down (Fig. 5-22), the head is relatively featureless. In the cervical region, **branchial arches** are prominent (Figs. 5-20, *B, C* and 5-23). The body stalk still occupies a significant part of the ventral body wall, and cephalic to the body stalk, the heart and liver make prominent bulges in the contours of the ventral body wall. Pos-

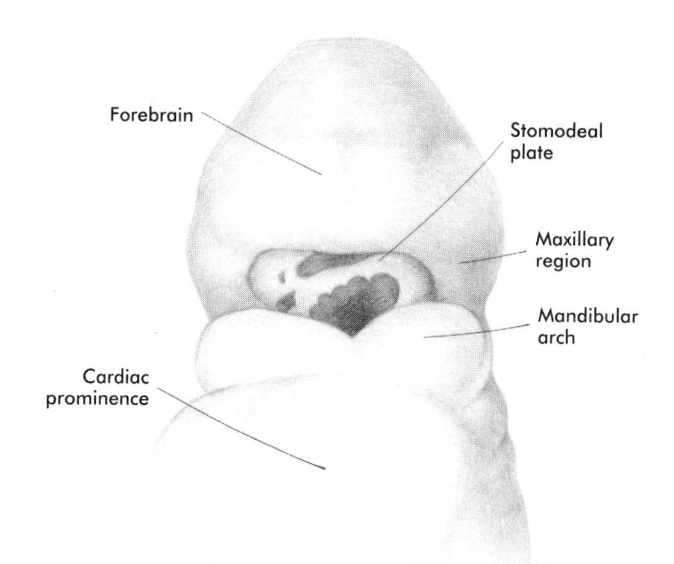

FIG. 5-22 Face of a human embryo during the fourth week showing the breakdown of the stomodeal plate.

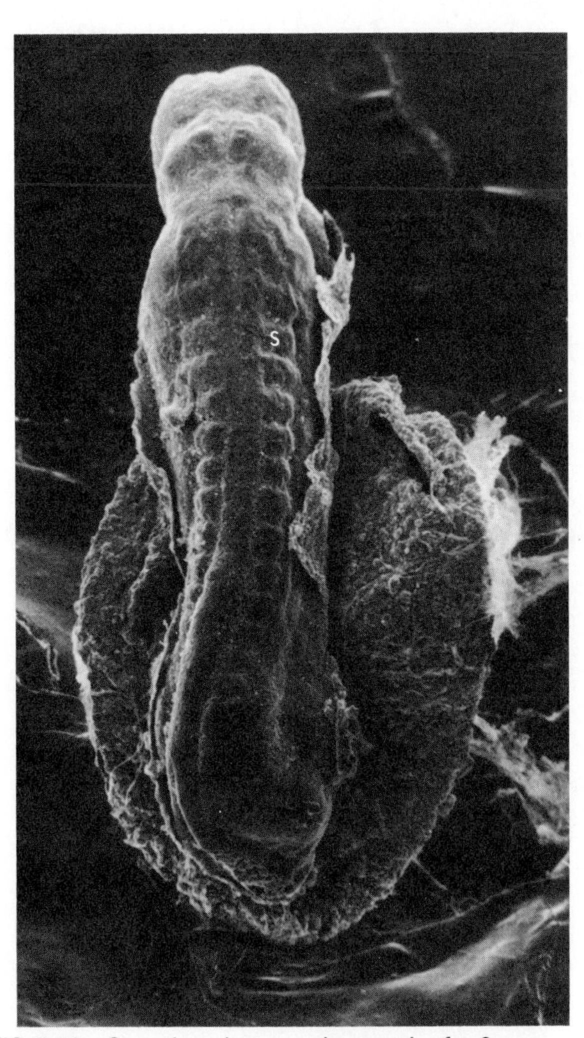

FIG. 5-21 Scanning electron micrograph of a 3-mm human embryo approximately 26 days old. *S,* Somite.
(From Jirásek JE: *Atlas of human prenatal morphogenesis,* Amsterdam, 1983, Martinus Nijhoff Publishers.)

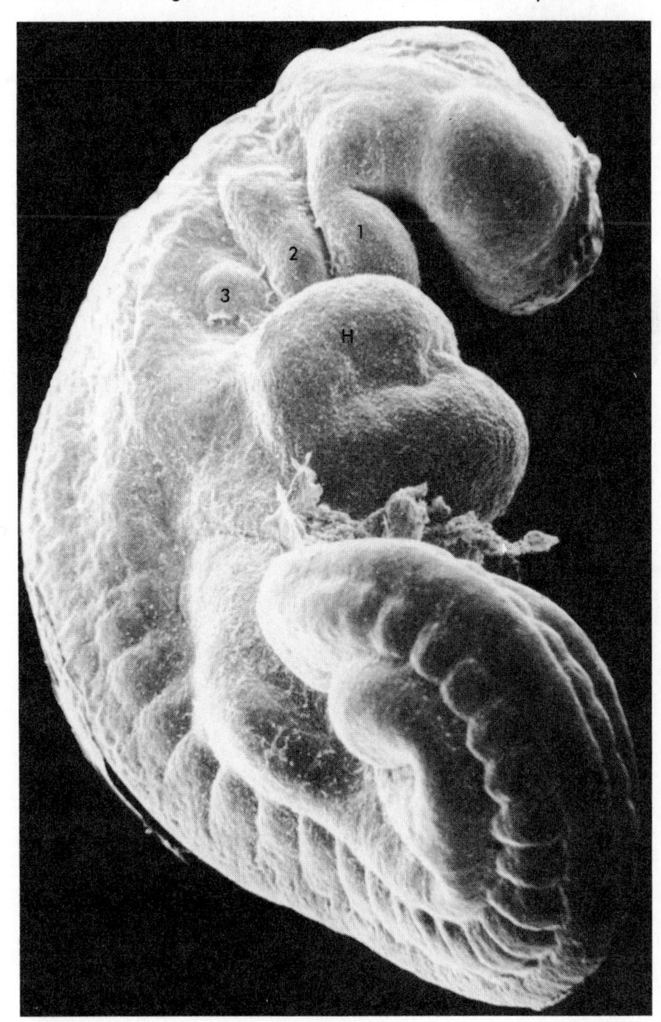

FIG. 5-23 Scanning electron micrograph of a 4 mm human embryo 30 days old. *1, 2, 3,* pharyngeal arches; *H,* heart.
(From Jirásek JE: *Atlas of human prenatal morphogenesis,* Amsterdam, 1983, Martinus Nijhoff Publishers.)

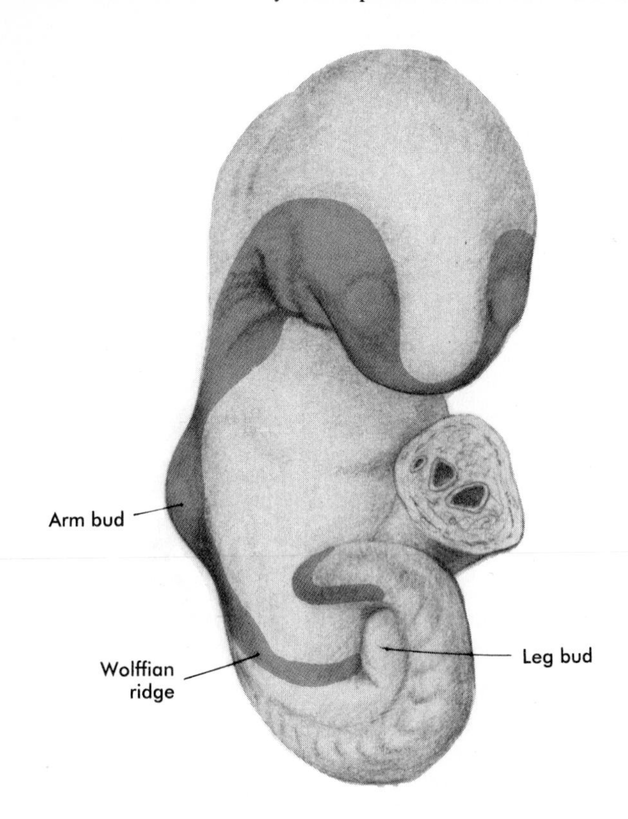

FIG. 5-24 Ventrolateral view of a 30-somite (4.2 mm) human embryo showing the thickened ectodermal ring *(gray)*. The portion of the ring between the upper and lower limb buds is the wolffian ridge.
(Based on studies by O'Rahilly and Gardner [1975].)

terior to the body stalk, the body tapers to a somewhat spiraled tail, which is prominent in embryos of this age.

Another prominent but little understood feature of embryos of this age is a ring of thickened ectoderm that encircles the lateral aspect of the body (Fig. 5-24). Its function is not well understood, but it spans the primordia of many structures (e.g., nose, eye, inner ear, branchial arches, limbs) that require tissue interactions for their early development. What role the thickened ectoderm plays in early organogenesis remains to be determined.

The Circulatory System

At 4 weeks of age, the embryo has a functioning two-chambered heart and a blood vascular system that consists of three separate circulatory arcs (Fig. 5-25). The first, the

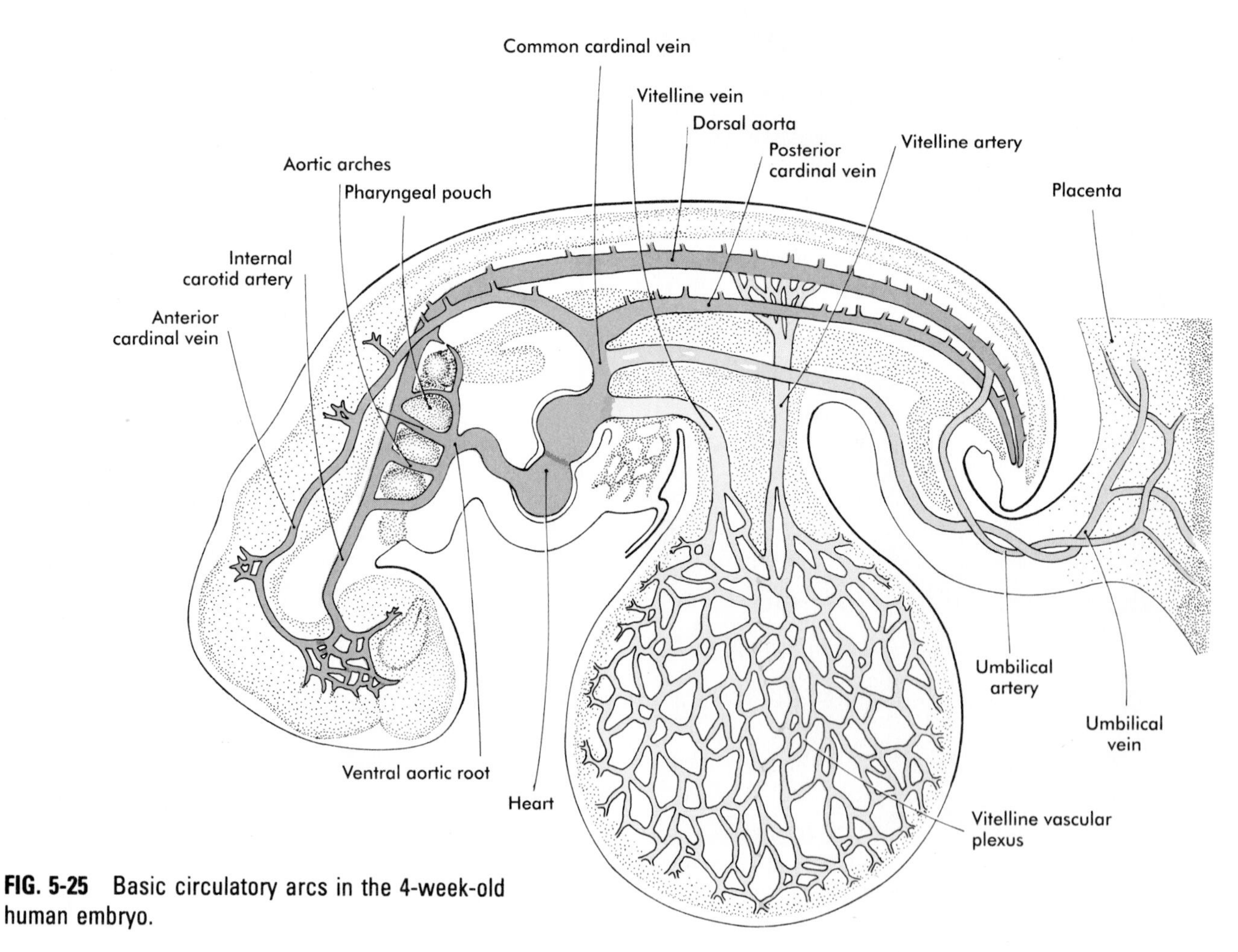

FIG. 5-25 Basic circulatory arcs in the 4-week-old human embryo.

intraembryonic circulatory arc, is organized in a similar manner to that of a fish. A ventral aortic outflow tract of the heart splits into a series of aortic arches passing through the branchial arches around the pharynx and then collecting into a cephalically paired dorsal aorta that distributes blood throughout the body. A system of cardinal veins collects the blood and returns it to the heart via a common inflow tract.

The second arc, commonly called the **vitelline** or **omphalomesenteric arc**, is principally an extraembryonic circulatory loop that supplies the yolk sac (see Fig. 5-25). The third circulatory arc, also extraembryonic, consists of the vessels associated with the allantois. In the human, this arc consists of the **umbilical vessels,** which course through the body stalk and spread in an elaborate network in the placenta and chorionic tissues. This set of vessels represents the real lifeline between the embryo and its mother. Although the two extraembryonic circulatory loops do not persist as such after birth, the intraembryonic portions of these arcs are retained as vessels or ligaments in the adult body.

SUMMARY

1. Evidence is increasing that the basic body plan of mammalian embryos is under the control of many of the same genes that have been identified to control morphogenesis in *Drosophila*. In *Drosophila,* the basic axes are fixed through the actions of maternal effect genes. A battery of segmentation genes (gap, pair-rule, and segment-polarity genes) are then activated. Two families of homeotic genes next confer a specific morphogenetic character to each body segment. Because of their regulative nature, mammalian embryos are not as rigidly controlled by genetic instructions as *Drosophila*.
2. The homeobox, a highly conserved region of about 180 base pairs, is found in almost all animals. The homeobox protein is a transcription factor. Homeobox-containing genes are arranged along the chromosome in a specific order and are expressed along the craniocaudal axis of the embryo in the same order. Activation of homeobox genes may involve interactions with other morphogenetically active agents, such as retinoic acid and TGF-β.
3. The response of dorsal ectodermal cells to primary induction is to thicken, forming a neural plate. Neurulation consists of lateral folding of the neural plate at hinge points to form a neural groove. Opposing sides of the thickened epithelium of the neural groove join to form a neural tube. The temporarily unclosed cranial and caudal ends of the neural tube are the anterior and posterior neuropores.
4. Cranially, the neural tube subdivides into a primitive three-part brain consisting of the prosencephalon, the mesencephalon, and the rhombencephalon. Much of the early brain also becomes subdivided into segments called neuromeres. Specific homeobox genes are expressed in a regular order in the neuromeres.
5. When the neural tube is completed, neural crest cells emigrate from the neural epithelium and spread through the body along well-defined paths. Secondary inductions acting on ectoderm in the cranial region result in the formation of several series of ectodermal placodes, which are the precursors of sense organs and sensory ganglia of cranial nerves.
6. The embryonic mesoderm is subdivided into three fundamental columns: the paraxial, intermediate, and lateral mesoderm. Paraxial mesoderm is the precursor tissue to the paired somites and somitomeres. The epithelial somites become subdivided into sclerotomes (precursors of vertebral bodies) and dermomyotomes, which in turn form dermatomes (dermal precursors) and myotomes (precursors of axial muscles). In further subdivisions, precursor cells of limb muscles are found in the lateral halves of the somites, and precursor cells of axial muscles are found in medial halves. The posterior half of one sclerotome joins with the anterior half of the next caudal somite to form a single vertebral body.
7. Intermediate mesoderm forms the organs of the urogenital system. The lateral plate mesoderm splits to form somatic mesoderm (associated with ectoderm) and splanchnic mesoderm (associated with endoderm). The space between becomes the coelom. The limb bud arises from lateral plate mesoderm, and extraembryonic mesoderm forms the body stalk.
8. Blood cells and blood vessels form initially from blood islands located in the mesodermal wall of the yolk sac. The heart, originating from a horseshoe-shaped region of splanchnic mesoderm anterior to the prechordal plate, forms two tubes on either side of the foregut. As the foregut takes shape, the two cardiac tubes come together to form a single tubular heart, which begins to beat around 22 days after fertilization.
9. The embryonic endoderm initially consists of the roof of the yolk sac. As the embryo undergoes lateral folding, the endodermal gut forms cranial and caudal tubes (foregut and hindgut), but the middle region (midgut) remains open to the yolk sac ventrally. As the tubular gut continues to take shape, the connection to the yolk sac becomes attenuated to form the yolk stalk. The future mouth (stomodeum) is separated from the foregut by an oropharyngeal membrane, and the hindgut is separated from the proctodeum by cloacal plate. A ventral evagination

from the hindgut is the allantois, which in many animals is an adaptation for removing urinary and respiratory wastes.

10. In the 4-week-old embryo, the circulatory system includes a functioning two-chambered heart and a blood vascular system that consists of three circulatory arcs. In addition to the intraembryonic circulation, the extraembryonic vitelline circulatory arc, which supplies the yolk sac, and the umbilical circulation, which supplies the placenta, are present.

REVIEW QUESTIONS

1. What is a homeobox?
2. What forces are involved in the folding of the neural plate to form the neural tube?
3. What role do neuromeres play in the formation of the central nervous system?
4. From what structures do the cells that form skeletal muscles arise?
5. Where do the first blood cells of the embryo form?
6. In the early embryo, does the gut open directly to the outside of the embryo?

REFERENCES

Alvarez IS, Schoenwolf GC: Expansion of surface epithelium provides the major extrinsic force for bending of the neural plate, *J Exp Zool* 261:340-348, 1992.

Bellairs R, Ede DA, Lash JW, eds: *Somites in developing embryos,* New York, 1986, Plenum.

Bergquist H: Studies on the cerebral tube in vertebrates: the neuromeres, *Acta Zool* 33:117-187, 1952.

Capecci M, ed: *Molecular genetics of early Drosophila and mouse development,* Cold Spring Harbor, New York, 1989, Cold Spring Harbor Laboratory.

DeRobertis EM, Oliver G, Wright CVE: Homeobox genes and the vertebrate body plan, *Sci Am* July, pp 46-52, 1990.

Dressler GR, Gruss P: Do multigene families regulate vertebrate development? *Trends Genet* 4:214-219, 1988.

French V and others: Mechanisms of segmentation, *Development* 104(suppl):1-254, 1980.

Gehring WJ: Homeotic genes, the homeobox, and the spatial organization of the embryo, *Harvey Lect* 81:153-172, 1987.

Graham A, Papalopulu N, Krumlauf R: The murine and *Drosophila* homeobox gene complexes have common features of organization and expression, *Cell* 57:367-378, 1989.

Jacobson AG: Somitomeres: mesodermal segments of vertebrate embryos, *Development* 104(suppl):209-220, 1988.

Jacobson AG, Sater AK: Features of embryonic induction, *Development* 104:341-359, 1988.

Kessel M, Gruss P: Murine developmental control genes, *Science* 249:374-379, 1990.

Keynes RJ, Stern CD: Segmentation and neural development in vertebrates, *Trends Neurosci* 8:220-223, 1985.

Keynes RJ, Stern CD: Mechanisms of vertebrate segmentation, *Development* 103:413-429, 1988.

Lash JW, Ostrovsky D: *On the formation of somites.* In Browder L, ed: *Developmental biology,* vol 2, New York, 1986, Plenum, pp 547-563.

Lumsden A: The cellular basis of segmentation in the developing hindbrain, *Trends Neurosci* 13:329-335, 1990.

McGinnis W, Krumlauf R: Homeobox genes and axial patterning, *Cell* 68:283-302, 1992.

McMahon AP: The *Wnt* family of developmental regulators, *Trends Genet* 8:236-242, 1992.

Meier S: Development of the chick embryo mesoblast: formation of the embryonic axis and establishment of the embryonic pattern, *Dev Biol* 73:24-45, 1979.

Meier S, Tam PPL: Metameric pattern development in the embryonic axis of the mouse. I. Differentiation of the cranial segments, *Differentiation* 21:95-108, 1982.

Morris-Kay G: Retinoic acid and development, *Pathobiology* 60:264-270, 1992.

Murtha MT, Leckman JF, Ruddle FH: Detection of homeobox genes in development and evolution, *Proc Natl Acad Sci USA* 88:10711-10715, 1991.

Nusse R, Varmus HE: *Wnt* genes, *Cell* 69:1073-1087, 1992.

O'Rahilly R, Müller F: The origin of the ectodermal ring in staged human embryos of the first 5 weeks, *Acta Anat* 122:145-157, 1985.

Ordahl CP, Le Douarin NM: Two myogenic lineages within the developing somite, *Development* 114:339-353, 1992.

Schindler JM: Basic developmental genetics and early embryonic development: what's all the excitement about? *J NIH Res* 2:49-55, 1990.

Schoenwolf GC: Histological and ultrastructural studies of secondary neurulation in mouse embryos, *Am J Anat* 169:361-376, 1984.

Schoenwolf GC: Mechanisms of neurulation: traditional viewpoint and recent advances, *Development* 109:243-270, 1990.

Scott MP: Vertebrate homeobox nomenclature, *Cell* 71:551-553, 1992.

Suzuki HR and others: Repeating developmental expression of G-Hox 7, a novel homeobox-containing gene in the chicken, *Dev Biol* 148:375-388, 1991.

Tam PPL, Meier S, Jacobson AG: Differentiation of the metameric pattern in the embryonic axis of the mouse. II. Somitomeric organization of the presomitic mesoderm, *Differentiation* 21:109-122, 1982.

Vaage S: The segmentation of the primitive neural tube in chick embryos *(Gallus domesticus), Adv Anat Embryol Cell Biol* 41(3):1-88, 1969.

Wright CV and others: Interference with function of a homeobox gene in *Xenopus* embryos produces malformations of the anterior spinal cord, *Cell* 59:81-93, 1989.

CHAPTER 6

Placenta and Extraembryonic Membranes

One of the most characteristic features of human embryonic development is the intimate relationship between the embryo and mother. The fertilized egg brings little with it except genetic material. To survive and grow during intrauterine life, the embryo must maintain an essentially parasitic relationship with the body of the mother for acquiring oxygen and nutrients and eliminating wastes. It must also avoid being rejected like a foreign body by the immune system of its maternal host. These exacting requirements are met by the placenta and extraembryonic membranes that surround the embryo and serve as the interface between the embryo and mother.

The tissues that make up the fetal/maternal interface (**placenta** and **chorion**) are derivatives of the **trophoblast,** which separates from the inner cell mass and surrounds the cellular precursors of the embryo proper even as the cleaving zygote travels down the uterine tube on its way to implanting into the uterine wall (see Fig. 3-16). Other extraembryonic tissues are derived from the inner cell mass. These include the **amnion** (an ectodermal derivative), which forms a protective fluid-filled capsule around the embryo; the **yolk sac** (an endodermal derivative), which in mammalian embryos no longer serves a primary nutritive function; the **allantois** (an endodermal derivative), which is associated with the removal of embryonic wastes; and the **extraembryonic mesoderm,** which forms the bulk of the umbilical cord, the connective tissue backing of the extraembryonic membranes, and the blood vessels that supply them.

EXTRAEMBRYONIC MEMBRANES

Amnion

The origin of the amniotic cavity within the ectoderm of the inner cell mass in the implanting embryo was described in Chapter 4 (see Figs. 3-16 and 4-2). As the early embryo undergoes cephalocaudal and lateral folding, the amniotic membrane surrounds the body of the embryo like a fluid-filled balloon (Fig. 6-1), allowing the embryo to be suspended in a liquid environment for the duration of pregnancy. The amniotic fluid serves as a buffer against mechanical injury to the fetus; in addition, it accommodates growth, allows normal fetal movements, and protects the fetus from adhesions.

The thin amniotic membrane consists of a single layer of extraembryonic ectodermal cells lined by a nonvascularized layer of extraembryonic mesoderm. Keeping pace with fetal growth, the amniotic cavity steadily expands until its fluid content reaches a maximum of nearly 1 L at weeks 33 to 34 of pregnancy (Fig. 6-2).

In many respects, amniotic fluid can be viewed as a dilute transudate of maternal plasma, but the origins and exchange dynamics of amniotic fluid are complex and not completely understood. There appear to be two phases in amniotic fluid production. The first phase encompasses the first 20 weeks of pregnancy, during which the composition of amniotic fluid is quite similar to that of fetal fluids. During this period the fetal skin is unkeratinized, and there is evidence that fluid and electrolytes are able to diffuse freely

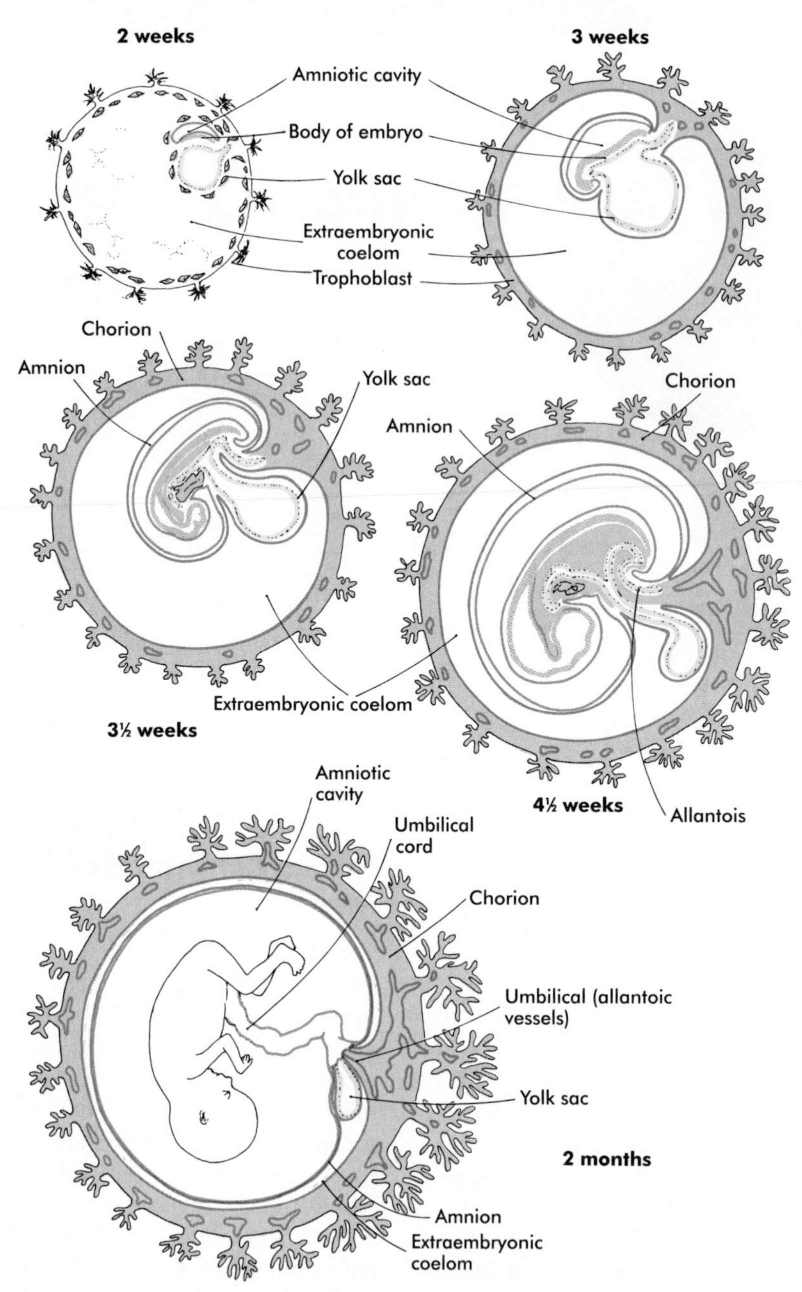

FIG. 6-1 Human embryos showing the relationships of the chorion and other extraembryonic membranes.
(After Patten.)

through the embryonic ectoderm of the skin. In addition, the amniotic membrane itself secretes fluid, and components of maternal serum pass through the amniotic membrane.

As pregnancy advances (especially after the 20th week, when the fetal epidermis begins to keratinize), there are changes in the source of amniotic fluid. There is not complete agreement on the sources (and their relative proportions) of amniotic fluid in the second half of pregnancy. Nonetheless, there are increasing contributions from fetal urine, filtration from maternal blood vessels in the chorion laeve (which is closely apposed to the amniotic membrane at this stage), and possibly filtration from fetal vessels in the umbilical cord and chorionic plate.

In the third trimester of pregnancy, the amniotic fluid turns over completely every 3 hours, and at term, the fluid exchange rate may approach 500 ml per hour. Although much of the amniotic fluid is exchanged across the amniotic membrane, fetal swallowing is an important mechanism in late pregnancy, with about 20 ml per hour of fluid being swallowed by the fetus. Swallowed amniotic fluid ultimately enters the fetal bloodstream after absorption

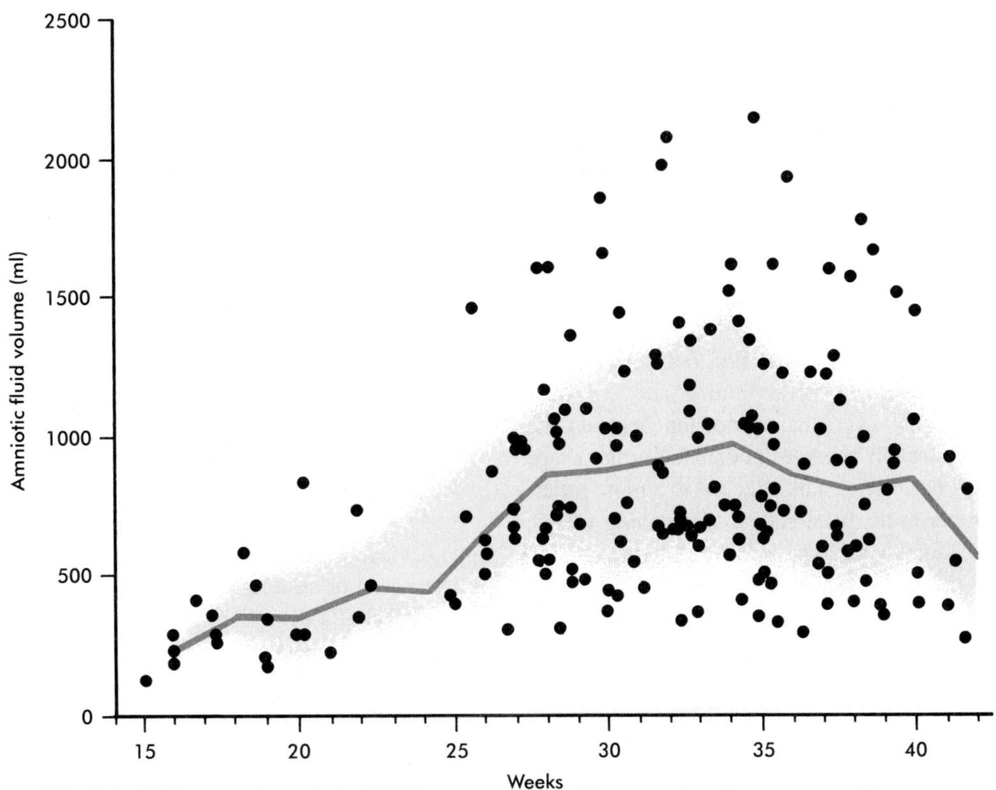

FIG. 6-2 Volumes of amniotic fluid in women at various weeks of pregnancy. The lined and colored area represents the mean ± standard deviation. Dots represent outlying values.

(Data from Queenan et al: *Am J Obstet Gynecol* 114:34-38,1972.)

through the gut wall. The ingested water can leave the fetal circulation through the placenta.

The normal amount of amniotic fluid at term is typically between 500 and 1000 ml. An excessive amount of amniotic fluid (over 2000 ml) is **hydramnios.** This condition is frequently associated with multiple pregnancies and **anencephaly** (a congenital anomaly characterized by gross defects of the head and often the inability to swallow [see Chapter 8]). Such circumstantial evidence supports the important role of fetal swallowing in the overall balance of amniotic fluid exchange. Too little amniotic fluid (less than 500 ml) is **oligohydramnios.** This condition is often associated with bilateral **renal agenesis** (absence of kidneys) and points to the role of fetal urinary excretion in amniotic fluid dynamics.

There are many components, both fetal and maternal, in amniotic fluid; for example, over 200 proteins of both maternal and fetal origins have been detected in amniotic fluid. With the analytical tools available, much can be learned about the condition of the fetus by examining the composition of amniotic fluid. **Amniocentesis** involves removing a small amount of amniotic fluid by inserting a needle through the mother's abdomen and into the amniotic cavity. Because of the small amount of amniotic fluid in early embryos, amniocentesis is usually not performed until the thirteenth or fourteenth week of pregnancy. Amniotic fluid has bacteriostatic properties, which may account for the low incidence of infections after amniocentesis is performed.

Fetal cells present in the fluid can be cultured and examined for a variety of chromosomal and metabolic defects. Recent techniques now permit examination of chromosomes in the cells immediately obtained instead of having to wait up to 2 to 3 weeks for cultured amniotic cells to proliferate to the point of being suitable for genetic analysis. In addition to the detection of chromosomal defects (e.g., trisomies), it is also possible to determine the sex of the fetus by direct chromosomal analysis. A high concentration of **α-fetoprotein** (a protein of the central nervous system) in amniotic fluid is a strong indicator of a neural tube defect. Fetal maturity can be assessed by determination of the concentration of creatinine or the **lecithin/sphingomyelin ratio** (which is a reflection of the maturity of the lungs). The severity of **erythroblastosis fetalis** (Rh disease) can also be assessed by examination of amniotic fluid.

Yolk Sac

The yolk sac, which is lined by extraembryonic endoderm, is formed ventral to the bilayered embryo at the time when the amnion appears dorsal to the embryonic disk (see Fig. 4-2). In contrast to birds and reptiles, the yolk sac of mammals is small and devoid of yolk. Although vestigial in terms of its original function of being a source of nutrition, the yolk sac remains vital to the embryo because of other functions that have become associated with it.

When it first appears, the yolk sac has the form of a hemisphere that is bounded at the equatorial region by the dorsal wall of the primitive gut (see Fig. 6-1). As the embryo grows and undergoes lateral folding and curvature along the craniocaudal axis, the connection between the yolk sac and forming gut becomes attenuated in the shape of a progressively narrowing stalk attached to a more spherical yolk sac proper at its distal end. In succeeding weeks, the yolk stalk becomes very long and attenuated as it is incorporated into the body of the umbilical cord. The yolk sac itself becomes situated near the chorionic plate of the placenta (Fig. 6-3).

The endoderm of the yolk sac is lined on the outside by well-vascularized extraembryonic mesoderm. Cells derived from each of these layers contribute vital components to the body of the embryo. During the third week, **primordial germ cells** become recognizable in the endodermal lining of the yolk sac (see Fig. 1-1). Soon these cells migrate into the wall of the gut and the dorsal mesentery as they make their way to the gonads, where they differentiate into oogonia or spermatogonia.

In the meantime, groups of mesodermal cells in the wall of the yolk sac become organized into **blood islands** (see Fig. 5-18), and many of the cells differentiate into primitive blood cells. **Extraembryonic hematopoiesis** continues in the yolk sac until about the sixth week, when blood-forming activity becomes transferred to intraembryonic sites, especially the liver.

As the tubular gut forms, the attachment site of the yolk stalk becomes progressively less prominent, until by 6 weeks it has effectively lost contact with the gut. In a small percentage of adults, traces of the yolk duct persist as a fibrous cord or an outpouching of the small intestine known as a **Meckel's diverticulum** (see Fig. 16-11, *A*). The yolk sac itself may persist throughout much of pregnancy, but it is not known to have a specific function in the fetal period. The proximal portions of the blood vessels of the yolk sac (the vitelline circulatory arc) persist as vessels that supply the midgut region.

FIG. 6-3 A 7-week-old human embryo surrounded by its amnion. The embryo was exposed by cutting open the chorion. The small sphere to the right of the embryo is the yolk sac.
(Carnegie embryo No. 8537A, Chester Reather, Baltimore, Md.)

Allantois

The allantois arises as an endodermally lined ventral outpocketing of the hindgut (see Fig. 6-1). In the human embryo, it is just a vestige of the large, saclike structure that is used by the embryos of many mammals, birds, and reptiles as a major respiratory organ and repository for urinary wastes. Like the yolk sac, the allantois in the human retains only a secondary function, in this case respiration. However, in the human this function is served by the blood vessels that differentiate from the mesodermal wall of the allantois. These vessels form the umbilical circulatory arc, consisting of the arteries and vein that supply the placenta (see Fig. 5-25). (The postnatal fate of these vessels is discussed in Chapter 19.)

The allantois proper, which consists of little more than a cord of endodermal cells, is embedded in the umbilical cord. Later in development, the proximal part of the allantois (called the **urachus**) is continuous with the forming urinary bladder (see Fig. 17-2). After birth, it becomes transformed into a dense fibrous cord (median umbilical ligament), which runs from the urinary bladder to the umbilical region (see Fig. 19-17).

CHORION AND PLACENTA

Formation of the placental complex represents a cooperative effort between the extraembryonic tissues of the embryo and the endometrial tissues of the mother. (Early stages of implantation of the embryo and the decidual reaction of the uterine lining are described in Chapter 3.) After implantation is complete, the original trophoblast surrounding the embryo has undergone differentiation into two layers, the inner **cytotrophoblast** and the outer **syncytiotrophoblast** (see Fig. 3-16, *D*). Lacunae in the rapidly expanding trophoblast have filled with maternal blood, and the connective tissue cells of the endometrium have undergone the decidual reaction (containing increased amounts of glycogen and lipids) in response to the trophoblastic invasion.

Formation of Chorionic Villi

In the early implanting embryo, the trophoblastic tissues have no consistent gross morphological features; consequently, this is called the period of the **previllous embryo.** Late in the second week, defined cytotrophoblastic projections called **primary villi** begin to take shape (see Fig. 4-2). Shortly thereafter, a mesenchymal core appears within an expanding villus, at which point it is properly called a **secondary villus** (Fig. 6-4). Surrounding the mesenchymal core of the secondary villus is a complete layer of cytotrophoblastic cells, and outside of that is the syncytiotrophoblast. By definition, the secondary villus becomes a **tertiary villus** when blood vessels penetrate its mesenchymal core and newly formed branches. This occurs toward the end of the third week of pregnancy. Although individual villi undergo considerable branching, most of them retain the same basic structural plan throughout pregnancy.

The terminal portion of a villus remains trophoblastic, consisting of a solid mass of cytotrophoblast called a **cytotrophoblastic cell column** and a relatively thin covering of syncytiotrophoblast over that. The villus is bathed in maternal blood. A further development of the tip of the villus occurs when the cytotrophoblastic cell column expands distally, penetrating the syncytiotrophoblastic layer (Fig. 6-5). These cytotrophoblastic cells abut directly on maternal decidual cells and spread over them to form a complete cellular layer known as the **cytotrophoblastic shell,** which surrounds the embryo complex. The villi that give off the cytotrophoblastic extensions are known as **anchoring villi** because they represent the real attachment points between the embryo complex and the maternal tissues.

It is important to understand the overall relationships of the various embryonic and maternal tissues at this stage of development (see Fig. 6-5). The embryo, attached by the **body stalk,** or **umbilical cord,** is effectively suspended in the **chorionic cavity.** The chorionic cavity is bounded by the **chorionic plate,** which consists of extraembryonic mesoderm overlain with trophoblast. The chorionic villi extend outward from the chorionic plate, and their trophoblastic covering is continuous with that of the chorionic plate. The villi and the outer surface of the chorionic plate are

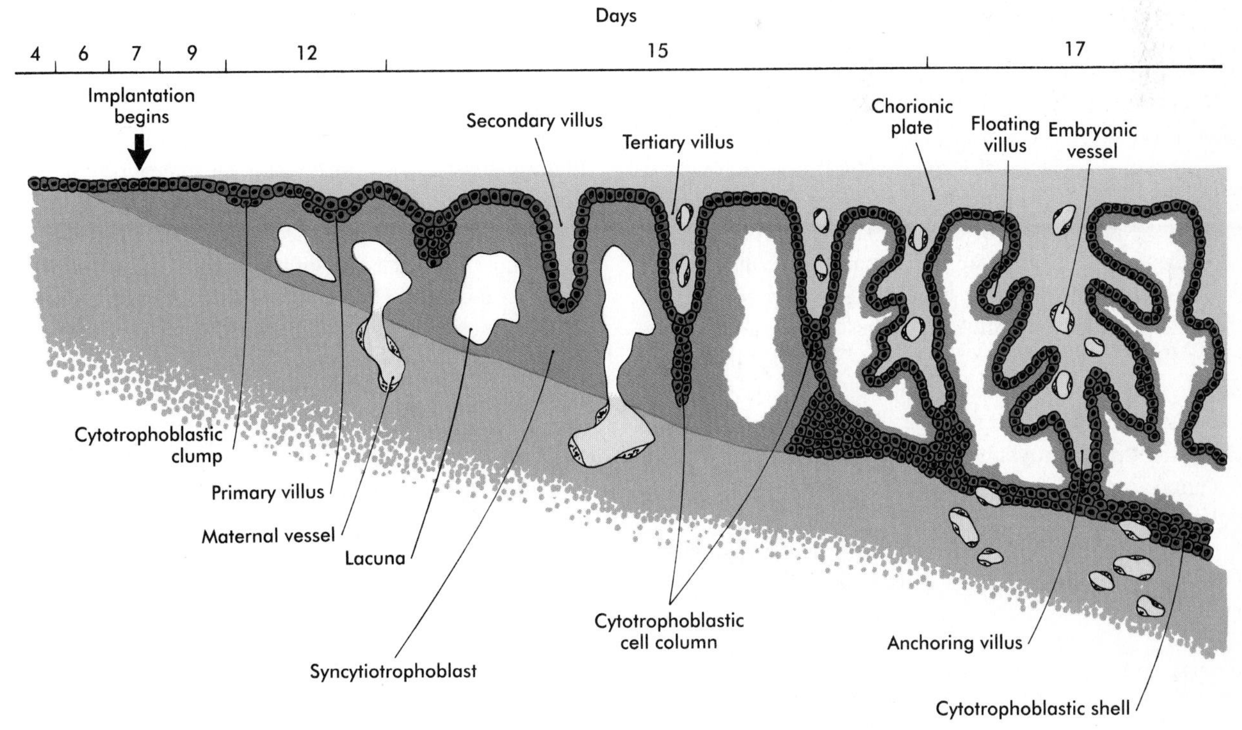

FIG. 6-4 Stages in the formation of a chorionic villus, starting with a cytotrophoblastic clump at the far left and progressing to an anchoring villus at the right.

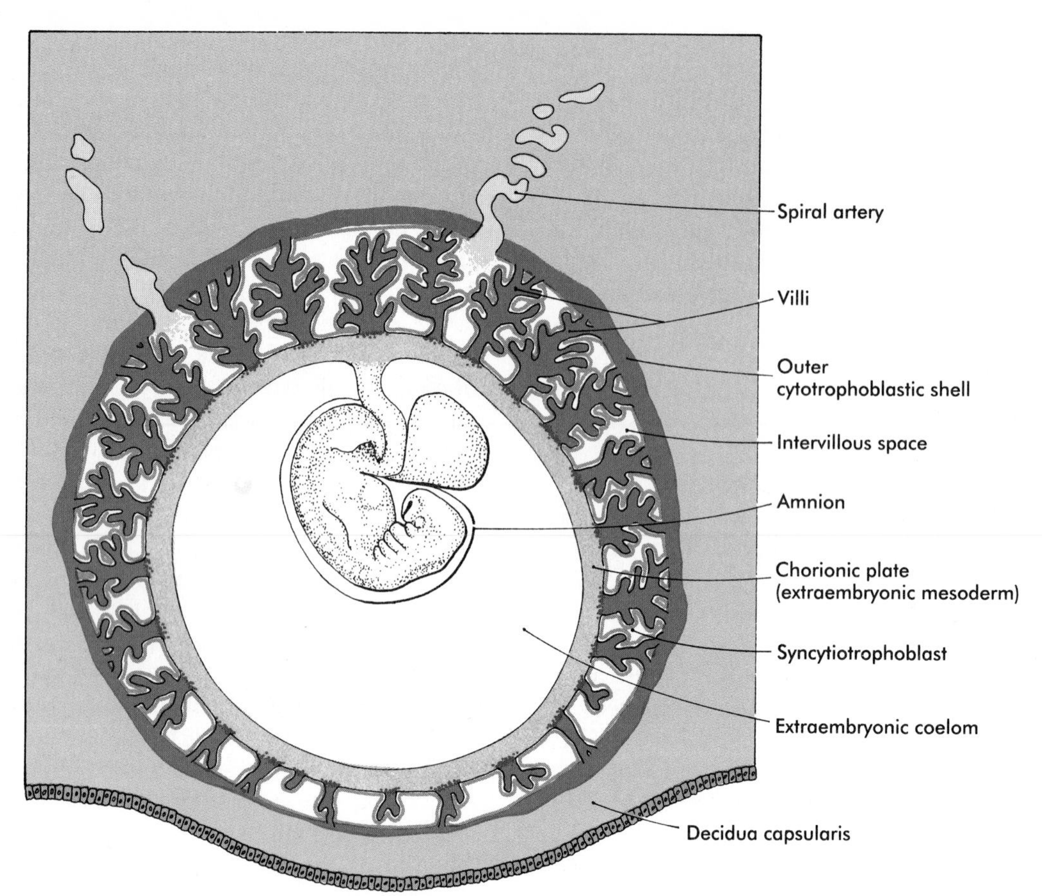

FIG. 6-5 Overall view of a 5-week-old embryo plus membranes showing the relationships of the chorionic plate, the villi, and the cytotrophoblastic shell.

bathed in a sea of continually exchanging maternal blood. Because of this, the human placenta is designated as the **hemochorial type.***

Although chorionic villi are structurally very complex, it is convenient to liken the basic structure of a villus complex to the root system of a plant. The anchoring villus is equivalent to the central tap root; by means of the cytotrophoblastic cell columns, it attaches the villus complex to the outer cytotrophoblastic shell. The unattached branches of the **floating villi** dangle freely in the maternal blood that fills the space between the chorionic plate and the outer cytotrophoblastic shell. All surfaces of the villi, the chorionic plate, and the cytotrophoblastic shell that are in contact with maternal blood are lined with a continuous layer of syncytiotrophoblast.

Maternal blood gains access to this syncytiotrophoblast-lined space through the open ends of the uterine spiral arteries, which pass through the cytotrophoblastic shell. The arteries were eroded by the invading trophoblast, but their lumens are invariably partially occupied by a plug of cytotrophoblastic cells, presumably an adaptation for controlling the flow of blood. Having left the spiral arteries under relatively high pressure, the maternal blood freely percolates throughout the intervillous spaces and bathes the surfaces of the villi. The blood is then picked up by the open ends of uterine veins, which also penetrate the cytotrophoblastic shell (see Fig. 6-9).

Gross Relations of Chorionic and Decidual Tissues

Within days after implantation of the embryo, the stromal cells of the endometrium undergo a striking transformation called the **decidual* reaction.** After the stromal cells swell as the result of accumulation of glycogen and lipid in their

*Other mammals have a variety of arrangements of tissue layers through which materials must pass to be exchanged between mother and fetus. For example, in an epitheliochorial placenta, found in pigs, the fetal component of the placenta (chorion) rests on the uterine epithelium instead of being directly bathed in maternal blood.

*Deciduum refers to tissues that are shed at birth. These include all the extraembryonic tissues plus the superficial layers of the endometrial connective tissue and epithelium.

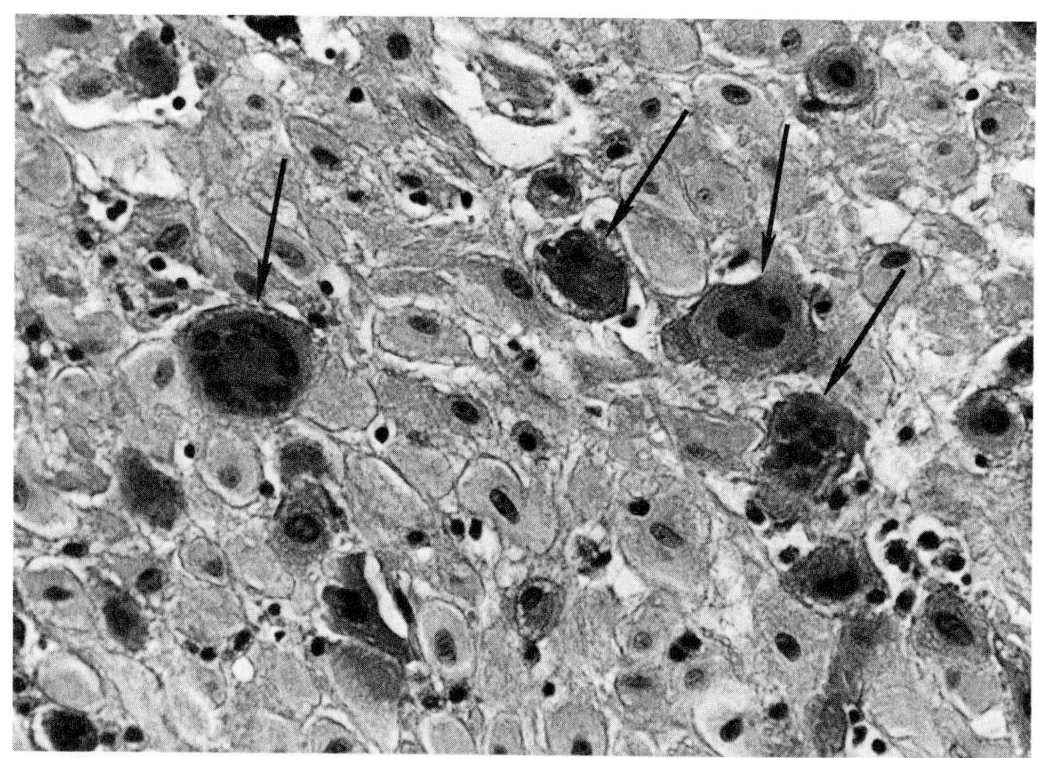

FIG. 6-6 Photomicrograph of decidual cells (light cells) in the endometrium during pregnancy. Hematoxylin and eosin stain. The arrows point to areas of syncytiotrophoblast within the maternal decidual tissue.
(From Naeye RL: *Disorders of the placenta, fetus and neonate,* St Louis, 1992, Mosby.)

cytoplasm, they are known as **decidual cells** (Fig. 6-6). The decidual reaction spreads throughout stromal cells in the superficial layers of the endometrium. The maternal decidua are given topographic names based on where they are located in relation to the embryo.

The decidual tissue that overlies the embryo and its chorionic vesicle is known as the **decidua capsularis,** whereas the decidua that lies between the chorionic vesicle and the uterine wall is called the **decidua basalis** (see Fig. 6-7). With continued growth of the embryo, the decidua basalis becomes incorporated into the maternal component of the definitive placenta. The remaining decidua, which consists of the decidualized endometrial tissue on the sides of the uterus not occupied by the embryo, is the **decidua parietalis.**

In human embryology, the chorion is defined as the layer consisting of the trophoblast plus the underlying extraembryonic mesoderm (see Fig. 6-1). The chorion forms a complete covering **(chorionic vesicle)** that surrounds the embryo, the amnion, yolk sac, and body stalk (Fig. 6-7). During the early period after implantation, primary and secondary villi project almost uniformly from the outer surface of the chorionic vesicle. The formation of tertiary villi, however, is asymmetrical, and the invasion of the cytotrophoblastic core of the primary villi by mesenchyme and embryonic blood vessels occurs preferentially in the primary villi located nearest the decidua basalis. As these villi continue to grow and branch, those located on the opposite side (the abembryonic pole) of the chorionic vesicle fail to keep up and eventually undergo atrophy as the growing embryo complex bulges into the uterine cavity. The region that contains the flourishing chorionic villi and that will ultimately become the placenta is known as the **chorion frondosum.** The remainder of the chorion, which ultimately becomes smooth, is called the **chorion laeve.**

The overall growth of the chorionic vesicle with its bulging into the uterine lumen pushes the decidua capsularis progressively farther from the endometrial blood vessels. By the end of the first trimester of pregnancy the decidua capsularis itself undergoes pronounced atrophy. Within the next month, portions of the atrophic decidua capsularis begin to disappear, leaving the chorion laeve in direct contact with the decidua parietalis on the opposite side of the uterus (Fig. 6-8). By midpregnancy, the chorion laeve has fused with the tissues of the decidua parietalis, thus effectively obliterating the original uterine cavity.

While the chorion laeve and decidua capsularis are undergoing progressive atrophy, the placenta takes shape in its definitive form and acts as the main site of exchange between the mother and embryo.

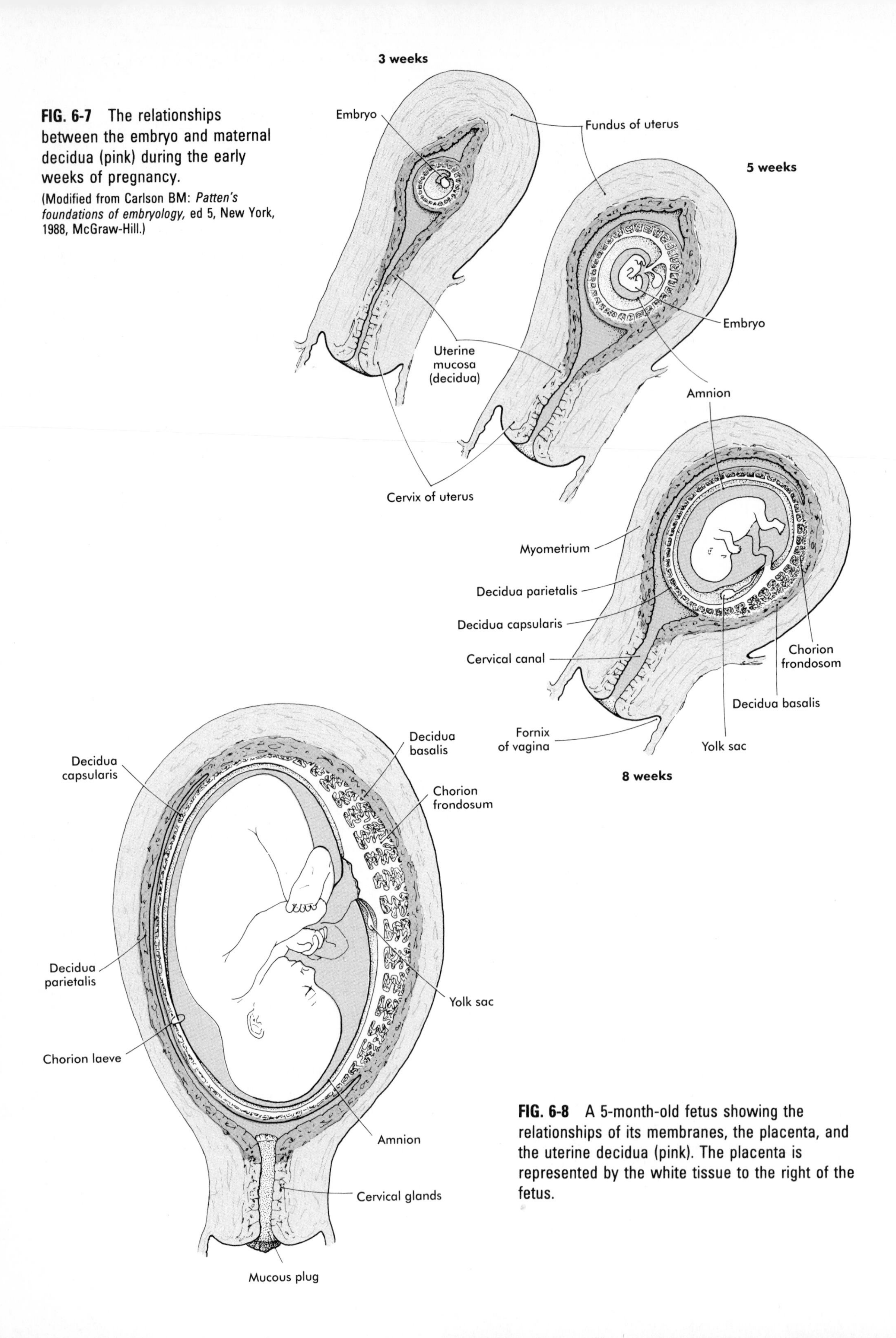

FIG. 6-7 The relationships between the embryo and maternal decidua (pink) during the early weeks of pregnancy.

(Modified from Carlson BM: *Patten's foundations of embryology,* ed 5, New York, 1988, McGraw-Hill.)

FIG. 6-8 A 5-month-old fetus showing the relationships of its membranes, the placenta, and the uterine decidua (pink). The placenta is represented by the white tissue to the right of the fetus.

Formation and Structure of the Mature Placenta

As the distinction between the chorion frondosum and chorion laeve becomes more prominent, the limits of the placenta proper can be defined. The placenta consists of a fetal and maternal component (Fig. 6-9). The fetal component is the part of the chorionic vesicle that is represented by the chorion frondosum. It consists of the wall of the chorion, called the **chorionic plate,** and the chorionic villi that arise from that region. The maternal component is represented by the decidua basalis, but covering the decidua basalis is the fetally derived outer cytotrophoblastic shell. The intervillous space between the fetal and maternal components of the placenta is occupied by freely circulating maternal blood. In keeping with its principal function as an organ-mediating exchange between the fetal and maternal circulatory systems, the overall structure of the placenta is organized to provide a very large surface area (over 10 m^2) for that exchange.

Structure of the Mature Placenta. The mature placenta is disklike in shape, with a thickness of 3 cm and a diameter of about 20 cm (Table 6-1). A typical placenta weighs about 500 g. The fetal side of the placenta is shiny due to the apposed amniotic membrane. From this aspect, the attachment of the umbilical cord to the chorionic plate and the large placental branches of the umbilical arteries and vein radiating from it are evident.

The maternal side of the placenta is dull and subdivided into as many as 35 lobes. The grooves between lobes are occupied by placental septae, which arise from the decidua basalis and extend toward the basal plate. Within a placental lobe are several **cotyledons,** each of which consists of a main stem villus and all its branches. The intervillous space in each lobe represents a nearly isolated compartment of the maternal circulation to the placenta.

The Umbilical Cord. The originally broad-based body stalk elongates and becomes relatively narrower as pregnancy progresses. The umbilical cord becomes the conduit for the umbilical vessels, which traverse its length between the fetus and placenta (Fig. 6-10). The umbilical vessels are embedded in a mucoid connective tissue that is often called **Wharton's jelly.**

The umbilical cord, which commonly attains a length of

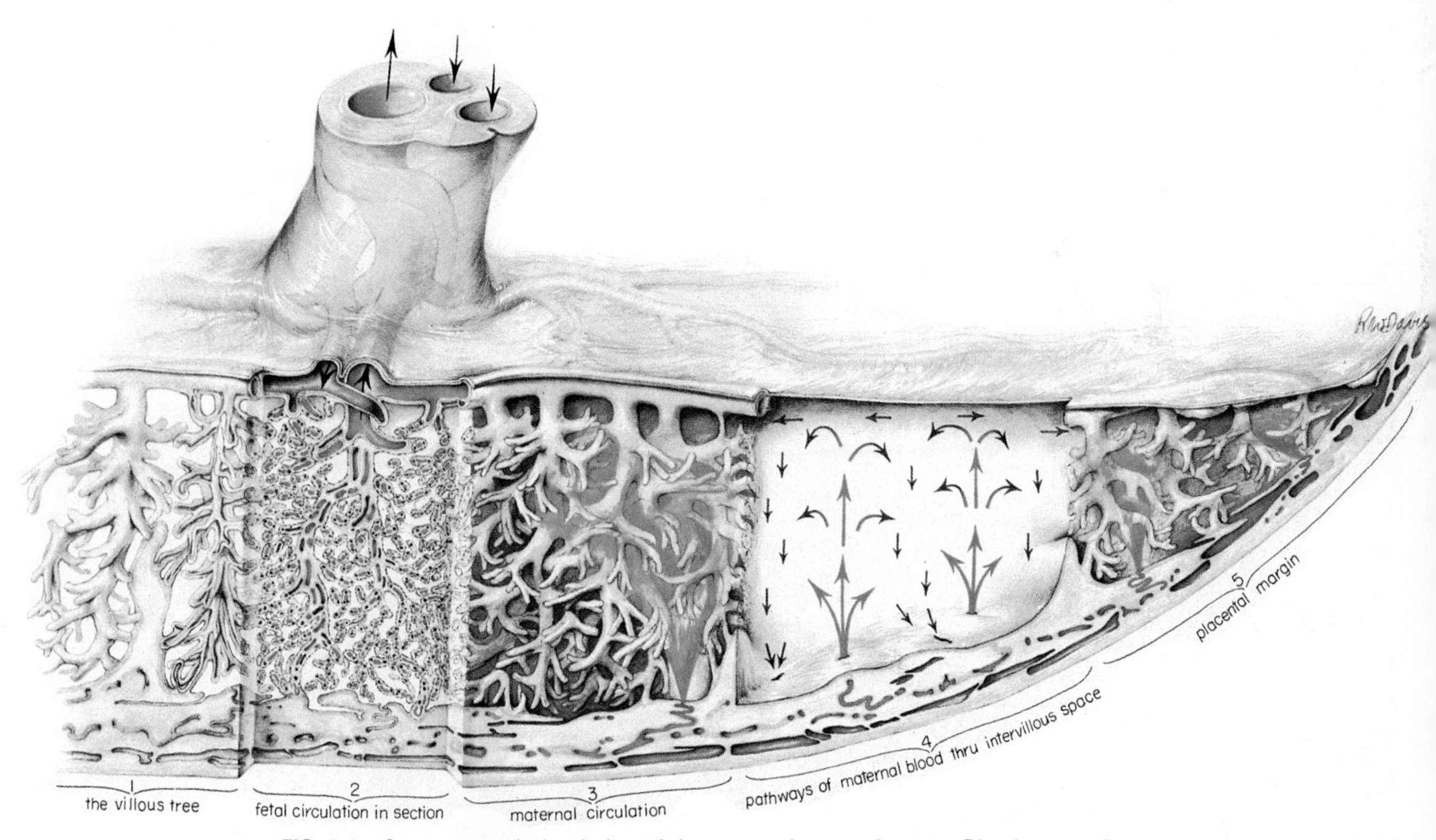

FIG. 6-9 Structure and circulation of the mature human placenta. Blood enters the intervillous spaces from the open ends of the uterine spiral arteries. After bathing the villi, the blood *(blue)* is drained via endometrial veins.

(From Bloom W, Fawcett DW: *Textbook of histology,* Philadelphia, 1986, Saunders.)

TABLE 6-1 The Developing Placenta

AGE OF EMBRYO (WEEKS AFTER FERTILIZATION)	PLACENTAL DIAMETER (mm)	PLACENTAL WEIGHT (g)	PLACENTAL THICKNESS (mm)	LENGTH OF UMBILICAL CORD (mm)
6	—	6	—	—
10	—	26	—	—
14	70	65	12	180
18	95	115	15	300
22	120	185	18	350
26	145	250	20	400
30	170	315	22	450
34	195	390	24	490
38	220	470	25	520

Modified from Kaufmann P, Scheffen I. In Polin R, Fox W, eds: *Fetal and neonatal physiology,* vol 1, Philadelphia, 1992, Saunders, p 48.

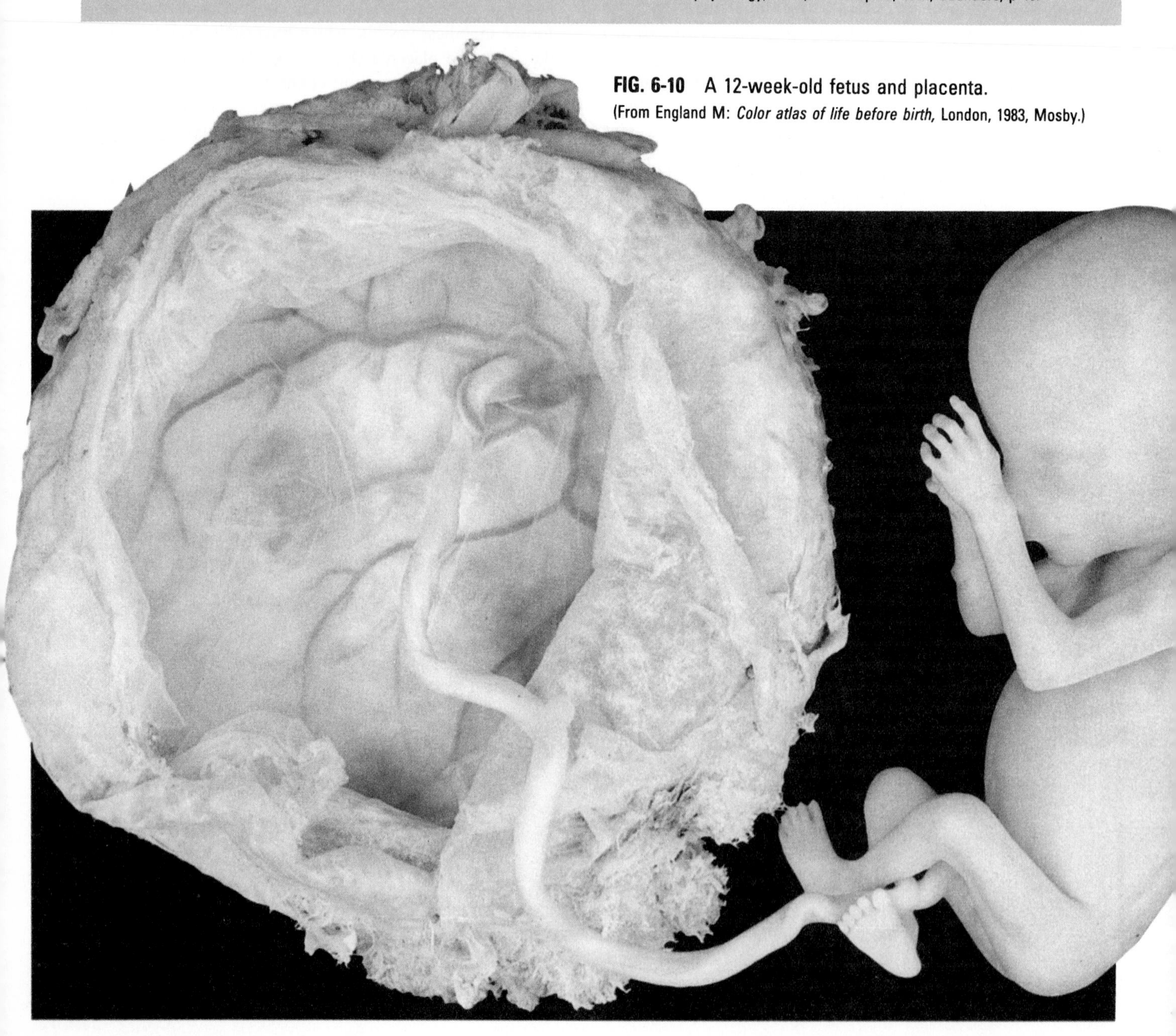

FIG. 6-10 A 12-week-old fetus and placenta.
(From England M: *Color atlas of life before birth,* London, 1983, Mosby.)

EMBRYO WEIGHT (g)/ PLACENTAL WEIGHT (g)	VILLOUS MASS (g)	TOTAL VILLOUS SURFACE AREA (cm^2)	DIFFUSION DISTANCE FROM MATERNAL TO FETAL CIRCULATION (μm)	MEAN TROPHOBLASTIC THICKNESS ON VILLI (μm)
0.18	5	830	55.9	15.4
0.65	18	3020	—	—
0.92	28	5440	40.2	9.6
2.17	63	14,800	27.7	9.9
3.03	102	28,100	21.6	7.4
4.00	135	42,200	—	—
4.92	191	72,200	20.6	6.9
5.90	234	101,000	11.7	5.2
7.23	273	125,000	4.8	4.1

50 to 60 cm by the end of pregnancy, is typically twisted many times. The twisting can be readily observed by gross examination of the umbilical blood vessels. In about 1% of full-term pregnancies, true knots occur in the umbilical cord. If they tighten as the result of fetal movements, they can cause anoxia and even death of the fetus.

Placental Circulation. Both the fetus and mother contribute to the placental circulation (see Fig. 6-9). The fetal circulation is contained in the system of umbilical and placental vessels. Fetal blood reaches the placenta through the two umbilical arteries, which ramify throughout the chorionic plate. Smaller branches from these arteries enter the chorionic villi and then break up into capillary networks in the terminal branches of the chorionic villi, where exchange of materials with the maternal blood occurs (see Fig. 6-14). From the villous capillary beds, the blood vessels consolidate into successively larger venous branches. These retrace their way through the chorionic plate into the large single umbilical vein and to the fetus.

In contrast to fetal circulation, which is totally contained within blood vessels, the maternal blood supply to the placenta is a free-flowing lake that is not bounded by vessel walls. As the result of the invasive activities of the trophoblast, roughly 80 to 100 spiral arteries of the endometrium open directly into the intervillous spaces and bathe the villi in about 150 ml of maternal blood, which is exchanged three to four times per minute.

The maternal blood enters the intervillous space under reduced pressure because of the partial cytotrophoblastic plugs that partially occlude the lumens of the spiral arteries. Nevertheless, the maternal blood pressure is sufficient to force the oxygenated maternal arterial blood to the bases of the villous trees at the chorionic plate. The overall pressure of the maternal placental blood is about 10 mm Hg in the relaxed uterus. From the chorionic plate, the blood percolates over the terminal villi as it returns to venous outflow pathways located in the decidual (maternal) plate of the placenta. An adequate flow of maternal blood to the placenta is vital to the growth and development of the fetus, and a reduced maternal blood supply to the placenta leads to a small fetus.

In the terminal villi, the fetal capillaries are located next to the trophoblastic surface to facilitate exchange between the fetal and maternal blood (Fig. 6-11). The placental barrier of the mature placenta consists of the syncytiotrophoblast, its basal lamina, the basal lamina of the fetal capillary, and the capillary endothelium. Often the two basal laminae seem to be consolidated. In younger embryos, a layer of cytotrophoblast is added to the placental barrier. (Placental transfer is described on p. 99.)

Structure of a Mature Chorionic Villus. Mature chorionic villi constitute a very complex mass of seemingly interwoven branches (Fig. 6-12). The core of a villus consists of blood vessels and mesenchyme that is similar in composition to the mesenchyme of the umbilical cord (see Fig. 6-11). Scattered among the mesenchymal cells are large **Hofbauer cells,** which probably fulfill a phagocytic function.

The villus core is covered by a continuous layer of syncytiotrophoblast, with minimum numbers of cytotrophoblastic cells beneath it. The surface of the syncytiotrophoblast is covered by immense numbers of microvilli (over 1 billion/cm^2 at term), which greatly increase the total surface area of the placenta (Fig. 6-13). The size and density of the microvilli are not constant but change with increasing age of the placenta and differing environmental conditions. For example, under conditions of poor maternal nutrition or oxygen transport, the microvilli increase in prominence. Poor adaptation of the microvilli to adverse conditions can lead to newborns with a low birth weight.

The trophoblastic surface is not homogeneous but rather seems to be arranged into territories. Among the many func-

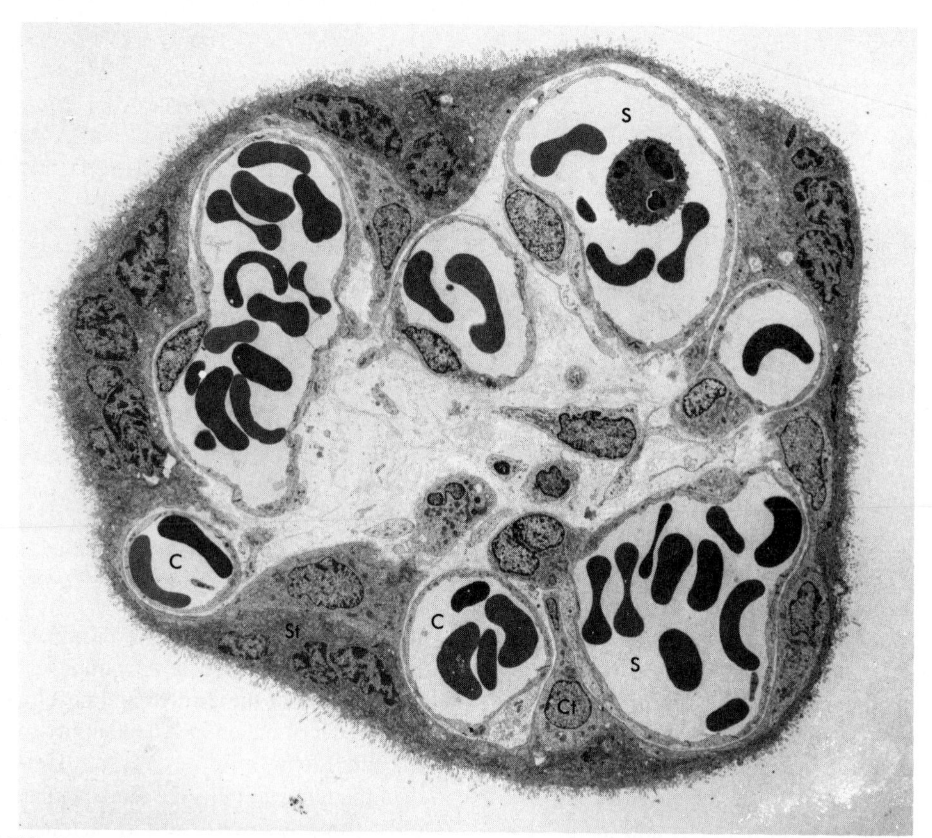

FIG. 6-11 Low-power transmission electron micrograph through a typical terminal villus of a human placenta. *C,* Capillary; *Ct,* cytotrophoblast; *S,* sinusoid (dilated capillary); *St,* syncytiotrophoblast.
(From Benirschke K, Kaufmann P: *Pathology of the human placenta,* ed 2, New York, 1990, Springer-Verlag.)

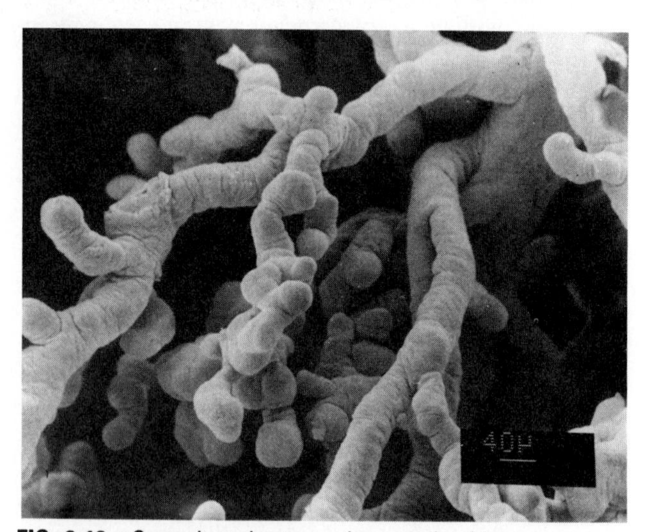

FIG. 6-12 Scanning electron micrograph of long, intermediate, knoblike terminal villi from a normal placenta near the termination of pregnancy.
(From Benirschke K, Kaufmann P: *Pathology of the human placenta,* ed 2, New York, 1990, Springer-Verlag.)

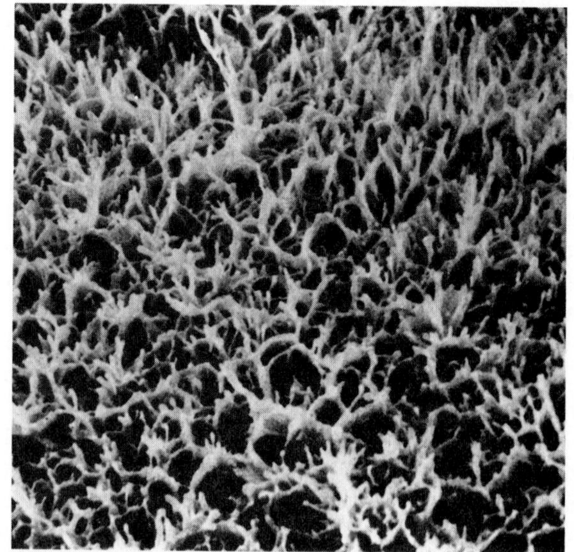

FIG. 6-13 Scanning electron micrograph of the surface of the syncytiotrophoblast of a human placenta in the twelfth week of pregnancy. The numerous microvilli increase the absorptive surface of the placenta. (×9000.)
(Courtesy Staffan Bergstrom, Uppsala, Sweden.)

tional components of the microvillous surface are (1) a wide variety of transport systems for substances ranging from ions to macromolecules, (2) hormone and growth factor receptors, (3) enzymes, and (4) numerous proteins with poorly understood functions. The placental surface is deficient or lacking in major histocompatibility antigens, which presumably plays a role in protecting maternal immune rejection of the fetus and its membranes. In keeping with its active role in both synthesis and transport, the syncytiotrophoblast is well supplied with a high density and wide variety of subcellular organelles.

Placental Physiology

Placental Transfer. The transport of substances both ways between the placenta and the maternal blood that bathes it is facilitated by the great surface area of the placenta, which expands from 5 m^2 at 28 weeks to almost 11 m^2 at term. Approximately 5% to 10% of the human placental surface consists of scattered areas where the barrier between fetal and maternal blood is extremely thin, measuring only a few microns. These areas, sometimes called **epithelial plates,** appear to be morphological adaptations designed to facilitate diffusion of substances between the fetal and maternal circulation (Fig. 6-14).

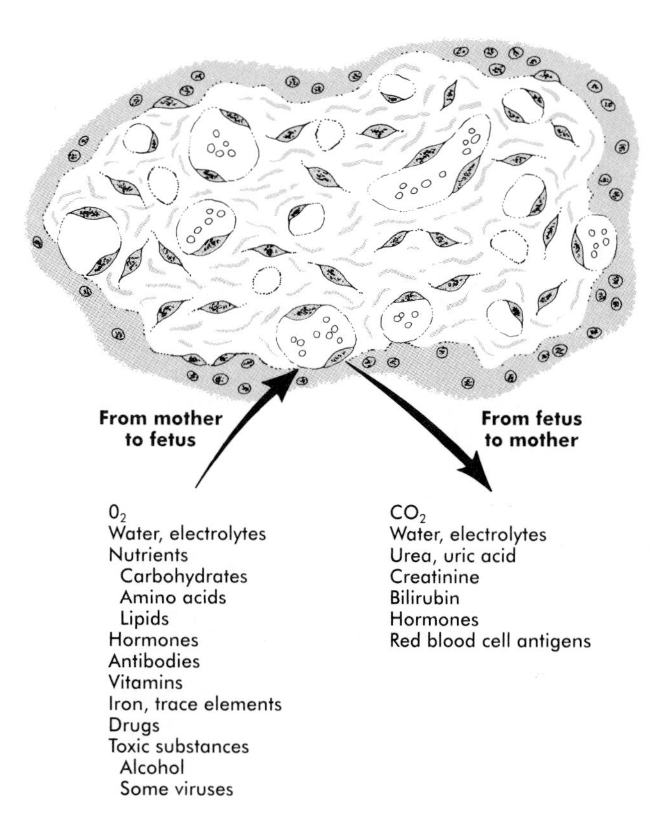

FIG. 6-14 The exchange of substances across the placenta between the fetal and maternal circulation.

The transfer of substances occurs both ways across the placenta. The bulk of the substances transferred from mother to fetus consists of oxygen and nutrients. The placenta represents the means of final elimination of carbon dioxide and other fetal waste materials into the maternal circulation. Under some circumstances, other substances, some of them harmful, can be transferred across the placenta.

Gases, principally oxygen from the mother and carbon dioxide from the fetus, readily cross the placental barrier by diffusion. The amount of exchange is limited more by blood flow than by the efficiency of diffusion. The placenta is also permeable to carbon monoxide and many inhalational anesthetics. The latter can interfere with the transition of the newborn to independent function (e.g., breathing) if used during childbirth.

Like gases, water and electrolytes are readily transferred across the placenta. The rates of transfer are modified by colloid osmotic pressure in the case of water and the function of ion channels. Fetal wastes (e.g., **urea, creatinine,** and **bilirubin**) are rapidly transferred across the placenta from the fetal circulation to the maternal blood bathing the villi.

The placenta is highly permeable to certain nutrients such as **glucose;** on the other hand, it is considerably less permeable to **fructose** and several common disaccharides. Amino acids are transported across the placenta through the action of specific receptors. A certain degree of transfer of maternal **free fatty acids** occurs, but more needs to be learned about the mechanism of transfer. Vitamins, especially water-soluble ones, are transferred from the maternal to the fetal circulation.

Steroid hormones cross the placental barrier from the maternal blood. Newborn males show evidence of the effects of exposure to maternal sex hormones. For example, the **prostatic utricle,** the vestigial rudiment of the uterine primordium (fused müllerian ducts [see Chapter 17]), is slightly enlarged in newborn males. Conversely, female fetuses exposed to testosterone or certain synthetic progestins (especially during the 1950s and 1960s before the effects were recognized) undergo masculinization of the external genitalia. Protein hormones are, in general, poorly transported across the placenta, although symptoms of maternal diabetes may be reduced during late pregnancy because of insulin produced by the fetus. Maternal thyroid hormone gains slow access to the fetus.

Some proteins are transferred very slowly through the placenta, mainly by means of pinocytosis (uptake by membrane-bound vesicles in the cells). Of considerable importance is the transfer of maternal antibodies, mainly of the IgG class. Because of its immature immune system, the fetus produces only small amounts of antibodies. The transfer of antibodies from the mother provides a passive immunity of the newborn to certain common childhood diseases, such as smallpox, diphtheria, and measles, until the immune system of the infant begins to function more efficiently.

Another maternal protein, **transferrin,** is important because, as its name implies, it carries iron to the fetus. The placental surface contains specific receptors for this protein. It appears that the iron is dissociated from its transferrin carrier at the placental surface and then actively transported into the fetal tissues.

Abnormal Placental Transfer. Unfortunately, the placenta is also permeable to substances that can be damaging to the embryo. Numerous maternally ingested drugs readily cross the placental barrier. Certain drugs can cause major birth defects if they reach the embryo during critical periods of morphogenesis. (Several classic examples of these are described in Chapter 7.) The placenta is highly permeable to alcohol, and excessive alcohol ingestion by the mother can produce **fetal alcohol syndrome** (see Chapter 8). The tragedy of newborns who are born addicted to heroin or crack cocaine is all too common in contemporary society.

In addition to drugs, certain infectious agents can penetrate the placental barrier and infect the fetus. Some (e.g., rubella virus) can cause birth defects if they infect the embryo at critical periods in development. Normally, bacteria cannot penetrate the placental barrier. Common viruses that can infect the fetus are rubella, cytomegalovirus, poliomyelitis, varicella, variola, and coxsackie viruses. The spirochete *Treponema pallidum,* which causes syphilis, can cause devastating fetal infections. The protozoan parasite *Toxoplasmosis gondii* can cross the placental barrier and cause birth defects.

Cellular Transfer and Rh Incompatibility. Small quantities of fetal blood cells often escape into the maternal circulation, either through small defects in the placental vasculature or through hemorrhage at the time of childbirth. If the fetal erythrocytes are positive for the Rh antigen and the mother is Rh negative, the presence of fetal erythrocytes in the maternal circulation can stimulate the formation of anti-Rh antibody by the immune system of the mother. The fetus in the first pregnancy is usually spared the effects of the maternal antibody (often because it has not formed in sufficient quantities), but in subsequent pregnancies, Rh-positive fetuses are attacked by the maternal anti-Rh antibodies, which make their way into the fetal bloodstream. This antibody causes hemolysis of the Rh-positive fetal erythrocytes and the fetus develops **erythroblastosis fetalis,** sometimes known as **hemolytic disease.** In severe cases the bilirubin released from the lysed red blood cells causes jaundice and brain damage in addition to anemia. When recognized, this condition is treated by exchange transfusions of Rh-negative donor blood into either the fetus or newborn. An indication of the severity of this condition can be gained by examining the amniotic fluid.

Placental Hormone Synthesis and Secretion

The placenta, and specifically the syncytiotrophoblast, is an important endocrine organ during much of the period of pregnancy. It produces both protein and steroid hormones.

The first protein hormone produced is **human chorionic gonadotropin (HCG),** which is responsible for maintaining the corpus luteum and its production of progesterone and estrogens. With synthesis beginning even before implantation, the presence of this hormone in maternal urine is the basis for many of the common tests for pregnancy. The production of HCG peaks at approximately the eighth week of gestation and then gradually declines. By the end of the first trimester, the placenta produces enough progesterone and estrogens so that pregnancy can be maintained even if the corpus luteum is surgically removed. The placenta can independently synthesize progesterone from acetate or cholesterol precursors, but it does not contain the complete enzymatic apparatus for the synthesis of estrogens. For estrogen to be synthesized, the placenta must operate in concert with the fetal adrenal gland and possibly the liver; these structures possess the enzymes that the placenta lacks.

Another placental protein hormone is **chorionic somatomammotropin,** sometimes called **human placental lactogen.** Similar in structure to human growth hormone, it influences growth, lactation, and lipid and carbohydrate metabolism. The placenta also produces small amounts of **chorionic thyrotropin** and **chorionic corticotropin.** When they are secreted into the maternal bloodstream, some placental hormones stimulate changes in the metabolism and cardiovascular function of the mother. These changes ensure that appropriate types and amounts of fundamental nutrients and substrates reach the placenta for transport to the fetus.

In certain respects, the placenta duplicates the multilevel control system that regulates hormone production in the postnatal body. Cells of the cytotrophoblast produce a homologue of gonadotropin-releasing hormone (GnRH), as is normally done by the hypothalamus. GnRH then passes into the syncytiotrophoblast, where it, along with certain opiate peptides and their receptors (which have been identified in the syncytiotrophoblast), stimulates the release of HCG from the syncytiotrophoblast. The opiate peptides and their receptors are also involved in the release of chorionic somatomammotropin from the syncytiotrophoblast. Finally, HCG appears to be involved in regulating the synthesis and release of placental steroids from the syncytiotrophoblast.

In addition to hormones, the placenta also produces a wide variety of other proteins that have principally been identified immunologically. The functions of the dozens of placental proteins that have been discovered are still very poorly understood.

Placental Immunology

One of the major mysteries of pregnancy is why the fetus and placenta, which are immunologically distinct from the mother, are not recognized as foreign tissue and rejected by the mother's immune system. (Immune rejection of foreign tissues normally occurs by the activation of cytotoxic lymphocytes, but humoral immune responses are also possible.) Despite considerable research, the answer to this question is still not known. Several broad explanations have been suggested to account for the unusual tolerance of the mother to the prolonged presence of the immunologically foreign embryo during pregnancy.

The first possibility is that the fetal tissues, especially those of the placenta, which constitute the direct interface between fetus and mother, do not present foreign antigens to the mother's immune system. To some extent this hypothesis is true, since neither the syncytiotrophoblast nor the nonvillous cytotrophoblast **(cytotrophoblastic shell)** expresses the two major classes of major histocompatibility (MHC) antigens that trigger the immune response of the host in the rejection of typical foreign tissue grafts (e.g., a kidney transplant). However, these antigens are present on cells of the fetus and in stromal tissues of the placenta. The expression of minor histocompatibility antigens (e.g., the HY antigen in male fetuses [see Chapter 17]) follows a similar pattern. Nevertheless, other minor antigens are expressed on trophoblastic tissues. In addition, due to breaks in the placental barrier, fetal red and white blood cells are frequently found circulating in the maternal blood. These cells should be capable of sensitizing the immune system of the mother.

A second major possibility is that the mother's immune system is somehow paralyzed during pregnancy so that it does not react to the fetal antigens to which it is exposed. This is not true because the mother is capable of mounting an immune response to infections or other foreign tissue grafts. There still remains the possibility of a selective repression of the immune response to fetal antigens (although the Rh incompatibility response shows that this is not universally the case).

A third possibility is that local decidual barriers prevent either immune recognition of the fetus by the mother or the reaching of competent immune cells from the mother to the fetus. Again, there is evidence for a functioning decidual immune barrier, but in a significant number of cases that barrier is known to be breached through trauma or disease.

Currently, studies are being directed toward conditions such as recurrent spontaneous abortion with the hope of finding further clues to the complex immunological interrelationships between the fetus and mother. What is abundantly clear is that this is not a simple relationship. Nevertheless, the solution to this problem may yield information that might be applied to the problem of reducing the host rejection of surgical tissue and organ transplants.

THE PLACENTA AFTER BIRTH

About 30 minutes after birth, the placenta, embryonic membranes, and remainder of the umbilical cord, along with much of the maternal decidua, are expelled from the uterus as the **afterbirth.** The fetal surface of the placenta is smooth, shiny, and grayish because of the amnion that covers the fetal side of the chorionic plate. The maternal surface, on the other hand, is a dull red and may be punctuated with blood clots. The maternal surface of the placenta must be examined carefully because if a cotyledon is missing and is retained in the uterine wall, it could be the cause of serious postpartum bleeding. Recognition of certain types of placental pathology can provide valuable clues to intrauterine factors that could affect the well-being of the newborn; details of these are beyond the scope of this text.

THE PLACENTA AND MEMBRANES IN MULTIPLE PREGNANCIES

Several different configurations of the placenta and extraembryonic membranes are possible in multiple pregnancies. Dizygotic twins or monozygotic twins resulting from complete separation of blastomeres very early in cleavage can have completely separate placentas and membranes if the two embryos implant in distant sites on the uterine wall (Fig. 6-15, *A*). In contrast, if the implantation sites are closer together, the placentas and chorions, (which were initially separate at implantation) can fuse, although the vascular systems of the two embryos remain separate (Fig. 6-15, *B*).

When monozygotic twins form by splitting of the inner cell mass in the blastocyst, it is usual to have a common placenta and a common chorion, but inside the chorion the twin embryos each develop within separate amnions (Fig. 6-15, *C*). In this case, there can be separate or fused vascular systems within the common placenta. When the vascular systems are fused, one twin may receive a much greater proportion of the placental blood flow than the other. This may result in mild to severe stunting of growth of the embryo that receives the lesser amount of blood from the placenta.

In conjoined twins and rarely in dizygotic twins with minimal separation of the inner cell mass, the embryos develop within a single amnion and chorion and have a common placenta with a common blood supply (Fig. 6-15, *D*). This and the previously described conditions can be readily determined by examination of the membranes of the afterbirth. At one time it was thought that it could be determined whether twins were monozygotic or dizygotic by simple examination of the membranes. Although in most cases the correct inference can be made, this method is by no means foolproof. Other methods ranging from simple observation of gender, eye color, and fingerprint patterns to determina-

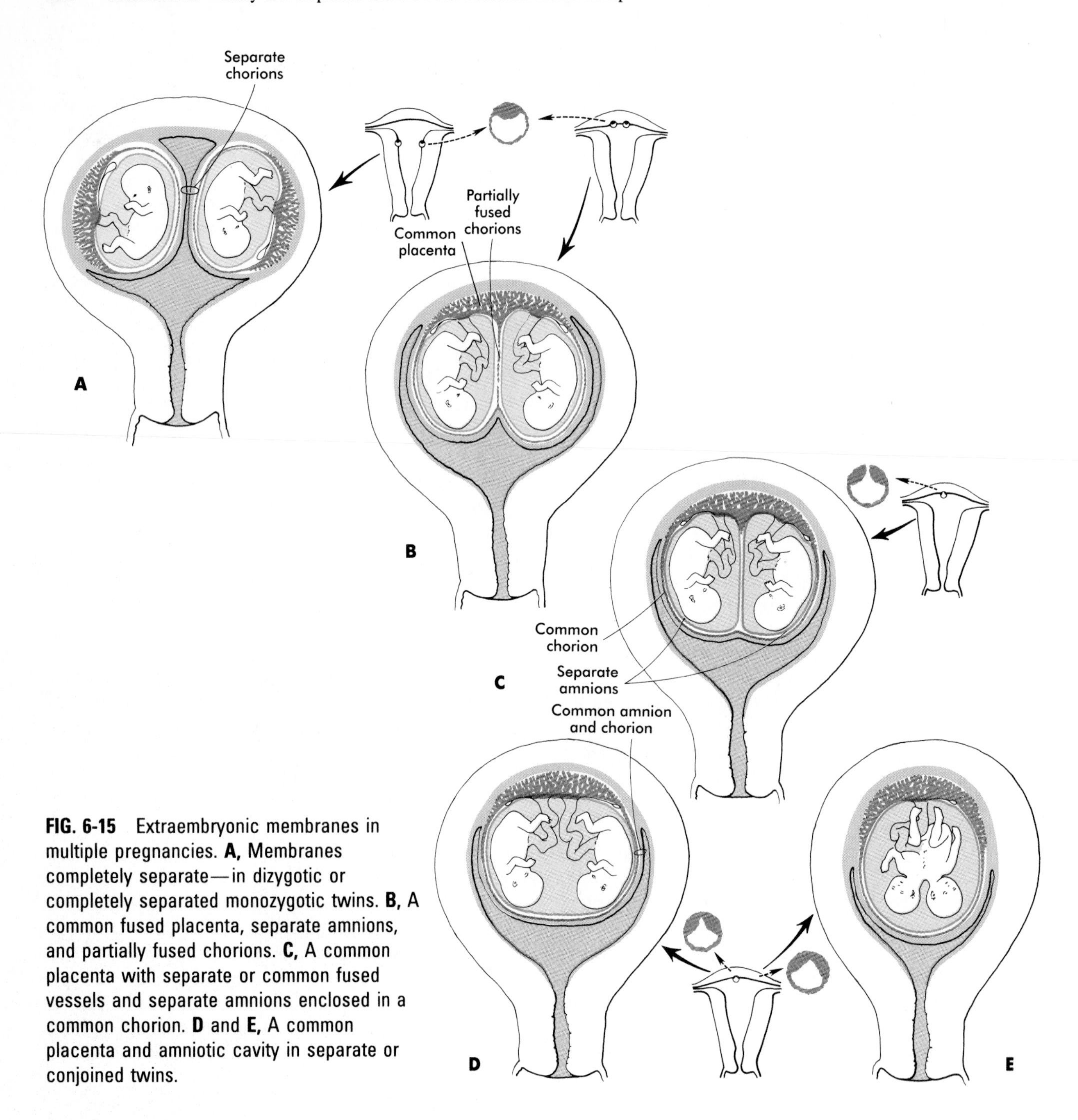

FIG. 6-15 Extraembryonic membranes in multiple pregnancies. **A,** Membranes completely separate—in dizygotic or completely separated monozygotic twins. **B,** A common fused placenta, separate amnions, and partially fused chorions. **C,** A common placenta with separate or common fused vessels and separate amnions enclosed in a common chorion. **D** and **E,** A common placenta and amniotic cavity in separate or conjoined twins.

tion of blood types or even DNA fingerprinting should be used for a definitive determination. In the current age of organ and cell transplantation, it can be vital to know whether twins are monozygotic in the event that one of the twins develops a condition that can be treated by a transplant.

PLACENTAL PATHOLOGY

Placental pathology covers a wide spectrum of conditions, ranging from abnormalities of implantation site to neoplasia to frank bacterial infections. Much can be learned about the past history and future prospects of a newborn by examination of the placenta. This section deals only with those aspects of placental pathology that are relevant to developmental mechanisms. Specific pathologies are well covered in compendia on placental pathology.

Abnormal Implantation Sites

Abnormal implantation sites within the uterine cavity are known as **placenta previa.** (Ectopic pregnancy is covered

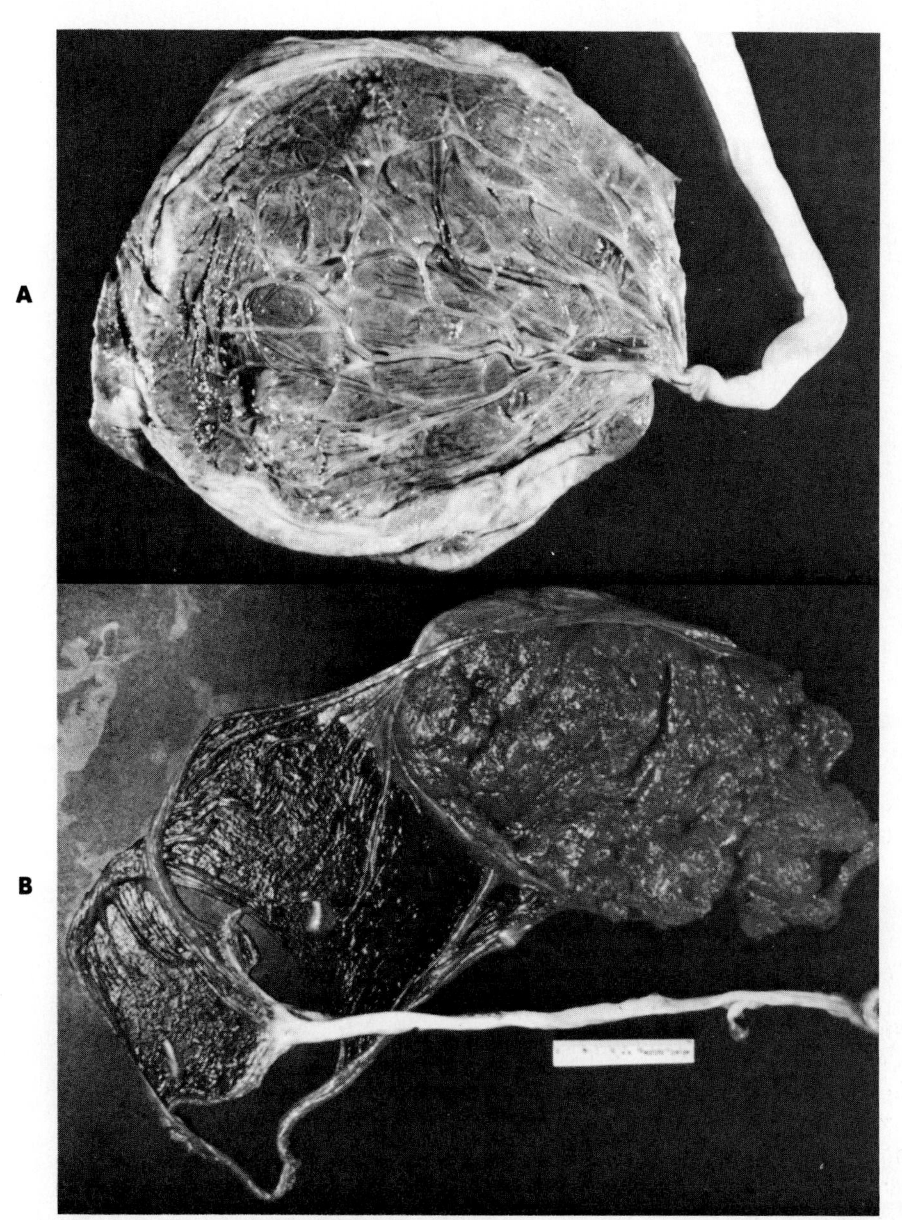

FIG. 6-16 Variations in placental shape. **A,** Marginal insertion of the umbilical cord. **B,** Velamentous insertion of umbilical cord.
(From Naeye RL: *Disorders of the placenta, fetus, and neonate,* St Louis, 1992, Mosby.)

Continued.

in Chapter 3.) When part of the placenta covers the cervical outlet of the uterine cavity, its presence is a mechanical obstacle in the birth canal. Additionally, hemorrhage, which can be fatal to the fetus or even the mother, is a common consequence of placenta previa due to premature separation of part of the placenta from the uterus.

Gross Placental Anomalies

Many variations in shape of the placenta have been described, but few of them appear to be of any functional significance. One involves marginal rather than central attachment of the umbilical cord (Fig. 6-16, *A*). If the umbilical cord attaches to the smooth membranes outside the boundaries of the placenta itself, the condition is known as a **velamentous insertion** of the umbilical cord (Fig. 6-16, *B*).

The placenta itself can be subdivided into **accessory lobes** (Fig. 6-16, *C*), or it can be completely divided into two parts, with smooth membrane between (Fig. 6-16, *D*).

Hydatidiform Mole

A **hydatidiform mole** is a noninvasive condition in which many of the chorionic villi are characterized by nodular swellings, giving them an appearance almost like bunches of grapes. Commonly, much of the villous surface of the

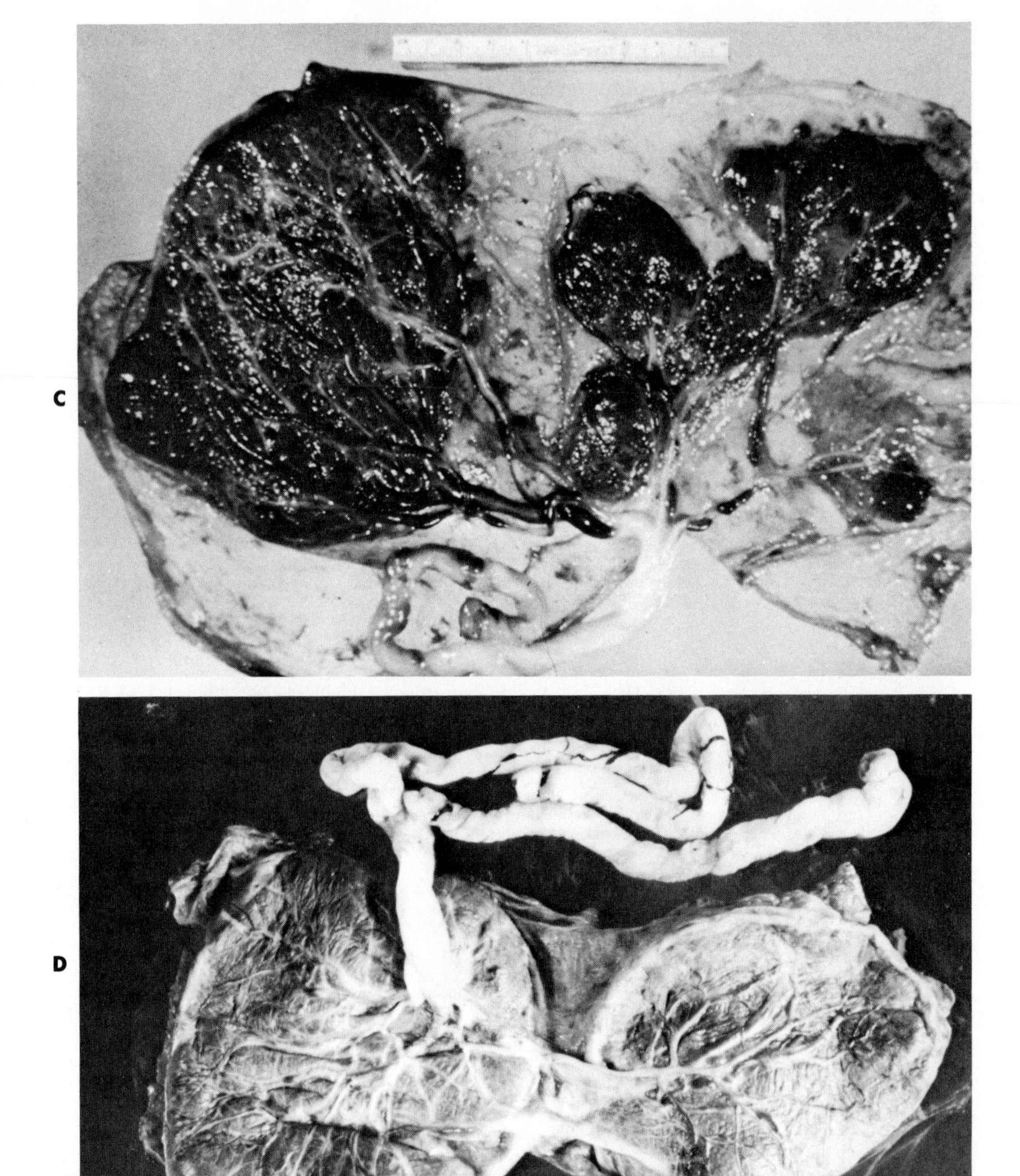

FIG 6-16, cont'd. C, A placenta with accessory (succenturiate) lobes. **D,** A completely bilobed placenta.

FIG. 6-17 **A,** Distended uterus containing a hydatiform mole. The ovaries *(top* and *bottom)* contain bilateral theca lutein cysts.
(Based on studies by Benirschke and Kaufmann [1990].)
B, View at greater magnification showing swollen villi. (Courtesy K. Benirschke, San Diego, Calif.)

placenta takes on this appearance; in addition, the embryo is either absent or not viable (Fig. 6-17). The villi show no evidence of vascularization.

Recent genetic analysis has determined that hydatidiform moles represent the results of paternal imprinting in cases where the female pronucleus of the egg does not participate in development (see Chapter 3). Instead, the chromosomal material is derived from two sperm that had penetrated the egg or by duplication of a single sperm pronucleus within the egg. The chromosomes of hydatidiform moles are paternally derived 46,XX, since the number of lethal genes in 46,YY embryos is not compatible with tissue survival.

Choriocarcinoma

Choriocarcinomas are malignant tumors derived from embryonic cytotrophoblast and syncytiotrophoblast. These tumors are highly invasive into the maternal decidual tissues and blood vessels. Like hydatidiform moles, most choriocarcinomas contain only paternally derived chromosomes and are thus products of paternal imprinting.

Chorionic Villus Biopsies

In recent years, biopsies of chorionic villi during the latter half of the second embryonic month have sometimes been performed instead of sampling amniotic fluid. Performed with the assistance of ultrasonography (see Fig. 19-17, *B*), chorionic villus biopsies are obtained for the analysis of possible chromosomal disorders or the diagnosis of certain metabolic disorders.

SUMMARY

1. The extraembryonic membranes consist of the chorion (the combination of trophoblast plus underlying extraembryonic mesoderm), the amnion, the yolk sac, and the allantois.
2. The amnion, a thin ectodermal membrane lined with mesoderm, grows to enclose the embryo like a balloon. It is filled with a clear fluid, which is generated from many sources such as the fetal skin, the amnion itself, the fetal kidneys, and possibly fetal vessels. At term the volume of amniotic fluid approaches 1 L. Amniotic fluid is removed by exchange across the amniotic membrane and by fetal swallowing.
3. The yolk sac is a ventral endodermally lined structure that does not serve a nutritive function in mammalian embryos. Mesodermal blood islands in the wall of the yolk sac form the first blood cells and vessels. Primordial germ cells are recognizable in the endodermal wall of the yolk sac.

4. The allantois is a small, endodermally lined diverticulum off the ventral side of the hindgut. It does not serve a direct function of respiration or storage of wastes in mammals, but through the associated umbilical vessels, these functions are carried out through the placenta.
5. Chorionic villi form as outward projections from the trophoblast. Primary villi consist of projections of trophoblast alone. When a mesenchymal core forms within a villus, it is a secondary villus, and when the mesenchyme becomes vascularized, the villus is a tertiary villus. As villi mature, the cytotrophoblast in some of them grows through the syncytiotrophoblast and makes contact with the maternal endometrial tissue. Cytotrophoblast continues to grow around the blood-filled space around the chorion to form a cytotrophoblastic shell, which is the direct interface between fetal and maternal tissues. Villi that make direct contact with maternal tissues are anchoring villi; villi that do not make such contact are floating villi. Because chorionic villi float in a pool of maternal blood, the human placenta is designated as a hemochorial placenta.
6. Stimulated by the implanting embryo, endometrial stromal cells undergo the decidual reaction. Maternal tissues that are lost at childbirth are collectively the decidua. The decidua basalis underlies the placenta; the decidua capsularis encircles the remainder of the chorion like a capsule; portions of the uterine wall not occupied by the fetal chorion are the decidua parietalis. As the fetal chorion matures, it becomes subdivided into a chorion laeve, in which the villi regress, and the chorion frondosum, the region of chorion nearest the basal tissues of the endometrium. The chorion frondosum ultimately develops into the placenta.
7. The mature placenta consists of the wall of the chorion (the chorionic plate) and numerous villi protruding from it. The fetal surface of the placenta is smooth and shiny due to the apposed amniotic membrane. The maternal surface is dull and lobulated, with cotyledons of numerous placental villi and their branches. The umbilical cord (formerly the body stalk) enters the middle of the placenta. Blood from the fetus reaches the placenta via the umbilical arteries. These branch out into numerous small vessels, terminating into capillary loops in the ends of the placental villi. There, oxygen, nutrients, and wastes are exchanged between fetal and maternal blood, which bathes the villi. Fetal blood returns to the body of the mature fetus via a single umbilical vein. Maternal blood exiting open-ended spiral arteries of the endometrium bathes the placental villi.
8. The transfer of substances from fetal to maternal blood must occur across the endothelium of the fetal capillaries, a basal lamina, and trophoblastic tissues before reaching the maternal blood. Transfer of substances is accomplished by both passive and active mechanisms. In addition to normal substances, alcohol, certain drugs, and some infectious agents can pass from the maternal blood into the fetal circulation and interfere with normal development. If a fetus is Rh positive and the mother is Rh negative, maternal anti-Rh antibodies can pass to the fetus to cause erythroblastosis fetalis.
9. The placenta produces a wide variety of hormones, many of which are normally synthesized in the hypothalamus and anterior pituitary gland. The first hormone released is human chorionic gonadotropin, which serves as the basis of many pregnancy tests. Other placental hormones are chorionic somatomammotropin (human placental lactogen), steroid hormones, and chorionic thyrotropin and corticotropin.
10. The fetal and placental tissues are immunologically different from those of the mother, but the placenta and fetus are not immunologically rejected. The reason is still not clear, but some explanations involve reduced antigenicity of the trophoblastic tissues, paralysis of the mother's immune system during pregnancy, and local immunological barriers between the fetus and mother.
11. The placenta is delivered about 30 minutes after the fetus as the afterbirth. Inspection of the placenta can reveal placental pathology, missing cotyledons, or the arrangement of membranes in multiple pregnancies. The last finding can help to determine whether a multiple birth is monozyotic in origin. Placental pathology includes abnormal gross shape, benign hydatidiform moles, and malignant choriocarcinomas.

REVIEW QUESTIONS

1. What congenital anomalies are associated with oligohydramnios and hydramnios and why?
2. Why is the human placenta designated a hemochorial type of placenta?
3. Through what layers must a molecule of oxygen pass to go from the maternal blood into the embryonic circulation?
4. What embryonic hormone has served as the basis for many standard pregnancy tests and why?
5. Why must a pregnant woman be very careful of what she eats and drinks?

REFERENCES

Ahmed MS, Cemerikic B, Agbas A: Properties and functions of human placental opioid system, *Life Sci* 50:83-97, 1991.

Aplin JD: Implantation, trophoblast differentiation and haemochorial placentation: mechanistic evidence in vivo and in vitro, *J Cell Sci* 99:681-692, 1991.

Benirschke K, Kaufmann P: *Pathology of the human placenta,* ed 2, New York, 1990, Springer-Verlag.

Billingham RE, Beer AE: Reproductive immunology: past, present and future, *Prospect Biol Med* 27:259-275, 1984.

Bohn H, Dati F, Lueben G: *Human trophoblast specific products other than hormones.* In Loke YW, Whyte A, eds: *Biology of trophoblast,* Amsterdam, 1983, Elsevier Science, pp 317-352.

Bohn H, Winckler W, Grundmann U: Immunochemically detected placental proteins and their biological functions, *Arch Gynecol Obstet* 249:107-118, 1991.

Boyd JD, Hamilton WJ: *The human placenta,* Cambridge, England, 1970, Heffer & Sons.

Chamberlain GVP, Wilkinson AW, eds: *Placental transfer,* Tunbridge Wells, Kent, England, 1979, Pitman Medical.

Cullen TS: *Embryology, anatomy and diseases of the umbilicus, together with diseases of the urachus,* Philadelphia, 1916, Saunders.

Dallaire L, Potier M: *Amniotic fluid.* In Milunsky A, ed: *Genetic disorders and the fetus,* New York, 1986, Plenum, pp 53-97.

Dearden L, Ockleford CD: *Structure of human trophoblast: correlation with function.* In Loke YW, Whyte A, eds: *Biology of trophoblast,* Amsterdam, 1983, Elsevier Science, pp 69-110.

Enders AC: Trophoblast differentiation during the transition from trophoblastic plate to lacunar stage of implantation in the rhesus monkey and human, *Am J Anat* 186:85-98, 1989.

Faber JJ, Thornburg KL, eds: *Placental physiology,* New York, 1983, Raven.

Foidart J-M and others: The human placenta becomes haemochorial at the 13th week of pregnancy, *Int J Dev Biol* 36:451-453, 1992.

Gruenwald P, ed: *The placenta and its maternal supply line,* Baltimore, 1975, University Park.

Kaufmann P: Basic morphology of the fetal and maternal circuits in the human placenta, *Contrib Gynecol Obstet* 13:5-17, 1985.

Kaufmann P, Scheffen I: *Placental development.* In Polin R, Fox W, eds: *Fetal and neonatal physiology,* vol 1, Philadelphia, 1992, Saunders, pp 47-56.

Knoll BJ: Gene expression in the human placental trophoblast: a model for developmental gene regulation, *Placenta* 13:311-327, 1992.

Lavrey JP, ed: *The human placenta: clinical perspectives,* Rockville, Md, 1987, Aspen.

Naeye RL: *Disorders of the placenta, fetus, and neonate,* St Louis, 1992, Mosby.

Ramsey EM: *The placenta: human and animal,* New York, 1982, Praeger.

Schneider H: Placental transport function, *Reprod Fertil Dev* 3:345-353, 1991.

Schneider H: The role of the placenta in nutrition of the human fetus, *Am J Obstet Gynecol* 164:967-973, 1991.

Schroeder J: Review article: transplacental passage of blood cells, *J Med Genet* 12:230-242, 1975.

Sibley CP, Boyd RDH: *Mechanisms of transfer across the human placenta.* In Polin R, Fox W, eds: *Fetal and neonatal physiology,* vol 1, Philadelphia, 1992, Saunders, pp 62-74.

Truman P, Ford HC: The brush border of the human term placenta, *Biochem Biophys Acta* 779:139-160, 1984.

CHAPTER 7

Manipulating Human Reproduction

For centuries, the manipulation of human reproduction consisted principally of contraception and abortion, the intent being to limit the number of offspring in the family or tribe. In some cases (e.g., the reduction of fertility associated with extended breast feeding), societal norms and nutritional necessity rather than conscious choice may have been the basis for unknowingly manipulating the reproductive function of a woman.

The past several decades have seen a virtual explosion in the ability to manipulate the reproductive process. In addition to developing an impressive armamentarium of methods of birth control, reproductive biologists have a variety of techniques for increasing fertility, storing gametes and embryos, fertilizing eggs in vitro, and raising embryos through surrogate mothers. Interestingly, many of the last techniques were first developed for use in domestic animals to economize or maximize the genetic influence of prize breeding stock. All these developments, whether applied to humans or animals, were based on the results of many years of research that has provided a detailed knowledge of the structural features and physiological controls of reproductive processes.

A new phase of reproductive manipulation features the cellular and molecular engineering of gametes and embryos. How far scientific capabilities and societal guidelines will allow these fields to develop and be applied to humans remains to be determined.

STRATEGIES OF CONTRACEPTION

It is beyond the scope of this book to present in detail the many specific methods of contraception. Nevertheless, they can be categorized according to several broad strategies (Table 7-1).

Contraception in Males

An often discussed strategy is to interfere with spermatogenesis. Many chemical and hormonal methods have been tried with varying degrees of success, but none is in wide use today. This is partly due to the problem of ensuring that all of the hundreds of millions of maturing spermatozoa are equally affected. Such methods are aimed at interfering with one of three levels of control:

1. Hypothalamic stimulation of the anterior pituitary via releasing factors: This is done with either analogues to the critical releasing factor luteinizing hormone–releasing hormone (LHRH) or through the use of steroids to cause a feedback inhibition of LHRH release.
2. Direct inhibition of follicle-stimulating hormone (FSH) production or release from the anterior pituitary by the use of inhibin (see Fig. 1-17) or immunological inhibition of FSH function: To date such methods have not been very successful.
3. Direct effects on spermatozoa or sperm-producing cells in the seminiferous tubules: One compound falling into this category is **gossypol,** a component of cotton seed oil that was found to cause severe **oligospermia** (reduction in number of sperm). A secondary effect of gossypol is greatly reduced sperm motility, possibly due to interference with the fructose in seminal fluid, which is the main energy source of mature spermatozoa.

Although increased temperature interferes with spermatogenesis, the home remedy of taking hot baths to reduce sperm production is extremely unreliable. Contraceptive methods directed toward the interference with sperm function are mostly still in developmental stages. Most methods are immunologically based, but ensuring that systemic autoimmunity does not occur is a major concern.

The most effective method of contraception in the male

TABLE 7-1 Commonly Used Contraceptive Methods in the United States

METHOD	ESTIMATED USE (%)	ACCIDENTAL PREGNANCY IN FIRST YEAR OF USE (%)	CHARACTERISTICS
Vasectomy	14	0.15	Single procedure, minimal health risks, surgically reversible (up to 50% success)
Tubal ligation	19	0.4	Single procedure, minimal health risks, limited surgical reversibility
Combined oral contraceptive pill (estrogen and progestin)	32	3	Cessation of ovulation, regulation of menstrual cycle, some side effects; regular usage to be effective, reversible on cessation of treatment
Oral contraceptive minipill (progestin only)	NA	5	Thickening of cervical mucus, causing atrophy of endometrium and possibly inhibiting ovulation; fewer side effects than combined oral contraceptives; reversible on cessation of treatment
Long-acting steroid implants (progestin only)	NA	0.2	Easy insertion, few side effects, stable blood levels of steroid, readily reversible
Intrauterine device (IUD)	3	6	Single procedure, relatively high frequency of side effects (including 1/2500 incidence of perforation); reversible on removal of device
Condom	17	12	Inexpensive, easy to use, protection against sexually transmitted diseases
Diaphragm	4-6	2-23	Training and care necessary to be effective, occasional allergic reactions to rubber or associated cream or jelly
Contraceptive sponge	3	18	Easy to obtain and use, occasional removal problems, higher risk of pregnancy than with most other reversible methods
Rhythm method	4	20	High risk of pregnancy, considerable knowledge and motivation necessary to be effective
Vaginal contraceptives (creams, jellies, foams)	2	21	Easy to obtain, relatively unreliable, use within minutes before intercourse necessary to be effective

Modified from Mastrovianni L, Donaldson PJ, Kane TT, eds: *Developing new contraceptives; obstacles and opportunities,* Washington, 1990, National Academy.

is to block sperm transport by surgically severing the ductus deferens **(vasectomy).** Vasectomy is among the most reliable permanent methods of contraception, and it has the advantage of being surgically reversible in a relatively high percentage of cases. On the other hand, vasectomy is associated with an increased risk of prostate cancer.

Probably the most common contraceptive method used by males is the **condom,** which functions in a purely mechanical sense to block the deposition of sperm in the vagina. Although not as reliable as other methods of contraception, the secondary function of condoms in reducing the spread of AIDS and other sexually transmitted diseases has resulted in a significant increase in their use. Early withdrawal during intercourse is an ancient and highly ineffective contraceptive technique.

Contraception and Birth Control in Females

The vast majority of contraceptive research and development has been conducted in the female. Contraceptive techniques have been targeted at virtually all levels of the female reproductive system from the hypothalamus to the gravid uterus.

Many systemic modes of contraception are designed to

prevent ovulation. One strategy is to administer analogues of the hypothalamic releasing factor gonadotropin-releasing hormone (GnRH), which prevents the release of the pituitary hormones required for normal follicular development in the ovary. The various **oral contraceptives** also act on the cells of the anterior pituitary but via a different feedback mechanism. Considerable attention is currently being directed toward the development of methods that permit the long-term sustained release of progestins and the local delivery of hormonally based contraceptive implants.

Tubal ligation is the female counterpart of vasectomy and is designed to create a block between the ovulated egg and sperm that have entered the female reproductive tract. Although normally considered a permanent form of sterilization, it is sometimes possible to surgically repair the ligated uterine tubes. However, the incidence of tubal pregnancies (see Chapter 6) is considerably increased after repair.

Another important contraceptive strategy is to prevent viable spermatozoa from entering the upper regions of the female reproductive tract. **Diaphragms, pessaries,** and **female condoms** provide purely mechanical interference, but diaphragms are normally used with spermicidal jellies. Contraceptive vaginal jellies or foams and sponges impregnated with spermicidal agents are designed to provide a hostile environment for spermatozoa in the upper reaches of the vagina. Because of the rapidity with which some spermatozoa enter the cervix, douching is an extremely unreliable method of preventing insemination no matter how soon after intercourse it is done.

The **rhythm method** involves the use of physiological timepoints (e.g., the menstrual period and the rise in basal body temperature at ovulation) in the woman as guides for "safe periods" to have intercourse. This method can be moderately effective, but its accuracy is considerably reduced in women with irregular menstrual cycles, which often occur at the beginning and end of a woman's reproductive period.

After fertilization, the preimplantation embryo remains extremely vulnerable. The **"morning after" pill,** with its high estrogen content, alters the endometrium so that implantation fails to occur. Similarly, **intrauterine devices (IUDs)** appear to function by preventing implantation, probably by chronic irritation of the endometrium. The effectiveness of other systemic contraceptive agents (principally hormonal and immunological agents currently being tested) may be due in part to their interfering with the normal hormonal preparations of the endometrium for implantation.

After implantation, birth control is accomplished by aborting the embryo. Over the centuries, many methods, ranging from a variety of mechanical means to pharmacological agents (e.g., ergot), have been developed. By far the most controversial agent (and one of the most effective) is the progesterone antagonist **RU 486.** This drug, which was developed during the early 1980s, functions by preferentially occupying the progesterone receptors in the nuclei of the normal target tissues. Progesterone receptors occupied by RU 486 do not undergo the normal conformational changes that occur after binding by progesterone. In this altered configuration, the receptors are unable to bind to the nuclear chromatin so that normal progesterone-induced transcription can occur. At the tissue level, this causes deterioration of the endometrium and the detachment of the embryo. Increased motility of the smooth muscle of the uterus, which may be accentuated by small doses of prostaglandins, results in expulsion of the embryo from the uterus.

TREATMENT OF INFERTILITY

The opposite of contraception is the treatment of infertility. The causes of infertility include malfunctions at many of the steps that contraceptive methods target. Various methods of treating infertility, particularly those involving surgery (e.g., reconstructing blocked or ligated uterine tubes), are beyond the scope of this text, but several of them have directly or indirectly permitted the development of some amazing techniques for manipulating human reproduction. Many of these are incorporated in the overall strategy for overcoming certain types of infertility by means of **in vitro fertilization** and **embryo transfer.** This type of manipulation is typically used with patients who are capable of forming gametes but in whom an often anatomical block prevents the eggs and spermatozoa from meeting or prevents the free-floating embryo from reaching or implanting in the uterus.

In Vitro Fertilization and Embryo Transfer

The relevant techniques involve (1) stimulating gamete production, (2) obtaining male and female gametes, (3) storing gametes, (4) fertilizing eggs, (5) performing an in vitro culture of cleaving embryos, (6) preserving embryos, and (7) introducing embryos into the uterus (Fig. 7-1).

Stimulation of Gamete Production. Most contemporary efforts attempt to stimulate ovulation by altering existing hormonal relationships. For women who are **anovulatory** (do not ovulate), these techniques alone may be sufficient to allow conception. Another group consists of women who ovulate normally but are candidates for in vitro fertilization and embryo transfer because of defects of the uterine tubes.

Three types of therapy have been commonly used to stimulate gamete production. The first involves the administration of **clomiphene citrate,** a nonsteroidal **antiestrogen** that competes with estrogens for binding sites in the pituitary and possibly the hypothalamus and ovaries. Competition acts to suppress the normal negative feedback by estrogens on the pituitary, resulting in an increase in the

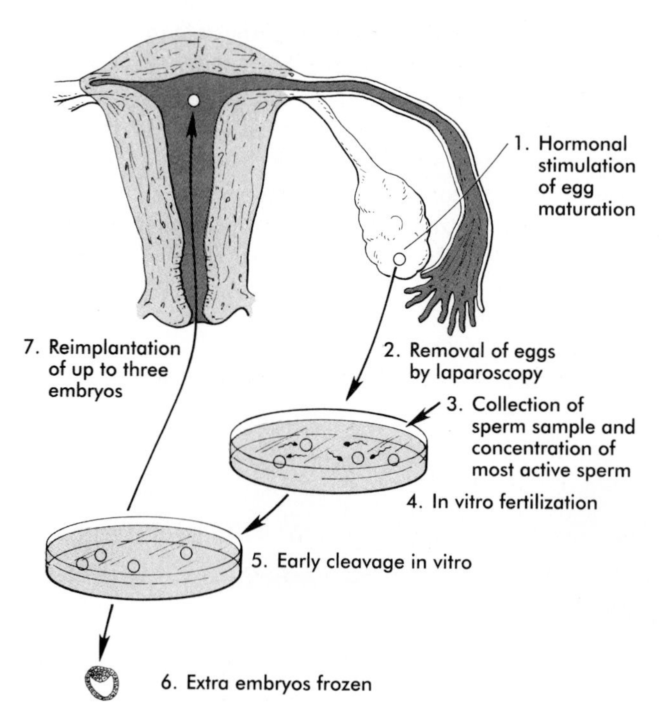

FIG. 7-1 A typical in vitro fertilization and embryo transfer procedure in humans.

serum concentrations of luteinizing hormone (LH) and FSH. This typically results in multiple ovulation, a desired outcome for artificial fertilization and embryo transfer procedures because fertilizing more than one egg at a time is more efficient. Sometimes, however, a woman who has used clomiphene for the induction of ovulation produces multiple offspring, with a number of cases of quintuplet to septuplet births having been recorded.

Another standard method of inducing ovulation is the application of **human menopausal gonadotropins (hMGs),** either alone or with clomiphene or FSH. An advantage of this method over clomiphene alone is a greater predictability of the ovulation response, but this is attained at a considerably greater financial cost to the patient than clomiphene therapy.

More recently, the pulsatile administration of GnRH has been used to induce ovulation. This technique, which in many respects is more physiological (sometimes referred to as the application of an artificial hypothalamus), produces results in concert with the patient's own gonadotrophic hormones and typically results in a lower incidence of multiple births. Pulsatile GnRH is also somewhat less expensive and requires less monitoring than hMG therapy. Further technical comparisons of these methods of stimulating ovulation can be found in clinical textbooks.

Obtaining Gametes. For artificial insemination in vivo or artificial fertilization in vitro, spermatozoa are typically collected by masturbation. The collection of eggs involves technological assistance. Today, ongoing monitoring of the course of induced ovulation is accomplished by imaging techniques, especially ultrasound.

Actual recovery of oocytes involves their aspiration from the ripe follicles. This has traditionally been done with the concurrent use of **laparoscopy,** which involves inserting a viewing instrument **(laparoscope)** through a small slit in the woman's abdominal wall, and an aspiration needle, which is inserted gently into each mature follicle. Once identified, the oocyte is gently sucked into the aspiration needle and is then placed into culture medium in preparation for fertilization in vitro. In recent years, ultrasound instead of the laparoscope has sometimes been used to guide the path of the aspiration needle. Although the use of ultrasound in oocyte retrieval has been increasing, both laparoscopy and ultrasound will likely continue to have their place, depending on the clinic and clinical circumstances.

Storing Gametes. Although eggs and sperm are usually placed together shortly after they are obtained, in some circumstances, the gametes (especially spermatozoa) are stored for various periods before their use. One example is sperm banks for donors other than the husband. In other cases (e.g., before a person undergoes certain operations that can interfere with spermatogenesis or ejaculatory function or before military personnel are involved in a war), individuals may bank some of their own sperm as insurance of genetic continuity. With contemporary techniques (e.g., bringing glycerinated preparations of spermatozoa down to the temperature of liquid nitrogen [−196° C]), spermatozoa can be kept for years without losing their normal fertilizing power. The freezing of eggs is possible but much more problematic.

In Vitro Fertilization and Embryo Culture. The three ingredients for successful in vitro fertilization are mature eggs; normal, active spermatozoa; and an appropriate culture environment. After considerable testing, culture media and gaseous environments that foster fertilization and embryo growth have been refined to the point of being routine. The formulas can be found in many manuals on in vitro fertilization.

One of the most important factors in obtaining successful in vitro fertilization is having oocytes that are properly mature. The eggs aspirated from a woman are sometimes of different stages of maturity. Immature eggs are cultured for a short time to become more fertilizable. The aspirated eggs are surrounded by the zona pellucida, the corona radiata, and a varying amount of cumulus oophorus tissue. Removing excess cumulus tissue sometimes facilitates sperm penetration.

Spermatozoa, either fresh or frozen, are prepared by separating them as much as possible from the seminal fluid. Seminal fluid reduces their fertilizing capacity, in part because it contains decapacitating factors. After capacitation, which in the human can be accomplished by exposing spermatozoa to certain ionic solutions, a defined number of spermatozoa are added to the culture in concentrations of from 10,000 to 500,000/ml. Their movements and penetration of the egg membranes can be visually monitored under favorable circumstances. In a number of studies, frozen spermatozoa have shown a slight edge in fertilizing capacity over fresh ones. Rates of fertilization in vitro vary from one center to another, but 75% represents a realistic average.

In cases of infertility due to **oligospermia** (too few spermatozoa) or excessively high percentages of abnormal sperm cells, multiple ejaculates may be obtained over an extended period. These are frozen and pooled to obtain adequate numbers of viable sperm. In some cases, small numbers of spermatozoa are microinjected into the **perivitelline space** inside the zona pellucida. Although this procedure

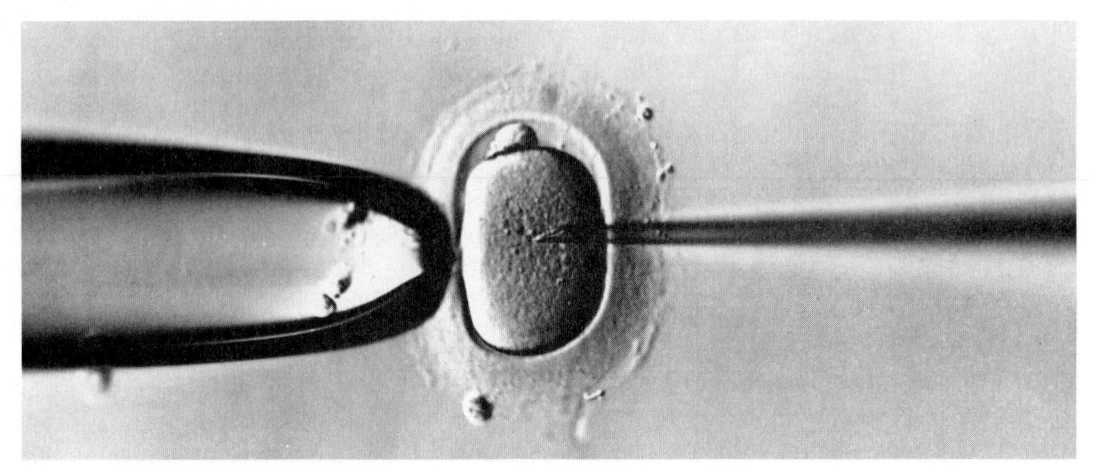

FIG. 7-2 The microinjection of a spermatozoon into a human oocyte. The micropipette containing the spermaozoon is entering the oocyte from the right side.
(From Veeck LL: *Atlas of the human oocyte and early conceptus,* vol 2, Baltimore, 1991, Williams & Wilkins.)

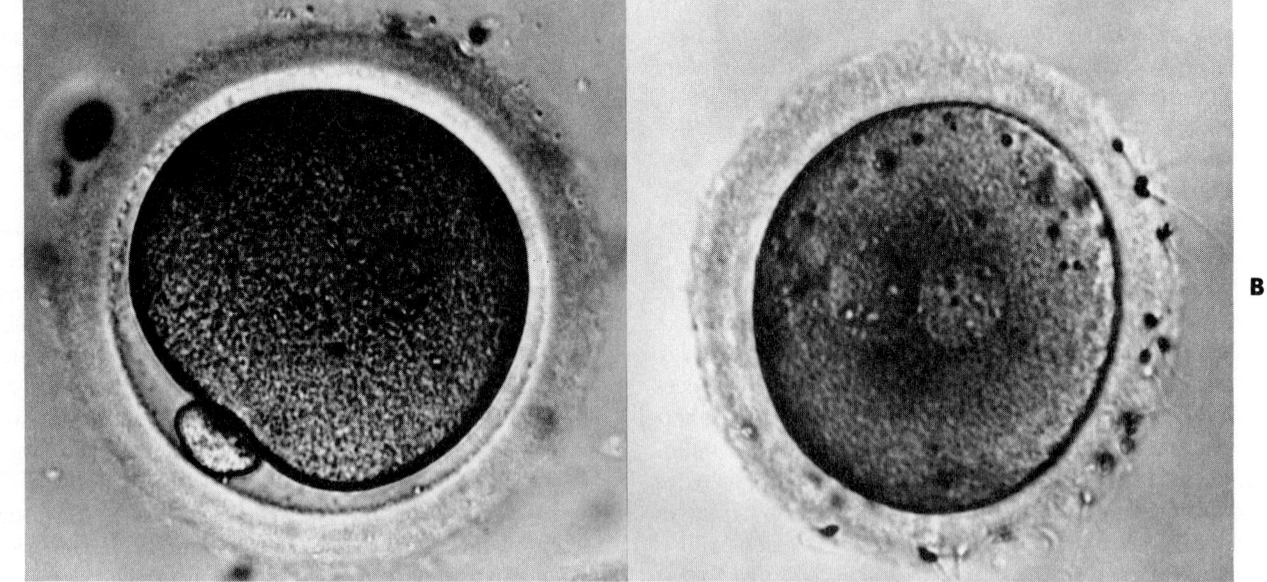

FIG. 7-3 **A,** Photomicrograph of a mature human oocyte arrested at metaphase II. This oocyte will be fertilized in vitro. **B,** Photomicrograph of a human oocyte newly fertilized in vitro. Two pronuclei are visible.
(From Veeck LL: *Atlas of the human oocyte and early conceptus,* vol 2, Baltimore, 1991, Williams & Wilkins.)

can compensate for very small numbers of viable spermatozoa, it introduces the increased risk of polyspermy because the normal gating function of the zona (which acts as a slow block to fertilization [see Chapter 2]) is bypassed. The most recent variant on in vitro fertilization techniques is direct injection of a spermatozoon into an oocyte (Fig. 7-2). This technique has been used in cases of severe sperm impairment.

The initial success of in vitro fertilization is determined the next day by examination of the egg. If two pronuclei are evident (Fig. 7-3), fertilization is assumed to have taken place.

Fortunately, the cleavage in vitro of human embryos is more successful than that of most other mammalian species. The embryos are usually allowed to develop to the 2- to 8-cell stage before they are considered ready to implant into the uterus.

Typically, all the eggs obtained from the multiple ovulations of the woman are fertilized in vitro during the same period. Very practical reasons exist for doing this. One is that because of the low success rate of embryo transfer, implanting more than one embryo (commonly up to three) into the uterus at a time is advisable. Another reason is financial and also relates to the low success rate of embryo transfer. Embryos other than those used during the initial procedure are stored for future use if the first embryo transfer proves to be unsuccessful. Such stockpiling saves a great deal of time and thousands of dollars for the patient.

Embryo Preservation. The embryos preserved for potential future use are usually in the 2- to 8-cell stage and are treated with cryoprotectants (usually glycerol or dimethyl sulfoxide) to reduce ice crystal damage. They are slowly subjected to very low temperatures (usually below −100° C) to halt all metabolic activity. Although blastocysts tolerate freezing better (possibly because their cells are smaller and less susceptible to damage), success rates with the implantation of frozen blastocysts are less than with earlier cleavage stages. The length of time frozen embryos should be kept and the procedure for handling them if the first implantation attempt is successful is a question with technical and ethical aspects.

Embryo Transfer into the Mother. Transfer of the embryo into the mother is technically simple, yet this is the step in the entire operation that is subject to the greatest failure rate. Typically only 10% to 25% of the embryo transfer attempts result in a viable pregnancy.

Embryo transfer is commonly performed by introducing a catheter through the cervix into the uterine cavity and then expelling the embryo or embryos from the catheter. The patient remains quiet, preferably lying down for several hours following embryo transfer.

The reasons for the low percentage of success of embryo transfers are poorly understood, but the percentage of completed pregnancies after normal fertilization in vivo is also likely to be only about one-third. Some possible reasons that have been advanced are (1) a low percentage of continued development of embryos maintained in vitro, (2) unsuccessful implantation possibly due to unpreparedness of the endometrium, and (3) later intrinsic failure of the embryo. If normal implantation does occur, the remainder of the pregnancy is typically uneventful and is followed by a normal childbirth.

Gamete Intrafallopian Transfer

Certain types of infertility are due to factors such as hostile cervical mucus and pathology or anatomical abnormalities of the upper end of the uterine tubes. A somewhat simpler method for dealing with these conditions is to introduce both male and female gametes directly into the lower end of the uterine tube (often at the junction of its isthmic and ampullary portions). Fertilization occurs within the tube, and the early events of embryogenesis occur naturally. Although relatively recent, in some clinics the method of **gamete intrafallopian transfer (GIFT)** has resulted in slightly higher percentages of pregnancies than the standard in vitro fertilization and embryo transfer methods.

A variant on this technique is **ZIFT (zygote intrafallopian transfer).** In this case a cleaving embryo that has been produced by in vitro fertilization is implanted into the uterine tube.

Surrogacy

There are some circumstances in which a woman can produce fertile eggs but cannot be pregnant. One example is a woman whose uterus has been removed but who still possesses functioning ovaries. One option in this case is in vitro fertilization and embryo transfer, but the embryo is transferred into the uterus of another woman **(surrogate mother).** From the biological perspective, this procedure differs little from embryo transfer into the uterus of the biological mother, but it introduces a host of social, ethical, and legal issues.

Ethical and Legal Issues

The many new and often unconventional procedures involved in reproductive technology have raised a number of ethical issues, both prospectively and retrospectively. Health care workers must be sensitive to the fact that opinions are sometimes divided regarding the moral and legal basis for performing certain manipulations. Some of these are based on religious beliefs; others reflect societal norms, sometimes established thousands of years ago in certain cultures.

Probably the most difficult issue is the question of when human life begins. This question is the basis of the abortion controversy and is of direct relevance to the issue of the disposition of extra frozen embryos in in vitro fertilization programs. There is a wide spectrum of opinions regarding the time of origin of human life, typically ranging from the time of conception to the time when the mother first feels quickening of the fetus in the uterus. Another opinion holds that with the continuity of germ plasm and all the preparative events that occur during oogenesis (e.g., the redistribution of chromosomes during meiosis, the storing of maternal genetic messages in the preovulatory oocyte), fertilization is just another event in a long continuous series that ultimately results in the formation of a biologically independent being.

Other facets of the in vitro fertilization program, such as the collection of spermatozoa by masturbation and artificially bypassing any component of the normal course taken by an ovulated egg, run counter to the teachings of a number of religions and cultures. Similar objections may apply to artificial insemination, either by spouse or donor. These fall into the general category of "unnatural" reproductive practices.

Surrogate motherhood has raised a number of issues, including the psychological bonding of the surrogate mother with the fetus and the morality of paying someone to assume the risks of pregnancy and childbirth. In a number of cases a sister, mother, or even grandmother has acted as a surrogate for a family member. In the legal arena, a number of pending court cases involve a surrogate mother who has refused to give up custody of a baby to the biological parents. Biological parents may also refuse to take custody of an infant borne of a surrogate mother if that infant is defective or of the undesired gender.

Another concrete example of the new legal issues raised by reproductive technology is the case of a wealthy couple who had undergone the in vitro fertilization and embryo transfer procedure and were then both killed in a plane crash. Several frozen embryos remained at the clinic. What are the legal rights of the embryos in relation to what could be a large inheritance?

A major ethical issue that has often been raised is the great investment of money, people, and technology for a technique designed to bring more children into an already overpopulated world. Valid arguments exist regarding this and most of these questions. If any truism exists, it is that the advancing technology inevitably interfaces with the moral and ethical structure of the society in which it is performed. Open dialogue by all parties concerned is clearly essential.

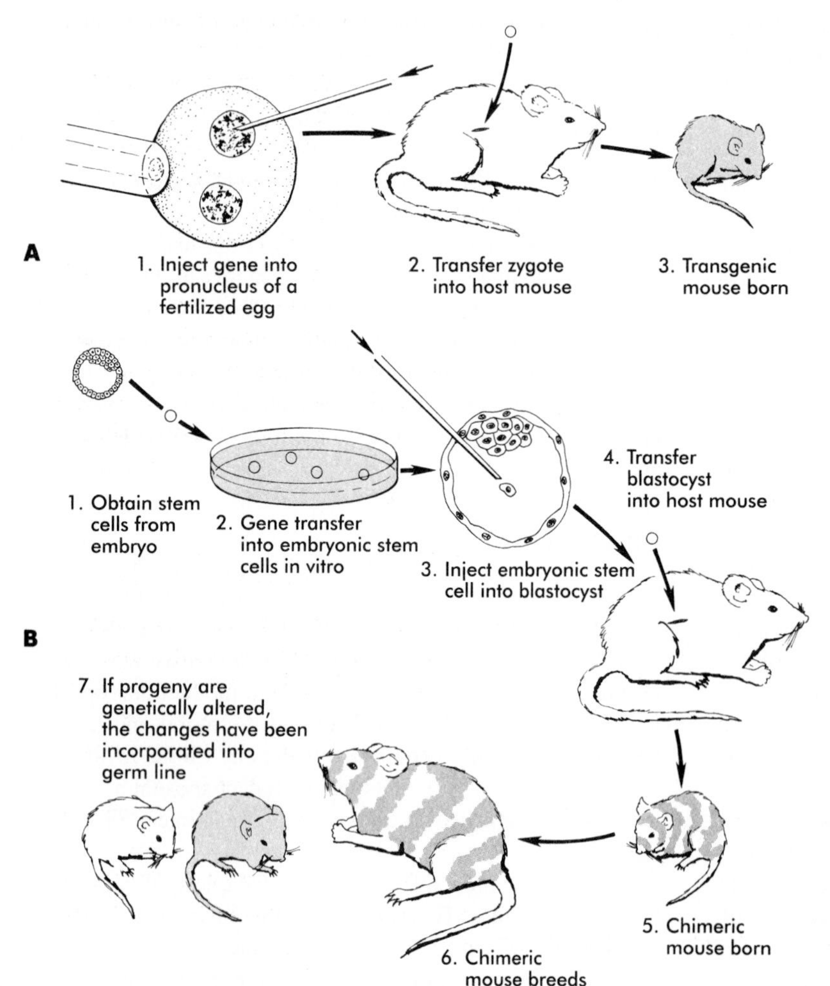

FIG. 7-4 **A,** The procedure for creating transgenic mice. **B,** The procedure for inserting genes into mice by first introducing them into embryonic stem cells and then inserting the transfected stem cells into an otherwise normal blastocyst.

CELLULAR AND GENETIC MANIPULATION OF EMBRYOS

On the basis of experiments performed on lower vertebrates, a number of possible scenarios resulting from the misuse of reproductive technology have been proposed. Armies of genetically identical drones cloned by a ruthless dictator or a wealthy man who hires a scientist to produce a copy of him have excited the popular imagination. Such fantasies have resulted from extrapolation of the sometimes successful nuclear transplantation experiments in amphibians. Adult frogs have been produced by inserting nuclei taken from larval or adult frog tissues into irradiated frog eggs. Mammals have proved to be resistant to such attempts, and several reports of positive results have later been recanted. Nevertheless, it cannot be categorically stated that successful cloning of mammals is not possible.

Of greater relevance today is the production of **transgenic animals** by injecting foreign genetic material directly into a pronucleus of a newly fertilized egg. With the early techniques, the transfered genes were essentially randomly inserted into the DNA of the host. Under these circumstances, expression of these genes was extremely variable for a number of reasons (e.g., the presence of improper regulatory sequences for the genes in question). Researchers can now target the inserted genes with a remarkable degree of precision. In experimental animals, particularly mice, gene transfer has been used to treat genetic defects and to introduce defective genes, thus creating animal models for certain human hereditary diseases. Transferred genes in sheep are being used to produce large quantities of human proteins. If the genes are expressed in blood or milk, commercially useful quantities can be obtained in an efficient manner throughout the life span of the transgenic animal.

An even newer technique is to isolate **embryonic stem cells** (i.e., cells of the early embryo that still retain the potential to develop into any cell type in the adult) and then perform targeted gene transfer into these cells. Once it is determined that the appropriate genes can be expressed or disrupted **(gene knockout),** the embryonic stem cells are microinjected into the blastocyst (Fig. 7-4). The **transfected** stem cell is incorporated into the inner cell mass of the host, and its progeny form parts of the now mosaic embryo. If the stem cell progeny become gametes, permanent lines expressing the transferred gene can be produced.

Methods like these can be used for studies of gene function, for gene therapy of defective genes, and for introducing defective genes into experimental animals to model specific diseases. To date they have been employed on experimental animals, especially mice and sheep. Although the application of transgenic techniques to human embryos has not been reported, the technology itself is not the main limiting factor. The biomedical community has dealt very conservatively with the issue of genetic manipulations in humans that could result in the transmission of introduced hereditary traits from one generation to the next.

SUMMARY

1. Contraceptive strategies in the male involve interference with spermatogenesis by blocking hypothalamic or pituitary hormones or by directly affecting sperm-producing cells or spermatozoa and blocking access to the female reproductive tract by vasectomy or condoms.
2. Contraception and birth control in the female are accomplished at several levels, including hormonal interference with ovulation, endometrial function, or the consistency of cervical mucus; physical separation between the egg and sperm (e.g., tubal ligation, diaphragm); killing of spermatozoa with spermicides; interference with implantation (e.g., intrauterine devices); and ejection of the implanted embryo (e.g., RU 486).
3. The treatment of infertility by in vitro fertilization and embryo transfer is a multistage process involving stimulating gamete production by drugs such as clomiphene citrate, obtaining gametes by laparoscopic techniques in the female and by masturbation in the male, storing gametes by freezing, performing in vitro fertilization and culture of embryos to the blastocyst stage, preserving the embryo, and transferring the embryo into the mother.
4. Other techniques used for the treatment of infertility are GIFT—the transfer of gametes directly into the uterine tube—and ZIFT—the transfer of zygotes into the uterine tube. These techniques can be used with surrogate and biological mothers.
5. New cellular and molecular techniques make it possible to directly manipulate the genetic nature of embryos through the production of transgenic embryos, the altering of genetic stem cells, and the introduction of such cells into host blastocysts.
6. Many reproductive techniques raise numerous ethical and legal issues. These issues are based on societal, religious, and personal values, as well as scientific uncertainty in many cases.

REVIEW QUESTIONS

1. What methods of birth control operate by preventing implantation or expelling a newly implanted embryo?
2. A woman gives birth to septuplets. What is the likely reason for the multiple births?
3. When multiple oocytes obtained by laparoscopy are fertilized in vitro, why are up to three embryos implanted into the woman's uterine tube and why are the other embryos commonly frozen?
4. Why do some reproductive technology centers insert spermatozoa under the zona pellucida or even directly into the oocyte?

REFERENCES

American Fertility Society: Guidelines for human embryology and andrology laboratories, *Fertil Steril* 1(suppl 58):1S-16S, 1992.

Austin CR: *Human embryos: the debate on assisted reproduction,* Oxford, England, 1989, Oxford University.

Avrech OM and others: Mifepristone (RU 486) alone or in combination with a prostaglandin analogue for termination of early pregnancy: a review, *Fertil Steril* 56:385-393, 1991.

Fishel S, Symonds EM, eds: *In vitro fertilization: past, present and future,* Oxford, England, 1986, IRL Press.

In vitro fertilization-embryo transfer (IVF-ET) in the United States: 1990 results from the IVF-ET registry, *Fertil Steril* 57:15-24, 1992.

Ivani KA, Seidel GE: *Micromanipulation of mammalian gametes.* In Dunbar BS, O'Rand MG, eds: *A comparative overview of mammalian fertilization,* New York, 1991, Plenum, pp 403-421.

Jones HW, Schrader C, eds: *In vitro fertilization and other assisted reproduction, Ann NY Acad Sci* 541:1-775, 1988.

Jones HW and others, eds: *In vitro fertilization—Norfolk,* Baltimore, 1986, Williams & Wilkins.

Keller DW, Strickler RC, Warren JC: *Clinical infertility,* Norwalk, Conn, 1984, Appleton-Century-Crofts.

Mastroianni L, Donaldson PJ, Kane TT, eds: *Developing new contraceptives: obstacles and opportunities,* National Research Council, Institute of Medicine, Washington, 1990, National Academy Press.

McKinnell RG: *Cloning of frogs, mice, and other animals,* Minneapolis, 1985, University of Minnesota Press.

Palermo G and others: Pregnancies after intracytoplasmic injection of single spermatozoon into an oocyte, *Lancet* 340:17-18, 1992.

Robertson JA: Ethical and legal issues in human egg donation, *Fertil Steril* 52:353-363, 1989.

Sathananthan AH, Trounson AO, Wood C: *Atlas of fine structure of human sperm penetration, eggs and embryos cultured in vitro,* New York, 1986, Praeger Scientific.

Seppala M, Edwards RG, eds: In vitro fertilization and embryo transfer, *Ann NY Acad Sci* 442:1-619, 1985.

Speroff L, Darney P: *A clinical guide to contraception,* Baltimore, 1992, Williams & Wilkins.

Talbert LM: The assisted reproductive technologies: a historical overview, *Arch Pathol Lab Med* 116:320-322, 1992.

Travis J: Scoring a technical knockout in mice, *Science* 256:1392-1394, 1992.

Trounson A, Wood C, eds: *In vitro fertilization and embryo transfer,* Edinburgh, 1984, Churchill Livingstone.

Ulmann A, Teutsch G, Philibert D: RU 486, *Sci Am* 262:42-48, 1990.

Veeck LL: *Atlas of the human oocyte and early conceptus,* vols 1 and 2, Baltimore, 1986, 1991, Williams & Wilkins.

Wagner EF: On transferring genes into stem cells and mice, *EMBO J* 9:3025-3032, 1990.

Warwick BL, Berry RO: Inter-generic and intra-specific embryo transfers in sheep and goats, *J Hered* 40:297-303, 1949.

Wolf DP: *Fertilization in man.* In Dunbar BS, O'Rand MG, eds: *A comparative overview of mammalian fertilization,* New York, 1991, Plenum, pp 385-400.

Wood C, Trounson A, eds: *Clinical in vitro fertilization,* ed 2, London, 1989, Springer-Verlag.

CHAPTER 8

Developmental Disorders—Causes, Mechanisms, and Patterns

Congenital malformations have attracted attention since the dawn of human history. When seen in either humans or animals, malformations were often interpreted as omens of good or evil. Because of the great significance attached to congenital malformations, they were frequently represented in folk art, either as sculptures or paintings. As far back as the classical Greek period, people speculated that maternal impressions during pregnancy (e.g., being frightened by an animal) caused development to go awry. In other cultures, women who gave birth to malformed infants were assumed to have had dealings with the devil or other evil spirits.

Earlier representations of malformed infants were often remarkable in their anatomical accuracy, and it is often possible to diagnose specific conditions or syndromes from the ancient art (Fig. 8-1, *A*). By the time of the Middle Ages, however, representations of malformations were much more imaginative, with hybrids of humans and other animals often represented (Fig. 8-1, *B*).

Among the first applications of scientific thought to the problem of congenital malformations were those of the sixteenth-century French surgeon, Ambrose Paré, who suggested a role of hereditary factors and mechanical influences such as intrauterine compression in the genesis of birth defects. Less than a century later, William Harvey, who is also credited with first describing the circulation of blood, elaborated the concept of developmental arrest and further refined thinking on mechanical causes of birth defects.

In the early ninteenth century, Etienne Geoffroy de St. Hilaire coined the term **teratology,** which literally means "the study of monsters," as a descriptor for the newly emerging study of congenital malformations. Late in the ninteenth century, the scientific study of teratology was put on a firm foundation with the publication of several encyclopedic treatises that exhaustively covered the anatomical aspects of the recognized congenital malformations.

Following the flowering of experimental embryology and genetics in the early twentieth century, laboratory researchers began to produce specific recognizable congenital anomalies by means of defined experimental genetic or laboratory manipulations on laboratory animals. This led to the demystification of congenital anomalies and to a search for rational scientific explanations for birth defects. Nevertheless, old beliefs are very tenacious, and even today patients may adhere to traditional beliefs.

The first of two major milestones in human teratology occurred in 1941, when Gregg in Australia recognized that the **rubella** virus was a cause of a recognizable **syndrome** of abnormal development, consisting of defects in the eyes, ears, and heart. About 20 years later, the tragic story of **thalidomide** sensitized the medical community to the potential danger of certain drugs and other environmental **teratogens** (agents that produce birth defects) to the developing embryo. Thalidomide is a very effective sedative that was widely used in West Germany, Australia, and other countries during the late 1950s. Soon physicians began to see infants born with extremely rare birth defects. One example is **phocomelia** (which means "seal limb"), a condition in which the hands and feet seem to arise almost directly from the shoulder and hip (Fig. 8-2). Another is **amelia,** in which a limb is entirely missing. Only after some careful epidemiological detective work that involved collecting individual case reports and sorting the drugs that had been taken by mothers during the early period of their pregnancies was it possible to identify thalidomide as the certain cause. With the intense investigations that followed the thalidomide disaster, modern teratology came of age. It is remarkable, however, that despite much effort, causes for most congenital malformations are still unknown.

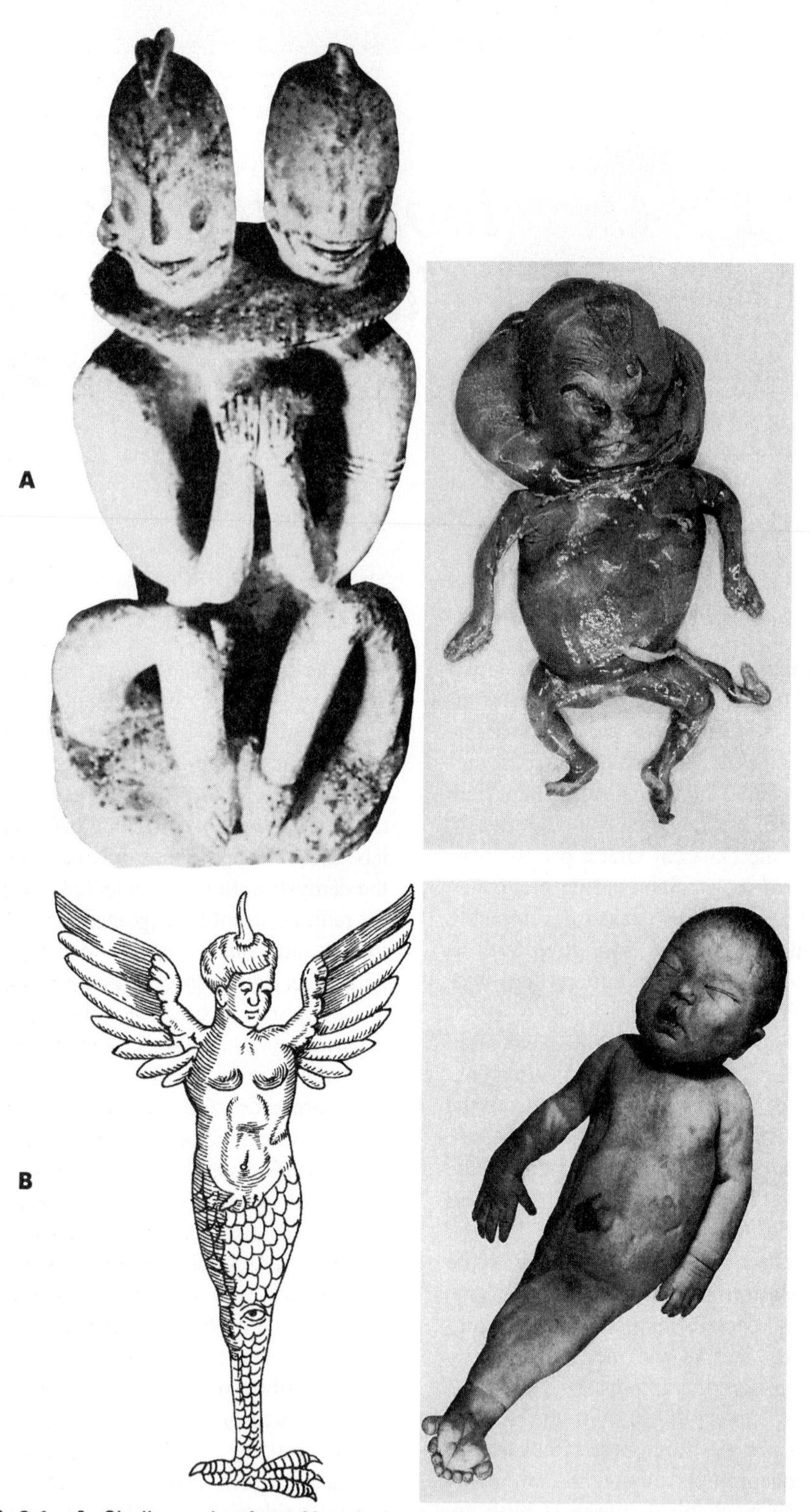

FIG. 8-1 **A,** Chalk carving from New Ireland in the South Pacific showing dicephalic, dibrachic conjoined twins *(left)* (From Brodsky I: *Med J Aust 1:417-420, 1943.*) Note also the "collar" beneath the heads, which is a representation of the malformation cystic hygroma colli *(right)*

(Courtesy Mason Barr, Ann Arbor, Mich.)

B, The bird-boy of Paré (about 1520) *(left).* A stillborn fetus with sirenomelia (fused legs) *(right).* Compare with the lower part of the bird-boy.

(Courtesy Mason Barr, Ann Arbor, Mich.)

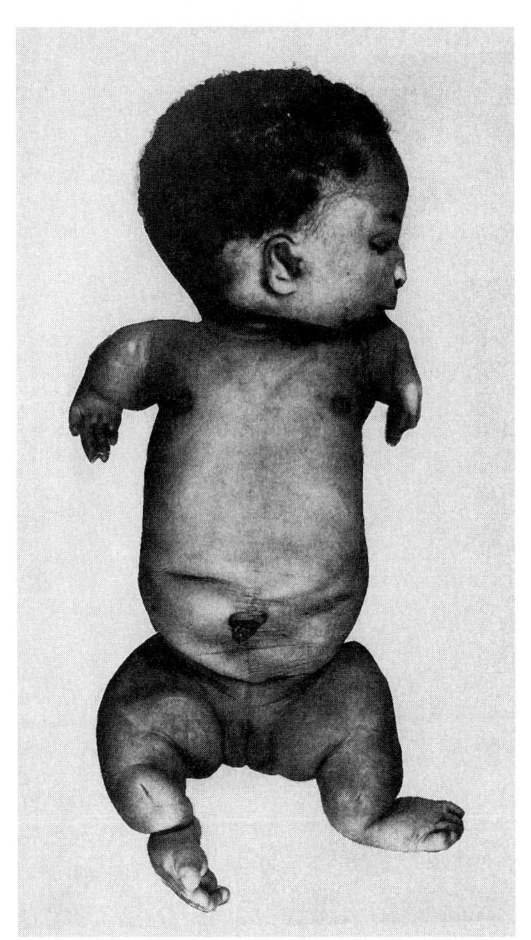

FIG. 8-2 Phocomelia in all four limbs. This fetus had not been exposed to thalidomide.
(Courtesy Mason Barr, Ann Arbor, Mich.)

GENERAL PRINCIPLES

According to most studies, approximately 2% to 3% of all living newborns show at least one recognizable congenital malformation. This percentage is doubled when anomalies that are diagnosed in children during the first few years after birth are considered. With the decline in infant mortality resulting from infectious diseases and nutritional problems, congenital malformations now rank high among the causes of infant mortality (currently over 20%), and an ever-increasing percentage of infants admitted to neonatology or pediatric units (up to 30%) can be attributed to various forms of genetic diseases or congenital defects.

Congenital defects range from enzyme deficiencies caused by single nucleotide substitutions in the DNA molecule to very complex associations of gross anatomical abnormalities. Although medical embryology textbooks traditionally cover principally structural defects—congenital malformations—there is a continuum between purely biochemical abnormalities and those that are manifested as abnormal structures. This continuum includes defects that constitute abnormal structure, function, metabolism, and behavior.

According to contemporary theory, the genesis of congenital defects can be viewed as an interaction between the genetic endowment of the embryo and the environment in which it develops. The basic information is encoded in the genes, but as the genetic instructions unfold, the developing structures or organs are subjected to microenvironmental or macroenvironmental influences that are either compatible or interfere with normal development. In the case of genetically based malformations or anomalies based on chromosomal aberrations, the defect is intrinsic and commonly expressed even in a normal environment. Purely environmental causes interfere with embryological processes in the face of a normal genotype. However, in other cases, environment and genetics interact. Penetrance (the degree of manifestation) of an abnormal gene or expression of one component of a genetically multifactorial cascade can sometimes be profoundly affected by environmental conditions.

One of the first clear-cut examples of the interactions between genetics and environment was provided in the 1950s by the experiments of Fraser on cleft palate formation in mice. In the presence of cortisone (the teratogen), 100% of the embryos of the A/J strain of mice developed cleft palate, whereas only about 20% of offspring of the C57BL strain were born with the anomaly. With crosses of the two strains, the incidence of cleft palate was approximately 40%. These results were ultimately shown to be related to strain-specific differences in the growth rate of the palatal shelves and to the width of the head at specific days of embryogenesis.

A number of factors are associated with various types of congenital malformations. At present, they are understood more at the level of statistical associations than as points of interference with specific developmental controls, but they are important clues to why development can go wrong. Among the factors associated with increased incidences of congenital malformations are (1) parental age, (2) season of the year, (3) the country of residence, (4) race, and (5) familial tendencies.

There are well-known correlations between parental age and the incidence of certain malformations, a classic one being the increase in incidence of **Down syndrome** in women over 35 years of age. Other conditions are related to paternal age (Fig. 8-3).

Other types of anomalies have a higher incidence among infants born at certain seasons of the year. For example, **anencephaly** (Fig. 8-4) occurs more frequently in January. Recognizing that the primary factors leading to anencephaly occur during the first month of embryonic life, researchers must seek the potential environmental causes that are more prevalent in April.

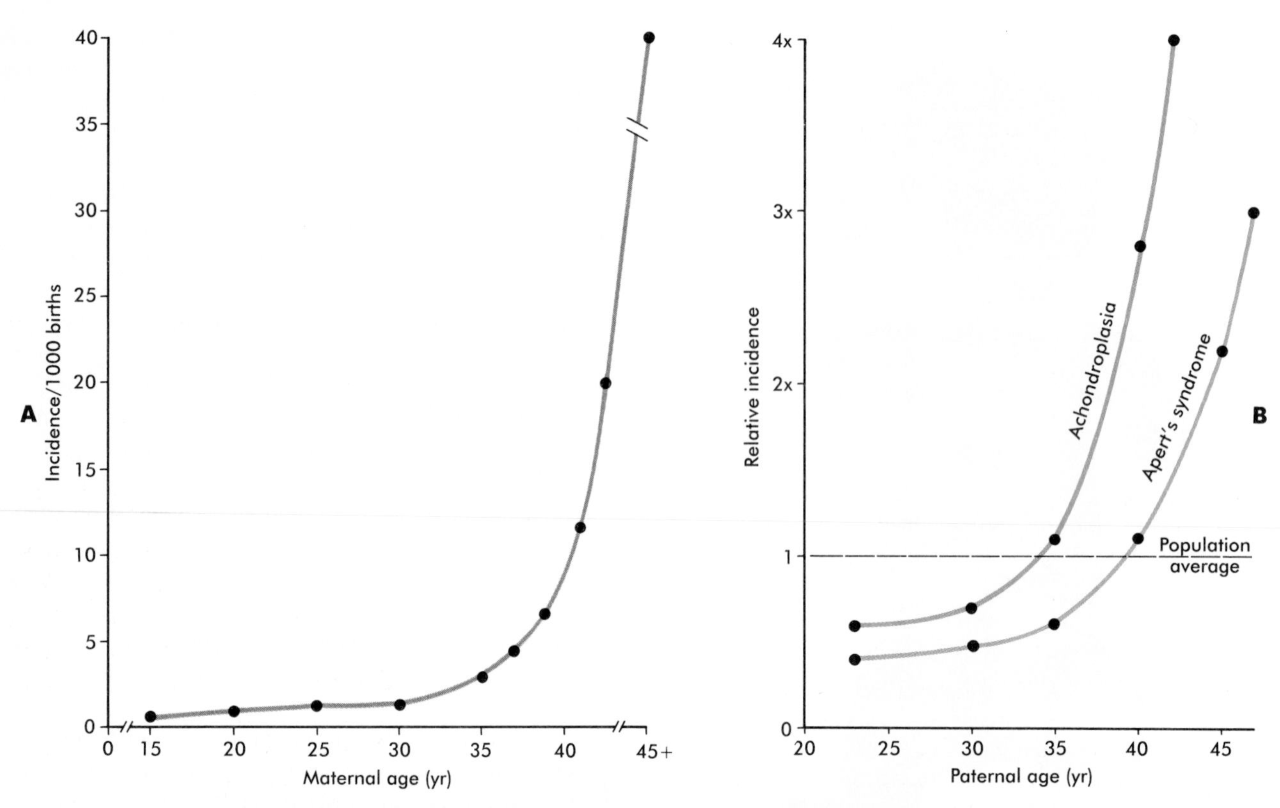

FIG. 8-3 The increased incidence of **(A)** Down syndrome with increasing maternal age and **(B)** achondroplasia and Apert's syndrome with increasing paternal age. Apert's syndrome (acrocephalosyndactyly) is characterized by a towering skull and laterally fused digits.

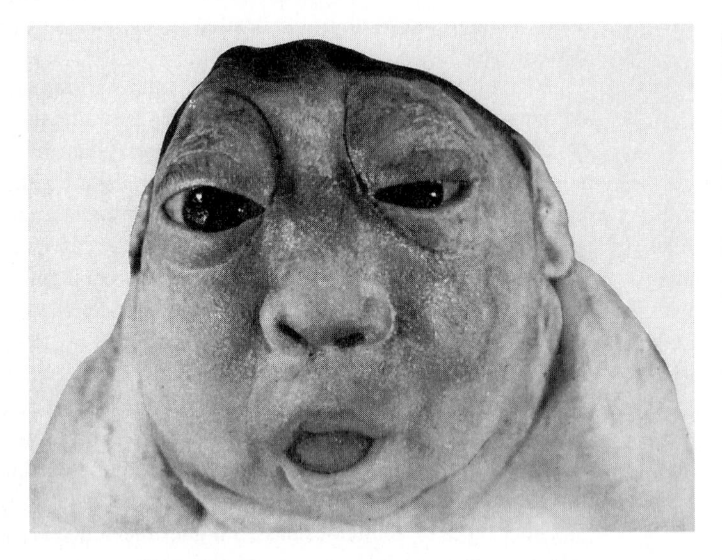

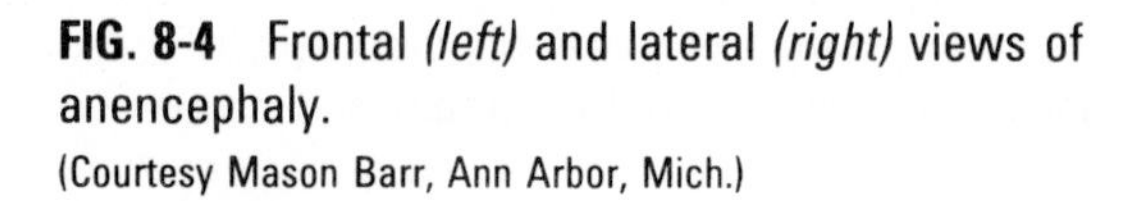

FIG. 8-4 Frontal *(left)* and lateral *(right)* views of anencephaly.
(Courtesy Mason Barr, Ann Arbor, Mich.)

TABLE 8-1 Incidence of Neural Tube Defects

SITE	INCIDENCE/1000 LIVE BIRTHS
India	0.6
Ireland	10
United States	1
Worldwide	2.6

The relationship between the country of residence and an increased incidence of specific malformations can be related to a variety of factors, ranging from racial tendencies, local environmental factors, and even governmental policies. A classic example of the last is the incidence of severely malformed infants due to exposure to thalidomide. These cases were concentrated in West Germany and Australia because the drug was commonly sold in these locations. Because thalidomide was not approved by the FDA (Food and Drug Administration), the United States was spared from this epidemic of birth defects. Another classic example of the influence of country as a factor in the incidence of malformations is seen in neural tube defects (Table 8-1). The reason neural tube defects (especially anencephaly) are so common in Ireland has been the topic of much speculation. One of the prevalent theories was that it may be due to the mother's eating old potatoes during the late winter months. More recent epidemiological studies have shown a high correlation between folic acid deficiency and an increased incidence of anencephaly and neural tube defects.

Race is a factor in many congenital malformations and a variety of diseases. In humans as well as mice, there are racial differences in the incidence of cleft palate. The incidence of cleft palate among whites is twice as high as it is among African Americans and twice as high among Orientals as among whites.

A number of malformations, particularly those that have a genetic basis, are found more frequently with certain families, especially if there is any degree of consanguinity in the marriages over the generations. A good example is the increased occurrence of extra digits among some families within the American Amish community.

Periods of Susceptibility to Abnormal Development

It is well known that at certain critical periods, embryos are more susceptible to agents or factors causing abnormal development than at others. The results of many investigations have allowed the following generalization: Insults to the embryo during the first 3 weeks of embryogenesis (the early period before organogenesis begins) are unlikely to result in defective development because they either kill the embryo or are compensated for by the powerful regulatory properties of the early embryo. The period of maximal susceptibility to abnormal development occurs between weeks 3 and 8, which is the period when most of the major organs and body regions are first being established. Major structural anomalies are unlikely to occur after the eighth week of pregnancy because by this point, most organs have become well established. Anomalies arising from the third to the ninth month of pregnancy tend to be functional (e.g., mental retardation) or involve disturbances in the growth of already-formed body parts. Such a simplified view of susceptible periods does not, however, take into account the possibility that a teratogen or some other harmful influence might be applied at an early stage of development but not be expressed as a developmental disturbance until later during embryogenesis. On the other hand, certain other influences (e.g., intrauterine diseases, toxins), may result in the destruction of all or parts of structures that have already been formed.

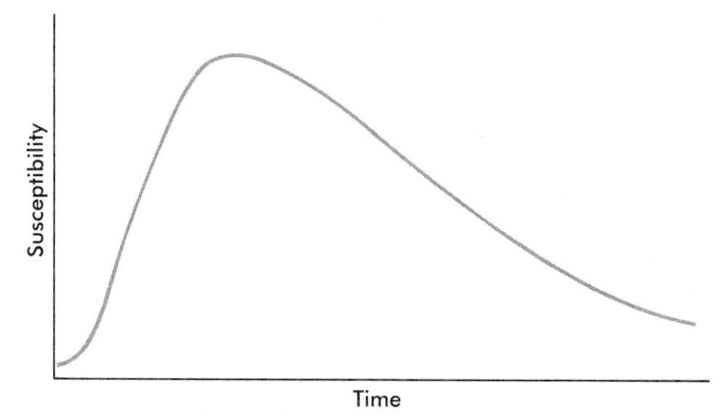

FIG. 8-5 Generalized susceptibility curve to teratogenic influences by a single organ.

Typically, a developing organ has a curve of susceptibility to teratogenic influences similar to that illustrated in Fig. 8-5. Before the critical period, exposure to a known teratogen has little influence on development. During the first days of the critical period, the susceptibility, measured as incidence of malformation, increases sharply and then declines over a much longer period.

Different organs have different periods of susceptibility during embryogenesis (Fig. 8-6). Organs that form the earliest, (e.g., the heart) tend to be sensitive to the effects of teratogens earlier than those that form later (e.g., external genitalia). Some very complex organs, especially the brain and major sense organs, show prolonged periods of high susceptibility to the disruption of normal development.

Not all teratogenic influences act on the same developmental periods (Table 8-2). Some cause anomalies if the embryo is exposed to them early in development, but they are innocuous at later periods of pregnancy. Others affect only later developmental periods. A good example of the former is thalidomide, which has a very narrow and well-defined danger zone. In contrast, tetracycline, which affects bony structures and teeth by staining them, can only exert its effects after hard skeletal structures in the fetus have been formed.

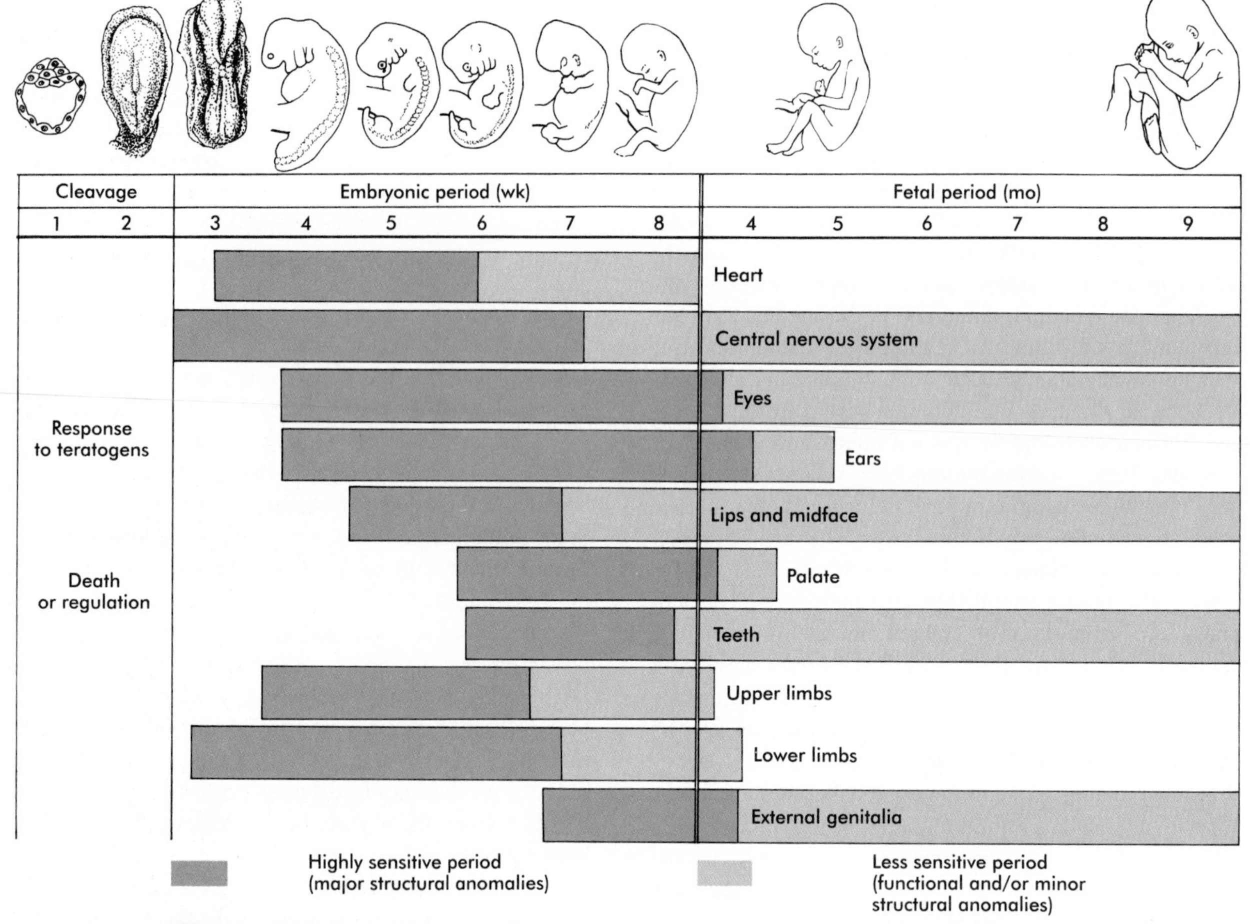

FIG. 8-6 Periods and degrees of susceptibility of embryonic organs to teratogens.

TABLE 8-2 Developmental Times at Which Various Human Teratogens Exert Their Effects

TERATOGENS	CRITICAL PERIODS (GESTATIONAL DAYS)	COMMON MALFORMATIONS
Rubella virus	0-60	Cataract or heart malformations
	0-120+	Deafness
Thalidomide	21-40	Reduction defects of limbs
Androgenic steroids	Earlier than 90	Clitoral hypertrophy and labial fusion
	Later than 90	Clitoral hypertrophy only
Coumadin anticoagulants	Earlier than 100	Nasal hypoplasia
	Later than 100	Possible mental retardation
Radioiodine therapy	Later than 65-70	Fetal thyroidectomy
Tetracycline	Later than 120	Staining of dental enamel in primary teeth
	Later than 250	Staining of crowns of permanent teeth

Modified from Persaud TVN, Chudley AE, Skalko RG, eds: *Basic concepts in teratology,* New Ycrk, 1985, Liss.

Patterns of Abnormal Development

Although isolated structural or biochemical defects are not rare, it is common to find multiple abnormalities in the same individual. This can be due to a number of reasons. One possibility is that a single teratogen acted on the primordia of several organs during susceptible periods of development. Another is that a genetic or chromosomal defect spanned genes affecting a variety of structures or that a single metabolic defect affected different developing structures in different ways.

CAUSES OF MALFORMATIONS

Despite considerable research over the past 50 years, the cause of at least 50% of human congenital malformations remains unknown (Fig. 8-7). Of the other 50%, roughly 25% are genetically based (chromosomal defects or mutants based on mendelian genetics), and less than 10% are attributed to environmental factors or physical or chemical teratogens. Multifactorial causes make up the rest.

Genetic Factors

Genetically based malformations can be due to abnormalities of chromosomal division or mutations of genes. Chromosomal abnormalities are usually classified as structural or numerical errors. These arise during cell division, especially meiosis. Numerical errors of chromosomes result in **aneuploidy,** defined as a total number of chromosomes other than the normal 46.

Abnormal Chromosome Numbers

Polyploidy. Polyploidy is the condition in which the chromosomal number is a multiple of the haploid number (23) of chromosomes other than 2. In the vast majority of cases, polyploid embryos abort spontaneously early in pregnancy. In fact, a high percentage of spontaneously aborted fetuses show major chromosomal abnormalities. Causes for polyploidy, especially triploidy, are likely to be either the fertilization of an egg by more than one sperm or the lack of separation of a polar body during meiosis.

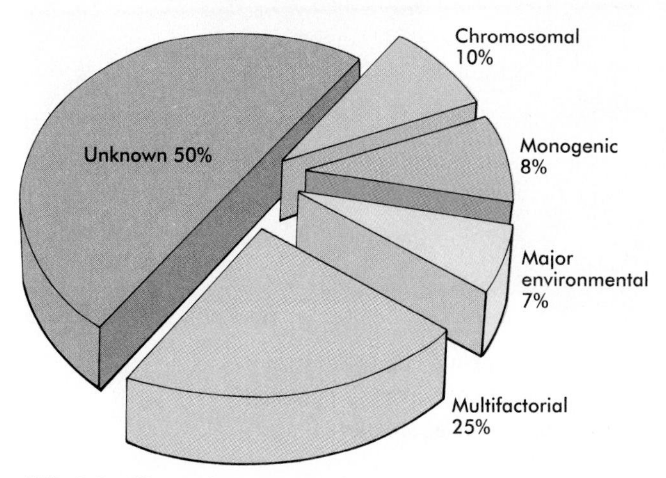

FIG. 8-7 The major causes of congenital malformations.

(Data from Persaud TVN, Chudley AE, Skalko RG, eds: *Basic concepts in teratology,* New York, 1985, Liss.)

Monosomy and trisomy. Monosomy (the lack of one member of a chromosome pair) and **trisomy** (a triplet instead of the normal chromosome pair) are typically the result of nondisjunction during meiosis (see Fig. 1-7). When this happens, one gamete shows monosomy, and the other shows trisomy of the same chromosome.

In most cases, embryos with monosomy of the autosomes or sex chromosomes are not viable. However, some individuals with monosomy of the sex chromosomes (45, XO genotype) can survive (Fig. 8-8). Such individuals,

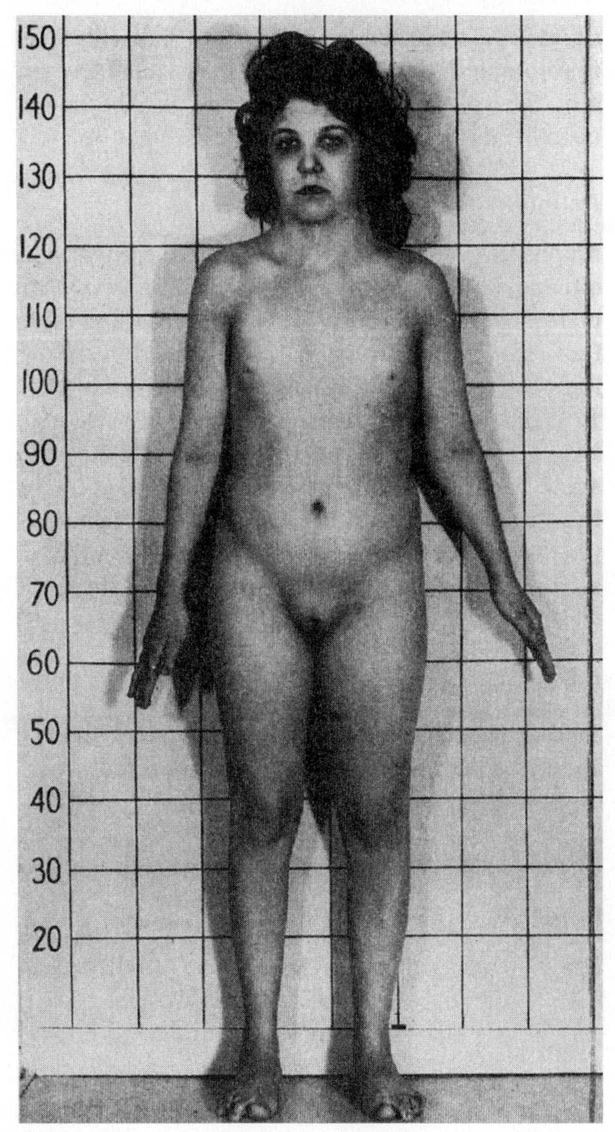

FIG. 8-8 An adult female with Turner syndrome. Note the short stature, webbed neck, and infantile sexual characteristics.

(From Connor J, Ferguson-Smith M: *Essential medical genetics,* ed 2, Oxford, England, 1987, Blackwell Scientific.)

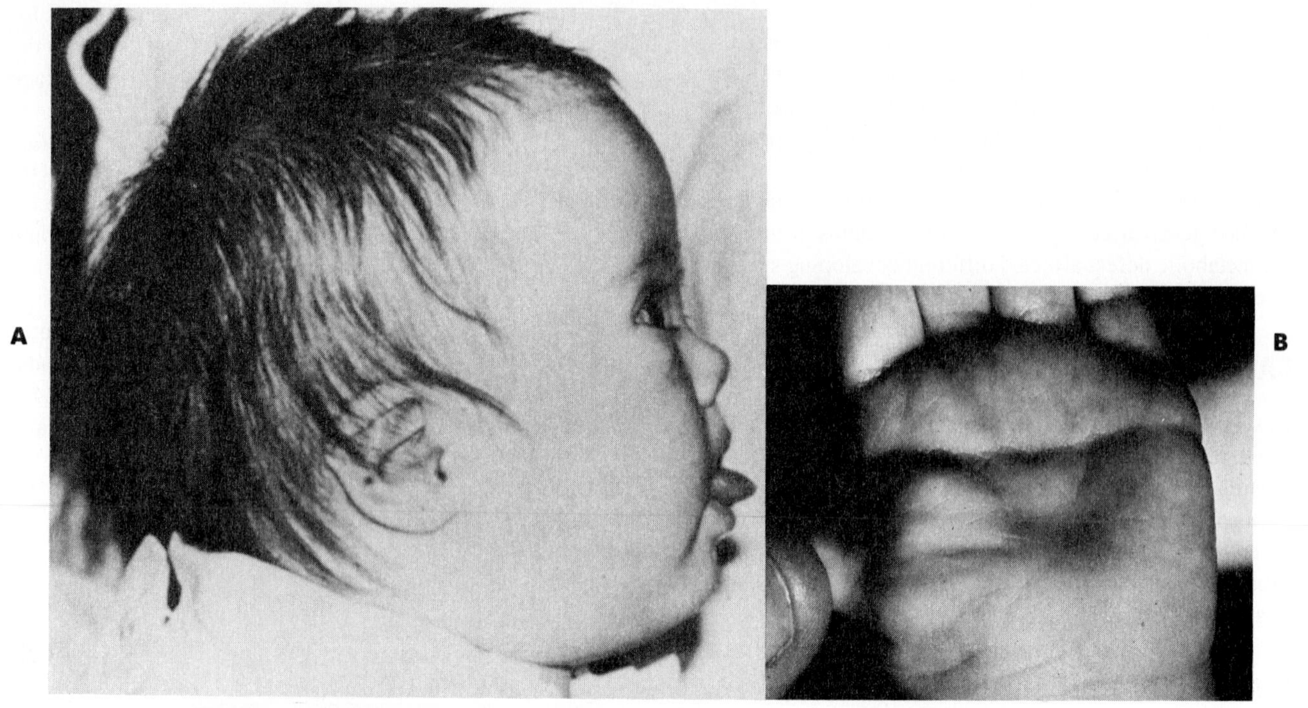

FIG. 8-9 **A,** Profile of a child with Down syndrome. Note the flat profile, protruding tongue, saddle-shaped bridge of nose, and low-set ears.
(From Garver K, Marchese S: *Genetic counseling for clinicians,* Chicago, 1986, Mosby.)
B, Hand of an infant with Down syndrome, showing the prominent simian crease that crosses the entire palm.
(From Fanaroff A, Martin RJ: *Neonatal-perinatal medicine,* ed 5, St Louis, 1992, Mosby.)

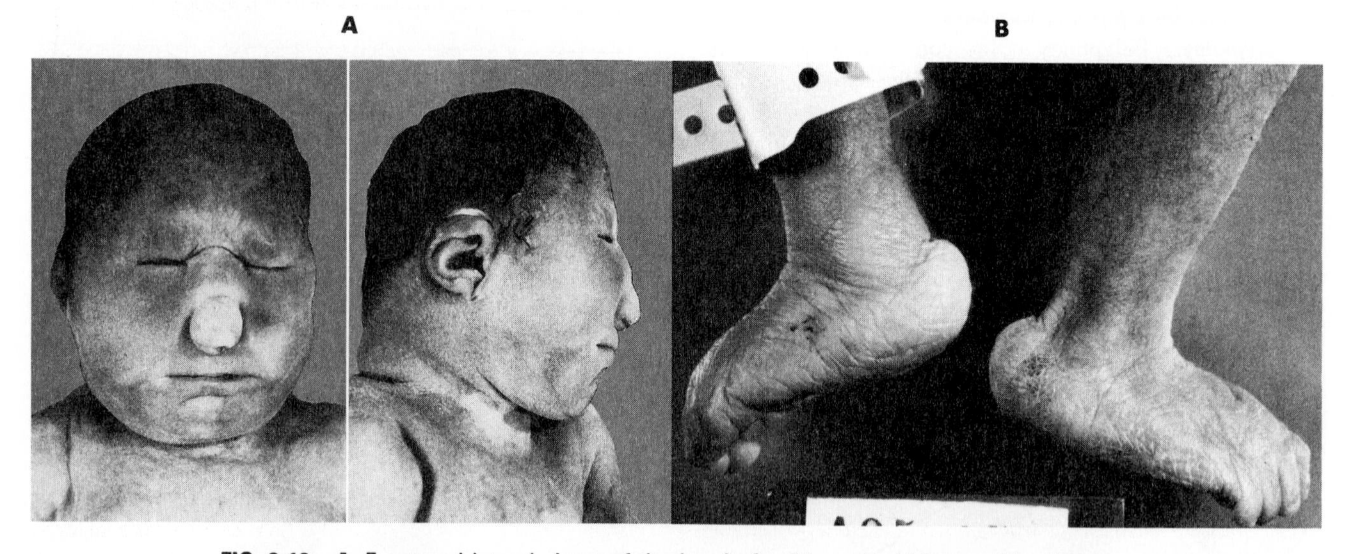

FIG. 8-10 **A,** Front and lateral views of the head of a 34-week-old fetus with trisomy 13. This fetus shows pronounced cebocephaly, with a keel-shaped head, a flattened nose, abnormal ears, and a reduction of forebrain and upper facial structures.
(Courtesy Mason Barr, Ann Arbor, Mich.)
B, Rocker bottom feet from a fetus with trisomy 18. Note the prominent heels and somewhat convex profile of the soles of the feet.
(From Nyberg D, Mahony B, Pretorious D: *Diagnostic ultrasound of fetal anomalies,* St Louis, 1990, Mosby.)

TABLE 8-3 Variations in Numbers of Sex Chromosomes

SEX CHROMOSOME COMPLEMENT	INCIDENCE	PHENOTYPE	CLINICAL FACTORS
XO	1:3000	Immature female	Turner syndrome: short stature, webbed neck, high arched palate
XX		Female	Normal
XY		Male	Normal
XXY	1:1000	Male	Klinefelter syndrome: small testes, infertile, often tall with long limbs
XYY	1:1000	Male	Tall, normal appearance, reputed difficulty with impulsive behavior
XXX	1:1000	Female	Normal appearance, mentally retarded (up to one third of cases), fertile (in many cases)

who are said to have **Turner syndrome,** exhibit a female phenotype, but the gonads are sterile.

Three autosomal trisomies produce infants with characteristic associations of anomalies. The best known is **trisomy 21,** also called *Down syndrome*. Individuals with Down syndrome are typically mentally retarded and have a characteristic broad face with a flat nasal bridge, wide set eyes, and prominent epicanthic folds. The hands are also broad, and the palmar surface is marked by a characteristic transverse **simian crease** (Fig. 8-9). Individuals with Down syndrome are prone to the early appearance of Alzheimer's disease and typically have a shortened life span.

Other trisomies of chromosomes 13 and 18 result in severely malformed fetuses, many of which do not survive to birth. Both **trisomy 13** and **trisomy 18** infants show severe mental retardation and other defects of the central nervous system. Cleft lip and cleft palate are common. Polydactyly is often seen in trisomy 13, and infants with both syndromes exhibit other anomalies of the extremities such as **"rocker bottom feet",** meaning a rounding under and protrusion of the heels (Fig. 8-10). Most infants born with trisomy 13 or 18 die within the first month or two after birth.

Abnormal numbers of the sex chromosomes are relatively common and can be detected by examination of the sex chromatin (X chromosome) or the fluorescence reactions of the Y chromosomes. Some of the various types of deletions and duplications of the sex chromosomes are summarized in Table 8-3.

Abnormal Chromosome Structure. A variety of abnormalities of chromosome structure can give rise to malformations in development. Some chromosomal abnormalities are the result of chromosome breakage induced by environmental factors such as radiation and certain chemical teratogens. This type of structural error is usually unique to a given individual and not transmitted to succeeding generations.

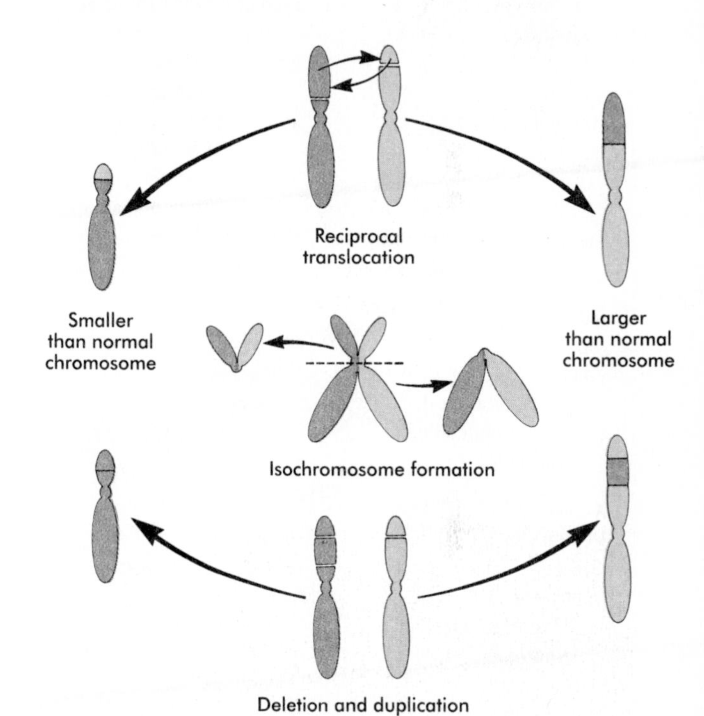

FIG. 8-11 Different types of structural errors of chromosomes.

Other types of structural abnormalities of chromosomes are generated during meiosis and, if present in the germ cells, can be inherited. Common types of errors in chromosome structure are **reciprocal translocations, isochromosome formation,** and **deletions,** or **duplications** (Fig. 8-11). One well-defined congenital malformation resulting from a deletion in the short arm of chromosome 5 is the **cri du chat syndrome.** Infants with this syndrome are severely retarded, have microcephaly, and make a cry that sounds like the mewing of a cat.

Genetic Mutations. Many genetic mutations are expressed as morphological abnormalities. These can be dominant or recessive genes of either the autosomes or the sex chromosomes. For some of these conditions (e.g., hemophilia, Lesch-Nyhan syndrome, muscular dystrophy, cystic fibrosis), the molecular or biochemical lesion has been identified, but the manner in which these defects are translated into abnormal development is not clear. Many of these conditions are extensively treated in textbooks of human genetics, and they are only listed here (Table 8-4).

TABLE 8-4 Genetic Mutations Leading to Abnormal Development

CONDITION	CHARACTERISTICS
AUTOSOMAL DOMINANT	
Achondroplasia	Dwarfism mainly due to shortening of limbs
Aniridia	Absence of iris (usually not complete)
Crouzon syndrome (craniofacial dysostosis)	Premature closure of certain cranial sutures, leading to flat face and towering skull
Neurofibromatosis	Multiple neural crest–derived tumors on skin, abnormal pigment areas on skin
Polycystic kidney disease (adult onset, type III)	Numerous cysts in kidneys
AUTOSOMAL RECESSIVE	
Albinism	Absence of pigmentation
Polycystic kidney disease (perinatal type I)	Numerous cysts in kidneys
Congenital phocomelia syndrome	Limb deformities
X-LINKED RECESSIVE	
Hemophilia	Defective blood clotting
Hydrocephalus	Enlargement of cranium
Ichthyosis	Scaly skin
Testicular feminization syndrome	Female phenotype due to inability to respond to testosterone

Environmental Factors

A variety of environmental factors are linked with birth defects. These range from chemical teratogens and hormones to maternal infections and nutritional factors. Although the list of suspected teratogenic factors is long, relatively few are unquestionably teratogenic in humans.

Maternal Infections. After the recognition in 1941 that rubella was the cause of a spectrum of developmental anomalies, several other maternal diseases have been implicated as direct causes of birth defects. With infectious diseases, however, it is important to distinguish between those that cause malformations by interfering with early stages in the development of organs and structures from those that interfere by destroying structures already formed. The same pathogenic organism can cause lesions by interference with embryonic processes or by destruction of differentiated tissues, depending on when the organism attacks the embryo.

Most infectious diseases that cause birth defects are viral, with **toxoplasmosis** (caused by the protozoan *Toxoplasma gondii*) and syphilis (caused by the spirochete *Treponema pallidum*) being notable exceptions. (A summary of the infectious diseases known to cause birth defects in humans is given in Table 8-5).

The time of infection is very important in relation to the types of effects on the embryo. Rubella causes a high percentage of malformations during the first trimester, whereas cytomegalovirus infections usually kill the embryo during the first trimester. The agents of both syphilis and toxoplasmosis cross the placental barrier during the fetal period and to a large extent cause malformations by the destruction of existing tissues.

TABLE 8-5 Infectious Diseases That Can Cause Birth Defects

INFECTIOUS AGENT	DISEASE	CONGENITAL DEFECTS
VIRUSES		
Rubella virus	German measles	Cataracts, deafness, cardiovascular defects, fetal growth retardation
Cytomegalovirus	Cytomegalic inclusion disease	Microcephaly, microphthalmia, cerebral calcification, intrauterine growth retardation
SPIROCHETES		
Treponema pallidum (syphilis)	Syphilis	Dental anomalies, deafness, mental retardation, skin and bone lesions, meningitis
PROTOZOA		
Toxoplasma gondii	Toxoplasmosis	Microcephaly, hydrocephaly, cerebral calcification, microphthalmia, mental retardation, prematurity

Chemical Teratogens. Many substances are known to be teratogenic in animals or are associated with birth defects in humans, but for only a relatively small number is there convincing evidence that links the substance directly to congenital malformations in humans (Table 8-6). This makes testing drugs for teratogenicity difficult, since what can cause a high incidence of severe defects in animal fetuses (e.g., cortisone and cleft palate in mice) may not cause malformations in other species of animals or in humans. Conversely, the classic teratogen thalidomide is highly teratogenic in humans, rabbits, and some primates but not in commonly used laboratory rodents.

Folic Acid Antagonists. At one time, folic acid antagonists, which are known to be highly embryolethal, were used in clinical trials as **abortifacients** (agents causing abortion). Although three fourths of the pregnancies were terminated, almost a fourth of the embryos that survived to term were severely malformed. A classic example of an embryotoxic folic acid antagonist is aminopterin, which produces multiple severe anomalies such as anencephaly, growth retardation, cleft lip and palate, hydrocephaly, hypoplastic mandible, and low-set ears.

Androgenic Hormones. Administration of androgenic hormones to pregnant women to either treat tumors or prevent threatened abortion resulted in the birth of hundreds of female infants with varying degrees of masculinization of the external genitalia. The anomalies consisted of clitoral hypertrophy and often varying amounts of fusion of the genital folds to form a scrotumlike structure (Fig. 8-12).

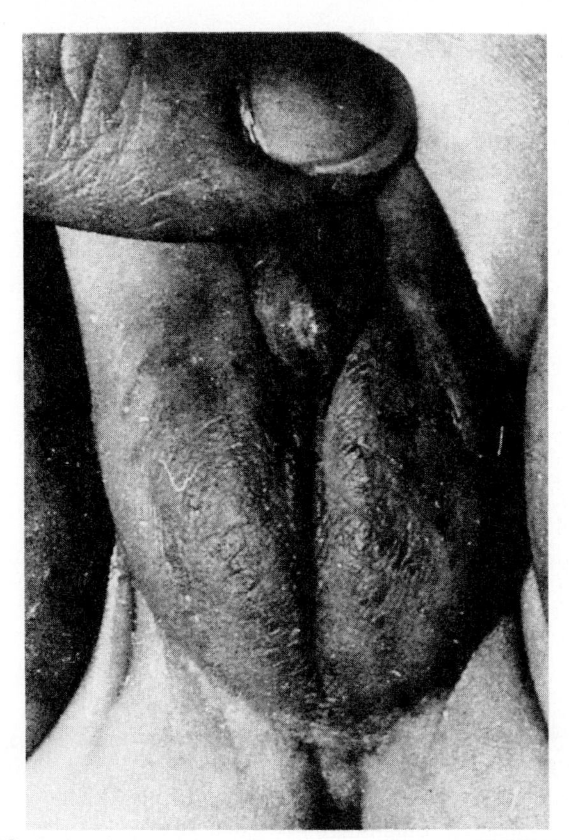

FIG. 8-12 Ambiguous external genitalia in a newborn with pseudohermaphroditism.

(From Reed GB, Claireaux AE: *Diseases of the fetus and newborn,* St Louis, 1989, Mosby.)

TABLE 8-6 Chemical Teratogens in Humans

AGENT	EFFECTS
Alcohol	Growth and mental retardation, microcephaly, various malformations of face and trunk
Androgens	Masculinization of females, accelerated genital development in males
Anticoagulants (warfarin, dicumarol)	Skeletal abnormalities; broad hands with short fingers; nasal hypoplasia; anomalies of eye, neck, central nervous system
Antithyroid drugs (e.g., propylthiouracil, iodide)	Fetal goiter, hypothyroidism
Chemotherapeutic agents (methotrexate, aminopterin)	Variety of major anomalies throughout body
Diethylstilbestrol	Cervical and uterine abnormalities
Lithium	Heart anomalies
Organic mercury	Mental retardation, cerebral atrophy, spasticity, blindness
Phenytoin (Dilantin)	Mental retardation, poor growth, microcephaly, dysmorphic face, hypoplasia of digits and nails
Isotretinoin (Accutane)	Craniofacial defects, cleft palate, ear and eye deformities, nervous system defects
Streptomycin	Hearing loss, auditory nerve damage
Tetracycline	Hypoplasia and staining of tooth enamel, staining of bones
Thalidomide	Limb defects, ear defects, cardiovascular anomalies
Trimethadione and paramethadione	Cleft lip/palate, microcephaly, eye defects, cardiac defects, mental retardation
Valproic acid	Neural tube defects

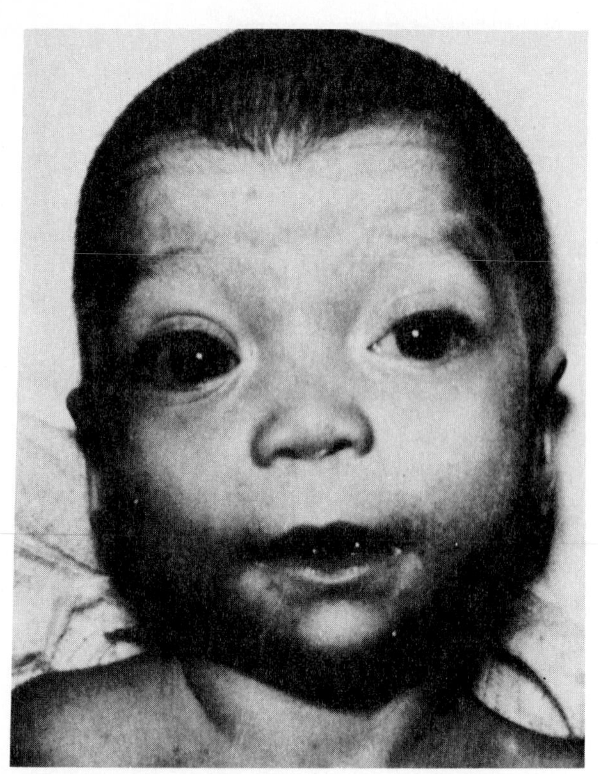

FIG. 8-13 Face of an infant with fetal hydantoin syndrome. This infant has prominent eyes, hypertelorism (increased space between the eyes), micrognathia, and microcephaly.

(From Wigglesworth JS, Singer DB: *Textbook of fetal and perinatal pathology,* 2 vols, Oxford, England, 1991, Blackwell Scientific.)

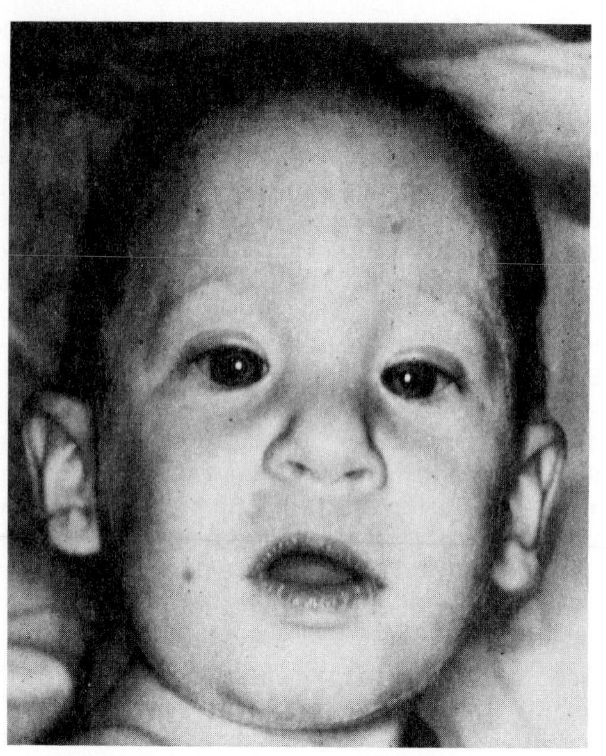

FIG. 8-14 Face of an infant with fetal alcohol syndrome showing a long and thin upper lip, shortened and upwardly slanting palpebral fissures, epicanthic folds, and mild hirsutism.

(From Wigglesworth JS, Singer DB: *Textbook of fetal and perinatal pathology,* 2 vols, Oxford, England, 1991, Blackwell Scientific.)

Anticonvulsants. Several commonly used anticonvulsants are known or strongly suspected to be teratogenic. Diphenylhydantoin produces a "fetal hydantoin syndrome" of anomalies, including growth anomalies, craniofacial defects, nail and digital hypoplasia, and mental retardation (Fig. 8-13). Trimethadone also produces a syndrome of anomalies involving low-set ears, cleft lip and palate, and skeletal and cardiac anomalies.

Sedatives and Tranquilizers. Thalidomide is highly teratogenic when administered even as little as one time during a very narrow window of pregnancy, especially between days 20 and 35. This represents the period when the primordia of most major organ systems are being established. The most characteristic lesions produced are gross malformations of limbs, but the thalidomide syndrome also includes malformations of the cardiovascular system, absence of the ears, and assorted malformations of the urinary system, gastrointestinal system, and face. Lithium carbonate, a commonly used agent for certain psychoses, is known to cause malformations of the heart and great vessels if administered early during pregnancy.

Antineoplastic Agents. Several antineoplastic agents are highly teratogenic, in large part because they are designed to kill or incapacitate rapidly dividing cells. Among these is aminopterin. Methotrexate and the combination of busulfan and 6-mercaptopurine cause severe anomalies of multiple organ systems. The use of these drugs during pregnancy is a difficult medical decision that must consider the lives of both the mother and fetus.

Alcohol. Accumulated evidence now leaves little doubt that maternal consumption of alcohol during pregnancy can lead to a well-defined constellation of developmental abnormalities that includes poor postnatal growth rate, mental retardation, heart defects, and hypoplasia of facial structures (Fig. 8-14). This is now popularly known as **fetal alcohol syndrome.**

Retinoic Acid (Vitamin A). Derivatives of retinoic acid are used in the treatment of acne, but in recent years it has been established that retinoic acid acts as a potent teratogen when taken orally, especially during the period of organogenesis. Major defects produced by retinoic acid are neural tube anomalies, cleft palate, hypoplasia of structures of the lower face, and heart and thymic defects. In view of the increasing recognition that retinoic acid or its metabolites may play a damaging role in pattern formation during early development, extreme caution with the use of vitamin A in doses above those needed for basic nutritional requirements is recommended.

Antibiotics. The use of two antibiotics during pregnancy is associated with birth defects. Streptomycin in high doses can cause inner ear deafness. Tetracycline given to the mother during late pregnancy crosses the placental barrier and seeks out sites of active calcification in the teeth and bones of the fetus. Tetracycline deposits cause a yellowish discoloration of teeth and bones and, in high doses, can interfere with enamel formation.

Other Drugs. A number of other drugs, such as the anticoagulant warfarin, are known to be teratogenic, and others are strongly suspected. However, firm proof of a drug's teratogenicity in humans is not easy to obtain. Several drugs, such as agent orange and some of the social drugs (e.g., LSD, marijuana), have often been claimed to cause birth defects, but the evidence to date is not entirely convincing.

Physical Factors

Ionizing Radiation. Ionizing radiation is a potent teratogen, and the response is both dose dependent and related to the stage at which the embryo is irradiated. In addition to numerous animal studies, there is direct human experience based on survivors of the Japanese atomic bomb blasts and pregnant women who were given large doses of irradiation (up to several thousand rads) for therapeutic reasons. There is no evidence that doses of irradiation at diagnostic levels (only a few millirads) pose a significant threat to the embryo. Nevertheless, because ionizing radiation can cause breaks in DNA and is also known to cause mutations, it is prudent for a woman who is pregnant to avoid exposure to radiation if possible, although the dose in a diagnostic x-ray examination is so small that the risk is minimal.

Although ionizing radiation can cause a variety of anomalies in embryos (e.g., cleft palate, microcephaly, and malformations of the viscera, limbs, and skeleton), defects of the central nervous system are very prominent in irradiated embryos. The spectrum runs from spina bifida to mental retardation.

Other Physical Factors. A number of studies on the teratogenic effects of extremes of temperature and different concentrations of atmospheric gases have been conducted on experimental animals, but the evidence relating any of these factors to human malformations is still equivocal. One exception is the effect of excess concentrations of oxygen on prematurely born infants. When this was a common practice, **retrolental fibroplasia** developed in over 10% of premature infants weighing less than 3 pounds and in about 1% of those weighing between 3 and 5 pounds. When this was recognized, the practice of maintaining high concentrations of oxygen in incubators ceased, and this problem is now of only historical interest.

Maternal Factors

A number of maternal factors have been implicated in the genesis of congenital malformations. **Maternal diabetes** is frequently associated with a fetus with a large birth weight and stillbirths. Structural anomalies occur several times more frequently in infants of diabetic mothers than among the general population. Although there is a correlation between the duration and severity of the mother's disease and the effects on the fetus, the specific cause of interference with development has not been identified.

In general, maternal nutrition does not seem to be a significant factor in the production of anomalies, but if the mother is severely deficient in iodine, the newborn is likely to show the symptoms of **cretinism** (growth retardation, mental retardation, short and broad hands, short fingers, dry skin, and difficulty in breathing).

There is now considerable evidence that **heavy smoking** by a pregnant woman leads to an increased risk of low birth weight and a low rate of growth after birth.

Mechanical Factors

Although mechanical factors have been implicated in the genesis of congenital malformations for centuries, only in recent years has it been possible to relate specific malformations to mechanical causes. A number of the most common anomalies such as **clubfoot, congenital hip dislocations,** and even certain deformations of the skull can be attributed in large measure to abnormal intrauterine pressures imposed on the fetus. This can often be related to uterine malformations or a reduced amount of amniotic fluid **(oligohydramnios).**

Amniotic bands constricting digits or extremities of the fetus have been implicated as causes of intrauterine amputations (Fig. 8-15). These bands presumably form as the result of tears to the extraembryonic membranes earlier during pregnancy.

DEVELOPMENTAL DISTURBANCES THAT CAN RESULT IN MALFORMATIONS

Duplications

The classic example of a duplication is identical twinning. Under normal circumstances, both members of the twin pair are completely normal, but rarely the duplication is not complete and **conjoined twins** result (see Figs. 3-13 and 3-14). Twins can be conjoined at almost any site and to any degree. With modern surgical techniques, it is now possible to separate members of some conjoined pairs. A type of conjoined twinning is the condition of **parasitic twinning,** in which one member of the pair is relatively normal but the other is represented by a much smaller body, often consisting of just the torso and limbs, attached to an area such as the mouth or lower abdomen of the host twin (see Fig. 3-15).

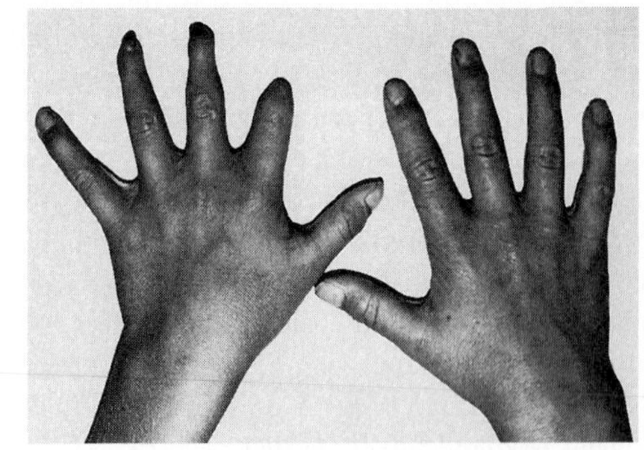

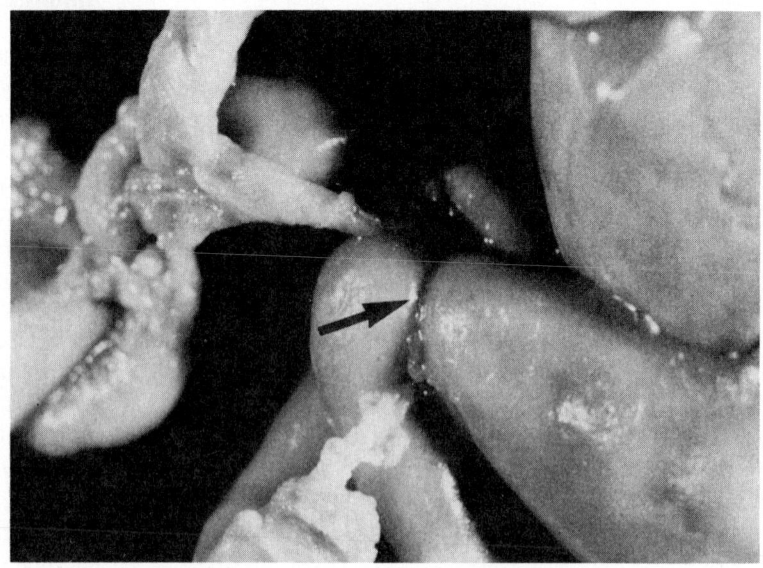

FIG. 8-15 **A,** Digital amputations, presumably caused by amniotic bands, on the left hand.
(Courtesy Mason Barr, Ann Arbor, Mich.)
B, Amniotic bands involving the umbilical cord and limbs of a fetus. The arrow shows a constriction ring around the thigh.
(From Wigglesworth JS, Singer DB: *Textbook of fetal and perinatal pathology,* 2 vols, Oxford, England, 1991, Blackwell Scientific.)

In a high percentage of conjoined twins, one member of the pair has reversed symmetry in relation to the other (see Fig. 3-14). Over a century ago, the biologist Oscar Bateson compiled a large number of instances of reversals of symmetry in duplicated structures throughout the animal kingdom. His recognition of this phenomenon is now called **Bateson's rule.** (An example of this rule as it applies to a limb duplication is illustrated in Fig. 11-9.)

In rare instances (approximately 1 in 10,000 births), an otherwise normal individual is found to have a partial or complete reversal of symmetry of the internal organs. The cause of this condition, called **situs inversus,** is difficult to determine. One possibility is a genetic mutation (e.g., strains of snails with reversed coiling of the shells are not uncommon); another is that the individual with situs inversus is the only surviving member of a duplicated pair and the one that had a reversal of symmetry.

Faulty Inductive Tissue Interactions

Absent or faulty induction early in development (e.g., induction of the central nervous system) is incompatible with life, but there are malformations consistent with disturbances in later inductions. Absence of the lens **(aphakia)** or a kidney **(renal agenesis)** could be due to an absent or abnormal inductive interaction.

Absence of Normal Cell Death

Genetically or **epigenetically** (environmental influences imposed on the genetic background) controlled cell death is an important mechanism in sculpting a number of regions of the body. The absence of normal interdigital cell death has been implicated in **syndactyly** (webbed digits) and abnormal persistence of the tail (see Fig. 10-22, *B*). The latter phenomenon has sometimes been considered as an example of **ativism** (the persistence of phylogenetically primitive structures).

Failure of Tube Formation

The formation of a tube from an epithelial sheet is a fundamental developmental mechanism. A classic case of failure of tube formation is seen in the family of spina bifida anomalies, which are based on the incomplete fusion of the neural tube. (Some of the possible mechanisms involved in normal formation of the neural tube are given in Chapter 12.)

Disturbances in Tissue Resorption

Some structures that are present in the early embryo must be resorbed for subsequent development to proceed normally. Good examples of this are the membranes that cover the future oral and anal openings. These membranes are composed of opposing sheets of ectoderm and endoderm, but if mesodermal cells become interposed between the two and this tissue becomes vascularized, breakdown typically fails to occur. **Anal atresia** (see Chapter 16) is a common anomaly of this type (see Fig. 16-15).

Failure of Migration

Migration is an important developmental phenomenon that occurs at the level of cells or entire organs. The neural crest is a classic example of massive migrations at the cellular level, and disturbances in migration can cause abnormalities in any of the structures for which the neural crest is a precursor (e.g., thymus, outflow tracts of the heart, adrenal medulla). At the organ level, the kidneys undertake a prominent migration into the abdominal cavity from their origin in the pelvic region, and the testes migrate from the abdominal cavity into the scrotum. **Pelvic kidneys** and undescended testes (**cryptorchidism** [see Chapter 17]) are not rare.

Developmental Arrest

Early in the history of teratology, some malformations were recognized as the persistence of structures in a state that was normal at an earlier stage of development. Many of the patterns of **cleft lip** and **cleft palate** (see Figs. 15-17 and 15-18), are examples of developmental arrest, although it is incorrect to assume that development has been totally arrested since the sixth to eighth weeks of embryogenesis.

Another example of the persistence of an earlier stage in development is a **thyroglossal duct** (see Fig. 15-29), in which persisting epithelial cells mark the path of the thyroid gland as it migrates from the base of the tongue to its normal position.

Destruction of Formed Structures

A number of teratogenic diseases or chemicals produce malformations by the destruction of structures already present. If the structure is in the early primordial stage, any tissues that the primordium would normally give rise to are missing or malformed. Interference with the blood supply of a structure can cause unusual patterns of malformations. For example, in **phocomelia** (see Fig. 8-2), vascular breakdown could destroy the primordia of the upper limb segments, but the cells that would give rise to the hands or feet could be spared.

Failure to Fuse or Merge

If two structures such as the palatal shelves fail to meet at the critical time, they are likely to remain separate. Similarly, the relative displacements of mesenchyme **(merging)** that are involved in the shaping of the lower jaw may not occur on schedule or in adequate amounts. This accounts for some malformations of the lower face.

Hypoplasia and Hyperplasia

The normal formation of most organs and complex structures requires a precise background of cellular proliferation. If cellular proliferation in a forming organ is abnormal, the structure can become too small **(hypoplastic)** or too large **(hyperplastic).** Even relatively minor growth disturbances can cause severe problems in complex regions like the face. Occasionally, **gigantism** of a structure such as a digit or whole limb occurs. The mechanism underlying this excessive growth remains obscure.

Defective Fields

Proper morphogenesis of many regions of the body is under the control of poorly understood morphogenetic fields. These regions of the body are under the control of an overall developmental blueprint. Disturbances in the boundaries or overall controls of fields can sometimes give rise to massive anomalies. One example is the fusion of lower limb fields, probably associated with a larger defect in the field controlling development of the caudal region of the body. This mermaidlike anomaly is called **sirenomelia** (see Fig. 8-1, *B*).

Effects Secondary to Other Developmental Disturbances

Because so much of normal development involves tight interlocking of individual processes or building on completed structures, it is not surprising that a number of malformations are secondary manifestations of other disturbed embryonic processes. There are a number of examples in craniofacial development. Some cases of cleft palate have been attributed to a widening of the cranial base such that the palatal shelves, which may have been perfectly normal, are unable to make midline contact.

The single or widely separated tubular probosces that appear in certain major facial anomalies such as **cyclopia** (Fig. 8-16) are very difficult to explain unless it is understood that one of several primary defects, whether too much or too little tissue of the midface, prevented the two nasal primordia from joining in the midline.

Germ Layer Defects

An understanding of normal development can explain the basis for a seemingly diverse set of anomalies. Ectodermal dysplasias, which are based on abnormalities in the ectodermal germ layer, can include malformations as diverse as thin hair, poorly formed teeth, short stature, dry and

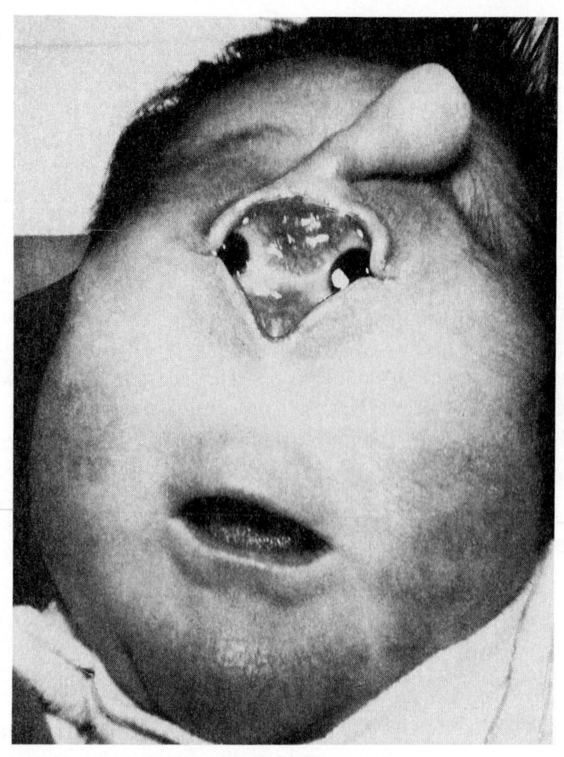

FIG. 8-16 Cyclopia in a newborn. Note the fleshy proboscis above the partially fused eye.
(Courtesy Mason Barr.)

scaly skin, and hypoplastic nails. Other syndromes with diverse phenotypic abnormalities are related to defects of the neural crest (see Chapter 13).

THE DIAGNOSIS AND TREATMENT OF BIRTH DEFECTS

Only a few decades ago, birth defects were diagnosed only after the fact, and sometimes it was years postnatally before certain defects could be discovered and treated. Although this can still happen today, technological changes have permitted the earlier diagnosis and treatment of certain congenital malformations.

One of the first advances was the technology associated with karyotyping and sex chromosome analysis. Initially these techniques were applied to postnatal individuals to diagnose conditions based on abnormalities in chromosome number or structure. After the development of amniocentesis (the removal of samples of amniotic fluid during early pregnancy), chromosomal analysis could be applied to cells present in the amniotic fluid. This was particularly useful in the diagnosis of Down syndrome, but it also permitted the prenatal diagnosis of the gender of the infant. Biochemical analysis of amniotic fluid has permitted the diagnosis of a number of inborn errors of metabolism and neural tube defects (the latter through the detection of **S-100 protein,** which leaks through the open neural tube into the amniotic fluid).

More recently, techniques have been developed for the direct sampling of tissue from the chorionic villi. The risk-to-benefit ratio of this technique is still being debated.

With the development of modern imaging techniques such as ultrasound, computer tomography, and magnetic resonance imaging (MRI), visualization of fetal morphology is now possible (see Figs. 19-10 to 19-13). These images can now serve as a direct guide to surgeons, who are attempting to correct certain malformations by intrauterine surgery. Because fetuses typically heal surgical wounds without scarring, there are distinct advantages to fetal corrective surgery (see Chapter 19).

SUMMARY

1. Developmental disorders have been recognized for centuries, but a direct connection between environmental teratogens and human birth defects was not demonstrated until 1941.
2. Abnormal development is often the result of environmental influences imposed on genetic susceptibility. Factors involved in abnormal development include age, race, country, nutrition, and time of year. The study of abnormal development is teratology, and an agent that causes abnormal development is a teratogen.
3. Genetic factors cause a significant number of birth defects. Abnormal chromosome numbers are associated with prenatal death and syndromes of abnormal structures. Common causes of abnormalities are monosomies and trisomies, which are commonly the result of nondisjunction during meiosis. Other malformations are based on abnormalities of chromosome structure. Certain malformations are based on genetic mutations.
4. Among the environmental factors leading to defective development are maternal infections, chemical teratogens, physical factors such as ionizing radiation, maternal factors, and mechanical factors.
5. A variety of disturbed developmental mechanisms may be involved in the production of a given congenital malformation. These include (a) duplications, (b) faulty inductive tissue interactions, (c) absence of normal cell death, (d) failure of tube formation, (e) disturbances in tissue resorption, (f) failure of migration, (g) developmental arrest, (h) destruction of an already formed structure, (i) failure to fuse or merge, (j) hypoplasia or hyperplasia, (k) defective fields, (l) effects secondary to other developmental disturbances and (m) germ layer defects.
6. With technological developments, it is now possible to diagnose increasing numbers of birth defects in

utero. Some of the diagnostic techniques are karyotyping and sex chromosome analysis on cells obtained from amniotic fluid, biochemical analysis of amniotic fluid, biochemical and molecular analysis of cells obtained from amniotic fluid of chorinic villus sampling, and imaging techniques, especially ultrasonography. There have been a few attempts to correct malformations by surgery in utero.

REVIEW QUESTIONS

1. A woman who was in a car accident and suffered abdominal bruising during the fourth month of pregnancy gave birth to an infant with a cleft palate. She sued the driver of the other car for expenses associated with treatment of the birth defect, claiming that it was caused by the accident. You are asked to be a witness for the defense. What is your case?
2. A woman who took a new sedative during the second month of pregnancy felt somewhat nauseated after ingestion of the drug and stopped taking it after a couple of weeks. She gave a birth to an infant who had a septal defect of the heart and sued the manufacturer of the drug, saying that the defect was caused by the drug that made her nauseous. You are asked to be a witness for the manufacturer. What is your case?
3. What is a likely cause of a badly turned-in ankle in a newborn?
4. A 3-year-old child is much smaller than normal, has sparse hair, and has irregular teeth. What is a likely basis for this constellation of defects?

REFERENCES

Buyse ML, ed: *Birth defects encyclopedia,* Dover, Mass, 1990, Centre for Birth Defects Information Services.

Coles CD: *Prenatal alcohol exposure and human development.* In Miller M, ed: *Development of the central nervous system: effects of alcohol and opiates,* New York, 1992, Wiley-Liss, pp 9-36.

Czeizel AE, Dudas I: Prevention of the first occurrence of neural-tube defects by periconceptional vitamin supplementation, *N Engl J Med* 327:1832-1835, 1992.

Frazer C: Of mice and children: reminiscences of a teratogeneticist, *Issues Rev Teratol* 5:1-75, 1990.

Hansen DK: The embryotoxicity of phenytoin: an update on possible mechanisms, *Proc Soc Exp Biol Med* 197:361-368, 1991.

Jones KL: *Smith's recognizable patterns of human malformation,* ed 4, Philadelphia, 1988, Saunders.

Larson JW, Grunsdale K: Acog technical bulletin number 84—February 1985: teratology, *Teratology* 32:493-496, 1985.

Miller MM: *Effects of prenatal exposure to ethanol on cell proliferation and neuronal migration.* In Miller M, ed: *Development of the central nervous system: effects of alcohol and opiates,* New York, 1992, Wiley-Liss, pp 47-69.

Naeye RL: *Disorders of the placenta, fetus, and neonate,* St Louis, 1992, Mosby.

Nishimura H, Okamoto N: *Sequential atlas of human congenital malformations,* Baltimore, 1976, University Park.

Pennington SN: Molecular changes associated with ethanol-induced growth suppression in the chick embryo, *Alcohol Clin Exp Res* 14:832-837, 1990.

Persaud TVN, Chudley AE, Skalko RG, eds: *Basic concepts in teratology,* New York, 1985, Liss.

Reed GB, Claireaux AE, Bain AD: *Diseases of the fetus and newborn,* St Louis, 1989, Mosby.

Saxén L, Rapola J: *Congenital defects,* New York, 1969, Holt, Rinehart and Winston.

Sulik KK, Alles AJ: *Teratogenicity of the retinoids.* In Saurat J-H, ed: *Retinoids: 10 years on,* Basel, Switzerland, 1991, Karger, pp 282-295.

Volpe JJ: Effect of cocaine use on the fetus, *N Engl J Med* 327:399-407, 1992.

Warkany J: *Congenital malformations,* Chicago, 1971, Mosby.

Werler MM and others: Maternal vitamin A supplementation in relation to selected birth defects, *Teratology* 42:497-503, 1990.

Wigglesworth JS, Singer DB: *Textbook of fetal and perinatal pathology,* 2 vols, Oxford, England, 1991, Blackwell Scientific.

Willis, RA: *The borderland of embryology and pathology,* ed 2, London, 1962, Butterworths.

Wilson GN: Genomics of human dysmorphogenesis, *Am J Med Genet* 42:187-196, 1992.

Wilson JG, Fraser FC, eds: *Handbook of teratology,* vols 1-4, New York, 1977, Plenum.

Yen IH and others: The changing epidemiology of neural tube defects, *Am J Dis Control* 146:857-861, 1992.

CHAPTER 9

Differentiation and the Generation of Cell Diversity

During the course of embryonic development, the fertilized egg undergoes many rounds of cell division, leading to not only almost exponential increase in the number of cells but also the formation of different types of cells. The formation of both the intraembryonic and extraembryonic germ layers (e.g., see Fig. 4-1) has already been outlined. Through various events of determination and restriction (see Fig. 4-12), the embryo ultimately produces more than 200 distinct cell types. This chapter first outlines some fundamental principles that are important in understanding the generation of cell diversity and then analyzes the developmental processes involved in the formation of several important types of cells.

GENERAL PRINCIPLES

Constancy of the Genome

As the result of a great deal of research conducted over the past century, contemporary developmental biology operates under the premise that virtually all cells of vertebrates possess a complete complement of genetic information at all stages of development. The classic basis for this assertion is the type of experiment that tested the developmental potential of individual nuclei taken from cells at various stages of development.

Early experiments involved separating 2- and 4-cell embryos into individual blastomeres and observing that if conditions were right, each blastomere could form a complete embryo. The formation of monozygotic twins from the separation of blastomeres at the 2-cell stage of cleavage is a rare but natural experiment that demonstrates the same point.

The great German embryologist, Spemann, constricted an amphibian zygote with a human hair to produce a lobe of cytoplasm that did not contain a nucleus (Fig. 9-1). The nucleated part of the zygote underwent several cleavage divisions before one of the nuclei was allowed to pass into the formerly nonnucleated lobe of cytoplasm. Even though this nucleus came from one cell of a multicelled embryo, it was still capable of directing the development of an entire embryo.

Later, the same principle was illustrated by the more refined technique of nuclear transplantation, in which the nucleus is removed from a donor cell with a micropipette and then inserted into a donor egg from which the nucleus has been inactivated (Fig. 9-2). Nuclei from cleaving embryos regularly produce normal embryos, but in rare cases even nuclei derived from adult frogs have been shown to be capable of producing normal embryos.

More contemporary demonstrations of the broad genetic potential of mature cells are experiments in which genes characteristic of completely different cell types can be expressed after a cell is appropriately stimulated. One involves the technique of cell fusion. For example, if a human nonmuscle cell is fused with a mouse myotube to form what is commonly called a **heterokaryon,** the human chromosomes form gene products characteristic of human muscle (Fig. 9-3, *A*.) Similarly, if cells cloned from a line of cultured mouse fibroblastic cells (10T1/2 cells) are exposed to the drug 5-azacytidine, they can become transformed into skeletal muscle cells, fat cells, or cartilage cells (Fig. 9-3, *B*). Experiments of this type demonstrate the presence of normally unexpressed genes in a variety of mature cell types.

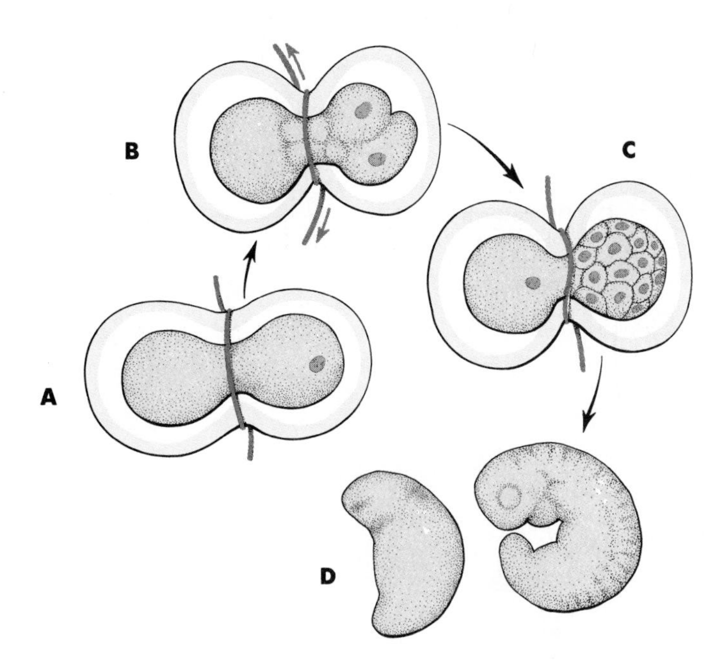

FIG. 9-1 Spemann's constriction experiment. **A,** A recently fertilized amphibian egg is partially constricted with a hair loop so that the cytoplasm on one side does not contain a nucleus. **B,** Cleavage begins on the nucleated side. **C,** After several cleavage divisions, the hair loop is slightly loosened, allowing a nucleus to migrate into the previously nonnucleated segment of cytoplasm. **D,** Embryos develop from both halves of the embryo, although development of the embryo derived from the originally enucleate cytoplasm and a cleavage stage nucleus is slightly delayed.

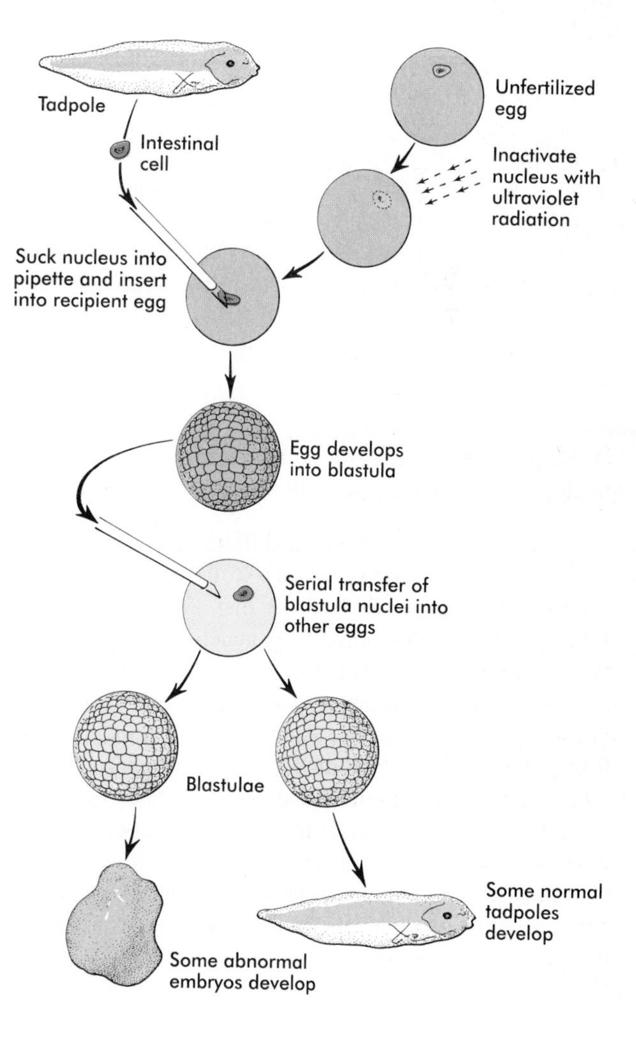

FIG. 9-2 Flow chart of a nuclear transplantation experiment, showing that it is possible to obtain normal embryos from nuclei derived from intestinal cells of tadpoles of *Xenopus.* Only a small percentage of the embryos derived from such nuclear transplants are normal. In the others, development is arrested or highly abnormal.

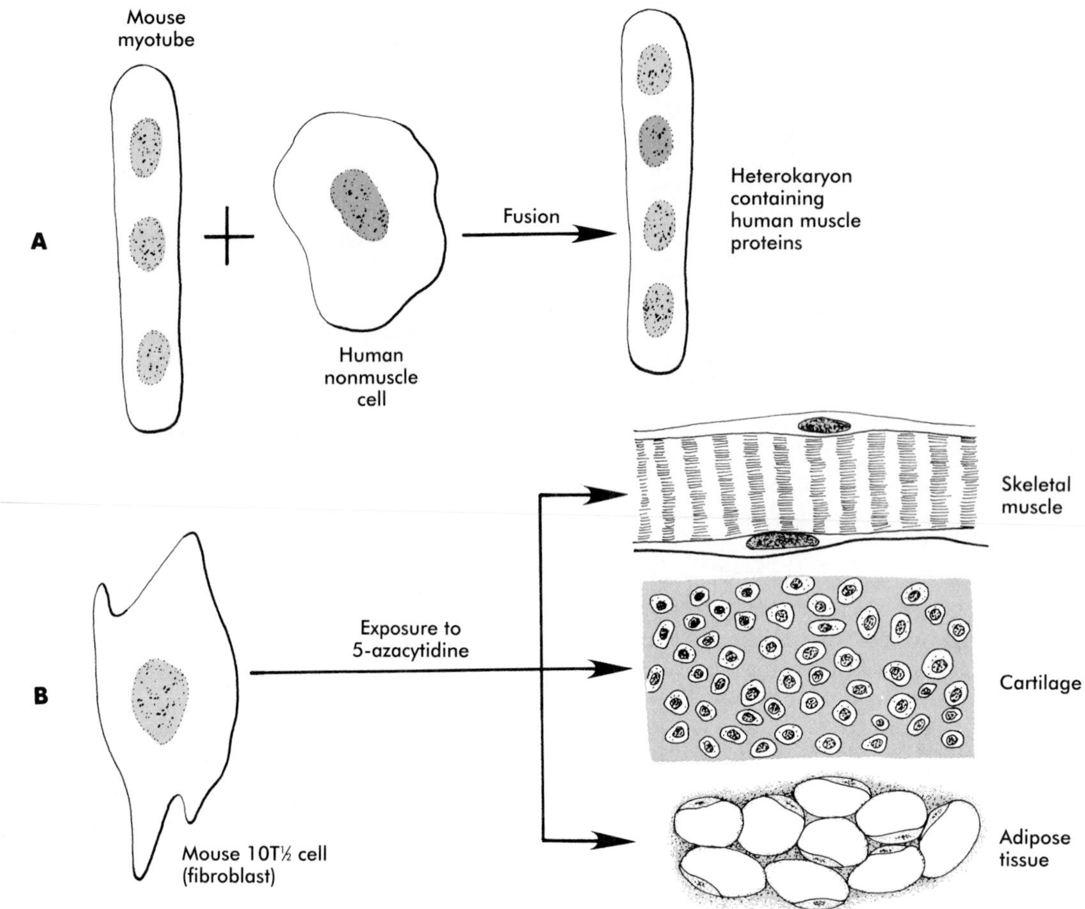

FIG. 9-3 Examples of eliciting the expression of normally quiescent genes in mammalian cells. **A,** When a heterokaryon is formed between mouse myotubes and human nonmuscle cells, human muscle proteins are found in the heterokaryon. **B,** After exposure to 5-azacytidine, fibroblasts from a standard tissue culture line differentiate into skeletal muscle, cartilage, or adipose tissue.

Lineage Versus Environmental Factors in the Generation of Cell Diversity

Cells in an embryo can become different from one another by following instructions intrinsic to the cells. For this to occur, it is necessary to postulate that after a cell division (1) a fundamental difference in the genetic material between the two daughter cells exists or (2) some substance in the cytoplasm that can influence gene expression is distributed unequally between the two daughter cells. Implicit in the **lineage model** is the assumption that the embryonic cells are never truly uncommitted to a particular fate. To change its fate, a cell must rely on intrinsic information supplied to it by its precursor cell. In most lineage models, cells change from their current phenotype to another type as the result of a mitotic division that permits the expression of a new portion of the genome. A major phenotypic change during an intermitotic period is unlikely.

According to the **environmental model** of generation of cell diversity, the fate of a cell is not fixed until it has been exposed to an external signal, for example, an induction or a hormonal influence. This type of signal causes a major change in the pattern of gene expression and ultimately alters the functional or morphological phenotype of the cell.

Although clearly defined examples of both models of cellular diversification have been recognized for decades, they were originally thought to be almost mutually exclusive. More recent research has demonstrated that many strictly defined cell lineages can be influenced by the local environment. In higher vertebrates, groups of cells that were once thought to arise purely through inductive influences are now known to be descended from well-defined lineages that can sometimes be traced to the early embryonic period.

Restriction, Determination, and Differentiation

In Chapter 4, the terms **restriction** and **determination** were introduced in relation to the induction of the neural plate. At various points in their life history, most cells pass through certain decision points that limit their subsequent developmental options. The commitment of cells during cleavage to become either inner cell mass or trophoblast and the segregation of embryonic cells into the three germ layers are early restriction events in the mammalian embryo (see Fig. 4-12). When a cell has passed its last restriction point and its fate is fixed (e.g., becoming a cartilage or muscle cell), the cell is *determined.* The determined cell may still pass through a number of developmental stages before it attains its mature state, but it is normally not able to jump onto another developmental track. For example, a determined muscle cell usually cannot be transformed into cartilage, or vice versa. The process by which a determined cell becomes structurally and functionally specialized is commonly called **differentiation.** Often there are well-defined stages in the differentiation of a cell.

Developmental Isoforms

One of the common themes of embryonic development is the orderly replacement of one cell or molecule by another that is quite similar but better adapted for the conditions prevailing at that particular time. The successive units in such a series of replacements are called **isoforms.** Although the term was first applied to molecules, it has become apparent that there are also cellular and even organ-level isoforms.

The development of red blood cells **(erythrocytes)** illustrates the isoform concept at both the cellular and molecular levels. Embryonic erythrocytes first arise in the yolk sac (see Fig. 9-12) and enter the circulating blood as nucleated cells. Later, another population of nonnucleated erythrocytes arises in the liver as a different cellular isoform. Within the erythrocytes, the succession of hemoglobin molecules from embryonic to fetal to adult isoforms (see Fig. 9-15) illustrates the adaptations of molecules, all designed to bind and transport oxygen, to changing conditions in the embryo.

There are several clear-cut examples of isoform transitions at the tissue and organ level as well. The long bones are initially composed entirely of hyaline cartilage, but soon the cartilage is replaced with true bony material (see Chapter 10). Even the replacement of deciduous (baby) teeth by permanent teeth is an isoform transition. Among the organs, one of the most complex isoform transitions is illustrated by the succession of different types of kidneys during embryonic development (see Chapter 17). Even with the different kidney isoforms, the principal function of the organ—the removal of body wastes—is maintained, although the functional mechanisms vary according to the circumstances of the embryo.

Molecular Mechanisms Regulating the Diversification and Differentiation of Cells*

The expression or control of expression of many genes (e.g., those for histones, ribosomal components, tRNAs, and a number of common enzymes) does not appear to differ greatly among most types of cells. Other genes, especially those that are uniquely expressed in specific cell types, are subject to a variety of control mechanisms that can act at levels from altering the DNA molecule itself to modification of a newly formed polypeptide chain (Fig. 9-4).

A typical gene coding for a mRNA molecule is illustrated in Fig. 9-5. Upstream from the gene itself are two **promoters (TATA box** and **CCAAT box). RNA polymerase II,** which is responsible for the transcription of mRNAs, binds to the nucleotides of the TATA box, which is located about 30 nucleotides from the transcription initiation site. The polymerase then moves along the DNA molecule, causing the helix to unwind and allowing the mRNA molecule to be synthesized behind it. The CCAAT box and sometimes another promoter element, called the **GC box,** are located upstream from the TATA box and are able to modify the efficiency of transcription.

In addition to the promoters, there may be other gene-specific regulatory elements called **enhancers** and **suppressors.** Enhancers can be located at various distances from the coding region of the gene in either the upstream or downstream direction.

One of the most exciting aspects of genetic control is the identification of increasing numbers of DNA-binding proteins called **transcription factors.** Transcription factors bind to specific sequences in the DNA molecule (e.g., on individual enhancers or promoters), and by doing so they regulate the transcription of the gene by RNA polymerase II. This regulation can take the form of either increasing or inhibiting the level of transcription. The large number of regulatory elements and transcription factors found in the nuclei of vertebrate cells allows for very precise control of gene expression.

Although individual transcription factors constitute a large class of proteins, many of them seem to operate on the basis of several general classes of molecular activity. One is the **helix-loop-helix protein,** which contains a short stretch of amino acids in which two α helices are separated by an amino acid loop. This region, with an adjacent basic region, allows the regulatory protein to bind to specific DNA sequences. This configuration is common in a number of the transcription factors that regulate myogenesis.

*Fully describing the myriad ways in which the regulation of gene expression provides the basis for both the diversification and differentiation of cells is beyond the scope of this text. At various places within the text, specific details are presented as the differentiation of certain cell and tissue types is discussed. This section outlines some general features of gene expression and its regulation.

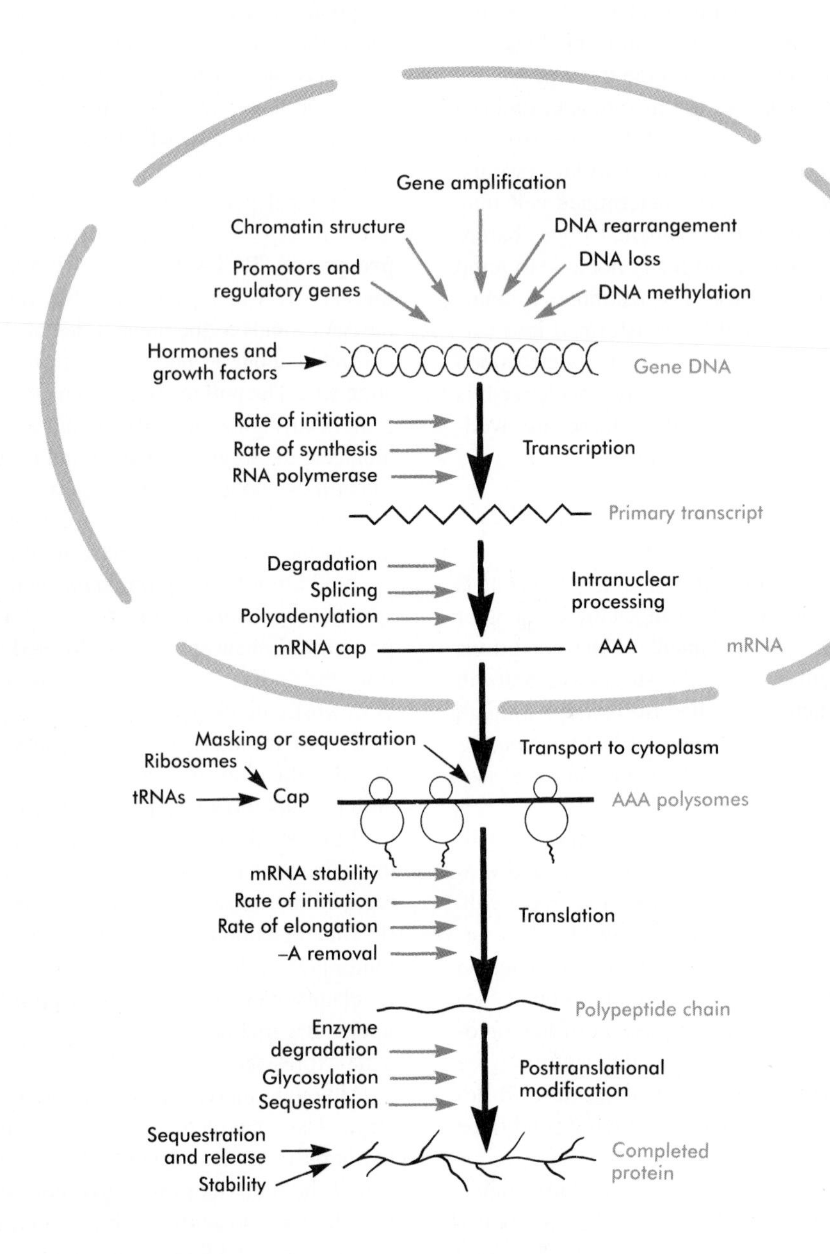

FIG. 9-4 Levels of gene control in a typical mammalian cell.

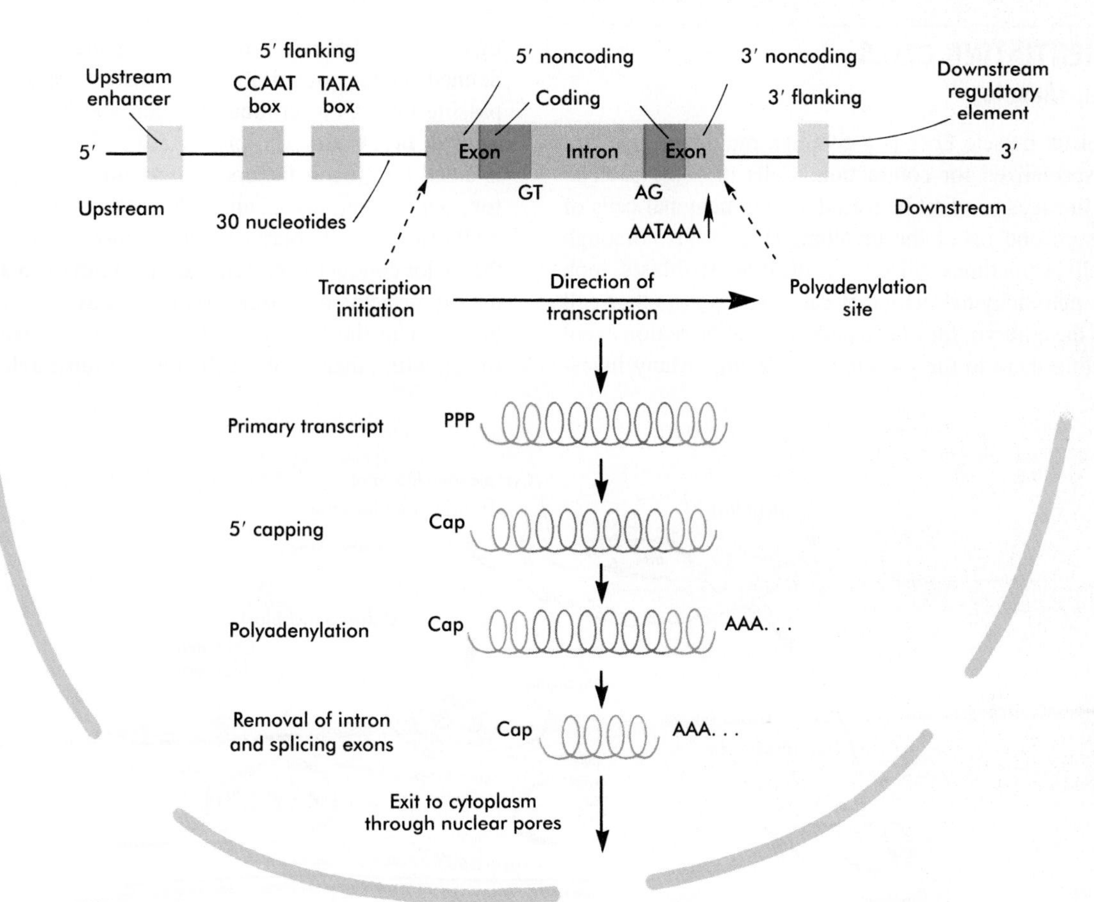

FIG. 9-5 The structure of a typical mammalian gene and the intranuclear processing of its mRNA.

Another family of regulatory proteins is the **zinc finger proteins.** In these proteins, regularly placed cystine and histidine units are bound by zinc ions in a manner that causes the polypeptide chain to pucker into fingerlike structures (Fig. 9-6). The "fingers" can be inserted into specific regions in the DNA helix. Examples of zinc finger proteins are steroid receptors.

Homeobox sequences are highly conserved regulatory elements of the helix-loop-helix variety. After the initial discovery of their DNA sequences in the homeotic genes, such as *antennapedia* in *Drosophila,* they have been found to be widespread throughout much of the animal kingdom, including humans. Much remains to be learned about the mechanisms by which they regulate gene expression.

Aside from transcriptional control, there are many other ways of regulating gene expression by either modifying the structure of the gene product or by influencing the amount and/or timing of expression (see Fig. 9-4).

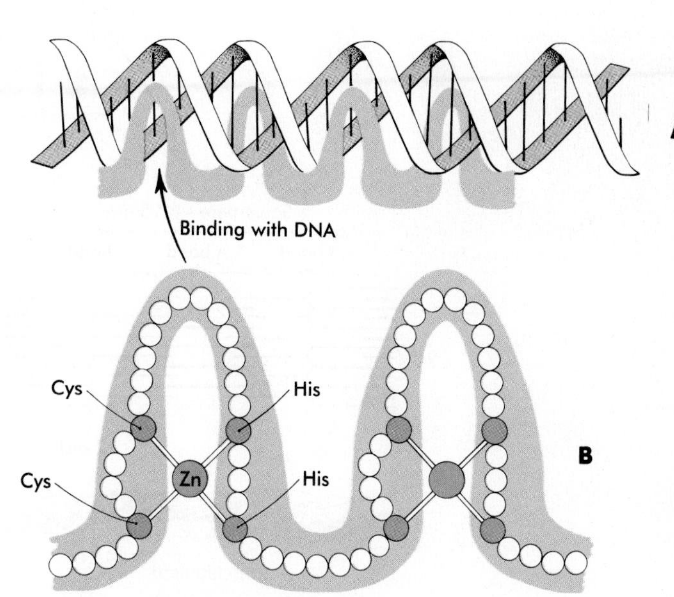

FIG. 9-6 **A,** The binding to DNA of a zinc finger. **B,** The structure of a zinc finger DNA-binding sequence.

DIFFERENTIATING CELLS

Skeletal Muscle

The skeletal muscle fiber is a complex multinucleated cell that is specialized for contraction. Cells of some skeletal muscle lineages can first be traced to mesenchymal cells of the dorsomedial lip of the myotome (Fig. 9-7). Although these cells, sometimes called **presumptive myoblasts,** look like the mesenchymal cells that can give rise to other cell types in the embryo, they have undergone a restriction event committing them to the muscle-forming line. Many investigators feel that a committed myogenic cell undergoes a defined number of mitotic divisions, perhaps four, before passing through a terminal division and becoming a **postmitotic myoblast.** The proliferative phase of myogenesis is aided by growth factors such as **fibroblast growth factor,** which keep myogenic cells in the mitotic cycle.

Postmitotic myoblasts begin to transcribe the mRNAs for the major contractile proteins **actin** and **myosin** and the regulatory proteins of muscle contraction as well, but the major event in the life cycle of a postmitotic myoblast is its fusion with other similar cells into a **multinucleated myo-**

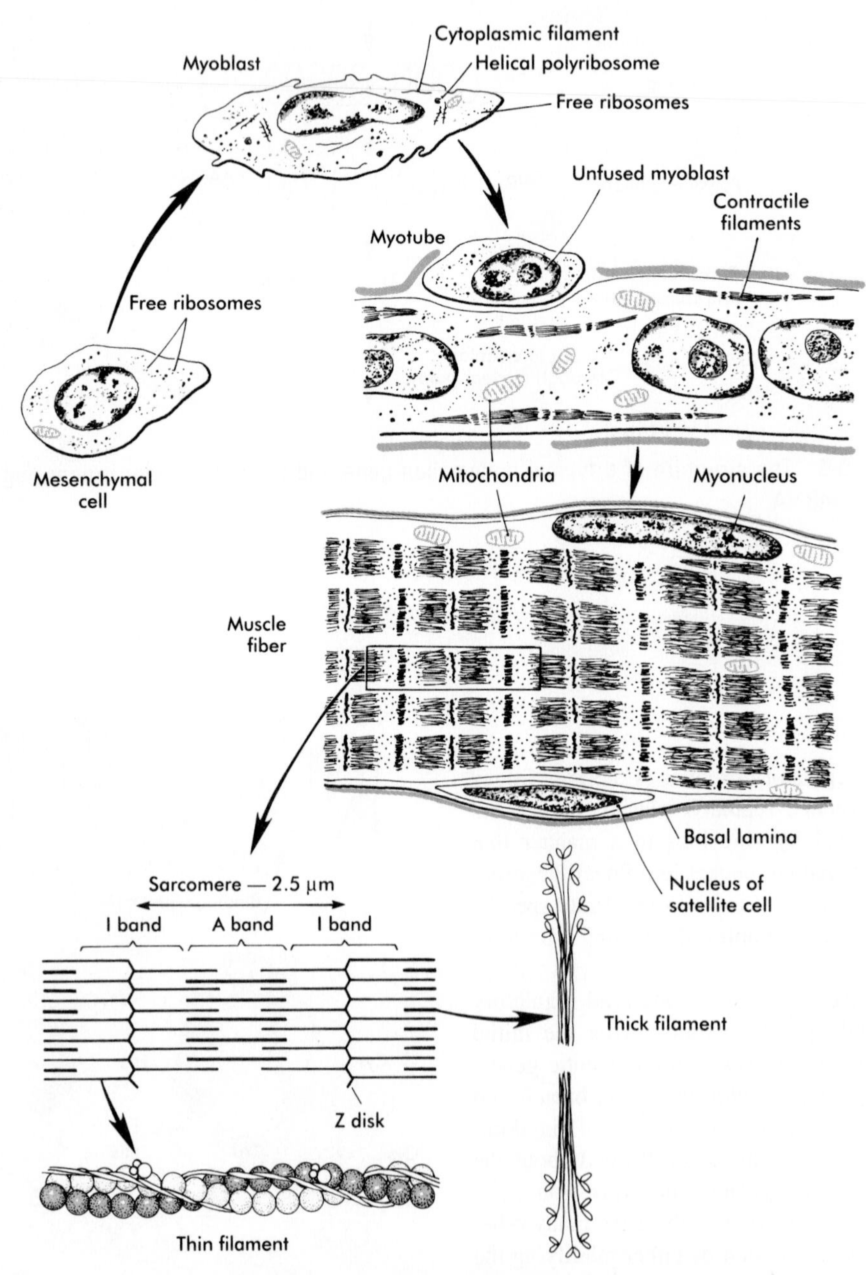

FIG. 9-7 Stages in the morphological differentiation of a skeletal muscle fiber. Important subcellular elements in a muscle fiber are also shown.

tube (see Fig. 9-7). The fusion of myoblasts is a precise process involving their lining up and adhering by Ca^{++}-mediated recognition mechanisms and the ultimate union of their plasma membranes.

Myotubes are intensively involved in mRNA and protein synthesis. In addition to forming actin and myosin, they synthesize a wide variety of other proteins, including the regulatory proteins of muscle contraction—**troponin** and **tropomyosin.** These proteins assemble into myofibrils, which are precisely arranged aggregates of functional contractile units called **sarcomeres.** As the myotubes fill with myofibrils, their nuclei, which had been arranged in regular central chains, migrate to the periphery of the myotube. At this stage the myotube is considered to be a **muscle fiber,** the final stage in the differentiation of the skeletal muscle cell.

The development of a muscle fiber is not complete, however, with the peripheral migration of the nuclei of the myotube. The nuclei **(myonuclei)** of a multinucleated muscle fiber are no longer able to proliferate, but the muscle fiber must continue to grow in proportion to the rapid growth of the fetus and the later infant. Muscle fiber growth is accomplished by means of a population of myogenic cells called **satellite cells,** which take up positions between the muscle fiber and the basal lamina in which each muscle fiber encases itself (see Fig. 9-7). Operating under a poorly understood control mechanism, satellite cells divide slowly during the growth of an individual. Some of the daughter cells fuse with the muscle fiber so that the muscle fiber contains an adequate number of nuclei to direct the continuing synthesis of contractile proteins required by the muscle fiber.

A typical muscle is not composed of homogeneous muscle fibers. Instead, there are usually several types of muscle fibers that are distinguished by their contractile properties and morphology and by possessing different isozymic forms of the contractile proteins. For the purposes of this text, muscle fiber types are divided into fast and slow categories.

Muscle Transcription Factors. Myogenesis begins with a restriction event that channels a population of mesenchymal cells into a lineage committed (determined) to forming muscle cells. The molecular basis for this commitment is the action of members of a family of transcription factors that, acting as master genetic regulators, turn on muscle-specific genes in the premuscle mesenchymal cells. This family of regulators, called the **MyoD family,** consists of several proteins with homologous basic and helix-loop-helix regions (Fig. 9-8).

As with many helix-loop-helix proteins, regulatory proteins of the MyoD family form dimers and bind to a specific DNA sequence (CA—TG) called the **E box** in the enhancer region of muscle-specific genes. The myogenic specificity of these proteins is encoded in the basic region.

The regulatory activities of MyoD and other members of that family are regulated by other regulatory proteins,

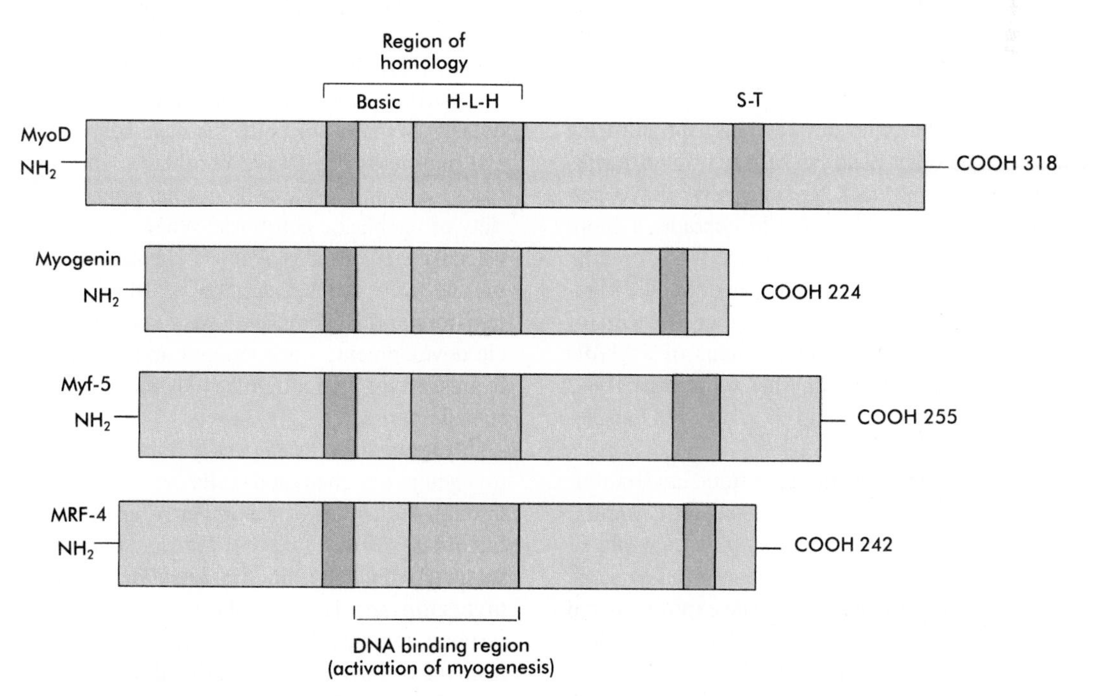

FIG. 9-8 Structural comparison of several myogenic regulatory factors. *H-L-H,* Homologous helix-loop-helix regions; *S-T,* homologous serine/threonine-rich region.

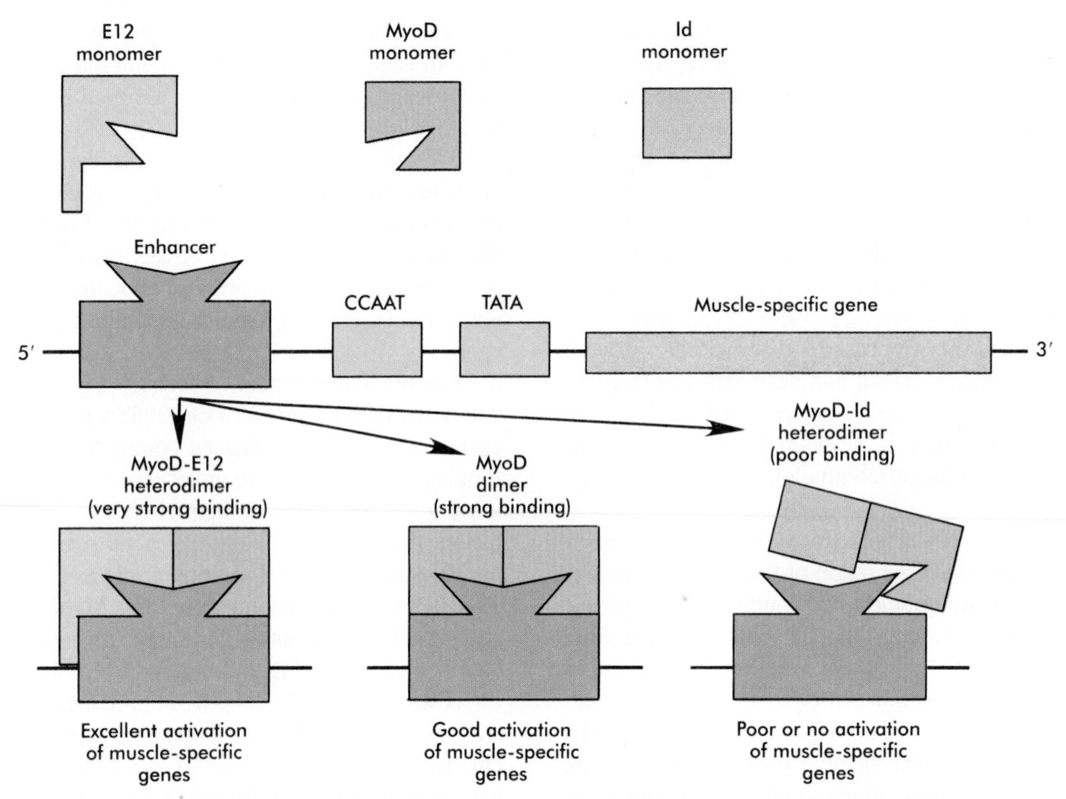

FIG. 9-9 MyoD regulation of early myogenesis, showing interactions between MyoD and a transcriptional activator *(E12)* and a transcriptional inhibitor *(Id)*.

which can modify its activities (Fig. 9-9). For example, many cells contain a **transcriptional activator** designated **E12.** When a molecule of E12 forms a **heterodimer** with a molecule of MyoD, the complex binds more tightly to the muscle-enhancer region of DNA than a pure MyoD dimer. This increases the efficiency of transcription of the muscle genes. On the other hand, a **transcriptional inhibitor** called **Id** (inhibitor of DNA binding) may form a heterodimer with a molecule of MyoD. Id contains a loop-helix-loop but no basic region, which is the DNA-binding part of the molecule. The Id molecule has a greater binding affinity for a MyoD molecule than another molecule of MyoD and can thus displace one of the units of a MyoD dimer, resulting in more Id-MyoD heterodimers. These bind very poorly to DNA and often fail to activate muscle-specific genes.

Because each of the regulators, activators, and inhibitors is itself a protein, their formation is subject to similar positive and negative controls. This complex example of the regulation of the first step in myogenesis gives some idea of the multiple levels of control of gene expression and stages of cell differentiation in mammals. Although molecular aspects of the control of myogenesis are better understood than those underlying the differentiation of most cells, it is unrealistic to think that similar sets of interlinked regulatory mechanisms do not operate in the differentiation of most cell types.

The muscle transcription factors of the MyoD family are so powerful that if they are introduced into a number of other differentiated cell types, these cells convert into myoblasts. Some of the cells that have been converted to skeletal muscle by MyoD are fibroblasts, adipocytes, chondrocytes, smooth muscle cells, retinal pigment cells, and a variety of neoplastic cell lines. When one fused cell of an artificially created heterokaryon becomes transformed into a muscle cell, the appearance of MyoD accompanies the transformation. MyoD is highly specific for skeletal muscle development, since smooth and cardiac muscle do not express MyoD, although they share some common muscle-specific genes.

Myogenesis. In the early embryo, proliferating premyogenic mesenchymal cells are kept in the cell cycle through the action of a number of growth factors, such as **fibroblast growth factor (FGF)** and **transforming growth factor-β (TGF-β).** In the myotome of the somite, first **myogenin** and later MyoD are expressed in the muscle-forming regions. MyoD removes the mesenchymal cells from the cell cycle by preventing their entry into the S phase and also stimulates their differentiation by turning on muscle-specific genes. Other growth factors, such as **in-**

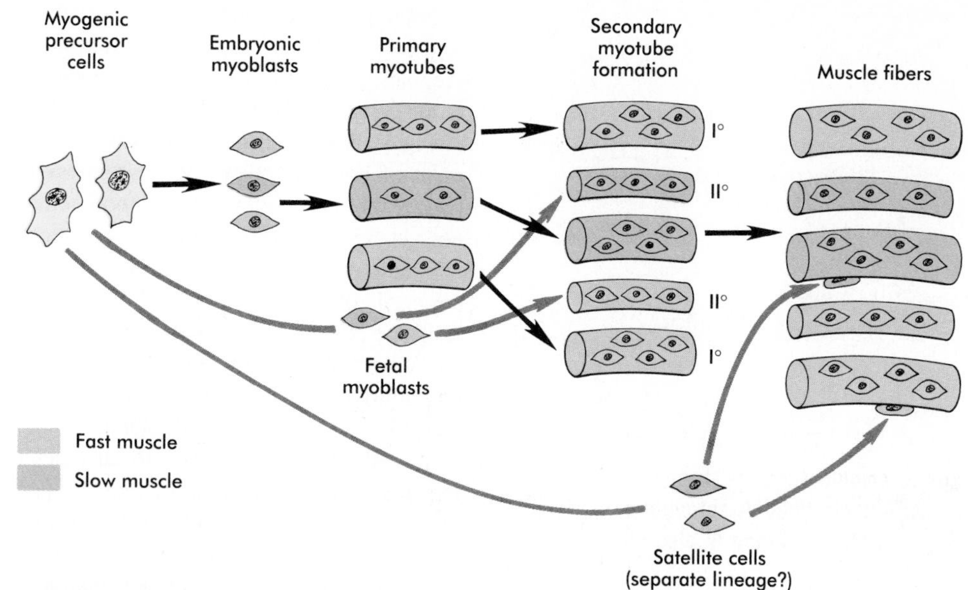

FIG. 9-10 Stages in the formation of primary and secondary muscle fibers. A family of embryonic myoblasts contributes to the formation of the primary myotubes, and fetal myoblasts contribute to secondary myotubes. Doubts remain about the origin of satellite cells.

sulinlike growth factor (IGF), are also involved in promoting muscle differentiation. These influences convert the undifferentiated myogenic cells into committed myoblasts.

At one time it was thought that all myoblasts were essentially identical and that their different characteristics (e.g., fast or slow) were imposed on them by their motor innervation. Recent research, however, has shown that in birds and several species of mammals, there may be distinct populations of fast and slow muscle cells as early as the myoblast stage, well before nerve fibers reach the developing muscles. The molecular basis for the early differentiation of distinct phenotypes in myoblasts is not understood.

Not only are there fast and slow myoblasts, but there are also early and late cellular isoforms of myoblasts, which have different requirements for serum factors and nerve interactions in their differentiation. When the early myoblasts fuse into myotubes, they give rise to **primary muscle fibers,** which form the initial basis for an embryonic muscle. Subsequently, smaller **secondary muscle fibers** arising from late myoblasts form along the primary muscle fibers (Fig. 9-10). A primary muscle fiber and its associated secondary muscle fibers are initially contained within a common basal lamina and are electrically coupled. These muscle fibers actively synthesize a wide array of contractile proteins.

Early in their life history, embryonic muscle fibers are innervated by motor neurons. Although it has long been assumed that fast and slow motoneurons impose their own functional characteristics on the developing muscle fibers, it now appears that they may select muscle fibers of a compatible type through information contained on the cell surfaces. Initially, a motor nerve fiber may terminate on both fast and slow muscle fibers, but ultimately inappropriate connections are broken, and fast nerve fibers innervate only fast muscle fibers and slow nerves innervate only slow muscle fibers. If the current data hold, nerve fibers will be viewed principally as helping muscle fibers maintain their state of differentiation rather than determining qualitative differences between fast and slow muscle fibers.

The phenotypes of muscle fibers depend on the nature of the specific proteins that make up their contractile apparatus. There are qualitative differences between fast and slow muscle fibers in many of the contractile proteins, and within each type of muscle fiber, there is a succession of isoforms of the major proteins during embryonic development. (The isoform transitions of **myosin** in a developing muscle fiber are used as an example.)

The myosin molecule is a complex, consisting of two heavy chains and a series of four light chains (Fig. 9-11). Mature fast muscle fibers have one LC_1, two LC_2, and one LC_3 light chain subunits; slow muscle myosin contains two LC_1 and two LC_2 light chain subunits. In addition, there are fast and slow forms (MHC_f and MHC_s) of the **myosin heavy chain** subunits. The myosin molecules possess ATPase activity, and differences in activity may partly account for differences in the speed of contraction between fast and slow muscle fibers.

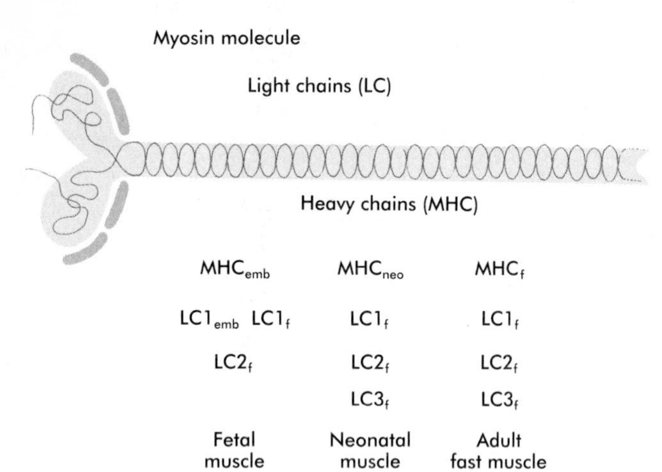

FIG. 9-11 Changes in myosin subunits during the development of a fast muscle fiber. A schematic representation of the myosin molecule is also shown.

The myosin molecule undergoes a succession of isoform transitions during development. From the fetal period to maturity, a series of three developmental isoforms of the myosin heavy chain (embryonic [MHC_{emb}], neonatal [MHC_{neo}], and adult fast [MHC_f]) passes through a fast muscle fiber. (Developmental changes in the light chain subunits are summarized in Fig. 9-11.) Other contractile proteins of muscle fibers (e.g., actin, troponin) pass through similar isoform transitions. After injury to muscle in the adult, the regenerating muscle fibers undergo sets of cellular and molecular isoform transitions that closely recapitulate those that occur during normal ontogenesis.

Contrary to former beliefs, the phenotype of skeletal muscle fibers is not irreversibly fixed. Muscle fibers possess a remarkable degree of plasticity. They respond to exercise by either undergoing hypertrophy or becoming more resistant to fatigue. On the other hand, they adapt to inactivity or denervation by becoming atrophic. All these changes are accompanied by a variety of changes in gene expression. Many other types of cells can also modify their phenotypes in response to changes in the environment, but the molecular changes are not always as striking as those seen in muscle fibers. On the other hand, other cells such as erythrocytes are terminally differentiated and normally cannot change their phenotype.

Red Blood Cells

The red blood cell **(erythrocyte)** is unquestionably one of the most simple cells in the body. However, the life history of an erythrocyte is remarkably complex, and this type of cell demonstrates a variety of very sensitive control mechanisms that operate at different times and different levels. Successions of developmental isoforms occur at the cellular and molecular levels. Isoform transitions are adaptive mechanisms that meet the physiological needs of the embryo.

Sites of Hematopoiesis. Blood cell formation (**hematopoiesis**) begins in the yolk sac during the third week of pregnancy. At this time the embryo has attained a size that is too large for the distribution of oxygen to all tissues by diffusion alone. This necessitates the very early development of both the heart and vascular system. Because the tissues that normally produce blood cells in the adult have not yet begun to form, yolk sac hematopoiesis appears to be a temporary adaptation for accommodating the immediate needs of the embryo.

There is some evidence that an inductive interaction with the underlying yolk sac endoderm stimulates the formation of primitive **blood islands** in the mesoderm of the yolk sac (see Fig. 5-18). The blood islands contain **pluripotential stem cells** that can give rise to all types of cells found in the blood. The way some of their progeny are channeled into forming red blood cells is discussed below.

The erythrocytes produced in the yolk sac are large, nucleated cells that enter the bloodstream just before the primitive heart tube begins to beat at about 22 days gestation. For the first 6 weeks, the circulating erythrocytes are almost entirely yolk-sac derived, but during that time, preparations for the next stage of hematopoiesis are taking place. At roughly 4 weeks, the primordium of the liver arises from the embryonic foregut, and by 5 to 6 weeks, sites of hematopoiesis become prominent within the liver. In mammals, including the human, stem cells derived from the yolk sac seed the liver and represent the origin of the hematopoietic cells there. In mouse embryos, removal of the yolk sac from a presomite embryo is followed by the absence of hematopoiesis in the embryo itself.

The erythrocytes produced by the liver are quite different from those derived from the yolk sac. Although still considerably larger than normal adult red blood cells, liver-derived erythrocytes are nonnucleated and contain different types of hemoglobin. By 6 to 8 weeks gestation, the liver replaces the yolk sac as the main source of blood cells. Although the liver continues to produce red blood cells until the early neonatal period, its contribution begins to decline in the sixth month of pregnancy. At this time the formation of blood cells shifts to the bone marrow, the definitive site of adult hematopoiesis. Although the spleen has traditionally been thought to play a role in embryonic hematopoiesis, recent evidence suggests that the red blood cells found in the spleen may be derived from the liver and that active **erythropoiesis** (the formation of red blood cells) may not occur there.

Cellular Aspects of Hematopoiesis. The first hematopoietic stem cells that arise in the embryo are truly pluripotent in that they can give rise to all the cell types found in the blood (Fig. 9-12). These pluripotent stem cells, sometimes called **hemocytoblasts,** have a great proliferative ability. They produce vast numbers of progeny, most of which are cells at the next stage of differentiation, but they also produce small numbers of their original stem cell type, which act as a reserve capable of replenishing indi-

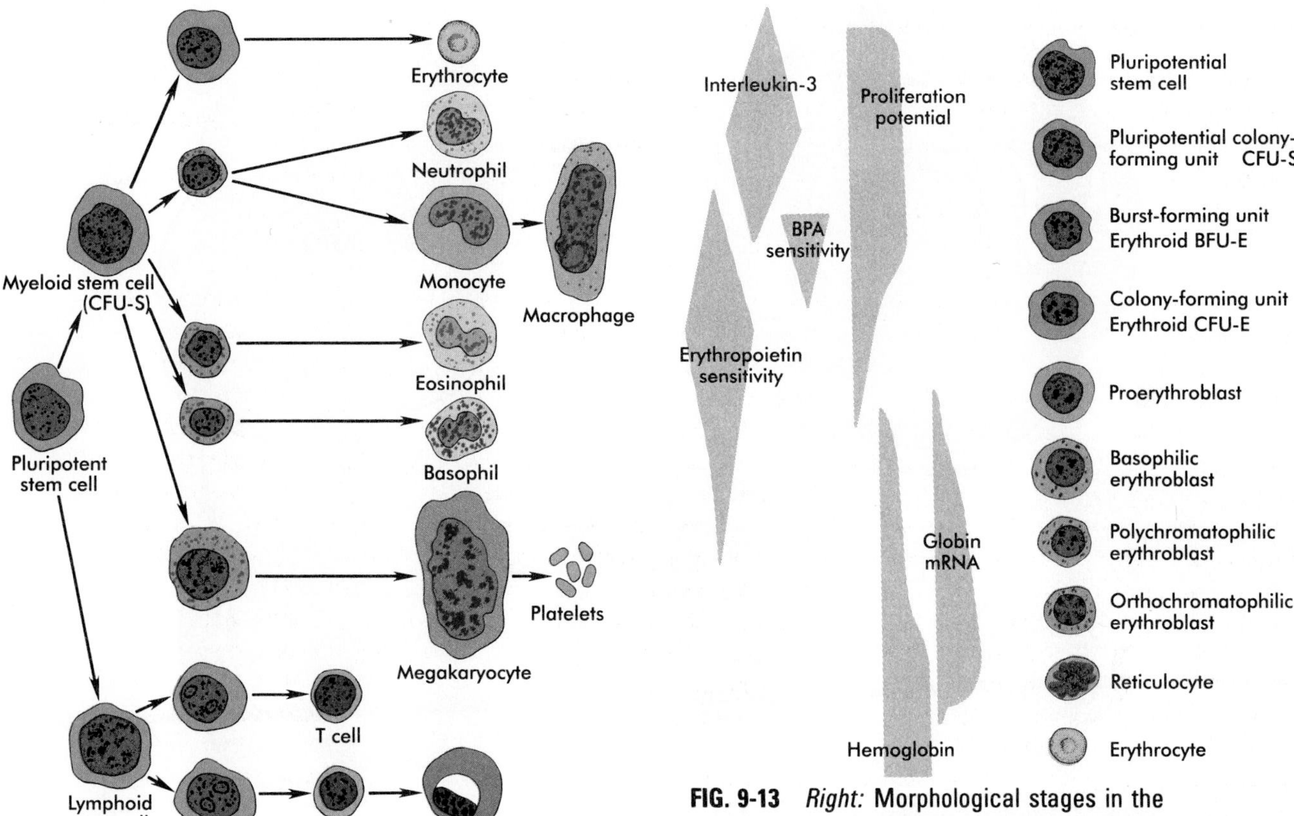

FIG. 9-12 Major cell lineages during hematopoiesis. *Right:* Mature blood cells.

FIG. 9-13 *Right:* Morphological stages in the differentiation of a red blood cell from a pluripotential stem cell. *Left:* Molecular correlates of differentiation. The thickness of the gray background is proportional to the amount at the corresponding stages of erythropoiesis.

vidual lines of cells should the need arise. Very early in development, the line of active blood-forming cells subdivides into two separate lineages. **Lymphoid stem cells** ultimately form the two lines of lymphocytes—the **B lymphocytes** (which are responsible for antibody production) and the **T lymphocytes** (which are responsible for cellular immune reactions). **Myeloid stem cells** are precursors to the other lines of blood cells—erythrocytes, the granulocytes (neutrophils, eosinophils, and basophils), monocytes, and platelets. The second generation stem cells (lymphoid and myeloid) are still pluripotential, although their developmental potency is restricted, since neither can form the progeny of the other type.

Stemming from their behavior in certain experimental situations, the hematopoietic stem cells are often called **colony-forming units (CFUs).** The first generation stem cell is called the *CFU-ML* because it can give rise to both myeloid and lymphoid lines of cells. Stem cells of the second generation are called *CFU-L* (*L*, lymphocytes) and *CFU-S* (S, spleen [determined from experiments in which stem cell differentiation was studied in irradiated spleens]). In all cases except one, the progeny of the CFU-L and CFU-S are **committed stem cells,** which are capable of forming only one type of mature blood cell. For each lineage the forming cell types must pass through several stages of differentiation before they attain their mature phenotype.

What controls the diversification of stem cells into specific cell lines? Experiments begun in the 1970s provided evidence for the existence of specific **colony-stimulating factors (CSFs)** for each line of blood cell. CSFs are diffusible proteins that stimulate the proliferation of hematopoietic stem cells. Some CSFs act on a number of types of stem cells; others stimulate only one type. Although much remains to be learned about the sites of origin and modes of action of CSFs, many appear to be produced locally in stromal cells of the bone marrow, and some may be stored on the local extracellular matrix. CSFs are bound by small numbers of surface receptors on their target stem cells. Functionally, CSFs represent mechanisms for stimulating the expansion of specific types of blood cells when the need arises. Recognition of the existence of CSFs has prompted considerable interest in their clinical application to conditions characterized by a deficiency of white blood cells (leukopenia) (e.g., AIDS or the leukopenia frequently resulting from cancer chemotherapy).

Erythropoiesis. The erythrocytic lineage represents one line of descent from the CFU-S cells. Although the erythroid progenitor cells are restricted to forming only red blood cells, there are many generations of precursor cells (Fig. 9-13). Some of these are just beginning to be recog-

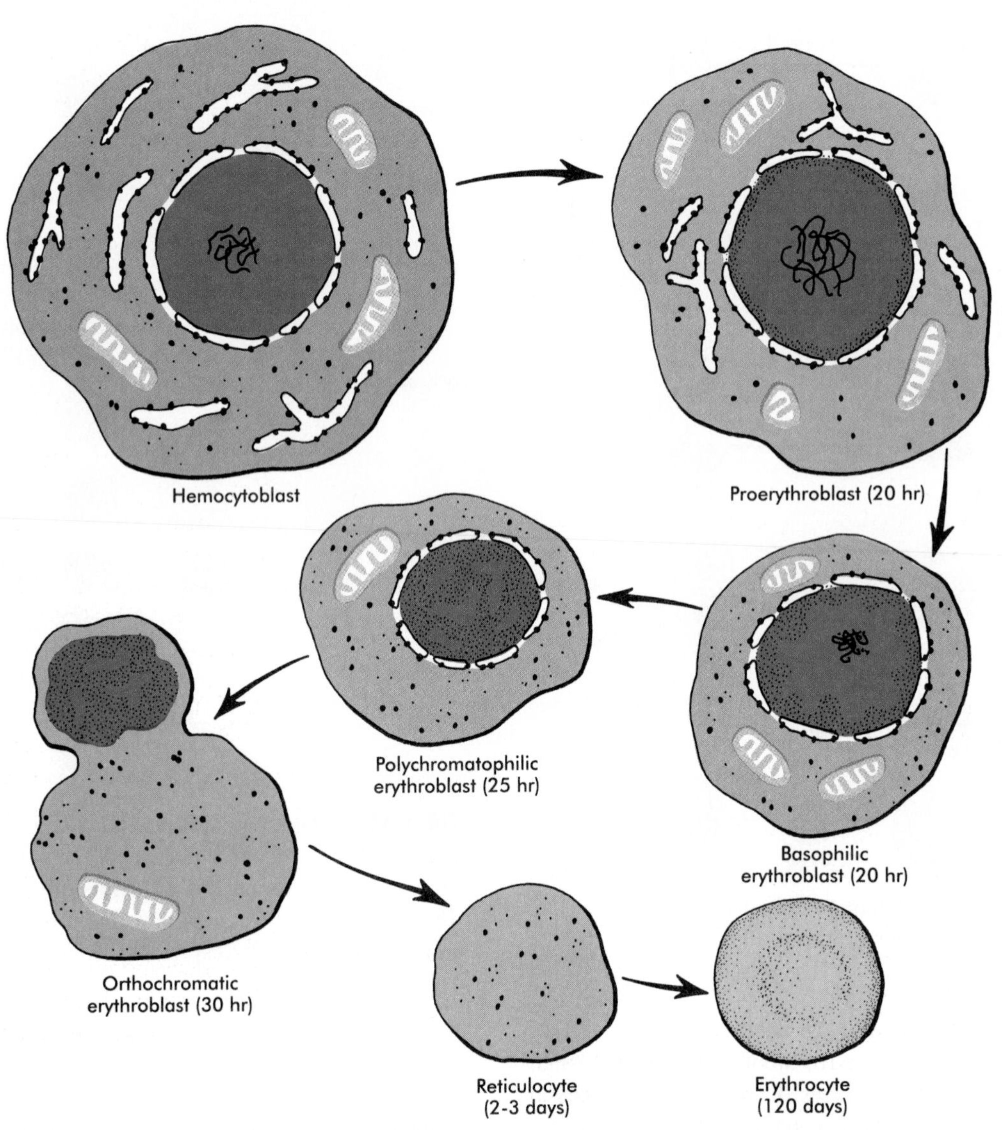

FIG. 9-14 Structural features of erythropoiesis. In successive stages, cytoplasmic basophilia decreases and the concentration of hemoglobin increases in the cells.

TABLE 9-1 Comparison between Morphologically Undifferentiated and Differentiated Cells

CHARACTERISTIC	UNDIFFERENTIATED CELLS	DIFFERENTIATED CELLS
Nuclear size	Larger	Smaller
Nucleocytoplasmic ratio	High	Low
Nuclear chromatin	Predominantly euchromatin	Predominantly heterochromatin
Nucleolus	Prominent	Less prominent
Cytoplasmic staining	Basophilic	Acidophilic
Ribosomes	Numerous	Less numerous
RNA synthesis	Greater	Lesser
Mitotic activity	Greater	Lesser
Metabolism	Generalized	Specialized
Extracellular matrix	Little	Often more prominent

nized, and the functions of many of the stages of precursor cells are just beginning to be understood.

The earliest stages of erythropoiesis are recognized by the behavior of the precursor cells in culture rather than by morphological or biochemical differences. These are called **erythroid burst-forming units (BFU-e)** and **erythroid colony-forming units (CFU-e).** Each responds to different stimulatory factors. The pluripotent CFU-S precursors respond to **interleukin-3,** a product of macrophages in adult bone marrow. A hormone designated as burst-promoting activity stimulates mitosis of the BFU-e precursors. A CFU-e, which has a lesser proliferative capacity than a BFU-e cell, requires the presence of **erythropoietin** as a stimulatory factor. Erythropoietin is a glycoprotein that stimulates the synthesis of the mRNA for globin and is first produced in the fetal liver. Later in development, synthesis shifts to the kidney, which remains the site of erythropoietin production in the adult. Under conditions of hypoxia (e.g., from blood loss or high altitudes) the production of erythropoietin by the kidneys increases, stimulating the production of more red blood cells to compensate for the increased need. In adult erythropoiesis, the CFU-e stage seems to be the one most responsive to environmental influences. The placenta appears to be impervious to erythropoietin, thereby insulating the embryo from changes in erythropoietin levels of the mother as well as eliminating the influence of fetal erythropoietin on the blood-forming apparatus of the mother.

One or two generations after the CFU-e stage, successive generations of erythrocyte precursor cells can be recognized by their morphology. The first recognizable stage is the **proerythroblast** (Fig. 9-14), a large, highly basophilic cell that has not produced sufficient hemoglobin to be detected by cytochemical analysis. These cells have a large nucleolus, much uncondensed nuclear chromatin, numerous ribosomes, and a high concentration of globin mRNAs. These are classic cytological characteristics of an undifferentiated cell (Table 9-1).

Succeeding stages of erythroid differentiation **(basophilic, polychromatophilic,** and **orthochromatic erythroblasts)** are characterized by a progressive change in the balance between the accumulation of newly synthesized hemoglobin and the decline of first the RNA-producing machinery and later the protein-synthesizing apparatus. The overall size of the cell decreases, and the nucleus becomes increasingly **pycnotic** (smaller with more condensed chromatin) until it is finally extruded at the stage of the orthochromatic erythrocyte. After the loss of the nucleus and most cytoplasmic organelles, the immature red blood cell, which still contains a small number of polysomes, is a reticulocyte. Reticulocytes are released into the bloodstream, where they continue to produce small amounts of hemoglobin for a day or two.

The final stage of erythropoiesis is the mature erythrocyte, which is a terminally differentiated cell because of the loss of its nucleus and most of its cytoplasmic organelles. Erythrocytes in embryos are larger than their adult counterparts and have a shorter lifespan (50 to 70 days in the fetus versus 120 days in adults).

Hemoglobin Synthesis and Its Control. Not only the red blood cells but also the hemoglobins within them undergo isoform transitions during embryonic development. The hemoglobin molecule is a complex composed of heme and four globin chains—two alpha (α) chains and two beta (β) chains. Both the α and β subunits are products of genes located on chromosomes 16 and 11, respectively (Fig. 9-15). Different isozymic forms of the subunits are encoded linearly on these chromosomes.

During the period of yolk sac hematopoiesis, embryonic globin isoforms are produced. The earliest embryonic hemoglobin, sometimes called Gower 1, is composed of two zeta (ζ) (α-type) and two epsilon (ϵ) (β-type) chains. After passing through a couple of transitional forms (Table 9-2), hemoglobin synthesis enters a fetal stage by 12 weeks, which corresponds to the shift in the site of erythropoiesis from the yolk sac to the liver. Fetal hemoglobin consists of two adult type α chains, which form very early in embryogenesis, and two γ chains, the major fetal isoform of the β chain. Fetal hemoglobin is the predominant form during the remainder of pregnancy. The main adaptive value of the fetal isoform of hemoglobin is that it has a higher affinity for oxygen than does the adult form. This is advantageous to the fetus, which depends on the oxygen concentration of the maternal blood. Starting at about 30 weeks gestation, there is a gradual switch from the fetal to the adult type of hemoglobin, with $\alpha_2\beta_2$ being the predominant type. A minor but functionally similar variant is $\alpha_2\zeta_2$.

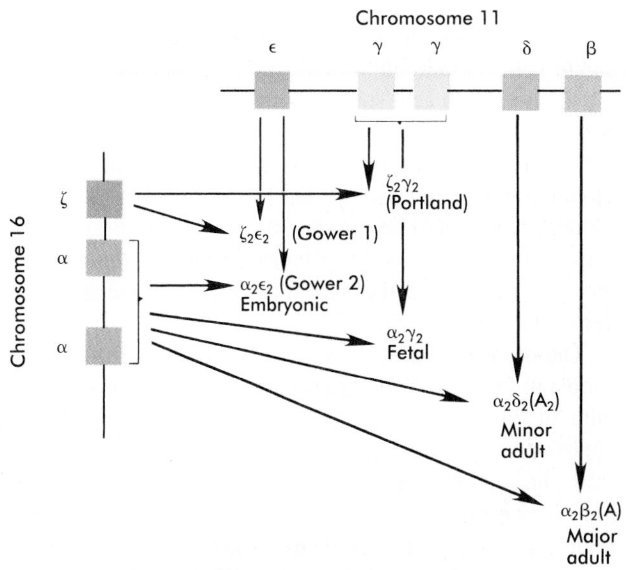

FIG. 9-15 Organization of hemoglobin genes along chromosomes 11 and 16 and their sequential activation during embryonic development.

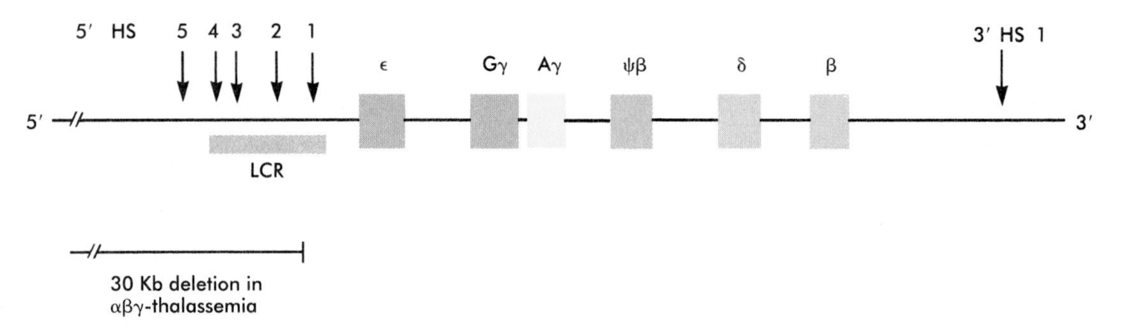

FIG. 9-16 Organization of the human β-globin gene cluster. *LCR,* Locus control region; *HS,* hypersensitive site.

TABLE 9-2 Developmental Isoforms of Human Hemoglobin

DEVELOPMENTAL STAGE	HEMOGLOBIN TYPE	GLOBIN-CHAIN COMPOSITION
Embryo	Gower 1	$\zeta_2\epsilon_2$
Embryo	Gower 2	$\alpha_2\epsilon_2$
Embryo	Portland	$\zeta_2\gamma_2$
Embryo to fetus	Fetal	$\alpha_2\gamma_2$
Fetus to adult	A (adult)	$\alpha_2\beta_2$
Adult	A_2	$\alpha_2\delta_2$
Adult	Fetal	$\alpha_2\gamma_2$*

Adapted from Brown MS: In Stockman J, Pochedly C, eds: *Developmental and neonatal hematology,* New York, 1988, Raven.
*The fetal hemoglobin expressed in adults differs from true fetal hemoglobin by an amino acid substitution at the 136 position of the γ chain.

A small number of clinically normal individuals retain a fetal type of hemoglobin ($\alpha_2\gamma_2$, the γ chain being slightly different from the true fetal γ chain) into adulthood. This finding has raised the hopes of finding a molecular solution for certain blood diseases such as the **thalassemias** and **sickle cell anemia,** in which abnormal function is correlated with amino acid substitutions in the β chain. Because the γ chains of these patients are typically normal, if a method of reversing the switch from fetal to adult hemoglobin could be devised so that fetal hemoglobin is again produced, it should be possible to bypass the disease process by genetic engineering. Numerous laboratories are now trying to understand the way "the switch" in the β-globin gene complex functions.

The organization of the human β-globin gene cluster is shown in Fig. 9-16. The genes coding for the isozymic subunits of the β-globin chains are arranged in order of their appearance in ontogeny in a 5′ to 3′ fashion. Upstream (5′) from the ε-globin gene is a roughly 30 Rb (ribobase) region hypersensitive to the action of DNAse I. Within that region, which is called the **locus control region (LCR)** are five individual **hypersensitive sites (HS).** At the 3′ end of the globin gene cluster is another single hypersensitive site. The importance of these sites is illustrated by a patient who was afflicted with **γ δ β-thalassemia,** in which the entire β-globin gene cluster was intact but none of the genes were functional. The molecular defect in this patient was a deletion that had removed the 5′ HS 2 to 5 sites from the LCR (see Fig. 9-16). The importance of the LCR has also been shown by experiments involving the use of transgenic mice. If the inserted globin **transgenes** are associated with the LCR region, the globins are strongly expressed in the mice. If the LCR is not included in the gene construct, globin expression is poor.

The mechanism by which the LCR regulates globin synthesis is not clear. There is considerable evidence that this region is involved in keeping the chromatin of the appropriate region in an open state and also acts as an enhancer of transcription, but the main question is what controls the sequential expression of the β-globin isoforms. There is some evidence that an active gene in the complex competes more effectively for interaction with the LCR than do the genes located downstream. When a gene farther downstream is activated, silencers act to prevent transcription from isoform genes located upstream to the currently active gene. Other recent research involving transgenic mice containing the human A γ-globin gene linked to the LCR has shown that this gene is silenced in the adult mouse, even in the absence of the β gene. This suggests that other factors than pure competition between the γ and the β genes

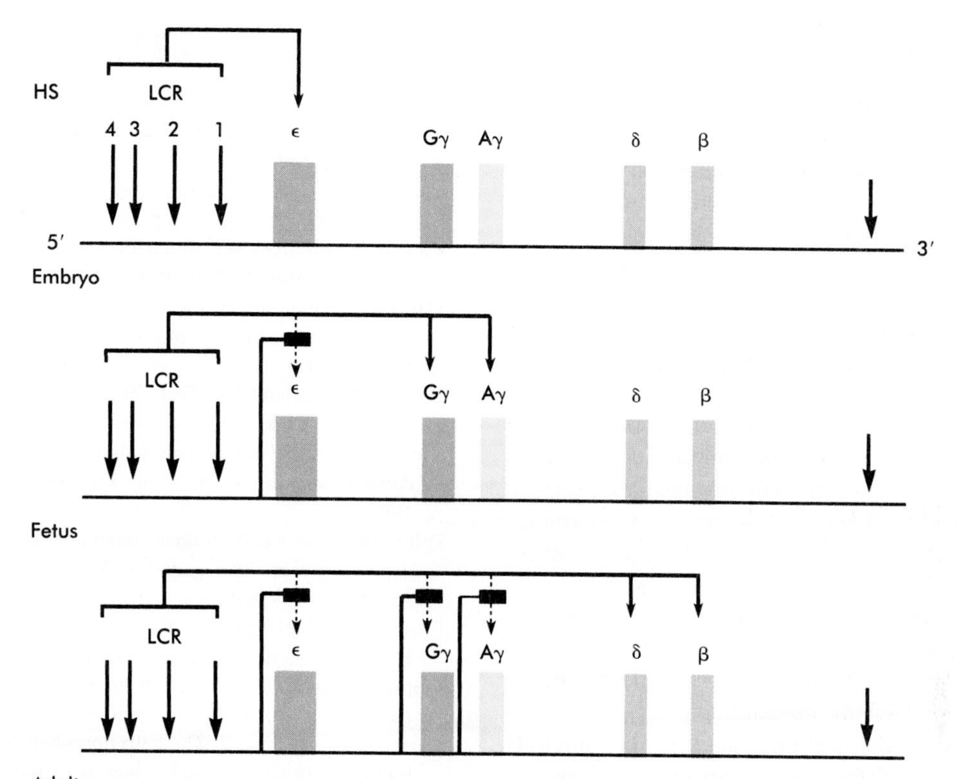

FIG. 9-17 A model for stage-specific regulation of β-globin gene expression. For each stage of development the solid arrows represent the activation of genes by the LCR (locus control region). The dashed arrows intersected by the dark rectangle represent stage-specific factors that silence the gene to which the arrow points.

are involved in the γ to β switch mechanism. (One current model of hemoglobin gene regulation is shown in Fig. 9-17.) The major unanswered question is exactly what mechanism underlies the transition from transcription of one isoform of the β-globin to the next more mature state. From the clinical standpoint, the critical question is how to stimulate the reexpression of the γ (fetal) isoform if the adult β isoform is defective.

SUMMARY

1. Experiments conducted over many years have shown that almost all mammalian cells contain a complete genome, but only a small portion of the genetic material is expressed at any time.
2. In the embryo, cells differentiate through several mechanisms. According to lineage models, a cell and its progeny contain sufficient intrinsic information to determine their phenotypic character. According to environmental models, cells respond to external signals and differentiate accordingly. A third option is that cell differentiation is controlled by both intrinsic and extrinsic information. Differentiation is the process by which a determined cell becomes structurally and functionally specialized.
3. Developmental isoforms are successive units in a series of replacements in which one molecule or structure takes over from one of a similar character. Each isoform is adapted to slightly different needs during development.
4. Gene regulation, which underlies differentiation, involves a variety of promoters, enhancers, and suppressors. Transcription factors, which bind to specific sequences on DNA (e.g., on enhancers or promoters), are important in gene regulation. Several classes of transcription factors have been identified. Some of these are helix-loop-helix proteins, zinc finger proteins, and homeobox sequences. Gene regulation can be accomplished at a number of levels past that of transcription.
5. Skeletal muscle undergoes a sequence of differentiation from mononuclear myoblasts, their fusion to form multinucleated myotubes, and final maturation into skeletal muscle fibers. Mononucleated reserve cells (satellite cells) can proliferate and fuse to growing or mechanically stressed muscle fibers.
6. Several transcription factors of the MyoD family can stimulate cells to differentiate into myoblasts. Other regulatory factors can activate (E12) or inhibit (Id) the activities of muscle transcription factors. Early

myoblasts are kept in the cell cycle by growth factors, such as fibroblast growth factor and transforming growth factor-β. Other growth factors promote muscle differentiation.

7. The first muscle fibers to form are primary muscle fibers. Secondary muscle fibers form around them. Innervation by motor nerve fibers is necessary for the full differentiation of muscle fibers. During the differentiative process, several sets of isoforms of myosin subunits and other contractile proteins appear in sequence in the muscle fibers.
8. Blood formation begins in blood islands in the wall of the yolk sac. The site of hematopoiesis later shifts to the liver and bone marrow. Both red and white blood cells pass through many intermediate stages before maturing. Some early stages (colony-forming units) are responsive to specific growth factors such as erythropoietin.
9. During the course of their differentiation, red blood cells pass through the following stages: proerythroblast (containing much hemoglobin mRNA but little hemoglobin), basophilic, polychromatophilic, and orthochromatic erythroblasts (which contain increasing amounts of hemoglobin). The nucleus is extruded at the orthochromatic erythroblast stage, and the cell is released into the blood as a reticulocyte, which soon matures into an erythrocyte.
10. Hemoglobin undergoes several isoform transitions during blood development. The corresponding stages of cellular differentiation are the embryonic, fetal, and adult stages, and each is characterized by a specific type of hemoglobin with unique arrangements of the four polypeptide chains that are part of the molecule.
11. Hemoglobin synthesis is highly regulated at the genetic level. Hypersensitive sites in the locus control region are important in the expression of specific constituent polypeptide chains of the hemoglobin molecule. Regulation of globin gene expression is still not completely understood.

REVIEW QUESTIONS

1. A teratoma, which arises from a single cell, contains many different types of tissues, including hair, muscle, and cartilage. Under certain circumstances, true bone can form within a muscle. Tissue culture cells treated with 5-azacytidine can form skeletal muscle, cartilage, or fat cells. What major point is illustrated by these examples?
2. What are two cellular-level developmental isoforms?
3. Many molecular geneticists are looking for regulatory factors that control the expression of specific polypeptide components of hemoglobin. Why is this important?

REFERENCES

Behringer RR and others: Human gamma- to beta-globin gene switching in transgenic mice, *Genes Dev* 4:379-389, 1990.

Brown MS: *Fetal and neonatal erythropoiesis.* In Stockman JA, Pochedly C, eds: *Developmental and neonatal hematology,* New York, 1988, Raven pp 39-56.

Caplan AI, Fiszman MY, Eppenberger HME: Molecular and cell isoforms during development, *Science* 221:921-927, 1983.

Davis RL, Weintraub H, Lassar AB: Expression of a single transfected cDNA converts fibroblasts to myoblasts, *Cell* 51:987-1000, 1987.

Dexter TM, Garland JM, Testa NG, eds: *Colony stimulating factors,* New York, 1990, Dekker.

Dieterlen-Lievre F: Embryonic chimeras and hemopoietic system development, *Bone Marrow Transplant* 9(suppl 1):30-35, 1992.

Dillon N, Grosveld F: Human gamma-globin genes silenced independently of other genes in the beta-globin locus, *Nature* 350:252-254, 1991.

Epstein HF, Fischman DA: Molecular analysis of protein assembly in muscle development, *Science* 251:1039-1044, 1991.

Florini JR, Magri KA: Effects of growth factors on myogenic differentiation, *Am J Physiol* 256:C701-C711, 1989.

Garrell J, Campuzano S: The helix-loop-helix domain: a common motif for bristles, muscles and sex, *Bioessays* 13:493-498, 1991.

Grieshammer U, Sassoon D, Rosenthal N: A transgene target for positional regulators marks early rostrocaudal specification of myogenic lineages, *Cell* 69:79-93, 1992.

Holliday R: A different kind of inheritance, *Sci Am* Jun 1989, pp 60-73.

Holtzer H, Biehl J, Holtzer S: *Induction-dependent and lineage-dependent models for cell diversification are mutually exclusive.* In Evans AE and others, eds: *Advances in neuroblastoma research,* New York, 1985, Liss, pp 3-11.

Kedes LH, Stockdale FE, eds: *Cellular and molecular biology of muscle development,* New York, 1989, Liss.

Ley TJ: The pharmacology of hemoglobin switching: of mice and men, *Blood* 77:1146-1152, 1991.

Miller JB: Myoblasts, myosins, MyoDs, and the diversification of muscle fibers, *Neuromuscular Disorders* 1:7-17, 1991.

Nienhuis AW, Benz EJ: Regulation of hemoglobin synthesis during the development of the red cell, *New Engl J Med* 297:1318-1328, 1371-1381, 1430-1436, 1977.

Olson EN: MyoD family: a paradigm for development? *Genes Dev* 4:1454-1461, 1990.

Olson EN: Regulation of muscle transcription by the MyoD family, *Circ Res* 72:1-6, 1993.

Olson EN and others: Molecular control of myogenesis: antagonism between growth and differentiation, *Mol Cell Biochem* 104:7-13, 1991.

Orkin SH: Globin gene regulation and switching: circa 1990, *Cell* 63:665-672, 1990.

Townes TM, Behringer RR: Human globin locus activation region (LAR): role in temporal control, *Trends Genet* 6:219-223, 1990.

Weintraub H and others: The myoD gene family: nodal point during specification of the muscle cell lineage, *Science* 251:761-766, 1991.

P[illegible] TWO

Development of the Body Systems

CHAPTER 10

Integumentary, Skeletal, and Muscular Systems

Just as individual cells undergo cytodifferentiation, tissues undergo characteristic sequences of developmental changes that are often lumped under the term **histogenesis.** This chapter outlines the histogenesis of the integumentary system, the skeleton, and the muscles.

THE INTEGUMENTARY SYSTEM

The skin, consisting of the epidermis and dermis, is one of the largest structures in the body. The epidermis represents the interface between the body and its external environment, and its structure is well adapted for local functional requirements. Simple inspection of areas such as the scalp and palm of the hand shows that the structure of the integument varies from one part of the body to another. Many of these local variations are the result of inductive interactions between the epidermis and the underlying dermis.

The Epidermis

Structural Development. The outer layer of the skin begins as a single layer of ectodermal cells (Fig. 10-1, *A*). As development progresses, the ectoderm becomes multilayered and regional differences in structure and the rate of development become apparent.

The first stage in epidermal layering is the formation of a thin outer layer of flattened cells known as the **periderm** at the end of the first month of gestation (Fig. 10-1, *B*). Cells of the periderm, which is present in the epidermis of all amniote embryos, appear to be involved in the exchange of water, sodium, and possibly glucose between the amniotic fluid and the epidermis.

By the third month the epidermis becomes a three-layered structure, with a mitotically active **basal** (or **germinative**) **layer,** an **intermediate layer** of cells that represent progeny of the dividing cells of the basal layer, and a superficial layer of peridermal cells bearing characteristic surface blebs (Fig. 10-2). Peridermal cells contain large amounts of glycogen, but its function remains uncertain.

During the sixth month, the epidermis beneath the periderm undergoes differentiation into the definitive layers characteristic of the postnatal epidermis. At this point, the peridermal cells are sloughed into the amniotic fluid, and the epidermis becomes a barrier between the fetus and the outside environment instead of a participant in exchange between the two. The change in function of the fetal epidermis may have adaptive value, since it occurs at about the time when urinary wastes begin to accumulate in the amniotic fluid.

Immigrant Cells in the Epidermis. Despite its homogeneous histological appearance, the epidermis is really a cellular mosaic, with contributions from cells derived from other regions of the body. These cells play important roles in the function of the skin.

Early in the second month, **melanoblasts** derived from the neural crest migrate into the embryonic dermis; slightly later, they migrate into the epidermis. Although melanoblasts can be recognized early by staining with a monoclonal antibody (HMB-45, which reacts with a cytoplasmic antigen common to melanoblasts and **melanomas** [pigment cell tumors]), these cells do not begin to produce recognizable amounts of pigment until midpregnancy. This occurs earlier in heavily pigmented individuals than in light-complected whites. The differentiation of melanoblasts into mature **melanocytes** involves the formation of pigment granules called **melanosomes** from **premelanosomes.**

The number of pigment cells in the skin does not differ greatly among the various races, but the melanocytes of dark-skinned individuals contain more pigment granules per cell. **Albinism** is a genetic trait characterized by the lack

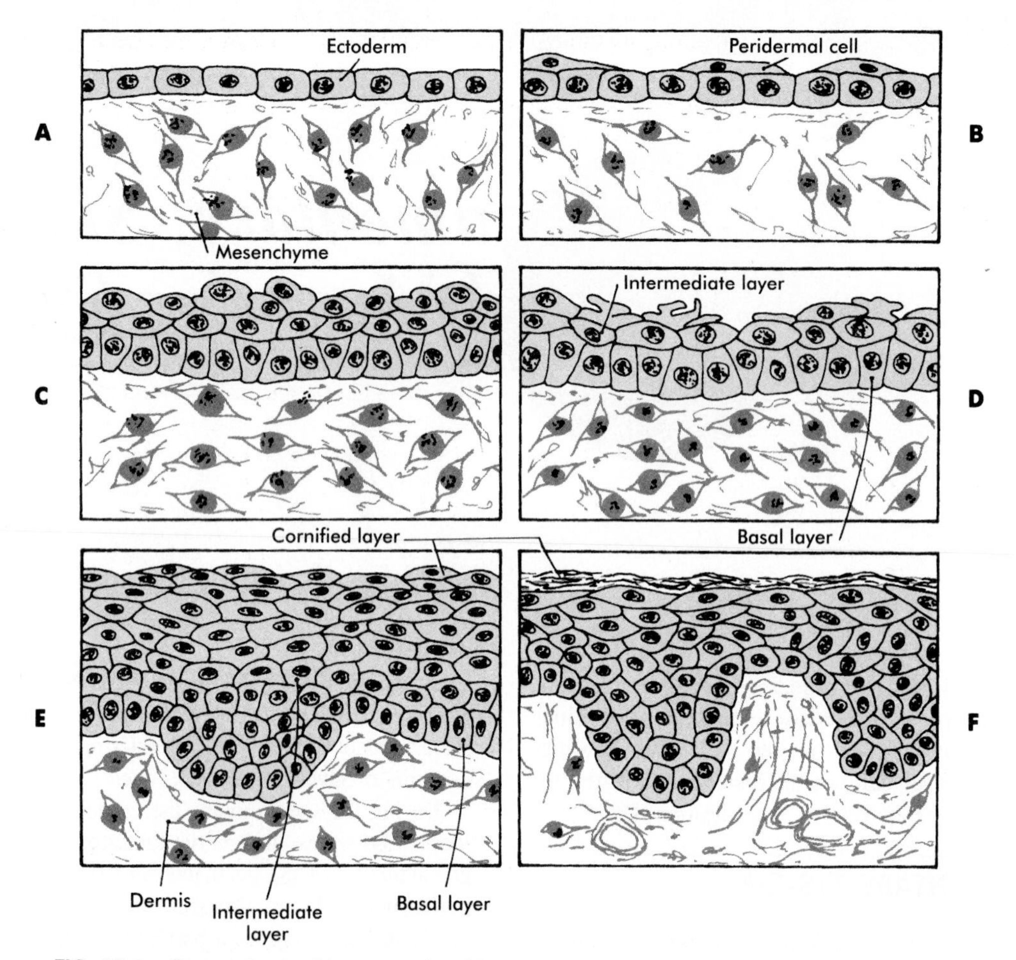

FIG. 10-1 Stages in the histogenesis of human skin. **A,** At 1 month. **B,** At 2 months. **C,** At 2½ months. **D,** At 4 months. **E,** At 6 months. **F,** Postnatal skin.

(Modified from Carlson B: *Patten's foundations of embryology,* ed 5, New York, 1988, McGraw-Hill.)

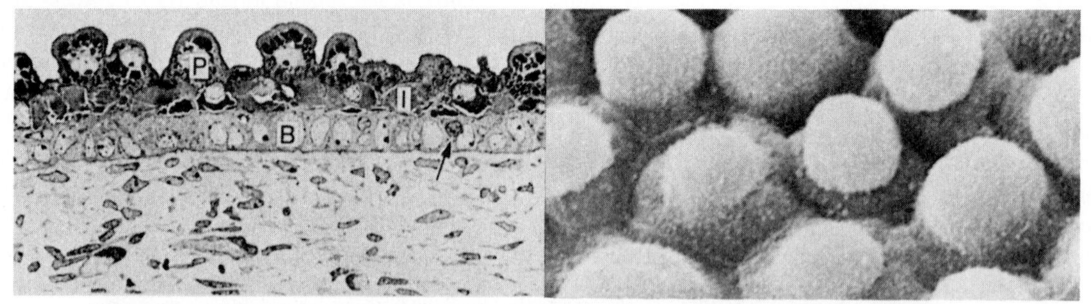

FIG. 10-2 Light micrograph (left) and scanning electron micrograph *(right)* of the epidermis of a 10-week-old human embryo. The prominent surface blebs seen in the scanning micrograph are represented by the irregular surface of the periderm *(P)* in the light micrograph. The arrow in the upper figure points to a melanocyte in the basal layer *(B)* of the epidermis. *I,* Intermediate layer of epidermis.

(From Sybert VP, Holbrook KA. In Reed G, Claireaux A, Bain A: *Diseases of the fetus and newborn,* St Louis, 1989, Mosby.)

of pigmentation, but albinos typically contain normal numbers of melanocytes in their skin. The melanocytes of albinos are typically unable to express pigmentation because they lack the enzyme **tyrosinase,** which is involved in the conversion of the amino acid tyrosine to **melanin.**

Late in the first trimester the epidermis is invaded by **Langerhans cells,** which arise from precursors in the bone marrow. These cells are peripheral components of the immune system and appear to be involved in the presentation of antigens; they cooperate with T lymphocytes (white blood cells involved in cellular immune responses) in the skin to initiate cell-mediated responses against foreign antigens. Although Langerhans cells are not readily evident in ordinary histological preparations, they can be distinguished through antibodies directed toward cell-specific surface antigens or by histochemical demonstration of their high membrane-bound ATPase activity. Langerhans cells are present in low numbers (about 65 cells/mm^2 of epidermis) during the first two trimesters of pregnancy, but then their numbers increase to 2% to 6% of the total number of epidermal cells in the adult.

A third cell type in the epidermis, the **Merkel cell,** is of uncertain origin. Some evidence suggests that Merkel cells may differentiate from epidermal cell precursors, but experiments on birds involving the grafting of marked neural crest tissues indicate that precursors of Merkel cells migrate into the limb from the neural crest. These cells, which appear in palmar and plantar epidermis as early as 8 to 12 weeks, are associated with free nerve terminals. They function as slow-adapting mechanoreceptors in the skin, but cytochemical evidence suggests that they may also function as neuroendocrine cells at some stage in their life history.

Epidermal Differentiation. Once the multilayered epidermis becomes established, a regular cellular organization and sequence of differentiation appears within it (Fig. 10-3). Relatively unspecialized cells of the basal layer **(stra-**

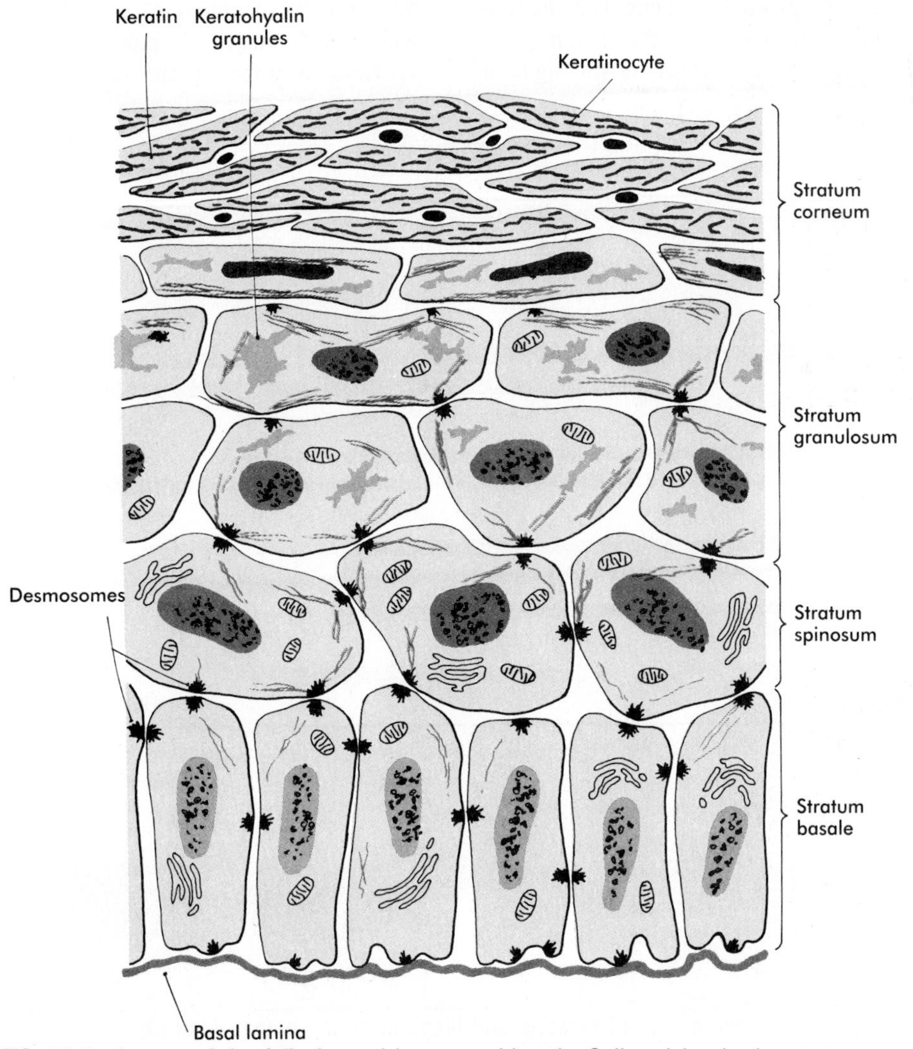

FIG. 10-3 Layers of the fully formed human epidermis. Cells arising in the stratum basale undergo terminal differentiation into keratinocytes as they move toward the surface.

(Modified from Carlson B: *Patten's foundations of embryology,* ed 5, New York, 1988, McGraw-Hill.)

tum basale) divide and contribute daughter cells to the next layer, the **stratum spinosum**. The movement of epidermal cells away from the basal layer is preceded by a loss of adhesiveness to basal lamina components (e.g., fibronectin, laminin, and collagen types I and IV). These cellular properties can be explained by the loss of several integrins (membrane proteins that mediate the attachment of cells to extracellular matrix molecules). Cells of the stratum spinosum produce prominent bundles of **keratin** filaments, which converge on the patchlike desmosomes binding the cells to one another.

Keratohyalin granules, another marker of epidermal differentiation, begin to appear in the cytoplasm of the outer, more mature cells of the stratum spinosum and are prominent components of the **stratum granulosum.** Recent research has shown that keratohyalin granules are actually composed of two types of protein aggregates—one histidine rich and one sulfur rich—closely associated with bundles of keratin filaments. Because of their high content of keratin, epidermal cells are given the generic name **keratinocytes.** As the keratinocytes move into the stratum granulosum, their nuclei begin to show characteristic signs of terminal differentiation, such as a flattened appearance, dense masses of nuclear chromatin, and early signs of breaking up of the nuclear membrane. In these cells the bundles of keratin become much more prominent, and the molecular weights of keratins that are synthesized are higher than in less mature keratinocytes.

As the cells move into the outer layer, the **stratum corneum,** they lose their nuclei and resemble very flattened bags densely packed with keratin filaments. The cells of this layer are interconnected by the histidine-rich protein **filaggrin,** which is derived from one of the granular components of keratohyalin. Depending on the region of the body surface, the cells of the stratum corneum accumulate to form approximately 15 to 20 layers of dead cells. In postnatal life, whether through friction or the degradation of the desmosomes and filaggrin, these cells are eventually shed (e.g., about 1300 cells/cm^2/hr in the human forearm) and commonly accumulate as house dust.

Biochemical work has correlated the expression of keratin proteins (members of a complex family of proteins) with specific stages of epidermal differentiation (Fig. 10-4). Keratins of the intermediate family of cellular filaments

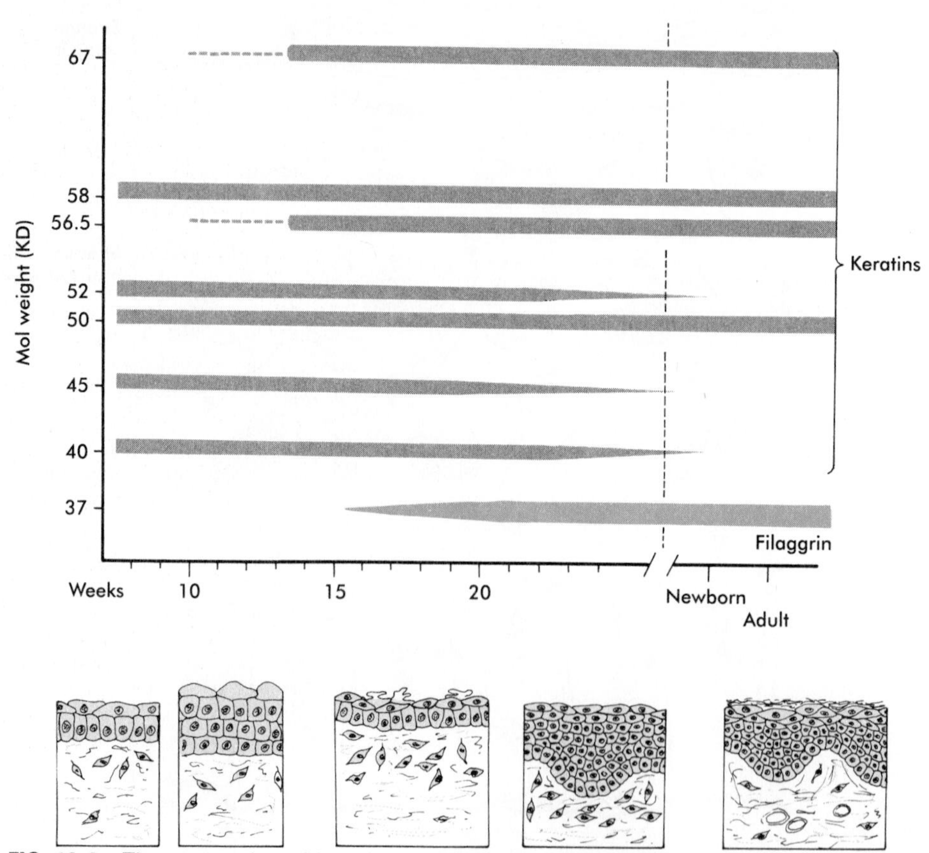

FIG. 10-4 The expression of keratins and filaggrin during human fetal skin development. The lower figures show approximate stages in skin development at the times indicated.

(Graph modified from Dale BA and others: *Cell Biol* 101:1257-1269, 1985.)

are first expressed in cells of the two-layered epidermis during the second month of pregnancy. Three of the keratins (40-, 45- and 52-KD) are characteristic of simple epithelia, and the other two (50- and 58-KD) are typically seen in stratified epithelia. As the epidermis begins to stratify (at 9 to 12 weeks gestation), the outer cells of the intermediate layer begin to express small amounts of 56.5- and 67-KD keratins, which are characteristic of keratinized epidermis. In succeeding weeks, the amounts of these two keratins increase greatly, whereas the prominence of the simple epithelial keratins (40-, 45- and 52-KD) declines during the late fetal period. The expression of filaggrin, the intercellular binding protein, is closely correlated with the later differentiation of the outer cornified layers of the fetal epidermis.

Proliferation of basal epidermal cells is under the control of a variety of growth factors. Some of these stimulate and others inhibit mitosis (box). Keratinocytes commonly spend about 4 weeks in their passage from the basal layer of the epidermis to ultimate desquamation, but in some skin diseases such as **psoriasis,** epidermal cell proliferation is poorly controlled and keratinocytes may be shed within a week after their generation.

One of the prominent features of the skin, particularly the thick skin of the palms and soles, is the presence of epidermal ridges and creases. On the tips of the digits, the ridges form loops and whorls in patterns that are unique to the individual. These patterns form the basis for the science of **dermatoglyphics,** in which the patterns constitute the basis for genetic analysis or criminal investigation.

Peptide Factors That Affect Keratinocyte Proliferation

Stimulators

Epidermal growth factor
Transforming growth factor-α
Insulin
Insulinlike growth factor I, II
Fibroblast growth factor, acidic
Fibroblast growth factor, basic
Interleukin-1/ETAF

Inhibitors

Transforming growth factor-β_1
Transforming growth factor-β_2
Interferon-α/β_1
Interferon-β_2
Interferon
Tumor necrosis factor

From Holbrook K: In Goldsmith LA, ed: *Physiology, biochemistry and molecular biology of the skin,* vol 1, New York, 1991, Oxford University.

Formation of epidermal ridges is closely associated with the earlier appearance of **volar pads** on the ventral surfaces of the fingers and toes (Fig. 10-5). Volar pools first form on the palms at about 6½ weeks, and by 7½ weeks they have formed on the fingers. The volar pads begin to regress by about 10½ weeks, but while they are present, they set

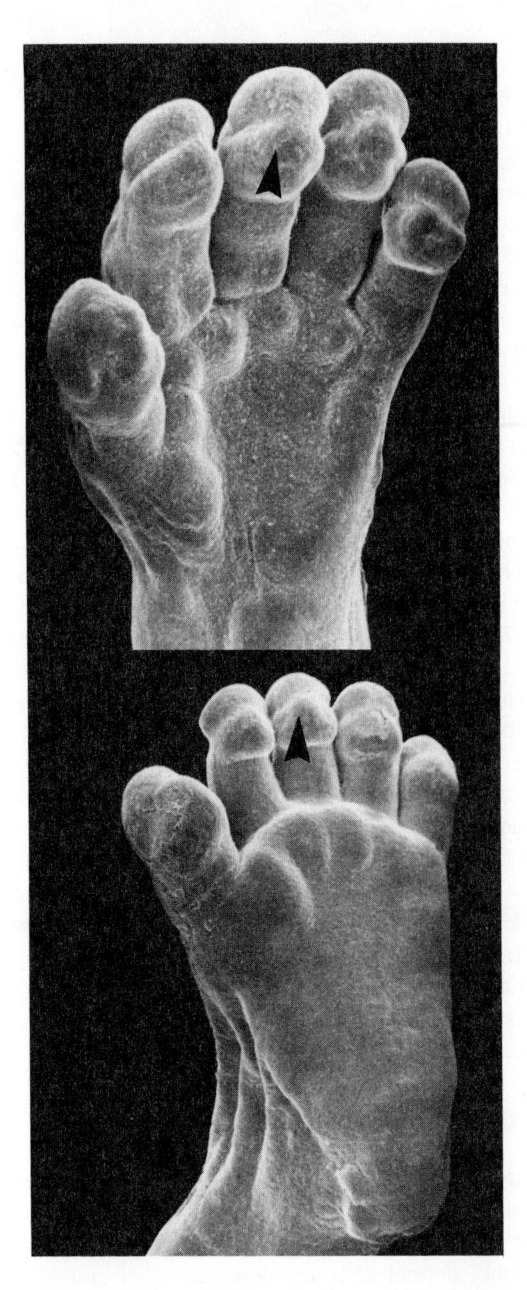

FIG. 10-5 Scanning electron micrographs of the ventral surfaces of **(A)** the hand and **(B)** the foot of a human embryo at the end of the second month. Volar pads are prominent near the tips of the digits *(arrowheads).*
(From Jirásek J: *Atlas of human prenatal morphogenesis,* Amsterdam, 1983, Martinus Nijhoff Publishers.)

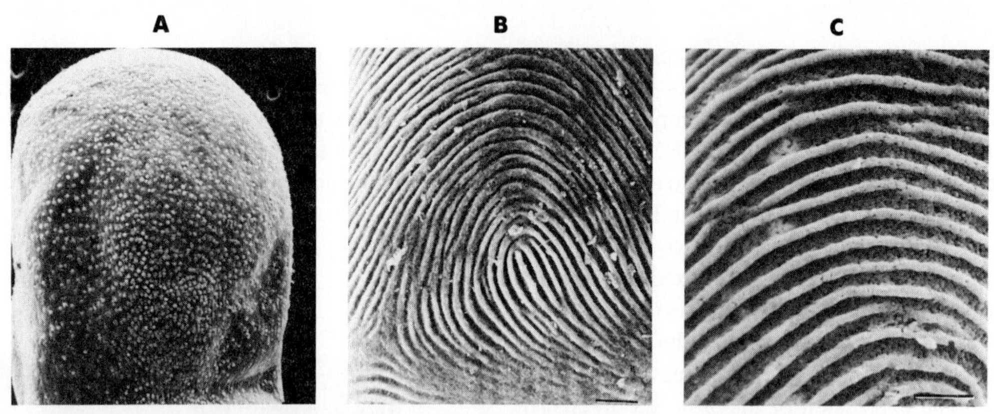

FIG. 10-6 Scanning electron micrographs of human digital palmar skin in a 14-week-old fetus. **A,** Low-power view of the palmar surface of a digit. **B,** Epidermal surface of the dermis of the fingertip showing primary dermal ridges. **C,** Basal surface of the epidermis showing the epidermal ridges. *Bars,* 100 μm.
(From Misumi Y, Akiyoshi T: *Am J Anat* 119:419, 1991.)

the stage for the formation of the epidermal ridges, which occurs between 11 and 17 weeks. Similar events in the foot are delayed approximately a week later than those in the hand.

The pattern of the epidermal ridges is correlated with the morphology of the volar pads when the ridges first form. If a volar pad is high and round, the epidermal ridges form a whorl; if the pad is low, an arch results. A pad of intermediate height results in a loop configuration of the digital epidermal ridges. The timing of ridge formation also appears to influence the morphology: early formation of ridges is associated with whorls and late formation with arches. The primary basis for dermatoglyphic patterns, however, is still not understood.

When the epidermal ridges first form, the tips of the digits are still smooth, and the fetal epidermis is covered with peridermal cells. Beneath the smooth surface, however, epidermal and dermal ridges begin to take shape (Fig. 10-6). Late in the fifth month of pregnancy the epidermal ridges become recognizable features of the surface landscape.

The Dermis

The **dermis** arises from mesodermal cells derived from the dermatome of the somites or other mesenchymal cells located just beneath the ectoderm. In the face and parts of the neck, dermal cells are descendants of cranial neural crest ectoderm (see Fig. 13-8).

The future dermis is initially represented by loosely aggregated mesenchymal cells that are highly interconnected by focal tight junctions on their cellular processes. These early dermal precursors secrete a watery intercellular matrix rich in glycogen and hyaluronic acid.

Early in the third month, the developing dermis undergoes a transition from the highly cellular embryonic form to a state characterized by the differentiation of the mesenchymal cells into fibroblasts and the formation of increasing amounts of a fibrous intercellular matrix. The principal types of fibers are types I and III collagen and elastic fibers. The dermis becomes highly vascularized, with an early capillary network transformed into layers of larger vessels. Shortly after the eighth week, sensory nerves growing into the dermis and epidermis help to complete reflex arcs allowing the fetus to respond to pressure and stroking.

Dermal/Epidermal Interactions. The transformation of simple ectoderm into a multilayered epidermis depends on continuing inductive interactions with the underlying dermis. Dermal/epidermal interactions are also the basis for the formation of a wide variety of epidermal appendages and the appearance of regional variations in the structure of the epidermis.

For example, early in development the epidermis covering the palms of the hands and soles of the feet becomes significantly thicker than that elsewhere in the body. These regions also do not produce hairs, whereas hairs of some sort, whether coarse or extremely fine, form in regular patterns from the epidermis throughout most of the rest of the body.

Tissue recombination experiments on a variety of vertebrate species have shown that the underlying dermis determines the course of development of the epidermis and its derivatives and that the ectoderm also influences the developmental course of the dermis. If the early ectodermal and mesenchymal components of the skin are dissociated and grown separately, the ectodermal component remains simple ectoderm without differentiating into a multilayered epidermis with appropriate epidermal appendages. Similarly,

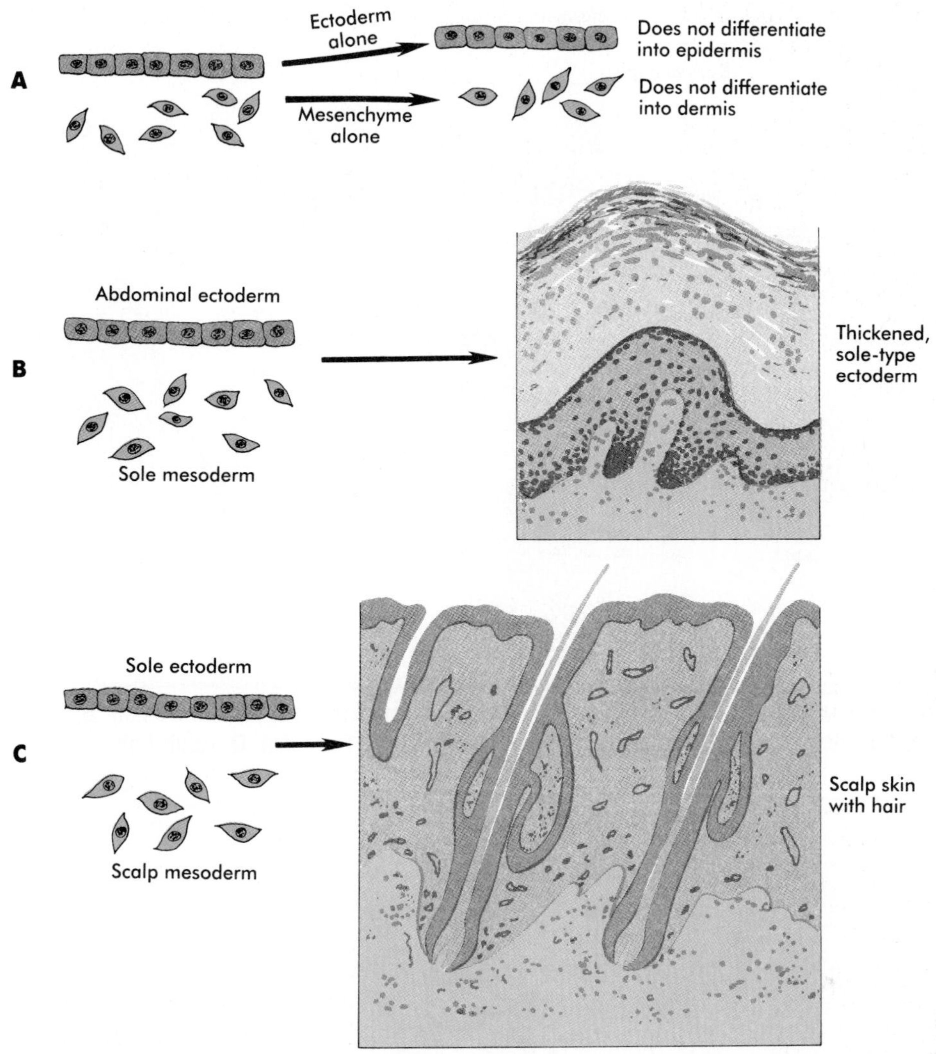

FIG. 10-7 Recombination experiments illustrating the importance of tissue interactions in the differentiation of the skin. **A,** When separated, ectoderm and underlying mesenchyme do not differentiate. Recombinations **(B)** and **(C)** show that the dermis determines the nature of the ectodermal differentiation.

isolated subectodermal mesenchyme retains its embryonic character without differentiating into dermis.

If ectoderm from one part of the body is combined with dermis from another area, the ectoderm differentiates into a regional pattern characteristic for that of the underlying dermis rather than that appropriate for the site of origin of the ectoderm (Fig. 10-7). Cross-species recombination experiments (see Fig. 11-11) have shown that even in distantly related animals, skin ectoderm and mesenchyme can respond to one another's inductive signals.

Epidermal Appendages

As the result of inductive influences of the dermis, the epidermis produces a wide variety of appendages such as hair, nails, sweat and sebaceous glands, mammary glands, and even the enamel component of teeth. (The development of teeth is discussed in Chapter 15.)

Hair. Hairs are specialized epidermal derivatives that arise as the result of inductive stimuli from the dermis. There are many types of hairs, ranging from the coarse hairs of the eyelashes and eyebrows to the barely visible hairs on the abdomen and back. Regional differences in morphology and patterns of distribution of the hairs are imposed on the epidermis by the underlying dermis.

Hair formation is first recognizable at about the twelfth week of pregnancy as regularly spaced epidermal downgrowths associated with small condensations of dermal cells called **dermal papillae** (Fig. 10-8). Under the continuing influence of the dermal papilla, the epidermal downgrowth

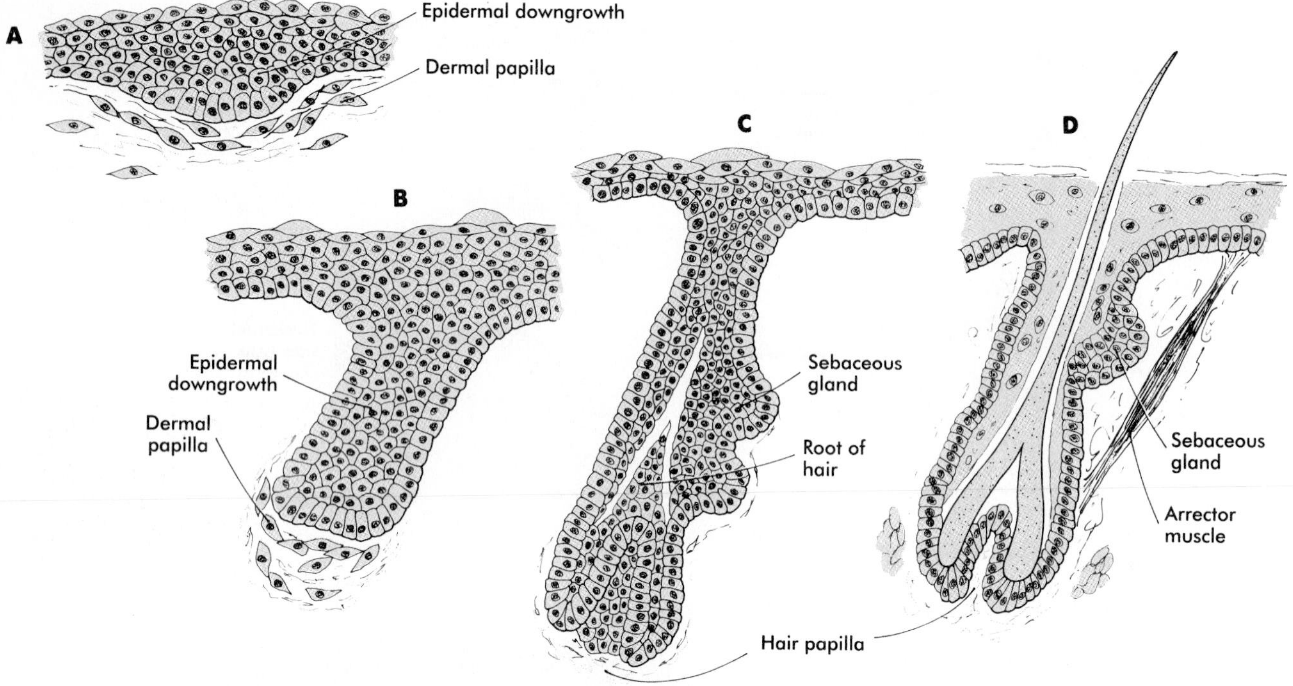

FIG. 10-8 Differentiation of a human hair follicle. **A,** Hair primordium (12 weeks). **B,** Early hair peg (15 to 16 weeks). **C,** Bulbous hair follicle (18 weeks). **D,** Adult hair.

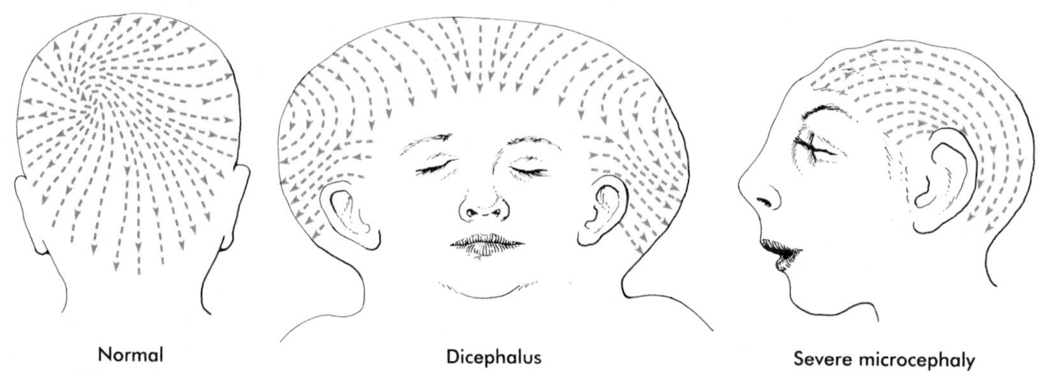

FIG. 10-9 Patterns of whorls of hair in normal and abnormal fetuses.

continues over the next few weeks and forms an early **hair peg.** In succeeding weeks the epidermal peg overgrows the dermal papilla, resulting in the shaping of an early **hair follicle.** At this stage the hair follicle still does not protrude beyond the outer surface of the epidermis, but in the portion of the follicle that penetrates deeply into the dermis, bulges in the epidermis presage the formation of **sebaceous glands,** which secrete an oily skin lubricant **(sebum)** and are the attachment site for the tiny **arrector pili muscle.** The arrector pili is a smooth muscle that lifts the hair to a near vertical position in a cold environment. In many animals, this increases the insulation properties of the hair.

Later differentiation of hairs is a leisurely process. Erupted hairs are first seen on the eyebrows shortly after 16 weeks. Within a couple of weeks, they cover the scalp. The eruption of hairs follows a cephalocaudal gradient over the body. During the later stages of hair formation, the hair bulb becomes infiltrated with melanocytes, which provide color to the hair. Starting around the fifth month, the epidermal cells of the hair shaft begin to undergo keratinization, forming firm granules of **trichohyalin,** which imparts hardness to the hair.

Products of the fetal sebaceous glands accumulate on the surface of the skin as **vernix caseosa.** This substance may serve as a protective coating for epidermis, which is continually exposed to the amniotic fluid. The first fetal hairs

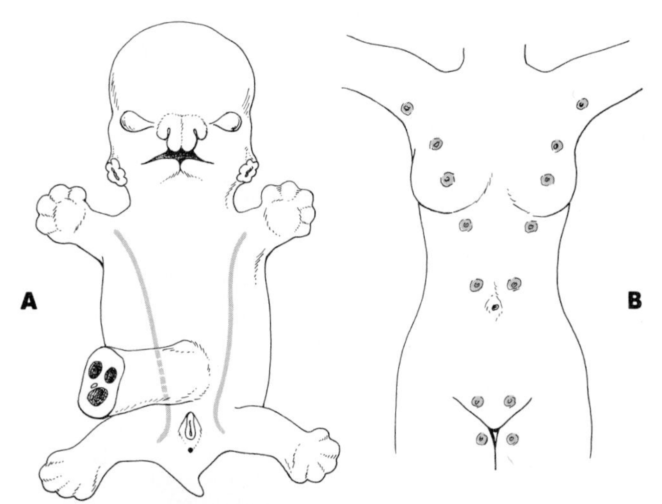

FIG. 10-10 **A,** The milk lines *(blue)* in a generalized mammalian embryo. Mammary glands form along these lines. **B,** Common sites of formation of supernumerary nipples or mammary glands along the course of the milk lines in the human.

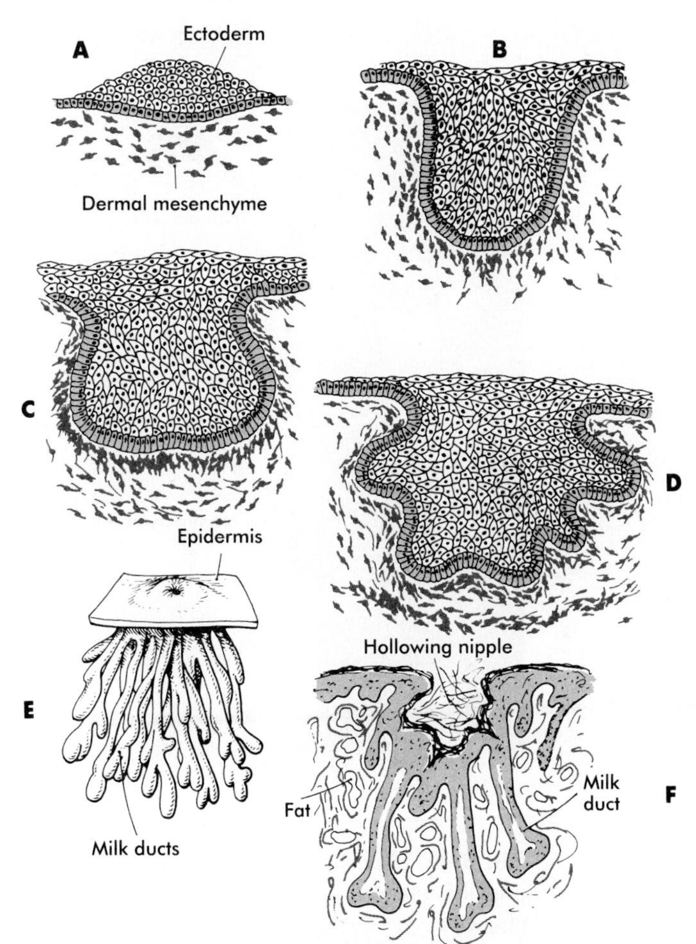

FIG. 10-11 Stages in the embryonic development of the human mammary gland. **A,** Sixth week. **B,** Seventh week. **C,** Tenth week. **D,** Fourth month. **E,** Sixth month. **F,** Eighth month.

are very fine in texture and are close together. Known as **lanugo,** they are most prominent during the seventh and eighth months. Lanugo hairs are typically shed just before birth and are replaced by coarser definitive hairs, which arise from newly formed follicles as tissue-level isoforms.

The pattern of epidermal appendages such as hairs has been shown experimentally to relate to patterns generated in the dermis. Other studies have compared patterns of scalp hairs between normal embryos and those with cranial malformations (Fig. 10-9) and have shown a correlation between whorls and the direction of hair growth and the tension on the epidermis at the time of formation of the hair follicles.

Mammary Glands. As with many glandular structures, the mammary glands arise as epithelial (in this case, ectodermal) downgrowths into mesenchyme in response to inductive influences by the mesenchyme. The first morphological evidence of mammary gland development is the appearance of two bands of ectodermal thickenings called **milk lines** running along the ventrolateral body walls in embryos of both sexes at about 6 weeks (Fig. 10-10, *A*). The craniocaudal level and the extent along the milk lines at which mammary tissue develops vary among species. Comparing the location of mammary tissue in cows (caudal), humans (in the pectoral region), and dogs (along the length of the milk line) demonstrates the wide variation in location and number of mammary glands. In humans, supernumerary mammary tissue or nipples can be found anywhere along the length of the original milk lines (10-10, *B*).

Mammary ductal epithelial downgrowths (Fig. 10-11) are associated with two types of mesoderm, fibroblastic and fatty. Experimental evidence suggests that inductive interactions with the fatty component of the connective tissue are responsible for the characteristic shaping of the mammary duct system. As with many developing glandular structures, the inductive message seems to be mediated to a great extent by the extracellular matrix of the connective tissue.

Although the mesoderm controls the branching pattern of the ductal epithelium, functional properties of the mammary ducts are intrinsic to the epithelial component. An experiment in which mouse mammary ectoderm was combined with salivary gland mesenchyme illustrates this point. The mammary ducts developed a branching pattern characteristic of salivary gland epithelium, but despite this, the mammary duct cells produced one of the milk proteins, **α-lactalbumin.**

In keeping with their role as secondary sexual characteristics, mammary glands are extremely responsive to the hormonal environment. This has been shown by experiments conducted on mice. In contrast to the continued

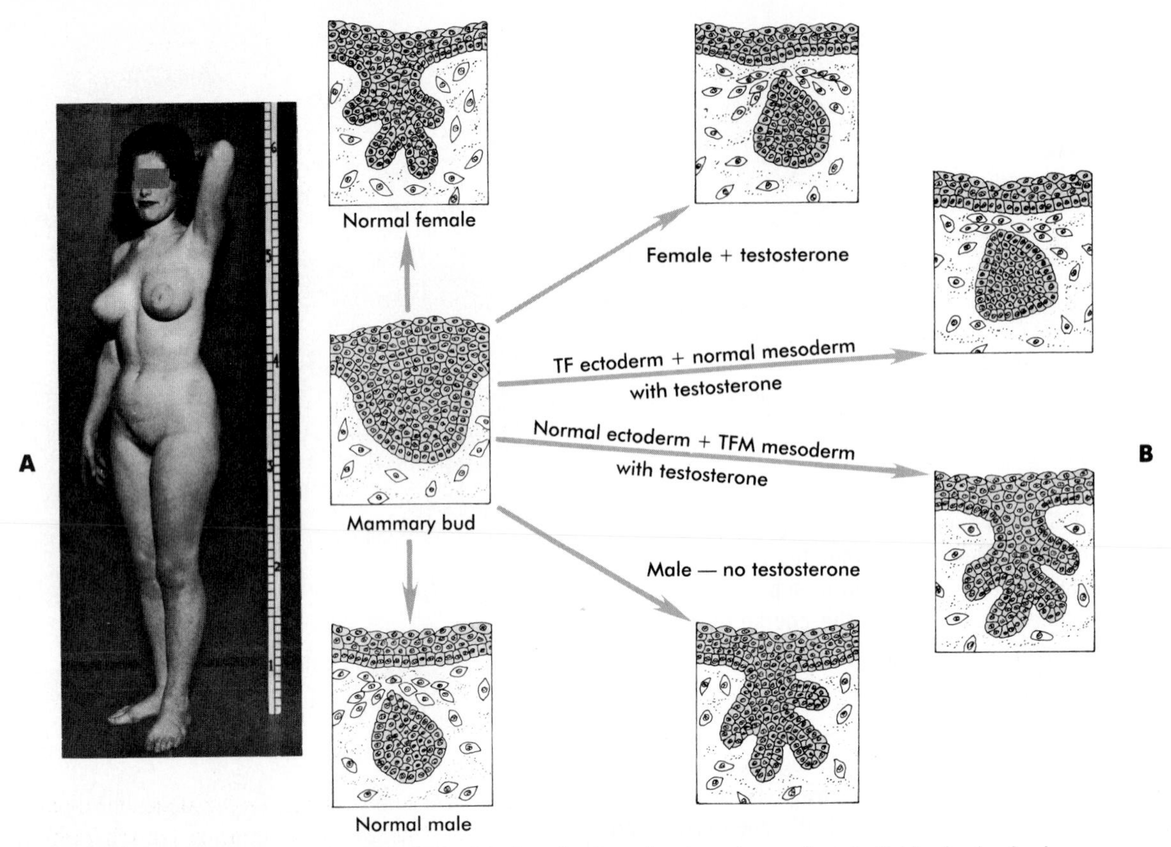

FIG. 10-12 A, Testicular feminization, showing the female phenotype of an individual who had primary amenorrhea. Examination of the gonads after removal revealed immature testicular tubules. (Based on studies by Kratochwil *J Embroyl Exp Morphol* 25:141, 1971.) (From Morris JM, Mahesh VB: *Am J Obstet Gynecol* 87:731,1963.)
B, Roles of genetic specificity and testosterone in the development of mouse mammary gland tissue. With normal female mammary tissue *(top center),* the addition of testosterone causes prospective duct tissue to detach and regress as in normal male development. Conversely, in the absence of testosterone, male ductal primordia *(bottom center)* assume a female configuration. In the testicular feminization *(TFM)* mutant, if normal mammary ectoderm is cultured with TFM mammary mesoderm in the presence of testosterone, mammary duct epithelium continues to develop *(lower right).* If normal male mammary mesoderm is combined with TFM ectoderm in the presence of testoterone, the normal male pattern of separation and regression of mammary duct epithelium occurs *(upper right),* showing that the genetic defect is expressed in TFM mesoderm.

downgrowth of ductal epithelium in females, the mammary ducts in male mice respond to the presence of testosterone by undergoing a rapid involution. Female mammary ducts react similarly if they are exposed to testosterone. Further analysis has shown that the effect of testosterone is mediated through the mammary mesenchyme rather than acting directly on the ductal epithelium. Conversely, if male mammary ducts are allowed to develop in the absence of testosterone, they assume a female morphology.

The role of the mesoderm and **testosterone receptors** is well illustrated in experiments involving mice with a genetic mutant, **androgen insensitivity syndrome.** This is the counterpart of a human condition called the **testicular feminization syndrome,** in which genetic males lack testosterone receptors. Despite high circulating levels of testosterone, these individuals develop female phenotypes, including typical female breast development (Fig. 10-12, *A*), because without receptors, the tissue cannot respond to the testosterone.

In vitro recombination experiments on mice with androgen insensitivity have been instrumental in understanding the role of the mesoderm in mediating the effects of testosterone on mammary duct development (Fig. 10-12, *B*). If mutant mammary ectoderm is combined with normal mesoderm in the presence of testosterone, the mammary ducts regress, but normal ectoderm combined with mutant mesoderm continues to form normal mammary ducts despite being exposed to high levels of testosterone. This shows that

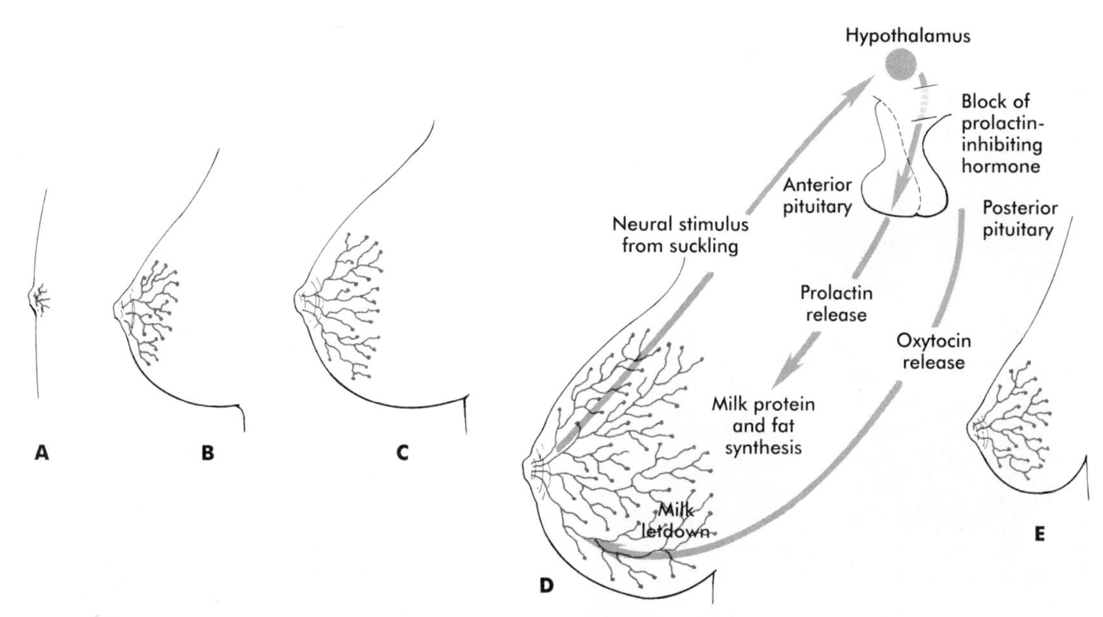

FIG. 10-13 Development of the mammary ducts and hormonal control of mammary gland development and function. **A,** Newborn. **B,** Young adult. **C,** Adult. **D,** Lactating adult. **E,** Postlactation.

the genetic defect in testicular feminization is expressed in the mesoderm.

The postnatal development of female mammary gland tissue is also highly responsive to its hormonal environment. The simple mammary duct system that was laid down in the embryo remains in an infantile condition until it is exposed to the changing hormonal environment at the onset of puberty (Fig. 10-13, *A*). Increasing levels of circulating estrogens and other less prominent hormonal changes stimulate proliferation of the mammary ducts and enlargement of the pad of fatty tissue that underlies it (Fig. 10-13, *B*). The next major change in the complete cycle of mammary tissue development occurs during pregnancy, although minor cyclic changes in mammary tissue are detectable in each menstrual cycle. During pregnancy, increased amounts of progesterone, along with prolactin and placental lactogen, stimulate the development of secretory alveoli at the ends of the branched ducts (Fig. 10-13, *C*). With continuing development of the alveoli, the epithelial cells build up increased numbers of the cytoplasmic organelles that are involved in protein synthesis and secretion.

Lactation involves a number of reciprocal short- and medium-term influences between the mammary glands and the brain. (These are summarized in Fig. 10-13, *D*). Stimulated by prolactin secretion from the anterior pituitary, the alveolar cells synthesize milk proteins (**casein** and α-lactalbumin) and lipids. In a rapid response to the suckling stimulus, the ejection of milk is triggered by the release of **oxytocin** by the posterior portion of the pituitary. Oxytocin causes the contraction of **myoepithelial cells,** which surround the alveoli. Suckling also causes an inhibition of release of luteinizing hormone–releasing hormone by the hypothalamus, resulting in the inhibition of ovulation and a natural form of birth control.

With cessation of nursing, reduced prolactin secretion and the inhibitory effects of nonejected milk in the mammary alveoli results in the cessation of milk production. The mammary alveoli regress, and the duct system of the mammary gland returns to the nonpregnant state (Fig. 10-13, *E*).

Abnormalities of Skin Development

In addition to those mentioned previously, several other types of anomalies affect the integumentary system. **Ectodermal dysplasia** is a germ layer defect that can affect a number of ectodermal derivatives depending on the type and severity of the condition. In addition to abnormalities of the epidermis itself, this syndrome can include the absence or abnormalities of hairs and teeth and short stature (due to anterior pituitary gland maldevelopment).

A number of relatively rare conditions are included among genetically transmitted disorders of keratinization. **Ichthyosis** is characterized by scaling and cracking of a hyperkeratinized epidermis. Disorders of sweat glands are commonly associated with this condition. A more severe autosomal recessive disorder is a **harlequin fetus,** in which epidermal platelike structures form, with deep cracks between the structures. Infants with this condition typically do not survive longer than a few weeks.

Angiomas of the skin (birthmarks) are vascular malformations characterized by localized red or purplish spots

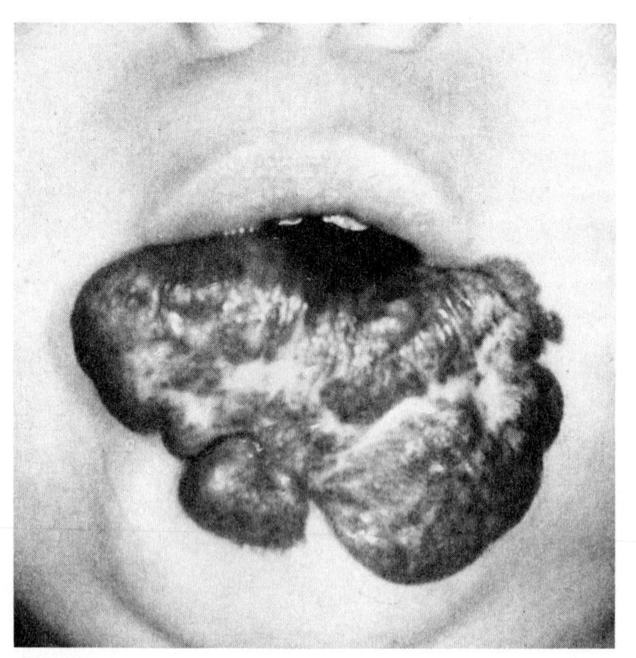

FIG. 10-14 Large hemangioma on the tongue.
(Courtesy A.R. Burdi, Ann Arbor, Mich.)

ranging in size from tiny dots to formations many inches in diameter. Angiomas consist of abnormally prominent plexuses of blood vessels in the dermis, and they may be raised above the level of the skin or a mucous membrane (Fig. 10-14).

THE SKELETON

Skeletal tissue is present in almost all regions of the body, and the individual skeletal elements are quite diverse in morphology and tissue architecture. However, there are some fundamental embryological commonalities despite this diversity.

All skeletal tissue arises from cells with a mesenchymal morphology, but the origins of the mesenchyme vary in different regions of the body. In the trunk the mesenchyme that gives rise to the segmented **axial skeleton** (i.e., that of the vertebral column and ribs) originates from the sclerotomal portion of the somites, whereas the appendicular skeleton (the bones of the limbs and their respective girdles) is derived from mesenchyme of the lateral plate mesoderm. All skeletal tissues of the trunk and appendages are mesodermal in origin.

The origins of the head skeleton are more complex. Some cranial bones (e.g., those making up the roof and much of the base of the skull) are mesodermal in origin, but the facial bones arise from mesenchyme derived from the ectodermal neural crest.

The deep skeletal elements (sometimes called the **endoskeleton**) typically first appear as cartilaginous models of the bones that will ultimately be formed (Fig. 10-15). At specific periods during embryogenesis, the cartilage is replaced by true bone through the process of **endochondral ossification.** In contrast, the superficial bones of the face and skull form by the direct ossification of mesenchymal cells without an intermediate cartilaginous stage **(intramembranous bone formation).** Microscopic details of both intramembranous and endochondral bone formation are presented in the standard histology texts and are not repeated here.

A common element of many mesenchymal cell precursors of skeletal elements is their migration or relative displacement from their site of origin to the area where the bone will ultimately form. The displacement can be relatively minor, such as the aggregation of cells from the sclerotome of the receding somite around the notochord to form the body **(centrum)** of a vertebra, or involve extensive migrations of cranial neural crest cells to their final destinations as membrane bones of the face.

To differentiate into defined skeletal elements, the mesenchymal precursor cells must typically interact with elements of their immediate environment—typically epithelia or their basal laminae—or components of the neighboring extracellular matrix. Details of the interactions vary among regions of the body. In the limb, for example, a continuous interaction between the **apical ectodermal ridge** (see Chapter 11) and the underlying limb bud mesoderm is involved in the specification of the limb skeleton. An inductive interaction between the sclerotome and notochord and/or neural tube initiates skeletogenesis of the vertebral column. In the head, preskeletal cells of the neural crest may receive information at levels ranging from the neural tube itself, to sites along their path of migration, to the region of their final destination. Inductive interactions between regions of the brain and the overlying mesoderm stimulate formation of the membrane bones of the cranial vault.

Regardless of the sources and types of environmental information that are used by the skeletal precursor cells, after they arrive at their final destinations, these cells typically undergo a local condensation before overt signs of differentiation occur. Condensation is the result of a number of incompletely understood influences. Evidence to date suggests that local cell proliferation and migration into areas where condensation is taking place may be less important than the effects of specific cell aggregation factors and local modifications of the extracellular matrix. Among the last may be the local synthesis and accumulation of **chondroitin sulfate** and **heparan sulfate proteoglycans** in sites of preskeletal cellular condensations.

The differentiation of mesenchyme into cartilage or bone follows the same course, regardless of the area of the body in which the process is taking place. Morphogenesis of the individual skeletal elements is often under complex sets of controls, ranging from information gained during the early interactions of mesenchymal precursor cells with epithelia to mechanical influences that can act at any time from the first stages of morphogenesis to late in postnatal life.

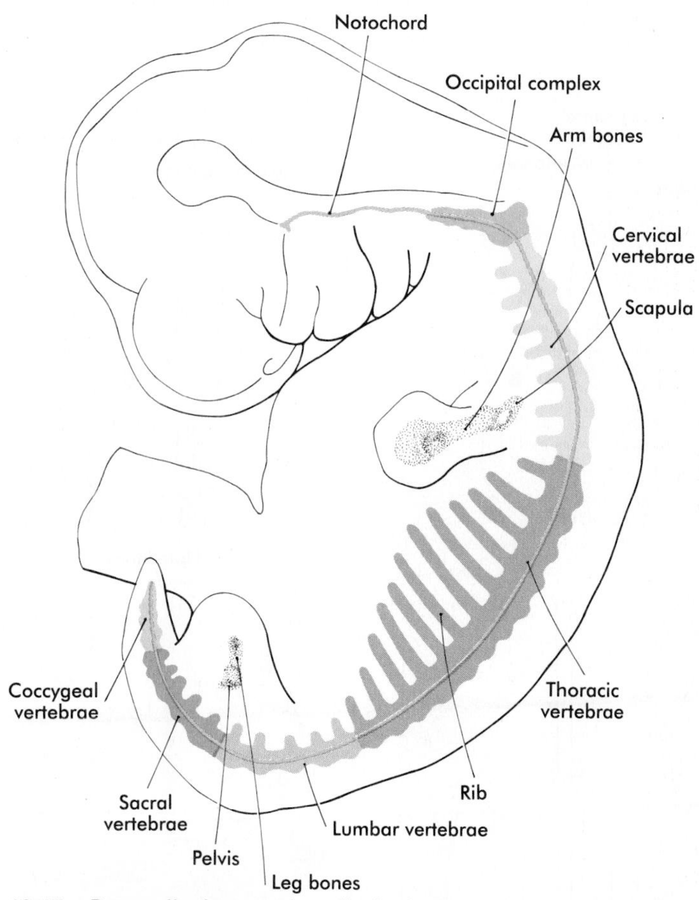

FIG. 10-15 Precartilaginous primordia in the 9-mm-long human embryo.

The Axial Skeleton

Vertebral Column and Ribs. The earliest stages in establishing the axial skeleton were introduced in Chapter 5. Formation of the axial skeleton, however, is more complex than the simple subdivision of the paraxial mesoderm into somites and the medial displacement of sclerotomal cells to form primordia of the vertebrae. Each vertebra has a complex and unique morphology specified by controls operating at several levels and developmental periods. Both the basic morphology of vertebral development and the true extent of the developmental controls are just beginning to be understood.

According to the traditional view of vertebral development (see Fig. 5-13), the sclerotomes split into cranial and caudal halves, and the densely packed caudal half of one sclerotome joins with the loosely packed cranial half of the next to form the centrum of a vertebra. Recent morphological research suggests that vertebral development is more complex than this model, and the scheme illustrated in Fig. 10-16 is becoming increasingly accepted.

The vertebral column is divided into several general areas (see Fig. 10-15): (1) an **occipital region,** which is incorporated into the bony structure of the base of the skull; (2) a **cervical region,** which includes the highly specialized **atlas** and **axis** that link the vertebral column to the skull; (3) the **thoracic region,** from which the true ribs arise; (4) the **lumbar region;** (5) a **sacral region,** in which the vertebrae are fused into a single **sacrum;** and (6) a **caudal region,** which represents the tail in most mammals and the rudimentary **coccyx** in humans. A typical vertebra arises from the fusion of several cartilaginous primordia. The centrum, which is derived from the sclerotomal positions of the paired somites, surrounds the notochord and serves as a bony floor for the spinal cord (Fig. 10-17). The independently arising **neural arches** fuse on either side with the centrum and above with other neural arches form a protective roof over the spinal cord. Incomplete closure of the bony roof results in a common anomaly called **spina bifida occulta** (see Fig. 12-38). The **costal process** forms the true ribs at the level of the thoracic vertebrae. At other levels along the vertebral column the costal processes become incorporated into the vertebrae proper.

There is increasing evidence that the fundamental regional characteristics of the vertebrae are specified by the actions of discrete combinations of homeobox-containing genes (Fig. 10-18). Expression of the *Hox* genes begins

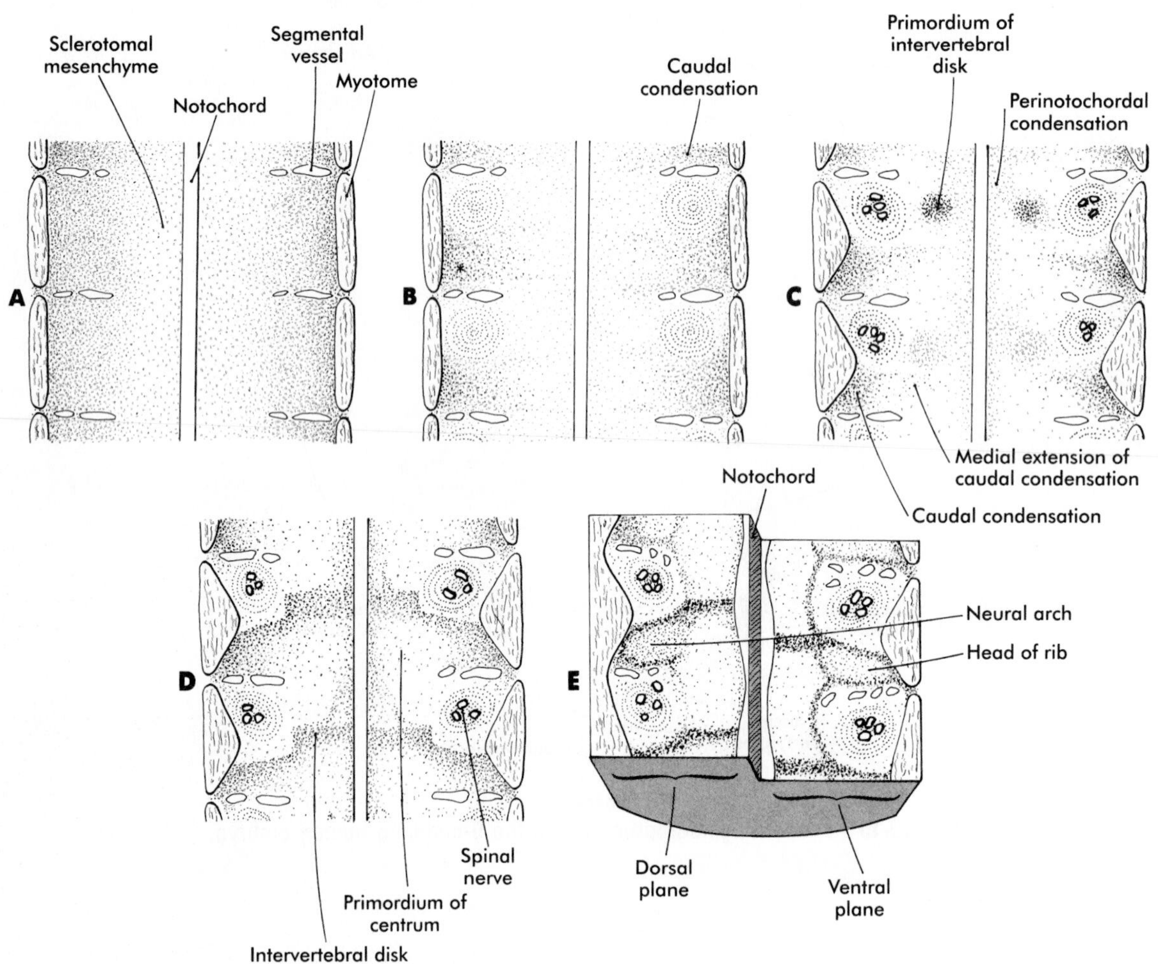

FIG. 10-16 Frontal sections through the developing vertebral column. **A,** Early stage, showing sclerotomal mesenchyme with a lesser density of cells around the notochord than laterally. The segmental vessels mark the division between adjacent somites. **B,** The lateral sclerotomal mesenchyme shows a band of greater density *(asterisk)* in the caudal half. The density of cells around the notochord is still low. **C,** A transverse band of densely packed cells in the cranial portion of the sclerotome represents the primordium of the intervertebral disk. Although a thin layer of densely packed cells surrounds the notochord, the bulk of the primordium of the centrum of the vertebra is a triangular mass of loosely packed cells. The condensed primordium of the neural arch and transverse process extends diagonally from just caudal to the intervertebral disk to the caudal edge of the lateral border of the sclerotome. **D,** All components described in **C** are better defined. The centrum arises from mesenchyme extending continuously from the caudal half of one somite to the cranial half of the next. (Note the position of the intersegmental vessels.) The spinal nerves pass through the cranial half of the sclerotome. **E,** The left side shows a dorsal level through the developing vertebrae, with the neural arch visible. The right shows a more ventral plane, with the head of the rib articulating with the centrum below the level of the neural arch and transverse process.

(Based on studies by Verbout [1985].)

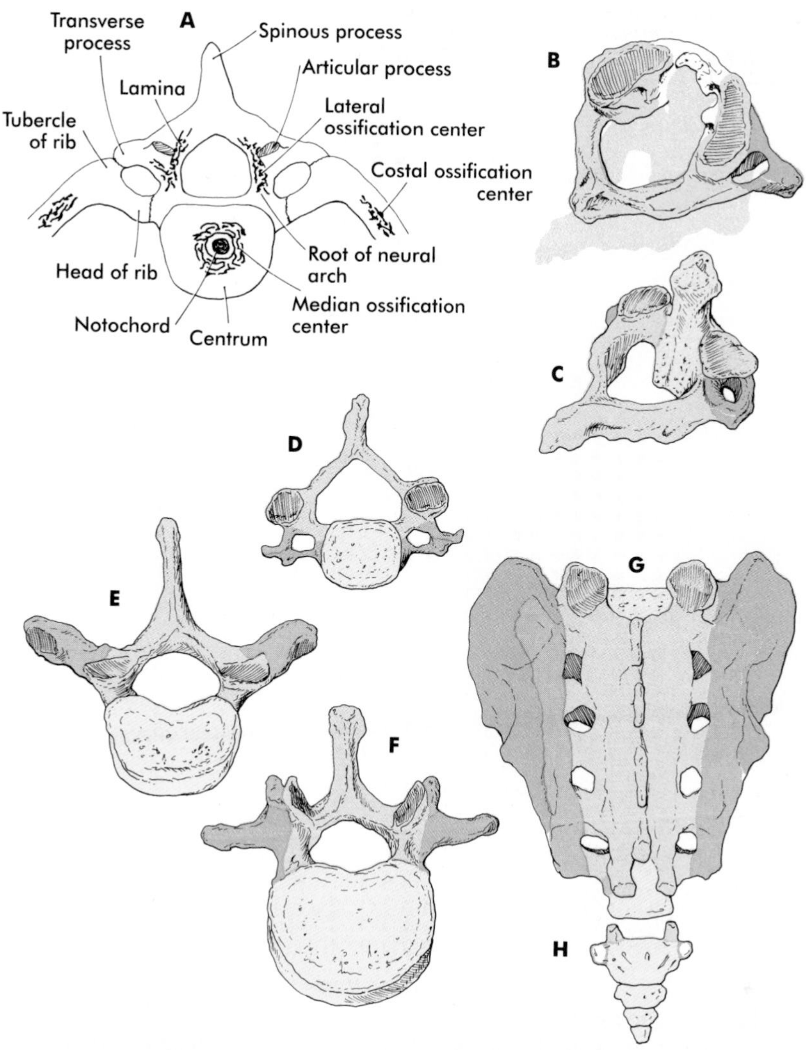

FIG. 10-17 **A,** Structure of a thoracic vertebra. **B** to **H,** Specific types of vertebrae, with homologous structures shown in the same color. **B,** Atlas, with axis shown in its normal position beneath. **C,** Axis. **D,** Cervical vertebra. **E,** Thoracic vertebra. **F,** Lumbar vertebra. **G,** Sacrum. **H,** Coccyx.

with the first appearance of the presomitic mesoderm and for most genes persists until chondrification begins in the primordia of the vertebrae. Formation of the normal segmental pattern along the craniocaudal axis of the vertebral column may be ensured by the fact that most vertebrae are specified by a unique combination of Hox genes. For example, in the mouse the atlas (C-1) is characterized by the expression of *Hox-2.9* (B1), *Hox-1.6* (A1), *Hox-1.5* (A3), and *Hox-4.2* (D4). The axis (C-2) is specified by these four, plus *Hox-1.4* (A4) and *Hox-2.6* (B4). Despite these apparent molecular safeguards, **retinoic acid** (vitamin A) can cause cranial or caudal level shifts in the overall segmental organizations of the vertebrae if applied at specific developmental periods. For example, if administered early, retinoic acid results in a cranial shift (the last cervical vertebra is transformed into the first thoracic vertebra), and later administration causes a caudal shift (e.g., thoracic vertebrae extend into the levels of the first two lumbar vertebrae). Such shifts in level are called **homeotic transformations** and are representative of the broad family of homeotic mutants described in Chapter 5. In about 5% of humans, there are minor variations in the number or proportions of vertebrae. The **Klippel-Feil syndrome,** sometimes called **brevicollis,** is characterized by a short neck with a reduced number of cervical vertebrae, a low hairline, and other assorted anomalies. These variations in vertebral organization are likely related to imbalances of the expression of key homeobox-containing gene products, but a direct relationship has not been established.

At a different level, other aspects of structural organization of the vertebrae result from local interactions with the notochord and spinal (sensory) ganglia. In the chick the

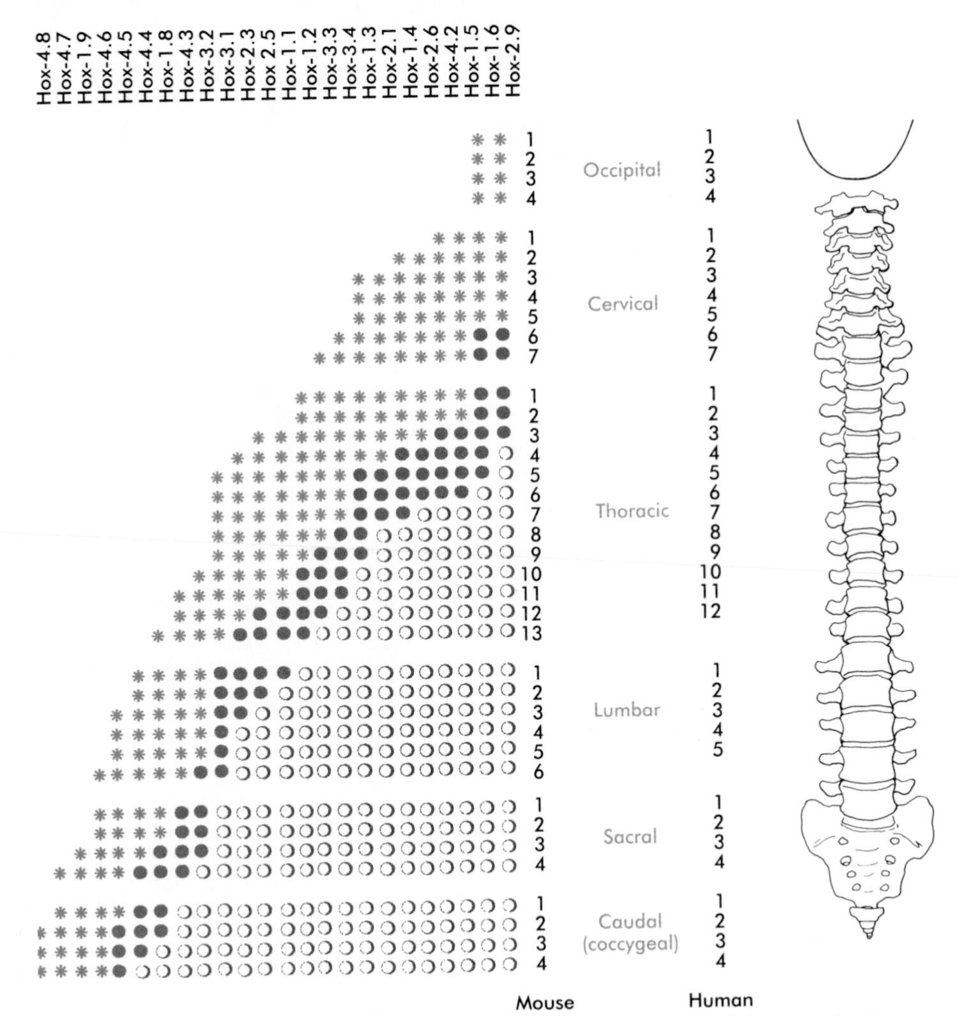

FIG. 10-18 *Hox* gene expression in relation to the development of the vertebral column in the mouse. Note that the vertebral column in the mouse *(left)* has one more thoracic and one more lumbar vertebra than does that of the human. Green asterisks indicate levels at which there is definite expression of the *Hox* gene indicated at the top of the column. Purple circles represent the caudal border where expression fades out. (To convert from older *Hox* gene terminology to the current recommended nomenclature, see Fig. 5-3.)

(Based on studies by Kessel and Gross [1990].)

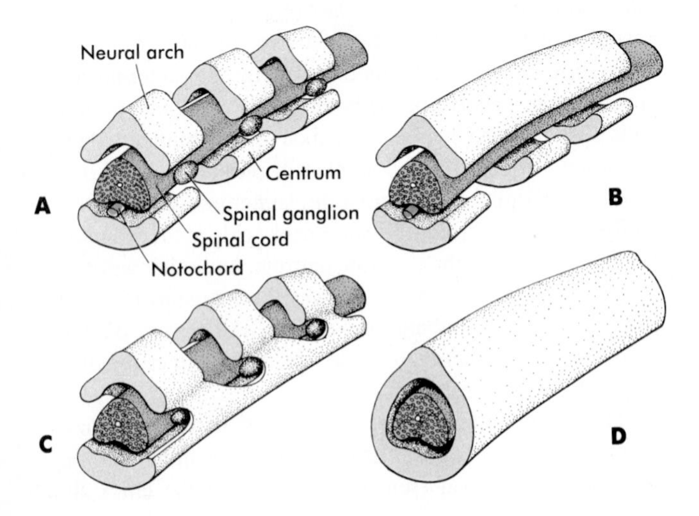

FIG. 10-19 Experiments involving the removal of the spinal ganglia and/or notochord on morphogenesis of the vertebral column of the chick embryo. **A,** Normal embryo. **B,** Excision of the spinal ganglia results in unsegmented neural arches, but the centra of the vertebrae are segmented. **C,** Excision of the notochord results in unsegmented centra, but the neural arches are segmented. **D,** Excision of both spinal ganglia and notochord results in an unsegmented vertebral cylinder around the spinal cord.

(Based on studies by Hall [1977].)

neural arch cartilage forms an unsegmented rod if the spinal ganglia are removed (Fig. 10-19, *B*). Similarly, the cartilaginous centra do not undergo segmentation if the notochord is removed. If both notochord and spinal ganglia are removed, the vertebral column forms as a continuous tube of unsegmented cartilage (Fig. 10-19, *D*).

Among the vertebrae, the axis and atlas have an unusual morphology and distinctive origin (Fig. 10-20). The centrum of the atlas is deficient, but the area of the centrum is penetrated by the protruding **odontoid process** of the axis. The odontoid process consists of three fused centra that are presumably equivalent: (1) a half-segment from the centrum of a transitional bone (the **proatlas**) not found in humans, (2) the centrum that should have belonged to the atlas, and (3) the normal centrum of the axis. This arrangement permits a greater rotation of the head about the cervical spine. When the ubiquitously expressed *Hox-1.1* (A7) transgene was introduced into the germline of mice, the cranial part of the vertebral column was posteriorized. The base of the occipital bone was transformed into an occipital vertebra (the proatlas), and the atlas was combined with its centrum, resulting in an axis that did not possess an odontoid process.

The ribs arise from zones of condensed mesenchymal cells lateral to the centrum (see Fig. 10-17). By the time ossification in the vertebrae begins, the ribs separate from the vertebrae. **Accessory ribs,** especially in the upper lumbar and lower cervical levels, are not uncommon, but estimates of incidence vary widely from one series to the next. Somewhat less than 1% may be a realistic estimate. These and other common rib anomalies **(forked** or **fused ribs)** are typically asymptomatic and are usually detected on x-ray examination.

The **sternum** arises as a pair of cartilaginous bands that converges at the ventral midline as the ventral body wall consolidates (Fig. 10-21). After the primordial sternal bands come together, they reveal their true segmental nature by secondarily subdividing into craniocaudal elements. Many of these ultimately fuse as they ossify to form a common unpaired body of the sternum. Several common anomalies of the sternum (e.g., **split xiphoid process**) are readily understood from its embryological development.

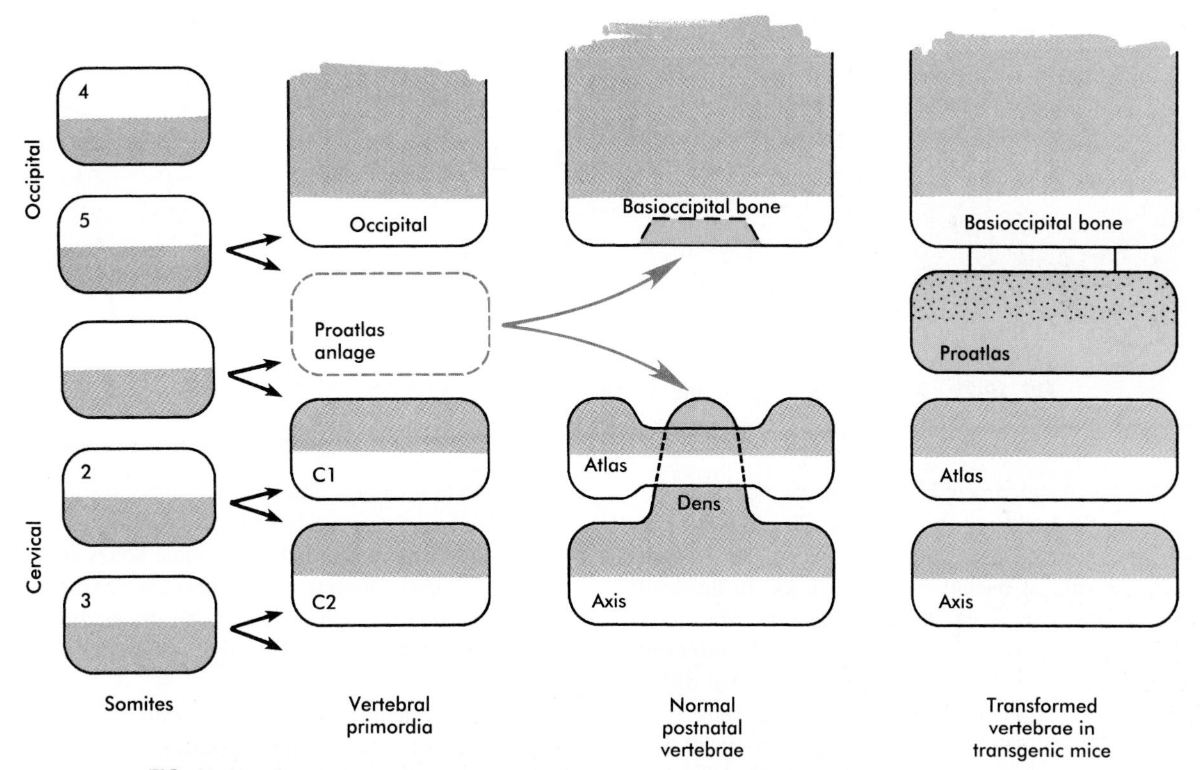

FIG. 10-20 Formation of the atlas and axis in normal and transgenic mice. In normal development, cells from a proatlas anlage contribute to the formation of the basioccipital bone and the dens of the axis. The normal atlas forms an anterior arch (only a transient structure in other vertebrae) instead of a centrum. The cells that would normally form the centrum at the level of the atlas instead fuse with the axis to form the dens of the axis. In mice containing the *Hox-1.1* (A7) transgene, a proatlas forms, and the atlas and axis have the form of typical vertebrae *(right column).*
(Based on studies by Kessel and others [1990].)

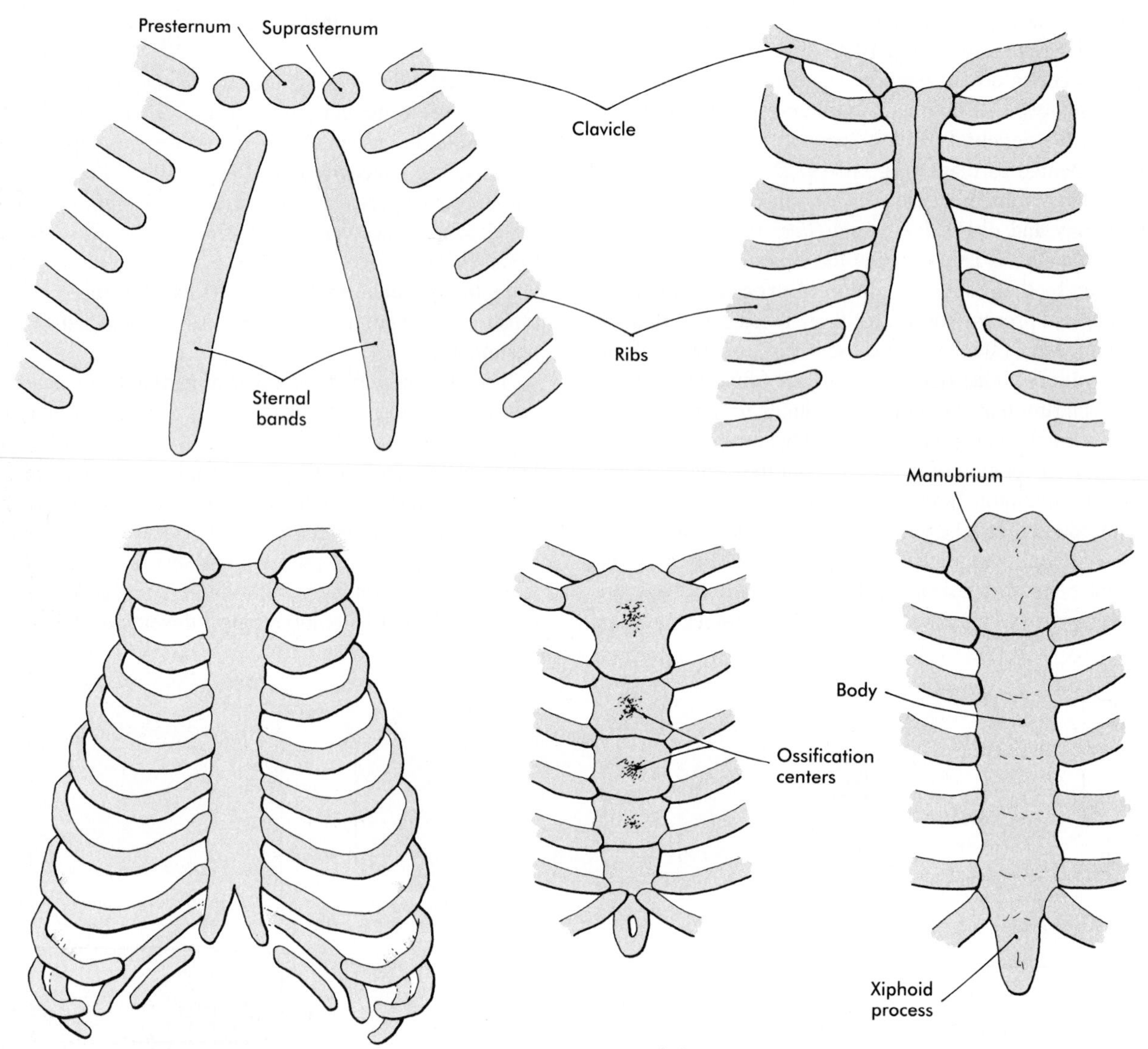

FIG. 10-21 Successive stages in the development of the sternum.

Another late development is the disappearance of the notochord from the bodies of the vertebrae. Between the vertebrae, the notochord expands into the condensed mesenchymal primordia of the intervertebral disks. In the adult the notochord persists as the **nucleus pulposus,** which constitutes the soft core of the disk. The bulk of the **intervertebral disk** consists of layers of fibrocartilage that differentiate from sclerotomally derived mesodermal cells.

The caudal end of the axial skeleton is represented by a well-defined, tail-like appendage during much of the second month (Fig. 10-22, *A*). During the third month, the tail normally regresses, largely through cell death and differential growth, to persist as the coccyx, but rarely a well-developed tail persists in newborns.

The Skull. The skull is a composite structure consisting of two major subdivisions—the **neurocranium,** which surrounds the brain, and the **viscerocranium,** which surrounds the oral cavity, pharynx, and upper respiratory passages. Each of these subdivisions, in turn, consists of two components, one in which the individual bones are first represented by cartilaginous models and are subsequently replaced by bone through endochondral ossification and another in which bone arises directly through the ossification of mesenchyme.

The phylogenetic and ontogenetic foundation of the skull is represented by the **chondrocranium,** which is the cartilaginous base of the neurocranium (Fig. 10-23, *A*). The fundamental pattern of the chondrocranium has been remarkably preserved in the course of phylogeny. It is initially represented by several sets of paired cartilages. One group (parachordals, hypophyseal cartilages, and trabeculae cranii) is closely related to midline structures. Caudal to the parachordal cartilages are four **occipital sclerotomes.** Along with the parachordal cartilages, the occipital sclero-

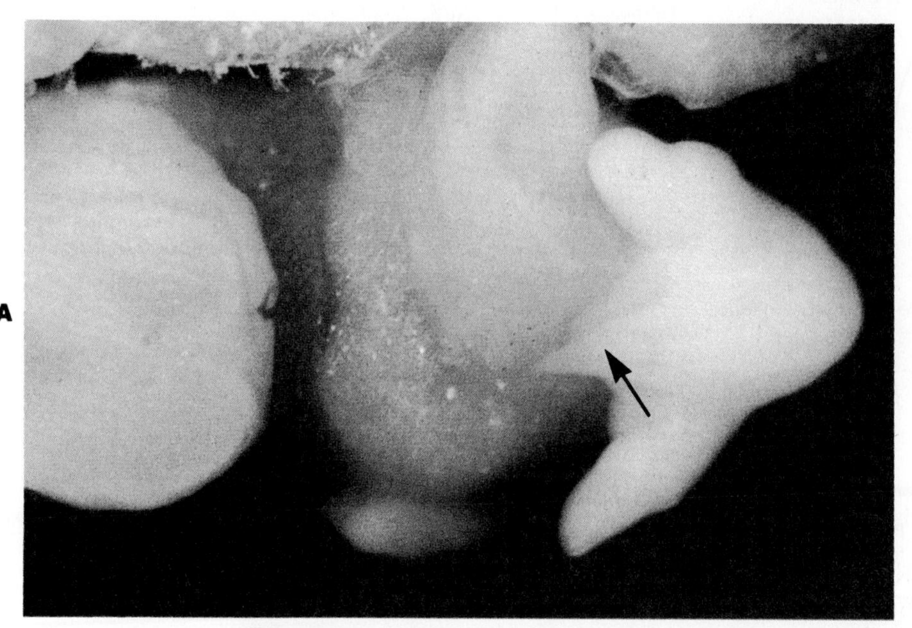

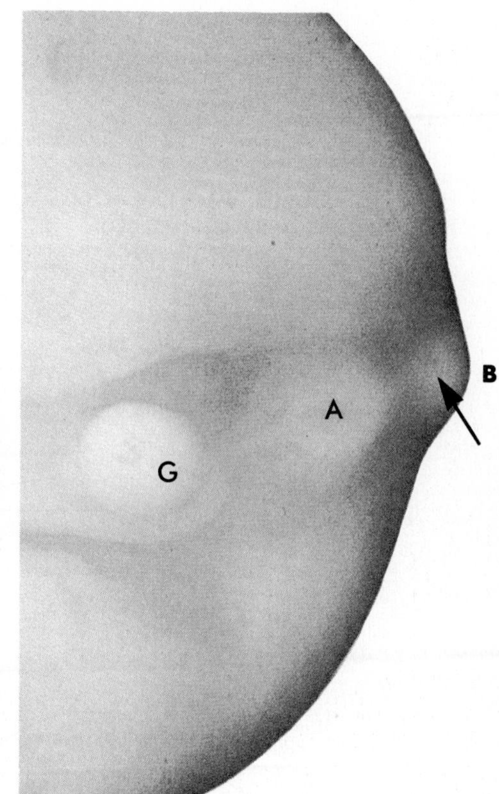

FIG. 10-22 **A,** Ventral view of a 5-week-old embryo, showing a prominent tail *(right).* **B,** Ventral view of the caudal region of a 10-week-old embryo, showing a coccygeal lump *(right)* where the tail is regressing. *A,* Anus; *G,* external genitalia.
(From England MA: *Color atlas of life before birth,* Chicago, 1983, Mosby.)

tomes, which are homologous with precursors of the vertebrae, fuse to form the base of the occipital bone. More laterally, the chondrocranium is represented by pairs of cartilage that are associated with epithelial primordia of the sense organs (olfactory organ, eyes, and auditory organ).

The individual primordial elements of the chondrocranium undergo several patterns of growth and fusion to form the structurally complex bones of the basicranium (the occipital, sphenoid, and temporal bones, as well as much of the deep bony support of the nasal cavity) (Fig. 10-23, *B*). In addition, some of these bones (e.g., the occipital and temporal bones) incorporate membranous components during their development, so in their final form, they are truly composite structures (Fig. 10-23, *D*). Other components of the neurocranium, such as the parietal and frontal bones, are purely membranous bones (see box on p. 173).

Virtually all the bones of the neurocranium arise as the result of an inductive influence of an epithelial structure on the neighboring mesenchyme. These interactions are typically matrix mediated, and immunocytochemical studies have shown the transient appearance of **type II collagen** (the principal collagenous component of cartilage) at the sites and times during which the interactions leading to the formation of the chondrocranium take place (Fig. 10-24). Other studies have shown that the type II collagen is produced by the epithelial component initiating the inductive interaction. In addition to type II collagen, a **cartilage-specific proteoglycan** also accumulates in areas of induction of chondrocranial elements. There is increasing evidence that epithelial elements in the skull not only induce the skeleton but also control its morphogenesis. This contrasts with morphogenetic control of the appendicular skeleton, which is determined by the mesoderm rather than the ectoderm of the limb bud.

Elements of the membranous neurocranium (the paired parietal and frontal bones and the interparietal part of the occipital bone) arise as flat, platelike aggregations of bony spicules (trabeculae) from mesenchyme that has been induced by specific parts of the developing brain. These bones remain separate structures during fetal development, and even at birth they are separated by connective tissue sutures. Intersections between sutures where more then two bones meet are occupied by broader areas of connective tissue called **fontanelles.** The most prominent fontanelles are the

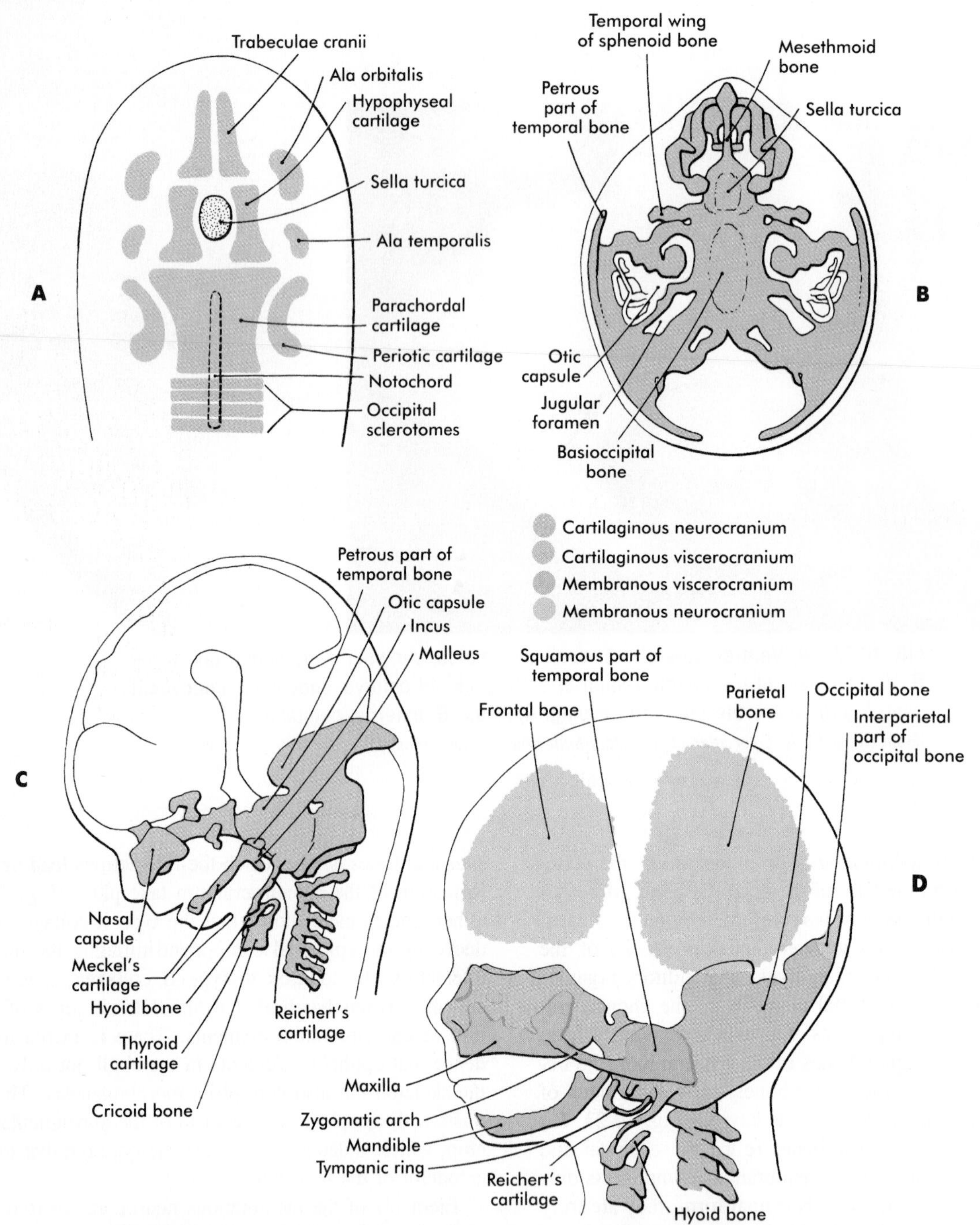

FIG. 10-23 The origins and development of the major skull bones. **A,** Basic skeletal elements of a 6-week-old embryo viewed from above. **B,** Chondrocranium of an 8-week-old embryo viewed from above. **C,** Lateral view of the embryo illustrated in **B. D,** Skull of a 3-month-old embryo.

(Modified from Carlson B: *Patten's foundations of embryology,* ed 5, New York, 1988, McGraw-Hill.)

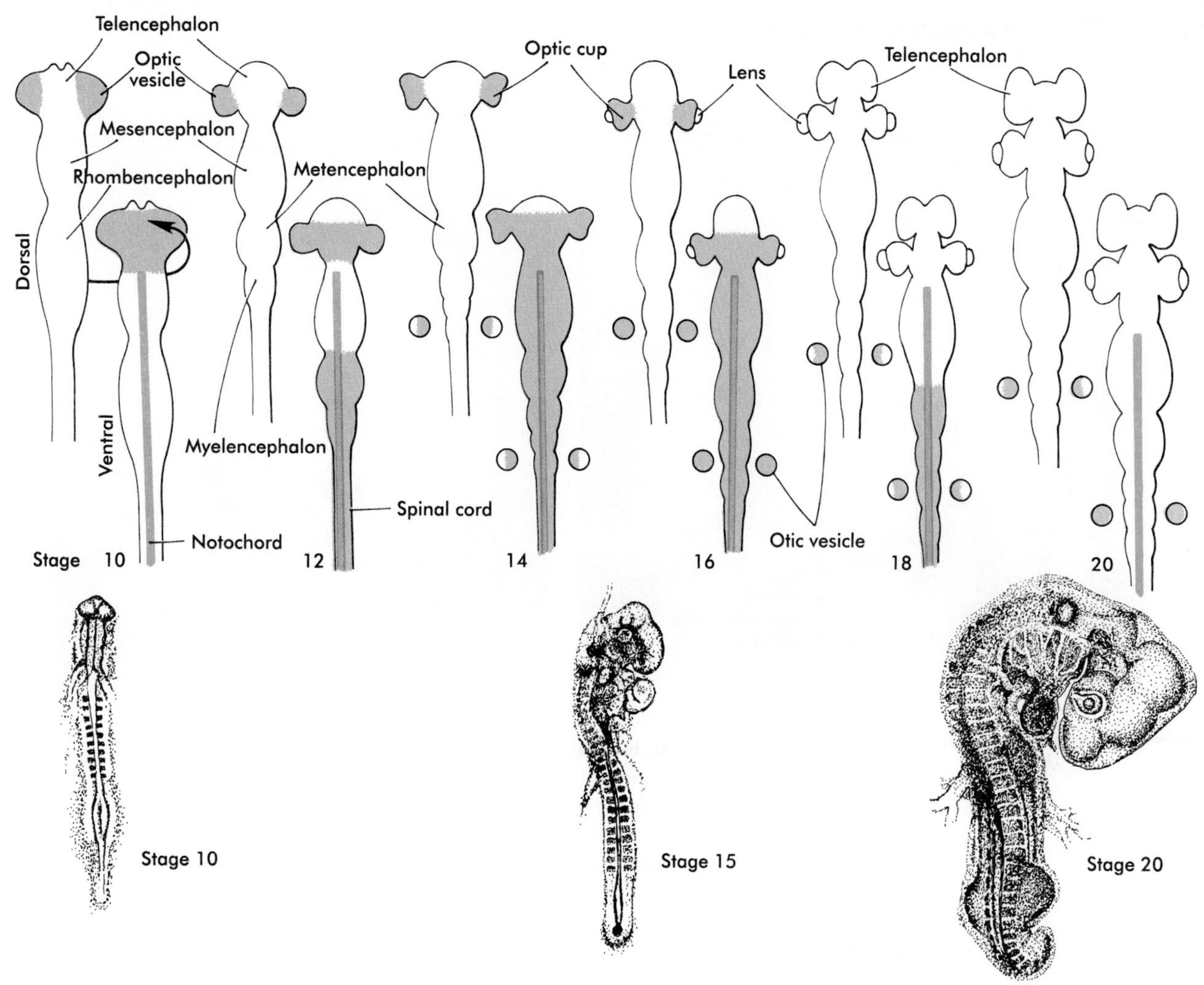

FIG. 10-24 Sites of transient type II collagen distribution (green) during early induction of the skull bones in the chick embryo.
(Based on studies by Thorogood [1988].)

Embryological Origins of Bones of the Cranium

Neurocranium

Chondrocranium
- Occipital
- Sphenoid
- Ethmoid
- Petrous and mastoid part of temporal

Membranous neurocranium
- Interparietal part of occipital
- Parietal
- Frontal
- Squamous part of temporal

Viscerocranium

Branchial arch I

Cartilaginous viscerocranium
- Meckel's cartilage
  - Malleus
  - Incus

Viscerocranium—cont'd

Branchial arch—cont'd

Membranous viscerocranium
- Maxillary process (superficial)
  - Squamous part of temporal
  - Zygomatic
  - Maxillary
  - Premaxillary
  - Nasal?
  - Lacrimal?
- Maxillary process (deep)
  - Palatine
  - Vomer
  - Pterygoid laminae
- Mandibular process
  - Mandible
  - Tympanic ring

Branchial arch II

Cartilaginous viscerocranium
- Reichert's cartilage
- Stapes
- Styloid process

A

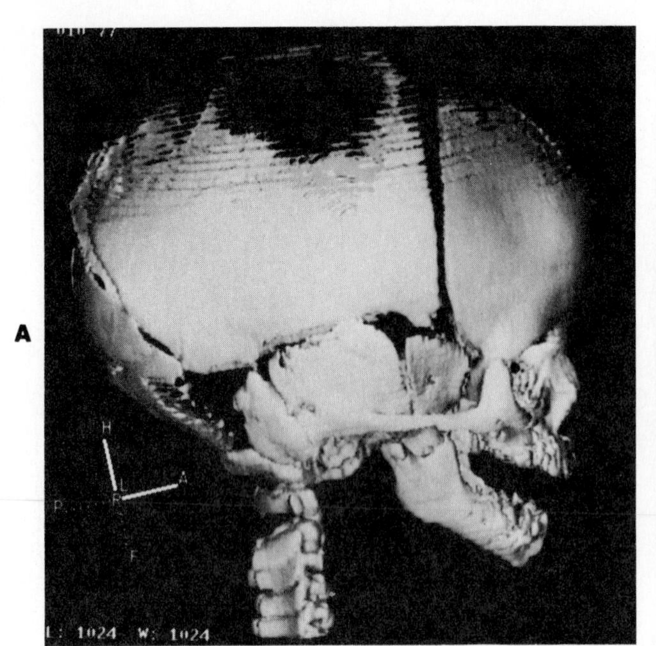

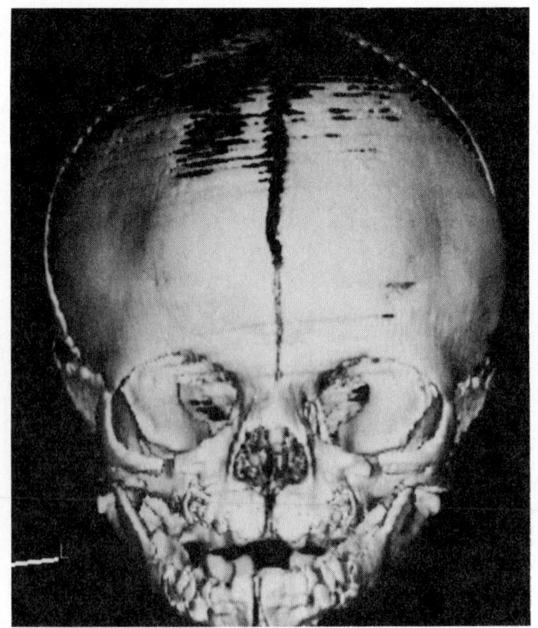

B

FIG. 10-25 High resolution computer tomography (CT) scans of the skull of a 34½-week-old fetus. **A,** Lateral view. **B,** Frontal view. The midline fissure in the forehead area is the metopic suture, which normally becomes obliterated after birth. The irregular black area above that is the anterior fontanelle, one of the "soft spots" in a newborn's head. In these images, the skeletal anatomy was reconstructed by two-dimensional algorithms followed by three-dimensional CT reconstructions. (Courtesy R.A. Levy, H. Maher, and A.R. Burdi, Ann Arbor, Mich.)

anterior fontanelle, located at the intersection of the two frontal and two parietal bones, and the **posterior fontanelle,** located at the intersection of the parietal bones and the single occipital bone (Fig. 10-25).

Like the neurocranium, the viscerocranium consists of two divisions: a **cartilaginous viscerocranium** and a **membranous viscerocranium.** In contrast to the neurocranium, the bones of the viscerocranium originate largely from neural crest–derived mesenchyme. Phylogenetically, the viscerocranium is related to the skeleton of the **branchial arches** (gill arches). Each branchial arch (more appropriately called *pharyngeal arch* in humans) is supported by a cartilaginous rod, which gives rise to a number of definitive skeletal elements (see the box on p. 173) (see Chapter 15). (Details of the organization and derivatives of the noncranial pharyngeal arch cartilages are discussed in Chapter 15 [see Fig. 15-22].)

The membranous viscerocranium consists of a series of bones associated with the upper and lower jaws and the region of the ear (see Fig. 10-23, *D*). These arise in association with the first arch cartilage **(Meckel's cartilage)** and take over some of the functions originally subserved by Meckel's cartilage, as well as a number of new ones.

A number of conditions are recognizable by gross deformities of the skull. Although many of these are true congenital malformations, others fall into the class of deformities that can be attributed to mechanical stress during intrauterine life or childbirth. Some malformations of the skull are secondary to disturbances in development of the brain. In this category are conditions such as **acrania** and **anencephaly** (see Fig. 8-4), which are associated with severe malformations of the brain; **microcephaly** (see Fig. 10-9), in which the size of the cranial vault accommodates to a very small brain; and **hydrocephaly** (see Fig. 12-34), a greatly enlarged cranial vault that represents the response of the skeleton of the head to an excessive buildup of cerebrospinal fluid.

One family of cranial malformations called **craniosynostosis** results from premature closure of certain sutures between major membrane bones of the neurocranium. Premature closure of the **sagittal suture** between the two parietal bones produces a long, keel-shaped skull referred to as **scaphocephaly** (Fig. 10-26). **Oxycephaly,** or turret skull, is the result of premature fusion of the **coronal suture,** located between the frontal and parietal bones. A dominant genetic condition, **Crouzon's syndrome,** presents a gross appearance quite similar to that of oxycephaly, but the malformation of the cranial vault is typically accompanied by malformations of the face, teeth, and ears and occasional malformations in other parts of the body (Fig. 10-27).

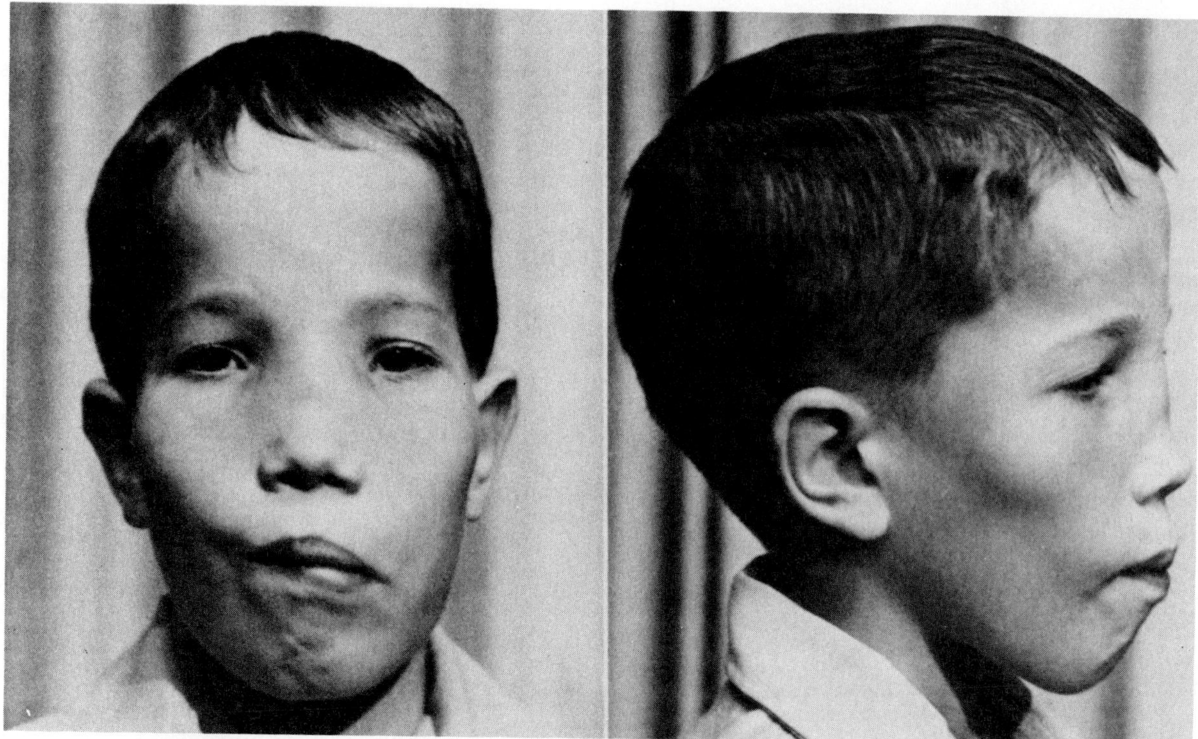

FIG. 10-26 Frontal and lateral views of a boy with a narrow and elongated scaphocephalic skull. Note the high forehead and flat bridge of the nose. This patient had an associated facial palsy and mixed deafness.
(From Goodman R, Gorlin R: *Atlas of the face in genetic disorders*, St Louis, 1977, Mosby.)

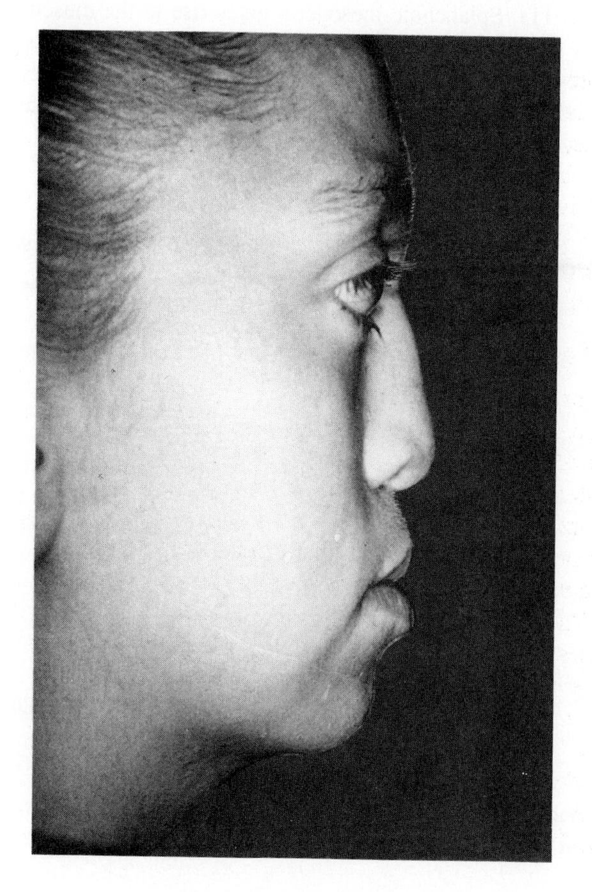

FIG. 10-27 Lateral view of the flattened face of an individual with Crouzon's syndrome.
(Courtesy A.R. Burdi, Ann Arbor, Mich.)

The Appendicular Skeleton

The **appendicular skeleton** consists of the bones of the limbs and limb girdles. There are fundamental differences in organization and developmental control between the axial and appendicular skeleton. The axial skeleton forms a protective casing around soft internal tissues (e.g., the brain, spinal cord, and pharynx), and the mesenchyme forming the bones is induced by the organs that the bones surround. In contrast, bones of the appendicular skeleton form a central supporting core of the limbs. Although interaction with an epithelium (the apical ectodermal ridge of the limb bud [see Chapter 11]) is required for the formation of skeletal elements in the limb, morphogenetic control of the limb is inherent in the mesoderm, with the epithelium playing only a stimulatory role.

All components of the appendicular skeleton begin as cartilaginous models, which convert to true bone by endochondral ossification later during embryogenesis. (Details of the formation of the appendicular skeleton are given in Chapter 11.)

THE MUSCULAR SYSTEM

Three types of musculature—skeletal, cardiac, and smooth—are formed during embryonic development. Virtually all skeletal musculature is derived from the paraxial mesoderm, specifically the somites or somitomeres (see Fig. 5-11). Splanchnic mesoderm gives rise to the musculature of the heart (cardiac muscle) and the smooth musculature of the gut and respiratory tracts (Table 10-1). Other smooth muscle, such as that of the blood vessels and the arrector pili muscles, is derived from local mesoderm.

Skeletal Muscles

Muscles of the Trunk and Limbs. There is increasing evidence that certain cells of the epiblast are determined to become myogenic cells even before the somites are completely formed. Among the evidence for this is the expression of the muscle regulatory protein **myf-5** by these cells (see Chapter 9).

Studies on muscle development have been greatly facilitated by the use of marked cells in transplantation experiments. For many decades the origin of the musculature was in question, with the somites and lateral plate mesoderm both being candidates. This issue was finally resolved by grafting tissue from quail embryos into homologous sites in chick embryos. The nuclei of quail cells contain a distinctive mass of dense chromatin, enabling researchers to distinguish quail from chick cells with great reliability (Fig. 10-28). If a putative precursor tissue is grafted from a quail into a chick embryo, the grafted quail tissue becomes well integrated into the chick host, and if cells migrate out from the graft, their pathway of migration into the host tissue can be clearly traced. Experiments involving this approach have been particularly useful in studies of muscle and the neural crest.

Quail/chick grafting experiments have clearly shown that the major groups of skeletal muscles in the trunk and limbs arise from myogenic precursors located in the somites (see Fig. 10-28). In the thorax, the intrinsic muscles of the back are derived from cells in the myotomes, whereas ventrolateral muscles (e.g., intercostal muscles) arise from epithelially organized ventral buds of the somites (Fig. 10-29). In the limb regions, myogenic cells migrate from the epithelium of the ventrolateral dermomyotome quite early dur-

TABLE 10-1 Embryologic Origins of the Major Classes of Muscle

EMBRYOLOGIC ORIGIN	DERIVED MUSCLE	INNERVATION
Somitomeres 1 through 3 and/or prechordal plate	Most extrinsic eye muscles	Cranial nerves III and IV
Somitomere 4	Jaw-closing muscles	Cranial nerve V (mandibular branch)
Somitomere 5	Lateral rectus of eye	Cranial nerve VI
Somitomere 6	Jaw-opening and other second-arch muscles	Cranial nerve VII
Somitomere 7	Third-arch branchial muscles	Cranial nerve IX
Somites 1 and 2	Intrinsic laryngeal muscles and pharyngeal muscles	Cranial nerve X
Occipital somites (1 through 7)	Muscles of tongue, larynx, and neck	Cranial nerves XI and XII, cranial cervical nerves
Trunk somites	Trunk muscles, diaphragm, limb muscles	Spinal nerves
Splanchnic mesoderm	Cardiac muscles	Autonomic
Splanchnic mesoderm	Smooth muscles of gut and respiratory tract	Autonomic
Local mesenchyme	Other smooth muscle: vascular, arrector pili muscles	Autonomic

From Carlson BM: *Patten's foundations of embryology*, ed 5, New York, 1988, McGraw-Hill.

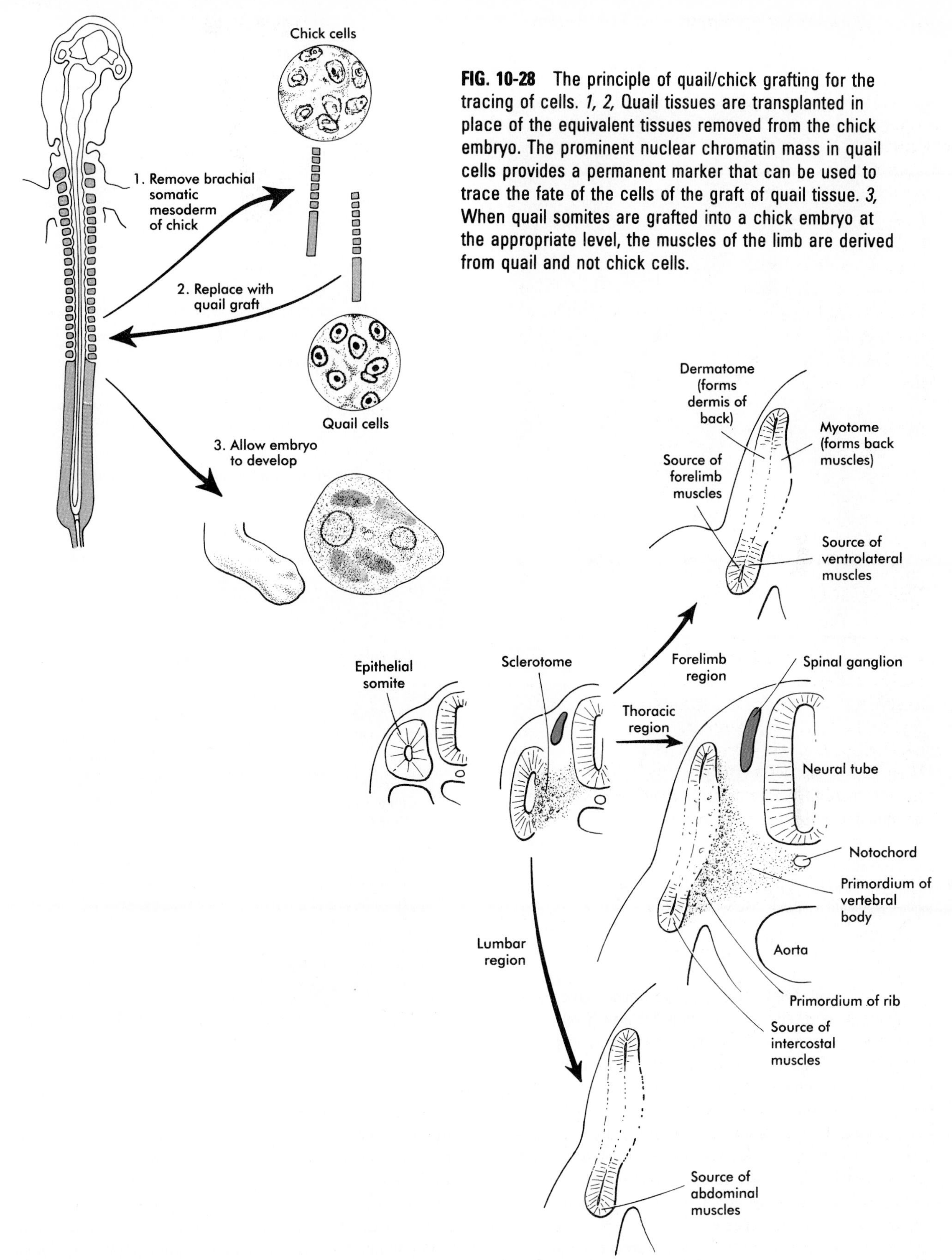

FIG. 10-28 The principle of quail/chick grafting for the tracing of cells. *1, 2,* Quail tissues are transplanted in place of the equivalent tissues removed from the chick embryo. The prominent nuclear chromatin mass in quail cells provides a permanent marker that can be used to trace the fate of the cells of the graft of quail tissue. *3,* When quail somites are grafted into a chick embryo at the appropriate level, the muscles of the limb are derived from quail and not chick cells.

FIG. 10-29 The origin of the trunk muscles from the somites in the forelimb, thoracic, and lumbar regions.
(Modified from Theiler K: *Adv Anat Embryol Cell Biol* 112:1-99, 1988.

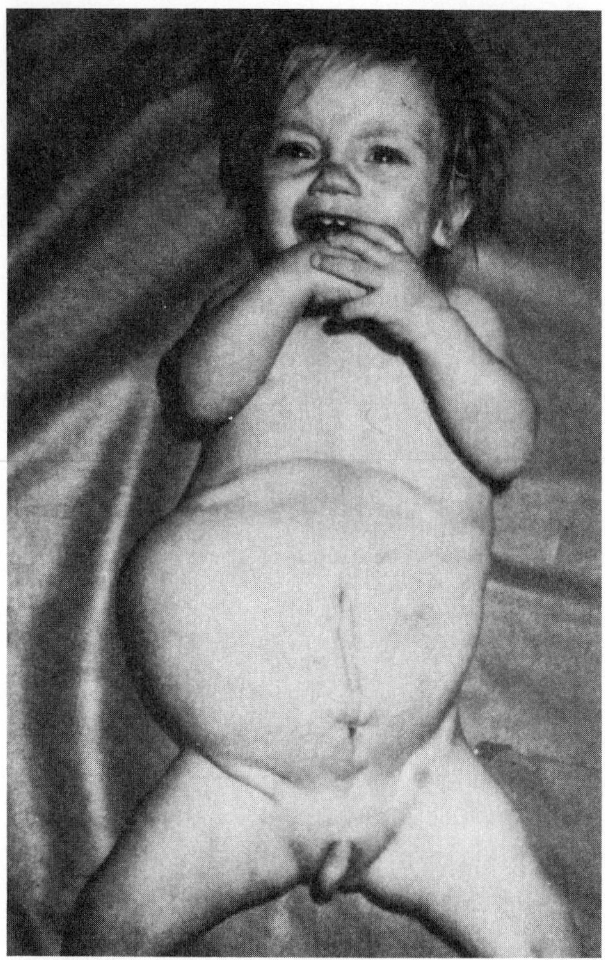

FIG. 10-30 A child with the prune belly syndrome. The abdominal musculature is absent or very hypoplastic. Urinary defects commonly accompany this defect. Note the characteristic dimples on the knees.
(From Wigglesworth J, Singer D: *Textbook of fetal and perinatal pathology,* Oxford, England, Blackwell Scientific.)

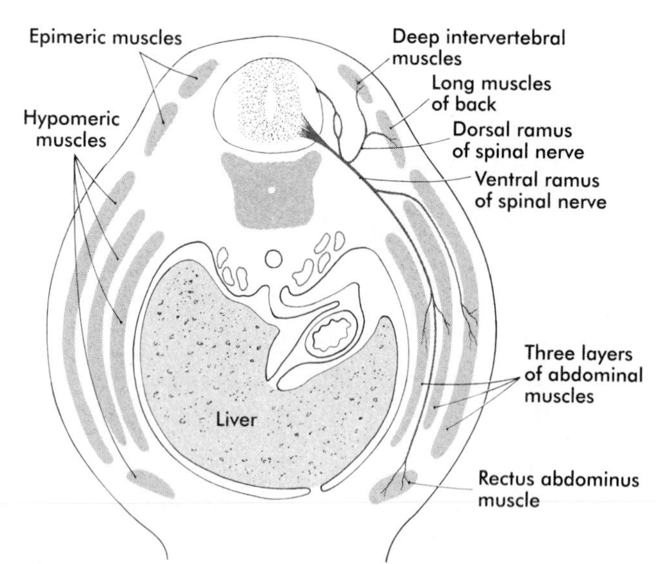

FIG. 10-31 The groups and layers of trunk muscles.

TABLE 10-2 Differences between Cellular Precursors of Axial and Limb Muscles

AXIAL MUSCLES	LIMB MUSCLES
Are located in medial half of somite	Are located in lateral half of somite
Differentiate largely in situ	Migrate into limb buds before differentiating
Differentiate initally as mononucleate myocytes	Differentiate initally as multinucleate myocytes
Express myogenic determination factors (Myf 5, MyoD 1) at or before the onset of myotome formation	Expression of myogenic determination genes is delayed until limb muscle masses begin to coalesce
Differentiation seems strongly influenced by neural tube and notochord	Migration and differentiation are little influenced by axial structures

ing development. Other grafting experiments have not supported the hypothesis that limb muscles arise from cells of the lateral plate mesoderm. More cranial myogenic cells originating in similar regions of the somites migrate into the developing tongue. At the lumbar levels, precursors of the abdominal muscles also move out of the epithelium of ventrolateral somitic buds. It is highly likely that the human anomaly the **prune belly syndrome,** which is characterized by the absence of the abdominal musculature (Fig. 10-30), will be found to be caused by a deficiency in this population of myogenic cells.

Recent experiments have shown different cellular behavior in areas of the myotomes adjacent to limb and nonlimb regions. In thoracic segments, cells of the dermatome surround the lateral edges of the myotome. This is followed by an increase in the number of myotubes formed in the myotome and the penetration of the primordia into the body wall. In contrast, at the levels of the limb buds, dermatome cells die before surrounding the early myotubes that form in the myotome. These myotubes do not increase significantly in number, nor do they move out from the myotomes to form separate muscle primordia.

Several experiments suggest that influences of the body regions surrounding the somites play some role in the early steps of release and morphogenesis of myogenic precursor cells from the somites. If a chick limb bud is removed, the dermatome cells do not die but rather surround the myotomes, allowing an increase in the number of myotubes that form in the myotomes. In contrast, when a limb bud is grafted onto a thoracic level, cells of the dermatome die and the underlying muscle primordia fail to mature. In another experimental approach, if somites or pieces of paraxial mesoderm are grafted in a rotated position with the medial edge facing laterally, myogenic cells migrate from the new lateral edge.

After their origin from the somites, the muscle primordia of the trunk and abdomen are organized into well-defined groups and layers (Fig. 10-31). (Later morphogenesis of the limb muscles is discussed in Chapter 11.)

Results of a number of experiments have demonstrated fundamental differences in cellular properties between the cellular precursors of limb muscles and axial muscles. These are summarized in Table 10-2.

Muscles of the Head and Cervical Region. The skeletal muscle of the head and neck is mesodermal in origin. Quail/chick grafting experiments have shown that the paraxial mesoderm, specifically the somitomeres, constitutes the main source of the cranial musculature, although some question still remains about the origin of the extraocular muscles. There is some evidence that at least some of the cells that make up the extraocular muscles pass by the prechordal plate of the early embryo. These newer findings have refuted early hypotheses of a special mode of origin for the "visceral" or pharyngeal arch musculature. Some questions still remain as to whether all aspects of myogenesis in the head are identical to those in the trunk, since a number of the craniofacial muscles have different phenotypic properties from trunk muscles (e.g., myosin isoforms and possibly elements of neuromuscular control of phenotype).

As with muscles in the trunk, muscles in the head and neck arise by the movement of myogenic cells away from the paraxial mesoderm through mesenchyme (either neural crest derived or mesodermal) on their way to their final destination. As in the trunk, the morphogenesis of muscles in the cranial region appears to be determined by information inherent in the connective tissues that ensheath the muscles. There is no early level specificity in the paraxial myogenic cells. This has been determined by grafting somites or somitomeres from one craniocaudal level to another. In these cases, myogenic cells that leave the grafted structures form muscles normal for the region into which they migrate rather than appropriate for the level of origin of the grafted somites.

Despite evidence that suggests the interchangeability of embryonic somites in muscle formation, other experimental data show the existence of well-defined properties of skeletal muscles (e.g., their ability to make connections with nerves from different axial levels and certain aspects of gene expression). These findings indicate a strong imprinting on the muscle fibers according to a pronounced rostrocaudal gradient. The time this form of axial specificity is imposed on the developing muscle fibers is not known.

Certain muscles of the head arise from the occipital somites in the manner of trunk muscles and undergo extensive migrations into the enlarging head. Their more caudal level of origin is evidenced by their innervation by the **hypoglossal nerve** (twelfth cranial nerve), which according to many comparative anatomists is a series of highly modified spinal nerves.

Anomalies of Skeletal Muscles. Variations and anomalies of skeletal muscles are common. Some, such as the absence of portions of the pectoralis major muscle, are associated with malformations of other structures. Further discussion of anomalies of specific muscles requires a level of anatomical knowledge beyond that assumed for this text.

Muscular dystrophy is a family of genetic diseases characterized by the repeated degeneration and regeneration of various groups of muscles during postnatal life. In Duchenne muscular dystrophy, which occurs in young boys, a membrane-associated protein called **dystrophin** is lacking from the muscle fibers. Although the function of dystrophin is still uncertain, its absence appears to make the muscle fibers more susceptible to damage when physically stressed.

Cardiac Muscle

Although a striated muscle, cardiac muscle differs from skeletal muscles in many aspects of its embryonic development. Derived from the splanchnic mesoderm of the early embryo, cardiac muscle cells arise from cells present in the myocardium. Differences between the differentiation of cardiac and skeletal muscle appear early, since MyoD and other common master regulators of skeletal muscle differentiation are not expressed in early cardiac muscle development.

Even early cardiac myoblasts contain relatively large numbers of myofibrils in their cytoplasm, and they are capable of undergoing pronounced contractions. In the embryo, the mononucleated cardiac myocytes face a difficult problem—the cells of the developing heart must continue to contract while the heart is increasing in mass. This functional requirement necessitates cardiac myocytes to undergo mitosis even though their cytoplasm contains many bundles of contractile filaments (Fig. 10-32). Cells of the body often lose their ability to divide when their cytoplasm contains structures characteristic of the differentiated state. Cardiac myocytes deal with this problem by partially disassembling their contractile filaments during mitosis. In contrast to skeletal muscle, cardiac myocytes do not undergo fusion but rather remain as individual cells, although occasionally they become binucleated. Cardiac myocytes keep in close structural and functional contact through the

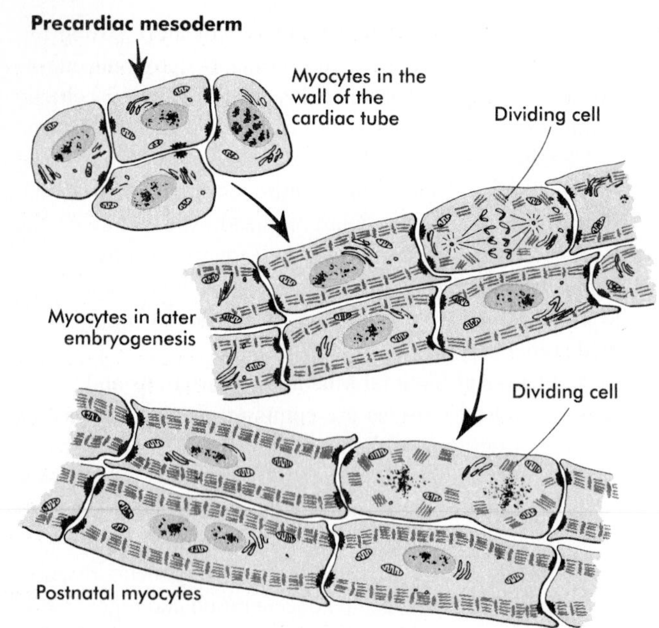

FIG. 10-32 Stages in the histogenesis of cardiac muscle. During mitosis, the contractile filaments undergo a partial disassembly.
(Modified from Rumyantsev P: *Cardiomyocytes in processes of reproduction, differentiation and regeneration* [in Russian], Leningrad, 1982, Nauka.)

use of **intercalated disks,** which join adjacent cells to one another.

Later in development a network of cardiac muscle cells undergoes an alternative pathway of differentiation characterized by increased size, a reduction in the concentration of myofibrils, and a greatly increased concentration of glycogen in the cytoplasm. These cells form the **conducting system,** parts of which are called **Purkinje fibers.** Purkinje fibers also express a different profile of contractile protein isoforms from either atrial or ventricular myocytes.

Smooth Muscle

As with cardiac muscle, much of the smooth muscle in the body arises from splanchnic mesoderm. Exceptions are the ciliary muscle and sphincter pupillae muscles of the eye, which are derived from neural crest ectoderm, and much of the vascular smooth muscle, which frequently arises from the local mesoderm. Very little is known about the morphology and mechanisms underlying the differentiation of smooth muscle cells.

SUMMARY

1. The epidermis starts as a single layer of ectoderm, to which a single superficial layer of peridermal cells is added. As other layers are added, three cell types migrate from other sources: (1) melanoblasts (pigment cells) from the neural crest, (2) Langerhans cells (immune cells) from precursors in the bone marrow, and (3) Merkel cells (mechanoreceptors), probably from the neural crest.
2. In the multilayered epidermis, unspecialized cells from the stratum basale differentiate as they move through the various layers toward the surface of the epidermis. The cells produce increasing amounts of intracellular keratins and filaggrin; the latter is involved in the interconnections of the keratinocytes, the final differentiated form of the epidermal cell.
3. The dermis arises from mesodermal cells derived from the dermatome of the somites. Dermal/epidermal interactions underlie the formation of epidermal appendages such as hairs. In mammary glands, hormonal influences are important in the further development of the duct system after the ductal epithelium is induced.
4. Skeletal tissue arises from mesenchyme of either mesodermal or neural crest origin. There are two major subdivisions of the skeleton—the axial skeleton of the trunk and the appendicular skeleton of the limbs.
5. The fundamental organization of the cranial components of the vertebral column is closely associated with homeobox-gene expression. Superimposed on this is the induction of many components of the axial skeleton by underlying ectodermal (usually neural) structures. Individual vertebrae are composite structures consisting of components derived from two adjoining somites.
6. The skull consists of two subdivisions—the neurocranium, which surrounds the brain, and the viscerocranium, which surrounds the oral cavity. The base of the neurocranium (chondrocranium) is initially represented by several sets of paired cartilage. These later become transformed into bone. Most bones surrounding the brain are formed by intramembranous bone, which differentiates directly from mesenchyme. The viscerocranium is also derived from cartilaginous and membranous components.
7. Skeletal muscles of the limbs and trunk arise from cellular precursors in the somites. The cranial musculature arises from the somitomeres. Dorsal and ventral muscles of the trunk arise from precursors located in different regions of the somites. The limb musculature also arises from cells in the ventrolateral regions of the somites. These cells migrate into the limb buds.
8. Cardiac muscle arises from splanchnic mesoderm. Cardiomyocytes differ from skeletal muscle cells in that they can divide mitotically after they are highly differentiated and contain contractile filaments.

REVIEW QUESTIONS

1. What component of the developing skin determines the nature of the hairs that form or the thickness of the epidermis in the fetus?
2. A male has two bilaterally symmetrical brownish spots about 8 mm in diameter located on the skin of the groin. What is one explanation for them?
3. Why is cranial bone typically not found over an area where part of the brain is missing?
4. How was it determined that the limb musculature arises from the somites?

REFERENCES

SKIN

Adams JC, Watt FM: Changes in keratinocyte adhesion during terminal differentiation: reduction in fibronectin binding precedes alpha$_5$beta$_1$ integrin loss from the cell surface, *Cell* 63:425-435, 1990.

Babler WJ: Embryologic development of epidermal ridges and their configurations, *Birth Defects* 27:95-112, 1991.

Duernberger H, Kratochwil K: Specificity of time interaction and origin of mesenchymal cells in the androgen response of the embryonic mammary gland, *Cell* 19:465-471, 1980.

Goldsmith LA, ed: *Physiology, biochemistry, and molecular biology of the skin,* ed 2, New York, 1991, Oxford University.

Halata Z, Grim M, Christ B: Origin of spinal cord meninges, sheaths of peripheral nerves, and cutaneous receptors, including Merkel cells, *Anat Embryol* 182:529-537, 1990.

Holbrook KA: *Structure and function of the developing human skin.* In Goldsmith LA, ed: *Physiology, biochemistry, and molecular biology of the skin,* ed 2, New York, 1991, Oxford University, pp 63-110.

Kimura S: Embryologic development of flexion creases, *Birth Defects* 27:113-129, 1991.

Krey AK and others: Morphogenesis and malformations of the skin, NICHD/NIADDK research workshop, *J Invest Dermatol* 88:464-473, 1987.

Sakakura T, Sakagami Y, Nishizuka Y: Dual origin of mesenchymal tissues participating in mouse mammary gland morphogenesis, *Dev Biol* 91:202-207, 1982.

Sengel P: *Morphogenesis of skin,* Cambridge, Mass, 1976, Cambridge University.

Topper YJ, Freeman CS: Multiple hormone interactions in the developmental biology of the mammary gland, *Physiol Rev* 60:1049-1106, 1980.

SKELETON

Balling R and others: *Development of the skeletal system, Ciba Found Symp* 165:132-143, 1992.

Bosma JF, ed: *Symposium on development of the basicranium,* DHEW Pub No (NIH) 76-989, Washington, DC, 1976, US Government Printing Office.

Gasser RF: Evidence that sclerotomal cells do not migrate medially during normal embryonic development of the rat, *Am J Anat* 154:509-524, 1979.

Hall BK: Cellular interactions during cartilage and bone development, *J Craniofac Genet Dev Biol* 11:238-250, 1991.

Hall BK: *The evolution of connective and skeletal tissues.* In Hinchliffe JR, Hurle JM, Summerbell D, eds: *Developmental patterning of the vertebrate limb,* New York, 1991, Plenum, pp 303-311.

Kessel M: Molecular coding of axial positions by hox genes, *Semin Dev Biol* 2:367-373, 1991.

Kessel M: Respecification of vertebral identities by retinoic acid, *Development* 115:487-501, 1992.

Kessel M, Balling R, Gruss P: Variations of cervical vertebrae after expression of a Hox-1.1 transgene in mice, *Cell* 61:301-308, 1990.

Schierhorn H: Ueber die Persistenz der embryonalen Schwanzknospe beim Menschen, *Anat Anz* 127:307-337, 1970.

Sensenig EC: The early development of the human vertebral column, *Carnegie Contr Embryol* 33:21-42, 1949.

Shinomura T, Kimata K: Precartilage condensation during skeletal pattern formation, *Dev Growth Differen* 32:243-248, 1991.

Theiler K: Vertebral malformations, *Adv Anat Embryol Cell Biol* 112:1-99, 1988.

Verbout AJ: The development of the vertebral column, *Adv Anat Embryol Cell Biol* 90:1-122, 1985.

MUSCLE

Christ B: *Entwicklung und Biologie des Bewegungsapparates.* In Staubesand J, ed: *Benninghoff Makroskopische und mikroskopisches Anatomie des Menschen,* Munich, 1985, Urban & Schwarzenberg, pp 167-180.

Christ B, Cihak R, eds: *Development and regeneration of skeletal muscles,* Basel, Switzerland, 1986, Karger.

Christ B, Jacob M, Jacob HJ: On the origin and development of the ventrolateral abdominal muscles in the avian embryo: an experimental and ultrastructural study, *Anat Embryol* 166:87-102, 1983.

Christ B and others: *The somite-muscle relationship in the avian embryo.* In Hinchliffe JR, Hurle JM, Summerbell D, eds: *Developmental patterning of the vertebrate limb,* New York, 1991, Plenum pp 265-271.

Cummins H: The topographic history of the volar pads (walking pads; Tastballen) in the human embryo, *Carnegie Contr Embryol* 113:103-126, 1929.

Kieny M and others: Origin and development of avian skeletal musculature, *Reprod Nutr Dev* 28(3B):673-686, 1988.

Noden DM: The embryonic origins of avian cephalic and cervical muscles and associated connective tissues, *Am J Anat* 168:257-276, 1983.

Ott, M-O and others: Early expression of the myogenic regulatory gene, myf-5, in precursor cells of skeletal muscle in the mouse embryo, *Development* 111:1097-1107, 1991.

Rong PM and others: The neural tube/notochord complex is necessary for vertebral but not limb and body wall striated muscle differentiation, *Development* 115:657-672, 1992.

Rumyantsev PP: *Cardiomyocytes in processes of reproduction, differentiation and regeneration* (in Russian), Leningrad, 1982, Nauka.

Sanes JR, Donoghue MJ, Merlie JP: *Positional differences among adult skeletal muscle fibers.* In Kelly AM, Blau HM, eds: *Neuromuscular development and disease,* New York, 1992, Raven, pp 195-209.

Sassoon D and others: Expression of two myogenic regulatory factors myogenin and MyoD1 during mouse embryogenesis, *Nature* 341:303-307, 1989.

CHAPTER 11

Limb Development

Limbs are remarkable structures that are designed almost solely for mechanical functions—motion and force. These properties are achieved through the coordinated development of a variety of tissue components. No single tissue in the limb takes shape without reference to the other tissues with which it is associated. The limb as a whole develops according to a master blueprint that appears to reveal itself sequentially with each successive stage in limb formation. Although some idea of many of the major themes in limb development is known, the nature of the master blueprint remains tantalizingly obscure.

One measure of the complexity and precision of limb development is demonstrated by the wide variety of limb anomalies. Virtually all are manifested as abnormalities of gross form rather than individual cellular function. Many of the underlying factors that control limb development cannot be seen but rather must be demonstrated by experimental means. This chapter outlines some of the classic methods of experimental embryology, how these methods can reveal the factors that guide normal limb development, and how their absence results in limb abnormalities.

THE INITIATION OF LIMB DEVELOPMENT

Limb development begins with the activation of a group of mesenchymal cells in the **lateral plate mesoderm** (Fig. 11-1). The initial stimulus for limb development is poorly understood. Interactions with a variety of neighboring tissues (e.g., gut endoderm, somites, pronephros) have been postulated, but a convincing explanation for the localization of the activation to the specific limb-forming regions has not been given.

The mesodermal limb primordia form at discrete locations beneath a broad band of thickened ectoderm that encircles the ventrolateral aspect of the embryo (see Fig. 5-24). In the earliest stages of limb development, the limb mesoderm is the prime mover. It influences the overlying ectoderm to be a functional part of an interacting mesodermal-ectodermal primordium that has sufficient developmental information to form a limb even if isolated from the rest of the body (a so-called **self-differentiating system**).

The primacy of the early limb mesoderm was demonstrated long ago by transplantation experiments on amphibian embryos. If early limb mesoderm is removed, a limb fails to form. However, if the same mesoderm is transplanted to the flank of an embryo, a supernumerary limb grows at that site. In contrast, if the ectoderm overlying the normal limb mesoderm is removed, new ectoderm heals the defect and a limb forms. If the original ectoderm that was removed is grafted to the flank, no limb forms. These experiments show that in *early* limb development, mesoderm is the primary bearer of the limb blueprint and ectoderm is only secondarily co-opted into the system.

Recently, a direct molecular marker of "limbness" has been found. The homeobox gene **X1HBox** (equivalent to the mammalian *Hox-3.3* [C6]) is expressed in future limb mesodermal cells well before there is any morphological evidence of limbness. It is not known how the expression of this gene product relates to cellular aspects of the initiation of limb development. In rare instances, individuals are born without one or sometimes all limbs **(amelia)** (Fig. 11-2). Although it is often impossible to pinpoint the exact cause of a congenital anomaly, it is likely that some cases of amelia can be attributed to the absence of expression of the appropriate homeobox or other limb-determining genes, which may be the result of genetic or environmental causes.

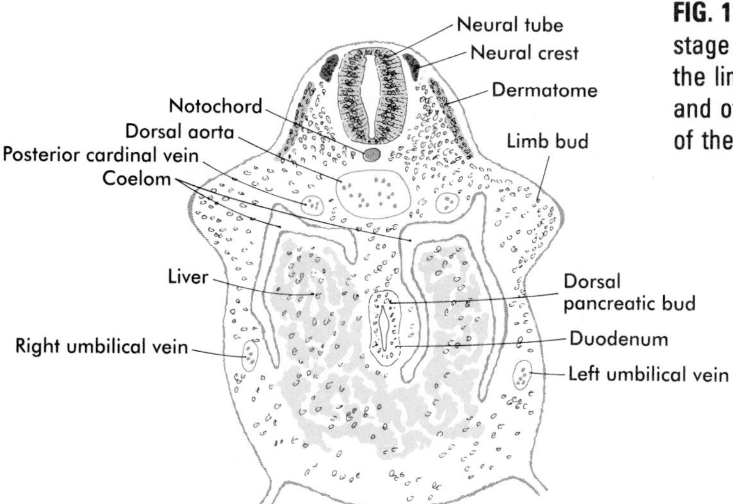

FIG. 11-1 Cross section through the trunk at an early stage of limb bud development showing the position of the limb bud in relation to that of the somite (dermatome) and other major structures. The limb bud is an outgrowth of the body wall (lateral plate mesoderm).

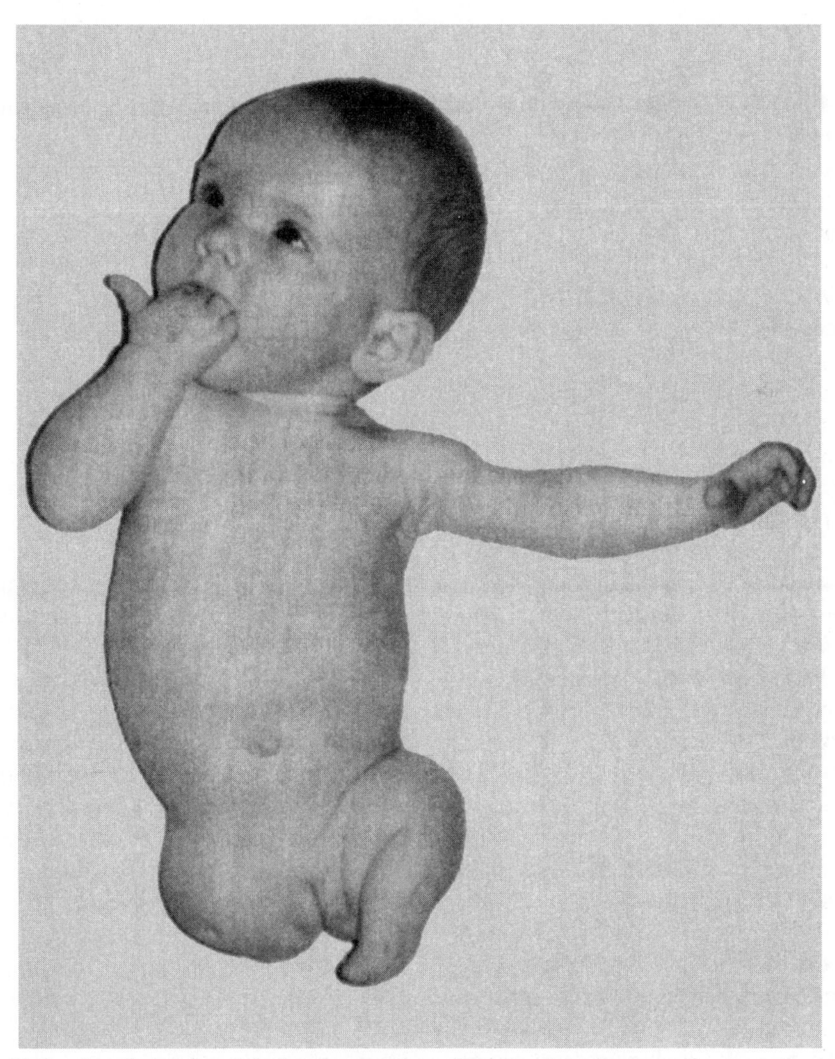

FIG. 11-2 Amelia of the right leg in an infant. Despite the absence of a foot, the left leg contains an upper and a lower leg segment.

From Connor JM, Ferguson-Smith MA: *Essential medical genetics,* ed 3, Oxford, England, 1991, Blackwell Scientific.)

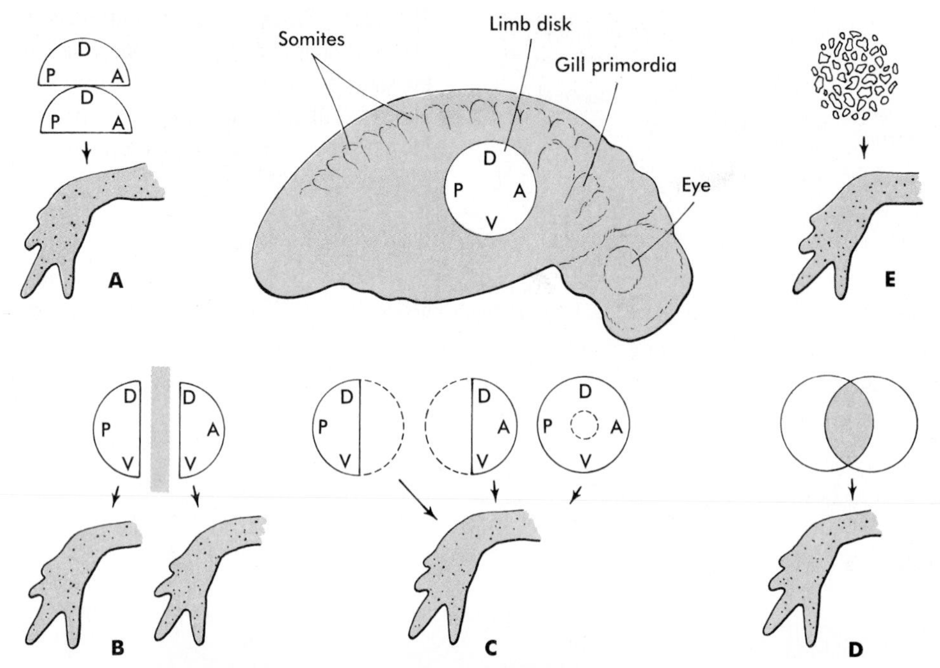

FIG. 11-3 Experiments demonstrating regulative properties of limb disks in amphibian embryos. **A,** Combining two identical halves of limb disks results in a single limb. **B,** Separation of two halves of a limb disk by a barrier results in each half forming a normal limb of the same polarity. **C,** After various types of tissue removal, the remaining limb tissue regulates to form a normal limb. **D,** Combining two disks results in the formation of a single normal limb. **E,** Mechanical disruption of a limb disk is followed by reorganization of the pieces and the formation of a normal limb. (Based on studies by Harrison and Swett.)

REGULATIVE PROPERTIES AND AXIAL DETERMINATION OF THE EARLY LIMB PRIMORDIUM

The early limb primordium is a highly regulative system, with properties very similar to those described for the cleaving embryo (see p. 40). These properties can be summarized with the following experiments (Fig. 11-3):

1. If part of a limb primordium is removed, the remainder reorganizes to form a complete limb.
2. If a limb primordium is split into two halves and these are prevented from fusing, each half gives rise to a complete limb (the twinning phenomenon).
3. If two equivalent halves of a limb primordium are juxtaposed, one complete limb forms.
4. If two equivalent limb disks are superimposed, they reorganize to form a single limb (see tetraparental embryos, p. 39).
5. In some species, disaggregated limb mesoderm can reorganize and form a complete limb.

The organization of the limb is commonly related to three linear axes based on the Cartesian coordinate system. The anteroposterior* axis runs from the first (anterior) to the fifth (posterior) digit. The back of the hand or foot is dorsal and the palm or sole is ventral. The proximodistal axis extends from the base of the limb to the digit tips.

Experiments involving transplantation and rotation of limb primordia in lower vertebrates have shown that these axes are fixed in a sequential order: anteroposterior → dorsoventral → proximodistal. Before all three axes are specified, a left limb primordium can be converted into a normal right limb simply by rotating it with respect to the normal body axes. These axes are important as reference points in several aspects of limb morphogenesis. Evidence indicates a similar sequence of axial specifications in certain other primordia, such as those of the retina and inner ear.

*Because of different conventions in the use of axial terms, some human embryologists would take exception to the axial terminology presented here. Specifically, according to strict human terminology, anterior means "ventral" and posterior means "dorsal." However, the axial terminology used in this chapter (anterior means "cranial" and posterior means "caudal") is so uniformly used in the experimental and comparative embryological literature that the student referring to the original literature in the field of limb development would find it very confusing to use human axial terminology.

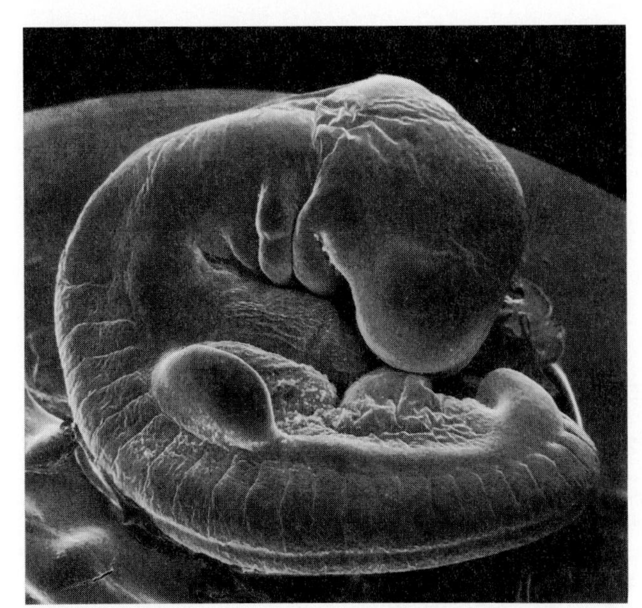

FIG. 11-4 Scanning electron micrograph of a 34-day-old human embryo (5 mm), with 34 pairs of somites. Toward the lower left, the right arm bud protrudes from the body. Pharyngeal arches are numbered.
(From Jirásek JE: *Atlas of human prenatal morphogenesis,* Amsterdam, 1983, Martinus Nijhoff Publishers.)

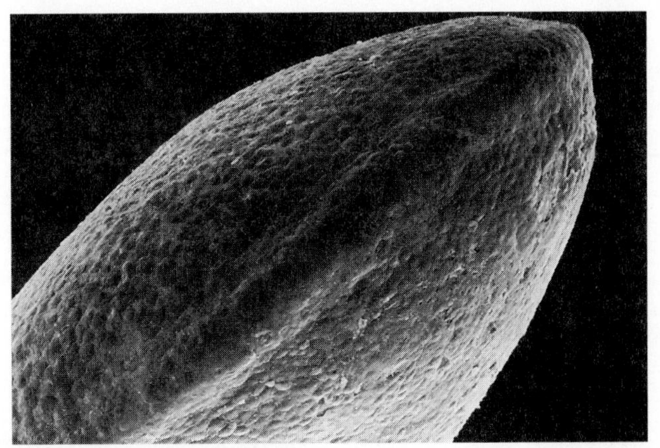

FIG. 11-5 Scanning electron micrograph of the flattened limb bud of a human embryo showing the prominent apical ectodermal ridge traversing the apical border.
(From Kelley RO, Fallon JF: *Dev Biol* 51:241-256, 1976.)

OUTGROWTH OF THE LIMB BUD

Shortly after its initial establishment, the limb primordium begins to bulge from the body wall (late in the first month for the human upper extremity [Fig. 11-4]). At this stage the limb bud consists of a mass of similar-looking mesodermal cells covered by a layer of ectoderm. Despite its apparently simple structure, the limb bud contains enough intrinsic information to guide its development, since if a mammalian limb bud is transplanted to another region of the body or is cultured in vitro, a recognizable limb forms.

A distinctive feature is the presence of a ridge of thickened ectoderm **(apical ectodermal ridge)** located along the anteroposterior plane of the apex of the limb bud (Fig. 11-5). During much of the time when the apical ectodermal ridge is present, the hand- and foot-forming regions of the developing limb bud are paddle shaped, with the apical ridge situated along the rim of the paddle (Fig. 11-6). Experiments have shown that the apical ectodermal ridge interacts with the underlying limb bud mesoderm to promote outgrowth of the developing limb. Other aspects of limb development such as **morphogenesis** (the development of form) are guided by information contained in the mesoderm.

This section outlines many of the ways in which the limb bud mesoderm and ectoderm interact to control limb development. Recognition of these developmental mechanisms is important in understanding the genesis of a number of limb malformations.

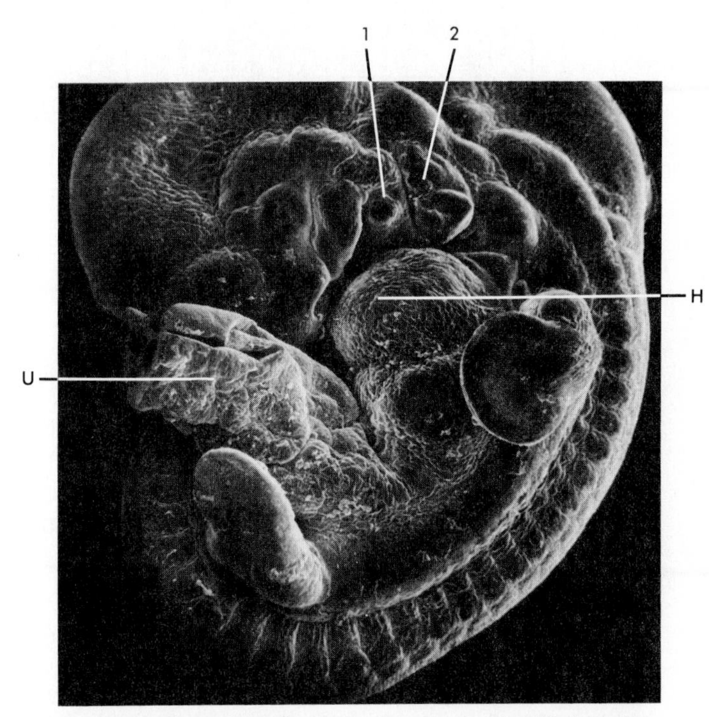

FIG. 11-6 Scanning electron micrograph of a 40-day-old human embryo (10 mm). The arm and leg buds are in the flattened paddle stage. *H,* Heart; *U,* umbilical cord; *1, 2,* pharyngeal arches 1 and 2.
(From Jirásek JE: *Atlas of human prenatal morphogenesis,* Amsterdam, 1983, Martinus Nijhoff Publishers.)

Apical Ectodermal Ridge

The human apical ectodermal ridge is a multilayered epithelial structure (Fig. 11-7) characterized by the presence of numerous gap junctions through which the cells are interconnected. A basal lamina is interposed between the apical ridge and the underlying mesodermal cells.

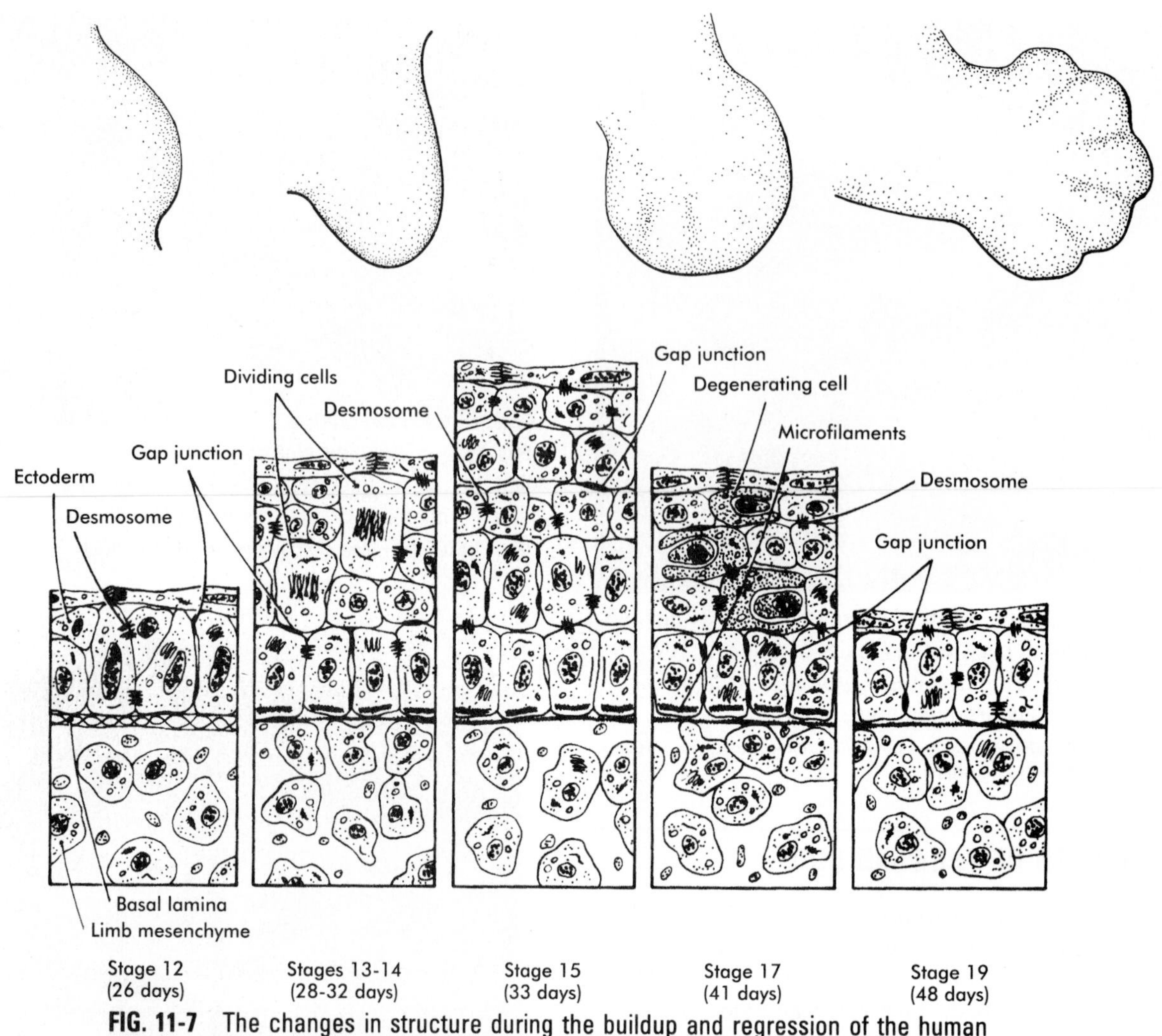

FIG. 11-7 The changes in structure during the buildup and regression of the human apical ectodermal ridge. Beneath the early ridge the basal lamina is double layered with cross-links. At later stages it is a single-layered structure.
(Modified from Kelley RO, Fallon JF: *Dev Biol* 51:241-256, 1976.)

Although the apical ectodermal ridge has been recognized morphologically for many years, its role in limb development was not understood until it was subjected to experimental analysis. Removal of the apical ridge results in an arrest of limb development, leading to distal truncation of the limb (Fig. 11-8). In the *limbless* mutant in chickens, early limb development is normal; later the apical ectodermal ridge disappears and further wing development ceases. If mutant ectoderm is placed over normal wing bud mesoderm, limb development is truncated, whereas combining mutant mesoderm with normal ectoderm results in more normal limb development, suggesting that the ectoderm is defective in this mutant. Conversely, the presence of an additional apical ectodermal ridge on a limb bud, whether experimentally transplanted or through genetic mutation, results in the formation of a supernumerary limb (Fig. 11-9).

How the apical ectodermal ridge promotes outgrowth of the limb bud is not well understood. The ridge keeps the distal-most mesenchymal cells of the limb bud in an undifferentiated state and appears to promote their proliferation. Evidence from in vitro studies suggests that the apical epidermis influences the behavior of the distal mesenchymal cells by affecting the properties of the extracellular matrix that surrounds them, but other influences may also be operating.

Recent experimental evidence suggests that the expression of two homeobox-containing genes in the chick (*GHox-7* and *GHox-8*) are involved as either signal or response elements in the apical ectodermal-mesodermal interaction that promotes limb outgrowth. *GHox-8* is expressed in the apical ectodermal ridge and in anterior limb bud mesoderm, whereas *GHox-7* is expressed only in the mesoderm of the limb bud, with a higher concentration in the anterior than that in the posterior mesoderm. In the *limbless* chick mutant, the expression of *GHox-7* is greatly reduced in distal limb mesoderm, suggesting that the apical ectodermal ridge emits a signal important for the expression of *GHox-7* in the mesoderm.

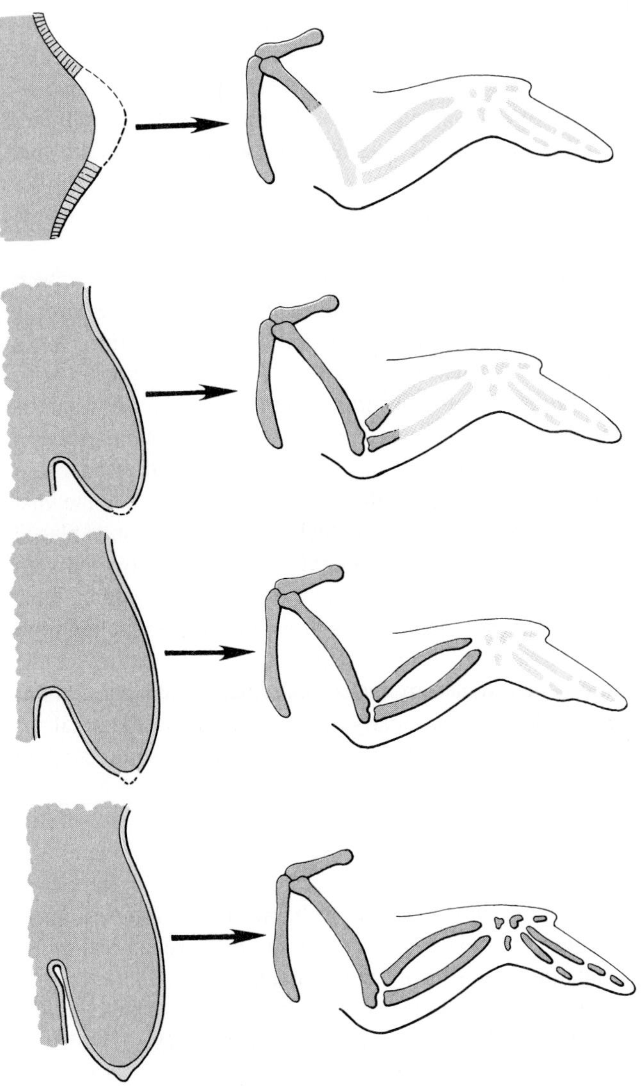

FIG. 11-8 The effect of removing the apical ectodermal ridge at successively later stages *(top three)* on development of the avian wing bud. The more mature the wing bud, the more skeletal elements form after apical ridge removal. Missing structures are shown in light gray. *Bottom:* Normal development of an untouched wing bud.
(Based on studies by Saunders.)

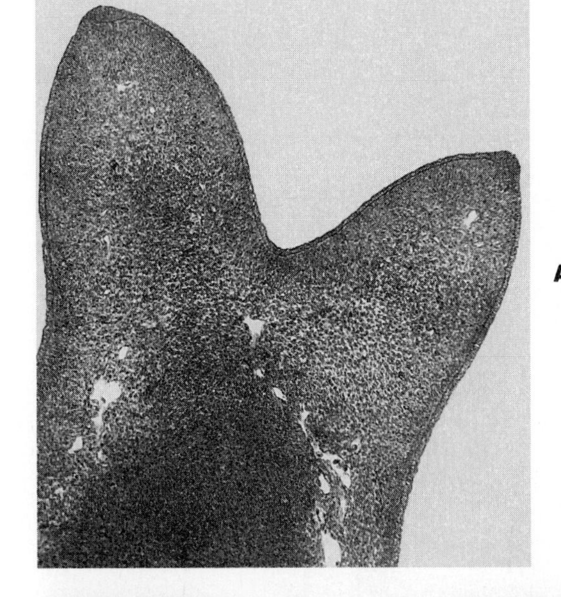

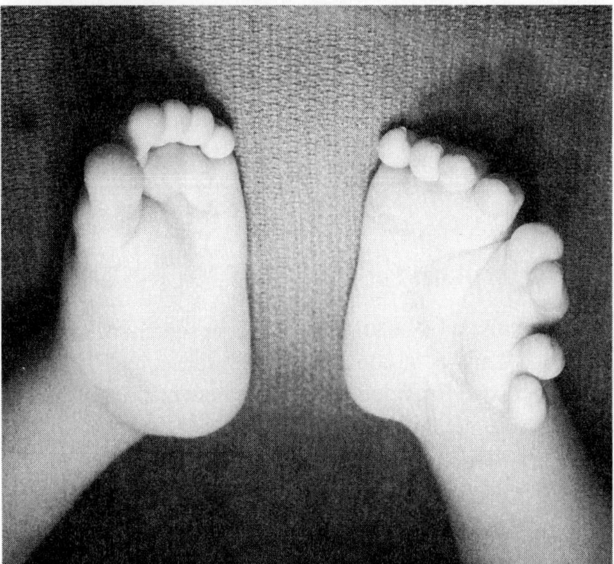

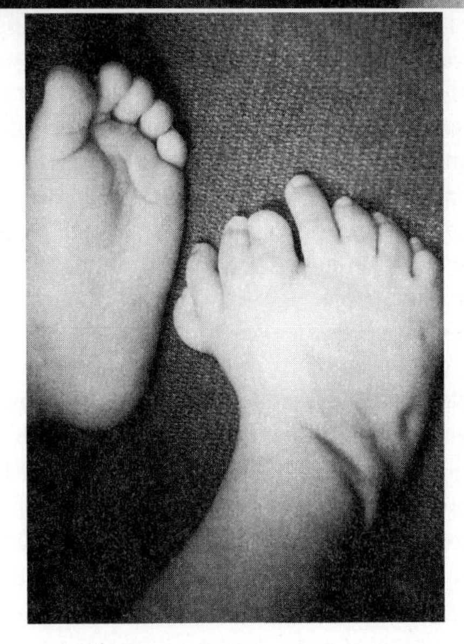

FIG. 11-9 A, Duplicated wing bud in a chick with eudiplopodia. Under the influence of a secondary apical ectodermal ridge, a supernumerary limb bud forms.
(Courtesy P. Goetinck, Boston, Mass.)
B, Diplopodia in the human. Dorsal and ventral views of the right foot.
(From Hootinck O and others: In Feinberg R and others, eds: *The development of the vascular system,* Basel, Switzerland, 1991, Karger.)

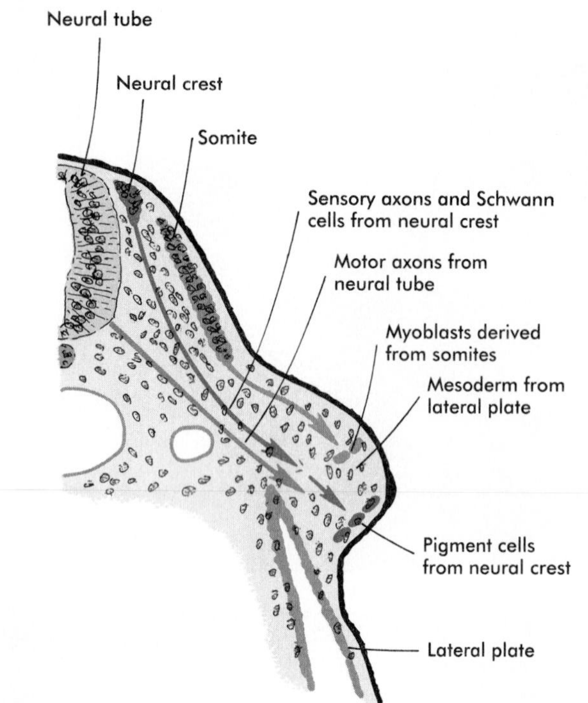

FIG. 11-10 The different types of cells that enter the limb bud.

Mesoderm of the Early Limb Bud

Structure and Composition. The mesoderm of the early limb bud consists of homogeneous mesenchymal cells supplied by a well-developed vascular network. The mesenchymal cells are embedded in a matrix consisting of a loose mesh of collagen fibers and ground substance, with hyaluronic acid and glycoproteins being prominent constituents of the latter. There are no nerves in the early limb bud.

It is not possible to distinguish different cell types within the early limb bud mesenchyme by their morphology alone. Nevertheless, mesenchymal cells from several sources are present (Fig. 11-10). Initially the limb bud mesenchyme consists exclusively of cells derived from the lateral plate mesoderm. These cells give rise to the skeleton, connective tissue, and blood vessels. Mesenchymal cells derived from the somites migrate into the limb bud as precursors of muscle cells. Another population of migrating cells is that from the neural crest, which ultimately form the Schwann cells of the nerves and pigment cells **(melanocytes).**

Mesodermal-Ectodermal Interactions and the Role of the Mesoderm in Limb Morphogenesis. Limb development occurs as the result of continuous interactions between the mesodermal and ectodermal components of the limb bud. As previously discussed, the ectoderm stimulates outgrowth of the limb bud, probably by promoting mitosis and preventing differentiation of the distal mesodermal cells of the limb bud. Although the apical ridge promotes outgrowth, its extent is determined by the mesoderm. If an apical ridge from an old limb bud is transplanted onto the mesoderm of a young wing bud, the limb grows normally until morphogenesis is complete. However, if old limb bud mesoderm is covered by young apical ectoderm, limb development ceases at a time appropriate for the age of the mesoderm and not that of the ectoderm.

Similar reciprocal transplantation experiments have been used to demonstrate that the overall shape of the limb is determined by the mesoderm and not the ectoderm. This is most dramatically represented by experiments done on birds because of the great differences in morphology between the extremities (Fig. 11-11). For example, if leg bud mesoderm in the chick embryo is covered with wing bud ectoderm, a normal leg covered with scales develops. In a somewhat more complex example, if chick leg bud ectoderm is placed over duck wing bud mesoderm, a duck wing covered with chicken feathers forms. Such experiments, which have sometimes involved mosaics of avian and mammalian limb bud components, show that the overall morphology of the limb is determined by the mesodermal component and not the ectoderm. Additionally, the nature of the ectodermal appendages, (hair in the case of mammals) is also dictated by the mesoderm. Cross-species grafting experiments, however, show that the ectodermal appendages that are formed are appropriate for the species from which the ectoderm was derived.

Polydactyly is a condition characterized by supernumerary digits and exists as a mutant in birds. Reciprocal transplantation experiments between mesoderm and ectoderm have shown that the defect is inherent in the mesoderm and not the ectoderm. Polydactyly in humans (Fig. 11-12) is typically inherited as a genetic recessive trait and is commonly found in populations such as certain American Amish communities, where the total genetic pool is relatively restricted.

Experimental evidence suggests that the limb mesoderm produces a still uncharacterized **maintenance factor** that preserves the integrity and function of the apical ectodermal ridge. Thus both mesoderm and ectoderm depend on influences exerted by the other for their own structure and function.

Cell Death and the Development of Digits. Although it may seem paradoxical, genetically programmed **cell death** is important in the development of a number of structures in the body. In the limb, it is prominently manifested in the future axillary region, between the radius and ulna, and in the interdigital spaces (Fig. 11-13). Experiments on avian embryos have shown that to a certain stage, mesodermal cells scheduled to die could be spared by transplanting them to areas in which cell death did not normally occur. However, after a certain time, a "death clock" was set (an example of determination) and the cells could no longer be rescued.

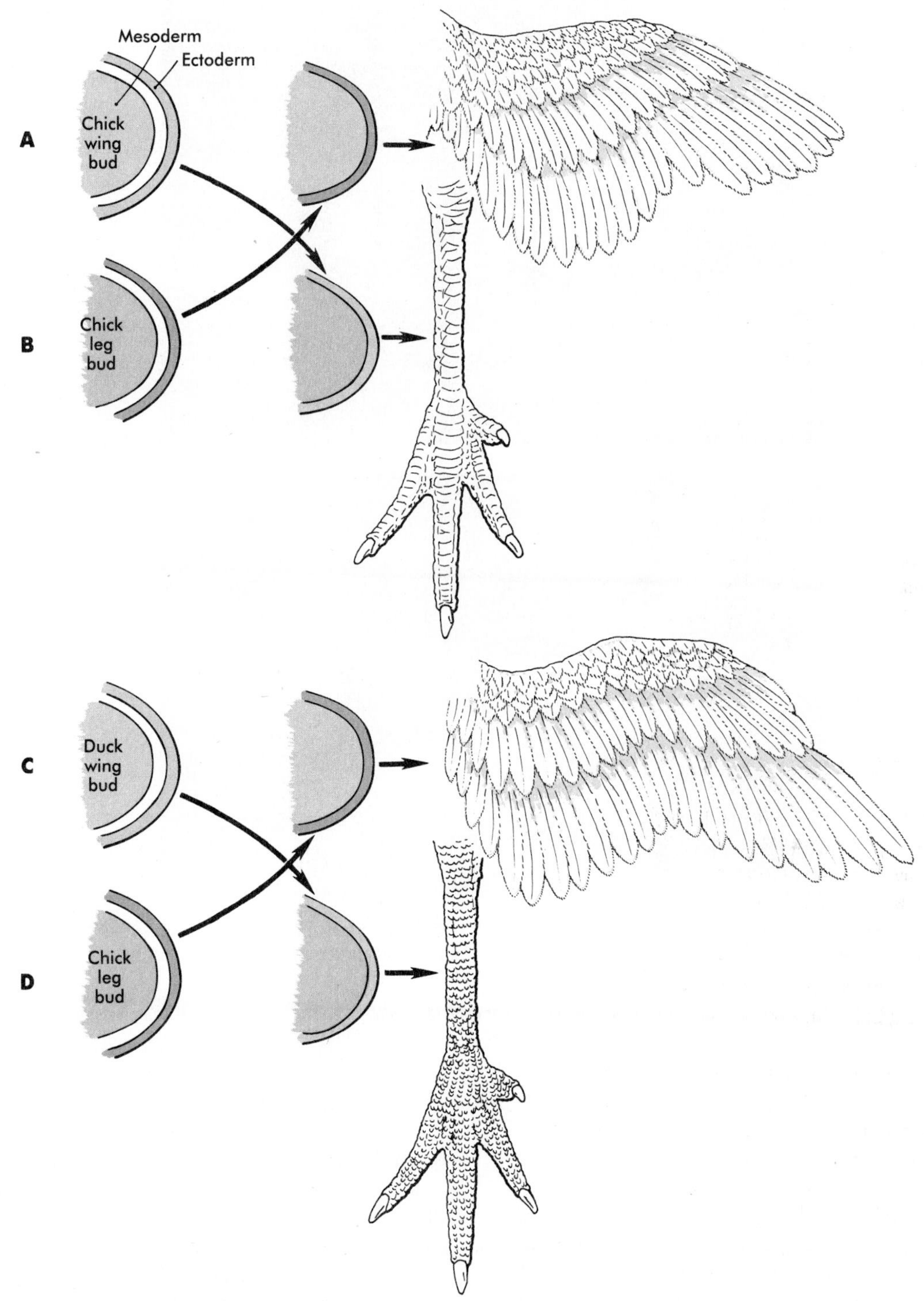

FIG. 11-11 Examples of recombination experiments involving limb bud mesoderm and ectoderm. **A,** Chick wing mesoderm covered by chick leg ectoderm produces a normal wing covered with feathers. **B,** Chick leg mesoderm covered with chick wing ectoderm produces a normal leg covered with scales. **C,** Duck wing mesoderm covered with chick leg ectoderm results in a duck wing covered with chick wing feathers. **D,** Chick leg mesoderm covered by duck wing ectoderm produces a chicken leg covered with duck scales.

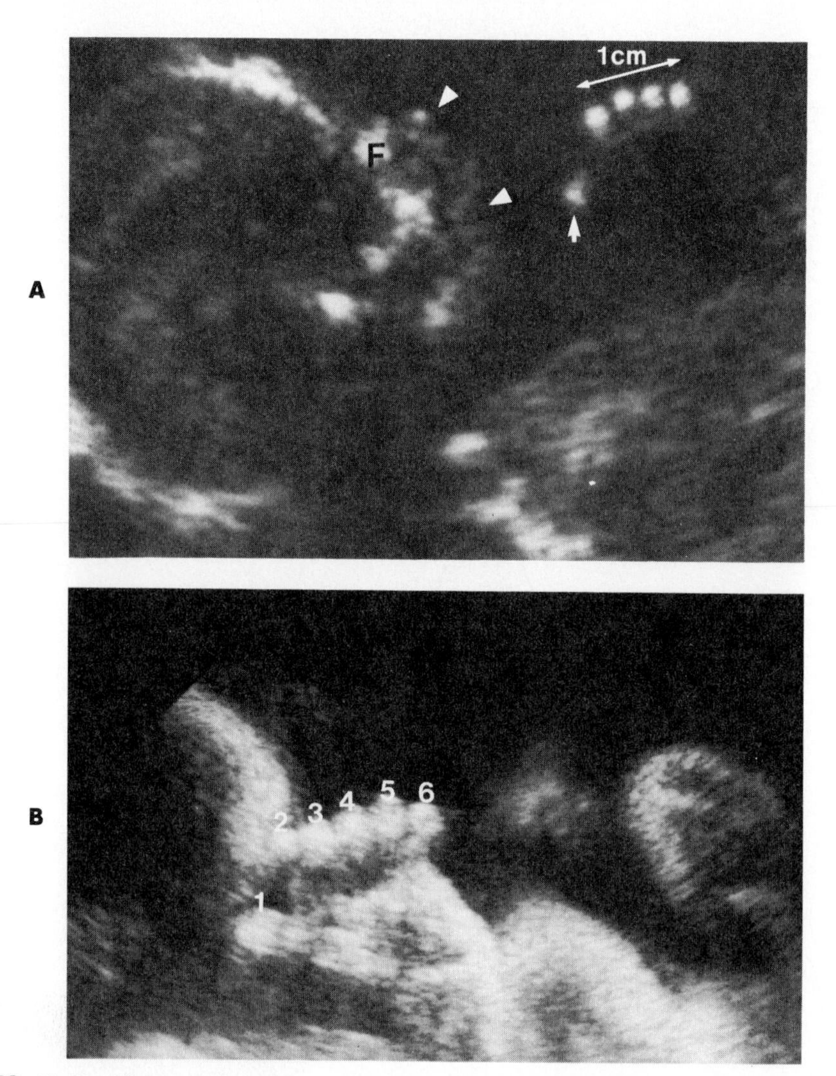

FIG. 11-12 Ultrasound images of **(A)** normal and **(B)** polydactylous (six digits) hands of human fetuses 16 and 31 weeks old. In both cases, the digits are imaged in cross section.

(**A** from Bowerman R: *Atlas of normal fetal ultrasonographic anatomy,* St Louis, 1992, Mosby; **B** from Nyberg D and others: *Diagnostic ultrasound of fetal anomalies,* St Louis, 1990, Mosby.)

As limb development proceeds and distal cell death sets in, changes also become apparent in the apical ectodermal ridge. Instead of remaining continuous around the entire apex of the limb, the ridge begins to break up, leaving intact segments of thickened ridge epithelium covering the digital rays. Between the digits the ridge regresses, possibly as the result of the cessation of epidermal maintenance factor by the dying mesodermal cells of the interdigital region (Fig. 11-13, *A*). In the chick embryo, areas of interdigital cell death are associated with the presence of retinoic acid receptors and binding protein (Fig. 11-13, *B*). *Hox-7* is also expressed in the same regions. As the digital primordia continue to grow outward, further cell death sculpts the interdigital spaces (Fig. 11-13, *C*). If interdigital cell death does not occur, a soft tissue web connects the digits on either side. This is the basis for the normal development of webbed feet on ducks and the abnormal formation of syndactyly (Fig. 11-14, *A*) in humans.

All human digits contain three phalangeal segments except for the first digits (thumb and great toe), which consist of only two segments. Some investigators have attributed the development of the diphalangeal first digits to the actions of a small zone of cell death that is thought to occur at the tip of the first digital primordium. On rare occasions an individual is born with a triphalangeal thumb (Fig. 11-14, *B*). This anomaly has been attributed to the absence of normally occurring cell death at the tip of the primordium of the thumb.

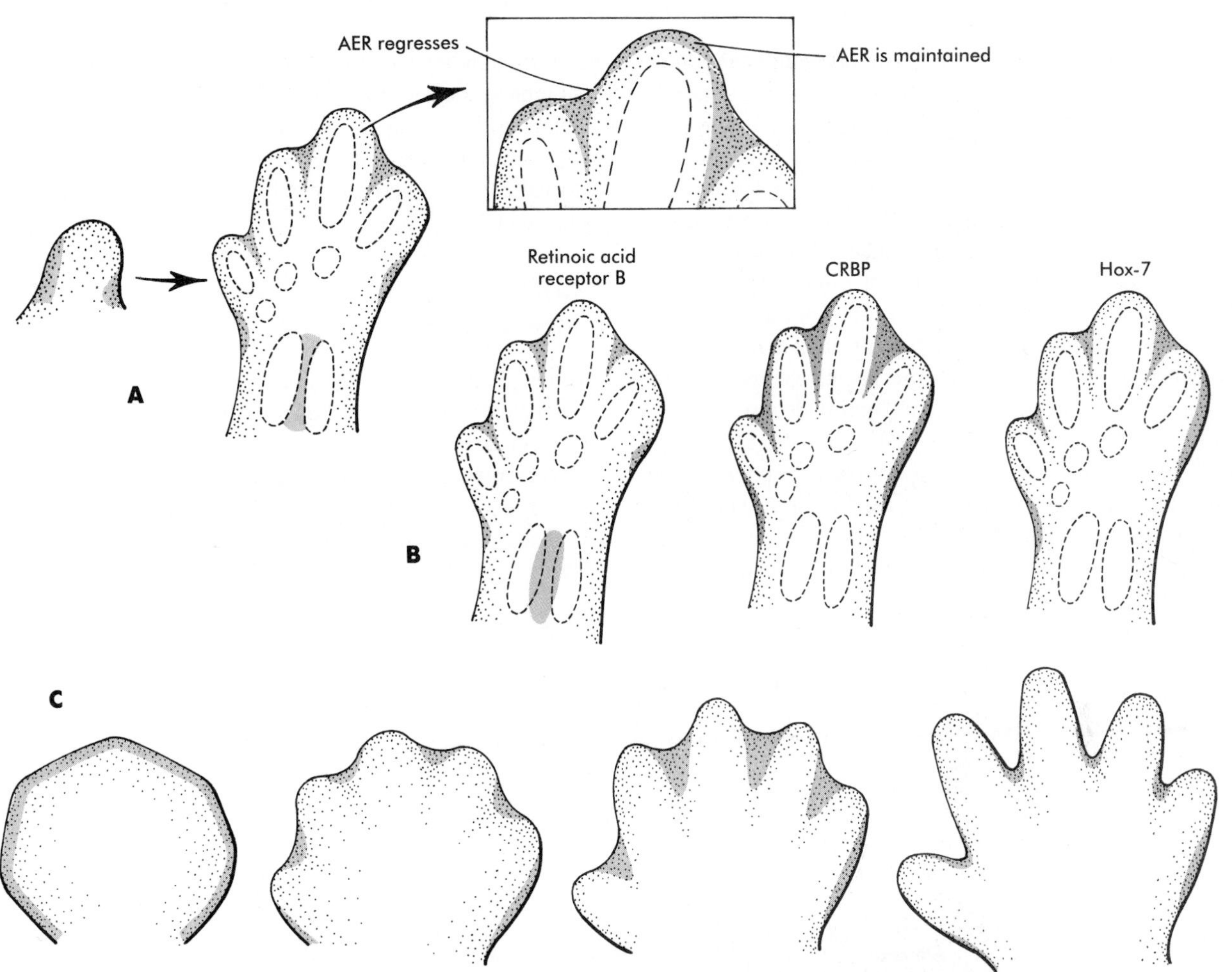

FIG. 11-13 Cell death in development of the hand and digits. **A,** Cell death in the chick limb bud. **B,** Gene expression in zones of cell death of the chick embryo. **C,** Cell death in the developing human hand. *CRBP,* Cellular retinoic acid binding protein.

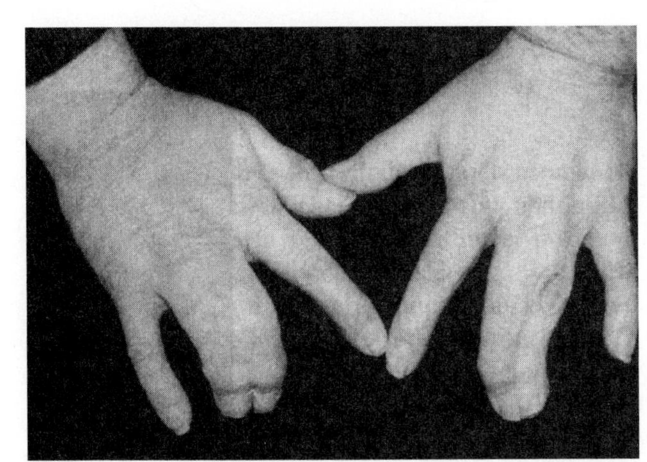
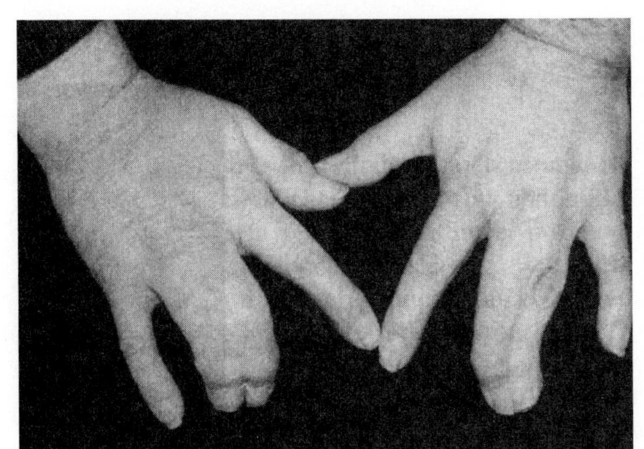

FIG. 11-14 **A,** Syndactyly in the human.

(From Connor J, Ferguson-Smith M: *Essential medical genetics,* ed 2, Oxford, England, 1987, Blackwell Medical.)

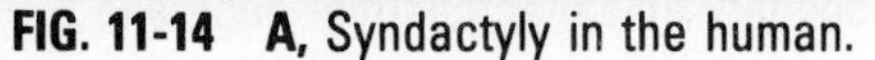

B, Triphalangeal thumb in a human fetus.

(Courtesy Mason Barr, Ann Arbor, Mich.)

The Zone of Polarizing Activity and Morphogenetic Signaling. During the course of experiments investigating morphogenetic cell death, researchers grafted mesodermal cells in the posterior base of the avian wing bud into the anterior margin. This manipulation resulted in the formation of a supernumerary wing, which was a mirror image of the normal wing (Fig. 11-15). Much subsequent experimentation has shown that the **zone of polarizing activity (ZPA)** acts as a biological signaler and appears to determine the organization of the limb along its anteroposterior axis. Cross-species grafting experiments have demonstrated that mammalian (including human) limb buds also contain a functional ZPA. A transplanted ZPA seems to act on the apical ectodermal ridge, eliciting a growth response from the mesenchymal cells just beneath the part of the ridge adjacent to the transplanted ZPA. As few as 50 cells from the ZPA can stimulate supernumerary limb formation.

The boundaries of the ZPA cannot be delineated by morphological means, and neither the nature of its signal nor the means by which the signal is transmitted have been defined. Recent experiments have shown that the effect of a grafted ZPA can be eliminated by blocking the function of gap junctions with an antibody.

Models of Morphogenetic Control of the Developing Limb

Although the control of limb morphogenesis is still not completely understood, several theoretical models have provided a basis for interpreting experimental results and for designing new experimental approaches directed toward identifying new factors that guide limb development.

Many of the current approaches toward the morphogenesis of limbs and other complex structures are based on the

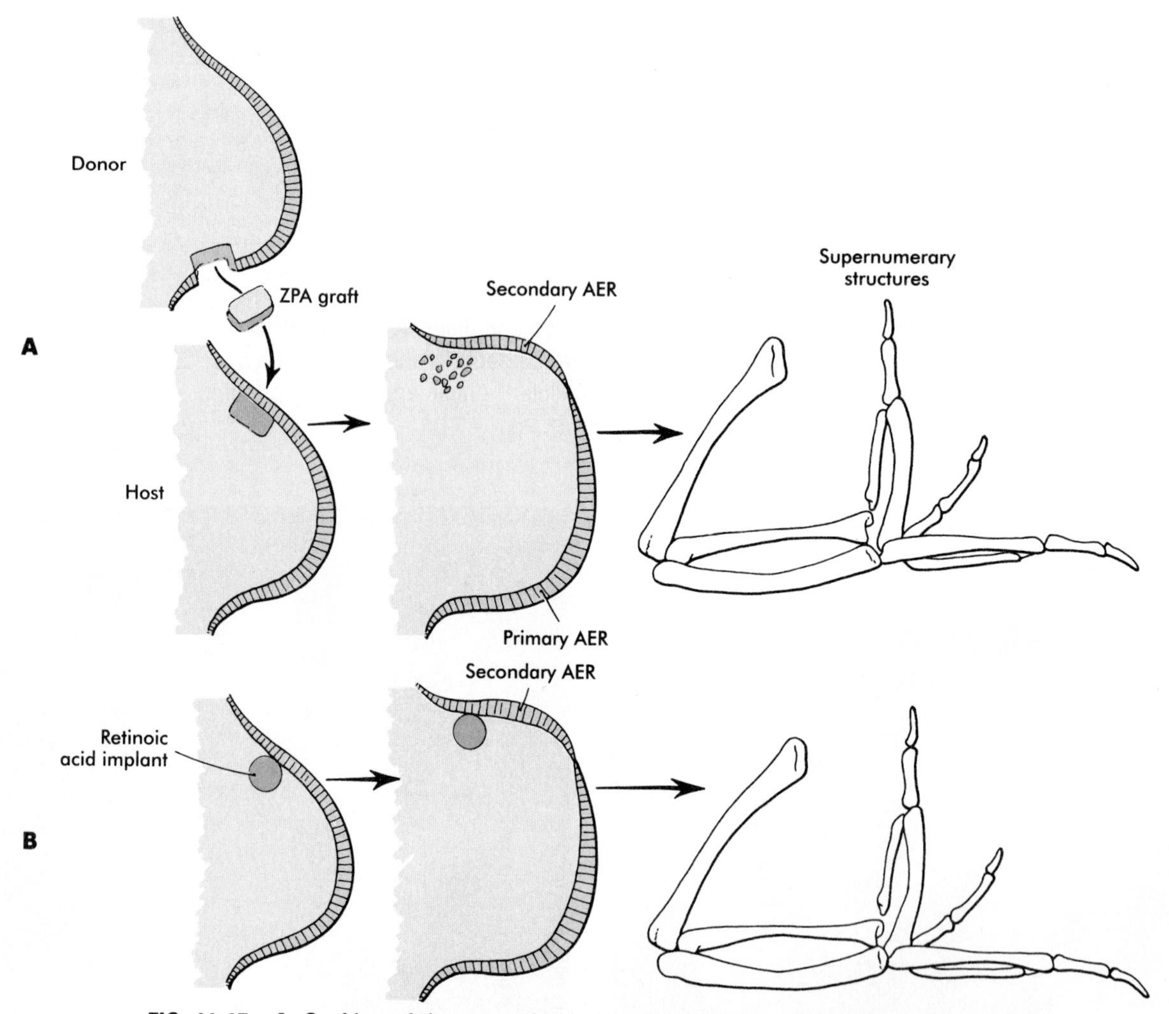

FIG. 11-15 **A,** Grafting of the zone of polarizing activity into the anterior border of the avian limb bud results in the formation of a secondary apical ectodermal ridge and a supernumerary limb. **B,** Implantation of a bead soaked in retinoic acid into the anterior border of the limb bud also stimulates the formation of a supernumerary limb.

theory of **positional information.** According to this concept, cells in the developing limb are exposed to positional cues that allow them to determine their relative position in the limb bud. The cells somehow process this information and then develop into structures appropriate to their relative positions. This theory implies that cells are morphogenetically interchangable (have a great regulative capacity) until they receive the appropriate positional information. In some systems, there is evidence that once they have been exposed to positional information the cells "remember" their original position (positional memory) even if they are transplanted elsewhere.

One commonly held hypothesis of morphogenetic control along the anteroposterior axis demonstrates the concept of positional information. As outlined previously, the ZPA is thought to control pattern formation along the anteroposterior axis. Many investigators hypothesize that the ZPA produces a diffusible **morphogen,** which forms a gradient from high to low concentration along the anteroposterior axis. The mesodermal cells along this axis sample and interpret the concentration of the ZPA morphogen (which represents the positional information that they receive). Those that are exposed to the highest concentrations of the putative morphogen organize into posterior structures, whereas those that see only low concentrations of the morphogen form anterior structures (Fig. 11-16). The formation of supernumerary limbs or digits as the result of grafts of ZPA cells into the anterior margin of the limb is attributed to the formation of a new gradient of ZPA morphogen with opposite polarity. This would also account for the supernumerary limb being a mirror image of the original limb.

Morphogenetic control along the proximodistal axis has been addressed by the **progress zone model.** According to this model the apical ectodermal ridge keeps a zone of mes-

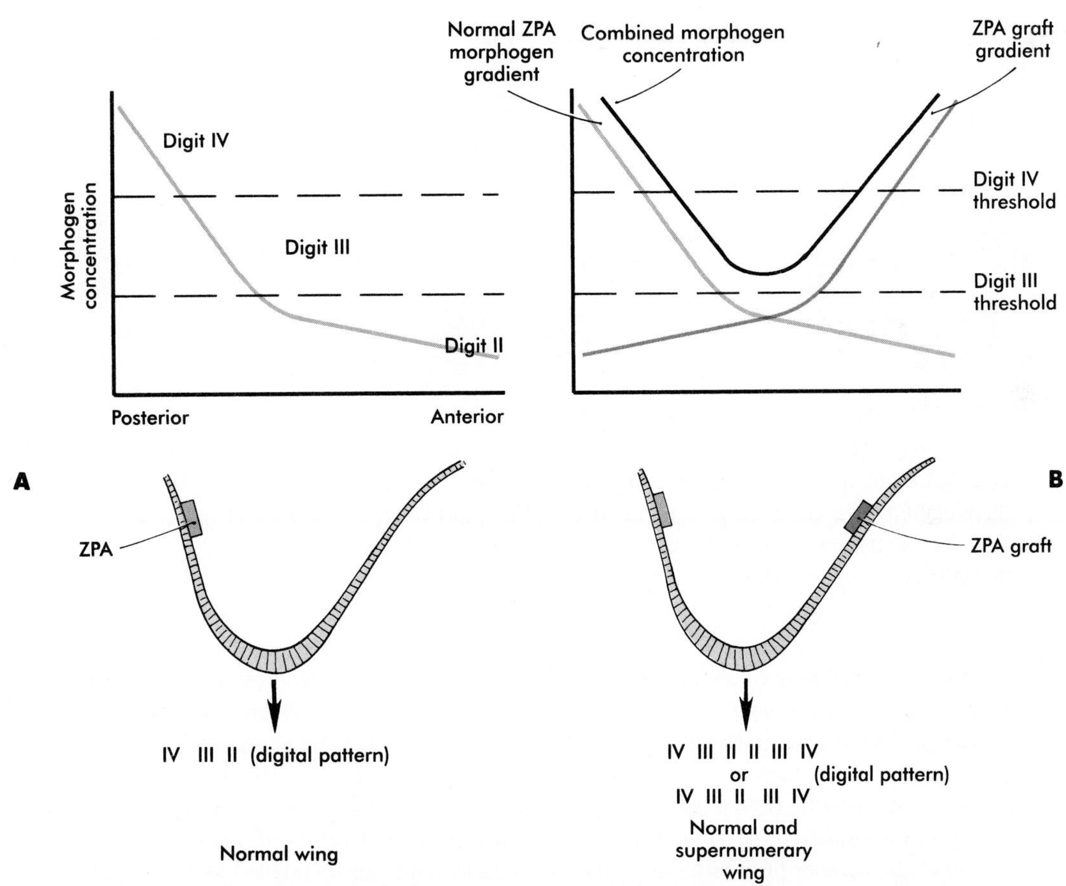

FIG. 11-16 A, The concentration gradient of the hypothetical morphogen released from the zone of polarizing activity. The lowest concentration *(below the lower dashed line)* is compatible with formation of digit II; the highest concentration *(above the upper dashed line)* is compatible with formation of digit IV. **B,** Explanation of the formation of duplicate limb structures after a ZPA graft. The additive level of the two morphogen gradients from the normal and grafted ZPAs determines the types of digits that form.

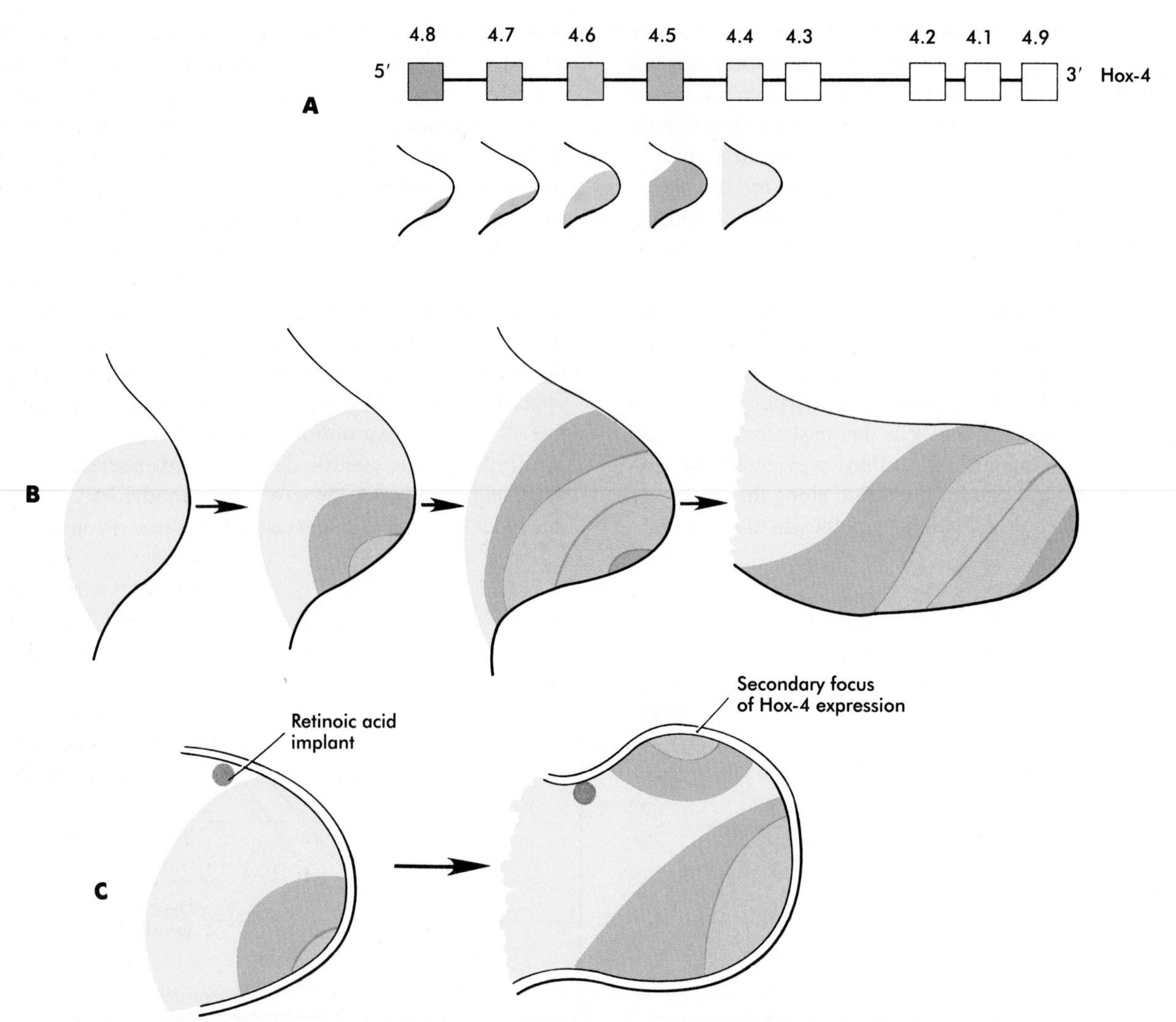

FIG. 11-17 *Hox-4 (Hox-D)* gene expression in the chick limb bud. **A,** The map of this gene family and the distribution of individual gene products. **B,** The development of the aggregate pattern of *Hox-4* gene expression over time in the normal limb bud. **C,** The development of a secondary focus of *Hox-4* gene expression in the area of supernumerary limb formation caused by an implant of retinoic acid.
(Based on studies by Tabin [1991].)

enchymal cells extending to a thickness of about 300 μm beneath it (the progress zone) in a proliferative and morphogenetically uninstructed state. As the limb bud elongates distally, cells formerly under the influence of the apical ridge escape from its influence. Having been exposed to an undefined form of positional information (possibly a period of time or a recognition of the number of mitotic divisions they have gone through), the mesenchymal cells that leave the progress zone have acquired positional values and are thus able to recognize their respective roles in the formation of proximodistal structures such as skeletal elements.

Evidence is increasing that expression of the homeobox-containing gene *GHox-7* is a property of cells in the progress zone. In the normal limb bud, expression of this gene is high in the progress zone, but it is weak or absent in more proximal mesoderm. If a piece of proximal mesoderm, which does not express *GHox-7,* is transplanted into the progress zone, the cells reexpress the gene. From earlier experimental work it is also known that the proximodistal positional values of a similarly transplanted piece of mesenchyme are reassigned as well. Thus there is a close correlation between the induction of *GHox-7* and a change in the positional values of cells.

In essentially all models based on positional information, the molecular and even cellular nature of positional information are poorly understood. Nevertheless, the predictive value of these models has allowed investigators to design new experimental approaches to limb development.

Retinoic Acid and Homeoboxes. For many years it has been recognized that **vitamin A** or its derivatives (retinoids) can cause malformations or disturbances in growth of avian and mammalian (including human) embryos. A major breakthrough in approaching mechanisms underlying limb development occurred when it was shown that exogenous retinoic acid applied to the anterior margin of the limb bud causes duplications virtually identical to those caused by ZPA grafts in the same location (see Fig. 11-15, *B*). This finding, along with the demonstration of a natural gradient of retinoic acid that has roughly the same slope as that of the postulated ZPA morphogen, led to the hypothesis that retinoic acid is the long-sought ZPA morphogen.

Subsequent experimentation and examination of the available data on amounts and distribution of cytoplasmic binding proteins and nuclear receptors for retinoic acid have not always been consistent with this hypothesis. In fact, evidence is increasing that one effect of retinoic acid on limb bud mesenchyme is to induce polarizing activity in the mesenchymal cells exposed to sufficient amounts of retinoic acid for a sufficient time. An intriguing relationship that is just beginning to be delineated is the sequential activation of certain homeobox genes after the application of exogenous retinoic acid.

As in several other regions of the body, the morphologically homogenous limb bud exhibits some striking patterns of expression of certain of the homeobox-containing genes. In fish and amphibians, the appearance of the *X1HBox 1* in a highly localized region of lateral plate mesoderm is the first indication of any sort that limb formation will take place. In embryos of birds and mammals, the human homologue of this gene, *Hox-3.3* (C6), appears along the anterior margin of the early limb bud.

Members of the **Hox-4** complex (earlier called *Hox-5* and now called *Hox-D*) show a striking pattern of sequential expression during limb development. The most downstream member of the complex (*Hox-4.4* [D9]) is first expressed in the posterior region of the early limb bud (Fig. 11-17, *A*). Subsequently, *Hox-4.4* (D9) through *Hox-4.8* (D13) are expressed in successive fronts extending from the initial site of expression of *Hox-4.4* (D9) toward the posterodistal margin of the limb bud. In normal limb development the expression domains of *Hox-4 (Hox-D)* and *Hox-3.3* (C6) almost appear to confront one another.

Of considerable interest is the pattern of *Hox* expression during the formation of supernumerary limbs induced by ZPA grafts or applications of retinoic acid (Fig. 11-17, *C*). The zone of expression of *Hox-3.3* (C6) expands to overlap with that of *Hox-4* (D), but even more striking is the appearance of a new zone of expression of the *Hox-4* (D) complex emanating from the region of the induced supernumerary limb.

Although the data collected do not provide a molecular explanation for limb morphogenesis, they show that the expression of *Hox-4* (D) is strongly correlated with the development of both normal and supernumerary limbs. The expression of *Hox-4.4* (D19) through *Hox-4.8* (D13) follows strict temporal and spatial gradients, the latter of which follows the order of the location of these genes on the chromosome. Members of the *Hox-4.4* (D9) through *Hox-4.8* (D13) complex can be expressed only if the previous member of the complex has been expressed; that is, the expression of *Hox-4.4* (D9) must precede that of *Hox-4.5* (D10) and so on. This may be an important factor in determining the order and types of the digits. In the developing avian wing, if supernumerary limbs are induced in such a way that the anterior-most digit in the normal limb is missing, the more upstream elements of the *Hox-4* (D) complex are correspondingly not expressed.

Application of retinoic acid to the anterior border of the chick limb bud results in striking changes in homebox gene expression as the supernumerary limb is being formed. Expression of *GHox-8,* which occurs in anterior mesoderm of the progress zone, is reduced, whereas a duplicate pattern of the *Hox-4* (D4) genes unfolds as the supernumerary limb takes shape. It is not certain whether retinoic acid directly activates or inactivates these homeobox genes. The complexity of the system is such that several unknown molecular steps might intervene between the accumulation of effective amounts of retinoic acid in the limb bud and the expression of homeobox genes.

DEVELOPMENT OF LIMB TISSUES

The morphogenetic events described take place largely during the early stages of limb development when the limb bud consists of a homogeneous-appearing mass of mesodermal cells covered by ectoderm. The differentiation and histogenesis of the specific tissue components of the limb are later developmental events that build on the morphogenetic blueprint already established.

Skeleton

The skeleton is the first major tissue of the limb to show signs of overt differentiation. Its gross morphology, whether normal or abnormal, closely reflects the major pattern-forming events that shape the limb as a whole. Formation of the skeleton can be first seen as a condensation of mesenchymal cells in the central core of the proximal part of the limb bud. Even before undergoing condensation, these cells are determined to form cartilage, and if they are transplanted to other sites or into culture, they differentiate only into cartilage. However, other mesenchymal cells that would normally form connective tissue retain the capacity to differentiate into cartilage if they are transplanted into the central region of the limb bud.

Little is known about the way the morphogenetic blueprint of the limb is translated into specific skeletal structure, in particular whether the cells respond to a set of glo-

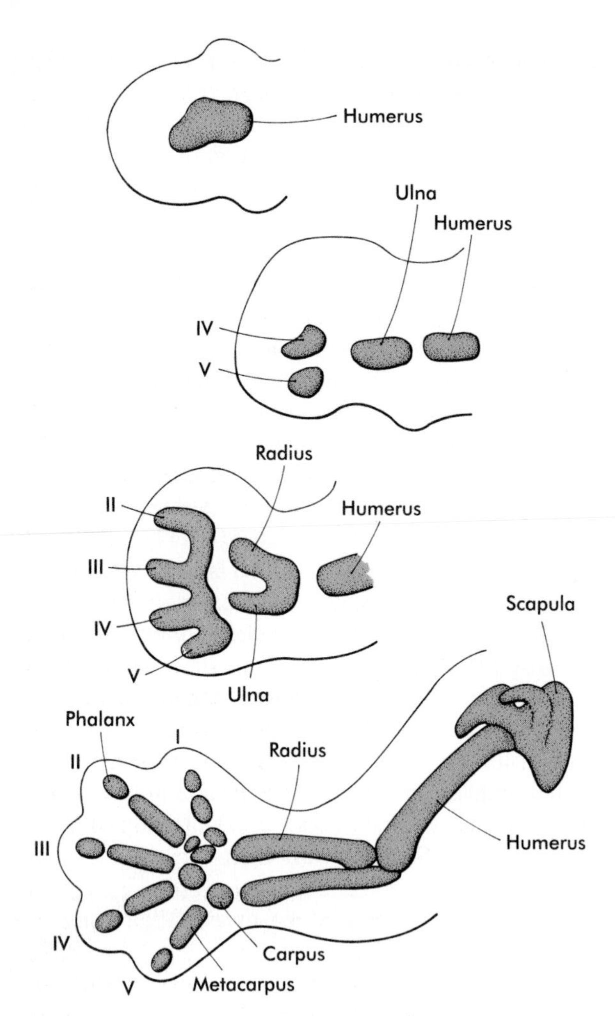

FIG. 11-18 Formation of the skeleton in the mammalian forelimb.

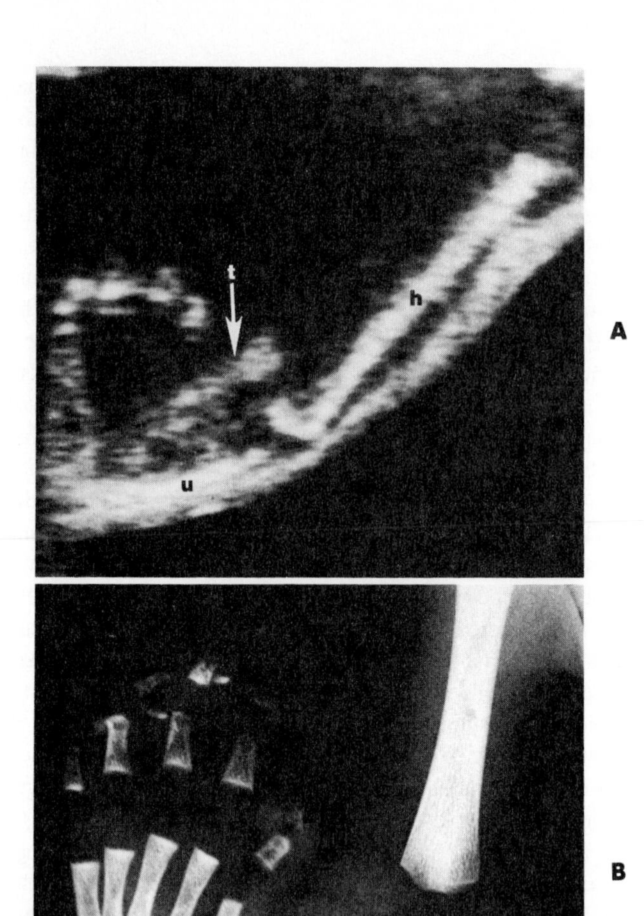

FIG. 11-19 Radial hemimelia (absent radius) in a 27-week-old fetus. **A,** Ultrasound image showing a thumb *(arrow)* but no radius. **B,** Postnatal x-ray image confirming the absent radius. *u,* Ulna; *h,* humerus.

(From Nyberg D, Mahony B, Pretorius D: *Diagnostic ultrasound of fetal anomalies,* St Louis, 1990, Mosby.)

bal instructions, to microenvironmental cues, or to both. It is now widely recognized that the ectoderm of the limb bud exerts an inhibitory effect on cartilage differentiation, so cartilage does not form in the region just beneath the ectoderm. In vitro studies suggest that the inhibition is mediated by a diffusible material secreted by the ectoderm. Whether this is the same mechanism that maintains the integrity of the progress zone remains to be determined.

Differentiation of the cartilaginous skeleton occurs in a proximodistal sequence, and in mammals the postaxial structures of the distal limb segments differentiate before the preaxial structures. For example, the sequence of formation of the digits is from the fifth to the first (Fig. 11-18). The postaxial skeleton of the arm is considered to be the ulna, digits 4 and 5, and the corresponding carpal elements. The radius, digits 1 through 3, and the corresponding carpal bones constitute the preaxial skeleton. Such terminology is often used to classify certain limb defects called **hemimelias,** in which many or all of the preaxial or postaxial components of the limb are missing (Fig. 11-19).

Musculature

The musculature of the limb is derived from myogenic cells that migrate into the very early limb bud from the somites. These cells, which are morphologically indistinguishable from the other mesenchymal cells, spread throughout the limb bud and keep pace with the elongation of the limb bud. According to experiments on avian embryos, however, myogenic cells do not extend into the distal 300 μm of the limb bud mesenchyme (roughly the extent of the progress zone).

Shortly after the condensations of the skeletal elements take shape, the myogenic cells themselves begin to coalesce into two common muscle masses, one the precursor of the flexor muscles and the other giving rise to the extensor muscles. The next stage in muscle formation is the splitting of the **common muscle masses** into anatomically recognizable precursors of the definitive muscles of the limb. Little is known about the mechanisms that guide the splitting of the common muscle masses. The fusion of myoblasts into early myotubes begins to occur during these early stages of muscle development.

Considerable evidence suggests that myogenic precursor cells do not possess intrinsic information guiding their morphogenesis. Rather, it appears that the myogenic cells follow the lead of connective tissue cells, which are the bearers and effectors of the morphogenetic information required to form anatomically correct muscles. Experiments in which the somites normally associated with a limb bud are removed and replaced by somites from elsewhere along the body axis have shown that the myogenic cells are morphogenetically neutral. Muscle morphogenesis is typically normal even though the muscle fiber precursors are derived from abnormal sources.

Through the removal of somites, the use of x-radiation, and the analysis of certain mutants such as *wingless,* it has been shown that the tendons and connective tissue components of a muscle are derived from the limb bud mesoderm, whereas the muscle fibers themselves are somite derived. In situations where myogenic cells are prevented from entering the limb bud, morphologically appropriate tendons still form, but these tendons are not attached to "muscles." In the same limbs, other connective tissue cells form a model of the muscle, even though it does not contain muscle fibers.

Depending on the specific muscle, migration, fusion, or displacement of muscle primordia may be involved in the genesis of the final form of the muscle. In one case, genetically programmed cell death, **apoptosis,** is responsible for the disappearance of an entire muscle layer (the **contrahentes muscle**) in the flexor side of the human hand. The myogenic cells differentiate to the myotube stage; they then accumulate with glycogen and soon degenerate. The contrahentes muscle layer is preserved in most of the great apes. The reason it degenerates in the human hand at such a late stage in its differentiation is not understood.

Although limb muscles assume their definitive form in the very early embryo, they must undergo considerable growth in both length and cross-sectional area to keep up with the overall growth of the embryo. This is accomplished by the division of the satellite cells (see Chapter 9) and the fusion of their progeny with the muscle fibers. The added satellite cell nuclei increase the potential of the muscle fiber to produce structural and contractile proteins, which increase the cross-sectional area of each muscle fiber. Accompanying this addition to the nuclear complement of the muscle fibers is their lengthening by adding additional sarcomeres, usually at the ends of the muscle fibers. The formation of new muscle fibers typically ceases at or shortly after birth. Although the muscles are capable of contracting in the early fetal period, their physiological properties continue to mature until after birth.

Innervation

Motor axons emanating from the spinal cord enter the limb bud at an early stage of development (during the fifth week) and begin to grow into the dorsal and ventral muscle masses before the latter have split up into primordia of individual muscles (Fig. 11-20). Tracing studies have shown a high degree of order in the projection of motor neurons into the limb. Neurons located in medial positions in the spinal cord send axons to the ventral muscle mass, whereas those located more laterally in the spinal cord supply the dorsal muscle mass. Similarly, a correlation exists between the craniocaudal position of neurons in the cord with the anteroposterior pattern of innervation of limb muscles within the common muscle masses. For example, the rostral-most neurons innervate the most anterior muscle primordia.

Local cues at the base of the limb bud seem to guide the entering pathways of nerve fibers into the limb bud. If a segment of the spinal cord opposite the area of limb bud outgrowth is reversed in the craniocaudal direction, the motor neurons change the direction of their outgrowth and enter the limb bud in their normal positions (Fig. 11-21). If larger segments of spinal cord are reversed and the neurons are at considerable distances from the level of the limb bud, their axons do not find their way to their normal locations in the limb bud. The muscles themselves apparently do not provide specific target cues to the ingrowing axons, since if muscle primordia are prevented from forming, the main patterns of innervation in the limb are still normal.

Sensory axons enter the limb bud after the motor axons and use them for guidance. Sensory nerve fibers do not enter muscles in the absence of motor nerve fibers. Similarly, neural crest cell precursors of Schwann cells lag slightly behind the outgrowth of motor axons into the limb bud. Cells of the neural crest surround both motor and sensory nerve fibers to form the coverings of the nerves in the limbs. By the time digits have formed in developing limbs, the basic elements of the gross pattern of innervation in the adult limb have been established.

Vasculature

The earliest vasculature of the limb bud is derived from endothelial cells arising from several segmental branches of the aorta and the cardinal veins and from **angioblasts** (endothelial cell precursors) endogenous to the limb bud mesoderm. Some evidence indicates that the latter contribute to the endothelium of arteries but not of veins. Initially, the limb vasculature consists of a fine capillary network, but soon some channels are preferentially enlarged, result-

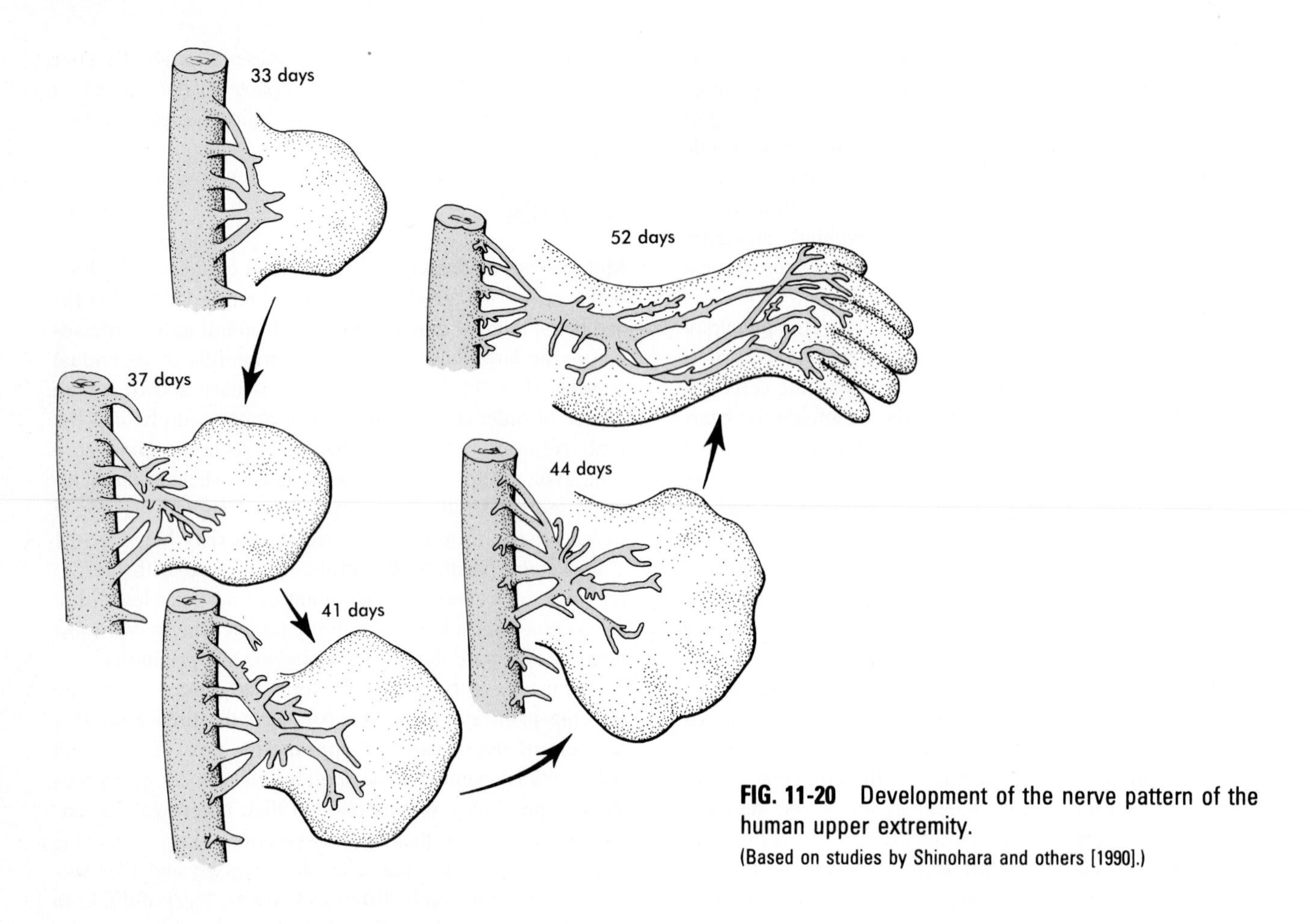

FIG. 11-20 Development of the nerve pattern of the human upper extremity.
(Based on studies by Shinohara and others [1990].)

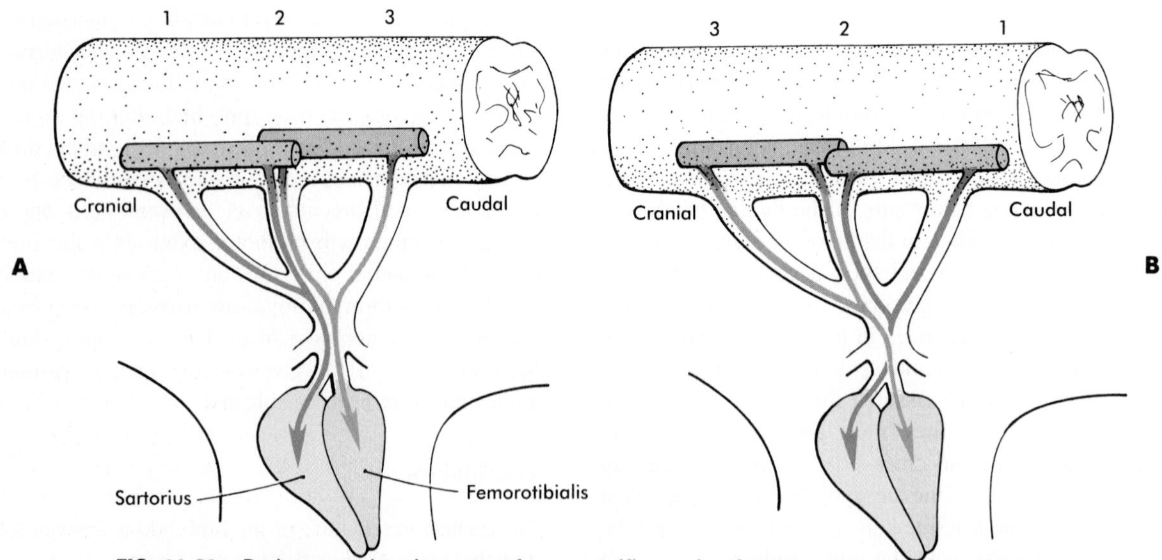

FIG. 11-21 Pathways taken by axons from specific pools of motoneurons in the spinal cord to limb muscles in the hindlimb of the chick embryo. **A,** Normal limb. **B,** After reversal of three segments of the embryonic spinal cord, axons originating from the spinal cord pass through abnormal pathways to innervate the muscles that they were originally destined to innervate.
(Modified from Brown M and others: *Essentials of neural development,* Cambridge Mass, 1990, Cambridge University.)

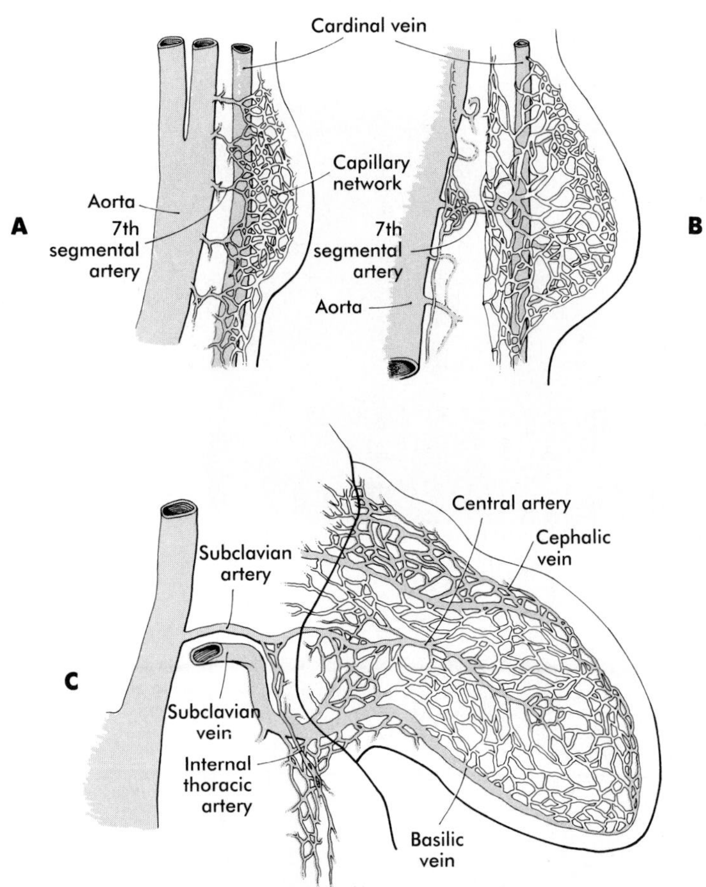

FIG. 11-22 Early stages in the development of the vascular pattern in the mammalian limb bud. **A,** Equivalent of a 4-week-old human embryo. **B,** Equivalent of 5-week-old human embryo. **C,** Equivalent of a 6-week-old human embryo.

ing in a large central artery that supplies blood to the limb bud (Fig. 11-22). From the central artery, the blood is distributed to the periphery via a mesh of capillaries and then collects into a **marginal sinus,** which is located beneath the apical ectodermal ridge. Blood in the marginal sinus drains into peripheral venous channels, which carry it away from the limb bud.

Even in the earliest limb bud there is a peripheral avascular zone of mesoderm within about 100 μm of the ectoderm of the limb bud (Fig. 11-23, *A*). The avascular region persists until the digits have begun to form. In the earliest stages of limb outgrowth, angioblasts are present in the avascular zone, but a short time later they are no longer present. Experimental studies have clearly shown that the proximity of ectoderm is inhibitory to vasculogenesis in the limb bud mesoderm. If the ectoderm is removed, vascular channels form to the surface of the limb bud mesoderm, and if a piece of ectoderm is placed into the deep limb mesoderm, an avascular zone forms around it (Fig. 11-23, *B*). According to some studies, degradation products of hyaluronic acid, which is secreted by the ectoderm, are the inhibitory agents.

Just before the skeleton begins to form, avascular zones appear in the areas where the cartilaginous models of the bones will take shape. Neither the stimuli for the disappearance of the blood vessels nor the fate of the endothelial cells that were present in these regions is understood at this time.

The pattern of the vascular channels changes constantly as the limb develops, mainly by the outgrowth of sprouts from existing channels, the regression of original channels, and the coalescence of the sprouts to form new ones. Such a mechanism accounts for the distal progression of the marginal sinus. With the establishment of the digital rays, the apical portion of the marginal sinus breaks up, but the proximal channels of the marginal sinus persist into adulthood as the **basilic** and **cephalic veins** of the arm.

Similar major changes take place in the arterial channels that course through the developing limb (Fig. 11-24). Branches arising from the **primary axial artery** ultimately take ascendency, especially in the forearm, leaving the original primary axial artery a relatively minor vessel (the **interosseous artery**) in the forearm.

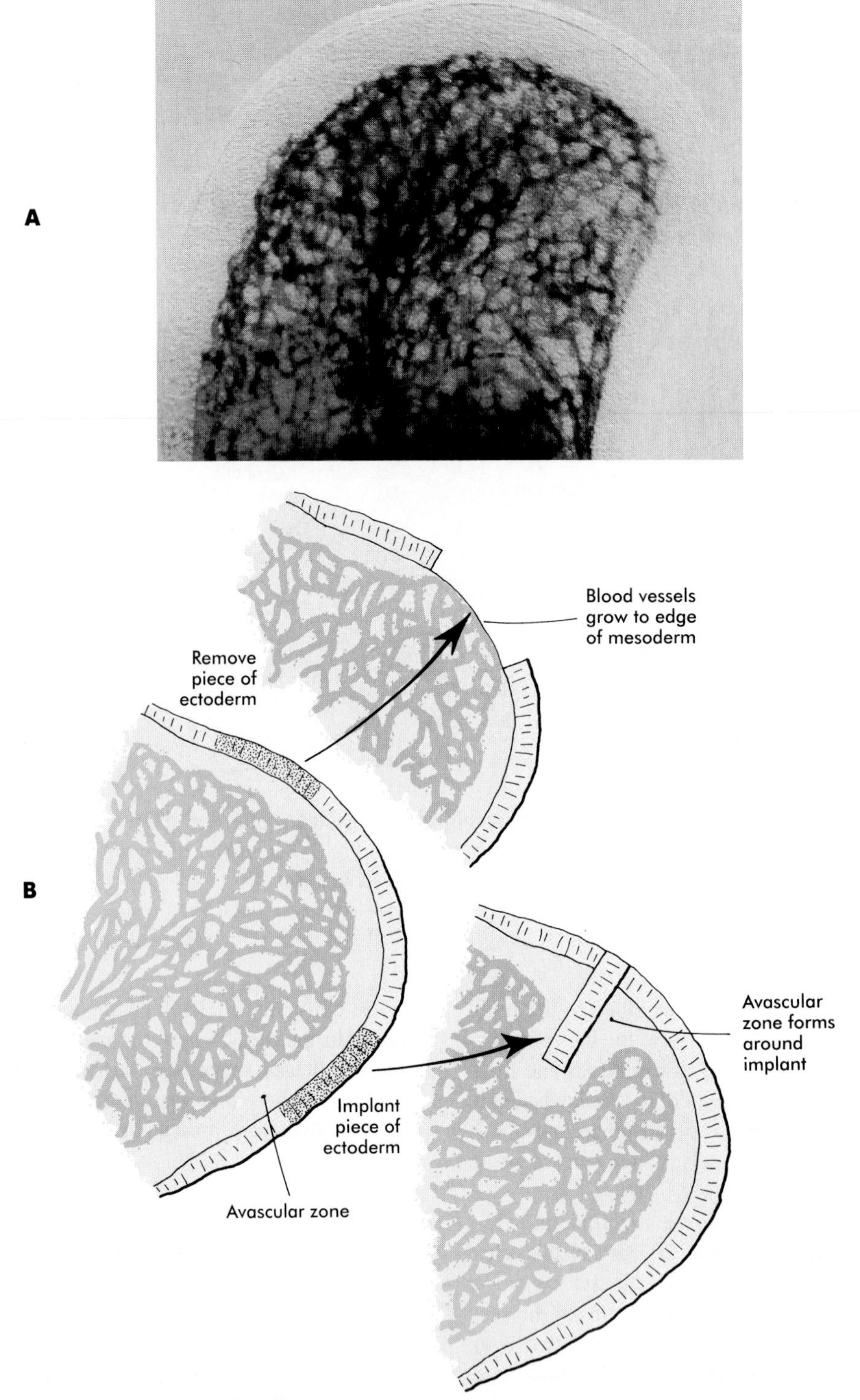

FIG. 11-23 A, Photomicrograph of a quail wing bud with ink-injected blood vessels. **B,** Experiments illustrating the inhibitory effect of limb ectoderm of vascularization of the subjacent mesoderm. *Left:* Normal limb bud with an avascular zone beneath the ectoderm. *Upper right:* After removal of a piece of ectoderm, capillaries grow to the edge of the mesoderm in the region of removal. *Lower right:* A zone of avascularity appears around a piece of implanted ectoderm.

(**A** courtesy R. Feinberg. Based on studies by Feinberg and Noden [1991].)

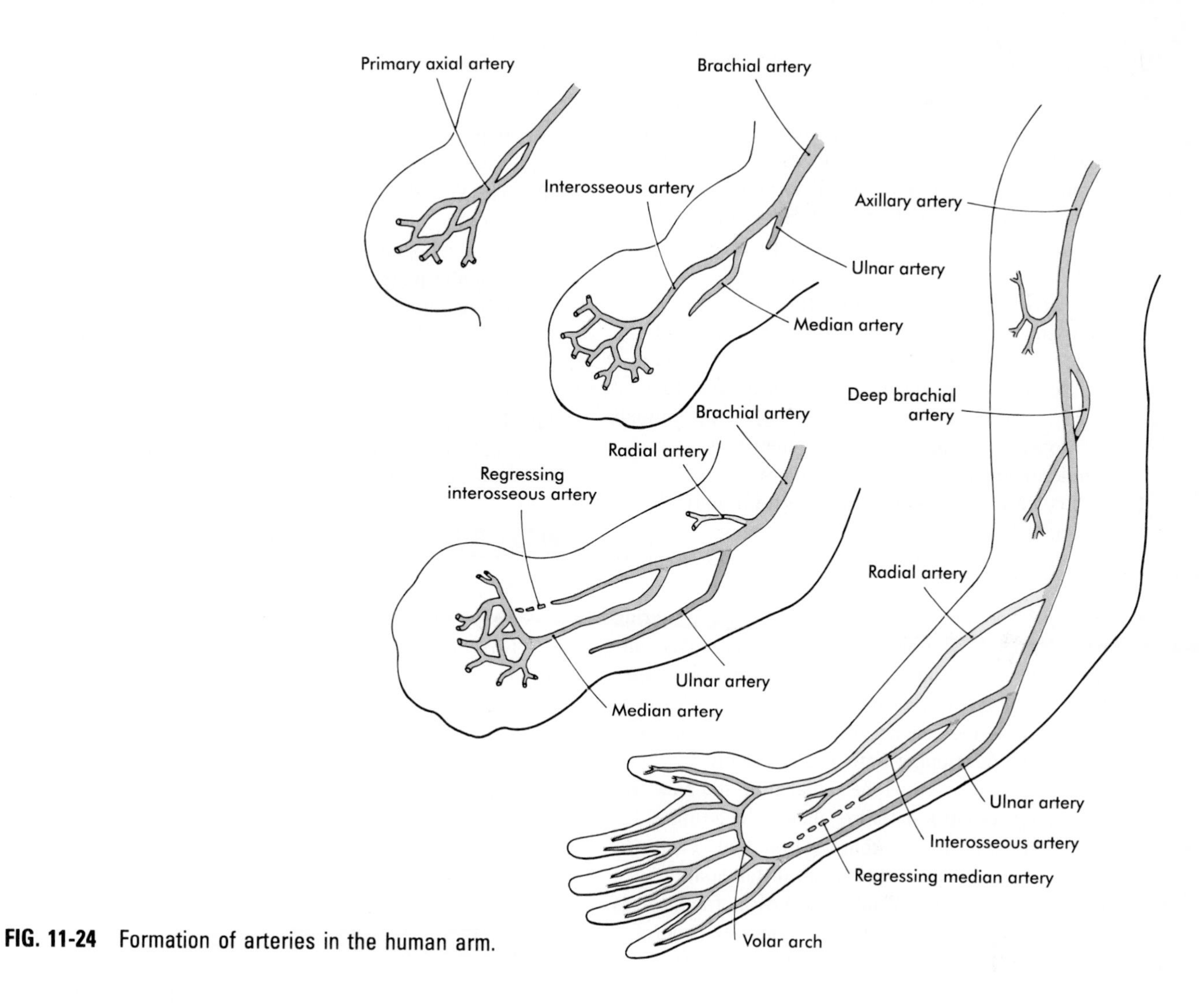

FIG. 11-24 Formation of arteries in the human arm.

LIMB ANOMALIES

Because they are so obvious, limb anomalies are frequently described in the literature. For many of these, however, the etiology remains unknown, especially when individual cases are considered. Some human anomalies can parallel those that can be produced by specific experimental manipulations. Many other limb malformations cannot be attributed to disturbances of specific mechanisms.

A number of limb anomalies have been attributed to vascular anomalies, but it is not always easy to know whether the vascular abnormalities preceded or accompanied the overall defective development in the limb. One of the persisting explanations for the proximal reduction defect in **phocomelia** (see Fig. 8-2), in which the distal limb structures are well formed, is secondary damage to the blood vessels in the proximal part of the limb bud after the pattern for distal hand or foot structures has been established.

Some of the most commonly encountered limb defects can be attributed to mechanical causes. **Intrauterine amputations** by amniotic bands, presumably caused by tears in the amnion, can result in the loss of parts of digits or even hands or feet (see Fig. 8-15). Other deformities, such as **clubfoot (talipes equinovarus)** and some cases of congenital dislocations, have been attributed to persistent mechanical pressures of the uterine wall on the fetus, particularly in cases of **oligohydramnios** (see Chapter 6).

Other limb anomalies are familial and have a genetic basis as either dominant or recessive traits. Although some of the anomalies such as amelia and polydactyly are known to exist as mutants in laboratory animals, the developmental defect underlying the genesis of a given human defect may not be the same as the one in the animal model; several defective mechanisms can give rise to the same phenotype.

A very rare limb deformity is **macromelia** (or **macrodactyly**), in which a limb or a digit is considerably enlarged over normal. Such abnormalities can be associated with neurofibromatosis, and the neural crest may be involved in this defect.

SUMMARY

1. Limbs arise from the lateral plate mesoderm and the overlying ectoderm. The early limb bud is a highly regulative system that can compensate for a variety of surgical disturbances and still form a normal limb. The axes of the limb are fixed in an anteroposterior, dorsoventral, and proximodistal sequence.
2. The early limb bud mesoderm stimulates the overlying ectoderm to form an apical ectodermal ridge that stimulates outgrowth of the limb through proliferation of the underlying mesodermal cells. The overall morphogenesis of the limb is determined by properties of the mesoderm, whereas the ectoderm acts in a more permissive sense.
3. Cell death is an important mechanism in normal limb development. Regions of programmed cell death include the axillary region and the interdigital spaces.
4. A zone of polarizing activity located in the posterior mesoderm acts as a biological signaler and plays an important role in anteroposterior organization of the limb, possibly by releasing a diffusible morphogen. According to the concept of positional information, cells in the developing limb are exposed to positional cues (such as the signal from the zone of polarizing activity) that allow them to determine their relative position within the limb bud. The cells then process this information and differentiate accordingly. Proximodistal control of morphogenesis may reside in the progress zone, a narrow band of mesoderm beneath the apical ectodermal ridge.
5. Retinoic acid exerts a profound effect on limb morphogenesis and can cause the formation of a supernumerary limb if applied to the anterior border of the limb bud. Expression of a variety of homeobox genes follows well-defined patterns in the normally developing limb. Some patterns of gene expression are profoundly altered in limbs treated with retinoic acid.
6. The skeleton of the limb arises from lateral plate mesoderm. The ectoderm of the limb bud inhibits cartilage formation in the mesoderm cells immediately beneath it. This could explain the reason the skeleton of the vertebrate limb forms in a central position.
7. Limb muscles arise from cells derived from somitic mesoderm. Myogenic cells first form dorsal and ventral common muscle masses, which later split into primordia of individual muscles. Morphogenetic control of muscles resides in the associated connective tissue rather than in the muscle cells themselves. Later stages in muscle development may involve cell death, fusion of muscle primordia, and the displacement of muscle primordia to other areas.
8. Nerves grow into the developing limb bud and become associated with the common muscle masses as they split into individual muscles. Local cues are important in guiding growing axons into the developing limb.
9. The vasculature of the limb bud arises from cells budding off the aorta and cardinal veins as well as from endogenous mesodermal cells. The early vascular pattern consists of a central artery, which drains into a peripheral marginal sinus and then into peripheral venous channels. Blood vessels do not form just beneath the ectoderm nor in the central cartilage-forming regions.
10. Limb anomalies can form as the result of genetic mutations, drug effects, disturbed tissue interactions, and purely mechanical effects.

REVIEW QUESTIONS

1. A baby whose mother underwent a chorionic villus sampling during pregnancy was born with the tips of two digits missing. What is a possible cause?
2. A woman who underwent amniocentesis during pregnancy gave birth to a child with a duplicated thumb. What is a possible cause?
3. If the somites opposite a limb-forming region are experimentally removed, the limbs form without muscles. Why?
4. A child is born with webbed fingers (syndactyly). What is the reason for this anomaly?

REFERENCES

Allen F, Tickle C, Warner A: The role of gap junctions in patterning of the chick limb bud, *Development* 108:623-634, 1990.

Christ B, Jacob HJ, Jacob M: Experimental analysis of the origin of the wing musculature in avian embryos, *Anat Embryol* 150:171-186, 1977.

Cihák R: Ontogenesis of the skeleton and intrinsic muscles of the human hand and foot, *Adv Anat Embryol Cell Biol* 46:1-194, 1972.

Coelho CND and others: Altered expression of the chicken homeobox-containing genes *GHox-7* and *GHox-8* in the limb buds of *limbness* mutant chick embryos, *Development* 113:1487-1493, 1991.

Coelho CND and others: Expression of the chicken homeobox-containing gene *GHox-8* during embryonic chick development, *Mech Dev* 34:143-154, 1991.

De Robertis EM, Morita EA, Cho KWY: Gradient fields and homeobox genes, *Development* 112:669-678, 1991.

Dolle P and others: Coordinate expression of the murine Hox-5 complex homeobox-containing genes during limb pattern formation, *Nature* 342:767-772, 1989.

Duboule D: The vertebrate limb: a model system to study the Hox/HOM gene network during development and evolution, *Bioessays* 14:375-384, 1992.

Dvorak L, Fallon JF: The *talpid²* chick limb has weak polarizing activity and can respond to retinoic acid and polarizing zone signal, *Dev Dynam* 193:40-48, 1992.

Dylevsky' I: Growth of the human embryonic hand, *Acta Univ Carol [Med Monogr]* 114:1-139, 1986.

Dylevsky' I: Connective tissue of the hand and foot, *Acta Univ Carol [Med Mongr]* 127:1-195, 1988.

Ede DA, Hinchliffe JR, Balls M, eds: *Vertebrate limb and somite morphogenesis,* Cambridge, Mass, 1977, Cambridge University.

Fallon JF, Caplan AI, eds: *Limb development and regeneration,* parts A and B, New York, 1983, Liss.

Feinberg RN: *Vascular development in the embryonic limb bud.* In Feinberg RN, Sherer GK, Auerbach R, eds: *The development of the vascular system,* Basel, Switzerland, 1991, Karger, pp 136-148.

Feinberg RN, Noden DM: Experimental analysis of blood vessel development in the avian wing bud, *Anat Rec* 231:136-144, 1991.

Grim M, Wachtler F: Muscle morphogenesis in the absence of myogenic cells, *Anat Embryol* 183:67-70, 1991.

Harrison RG: On relations of symmetry in transplanted limbs, *J Exp Zool* 32:1-136, 1921.

Hinchliffe JR, Johnson DR: *The development of the vertebrate limb,* Oxford, England, 1980, Clarendon.

Hornbruch A, Wolpert L: The spatial and temporal distribution of polarizing activity in the flank of the pre-limb-bud stages in the chick embryo, *Development* 111:725-731, 1991.

Izpisua-Belmonte J-C, Duboule D: Homeobox genes and pattern formation in the vertebrate limb, *Dev Biol* 152:26-36, 1992.

Izpisua-Belmonte J-C and others: Expression of the homeobox Hox-4 genes and the specification of position in chick wing development, *Nature* 350:585-589, 1991.

Kelley RO, Fallon JF: Ultrastructural analysis of the apical ectodermal ridge during morphogenesis. I. The human forelimb with special reference to gap junctions, *Dev Biol* 51:241-256, 1976.

Krabbenhoft KM, Fallon JF: *Talpid²* limb bud mesoderm does not express *GHox-8* and has an altered expression pattern of *GHox-7, Dev Dynam* 194:52-62, 1992.

Maden M and others: Spatial distribution of cellular protein binding to retinoic acid in the chick limb bud, *Nature* 335:733-735, 1988.

Maini PK, Solursh M: Cellular mechanisms of pattern formation in the developing limb, *Int Rev Cytol* 129:91-133, 1991.

Mrázková O: Blood vessel ontogeny in upper extremity of man as related to developing muscles, *Acta Univ Carol [Med Monogr]* 115:1-114, 1986.

Nohno T and others: Involvement of the *CHox-4* chicken homeobox genes in determination of anteroposterior axial polarity during limb development, *Cell* 64:1197-1205, 1991.

Nohno T and others: Differential expression of two *msh*-related homeobox genes *CHox-7* and *CHox-8* during chick limb development, *Biochem Biophys Res Commun* 182:121-128, 1992.

Robert B and others: The apical ectodermal ridge regulates *Hox-7* and *Hox-8* gene expression in developing limb buds, *Genes Dev* 5:2363-2374, 1991.

Ros MA and others: Apical ridge dependent and independent mesodermal domains of and *GHox-8* expression in chick limb buds, *Development* 116:811-818, 1992.

Rubin L, Saunders JW: Ectodermal-mesodermal interactions in the growth of limb buds in the chick embryo: constancy and temporary limits of the ectodermal induction, *Dev Biol* 28:94-112, 1972.

Sassoon D: Hox genes: a role for tissue development, *Am J Respir Cell Mol Biol* 7:1-2, 1992.

Saunders JW: The proximodistal sequence of origin on the parts of the chick wing and the role of the ectoderm, *J Exp Zool* 108:363-403, 1948.

Saunders JW, Gasseling MT: *Ectodermal-mesenchymal interactions in the origin of limb symmetry.* In Fleischmajer R, Billingham RE: *Epithelial-mesenchymial interactions,* Baltimore 1968, pp. 78-97, Williams & Wilkins.

Saunders JW, Gasseling MT, Saunders LC: Cellular death in morphogenesis of the avian wing, *Dev Biol* 5:147-178, 1962.

Seichert V: Significance of the differential growth, relative tissue shifts and the vascular bed in limb development, *Acta Univ Carol [Med Monogr]* 125:1-162, 1988.

Shinohara H and others: Development of innervation patterns in the upper limb of staged human embryos, *Acta Anat* 138:265-269, 1990.

Stephens TD: The wolffian ridge: history of a misconception, *Isis* 73:254-259, 1982.

Summerbell D, Lewis JW, Wolpert L: Positional information in chick limb morphogenesis, *Nature* 244:492-496, 1973.

Swett FH: Determination of limb-axes, *Q Rev Biol* 12:322-339, 1937.

Tabin CJ: Retinoids, homeoboxes and growth factors: toward molecular models for limb development, *Cell* 66:199-217, 1991.

Tabin CJ: Why we have (only) five fingers per hand: *Hox* genes and the evolution of paired limbs, *Development* 116:289-296, 1992.

Thaller C, Eichele G: Identification and spatial distribution of retinoids in the developing chick limb bud, *Nature* 327:625-628, 1987.

Thaller C, Eichele G: Isolation of 3,4-didehydroretinoic acid, a novel morphogenetic signal in the chick wing bud, *Nature* 345:815-819, 1990.

Tickle C: Retinoic acid and chick limb bud development, *Development* 1991 (suppl 1):113-122, 1991.

Uhthoff HK: *The embryology of the human locomotor system,* Berlin, 1990, Springer-Verlag.

Wolpert L: Positional information revisited, *Development* 107(suppl):3-12, 1989.

Wolpert L, Lewis J, Summerbell D: *Morphogenesis of the vertebrate limb.* In *Cell patterning,* London, 1975, Ciba Foundation Symposium 29, pp 95-130.

Yokouchi K and others: Chicken homeobox gene *Msx-1:* structure, expression in limb buds and effect of retinoic acid, *Development* 113:431-444, 1991.

Zwilling E: Limb morphogenesis, *Adv Morphogen* 1:301-330, 1961.

CHAPTER 12

The Nervous System

Many fundamental developmental processes are involved in the formation of the nervous system. Some of these dominate certain stages of embryogenesis; others occur only at limited times and in restricted locations. The major processes are the following:

1. **Induction,** including both primary induction of the nervous system by the underlying notochord and secondary inductions driven by neural tissues themselves
2. **Proliferation,** first as a response of the neuroectodermal cells to primary induction and later to build up critical numbers of cells for virtually all aspects of morphogenesis of the nervous system
3. **Pattern formation,** in which cells respond to genetic or environmental cues in forming the fundamental subdivisions of the nervous system
4. **Intercellular communication** and the adhesion of like cells
5. **Cell migration,** of which there are a variety of distinct patterns in the nervous system
6. **Cellular differentiation** of both neurons and glial cells
7. The formation of specific connections or **synapses** between cells
8. The **stabilization** or **elimination** of specific interneuronal connections, sometimes associated with massive episodes of cell death of unconnected neurons
9. The progressive **development of integrated patterns** of neuronal function, which results in coordinated reflex movements

EARLY ESTABLISHMENT OF THE NERVOUS SYSTEM

As described in Chapter 5, primary induction of the nervous system results in the formation of a thickened ectodermal **neural plate** overlying the notochord. Shortly thereafter, the neural plate begins to fold into the **neural tube** (see Fig. 5-7).

During these earliest stages in establishment of the nervous system, a striking change occurs in the distribution of **cell adhesion molecules (CAMs)** on the surfaces of the ectodermal cells. CAMs protrude from the plasma membrane. Cells of a similar type typically adhere by means of Ca^{++}, which bind CAMs of one cell to those of the other (see Fig. 4-13). In chick embryos, cells of the early epiblast express both **N-CAM** and **L-CAM** before primary induction takes place. As the neural plate takes shape after the primary inductive event, the profound changes that have occurred in the ectoderm are reflected not only in the shapes of the cells but in their expression of CAMs. Neuroepithelial cells lose their L-CAM and express only N-CAM, whereas the noninduced ectodermal cells lose N-CAM but retain L-CAM.

The neural tube, which is the structural manifestation of the earliest stages in establishing the nervous system, is a prominent structure. In the human, it dominates the cephalic end of the embryo (see Fig. 5-6). This chapter describes the way the neural tube develops into the major morphological and functional components of the nervous system.

EARLY CHANGES IN THE GROSS STRUCTURE OF THE NERVOUS SYSTEM

Closure of the neural tube first occurs in the region where the earliest somites appear; closure spreads both cranially and caudally (see Fig. 5-6). The unfused regions of the neural tube are known as the **cranial** and **caudal neuropores.** Even before the closure of the neuropores (24 days gestation for the cranial neuropore and 26 days gestation for the caudal neuropore), some fundamental subdivisions in the early nervous system can be distinguished. The future spi-

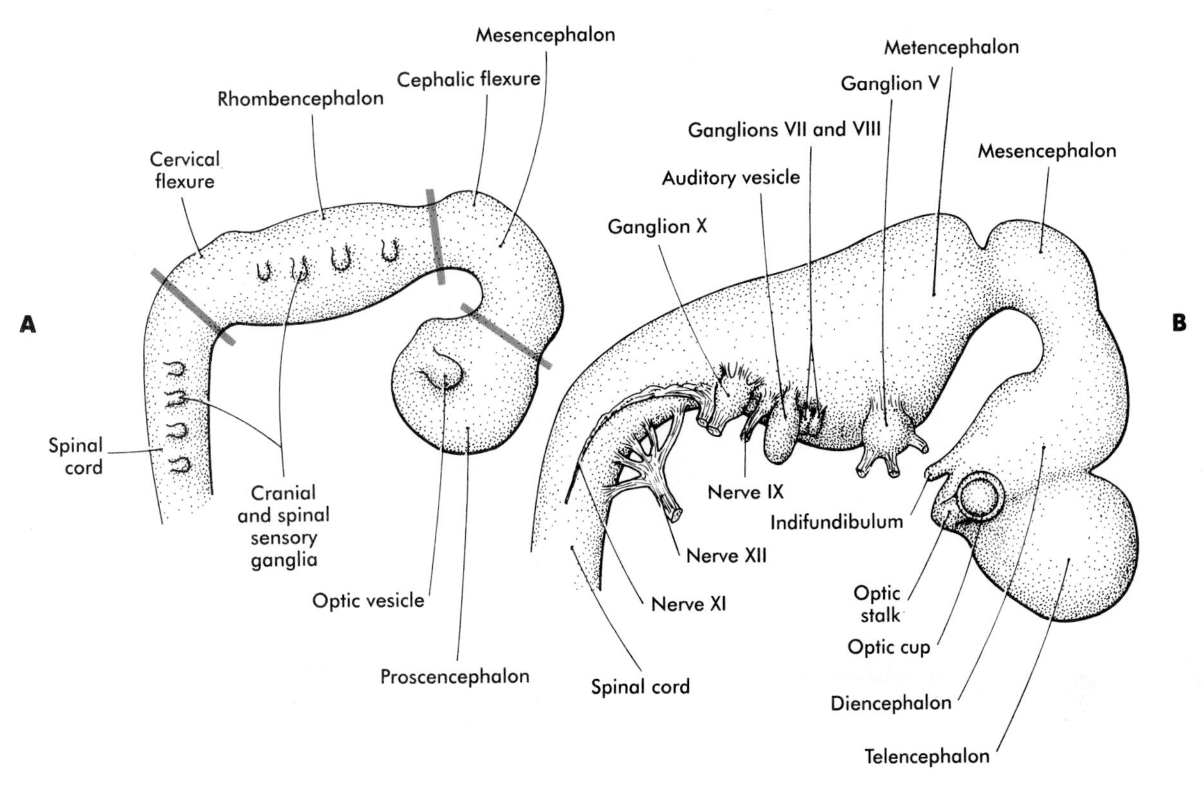

FIG. 12-1 Basic anatomy of the **(A)** three- and **(B)** five-part human brain.

nal cord and brain are recognizable, and within the brain the forebrain **(prosencephalon)**, midbrain **(mesencephalon)**, and hindbrain **(rhombencephalon)**, are distinguishable (Fig. 12-1).

A prominent force in shaping the early nervous system is the overall bending of the cephalic end of the embryo into a C shape. Associated with this bending is the appearance at the end of the third week of a prominent **cephalic flexure** of the brain at the level of the mesencephalon. Soon the brain almost doubles back on itself at the cephalic flexure. At the beginning of the fifth week, a **cervical flexure** appears at the boundary between the hindbrain and the spinal cord.

By the fifth week the original three-part brain has become further subdivided into five parts (Fig. 12-2). The prosencephalon gives rise to the **telencephalon** (endbrain), with prominent lateral outpocketings that will ultimately form the cerebral hemispheres, and a more caudal **diencephalon.** The diencephalon is readily recognizable because of the prominent lateral **optic vesicles** that extend from its lateral walls. The mesencephalon, which is sharply bent by the cephalic flexure, remains undivided and tubular in its overall structure. The roof of the rhombencephalon becomes very thin, and there are early indications of the subdivision of the rhombencephalon into a **metencephalon** and a more caudal **myelencephalon.** These five subdivisions of the early brain represent a fundamental organization that persists through adulthood. Many further structural and functional components give added layers of complexity to the brain over the next several weeks of embryonic life.

HISTOGENESIS WITHIN THE CENTRAL NERVOUS SYSTEM

Proliferation Within the Neural Tube

Shortly after primary induction, the thickening neural plate and early neural tube organize into a pseudostratified epithelium (Fig. 12-3). In this type of epithelium the nuclei appear to be located in several separate layers of cells, but the cytoplasm of the neuroepithelial cells extends from the basal to the apical surface. The nuclei undertake extensive shifts of position within the cytoplasm of the neuroepithelial cells.

The neuroepithelial cells are characterized by a high degree of mitotic activity, and the position of the nuclei within the neural tube and their stage in the mitotic cycle closely correlate (Fig. 12-4). DNA synthesis occurs in nuclei located near the external limiting membrane (the basal lamina surrounding the neural tube). As these nuclei prepare to go into mitosis, they migrate within the cytoplasm toward the lumen of the neural tube, where they undergo mitotic division. Nuclei of the resulting two daughter cells migrate up the cytoplasm toward the external limiting mem-

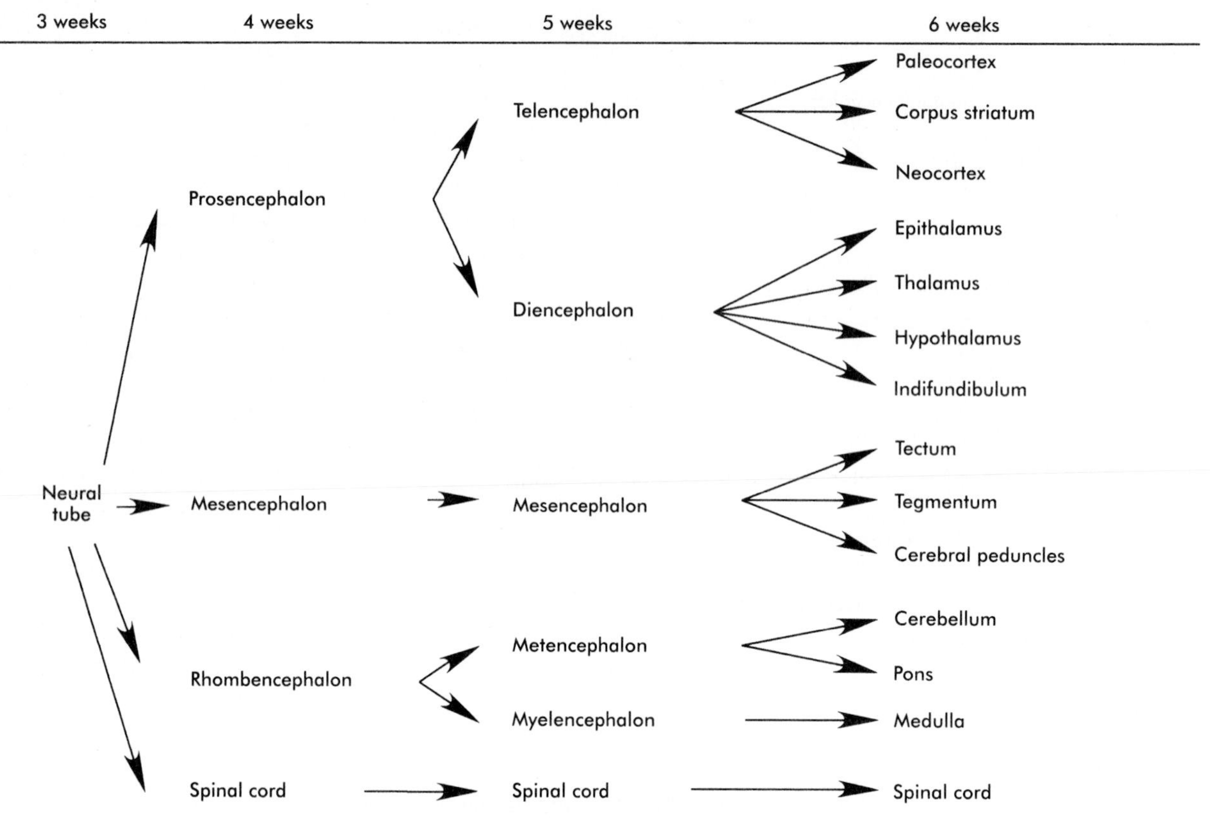

FIG. 12-2 Increasing levels of complexity of the developing human brain.

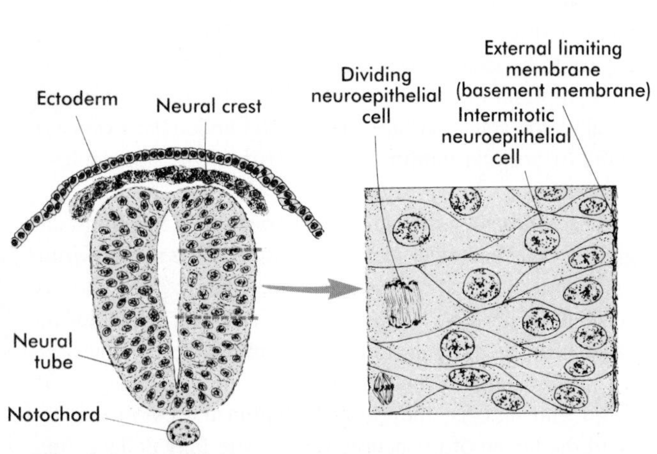

FIG. 12-3 *Left:* Cross section through the early neural tube. *Right:* Higher magnification of a segment of the wall of the neural tube.

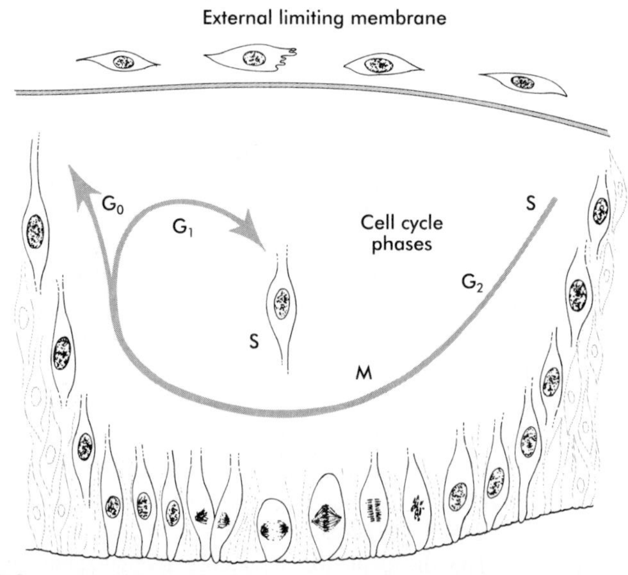

FIG. 12-4 Mitotic events in the early neural tube. In the pseudostratified cells of the early neural tube, the nuclei that synthesize DNA (S phase) are located near the external limiting membrane. The nuclei then move toward the inner margin of the neural tube, where mitosis occurs. Nuclei of daughter cells then move outward toward the external limiting membrane, where the cells containing them either undergo another round of DNA synthesis or differentiate into neuroblasts.

brane. At this point they have two options: (1) to undergo DNA synthesis again and feed back into the mitotic cycle or (2) to leave the mitotic cycle permanently and ultimately take up positions beneath the external limiting membrane as **neuroblasts.** The neuroblasts, cellular precursors of neurons, begin to produce cell processes that will ultimately become axons and dendrites.

Cell Lineages in Histogenesis of the Central Nervous System

Many schemes of cell lineages within the nervous system have been proposed. With the recent availability of specific antibody markers, clonal cell culture techniques, and in vivo labeling techniques, it is possible to piece together detailed lineage maps. Even with these contemporary tools, not all aspects of cell lineages are understood.

The origins of most cells found in the mature central nervous system can be traced to **multipotential stem cells** within the early neuroepithelium (Fig. 12-5). These cells undergo numerous mitotic divisions before maturing into **bipotential progenitor cells,** which give rise to either neuronal or glial progenitor cells.

The **neuronal progenitor cells** give rise to a series of neuroblasts. The earliest **bipolar neuroblasts** possess two slender cytoplasmic processes that contact both the external limiting membrane and the central luminal border of the neural tube. By retracting the inner process, a bipolar neuroblast loses contact with the inner luminal border in the process of becoming a **unipolar neuroblast.** The unipolar neuroblasts accumulate large masses of rough endoplasmic reticulum **(Nissl substance)** in their cytoplasm and then begin to send out several cytoplasmic processes. At this point they are known as **multipolar neuroblasts.** Their principal developmental activities are to send out axonal and dendritic processes and to make connections with other neurons or end organs. (Further details of the many patterns of specific neuronal differentiation are beyond the scope of this text but can be found in specialized reviews or monographs.)

The other major lineage stemming from the bipotential progenitor cells is the glial line. **Glial progenitor cells**

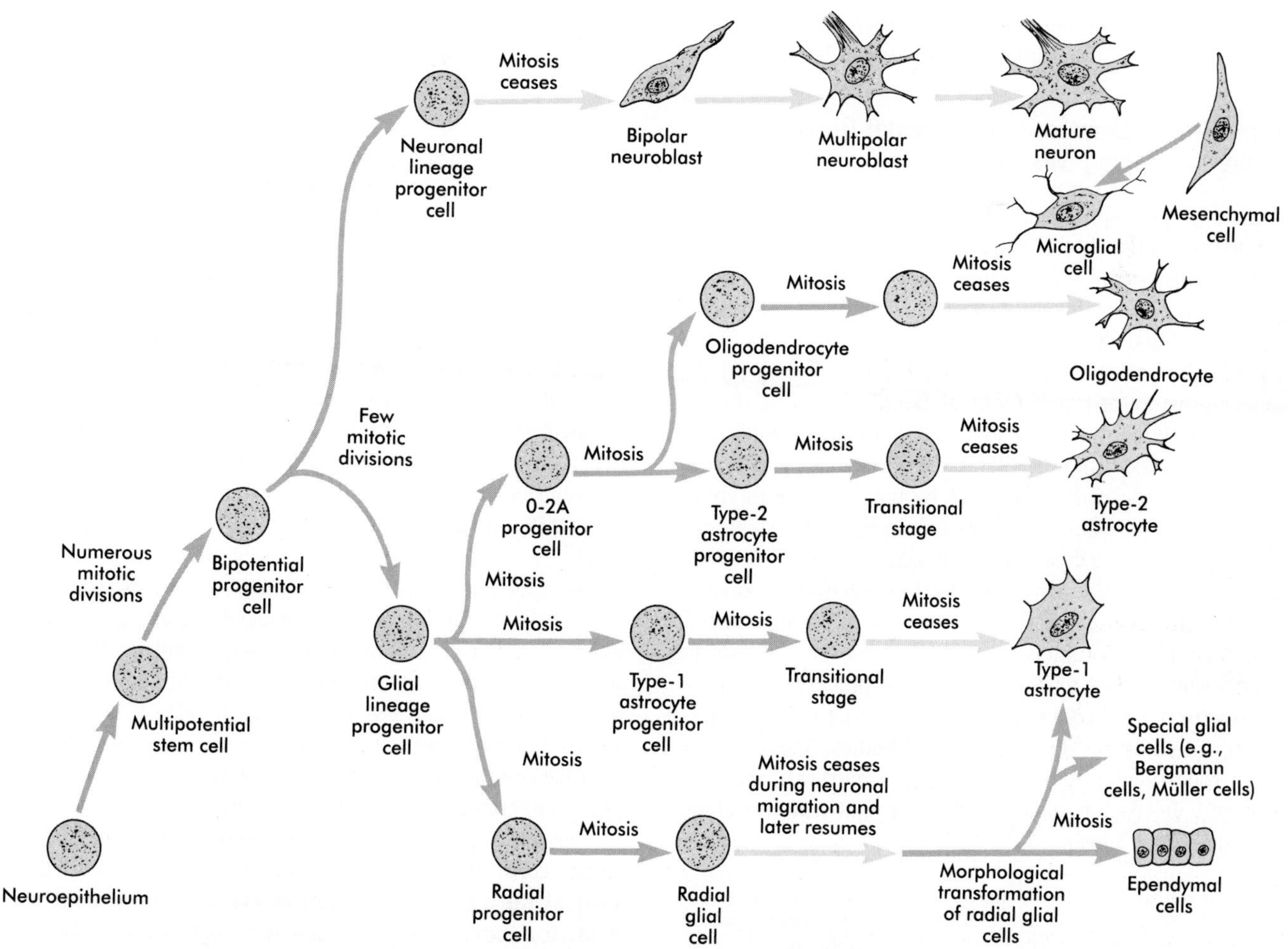

FIG. 12-5 Cell lineages in the developing central nervous system. (Based on studies by Cameron and Rakic [1991].)

continue to undergo mitosis, and their progeny split into several lines. One, the **O-2A progenitor cell** (see Fig. 12-5), is a precursor to two lines of glial cells that ultimately form the **oligodendrocytes*** and that give rise to **type-2 astrocytes.** Another glial lineage gives rise to **type-1 astrocytes.**

The third glial lineage has a more complex history. Radial progenitor cells give rise to **radial glial cells,** which act as guide wires in the brain for the migration of young neurons (see Fig. 12-19). When the neurons are migrating along the radial glial cells during midpregnancy, they inhibit the proliferation of these cells. After neuronal cell migration, the radial cells, now free from the inhibitory influence of the neurons, reenter the mitotic cycle. Their progeny can transform into a number of cell types. Some can seemingly cross lineage lines and differentiate into type-1 astrocytes (see Fig. 12-5). Other progeny differentiate into a variety of specialized glial cell types or even into **ependymal cells.** According to some authors, remaining neuroepithelial cells represent another source of ependymal cells.

Not all cells of the central nervous system originate in the neuroepithelium. **Microglial cells,** which serve a phagocytic function after damage to the brain, are mesodermally derived immigrant cells. Microglia are not found in the developing brain until it is penetrated by blood vessels.

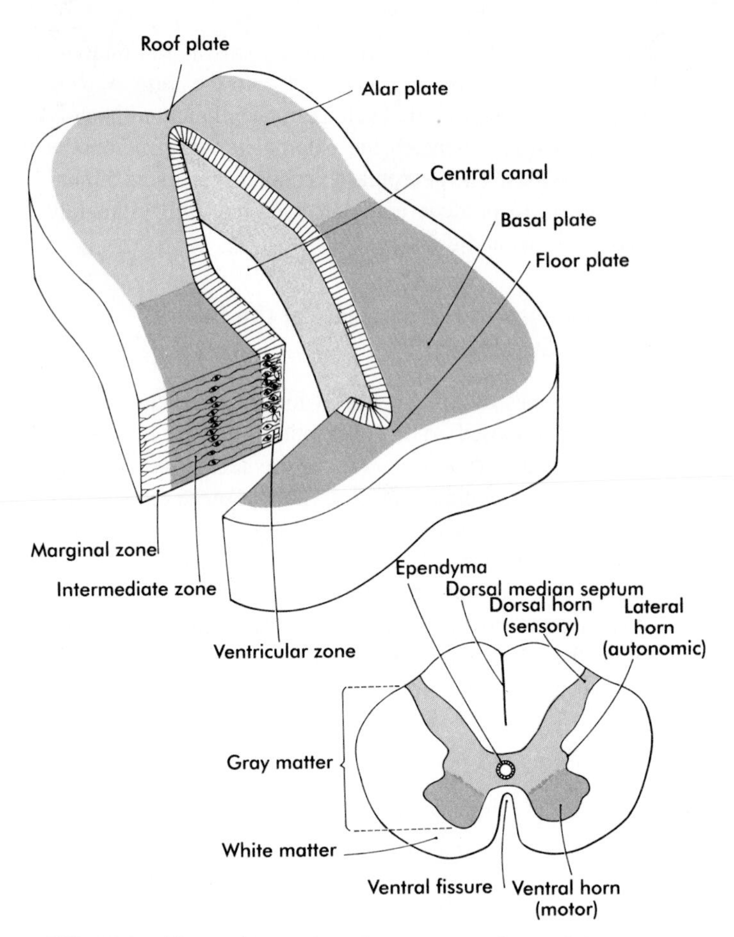

FIG. 12-6 The major regions in cross sections of the neural tube *(top)* and spinal cord *(bottom).*

Formation of Zones and Fundamental Organization of the Developing Neural Tube

The developing spinal cord is a useful prototype for studying the overall structural and functional features of the central nervous system because it preserves its fundamental organization through much of development. With the beginning of cellular differentiation in the neural tube, the neuroepithelium thickens and appears layered. The layer of cells closest to the lumen **(central canal)** of the neural tube remains epithelial and is called the **ventricular zone** (the **ependymal zone** in older literature). This zone, which still contains mitotic cells, ultimately becomes the **ependyma,** a columnar epithelium that lines the ventricular system and central canal of the central nervous system (Fig. 12-6). Farther from the ventricular zone is the **intermediate** (formerly called **mantle**) **zone,** which contains the cell bodies of the differentiating postmitotic neuroblasts. As the neuroblasts continue to produce axonal and dendritic processes, the processes form a peripheral **marginal zone** that contains neuronal processes but not neuronal cell bodies. The intermediate zone eventually becomes the **gray matter** of the central nervous system, whereas the marginal zone develops into the white matter.

*There is still not unanimity of opinion regarding the origin of oligodendrocytes. Although they are described as arising from precursor cells in the neural tube in this text, other embryologists feel that they arise from neural crest precursor cells.

As the spinal cord matures, the ventricular zone becomes the gray matter, in which the cell bodies of the neurons are located. The marginal zone is called the white matter because of the color imparted by the numerous tracts of myelinated nerve fibers in that layer (see Fig. 12-6). During development the proliferating progenitor cell populations in the ventricular zone become exhausted, and the remaining cells differentiate into the epithelium of the ependymal layer.

Once the basic layers in the spinal cord are established, a number of important topographical features can be recognized in cross sections of the cord. A **sulcus limitans** within the central canal divides the cord into a dorsal **alar plate** and a ventral **basal plate** on each side of the central canal. The right and left alar plates are connected dorsally over the central canal by a thin **roof plate,** and the two basal plates are connected ventrally by a **floor plate.**

The basal plate represents the motor component of the spinal cord. Axons arising from neurons located in the **ventral horn** of the gray matter exit the spinal cord as **ventral motor roots** of the spinal nerves (see Fig. 12-6). The gray matter of the alar plate, called the **dorsal horn,**

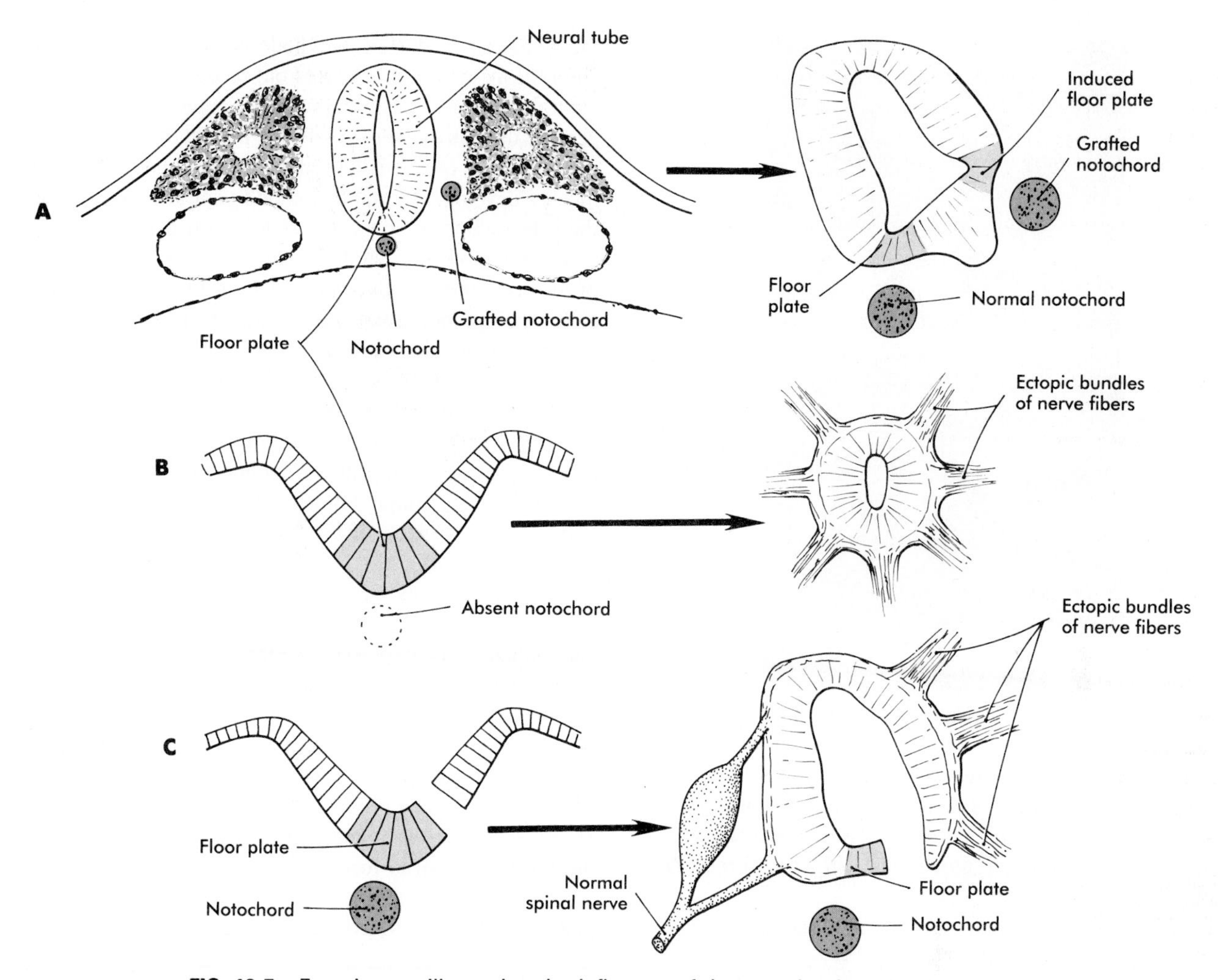

FIG. 12-7 Experiments illustrating the influence of the notochord on development of the floor plate and sites of exit of nerves from the spinal cord. **A,** Grafting an extra notochord near the neural tube induces a secondary floor plate. **B,** In the absence of a notochord, no floor plate forms and nerve fibers exit from multiple sites around the spinal cord. **C,** Slitting the neural tube on one side of the floor plate removes the wall of the neural tube from the influence of the notochord, allowing the disorganized exit of nerve fibers from that part of the spinal cord.
(Modified from Hirano S, Fuse S, Sohal GS: *Science* 251:310-313, 1991.)

is associated with sensory functions. Sensory axons from the spinal ganglia (neural crest derivatives) enter the spinal cord as dorsal roots and synapse with neurons in the dorsal horn. A small projection of gray matter between the dorsal and ventral horns contains cell bodies of autonomic neurons. This projection is called the **lateral horn** or sometimes the **intermediolateral gray column.**

The floor plate is far more than an anatomical connection between the right and left basal plates. Cells of the future floor plate are the first to differentiate in the neural plate after primary induction of the nervous system. Experimental work has demonstrated a specific inductive influence of the notochord on the neuroepithelial cells that overly it. Specific reactions are (1) characteristic changes in cell shape, (2) the appearance of unique surface antigens, and (3) the production of a potent diffusible factor acting as a chemoattractant to stimulate the growth of axons, which cross over to the other side of the developing spinal cord.

If an extra notochord is grafted along the lateral surface of the neural tube, the neuroepithelial cells closest to it acquire the properties of floor plate cells (Fig. 12-7). Conversely, if a segment of normal notochord is removed, the neuroepithelial cells overlying it do not acquire the properties of floor plate cells. Through its action on the floor plate, the notochord also exerts a profound effect on the organization of the dorsal and ventral roots that enter and leave

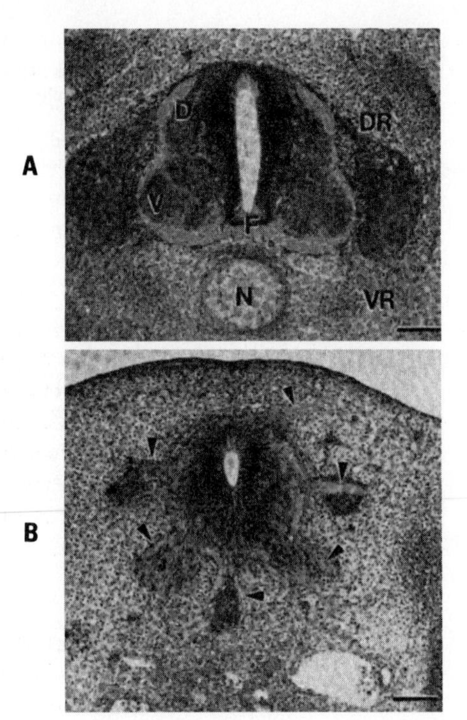

FIG. 12-8 **A,** Photomicrograph of a normal embryonic quail spinal cord. **B,** In an experiment in which the notochord was not permitted to form, the spinal cord is disorganized, with multiple exit sites of nerve fibers (see Fig. 12-7, *B*). Arrowheads indicate ectopic spinal nerves. *D,* Dorsal; *V,* ventral; *DR,* dorsal root; *VR,* ventral root; *N,* notochord; *F,* floor plate.
(From Hirano S, Fuse S, Sohol GS: *Science* 251:310-313, 1991.)

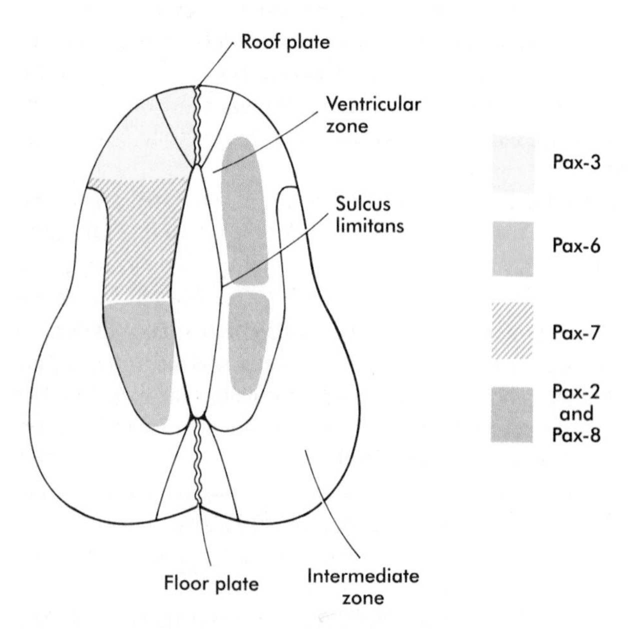

FIG. 12-9 Distribution of *Pax* gene products in a cross section of the embryonic spinal cord.
(Based on studies by Deutsch and Gruss [1991].)

the spinal cord. If Hensen's node is removed in the early embryo, the neural tube closes but recognizable dorsal and ventral roots are absent. Numerous ectopic nerve fibers appear in their place (Fig. 12-8). If the future floor plate is split, the side of the neural tube on which the notochord is located develops normal dorsal and ventral roots, whereas the side lacking these structures gives off ectopic nerves.

Recent molecular studies have shown discrete patterns of localization of products of a class of segmentation genes called **paired box genes** (in contrast to homeobox genes) in the early developing neural tube of mice. Expression of one family of paired box genes **(Pax-1** to **Pax-8)** is limited to specific regions in cross sections of the neural tube (Fig. 12-9). As with the products of homeobox genes, the proteins of *Pax* genes appear to function as transcription factors. Studies on simple brains of fish suggest a correlation between some aspects of *Pax* expression and the formation of specific axonal tracts in the central nervous system.

In addition to its other roles in induction of the neural tube and organization of the floor plate, the notochord can also influence the expression patterns of certain morphogenetically important genes. One such gene is **En-2** (engrailed). Normally, *engrailed* has a highly characteristic rostrocaudal pattern of expression in the mesencephalon and rostral part of the metencephalon, but within this region of expression the gene product is absent from the floor plate. However, when the notochord is removed, not only does the floor plate fail to form (see Fig. 12-7), but *En-2* is expressed in the ventral part of the neural tube where the floor plate would have formed.

PATTERN FORMATION AND SEGMENTATION IN THE CENTRAL NERVOUS SYSTEM

The organization of the caudal brain–forming regions into segments called **rhombomeres** was introduced in Chapter 5. Early in brain development a very close correlation exists between the expression of certain homeobox genes and specific rhombomeres (see Fig. 15-3). Other sets of homeobox genes expressed only in the forebrain and midbrain are also being identified. A relationship between homeobox gene expression and fundamental patterning processes within the central nervous system, as well as other components of the body, is widely assumed. As cellular differentiation begins in the hindbrain, the segmental nature of the neural tube again becomes evident when cellular behavior is considered.

The correspondence between the rhombomeres of the developing brain and other structures of the cranial and pharyngeal arch region is remarkable (see Chapter 15 and Fig. 15-2). The cranial nerves, which have a highly ordered pattern by which they supply structures derived from the pharyngeal arches and other structures in the head, have an equally highly ordered origin with respect to the rhombomeres (Fig. 12-10). For example, cranial nerve V inner-

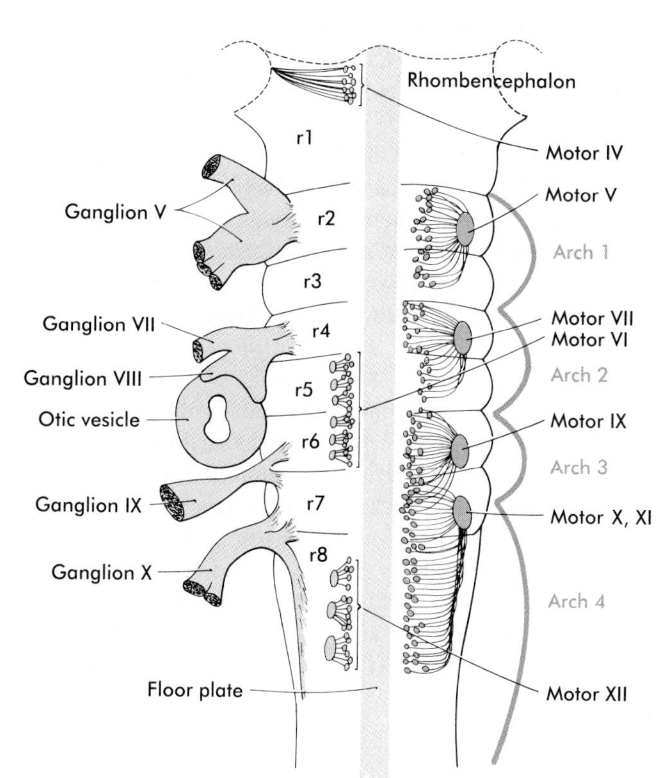

FIG. 12-10 Origin of cranial nerves in relation to rhombomeres in the developing brain.
(Based on studies by Lumsden and Keynes [1989].)

vates structures derived from the first pharyngeal arch. Cranial nerves VII and IX innervate second and third arch structures, respectively. In embryos of birds, the species studied most extensively, cell bodies of the motor components of cranial nerves V, VII, and IX are found exclusively in rhombomeres 2, 4, and 6. Dye injection studies have shown that progeny of a single neuroblast remain within the rhombomere containing the injected cell. Axons contributing to a cranial nerve extend laterally within the rhombomere and converge on a common exit site in the craniocaudal midpoint of the rhombomere. At a slightly later stage in development, motor neurons originating in the next more posterior rhombomere (3, 5, 7) extend axons laterally. However, before the axons reach the margin of the rhombomere, they cross into rhombomeres 2, 4, or 6 and converge on the motor axon exit site in the even-numbered rhombomere. Comparative studies indicate species differences in the relationship between rhombomeres and origins of the cranial nerves. Specific data on human embryos are not available.

The cell bodies (the collection of cell bodies of a single nerve is called a **nucleus**) of the cranial nerves that innervate the pharyngeal arches arise in register along the craniocaudal axis. The motor nuclei of other cranial nerves that innervate somatic structures (e.g., extraocular muscles, the tongue) arise in a different craniocaudal column along the hindbrain and do not occupy contiguous rhombomeres (see Fig. 12-10).

Direct and indirect evidence indicates that properties of the walls of the rhombomeres prevent axons from straying into inappropriate neighboring rhombomeres. One cellular property, which is also characteristic of regions of somites that restrict the movement of neural crest cells, is the ability of cells of the wall of the rhombomere to bind specific lectins. In apparent contradiction to the compartmentalization just described, processes growing from sensory neuroblasts and from nerves of a tract called the medial longitudinal fasciculus are free to cross rhombomeric boundaries. Blood vessels first enter the hindbrain in the region of the floor plate soon after the emergence of the motor axons and spread within the interrhombomeric junctions. The way the vascular branches recognize the boundaries of the rhombomeres is not known.

In contrast to the hindbrain, the pattern of nerves emanating from the spinal cord does not appear to be determined by craniocaudal compartmentalization within the cord. Rather, the segmented character of the spinal nerves is dictated by the somitic mesoderm along the neural tube. Outgrowing motor neurons from the spinal cord and migrating neural crest cells can easily penetrate the anterior mesoderm of the somite, but they appear to be repulsed by the posterior half of the somite. This results in a regular pattern of spinal nerve outgrowth, with one bilateral pair of spinal nerves per body segment. Rotating the early neural tube about its craniocaudal axis does not result in an abnormal pattern of spinal nerves. This further strengthens the viewpoint that the pattern of spinal nerves is not generated within the neural tube itself. Despite many studies on molecular differences between anterior and posterior somitic mesoderm, the exact molecular basis for the repulsion or attraction of spinal nerves by specific regions of the somites remains unknown.

There are many other aspects of pattern formation within the nervous system. The possibility of a causal relationship between the striking molecular patterning that is becoming more apparent along the neural tube and the ultimate functional morphology of the brain and spinal cord is likely but still unproved. Patterns of molecular expression can be correlated with subsequent morphological changes, but the way the molecular information is translated into specific morphology remains a mystery.

THE PERIPHERAL NERVOUS SYSTEM

Structural Organization of a Peripheral Nerve

The formation of a peripheral nerve begins with the outgrowth of axons from motor neuroblasts located in the intermediate zone of the basal plate (the future ventral horn of the gray matter) of the spinal cord (Fig. 12-11). On the

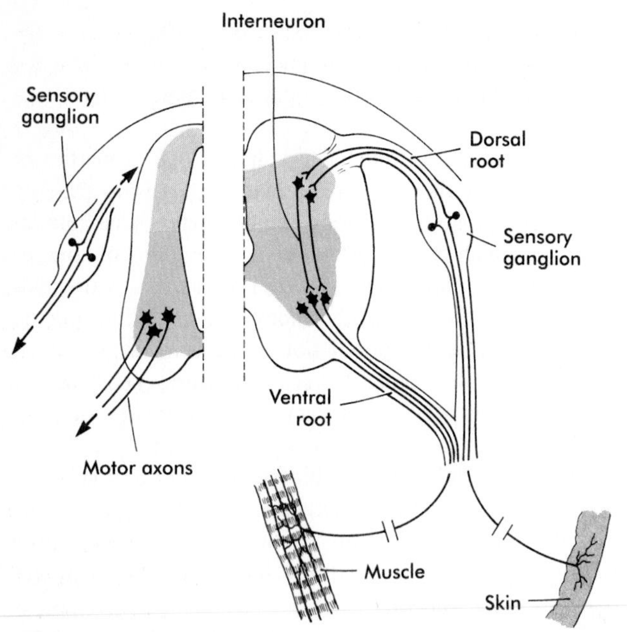

FIG. 12-11 Development of a peripheral nerve. *Left:* Early embryo. *Right:* Fetus.

dorsal aspect of the spinal cord, thin processes also begin to grow from neural crest–derived neuroblasts that have aggregated to form the spinal ganglia. **Dendrites,** which conduct impulses toward the nerve cell body, grow from the sensory neurons toward the periphery. **Axons,** which conduct impulses away from the cell body, penetrate the dorsolateral aspect of the spinal cord and terminate in the dorsal horn (the gray matter of the alar plate). Within the gray matter, short interneurons connect the terminations of the sensory axons to the motor neurons. These three connected neurons (motor, sensory, and interneuron) constitute a simple **reflex arc,** through which a sensory stimulus can be translated into a simple motor response. Autonomic nerve fibers are also associated with typical spinal nerves.

Within a peripheral nerve the neuronal processes can be myelinated or unmyelinated. At the cellular level, **myelin** is a multilayered spiral sheath consisting largely of phospholipid material that is formed by individual **Schwann cells** (neural crest derivatives) wrapping themselves many times around a nerve process like a jelly roll (Fig. 12-12). This wrapping serves as a form of insulation that determines

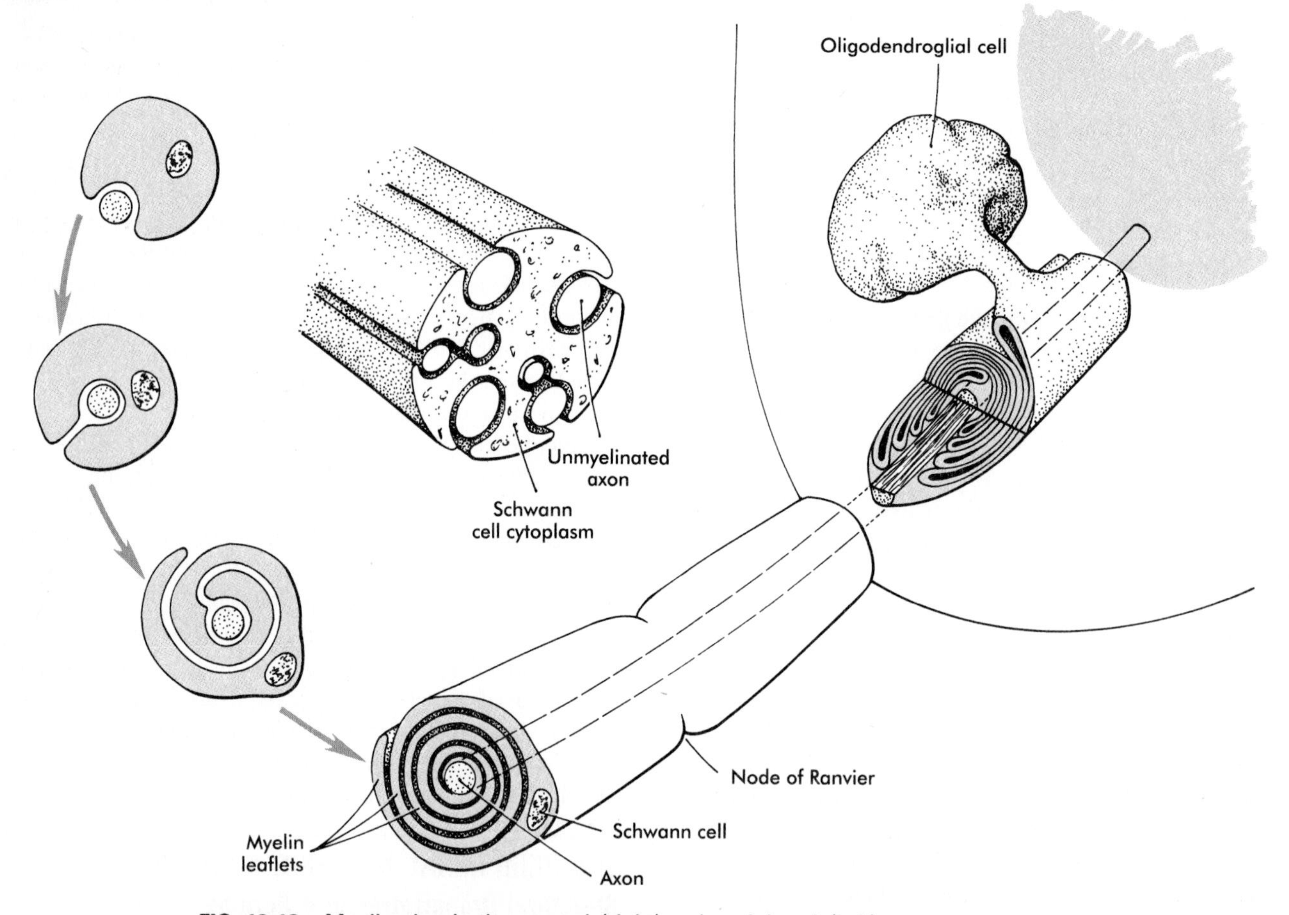

FIG. 12-12 Myelination in the central *(right)* and peripheral *(left)* nervous systems. Within the central nervous system, myelin is formed by oligodendroglial cells. In the peripheral nervous system, Schwann cells wrap around individual axons. The inset shows a segment through a region of unmyelinated nerve fibers embedded in the cytoplasm of a single Schwann cell.

to a great extent the character of the electrical impulse (action potential) traveling along the neuronal process. **Unmyelinated nerve fibers** are also embedded in the cytoplasm of Schwann cells, but they lack the characteristic spiral profiles of myelinated processes (Fig. 12-12).

Within the central nervous system the color of the white matter is the result of its high content of myelinated nerve fibers, whereas the gray matter contains unmyelinated fibers. Schwann cells are not present in the central nervous system; instead, myelination is accomplished by oligodendrocytes. Although one Schwann cell in a myelinated peripheral nerve fiber can wrap itself around only one axon or dendrite, a single oligodendrocyte can myelinate several nerve fibers in the central nervous system.

Patterns and Mechanisms of Neurite Outgrowth

The outgrowth of **neurites** (axons or dendrites) involves many factors both intrinsic and extrinsic to the neurite. Although similar in many respects, the outgrowth of axons and dendrites differs in fundamental ways.

An actively elongating neurite is capped by a **growth cone** (Fig. 12-13). Growth cones are characterized by an expanded region of cytoplasm with numerous spikelike projections called **filopodia.** In vitro and in vivo studies of living nerves show that the morphology of an active growth cone is in a constant state of flux, with filopodia regularly extending and retracting as if testing the local environment. Growth cones contain numerous cytoplasmic organelles, but much of the form and function of the filopodia depend on the large quantities of **actin** microfilaments that fill these processes. In the presence of agents that disrupt actin filaments, the filopodia retract and the growth cones cease to function normally.

Whether growth cones progress forward, remain static, or change directions depends in large measure on their interactions with the local environment. If the environment is favorable, a filopodium will remain extended and adhere to the substrate around it, whereas other filopodia on the same growth cone retract. Depending on the location of the adhering filopodia, the growth cone may lead the neurite to which it is attached straight ahead or change its direction of outgrowth.

Growth cones can respond to concentration gradients of diffusible substances (e.g., nerve growth factor) or to weak local electrical fields. They can also respond to fixed physical or chemical cues from the microenvironment immediately surrounding them. As previously discussed, the caudal half of the somite repulses the ingrowth of motor axons from the spinal cord and also neural crest cells into that area. On the other hand, extracellular matrix glycoproteins such as **fibronectin** and especially **laminin** strongly promote the adhesion and outgrowth of neurites. Integral membrane proteins on the neurites called **integrins** bind specifically to arginine-glycine-asparagine sequences on the glycoproteins and promote adhesion to the substrate containing these molecules.

Other molecules, such as **N-cadherin, N-CAM,** and **L1,**

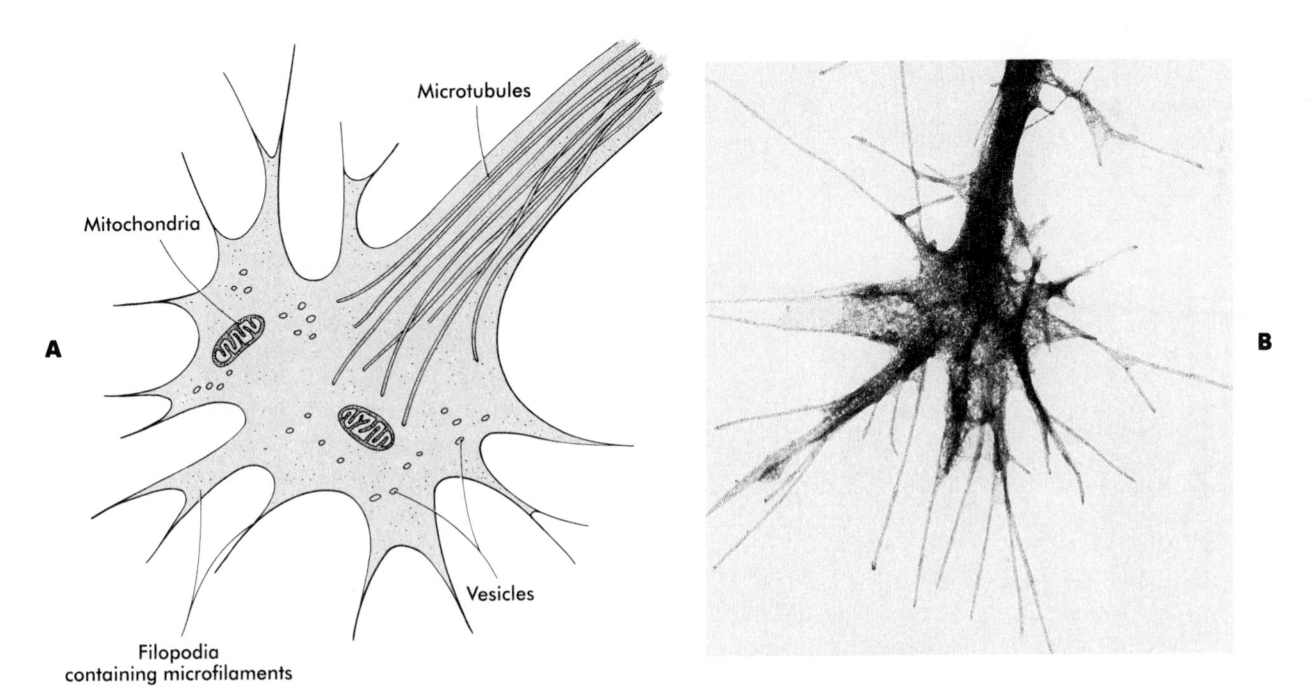

FIG. 12-13 **A,** A growth cone at the end of an elongating axon. **B,** High-voltage electron micrograph of a growth cone in culture.

(**A** from Landis S: *Annu Rev Physiol* 45:567, 1983; **B** courtesy K. Tosney, Ann Arbor, Mich.)

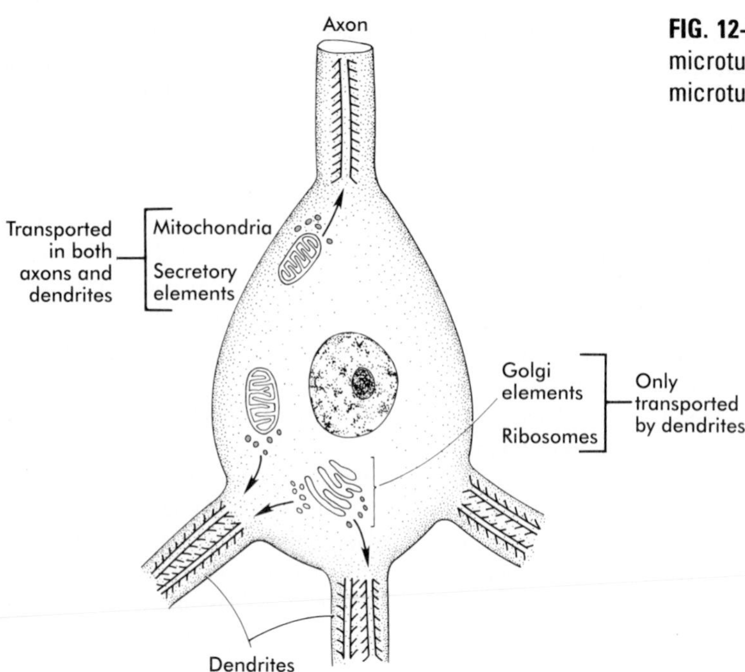

FIG. 12-14 Polarity in a developing neuron. In the axon, microtubules have only one polarity, but in dendrites, microtubules with opposite polarities are present.

are involved in intercellular adhesion at various stages of cell migration or neurite elongation. N-cadherin, which uses Ca^{++} as an ionic agent to bind two like molecules together, is heavily involved in the intercellular binding of cells in neuroepithelia. It also plays a role in the adhesion of parallel outgrowing neurites. In a peripheral nerve, one **pioneering axon** typically precedes the others in growing toward its target. Other axons then follow, forming **fascicles** (bundles) of axons. Fasciculation is facilitated by intercellular adhesion proteins such as L1, which help bind parallel nerve fibers. If antibodies to the L1 protein are administered to an area of neurite outgrowth, fasciculation is disrupted. N-CAM is present on the surfaces of most embryonic nerve processes and muscle fibers and is involved in the initiation of neuromuscular contacts. Antibodies to N-CAM interfere with the development of neuromuscular junctions in embryos. Outgrowing neurites interact with many other molecules, the full extent of which is just becoming apparent.

Although the growth cone can be thought of as the director of neurite outgrowth, other factors are important for the actual elongation of axons. Essential to the growth and maintenance of axons and dendrites is **axonal transport.** In this intracellular process, materials produced in the cell body of the neuron are carried to the ends of these neurites, which can be several feet long in humans.

The cytoskeletal backbone of an axon is an ordered array of microtubules and neurofilaments. **Microtubules** are long tubular polymers composed of **tubulin** subunits. As an axon extends from its cell body, tubulin subunits are transported down the axon and polymerize onto the distal end of the microtubule. The assembly of **neurofilaments** is organized in a similar polarized manner. The site of these cytoskeletal additions is close to the base of the growth cone, meaning that the axon elongates by being added to distally rather than being pushed out by addition to its proximal end near the neuronal cell body. A characteristic accompaniment of axonal growth is the production of large amounts of **growth-associated proteins (GAPs).** Particularly prominent among these is **GAP-43,** which serves as a substrate for protein kinase C and is concentrated in the growth cone.

Outgrowing axons and dendrites differ in several important ways. In contrast to axons, dendrites contain microtubules with polarity running in both directions (Fig. 12-14). Another prominent difference is the absence of GAP-43 protein in growing dendrites. In fact, one of the first signs of polarity of a developing neuron is the concentration of GAP-43 in the outgrowing axon and its disappearance from the dendritic processes.

Neurite/Target Relations during Development of a Peripheral Nerve

Developing neurites continue to elongate until they have contacted an appropriate end organ. In the case of motor neurons, that end organ is a developing muscle fiber. Dendrites of sensory neurons relate to a number of types of targets. The end of the neurite must first recognize its appropriate target, and then it must make a functional connection with it.

In the case of motor neurons, evidence that very specific cues guide individual nerves and axons to their muscle targets is increasing. Tracing and transplantation stud-

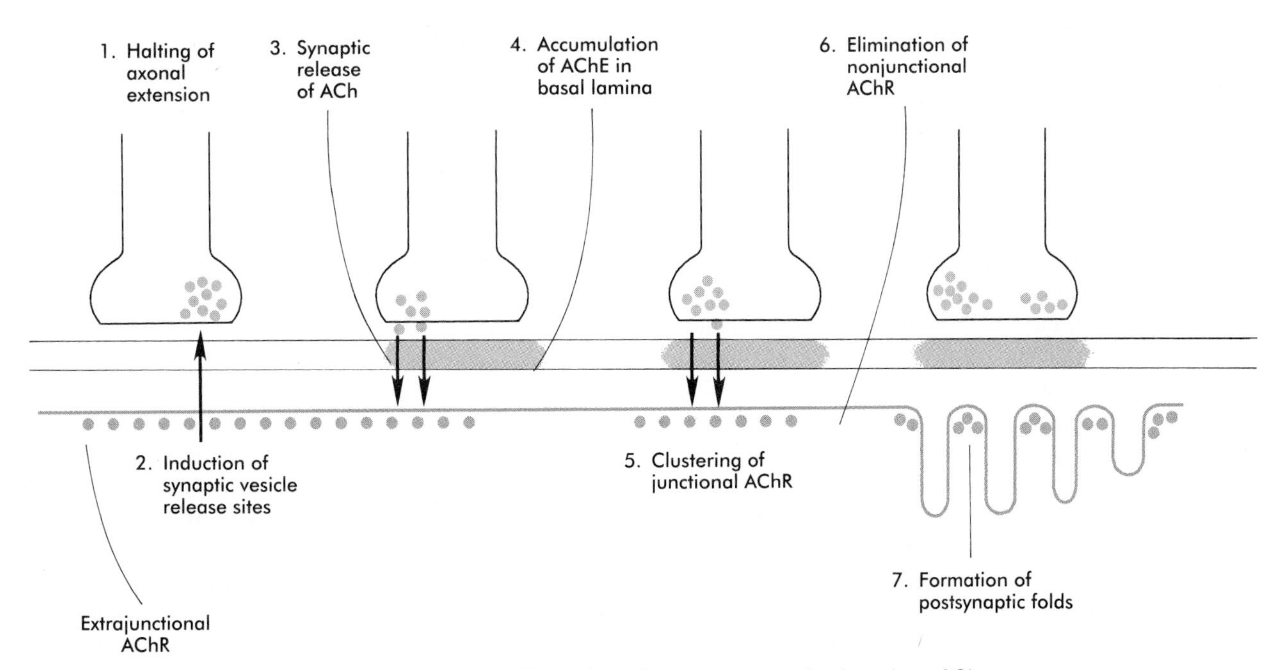

FIG. 12-15 Major steps in the formation of a neuromuscular junction. *ACh,* Acetylcholine; *AChE,* acetylcholinesterase; *AChR,* acetylcholine receptors.

ies have shown that outgrowing motor nerves to limbs supply the limb muscles in a well-defined order and that after minor positional dislocations, they will seek out the correct muscles (see Fig. 11-21). Recent evidence suggests that even at the level of neurons, "fast" axons are attracted to the precursors of fast muscle fibers and "slow" axons to slow muscle fibers. There are many similar examples of target specificity in dendrites in the peripheral nervous system and of both dendrites and axons in the central nervous system.

When a motor axon and a muscle fiber meet, a complex series of changes in both the nerve and muscle fibers marks the formation of a functional **synapse,** in this case called a **neuromuscular junction** (Fig. 12-15). The early changes consist of (1) the cessation of outgrowth of the axon, (2) preparation of the nerve terminal for the ultimate release of appropriate neurotransmitter molecules, and (3) modifications of the muscle fibers at the site of nerve contact so that the neural stimulus can be received and translated into a contractile stimulus. Both neural and muscular components of the neuromuscular junction are involved in stabilizing the morphology and functional properties of this highly specialized synapse.

One of the first signs of specialization at an incipient neuromuscular junction is the formation of synaptic vesicles, presumably due to some influence of the muscle fiber. These vesicles store and ultimately release the neurotransmitter substance **acetylcholine** from the nerve terminal (see Fig. 12-15). Before the developing muscle fiber is contacted by the motor neuron, **acetylcholine receptors** (nonjunctional type) are scattered throughout the length of the muscle fiber. After initial nerve contact, myonuclei in the vicinity of the neuromuscular junction produce junction-specific acetylcholine receptors that reside on nerve-induced postjunctional folds of the muscle fiber membrane, and the scattered nonjunctional receptors disappear. Between the nerve terminal and the postsnyaptic apparatus of the muscle fiber lies a basal lamina that contains molecules stabilizing the acetylcholine receptors at the neuromuscular junction and also **acetylcholinesterase,** an enzyme produced by the muscle fiber.

Factors Controlling Numbers and Kinds of Connections between Neurites and End Organs in the Peripheral Nervous System

At many stages in the formation of a peripheral nerve, interactions between the outgrowing neurites and the target structure influence the numbers and quality of either the nerve fibers or the targets. The existence of such mechanisms was demonstrated in the early 1900s by transplanting limb buds onto flank regions. The motor nerves and sensory ganglia that supplied the grafted limbs were substantially larger than the contralateral spinal nerves, which innervated only structures of the body wall. Examination of the spinal cord at the level of the transplant revealed larger ventral horns of gray matter containing more motor neurons than normal for levels of the spinal cord that supply only flank.

Addition experiments of this type cast light on normal

anatomical relations, which also show relatively larger volumes of gray matter and larger nerves at levels from which the normal limbs are innervated. Deletion experiments, in which a limb bud is removed before neural outgrowth, or the congenital absence of limbs results in deficient numbers of peripheral neurons and reduced volumes of gray matter in the affected regions.

Neuronal **cell death (apoptosis)** plays an important role in normal neural development. For example, when a muscle is first innervated, far more than the normal adult number of neurons supply it. At a critical time in development, massive numbers of neurons die. There seem to be a number of reasons for this seemingly paradoxical phenomenon: (1) Some axons fail to reach their normal target, and cell death is a way of eliminating them. (2) Cell death could be a way of reducing the size of the neuronal pool to something appropriate to the size of the target. (3) Similarly, cell death could compensate for a presynaptic input that is too small to accommodate the neurons in question. (4) Neuronal cell death may also be a means of eliminating connection errors between the neurons and their specific end organs, for example, muscle fibers in the case of motor nerves.

All of these reasons for neuronal cell death may be part of a general biological strategy that reduces superfluous initial connections to ensure that enough correct connections have been made. The other developmental strategy, which seems to be much less used, is to control the outgrowth and connection of neurites with their appropriate end organs so tightly that there is little room for error from the beginning. Because of the overall nature of mammalian development, such tight developmental controls would rob the embryo of the overall flexibility it needs to compensate for genetically or environmentally induced variations in other aspects of development.

The mechanisms by which innervated target structures prevent the death of the neurons that supply them are only beginning to be understood. A currently popular hypothesis is that the target cells release chemical **trophic factors** that neurites take up, usually by binding to specific receptors, and sustain. The classic example of a trophic factor is **nerve growth factor,** which sustains the outgrowth and prevents the death of sensory neurons. Even after 40 years of intensive study by many laboratories, the mechanism of action of nerve growth factor is not well understood. Several other well-characterized molecules are also candidates for trophic factors.

THE AUTONOMIC NERVOUS SYSTEM

The autonomic nervous system is the component of the peripheral nervous system that subserves many of the involuntary functions of the body, such as glandular activity and motility within the digestive system, heart rate, vascular tone, and sweat gland activity. It is divided into two major divisions, the sympathetic and parasympathetic nervous systems. Components of the **sympathetic nervous system** arise from the thoracolumbar levels of the spinal cord, whereas the **parasympathetic nervous system** has a widely separated dual origin from the cranial and sacral regions. Both components of the autonomic nervous system consist of two tiers of neurons—**preganglionic** and **postganglionic.** Postganglionic neurons are derivatives of the neural crest (see Chapter 13).

Sympathetic Nervous System

Preganglionic neurons of the sympathetic nervous system arise from the **intermediate horn** (visceroefferent column) of the gray matter in the spinal cord. At levels from the first thoracic to the second lumbar vertebrae, their myelinated axons grow from the cord through the ventral roots, paralleling the motor axons that supply the skeletal musculature (Fig. 12-16). Shortly after the dorsal and ventral roots of the spinal nerve join, the preganglionic sympathetic axons, which are derived from the neuroepithelium of the neural tube, leave the spinal nerve via a **white communicating ramus.** They soon enter one of a series of **sympathetic ganglia** to synapse with neural crest–derived postganglionic neurons.

The sympathetic ganglia, the bulk of which are organized as two chains running ventrolateral to the vertebral bodies, are laid down by neural crest cells that migrate from the closing neural tube along a special pathway (see Fig. 13-4). Once the **migrating sympathetic neuroblasts** have reached the site at which the **sympathetic chain ganglia** form, they spread both cranially and caudally until the extent of the chains approximates that seen in the adult. Some of the sympathetic neuroblasts migrate farther ventrally than the level of the chain ganglia to form a variety of other **collateral ganglia** (e.g., **celiac** and **mesenteric ganglia**), which occupy somewhat variable positions within the body cavity. The **adrenal medulla** can be broadly viewed as a highly modified sympathetic ganglion.

The outgrowing preganglionic sympathetic neurons either terminate within the chain ganglia or pass through on their way to more distant sympathetic ganglia to form synapses with the cell bodies of the second-order postganglionic sympathetic neuroblasts (see Fig. 12-16). Axons of some postganglionic neuroblasts, which are unmyelinated, leave the chain ganglia as a parallel group and reenter the nearest spinal nerve through the **gray communicating ramus.** Once in the spinal nerve, these axons continue to grow until they reach appropriate peripheral targets, such as sweat glands, arrector pili muscles, and walls of blood vessels. Axons of other postganglionic sympathetic neurons leave their respective ganglia as tangled **plexuses** of nerve fibers and grow toward other visceral targets.

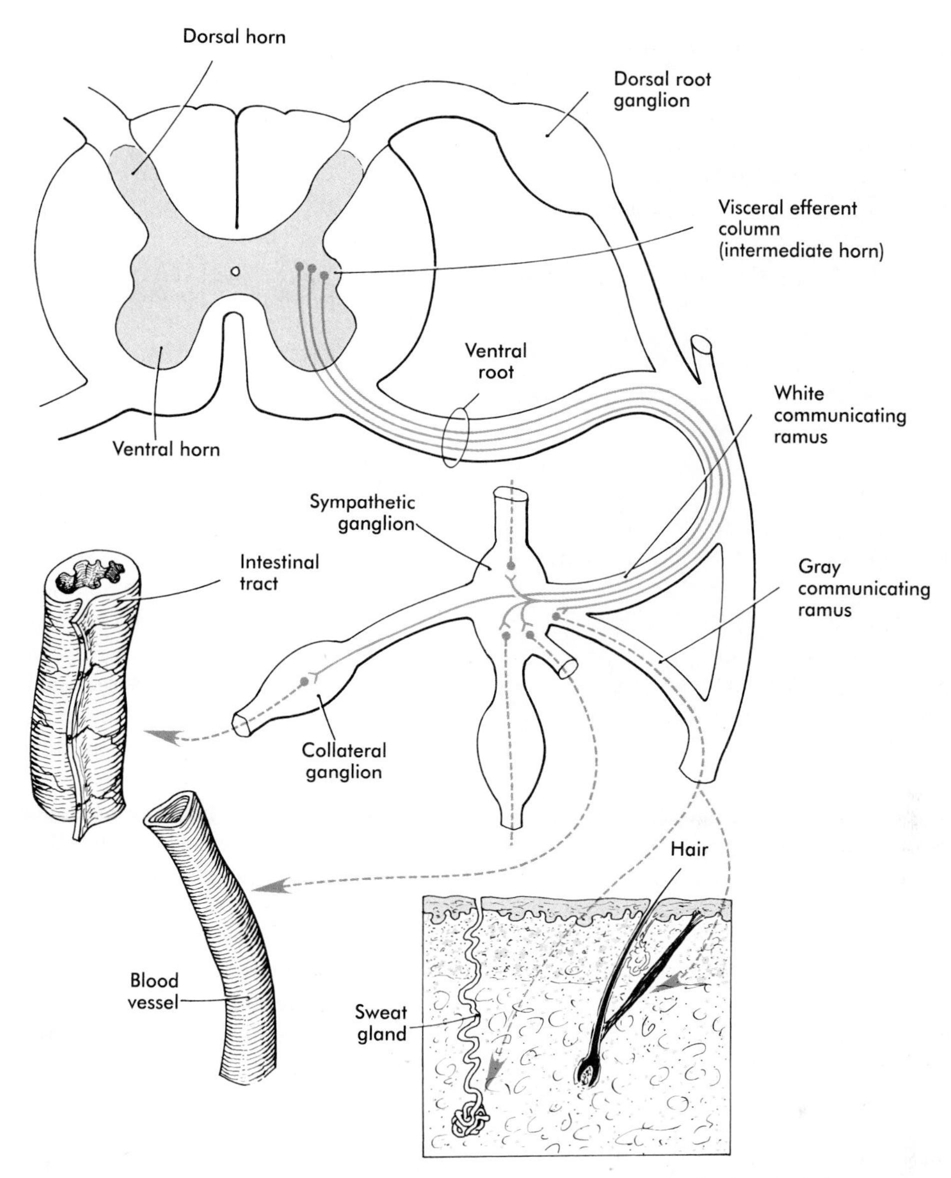

FIG. 12-16 The organization of the autonomic nervous system at the level of the thoracic spinal cord.

Parasympathetic Nervous System

Although also organized on a preganglionic and postganglionic basis, the parasympathetic nervous system has a distribution quite different from that of the sympathetic system. Like those of the sympathetic nervous system, preganglionic parasympathetic neurons originate in the visceroefferent column of the central nervous system. However, the levels of origin of these neuroblasts are in the midbrain and hindbrain (specifically associated with cranial nerves III, VII, IX, and X) and in the second to fourth sacral segments of the developing spinal cord. Axons from these preganglionic neuroblasts grow long distances before they meet the neural crest–derived postganglionic neurons. These are typically embedded in scattered small ganglia or plexuses in the walls of the organs that they innervate.

The neural crest precursors of the postganglionic neurons often undertake extensive migrations, for example, from the hindbrain to final locations in the walls of the intestines. The migratory properties of the neural crest precursors of parasympathetic neurons are impressive, but this population of cells also undergoes a tremendous expansion until the final number of enteric neurons approximates the number of neurons in the spinal cord. Evidence is increasing that factors in the gut wall stimulate the mitosis of the neural crest cells migrating there. A striking demonstration of the stimulatory powers of the gut is the ability of pieces

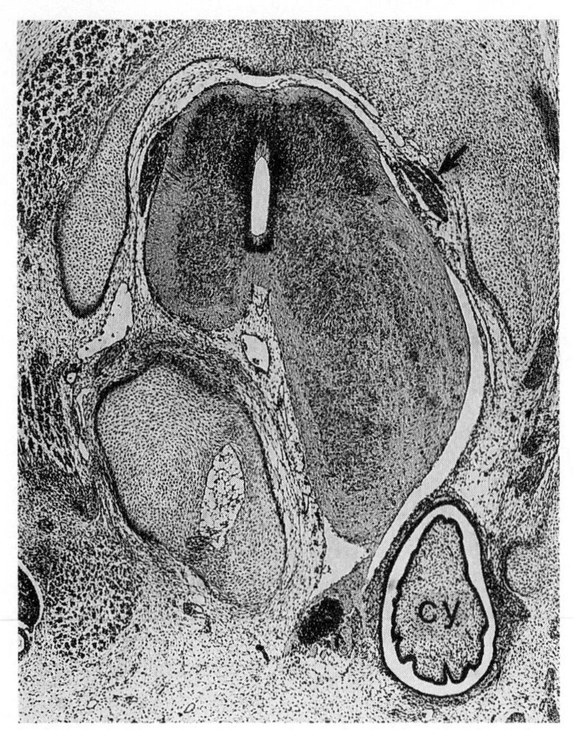

FIG. 12-17 Influence of the gut on growth of the neural tube. A graft of quail duodenum was placed between the neural tube and somites of a chick embryo host. The spinal cord on the side near the graft of gut (*cy,* cyst of donor endoderm) has greatly enlarged, causing secondary distortion of the musculoskeletal structures near it.

(From Rothman TP, Gershon MD, and others: *Dev Biol* 124:331-346, 1987.)

of gut wall transplanted along the neural tube to cause a great expansion of the region of the neural tube closest to the graft (Fig. 12-17).

The Differentiation of Autonomic Neurons

At least two major steps are involved in the differentiation of autonomic neurons. The first is the determination of certain migrating neural crest cells to differentiate into autonomic neurons instead of the other possible neural crest derivatives.

At early stages the neural crest cells have the option of becoming components of either the sympathetic or parasympathetic system. This was demonstrated by level-shift transplantations of neural crest cells in birds. For example, when cephalic neural crest, which would normally form parasympathetic neurons, was transplanted to the level of somites 18 to 24, the transplanted cells migrated and settled into the adrenal medulla as **chromaffin cells,** which are part of the sympathetic nervous system. Conversely, trunk neural crest cells transplanted into the region of the head often migrated into the lining of the gut and differentiated into postganglionic parasympathetic neurons.

A second major step in the differentiation of autonomic neurons involves the choice of the neurotransmitter that the neuron will use. Typically, parasympathetic postganglionic neurons are **cholinergic** (i.e., they use acetylcholine as a transmitter), whereas sympathetic neurons are **adrenergic** (noradrenergic) and use norepinephrine as a transmitter.

As they arrive at their final destinations, autonomic neurons are noradrenergic. They then undergo a phase resulting in the selection of the neurotransmitter substance that will characterize their mature state. Considerable experimental evidence suggests that the choice of transmitter proceeds independently of other concurrent events, such as axonal elongation and the innervation of specific target organs.

At surprisingly late stages in their development, autonomic neurons still retain flexibility in their choice of neurotransmitter. For example, sympathetic neurons in newborn rats are normally adrenergic, and if grown in standard in vitro culture conditions, they produce large amounts of norepinephrine and negligible amounts of acetylcholine. If the same neurons are cultured in a medium that has been conditioned by the presence of cardiac muscle cells, they undergo a functional conversion and produce large amounts of acetylcholine instead (Fig. 12-18).

An example of a natural transition of neurotransmitter phenotype from noradrenergic to cholinergic occurs in the sympathetic innervation of sweat glands in the rat. Neurotransmitter transitions depend on target-derived cues. An example of one such cue is **cholinergic differentiation factor,** a glycosylated basic 45-kD protein. This molecule, which is present in heart-conditioned medium, is one of a number of chemical environmental factors that can exert a strong influence on late phases of differentiation of autonomic neurons.

Congenital Aganglionic Megacolon (Hirschsprung's Disease)

If a newborn exhibits symptoms of complete constipation in the absence of any demonstrable physical obstruction, the cause is usually an absence of parasympathetic ganglia from the lower (sigmoid) colon and rectum. This condition, commonly called **Hirschsprung's disease,** is normally attributed to the absence of colonization of the wall of the lower colon by neural crest–derived parasympathetic neuronal precursors, presumably of sacral origin because of their distribution. In rare cases, greater parts of the colon are lacking in ganglia. In view of recent experimental findings of the role of the gut wall in promoting mitosis of parasympathetic neuronal precursors, the possibility that there is a deficit of the mitogenic stimulatory properties of the wall of the colon should be investigated.

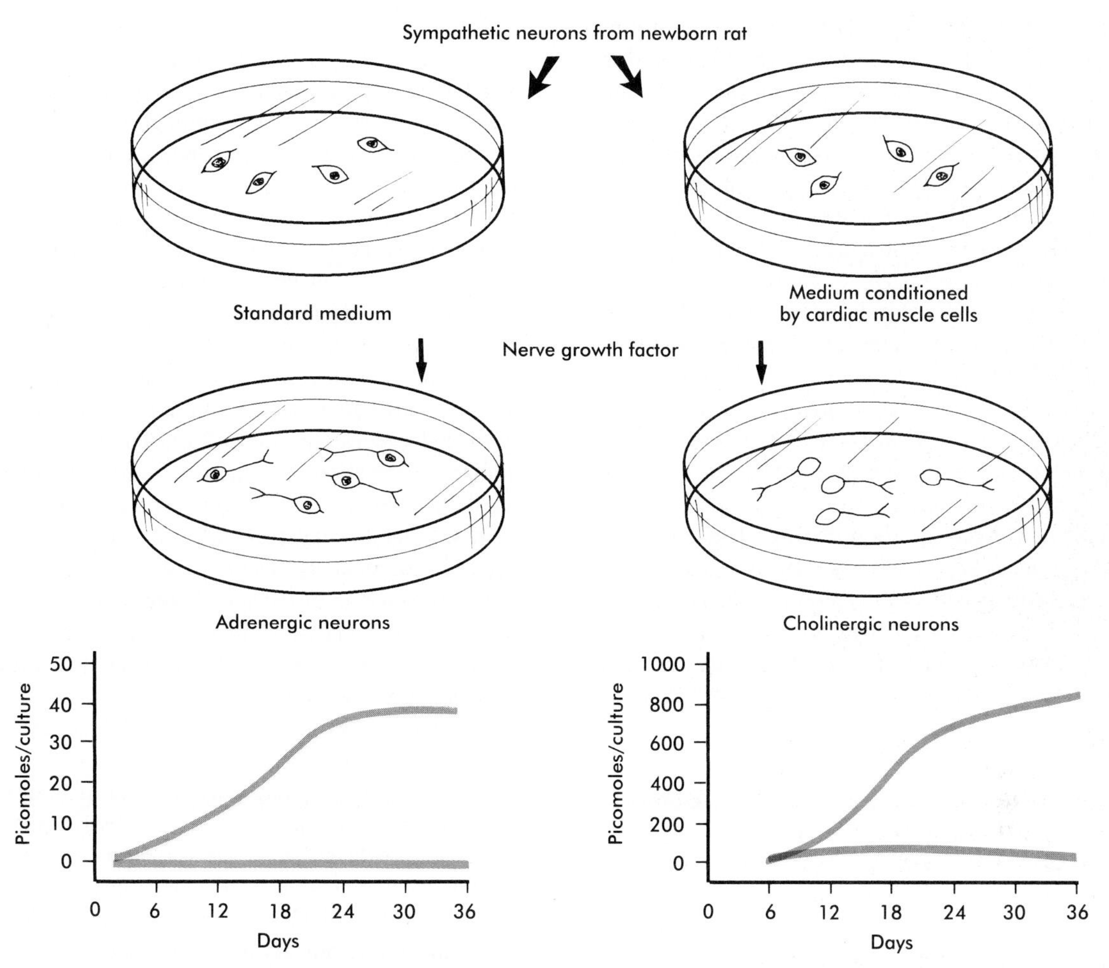

FIG. 12-18 Experiment illustrating the effect of the environment on the choice of transmitter by differentiating sympathetic neurons. In standard medium, they become adrenergic; in medium conditioned by cardiac muscles they become cholinergic. Levels of norepinephrine are in red; levels of acteylcholine are in blue.
(Based on studies by Patterson, *Sci Am* [1978].)

LATER STRUCTURAL CHANGES IN THE CENTRAL NERVOUS SYSTEM*

Histogenesis Within the Central Nervous System

A major difference between the brain and spinal cord is the organization of the gray and white matter. In the spinal cord the gray matter is centrally located, with white matter surrounding it (see Fig. 12-6). In many parts of the brain this arrangement is reversed, with a large core of white matter and layers of gray matter situated superficial to that.

One of the fundamental processes in histogenesis within the brain is cell migration. From their sites of origin close to the ventricles in the brain, neuroblasts migrate toward the periphery following set patterns. These patterns often result in a multilayering of the gray substance of the brain tissue. Key players in the migratory phenomenon are radial glial cells, which extend toward the periphery long processes radially from cell bodies located close to the ventricular lumen (Fig. 12-19). Young postmitotic neurons, which are typically simple bipolar cells, wrap themselves around the radial glial cells and use them as guides on their migrations from their sites of origin to the periphery.

In areas of brain cortex characterized by multiple layers of gray matter, the large neurons populating the innermost layer migrate first. The remainder of the layers of gray matter are formed by smaller neurons migrating through the first layer and other previously formed layers to set up a new layer of gray matter at the periphery. With this pattern of histogenesis, the outermost layer of neurons is the one

*The later changes in the central nervous system are so extensive that an exhaustive treatment of even one aspect, such as morphology, is well beyond the scope of this book. This section instead stresses fundamental aspects of the organization of the central nervous system and summarizes the major changes in organization of the brain and spinal cord.

that is formed last, and the innermost is the first-formed layer. In a mouse mutant called ***weaver,*** specific behavioral defects are related to abnormal function of the cerebellum. The morphological basis for this mutant is an abnormality of the radial glial cells in the cerebellum and a consequent abnormal migration of the cells that would normally form the granular layer of the cerebellar cortex.

Increasing evidence indicates that the seemingly featureless cerebral cortex is a matrix of discrete **columnar radial units,** which consist of radial glial cells and the neuroblasts that migrate along them. There may be as many as 200 million radial units in the human cerebral cortex. The radial units begin as proliferative units, with most cortical neurons generated between days 40 and 125. As with many aspects of neural differentiation, the number of radial units seems to be sensitive to their own neural input. For example, in cases of **congenital anophthalmia** (the absence of eyes), neural input from visual pathways to the area of the occipital cortex associated with vision is reduced. This results in both gross and microscopic abnormalities of the visual cortex, principally related to a reduced number of radial units in that region.

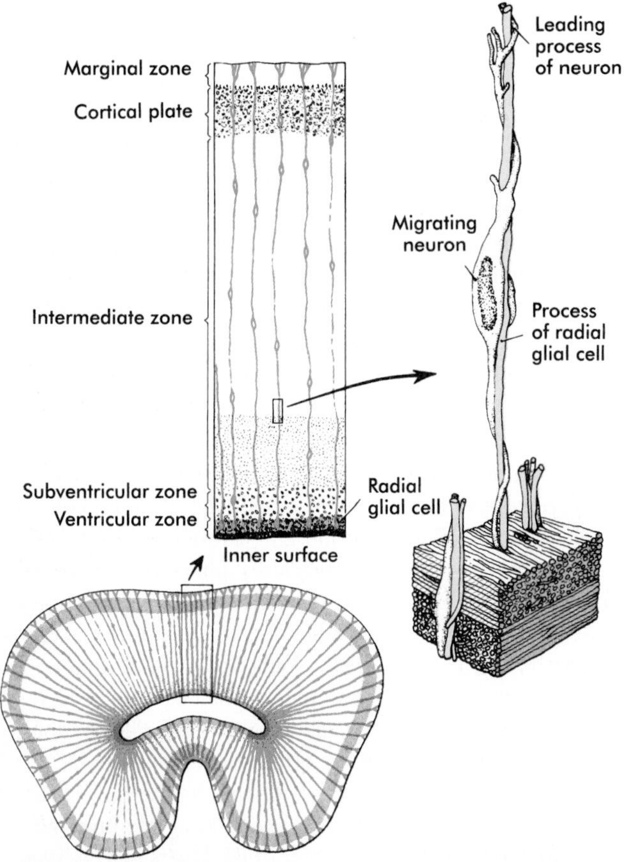

FIG. 12-19 Radial glial cells and their association with peripherally migrating neurons during development of the brain.
(Based on studies by Rakic [1975].)

Spinal Cord

The spinal cord is a structure where inputs from many peripheral sensory nerves are distributed as local reflex arcs or are channeled to the brain through tracts of axons. Here also, motor messages originating in the brain are distributed to appropriate peripheral locations via motor tracts and ventral (motor) roots of individual spinal nerves. Although a number of aspects of organization of the spinal cord were dealt with earlier in this chapter, some will be reviewed briefly because of their value in understanding the basic organization of the brain.

The early spinal cord is divided into alar and basal plate regions, which are precursors for the sensory and motor regions of the cord (see Fig. 12-6). The mature cord has a similar organization, but these regions are further subdivided into somatic and visceral components. Within the brain still another layer of input and output is added with "special" components. These are summarized in the box.

A gross change of the spinal cord that is of clinical significance is the relative shortening of the cord in relation to the vertebral column (Fig. 12-20). In the first trimester the spinal cord extends the entire length of the body, and the spinal nerves pass through the intervertebral spaces directly opposite their site of origin. In later months, growth of the posterior part of the body outstrips that of both the vertebral column and the spinal cord, but growth of the cord lags significantly behind that of the vertebral column. This disparity is barely apparent in the cranial and thoracic regions, but at birth the spinal cord terminates at the level of the third lumbar vertebra. By adulthood, the cord terminates at the second lumbar vertebra.

Functional Regions in the Spinal Cord and Brain

Alar Plate (Afferent or Sensory)

1. General somatic afferent—Sensory input from the skin, joints, and muscles
2. Special visceral afferent—Sensory input from the taste buds and pharynx
3. General visceral afferent—Sensory input from the viscera and heart

Basal Plate (Efferent—Motor or Autonomic)

1. General visceral efferent—Autonomic (two-neuron) links from the intermediate horn to the viscera
2. Special visceral efferent—Motor nerves to striated muscles of the branchial arches
3. General somatic efferent—Motor nerves to the striated muscles other than those of the pharyngeal arches

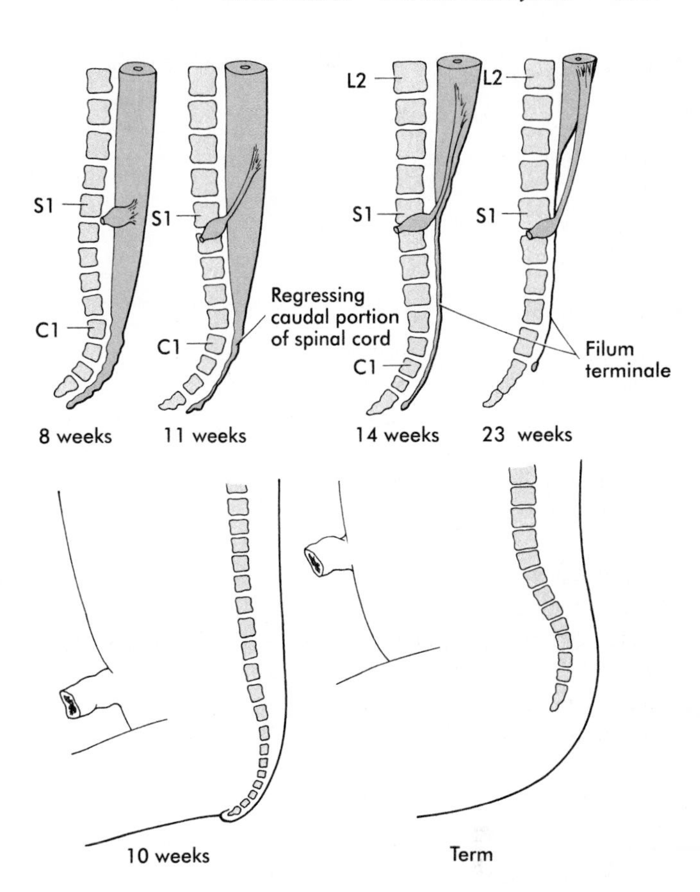

FIG. 12-20 Changes in the level of the end of the spinal cord in relation to bony landmarks in the vertebral column during fetal development. The lower illustrations show the development of the curvature of the spine.

The consequence of this disparity in growth is that the lumbar and sacral dorsal and ventral spinal nerve roots are considerably elongated to accommodate the increased distance between their point of origin and the appropriate intervertebral space. This gives them the collective appearance of a horse's tail, hence their name, **cauda equina.** A thin, filamentlike **filum terminale** extending from the end of the spinal cord proper to the base of the vertebral column marks the original excursion of the spinal cord. This arrangement is convenient for the clinician because the space below the termination of the cord is a safe place from which to withdraw cerebrospinal fluid for analysis.

Myelencephalon

The myelencephalon, the most caudal subdivision of the rhombencephalon (see Fig. 12-2), develops into the **medulla oblongata** of the adult brain (Figs. 12-1, *B*, and 12-21). It is in many respects a transitional structure between the brain and spinal cord, and the parallels between its functional organization and that of the spinal cord are readily apparent (Fig. 12-22). Much of the medulla serves as a conduit for tracts that link the brain with input and output nodes in the spinal cord, but in addition, it contains centers for the regulation of vital functions such as the heart beat and respiration.

The fundamental arrangement of alar and basal plates with an intervening sulcus limitans is retained almost unchanged in the myelencephalon. The major topographical change from the spinal cord is a pronounced expansion of the roof plate to form the characteristic thin roof overlying the expanded central canal, which in the myelencephalon is called the **fourth ventricle** (see Fig. 12-33). (Details of the ventricles and the coverings of the brain and spinal cord are presented later in this chapter.)

Special visceral **afferent** (leading toward the brain) and **efferent** (leading from the brain) columns of nuclei (aggregations of neuronal cell bodies in the brain) appear in the myelencephalon to accommodate structures derived from the branchial arches.

Metencephalon

The metencephalon, the more cranial subdivision of the rhombencephalon, consists of two main parts—the **pons,** which is directly continuous with the medulla, and the **cerebellum,** a phylogenetically newer and ontogenetically later appearing component of the brain (see Fig. 12-21).

As its name implies, the pons serves as a bridge that carries tracts of nerve fibers between higher brain centers and the spinal cord. Its fundamental organization remains like that of the myelencephalon, with three sets of afferent and

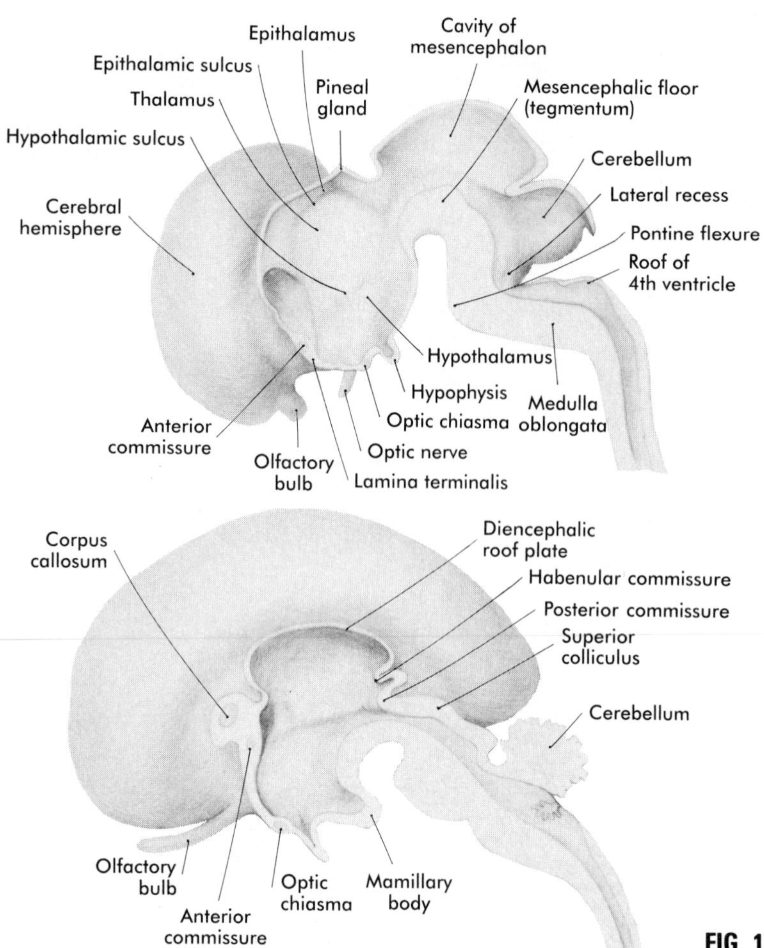

FIG. 12-21 The anatomy of the brain in 9- *(above)* and 16-week-old *(below)* human embryos.

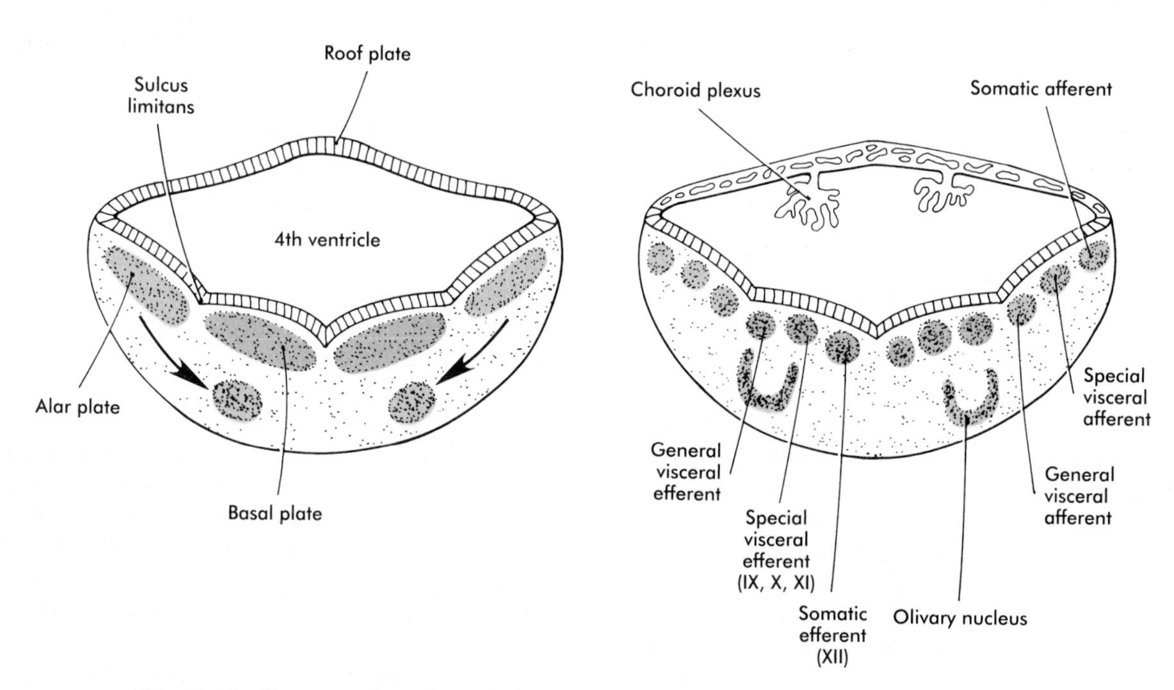

FIG. 12-22 Cross sections through the developing myelencephalon at early *(left)* and later *(right)* stages of embryonic development. Motor tracts (from the basal plate are shown in green; sensory tracts (from the alar plate) are orange.

(Modified from Sadler T: *Langman's medical embryology,* ed 6, 1990, Williams & Wilkins.)

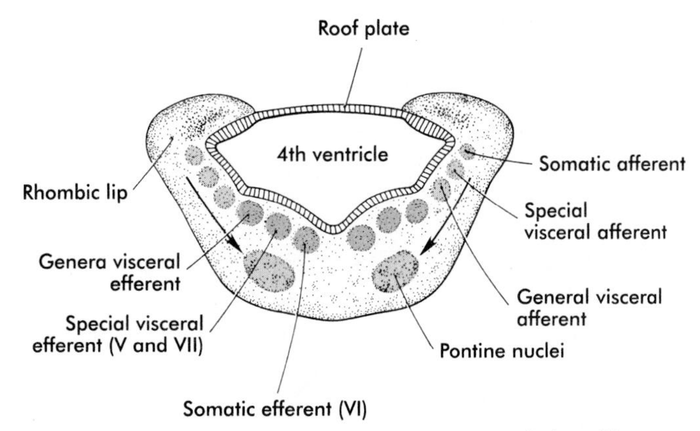

FIG. 12-23 Cross section through the developing metencephalon. Motor tracts are green; sensory tracts are orange.
(Modified from Sadler T: *Langman's medical embryology,* ed 6, 1990, Williams & Wilkins.)

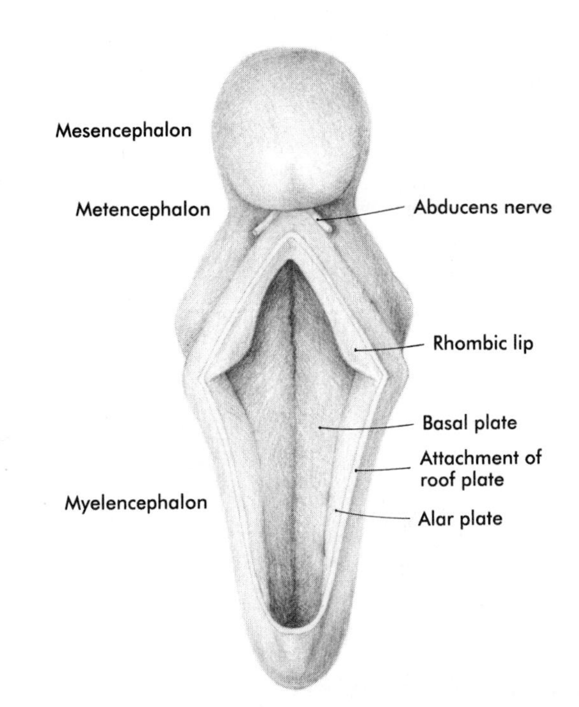

FIG. 12-24 Dorsal view of the midbrain and hindbrain of a 5-week-old embryo. The roof of the fourth ventricle has been opened.

efferent nuclei (Fig. 12-23). In addition to these, other special pontine nuclei, which originated from alar plate–derived neuroblasts, are present in the ventral white matter. The caudal part of the pons also has an expanded roof plate similar to that of the myelencephalon.

The cerebellum is both structurally and functionally complex, but phylogenetically it arose as a specialization of the vestibular system and was involved with balance. Other functions, such as orchestration of general coordination and involvement in auditory and visual reflexes, were later superimposed.

The future site of the cerebellum is first represented by the **rhombic lips** of the 5- to 6-week-old embryo (Fig. 12-24). The rhombic lips are located at the cranial edge of the thinned roof of the fourth ventricle, and they project partly into the ventricle. Until the end of the third month, the expansion of the rhombic lips is mainly inward, but thereafter the rapid growth in volume of the cerebellum is directed outward (Fig. 12-25).

As the volume of the developing cerebellum expands, the two lateral rhombic lips join in the midline, giving the early cerebellar primordium a dumbbell appearance. The

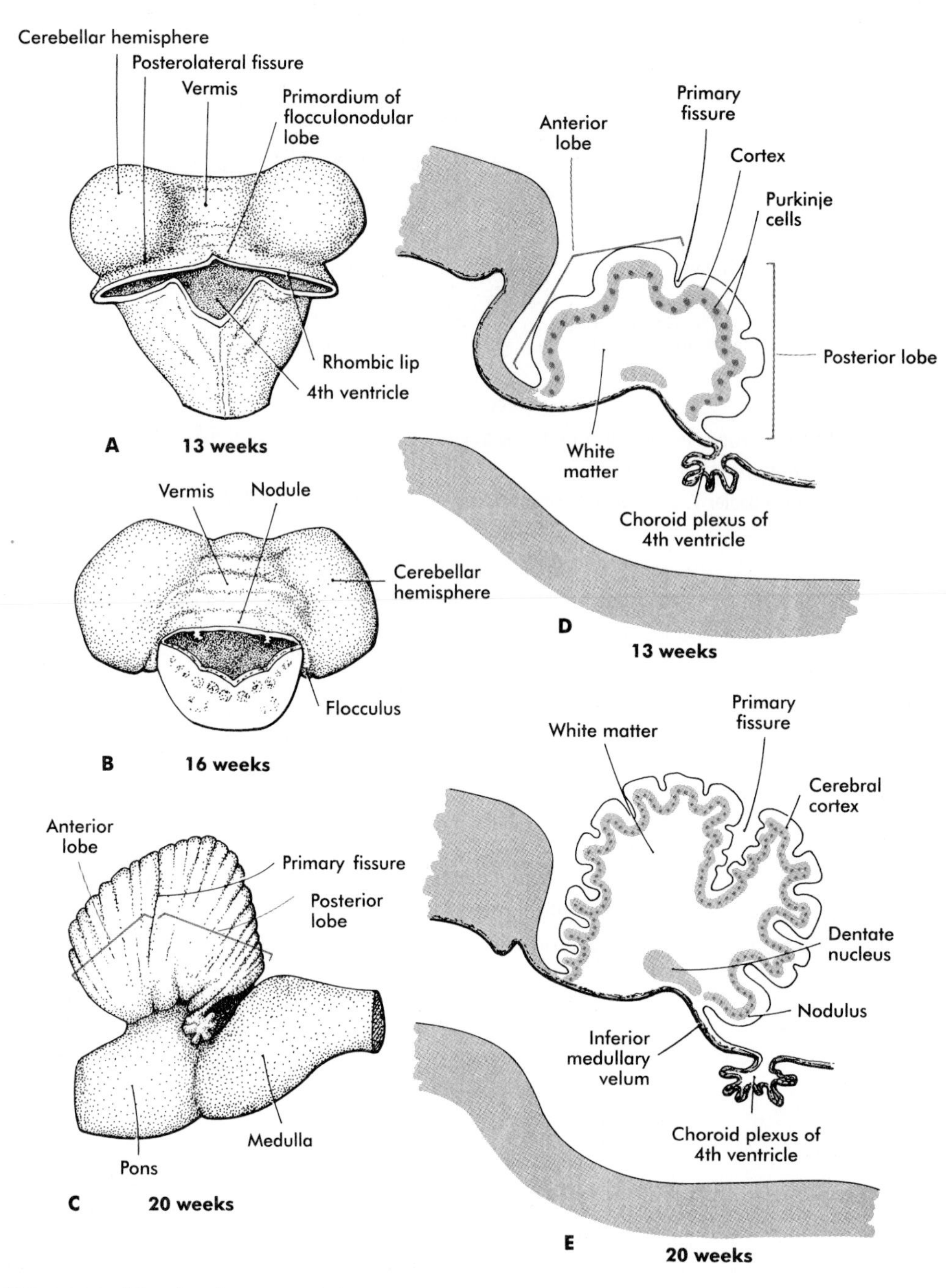

FIG. 12-25 The development of the cerebellum. **A** and **B,** Dorsal views. **C,** Lateral view. **D** and **E,** Sagittal sections.

cerebellum then enters a period of rapid development and external expansion. Internally, the cerebellum undergoes a complex process of histogenesis. (Generally, the processes underlying the cellular organization of the cerebellum can be appreciated by reviewing the section on histogenesis of the cerebral cortex [see Fig. 12-19].) Histogenesis of the cerebellum continues until well after birth. Many fibers emanating from the vast number of neurons generated in the cerebellum leave the cerebellum through a pair of massive **superior cerebellar peduncles,** which grow into the mesencephalon.

Mesencephalon

The mesencephalon, or **midbrain,** is structurally a relatively simple part of the brain in which the fundamental relationships between the basal and alar plates are essentially preserved (Fig. 12-26). In the region of the alar plates, neu-

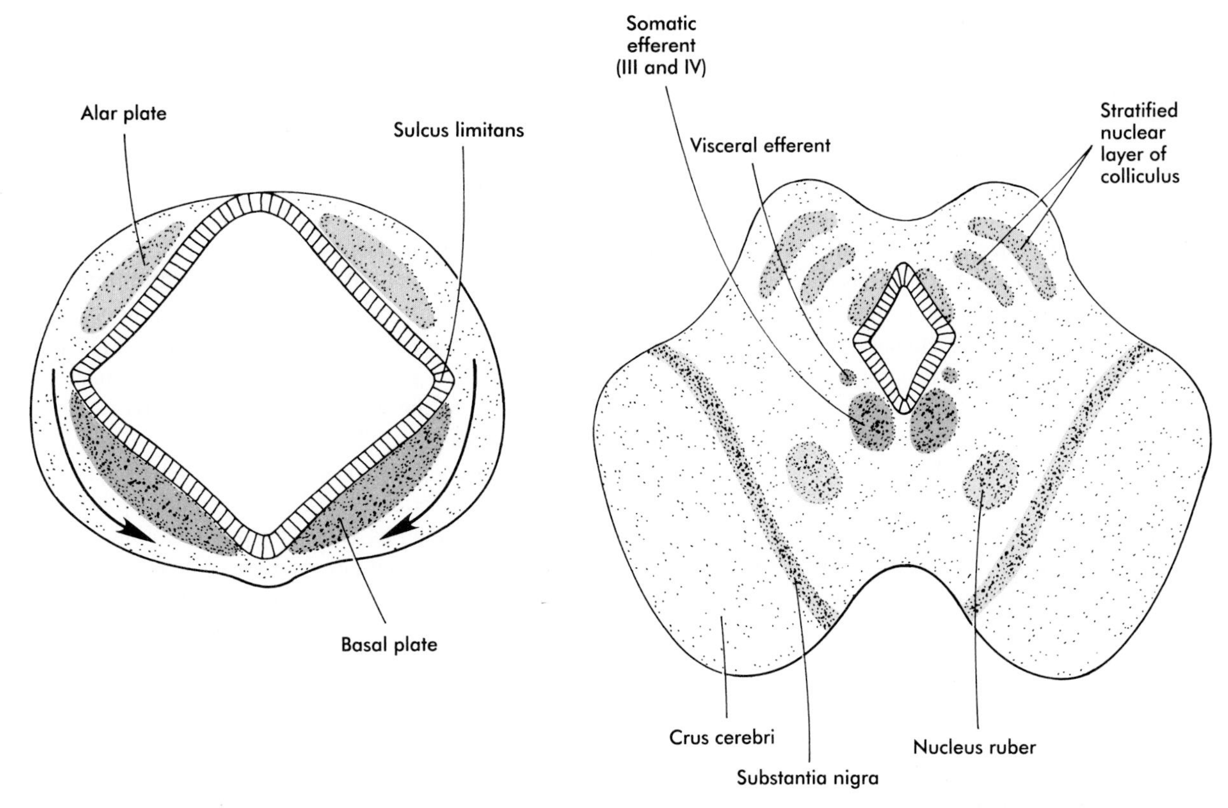

FIG. 12-26 Cross sections through the early and later developing mesencephalon. Motor tracts are green; sensory tracts are orange.
(Modified from Sadler T: *Langmen's medical embryology,* ed 6, 1990, Williams & Wilkins.)

roblasts migrate toward the roof **(tectum),** forming two prominent pairs of bulges collectively called the **corpora quadrigemina.** The caudal pair, called the **inferior colliculi,** are relatively simple in structure and are functionally part of the auditory system. The **superior colliculi** take on a more complex layered architecture through the migration patterns of the neuroblasts that give rise to it. The superior colliculi are an integral part of the visual system, and they serve as an important synaptic relay station between the optic nerve and the visual areas of the cerebral (occipital) cortex. Connections between the superior and inferior colliculi help to coordinate visual and auditory reflexes.

The primitive mesencephalic basal plates develop into the structurally less well-defined region called the **tegmentum.** In the tegmentum are located the somatic efferent nuclei of cranial nerves III and IV, which supply most of the extrinsic muscles of the eye. A small visceroefferent nucleus, the **Edinger-Westphal nucleus,** is responsible for innervation of the pupillary sphincter muscle of the eye. Two pairs of prominent nuclei of gray matter, the **nucleus ruber** (red nucleus) and **substantia nigra,** are still of uncertain origin.

The third major region of the mesencephalon is represented by prominent ventrolateral bulges of white matter called the **cerebral peduncles.** A number of the major descending fiber tracts pass through these structures on their way from the cerebral hemispheres to the spinal cord.

Diencephalon

Cranial to the mesencephalon the organization of the developing brain becomes so highly modified that it is difficult to relate later morphology to the fundamental alar plate/basal plate plan. In fact, it is widely believed that the forebrain structures (diencephalon and telencephalon) are highly modified derivatives of the alar plates and roof plate without significant representation by basal plates.

Development of the early diencephalon is characterized by the appearance of two pairs of prominent swellings on the lateral walls of the **third ventricle.** These swellings represent the greatly expanded central canal in this region (see Fig. 12-21). The largest pair of masses represents the developing **thalamus,** in which neural tracts from higher brain centers synapse with those of other regions of the brain and brainstem. Among the many thalamic nuclei are those that receive input from the auditory and visual systems and transmit them to the appropriate regions of the cerebral cortex. In later development the thalamic swellings may thicken to the point where they meet and fuse in the midline across the third ventricle. This connection is called the **massa intermedia.**

Ventral to the thalamus, the swellings of the **incipient hypothalamus** are separated from the thalamus by the **hypothalamic sulcus.** As mentioned earlier, the hypothalamus receives input from many areas of the central nervous system. It also acts as a master regulatory center, controlling many basic homeostatic functions such as sleep, temperature control, hunger, fluid and electrolyte balance, emotions, and rhythms of glandular secretion (e.g., of the pituitary). A number of its functions are neurosecretory; therefore the hypothalamus serves as a major interface between the neural integration of sensory information and the humoral environment of the body.

In early embryos (i.e., around 7 to 8 weeks gestational age), a pair of less prominent bulges dorsal to the thalamus mark the emergence of the **epithalamus** (see Fig. 12-21), a relatively poorly developed set of nuclei relating to masticatory and swallowing functions. The most caudal part of the diencephalic roof plate forms a small diverticulum that becomes the **epiphysis (pineal body),** a phylogenetically primitive gland that often serves as a light receptor. Under the influence of light/dark cycles, the pineal gland secretes (mainly at night) **melatonin,** a hormone that inhibits function of the pituitary-gonadal axis.

The **hypophysis** (pituitary gland) develops from two initially separate ectodermal primordia that secondarily unite. One of the primordia, called the **infundibular process,** forms as a ventral downgrowth from the floor of the diencephalon. The other primordium is **Rathke's pocket,** a midline outpocketing from the stomodeal ectoderm that extends toward the floor of the diencephalon as early as the fourth week (Fig. 12-27).

The infundibular process is intimately related to the hypothalamus (see Fig. 1-15), and certain hypothalamic neurosecretory neurons send their processes into the infundibular process, which ultimately becomes the **neural lobe of the hypophysis.** Throughout development the histological structure of the infundibulum retains a neural character.

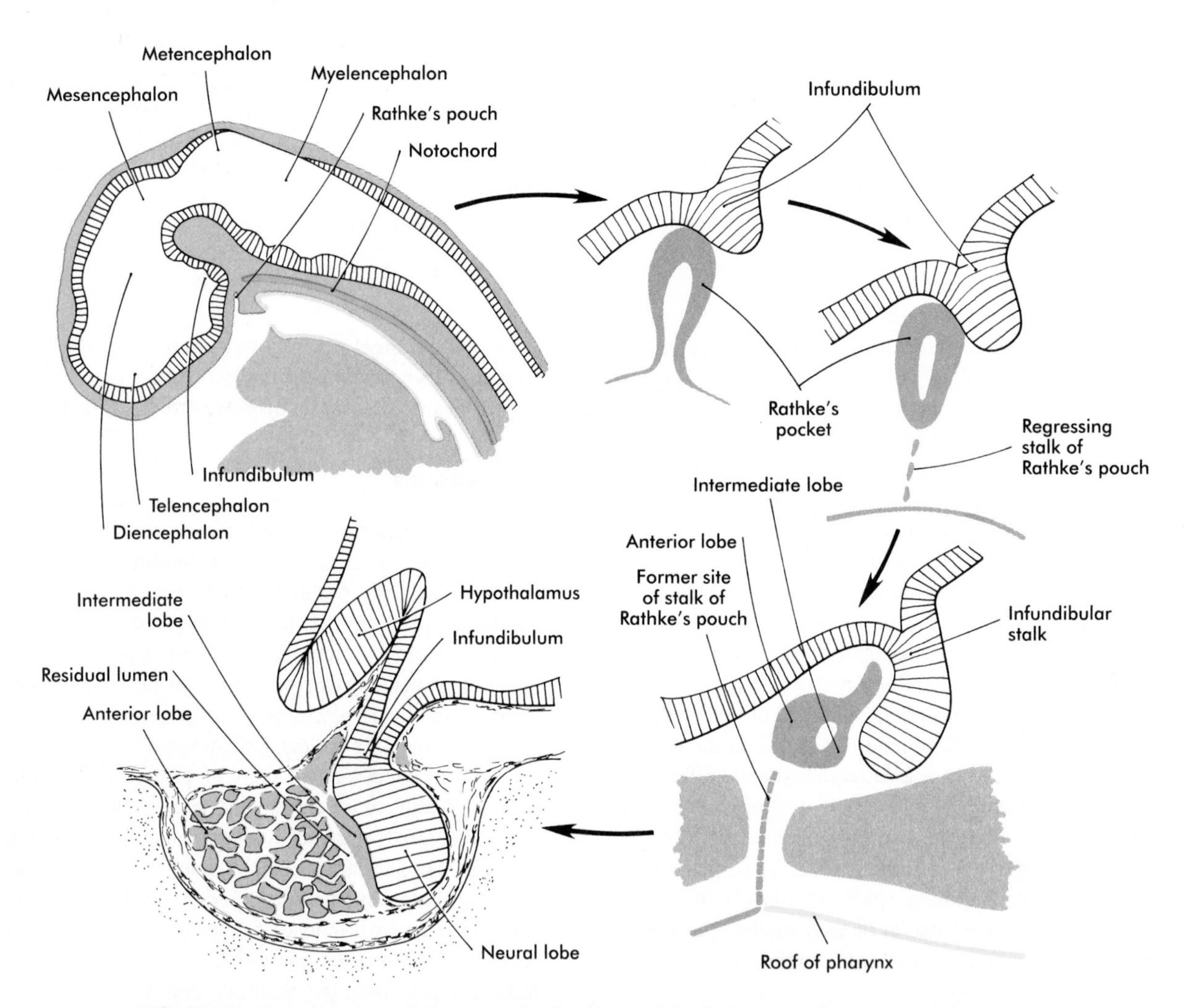

FIG. 12-27 Development of the hypophysis. *Upper left:* Reference diagram showing a sagittal section through a 4-week-old human embryo.

As development proceeds, Rathke's pocket elongates toward the infundibulum (see Fig. 12-27). While its blind end partially enfolds the infundibulum like a double-layered cup, the stalk of Rathke's pocket begins to regress. The outer wall of the cup thickens and assumes a glandular appearance in the course of its differentiation into the **pars distalis** (anterior lobe) of the hypophysis. The inner layer of the cup, which is closely adherent to the neural lobe, becomes the **pars intermedia.** It remains separated from the anterior lobe by a slitlike **residual lumen,** which represents all that remains of the original lumen of Rathke's pocket.

As pregnancy progresses, the hypophysis undergoes a phase of cytodifferentiation. Late in the fetal period, specific cell types begin to produce small amounts of hormones.

Although Rathke's pocket normally begins to lose its connections to the stomodeal epithelium by the end of the second month, portions of the tissue occasionally persist along the pathway of the elongating stalk. If the tissue is normal, it is called a **pharyngeal hypophysis.** Sometimes, however, the tissue rests become neoplastic and form hormone-secreting tumors called **craniopharyngiomas.**

The **optic cups** are major outpocketings of the diencephalic wall during early embryogenesis. (They and the optic nerves [cranial nerve II] are discussed in Chapter 14.)

Telencephalon

Development of the telencephalon is dominated by the tremendous expansion of the bilateral **telencephalic vesicles,** which ultimately become the cerebral hemispheres (see Fig. 12-21). The walls of the telencephalic vesicles surround the expanded lateral ventricles, which are outpocketings from the midline third ventricle located in the diencephalon (see Fig. 12-33). Although the cerebral hemispheres first appear as lateral structures, the dynamics of their growth cause them to approach toward the midline over the roofs of the diencephalon and mesencephalon (Fig. 12-28). The two cerebral hemispheres never actually meet in the dorsal midline because they are separated by a thin septum of connective tissue (part of the dura mater) known as the **falx cerebri.** Below this septum the two cerebral hemispheres are connected by the ependymal roof of the third ventricle.

Although the cerebral hemispheres expand greatly during the early months of pregnancy, their external surfaces remain smooth until the fourteenth week. With continued growth, the cerebral hemispheres undergo folding at several levels of organization. The most massive folding involves the large **temporal lobes,** which protrude laterally and rostrally from the caudal part of the cerebral hemispheres. From the fourth to the ninth month of pregnancy, the expanding temporal lobes and the frontal and parietal lobes completely cover areas of the cortex known as the

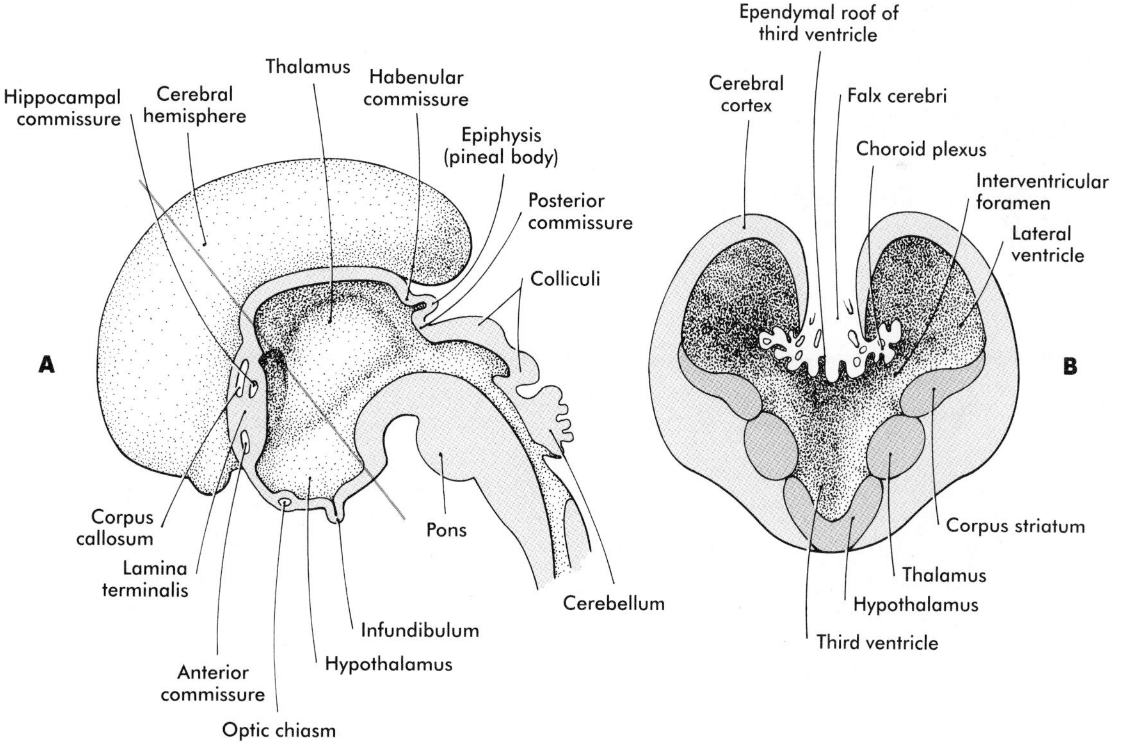

FIG. 12-28 Early formation of the cerebral hemispheres in a 10-week-old embryo. **A,** Sagittal section through the brain. **B,** Cross section through the level in **A.**
(Modified from Moore K: *The developing human,* ed 4, Philadelphia, 1988, Saunders.)

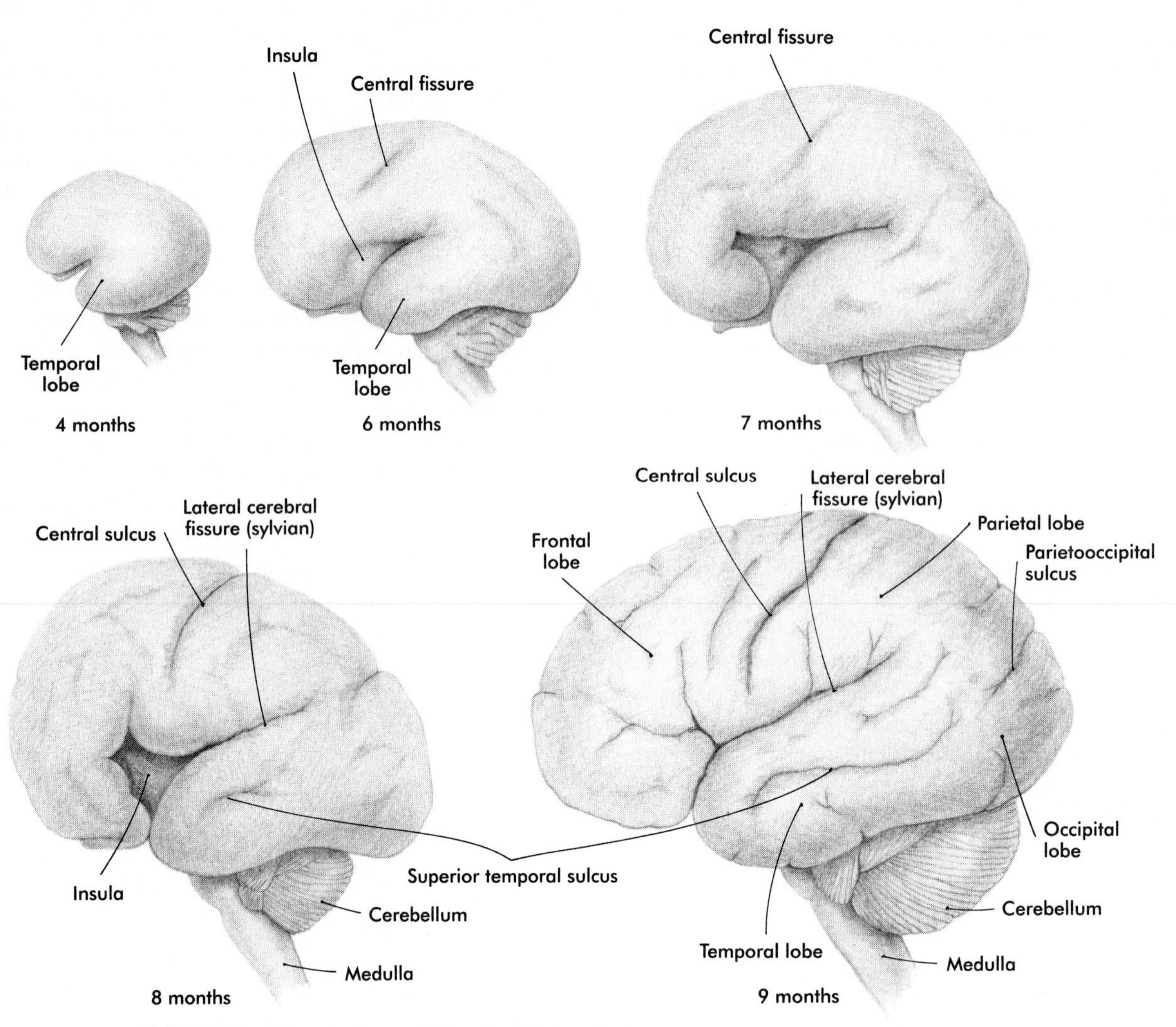

FIG. 12-29 Lateral views of the developing brain.

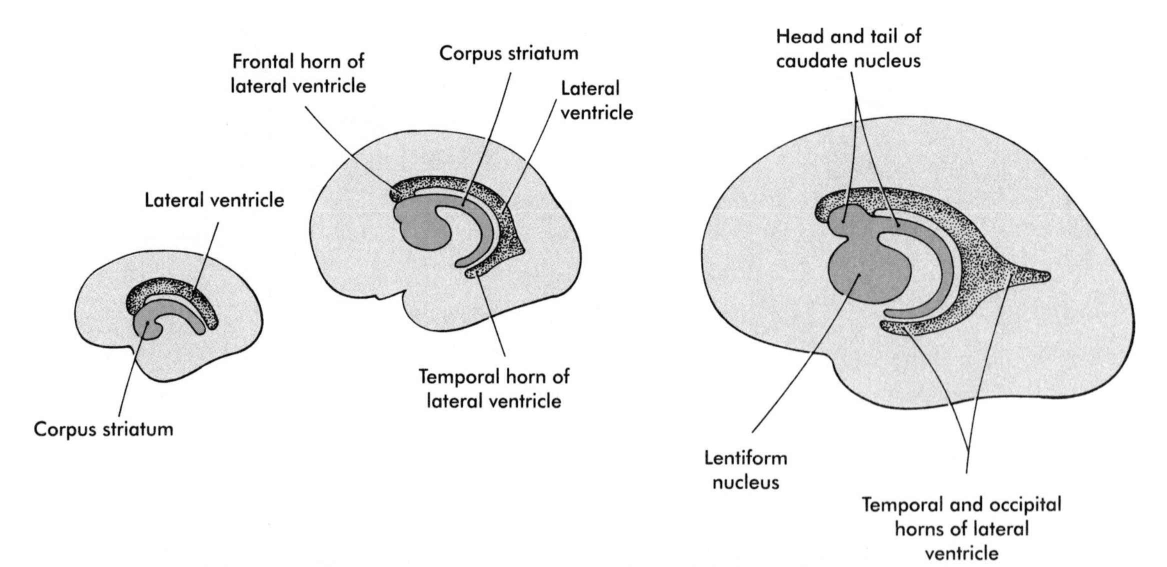

FIG. 12-30 Development of the corpus striatum and lateral ventricles.
(Modified from Moore K: *The developing human,* ed 4, Philadelphia, 1988, Saunders.)

insula (island) (Fig. 12-29). While these major changes in organization are occurring, other precursors of major surface landmarks of the definitive cerebral cortex are being sculpted. Several major sulci and fissures begin to appear as early as the sixth month. By the eighth month the **sulci** (grooves) and **gyri** (convolutions) that characterize the mature brain take shape.

Internally, the base of each telencephalic vesicle thickens to form the comma-shaped **corpus striatum** (Fig. 12-30). Located dorsal to the thalamus, the corpus striatum becomes more C shaped as development progresses. With histodifferentiation of the cerebral cortex, many fiber tracts converge on the area of the corpus striatum, which becomes subdivided into two major nuclei, the **lentiform nucleus** and the **caudate nucleus.** These structures, which are components of the complex aggregation of nuclei known as the **basal ganglia,** are involved in the unconscious control of muscle tone and complex body movements.

Aside from the telencephalic vesicles, the other major component of the early telencephalon is the **lamina terminalis,** which forms its median rostral wall (Figs. 12-31 and 12-33, *A*). Initially the two cerebral hemispheres develop

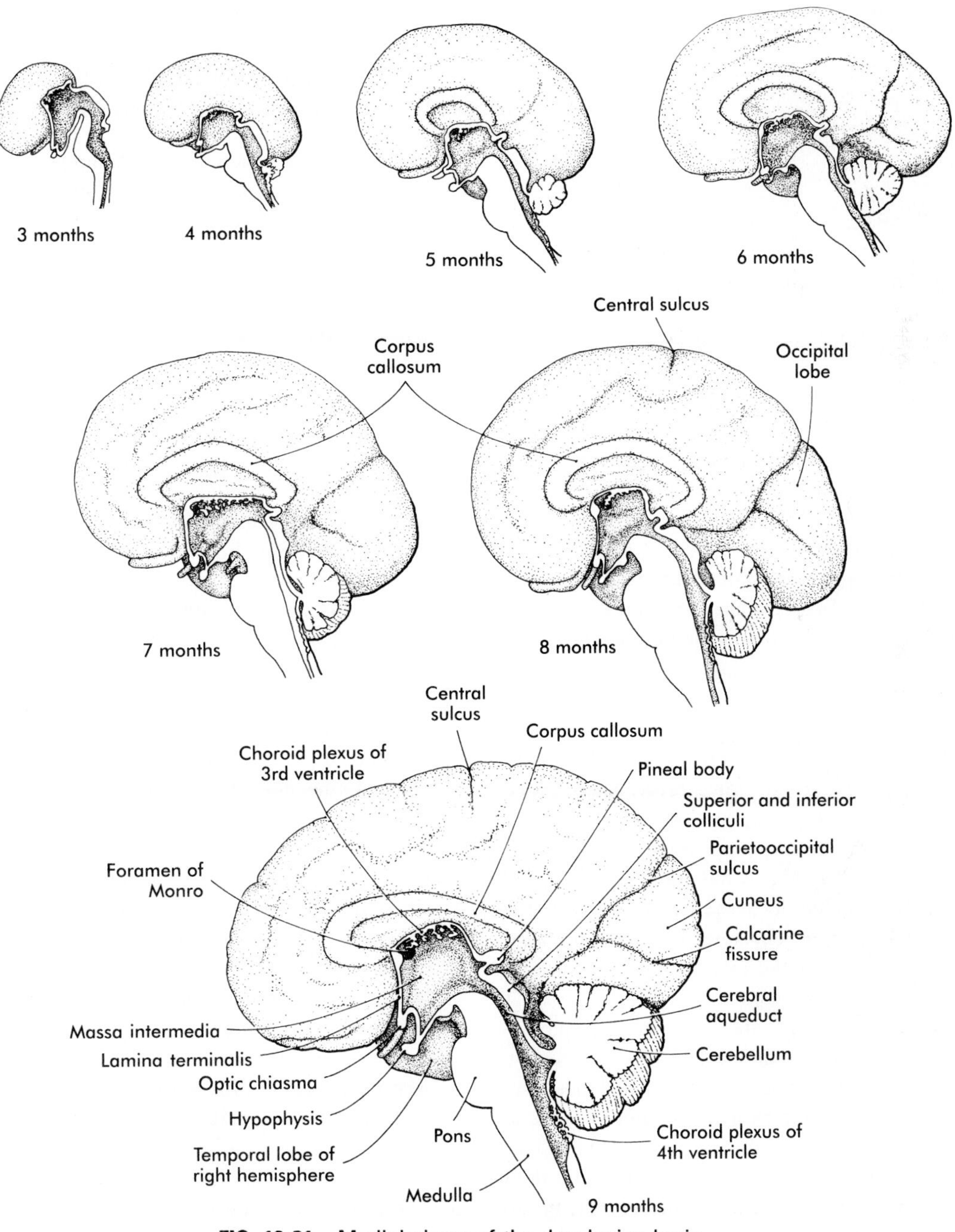

FIG. 12-31 Medial views of the developing brain.

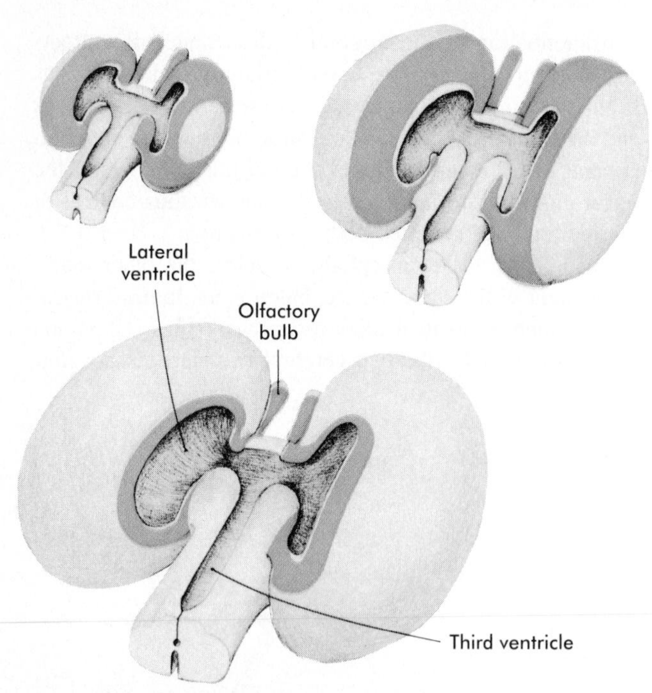

FIG. 12-32 The decrease in prominence of rhinencephalic areas *(green)* of the brain as the cerebrum expands.

separately, but toward the end of the first trimester of pregnancy, bundles of nerve fibers begin to cross from one cerebral hemisphere to the other. Many of these connections occur through the lamina terminalis.

The first set of connections to appear in the lamina terminalis becomes the **anterior commissure** (see Fig. 12-21, *B*), which connects olfactory areas from the two sides of the brain. The second connection is the **hippocampal commissure (fornix).** The third commissure to take shape in the lamina terminalis is the **corpus callosum,** the most important connection between the right and left halves of the brain. It initially forms as a small bundle in the lamina terminalis but expands greatly to form a broad band connecting a large part of the base of the cerebral hemispheres (see Fig. 12-31). Other commissures not related to the lamina terminalis are the **posterior** and **habenular commissures** (see Fig. 12-28), which are located close to the base of the pineal gland, and the **optic chiasma,** the region in the diencephalon where parts of the optic nerve fibers cross to the other side of the brain.

Neuroanatomists subdivide the telencephalon into several functional components that are based on the phylogenetic development of this region. The oldest and most primitive component is called the **rhinencephalon** (also the **archicortex** and **paleocortex**). As the name implies, it is heavily involved in olfaction. The morphologically dominant cerebral hemispheres are called the **neocortex.** In early development, much of the telencephalon is occupied by rhinencephalic areas (Fig. 12-32), but with the expansion of the cerebral hemispheres, the neocortex takes over as the component occupying most of the mass of the brain.

The **olfactory nerves** (cranial nerve I), arising from paired ectodermal placodes in the head, send fibers back into the **olfactory bulbs,** which are outgrowths from the rhinencephalon. Recent evidence strongly suggests that a subpopulation of cells from the olfactory placode migrates along the olfactory nerve into the brain, ultimately settling in the hypothalamus where they become the luteinizing hormone–releasing hormone (LHRH)-secreting cells. There are some potentially interesting behavioral implications in the placodal origin of cells that are extremely important in the control of reproduction.

VENTRICLES, MENINGES, AND CEREBROSPINAL FLUID FORMATION

The ventricular system of the brain represents an expansion of the central canal of the neural tube. As certain parts of the brain take shape, the central canal expands into well-defined **ventricles,** which are connected by thinner channels (Fig. 12-33). The ventricles are lined by ependymal epithelium and filled with clear **cerebrospinal fluid.** Cerebrospinal fluid is formed in specialized areas called **choroid plexuses,** which are located in specific regions in the roof of the third, fourth, and lateral ventricles. Choroid plexuses are highly vascularized structures that project into the ventricles (see Fig. 12-28, *B*) and secrete cerebrospinal fluid into the ventricular system.

Cerebrospinal fluid has a well-characterized circulatory path. As it forms, it flows from the lateral ventricles into the third and ultimately the fourth ventricle. Much of it then escapes through three small holes in the roof of the fourth ventricle and enters the **subarachnoid space** between two layers of meninges. Much of the fluid leaves the skull and bathes the spinal cord as a protective layer.

If an imbalance exists between the production and resorption of cerebrospinal fluid or if its circulation is blocked, the fluid may accumulate within the ventricular system of the brain and, through increased mechanical pressure, result in a massive enlargement of the ventricular system. This in turn causes a thinning of the walls of the brain and a pronounced increase in the diameter of the skull, a condition known as **hydrocephalus** (Fig. 12-34). The blockage of fluid can be due to a congenital **stenosis** (narrowing) of the narrow parts of the ventricular system, or it can be the result of certain fetal viral infections.

A specific malformation leading to hydrocephalus is the **Arnold-Chiari malformation,** in which a tonguelike overgrowth of the cerebellum herniates into the foramen magnum, thereby mechanically preventing the escape of cerebrospinal fluid from the skull. This condition is often associated with some form of **spina bifida** (see Fig. 12-37), which can bind the spinal cord. With the differential growth

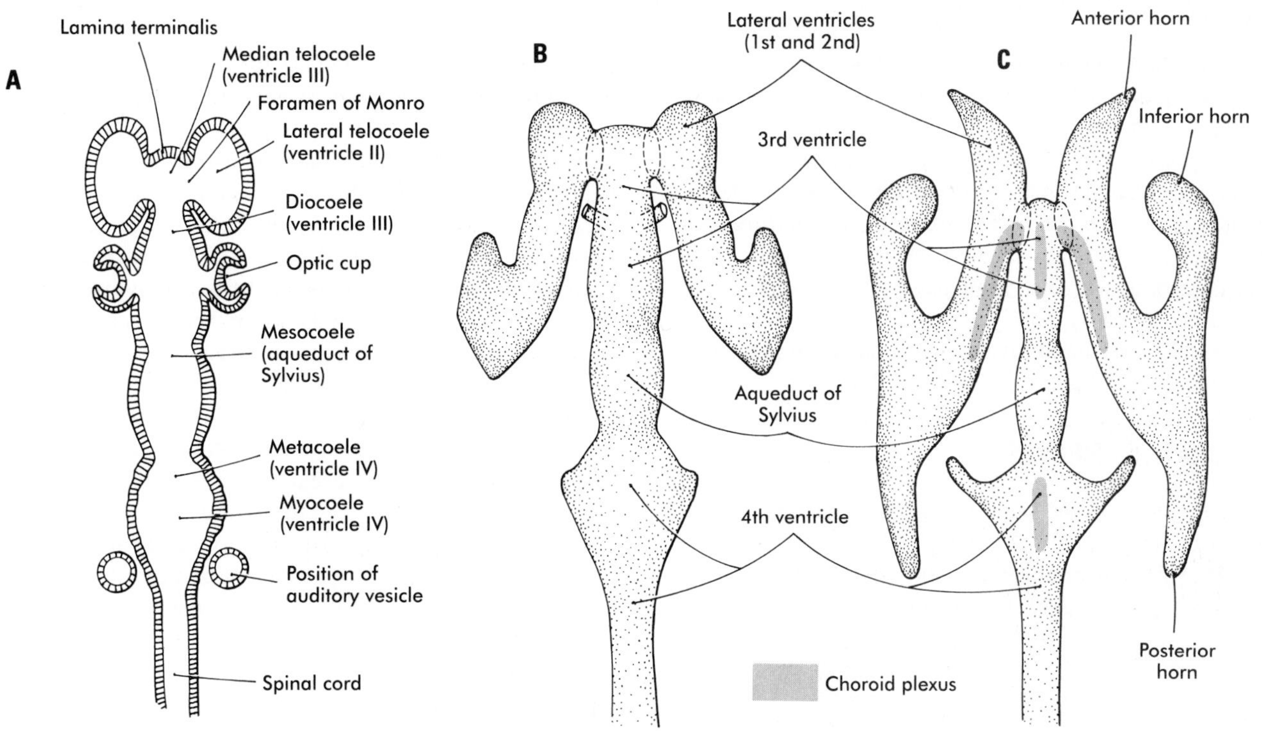

FIG. 12-33 Development of the ventricular system of the brain. **A,** Section from an early embryo. **B,** Ventricular system during expansion of the cerebral hemispheres. **C,** Postnatal morphology of the ventricular system.

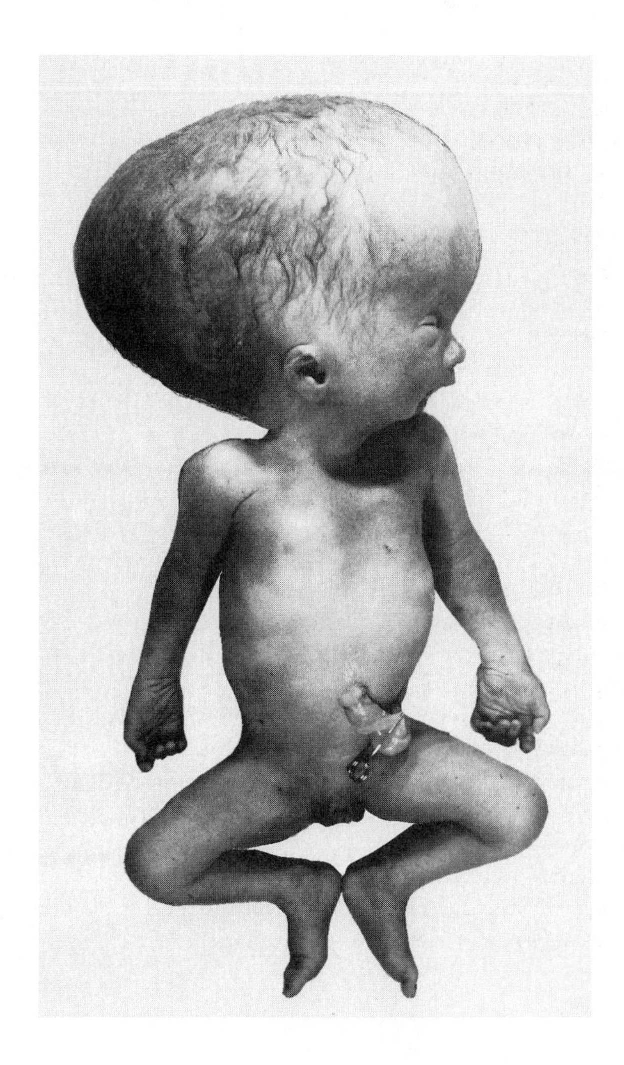

FIG. 12-34 Fetus with pronounced hydrocephalus.
(Courtesy Mason Barr, Ann Arbor, Mich.)

of the vertebral column, the spinal cord becomes stretched and may literally pull the brain into the foramen magnum, thus causing the blockage of fluid.

In the early fetal period, two layers of mesenchyme appear around the brain and spinal cord. The thick outer layer, which is of mesodermal origin, forms the tough **dura mater** as well as the membrane bones of the calvarium. A thin inner layer of neural crest origin later subdivides into a thin **pia mater,** which is closely apposed to the neural tissue, and a middle **arachnoid layer.** Spaces that form in the pia-arachnoid layer fill with cerebrospinal fluid.

THE CRANIAL NERVES

Although based on the same fundamental plan as the spinal nerves, the cranial nerves (Fig. 12-35) have lost their regular segmental arrangement and have become highly specialized (Table 12-1). One of the major differences is the tendency of many cranial nerves to be either sensory (dorsal root based) or motor (ventral root based).

The cranial nerves can be subdivided into several categories on the basis of their function and embryological origin. Cranial nerves I and II (olfactory and optic) are often regarded as extensions of brain tracts rather than true nerves. Cranial nerves III, IV, VI, and XII are pure motor nerves that appear to have evolved from primitive ventral roots. Nerves V, VII, IX, and X are mixed nerves with both motor and sensory components, and each nerve supplies derivatives of a different pharyngeal arch (Figs. 12-36 and 15-21). Although the motor components of these nerves are traditionally placed into a separate functional category (special visceral efferent), recent research shows that the branchial muscles are derived from somitomeres (see Chapter 10). Thus having a special category for the nerves supplying the branchial muscles may not be necessary.

The sensory components of the nerves supplying the

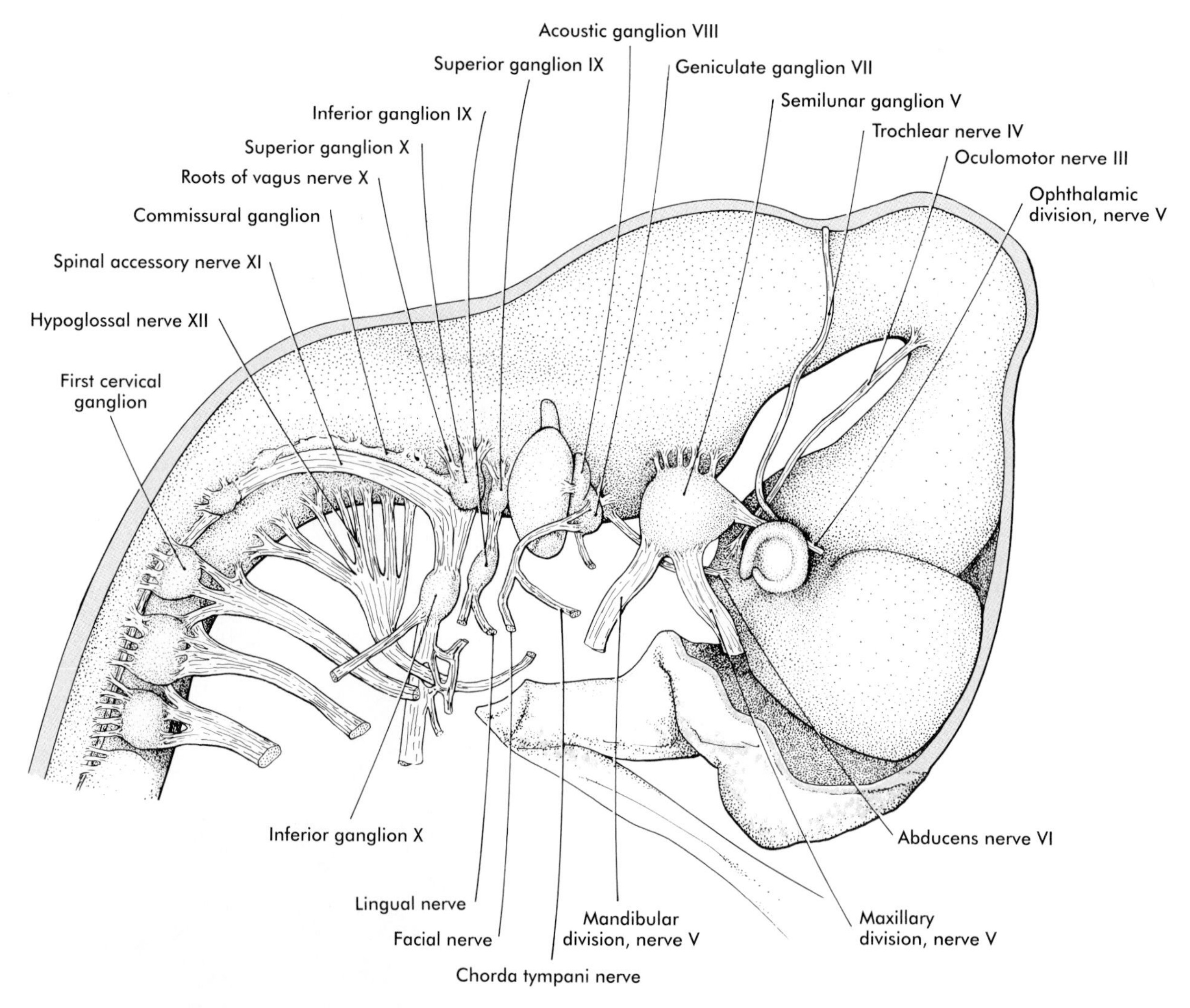

FIG. 12-35 Reconstruction of the brain and cranial nerves of a 12-mm-long pig embryo.

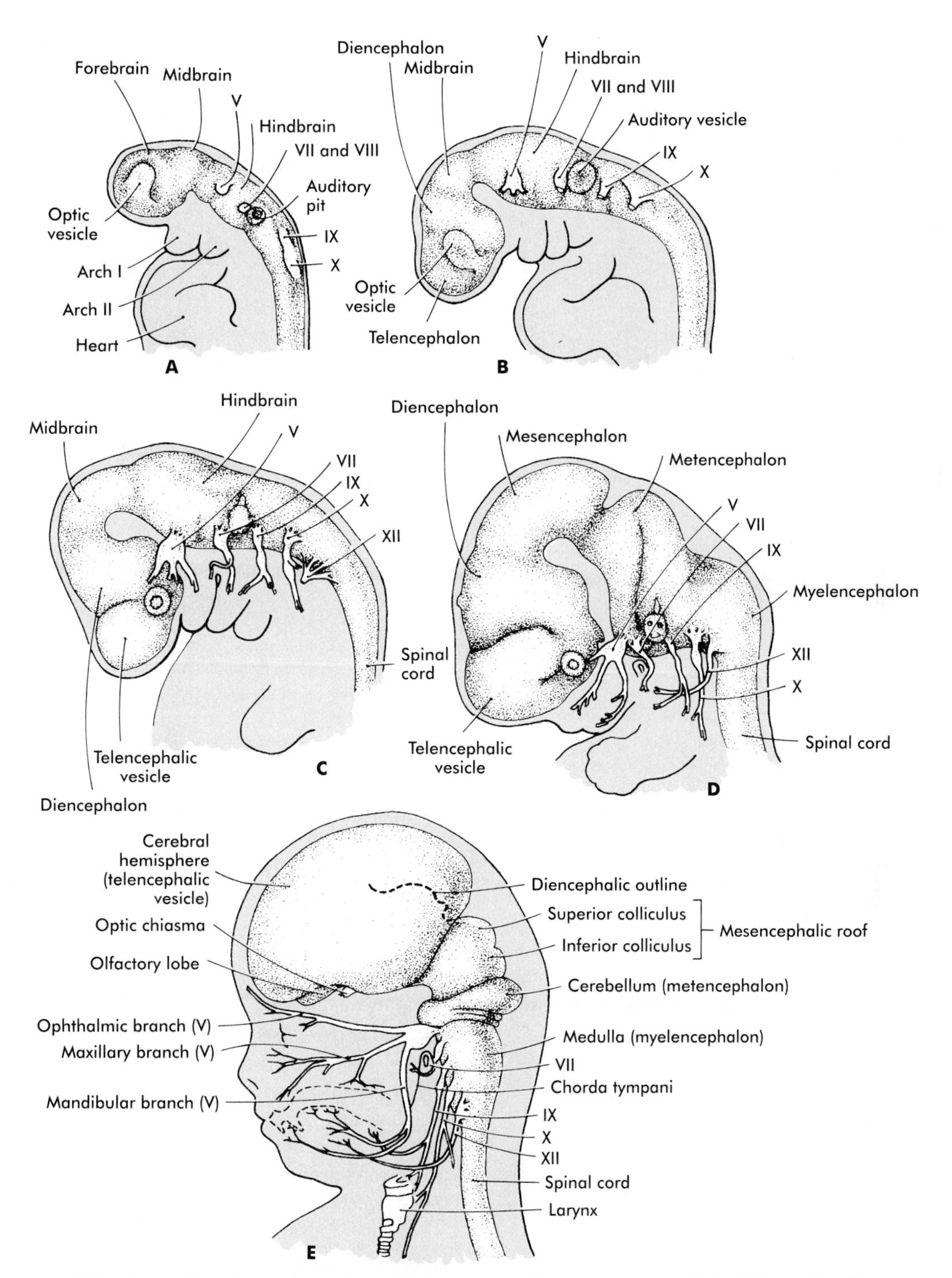

FIG. 12-36 Development of the cranial nerves in human embryos. **A,** At 3½ weeks. **B,** At 4 weeks. **C,** 5½ weeks. **D,** At 7 weeks. **E,** At 11 weeks.

TABLE 12-1 Cranial Nerves

CRANIAL NERVE	ASSOCIATED COMPONENT OF THE CENTRAL NERVOUS SYSTEM	FUNCTIONAL COMPONENTS	DISTRIBUTION
Olfactory (I)	Telencephalon	Special sensory (olfaction)	Olfactory area of the nose
Optic (II)	Diencephalon	Special sensory (vision)	Retina of the eye
Oculomotor (III)	Mesencephalon	Motor, autonomic (minor)	Intraocular and four extraocular muscles
Trochlear (IV)	Mesencephalon	Motor	Superior oblique muscle of the eye
Trigeminal (V)	Metencephalon	Sensory, motor (some)	Derivatives of branchial arch I
Abducens (VI)	Metencephalon	Motor	Lateral rectus muscle of the eye
Facial (VII)	Metencephalon/myelencephalon junction	Motor Sensory (some) Autonomic (minor)	Derivatives of branchial arch II
Auditory (VIII)	Metencephalon/myelencephalon junction	Special sensory (hearing, balance)	Inner ear
Glossopharyngeal (IX)	Myelencephalon	Sensory, motor (some)	Derivatives of branchial arch III
Vagus (X)	Myelencephalon	Sensory, motor, autonomic (major)	Derivatives of branchial arch IV
Accessory (XI)	Myelencephalon Spinal cord	Motor Autonomic (minor)	Gut, heart, visceral organs Some neck muscles
Hypoglossal (XII)	Myelencephalon	Motor	Tongue muscles

pharyngeal arches (V, VII, IX, and X) and the auditory nerve (VIII) have a multiple origin from both the neural crest and ectodermal placodes, which are located along the developing brain (see Fig. 5-9). These nerves have complex, often multiple sensory ganglia. Neurons in some parts of the ganglia are of neural crest origin, and those of other parts or ganglia arise from placodal ectoderm. (Ectodermal placodes are discussed in Chapter 14.)

CONGENITAL MALFORMATIONS OF THE NERVOUS SYSTEM

In an organ system as prominent and complex as the nervous system, it is not surprising that the brain and spinal cord are subject to a wide variety of congenital malformations. These range from severe structural anomalies resulting from incomplete closure of the neural tube to functional deficits caused by unknown factors acting late in pregnancy. A number of the closure defects can be diagnosed by the detection of elevated levels of **alpha-fetoprotein** in the amniotic fluid or by ultrasound scanning.

Defects in Closure of the Neural Tube

Failure of closure of the neural tube occurs most commonly in the regions of the anterior and posterior neuropore, but other locations are also possible. In this condition the spinal cord or brain in the affected area is splayed open, with the wall of the central canal or ventricular system constituting the outer surface. A closure defect of the spinal cord is called **rachischisis** and, in the brain, **cranioschisis.** Cranioschisis is incompatible with life. Rachischisis (Fig. 12-37) is associated with a wide variety of severe problems, including chronic infection, motor and sensory deficits, and disturbances in bladder function. These defects commonly accompany anencephaly (see Fig. 8-4), in which there is a massive deficiency of cranial structures.

Other Closure Defects

A defect in the formation of the bony covering overlying either the spinal cord or brain can result in a graded series of structural anomalies. In the spinal cord, the simplest defect is called **spina bifida occulta** (Fig. 12-38, *B*). The spi-

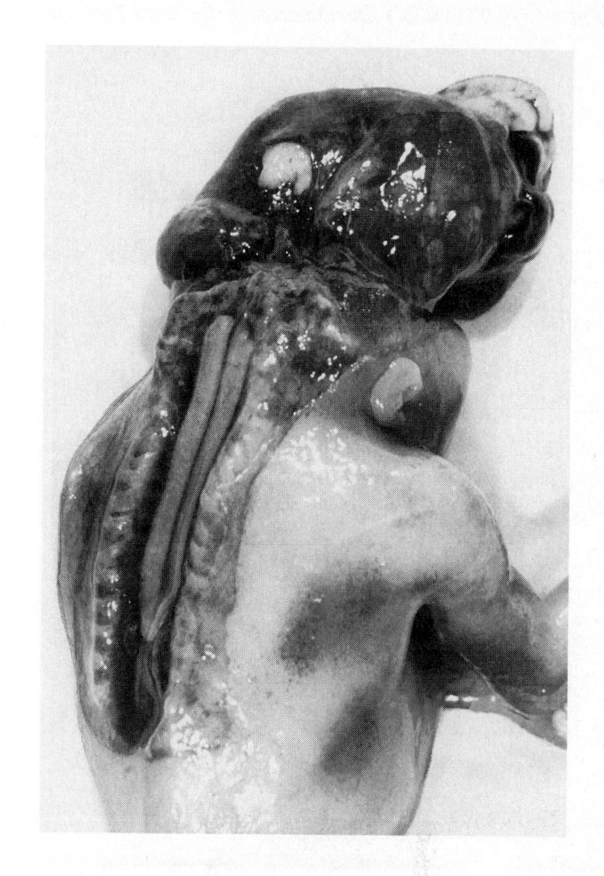

FIG. 12-37 Fetus with a severe case of rachischisis. The brain is not covered by cranial bones, and the light-colored spinal cord is totally exposed.
(Courtesy Mason Barr, Ann Arbor, Mich.)

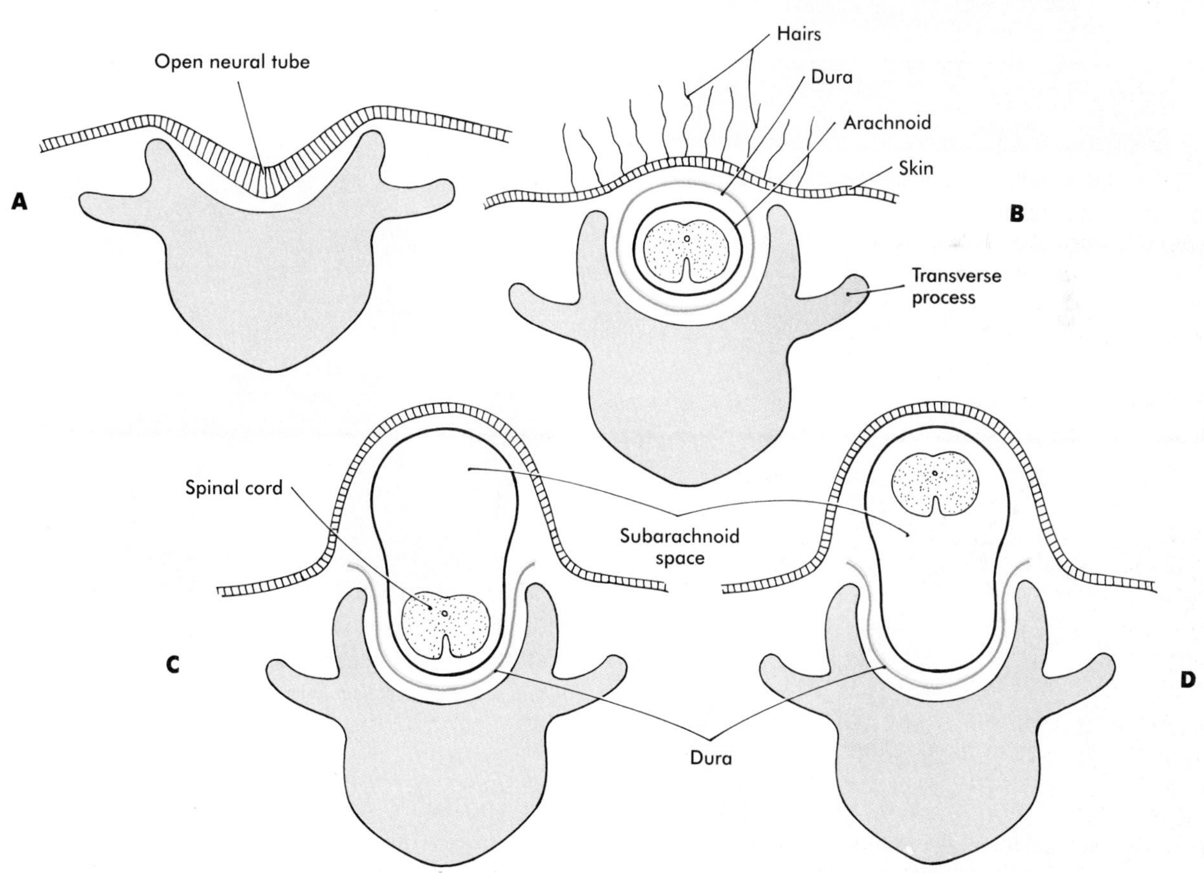

FIG. 12-38 Varities of closure defects of the spinal cord and vertebral column. **A,** Rachischisis. **B,** Spina bifida occulta, with hair growth over the defect. **C,** Meningocele. **D,** Myelomeningocele.

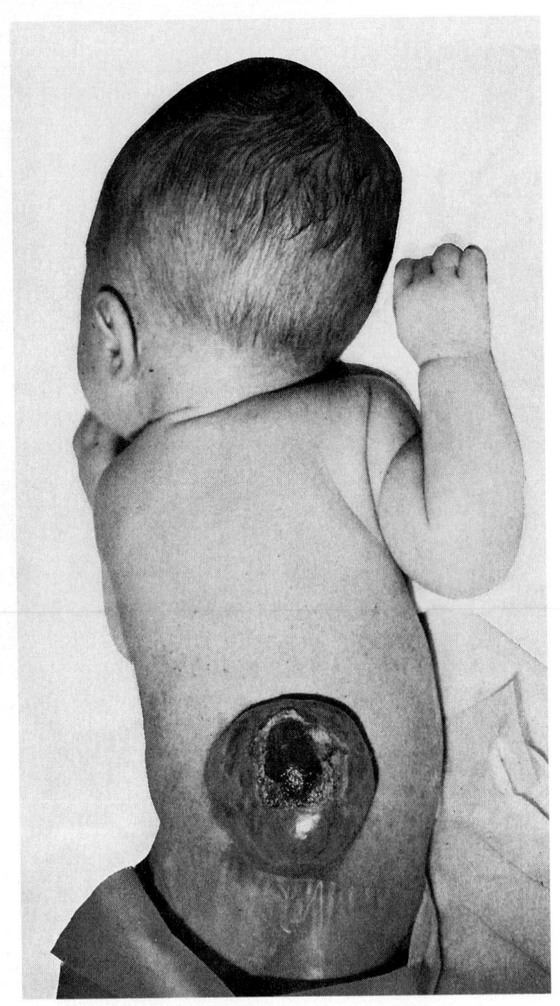

FIG. 12-39 Infant with a myelomeningocele and secondary hydrocephalus.
(Courtesy Mason Barr, Ann Arbor, Mich.)

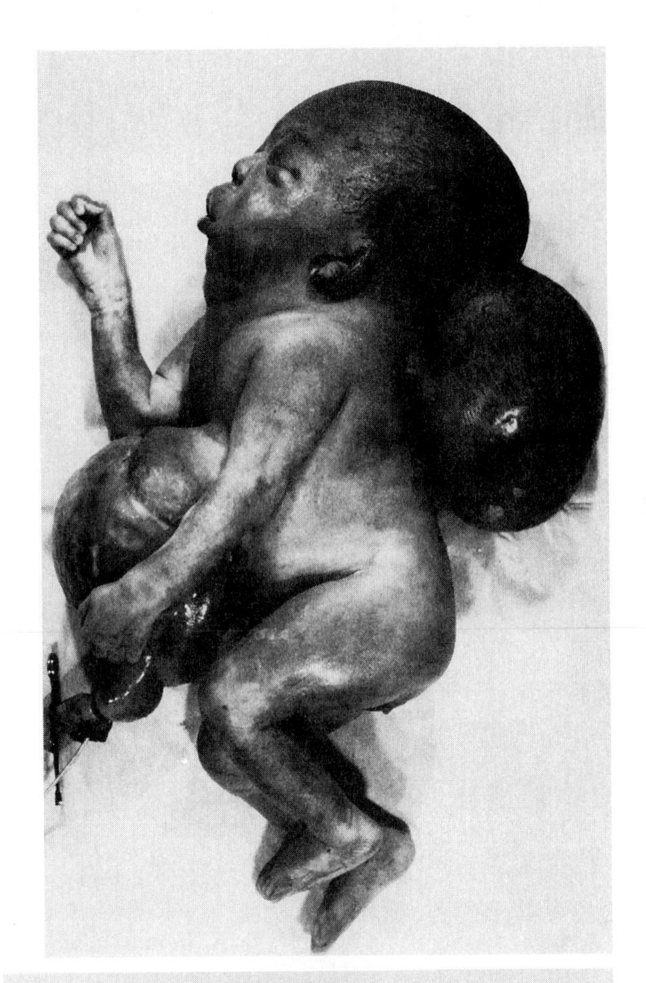

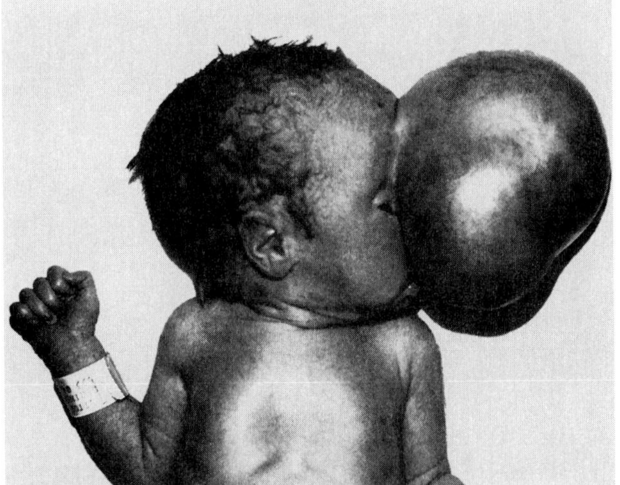

FIG. 12-41 Fetuses with **(A)** an occipital meningocele and **(B)** a frontal encephalocele.
(Courtesy Mason Barr, Ann Arbor, Mich.)

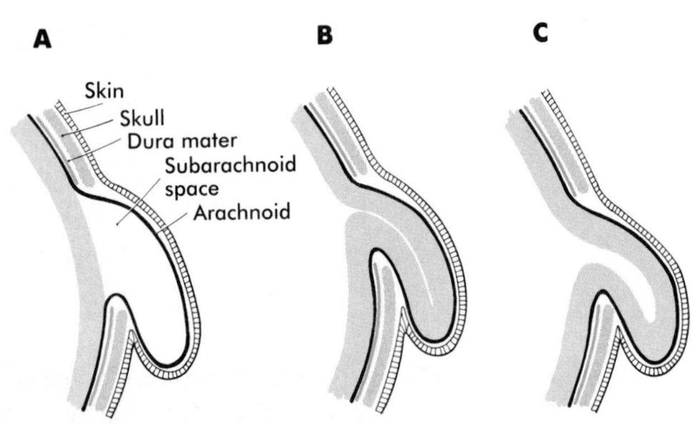

FIG. 12-40 Herniations in the cranial region. **A,** Meningocele. **B,** Meningoencephalocele. **C,** Meningohydroencephalocele.

nal cord and meninges remain in place, but the bony covering (neural arch) of one or more vertebrae is incomplete. Sometimes the defect goes unnoticed for many years. The site of the defect is often marked by a tuft of hair. The next most severe category of defect is a **meningocele,** in which the dura mater may be missing in the area of the defect and the arachnoid layer bulges prominently beneath the skin (Fig. 12-38, *C*). The spinal cord, however, remains in place, and neurological symptoms are often minor. The most severe condition is a **myelomeningocele,** in which the spinal cord bulges or is entirely displaced into the protruding subarachnoid space (Figs. 12-38, *D* and 12-39). Because of problems associated with displaced spinal roots, neurological problems are commonly associated with this condition.

A similar spectrum of anomalies is associated with cranial defects (Figs. 12-40 and 12-41). A **meningocele** is typically associated with a small defect in the skull, whereas brain tissue alone **(meningoencephalocele)** or brain tissue containing part of the ventricular system **(meningohydroencephalocoele)** may protrude through a larger opening in the skull. Depending on the nature of the protruding tissue, these malformations may be associated with neurological deficits. The mechanical circumstances may also lead to secondary hydrocephalus in some cases.

Microcephaly is a relatively uncommon condition characterized by underdevelopment of both the brain and the cranium (see Fig. 10-9). Although it can result from premature closure of the cranial sutures, in most cases its etiology is uncertain.

Many of the functional defects of the nervous system are poorly characterized, and their etiology is not understood. Studies on mice with genetically based defects of movement or behavior due to abnormalities of cell migration or histogenesis in certain regions of the brain suggest there is likely a parallel spectrum of human defects. **Mental retardation** is common and can be attributed to many causes, both genetic and environmental. The timing of the insult to the brain may be late in the fetal period.

THE DEVELOPMENT OF NEURAL FUNCTION

During the first 5 weeks of embryonic development, there is no gross behavioral evidence of neural function. Primitive reflex activity can be first elicited at the sixth week, when touching the perioral skin with a fine bristle is followed by contralateral flexion of the neck. Over the next 6 to 8 weeks, the region of skin sensitive to tactile stimulation spreads from the face to the palms of the hands, the upper chest, and by 12 weeks, the entire surface of the body except for the back and top of the head. As the sensitive areas of the skin expand, the nature of the reflexes elicited matures from generalized movements to specific responses of more localized body parts. There is a general craniocaudal sequence of appearance of reflex movements.

Starting at the end of the fourth month the fetus begins a pattern of periods of activity followed by times of inactivity. Many women first become aware of fetal movements at this time. Between the fourth and fifth month, the fetus becomes capable of gripping firmly onto a glass rod. Although weak protorespiratory movements are possible, they cannot be sustained. The sucking reflex appears during the sixth month. Striking changes in the brain wave patterns take place at about 28 weeks, a time when some prematurely born infants can sustain breathing and survive in an incubator.

Much of the behavioral development of fetuses after the sixth month has been learned by observation of the behavior of prematurely born infants. Behavioral changes during the last trimester are more subtle and often reflect not only the establishment and completion of neural circuits but their structural and functional maturation.

The development of functional circuitry can be illustrated by the spinal cord. Several stages of structural and functional maturation can be identified (Fig. 12-42). The first is a prereflex stage, which is characterized by the initial differentiation (including axonal and dendritic growth)

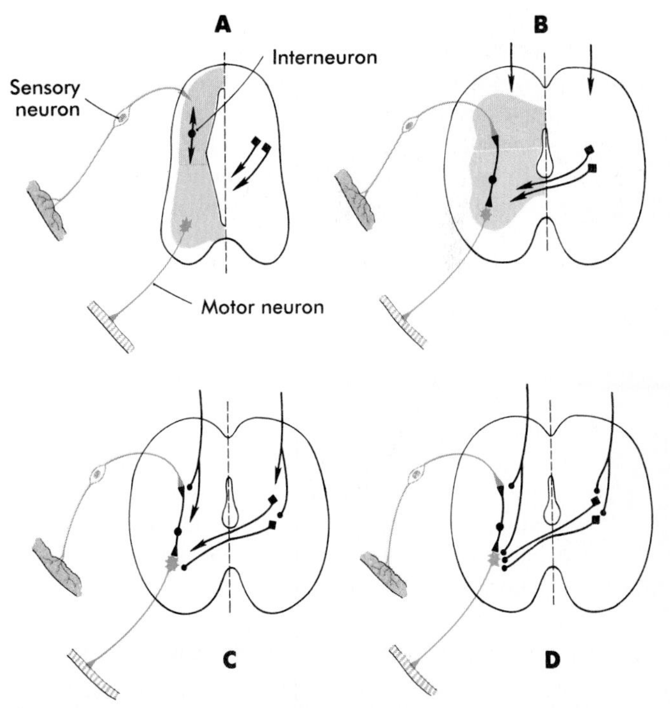

FIG. 12-42 Stages in the development of neural circuitry. **A,** Presynaptic stage. **B,** Closure of the primary reflex circuit. **C,** Connections with longitudinal and lateral inputs. **D,** Completion of circuits and myelination.
(Based on studies by Bodian [1970].)

of the neurons according to a well-defined sequence, starting with motor and followed by sensory and finally the interneurons that connect the two (see Figs. 12-11 and 12-42, *A*). The second stage consists of closure of the primary circuit, which allows the expression of local segmental reflexes. While the local circuit is being set up, other axons are growing down descending tracts in the spinal cord or are crossing from the other side of the cord. When these axons make contact with the components of the simple reflex that was established in the second stage, the anatomical basis for intersegmental and cross-cord reflexes is set up. Later in the fetal period, these more complex circuits are completed, and the tracts are myelinated by oligodendrocytes.

The functional maturation of individual tracts, as indicated by their myelination, takes place over a broad time span and is not completed until early adulthood. Particularly in early postnatal life, the maturation of functional tracts in the nervous system can be followed by clinical neurological examination.

Myelination begins in the peripheral nervous system, with motor roots becoming myelinated before sensory roots (which occurs in the second through fifth months). Myelination begins in the spinal cord at about 11 weeks and proceeds according to a craniocaudal gradient. During the third trimester, myelination begins to occur in the brain, but here, in contrast to the peripheral nervous system, myelination is first seen in sensory tracts (e.g., in the visual system). Myelination in complex association pathways in the cerebral cortex occurs after birth. In the **corticospinal tracts,** the main direct connection between the cerebral cortex and the motor nerves emanating from the spinal cord, myelination extends caudally only to the level of the medulla by 40 weeks. Myelination continues after birth, and its course can be appreciated by the increasing mobility of infants during their first year of life.

SUMMARY

1. While the neural tube is closing, its open ends are the cranial and caudal neuropores. The newly formed brain consists of three parts—the prosencephalon, the mesencephalon, and the rhombencephalon. The prosencephalon later subdivides into the telencephalon and the diencephalon, and the rhombencephalon forms the metencephalon and myelencephalon.
2. Within the neural tube, neuroepithelial cells undergo active mitotic proliferation. Their daughter cells form neuronal or glial progenitor cells. Among the glial cells, radial glial cells act as guide wires for the migration of neurons from their sites of origin to definite layers in the brain. Microglial cells arise from mesoderm.
3. The neural tube divides into ventricular, intermediate, and marginal zones. Neuroblasts in the intermediate zone (future gray matter) send out processes that collect principally in the marginal zone (future white matter). The neural tube is also divided into a dorsal alar plate and a ventral basal plate. The basal plate represents the motor component of the spinal cord, and the alar plate is largely sensory.
4. Much of the early brain is a highly segmented structure. This is reflected structurally in the rhombomeres and molecularly in the patterns of expression of homeobox genes. Neurons and their processes developing within the rhombomeres follow specific rules of behavior with respect to rhombomere boundaries. Nerve processes growing from the spinal cord react to external cues provided by the environment of the somites. Neurons and neural crest cells can readily penetrate the anterior but not the posterior mesoderm of the somite.
5. A peripheral nerve forms by the outgrowth of motor axons from the ventral horn of the spinal cord. The outgrowing axons are capped by a growth cone. This growing tip continually samples its immediate environment for cues that guide the amount and direction of axonal growth. The motor component of a peripheral nerve is joined by the sensory part, which is based on neural crest–derived cell bodies in dorsal root ganglia along the spinal cord. Axons and dendrites from the sensory cell bodies penetrate the spinal cord and also grow peripherally with the motor axons. Connections between the nerve and end organs are often mediated through trophic factors. Neurons that do not establish connections with peripheral end organs often die.
6. The autonomic nervous system consists of two components—the sympathetic and parasympathetic nervous system. Both components contain preganglionic neurons, which arise from the central nervous system, and postganglionic components, which are of neural crest origin. Typically, sympathetic neurons are adrenergic and parasympathetic neurons are cholinergic. However, the normal choice of transmitter can be overridden by environmental factors so that a sympathetic neuron, for example, can secrete acetylcholine.
7. The spinal cord functions as a pathway for organized tracts of nerve processes as well as an integration center for local reflexes. During the fetal period, growth in length of the spinal cord lags behind that of the vertebral column, pulling the nerve roots and leaving the spinal cord as a cauda equina.

8. Within the brain, the myelencephalon retains an organizational similarity to the spinal cord with respect to the tracts passing through, but centers that control respiration and heart rate also form at the site. The metencephalon contains two parts, the pons (which functions principally as a conduit) and the cerebellum (which integrates and coordinates many motor movements and sensory reflexes). In the cerebellum the gray matter forms on the outside. The ventral part of the mesencephalon is the region through which the major tracts of nerve processes that connect centers in the cerebral cortex with specific sites in the spinal cord pass. The dorsal part of the mesencephalon develops the superior and inferior colliculi, which are involved with the integration of visual and auditory signals, respectively.
9. Both the diencephalon and telencephalon represent modified alar plate regions. Many important nuclei and integrating centers develop in the diencephalon, among them the thalamus, hypothalalmus, neural hypophysis, and pineal body. The eyes also arise as outgrowths from the diencephalon. In humans, the telencephalon ultimately overgrows other parts of the brain. Like the cerebellum it is organized with the gray matter in layers outside the white matter. Neuroblasts migrate through the white matter to these layers by using radial glial cells as their guides.
10. Within the central nervous system, the central canal expands to form a series of ventricles in the brain. Specialized vascular plexuses form cerebrospinal fluid, which circulates throughout the central nervous system. Around the brain and spinal cord, two layers of mesenchyme form the meninges.
11. The cranial nerves are organized on the same fundamental plan as the spinal nerves, but they have lost their regular segmental pattern and have become highly specialized. Some are purely motor, others are purely sensory, and others are mixed.
12. Many congenital malformations of the nervous system are based on incomplete closure of the neural tube or associated skeletal structures. In the spinal cord, the spectrum of defects ranges from a widely open neural tube (rachischisis) to relatively minor defects in the neural arch over the cord (spina bifida occulta). A similar spectrum of defects is seen in the brain.
13. Neural function appears in concert with structural maturation of various components of the nervous system. The first reflex activity is seen in the sixth week. During successive weeks the reflex movements become more complex and spontaneous movements appear. Final functional maturation coincides with myelination of the tracts and is not completed until many years after birth.

REVIEW QUESTIONS

1. In an infant born with severe rachischisis of the lower spine, the head begins to increase in size. What is a likely explanation?
2. In the early days after birth, an infant does not pass fecal material and develops abdominal swelling. An anal opening is present. What is the likely condition?
3. What is the likely appearance of the spinal cord and brachial nerves in an infant who was born with the congenital absence of one arm (amelia)?

REFERENCES

Black IB: Stages of neurotransmitter development of autonomic neurons, *Science* 215:1198-1204, 1982.

Bodian D: *A model of synaptic and behavioral ontogeny.* In Quartan GC, Melnechuk T, Adelman G, eds: *The neurosciences: second study program,* New York, 1970, Rockefeller University, pp 129-140.

Breedlove SM: Sexual dimorphism in the vertebrate central nervous system, *J Neurosci* 12:4133-4142, 1992.

Brown MC, Hopkins WG, Keynes RJ: *Essentials of neural development,* Cambridge, Mass, 1991, Cambridge University.

Bunge R, Johnson M, Ross CD: Nature and nurture in development of the autonomic neuron, *Science* 199:1409-1416, 1978.

Cameron RS, Rakic P: Glial cell lineage in the cerebral cortex: a review and synthesis, *GLIA* 4:124-137, 1991.

Copp AJ and others: The embryonic development of mammalian neural tube defects, *Prog Neurobiol* 33:363-401, 1990.

Darnell DK, Schoenwolf GC, Ordahl CP: Changes in dorsoventral but not rostrocaudal regionalization of the chick neural tube in the absence of cranial notochord, as revealed by expression of *Engrailed-2, Dev Dynam* 193:389-396, 1992.

Deutsch U, Gruss P: Murine paired domain proteins as regulatory factors of embryonic development, *Semin Dev Biol* 2:413-424, 1991.

Easter SS, Barald KF, Carlson BM, eds: *From message to mind,* Sunderland, Mass, 1988, Sinauer Associates.

Ericson J and others: Early stages of motor neuron differentiation revealed by expression of homeobox gene *Islet-1, Science* 256:1555-1560, 1992.

Ferrer I and others: Cell death and removal in the cerebral cortex during development, *Prog Neurobiol* 39:1-43, 1992.

Hirano S, Fuse S, Sohal GS: The effect of the floor plate on pattern and polarity in the developing central nervous system, *Science* 251:310-313, 1991.

Hooker D: *The prenatal origin of behavior,* Lawrence, Kan, 1952, University of Kansas Press.

Jessen KR, Mirsky R: Schwann cell precursors and their development, *GLIA* 4:185-194, 1991.

Keynes RJ, Jaques KF, Cook GMW: Axon repulsion during peripheral nerve segmentation, *Development* 2(suppl):131-139, 1991.

Landis SC, Keefe D: Evidence for neurotransmitter plasticity in vivo: developmental changes in the properties of cholinergic sympathetic neurons, *Dev Biol* 98:349-372, 1983.

Landmesser LT, ed: *The assembly of the nervous system,* New York, 1989, Liss.

Landmesser LT: *Growth cone guidance in the avian limb: a search for cellular and molecular mechanisms.* In Letourneau PC, Kater SB, Macagno ER, eds: *The nerve growth cone,* New York, 1992, Raven, pp 373-385.

Lu S and others: Expression pattern of a murine homeobox gene, *Dbx,* displays extreme spatial restriction in embryonic forebrain and spinal cord, *Proc Natl Acad Sci USA* 89:8053-8057, 1992.

Lumsden A: Cell lineage restrictions in the chick embryo hindbrain, *Philos Trans R Soc Lond [Biol]* 331:281-286, 1991.

Lumsden A, Keynes R: Segmental patterns of neuronal development in the chick hindbrain, *Nature* 337:424-428, 1989.

Maden M and others: Domains of cellular retinoic acid-binding protein I (CRABP I) expression in the hindbrain and neural crest of the mouse embryo, *Mech Dev* 37:13-23, 1992.

Oppenheim RW: Cell death during development of the nervous system, *Annu Rev Neurosci* 14:453-501, 1991.

Oppenheim RW, Schwatrz LM, Shatz CJ: Neuronal death, a tradition of dying, *J Neurobiol* 23:1111-1115, 1992.

O'Rahilly R, Gardner E: The timing and sequence of events in the development of the human nervous system during the embryonic period proper, *Z Anat Entwickl-Gesch* 134:1-12, 1974.

Placzek M and others: Mesodermal control of neural cell identity: floor plate induction by the notochord, *Science* 250:985-988, 1990.

Placzek M and others: Orientation of commissural axons *in vitro* in response to a floor plate–derived chemoattractant, *Development* 110:19-30, 1990.

Purves D, Lichtman JW: *Principles of neural development,* Sunderland, Mass, 1985, Sinauer Associates.

Rakic P: Specification of cerebral cortical areas, *Science* 241:170-176, 1988.

Rao MS, Landis SC, Patterson PH: The cholinergic neuronal differentiation factor from heart cell conditioned medium is different from the cholinergic factors in sciatic nerve and spinal cord, *Dev Biol* 139:65-74, 1990.

Robinson SR, Smotherman WP: Fundamental motor patterns of the mammalian fetus, *J Neurobiol* 23:1574-1600, 1992.

Rothman TP and others: The effect of back-transplants of the embryonic gut wall on growth of the neural tube, *Dev Biol* 124:331-346, 1987.

Salinas PC, Nusse R: Regional expression of the *wnt-3* gene in the developing mouse forebrain in relationship to diencephalic neuromeres, *Mech Dev* 39:151-160, 1992.

Schwanzel-Fukuda M, Pfaff DW: Origin of luteinizing-releasing hormone neurons, *Nature* 338:161-164, 1989.

Simeone A and others: Nested expression domains of four homeobox genes in developing rostral brain, *Nature* 358:687-690, 1992.

Smith J: *Ontogeny of the autonimic nervous system.* In Gootman PM, ed: *Developmental neurobiology of the autonomic nervous system,* Clifton, NJ, 1986, Humana Press, pp 1-28.

Tosney KW: Cells and cell-interactions that guide motor axons in the developing chick embryo, *Bioessays* 13:17-23, 1991.

Vettivel S: Vertebral level of the termination of the spinal cord in human fetuses, *J Anat* 179:149-161, 1991.

Wray S, Gaunt P, Gainer H: Evidence that cells expressing luteinizing hormone-releasing hormone mRNA in the mouse are derived from progenitor cells in the olfactory placode, *Proc Natl Acad Sci USA* 86:8132-8136, 1989.

CHAPTER 13

The Neural Crest

Although its existence has been recognized for over a century, not until adequate methods of marking neural crest cells became available—first with isotopic labels and subsequently with stable biological markers, monoclonal antibodies, intracellular dyes, and genetic markers—did the neural crest become one of the most widely studied components of the vertebrate embryo. The vast majority of studies on the neural crest have been conducted on the avian embryo because of accessibility and the availability of specific markers. Research conducted on mammalian embryos suggests that except for relatively minor structural details, information learned from birds can be directly applied to mammalian embryos.

ORIGINS OF THE NEURAL CREST

The origins of the neural crest are still poorly understood. Before completion of the neural tube, future neural crest cells are located along the lateral border of the neural plate where it interfaces with the general cutaneous ectoderm. What causes these cells to become neural crest is unknown.

Neural crest cells break from the neural plate or neural tube by changing their shape and properties from those of typical neuroepithelial cells to those of mesenchymal cells. In the head region, incipient neural crest cells send out processes that penetrate the basal lamina underlying the neuroepithelium well before neural tube closure (Fig. 13-1). After the basal lamina is further degraded, the neural crest cells, which by this time have assumed a mesenchymal morphology, pass through the remnants of the basal lamina and embark on a remarkable series of migrations. Another significant change accompanying the epithelial-to-mesenchymal transformation of the neural crest cells is a loss of cell-to-cell adhesiveness. This is accompanied by the loss of cell adhesion molecules on the neural crest cells during their migratory phase. After neural crest cells have completed their migrations and differentiated into certain structures (e.g., spinal ganglia), cell adhesion molecules may be reexpressed.

In the trunk, neural crest cells do not leave the neuroepithelium until after the neural tube has formed. They do not, however, have to contend with penetrating a basal lamina because the dorsal part of the neural tube does not form a basal lamina until after emigration of the crest cells.

MIGRATIONS OF THE NEURAL CREST

After leaving the neuroepithelium, the neural crest cells are first in a relatively cell-free environment rich in extracellular matrix molecules (Fig. 13-2). In this environment, they undergo extensive migrations along several well-defined pathways. These migrations are determined by both intrinsic properties of the neural crest cells and features of the external environment encountered by the migrating cells.

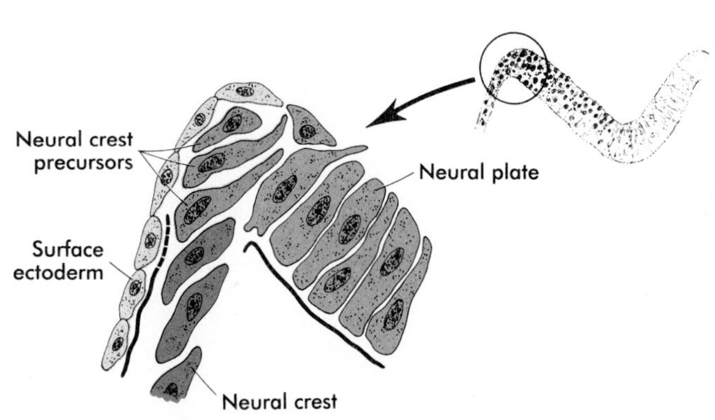

FIG. 13-1 The early migration of trunk neural crest cells from the lateral margin of the neural plate.

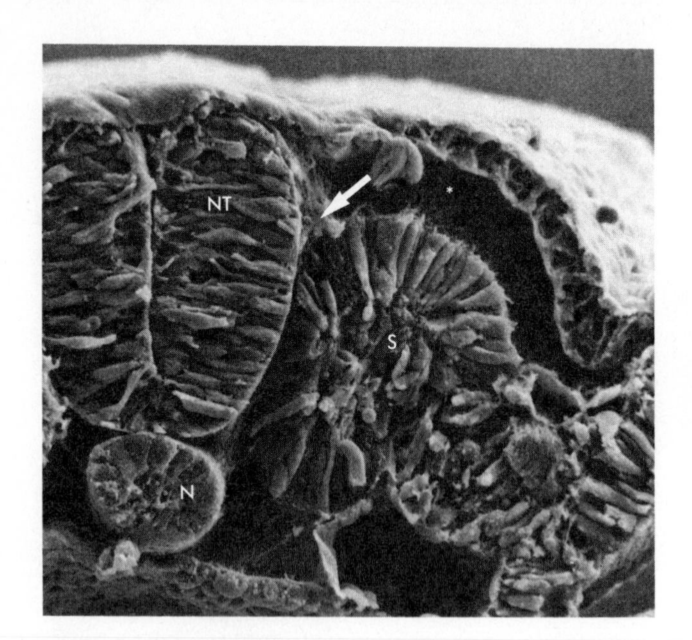

FIG. 13-2 Scanning electron micrograph of a chick embryo showing the early migration of neural crest cells *(arrow)* out of the neural tube *(NT)*. The subectodermal pathway of neural crest migration (*) is relatively cell free but contains a fine mesh of extracellular matrix molecules. *N*, Notochord; *S*, somite.
(Micrograph courtesy K. Tosney, Ann Arbor, Mich.)

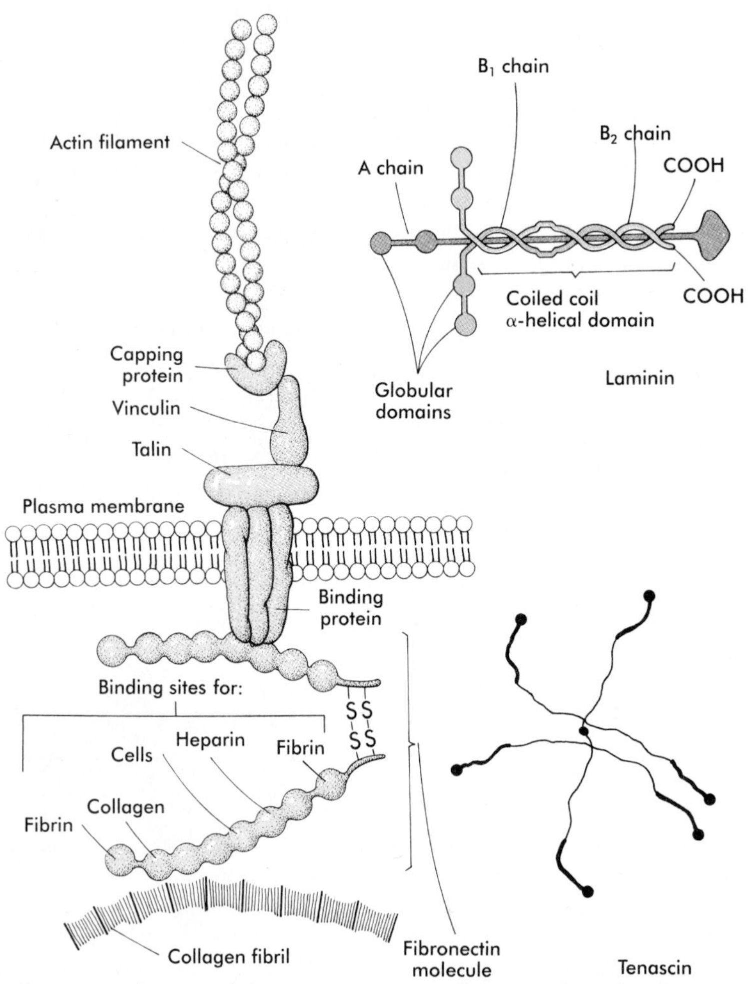

FIG. 13-3 Structure of some of the common extracellular matrix molecules.

Because of the lack of suitable markers, the pathways of migration of neural crest cells in mammals have still not been completely mapped. However, data suggest that the pathways are similar to those in the chick. The availability of stable cellular (e.g., chick-quail chimeras [see Fig. 10-28]) and monoclonal antibody markers have made the chick embryo the object of intense investigation pertaining to the migrations and differentiation of neural crest cells.

Only generalizations can be made about the specific controls underlying the migrations of neural crest cells. Permissive factors are substrates containing fibronectin, laminin, and certain types of collagen (e.g., type IV). Attachment to and migration over these substrate molecules is mediated by the family of attachment proteins called **integrins** (Fig. 13-3). Organized basal laminae (e.g., those of the surface ectoderm and ventral neural tube) act as barriers that guide migrating crest cells along their surfaces. Other extracellular matrix molecules, notably **chondroitin sulfate-rich proteoglycans** are not good substrates for migrating crest cells. Few neural crest cells are seen in areas that contain high concentrations of chondroitin sulfate (e.g., the sclerotome of the somites). The distribution of migrating neural crest cells in the somites illustrates this principle well. Neural crest cells enter only the anterior parts of the somites, where many of them settle to form the sensory ganglia. They do not penetrate the posterior regions of the somites, which have elevated concentrations of chondroitin sulfate. Neural crest cells in the trunk and the head follow different pathways of migration.

DIFFERENTIATION OF NEURAL CREST CELLS

Neural crest cells ultimately differentiate into an astonishing array of adult structures (Table 13-1). What controls their differentiation is one of the principal questions of neural crest biology. Two different hypotheses have been pro-

TABLE 13-1 Major Derivatives of the Neural Crest

	TRUNK CREST	CRANIAL CREST
Nervous system		
Sensory nervous system	Spinal ganglia	Ganglia of trigeminal nerve (V), facial nerve (VII), glossopharyngeal nerve (superior ganglion) (IX), vagus nerve, (jugular ganglion) (X)
Autonomic nervous system	Parasympathetic ganglia: pelvic plexus, visceral	Parasympathetic ganglia: ciliary, ethmoidal, sphenopalatine, submandibular, visceral
Nonneural cells	Satellite cells of sensory ganglia, Schwann cells of peripheral nerves, enteric glial cells	Satellite cells of sensory ganglia, Schwann cells of peripheral nerves, leptomeninges of prosencephalon and part of mesencephalon
Pigment cells	Melanocytes	Melanocytes
Endocrine and paraendocrine cells	Adrenal medulla, neurosecretory cells of heart and lungs	Carotid body (type I cells), parafollicular cells (thyroid)
Mesectodermal cells		
Skeleton	None	Cranial vault (squamosal and part of frontal), nasal and orbital, otic capsule (part), palate and maxillary, sphenoid (small contribution), trabeculae (part), visceral cartilages, external ear cartilage (part)
Connective tissue	None	Dermis and fat of skin; cornea of eye (fibroblasts of stroma and corneal endothelium); dental papilla (odontoblasts); connective tissue stroma of glands: thyroid, parathyroid, thymus, salivary, lacrimal; outflow tract (truncoconal region) of heart; cardiac semilunar valves; walls of aorta and aortic arch–derived arteries
Muscle	None	Ciliary muscles, dermal smooth muscles, vascular smooth muscle, minor skeletal muscle elements (?)

posed. One proposes that all neural crest cells are equal in developmental potential and that their ultimate differentiation is determined entirely by the environment through which they migrate and into which they finally settle. The other suggests that premigratory crest cells are already programmed for different developmental fates and that certain stem cells are favored while others are inhibited from further development during migration. Recent research indicates that the real answer can be found somewhere between these two positions.

Evidence is increasing of a correlation between the time of emigration of neural crest cells from the neural tube and their developmental potential. For example, the cells that first begin to migrate have the potential to differentiate into many different types of cells. Crest cells that begin to migrate later are capable of forming only derivatives characteristic of more dorsal locations (e.g., spinal ganglia) but not sympathetic neurons or adrenal medullary cells. Those that leave the neural tube last are restricted to the dorsalmost pathway of migration and can form only pigment cells.

Several experiments have shown that the fates of neural crest cells are not irreversibly fixed along a single pathway. One type of experiment involves transplantation of neural crest cells from one part of the body to another. For example, many neural crest cells from the trunk differentiate into sympathetic neurons that produce **norepinephrine** as the transmitter. In the cranial region, however, neural crest cells give rise to parasympathetic neurons, which produce **acetycholine.** If thoracic neural crest is transplanted into the head, some cells differentiate into cholinergic parasympathetic neurons instead of the adrenergic sympathetic neurons normally produced. Conversely, cranial neural crest cells grafted into the thoracic region respond to their new environment by forming adrenergic sympathetic neurons. A more striking example is the conversion of cells of the periocular neural crest mesenchyme, which in birds would normally form cartilage, into neurons if they are associated with embryonic hindgut tissue in vitro. Many of the regional influences on the differentiation of local populations of neural crest cells are now recognized to be interactions between the migrating neural crest cells and specific tissues that they encounter during migration. Some examples of tissue interactions that promote the differentiation of specific neural crest derivatives are given in Table 13-2.

The plasticity of differentiation of neural crest cells can be demonstrated by cloning single neural crest cells in culture. In the same medium and under apparently the same environmental conditions, the progeny of the single cloned cells frequently differentiate into neuronal and nonneuronal (e.g., pigment cell) phenotypes. Similarly, if individual neural crest cells are injected in vivo with a dye, over 50% of the injected cells give rise to progeny with two to four different phenotypes containing the dye.

TABLE 13-2 Environmental Factors Promoting Differentiation of Neural Crest Cells

NEURAL CREST DERIVATIVE	INTERACTING STRUCTURE
Bones of cranial vault	Brain
Bones of base of skull	Notochord, brain
Pharyngeal arch cartilages	Pharyngeal endoderm
Meckel's cartilage	Cranial ectoderm
Maxillary bone	Maxillary ectoderm
Mandible	Mandibular ectoderm
Palate	Palatal ectoderm
Otic capsule	Otic vesicle
Dentine of teeth	Oral ectoderm
Glandular stroma: thyroid, parathyroid, thymus, salivary	Local epithelium
Adrenal medullary chromaffin cells	Glucocorticoids secreted by adrenal cortex
Enteric neurons	Gut wall
Sympathetic neurons	Spinal cord, notochord, somites
Sensory neurons	Peripheral target tissue
Pigment cells	Extracellular matrix along pathway of migration

However, not all types of transformations among possible neural crest derivatives can occur. For example, crest cells from the trunk transplanted into the head cannot form cartilage or skeletal elements, although this is normal for cells of the cranial neural crest. Most experiments suggest that early neural crest cells segregate into intermediate lineages that preserve the option of differentiating into several but not all types of individual phenotypes. In the chick embryo, some neural crest cells are antigenically different from others even before they have left the neural tube.

A number of neural crest cells are bipotential, depending on signals from their local environment for cues to their final differentiation. One subline called the **sympathoadrenal lineage** forms adrenal medullary cells if exposed to adrenal glucocorticoid hormones. In contrast, if they are exposed first to **fibroblast growth factor (FGF)** and then to **nerve growth factor,** the same cells become sympathetic neurons. Similarly, cultured heart cells secrete a protein that converts postmitotic sympathetic neurons from an adrenergic (norepinephrine transmitter) phenotype to a cholinergic (acetylcholine-secreting) phenotype (see Fig. 12-18). During normal development the sympathetic neurons that innervate sweat glands are catecholaminergic until their axons actually contact the sweat glands. At that point they become cholinergic. With such a great variety of neural crest derivatives, other developmental switches of one functional cell type to another are likely to be discovered.

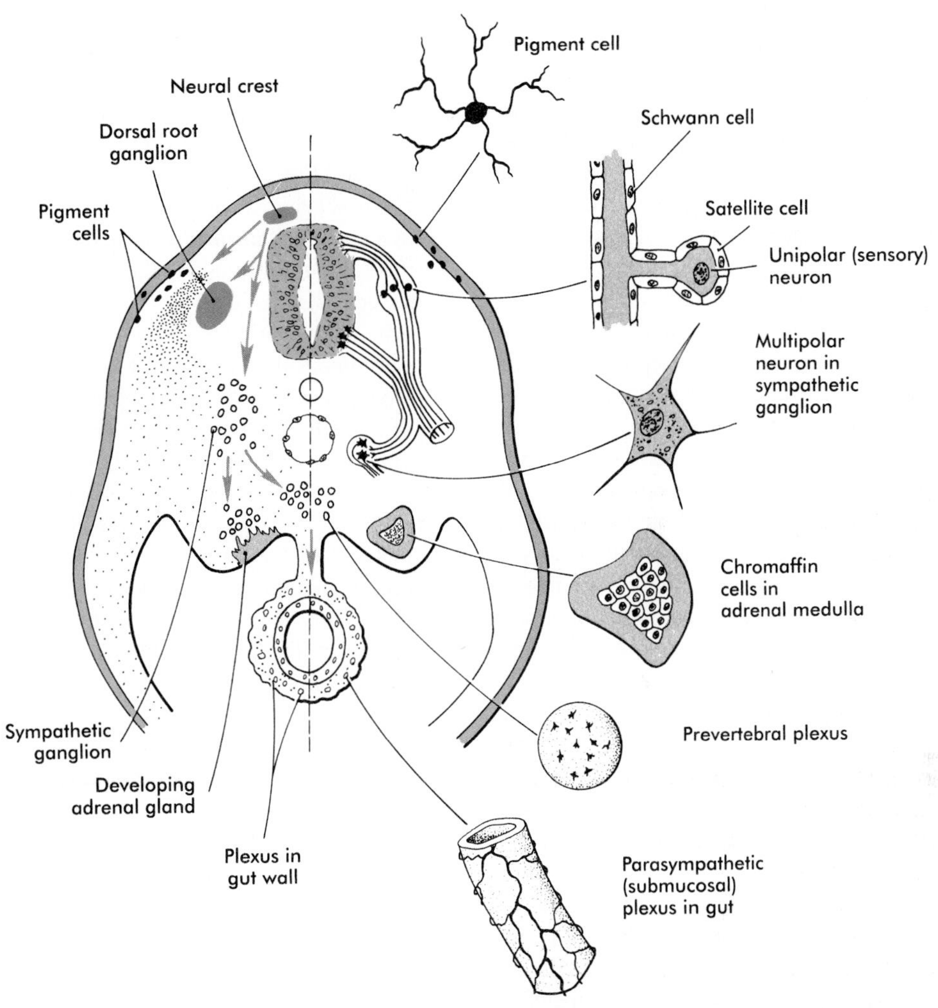

FIG. 13-4 Major neural crest migratory pathways and derivatives in the trunk. *Left:* Pathways in the early embryo. *Right:* Derivatives of the trunk neural crest.

THE TRUNK NEURAL CREST

The neural crest of the trunk extends from the level of the sixth somite to the most posterior extent of the neural crest. Within much of the trunk, three main migratory pathways of neural crest cells can be seen in cross section (Fig. 13-4). One is a dorsolateral pathway between the ectoderm and the somites. The cells that elect this pathway disperse beneath the ectoderm and ultimately enter the ectoderm as pigment cells **(melanocytes).**

The second pathway of migration is a ventral one, along which the neural crest cells initially move into the space between the somites and the neural tube. It continues just under the ventromedial surface of the somite until the cells reach the dorsal aorta. Cells that follow this branch belong to the **sympathoadrenal lineage.** These cells contribute to the formation of the adrenal medulla and elements of the sympathetic nervous system. A third ventrolateral pathway leads into the anterior halves of the somites. The cells that follow this route form the segmentally arranged sensory ganglia.

The sympathoadrenal lineage is derived from a committed sympathoadrenal progenitor cell that has already passed a number of restriction points so that it can no longer form sensory neurons, glia, or melanocytes. This progenitor cell gives rise to four types of cellular progeny: (1) adrenal chromaffin cells, (2) SIF (small intensely fluorescent) cells found in the sympathetic ganglia and carotid body, (3) adrenergic sympathetic neurons, and (4) a small population of cholinergic sympathetic neurons.

Somewhat further down this cellular lineage is a bipotential progenitor cell that can give rise to either adrenal chromaffin cells or sympathetic neurons. The bipotential progenitor cell already possesses some neuronal traits, but final differentiation depends on the environment surrounding these cells. In the presence of FGF in early sympathetic ganglia, these precursors differentiate into definitive sympathetic neurons. On the other hand, precursor cells in the forming adrenal medulla encounter glucocorticoids secreted by adrenal cortical cells. Under the hormonal influence, they lose their neuronal properties and differentiate into chromaffin cells. This differentiative choice, however, is not absolutely fixed, since chromaffin cells can be postnatally stimulated to **transdifferentiate** into neurons if they are exposed to nerve growth factor in vitro.

The entire length of the gut is populated by neural crest–derived parasympathetic neurons and associated cells, the enteric glia. These arise from neural crest cells in the cervical (vagal) and sacral levels and undertake extensive migrations along the developing gut. Considerable evidence now suggests that these cells are not committed to form gut-associated nervous tissue before leaving the spinal cord. If vagal crest is replaced by neural crest of the trunk, which normally does not give rise to gut-associated derivatives, the gut is colonized by the transplanted trunk-level neural crest cells. Evidence that the pathways of migration affect differentiation is seen in the neurotransmitters produced by these transplanted crest cells. Parasympathetic neurons differentiated from trunk cells produce serotonin but not catecholamines in the gut. If they had differentiated in their normal sites in the trunk, these neurons would have produced catecholamines and not serotonin.

Despite the strong influence of the environment of the gut on differentiation of neural crest cells exposed to its influences, neural crest cells retain a surprising degree of developmental flexibility. If crest-derived cells already in the gut of avian embryos are retransplanted into the trunk region of younger embryos, they seem to lose the memory of their former association with the gut. They enter the pathways (e.g., adrenal or peripheral nerve) common to trunk crest cells (except that they cannot enter pigment cell pathways) and differentiate accordingly.

Compared with the cranial neural crest, the trunk neural crest has a relatively limited range of differentiation options. The derivatives of the trunk neural crest are summarized in Table 13-1.

THE CRANIAL NEURAL CREST

The cranial neural crest is a major component of the cephalic end of the embryo. Comparative anatomical and developmental research suggests that the neural crest may represent the major morphological substrate for the evolution of the vertebrate head. Largely as the result of the availability of precise cellular marking methods, understanding of the cranial neural crest has increased dramatically in the past decade. The majority of studies on the cranial neural crest have been conducted on avian embryos. Research on mammalian embryos, however, suggests that the properties and role of the neural crest in mammalian cranial development are quite similar to those in birds.

In the mammalian head, neural crest cells leave the future brain well before closure of the neural folds (Fig. 13-5). Although the pathways of migration of the cranial neural crest in mammals are not nearly as well delineated as in birds, there nevertheless appear to be distinct but somewhat overlapping migratory territories in the embryonic mammalian head (Fig. 13-6). Because of the distribution of primary mesenchymal cells and extracellular matrix molecules in the mammalian head, the neural crest cells migrate in diffuse streams throughout the cranial mesenchyme to reach their final destinations.

There is remarkable specificity in the relationship among the origins of the neural crest in the hindbrain, its ultimate destination within the pharyngeal arches, and the expression of certain gene products (Figs. 13-7 and 15-5). The neural crest associated with rhombomere 2 migrates into and forms the bulk of the first pharyngeal arch, that of rhombomere 4 into the second arch, and that of rhombomere 6 into the third arch. The neural crest is not present lateral to rhombomeres 3 and 5. A close correlation exists between this pattern of migration and the expression of products of the *Hox-2 (Hox-B)* gene complex. *Hox-2.8* (B2), *Hox-2.7* (B3), and *Hox-2.6* (B4) products are expressed in a regular sequence in both the neural tube and the neural crest–derived mesenchyme of pharyngeal arches 2, 3, and 4. *Hox-2 (Hox-B)* is not expressed in rhombomere 2 or in the first branchial arch mesenchyme. Only after the pharyngeal arches become populated with neural crest cells does the ectoderm overlying the arches express a similar pattern of *Hox-2 (Hox-B)* gene products. These *Hox-2 (Hox-B)* genes may play a role in positionally specifying the neural crest cells with which they are associated. Furthermore, interactions between the neural crest cells and surface ectoderm of the pharyngeal arches may specify the ectoderm of the arches.

Another gene, *Krox 20,* also shows great specificity to neural crest in this region, since it is exposed lateral to all rhombomeres in the region except for 3 and 5, the ones that are lacking in neural crest cells. It is reasonable to expect that functional linkages rather than just positional correlations will soon be established between certain gene families and segmental specifications during early embryonic development.

In the posterior part of the pharynx, a circumpharyngeal crest passes behind the sixth pharyngeal arch. Ventral to the pharynx it sweeps cranially, providing the pathway through which the hypoglossal nerve (XII) and its associated skeletal muscle precursor cells pass. The muscles innervated by the hypoglossal nerve and the hypopharyngeal muscles are the only somite-derived skeletal muscles whose connective tissue cells originate in the neural crest.

A major difference between the cranial neural crest and that of the trunk is that the neural crest cells are patterned with level-specific instructions in the head, whereas those of the trunk do not appear to have imprinted level-specific instructions. Specifically, if neural crest cells that normally give rise to structures characteristic of branchial

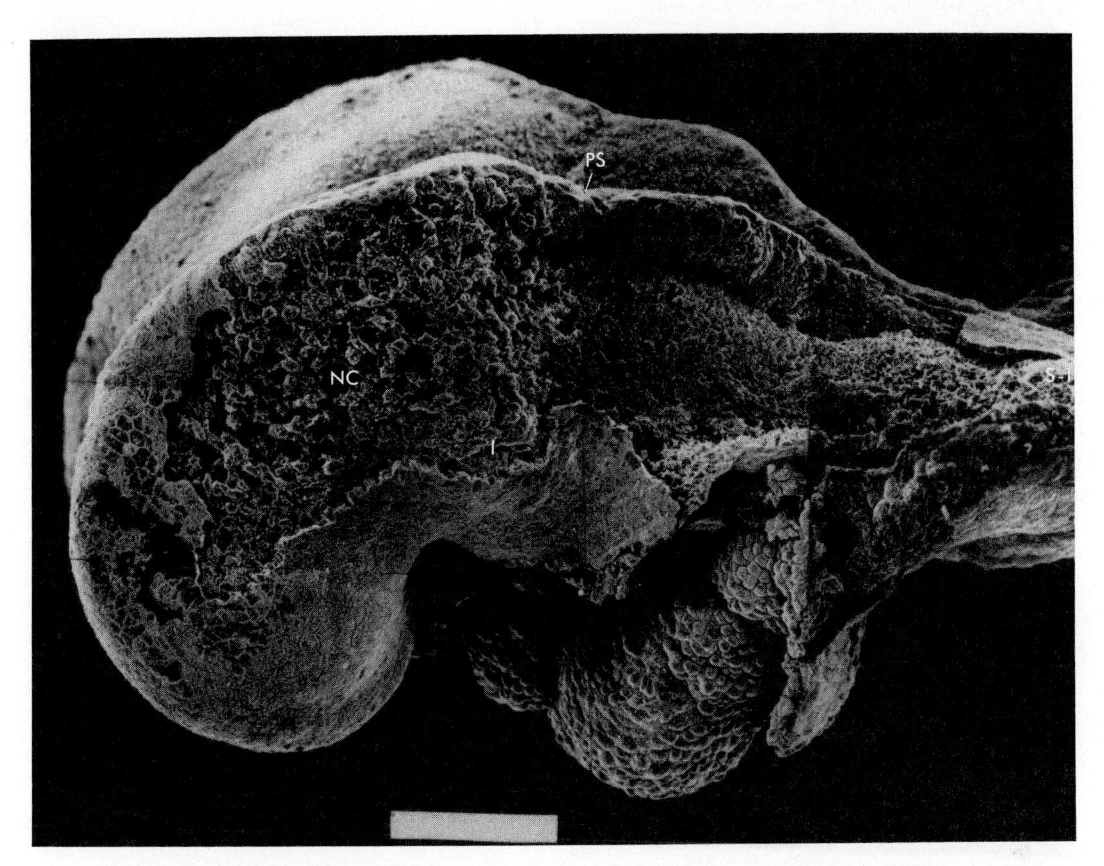

FIG. 13-5 Neural crest migration in the head of a 7-somite rat embryo. In this scanning electron micrograph, the ectoderm was removed from a large part of the side of the head, exposing migrating neural crest (*NC*) cells cranial (to the left) to the preotic sulcus (*PS*). Many of the cells are migrating toward the first pharyngeal arch (*I*). The area between the preotic sulcus and the first somite (*S-1*) is devoid of neural crest cells because in this region they have not begun to emigrate from the closing neural folds. The white bar at the bottom represents 100 μm.

(Based on studies from Tan and Morriss-Kay [1985].)

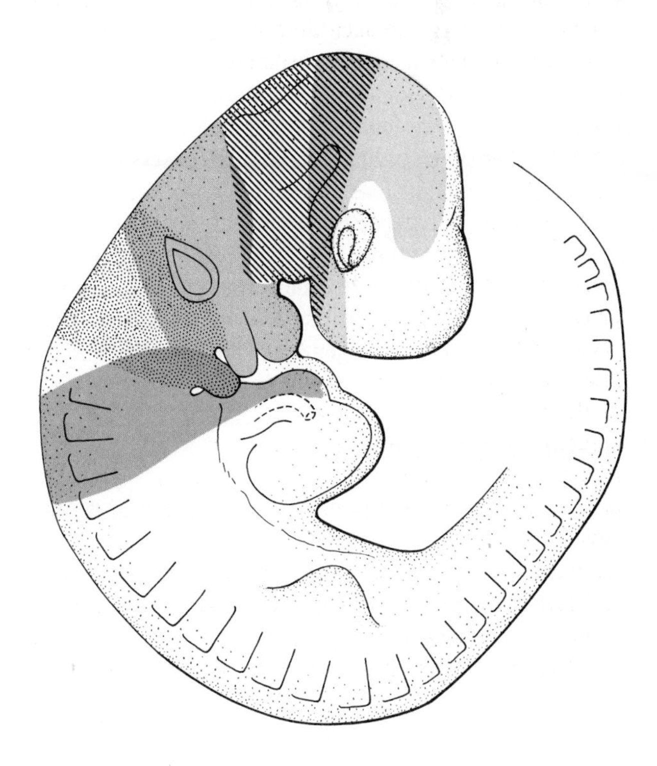

FIG. 13-6 Major cranial neural crest migration routes in the mammal.

(Based on studies by Morris-Kay and Tuckett [1991].)

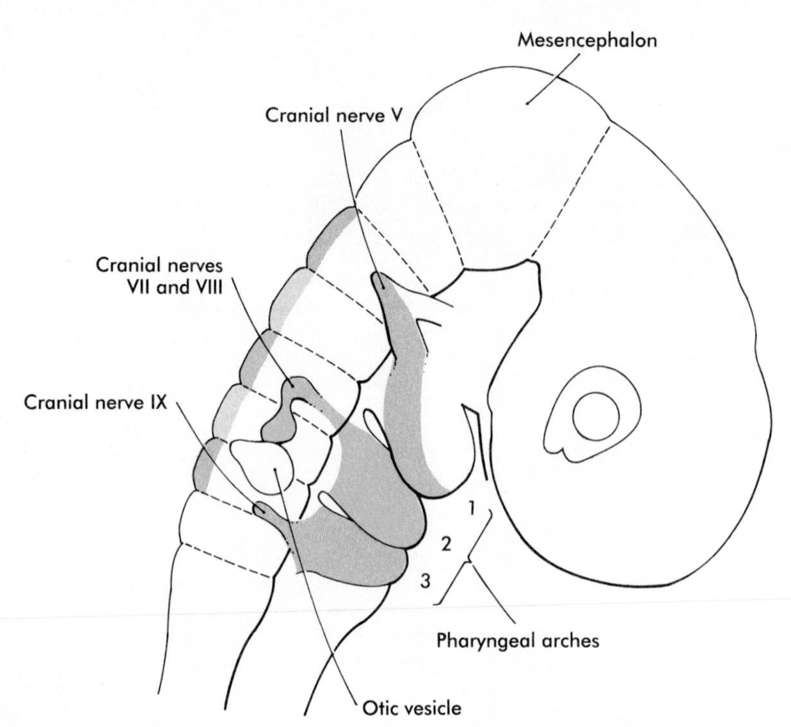

FIG. 13-7 Migration paths of neural crest cells from rhombomeres 2, 4, and 6 into the first three pharyngeal arches.

arch 1 are transplanted to the level of branchial arch 3, ectopic arch 1 structures (e.g., a supernumerary jaw) are formed at the level of arch 3, similar level shifts of transplanted neural crest do not result in the formation of abnormal structures.

In both avian and mammalian embryos a migration pathway is present from the occipital neural crest to the cardiac outflow tract. A disturbance in this area can result in cardiac septation defects **(aorticopulmonary septum)** as well as glandular and craniofacial malformations **(DiGeorge syndrome)**. Such a constellation of defects has been described in human embryos exposed to an excess of retinoic acid early in embryogenesis.

Cranial neural crest cells differentiate into a wide variety of cell and tissue types (see Table 13-1), including connective tissue and skeletal tissues. These tissues constitute much of the soft and hard tissues of the face (Fig. 13-8). (Specific details of morphogenesis of the head are presented in Chapter 15.)

NEUROCRISTOPATHIES

Because of the complex developmental history of the neural crest, a variety of congenital malformations are associated with its defective development. These have commonly been subdivided into two main categories—defects of migration or morphogenesis and tumors of neural crest tissues (box). Some of these defects involve only a single component of the neural crest; others affect multiple components and are recognized as syndromes.

Several syndromes or associations of defects are understandable only if the wide distribution of derivatives of the neural crest is recognized. For example, one association called **CHARGE** consists of *C*oloboma (see Chapter 14), *H*eart disease, *A*tresia of nasal choanae, *R*etardation of development, *G*enital hypoplasia in males, and anomalies of the *E*ar. The **Waardenburg syndrome** involves various combinations of pigmentation defects (commonly a white stripe in the hair and other pigment anomalies in the skin), deafness, cleft palate, and **ocular hypertelorism** (increased space between the eyes). The dominant gene for the Waardenburg syndrome and the human homologue of *Pax-3* (a paired homeobox gene) map to the same locus. It appears likely that the gene for Waardenburg syndrome and the *splotch* gene, which results in white spotting in mice, are homologous.

Neurofibromatosis (von Recklinghausen disease) is a common genetic disease that is manifest by multiple tumors of neural crest origin. Common features are **cafe au lait spots** (light-brown pigmented lesions) on the skin, multiple (often hundreds) **neurofibromas** (peripheral nerve tumors), occasional gigantism of a limb or digit, and a variety of other conditions. Neurofibromatosis occurs in approximately 1 of 3000 live births, and the gene (which has been recently cloned) is a very large one and subject to a high mutation rate.

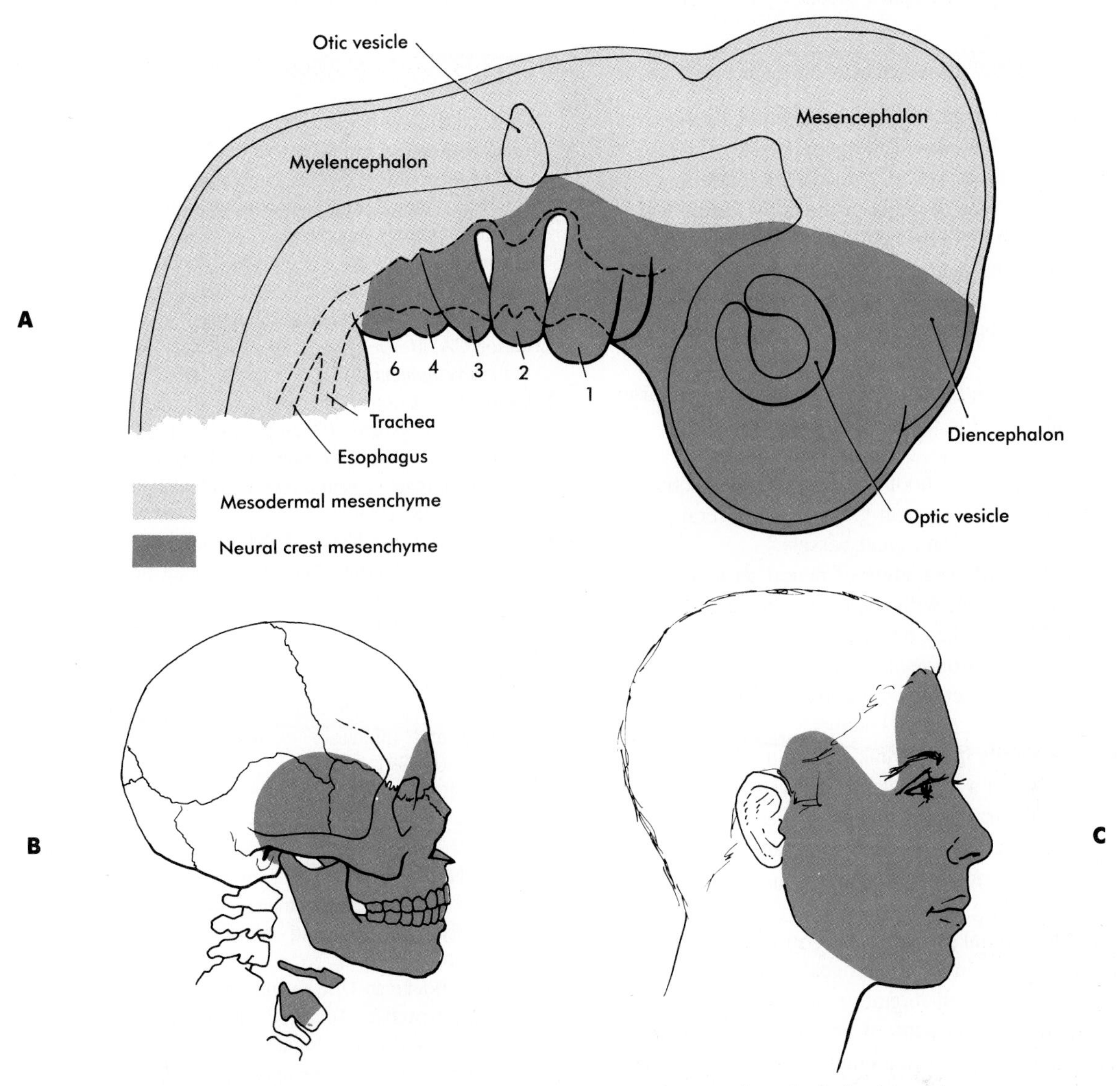

FIG. 13-8 Neural crest distribution in the human face and neck. **A,** In the early embryo. **B** and **C,** In the adult skeleton and dermis.

Major Neurocristopathies

Defects of Migration or Morphogenesis

Trunk neural crest

Hirschsprung's disease (aganglionic colon)

Cranial neural crest

Aorticopulmonary septation defects of heart

Anterior chamber defects of eye

Cleft lip and/or cleft palate

Frontonasal dysplasia

DiGeorge syndrome (hypoparathyroidism, thyroid deficiency, thymic dysplasia leading to immunodeficiency, defects in cardiac outflow tract and aortic arches)

Certain dental anomalies

Trunk and cranial neural crest

CHARGE association

Waardenburg syndrome

Tumors and Proliferation Defects

Pheochromocytoma—tumor of chromaffin tissue of adrenal medulla

Neuroblastoma—tumor of adrenal medulla and/or autonomic ganglia

Medullary carcinoma of thyroid—tumor of parafollicular (calcitonin-secreting) cells of thyroid

Carcinoid tumors—tumors of enterochromaffin cells of digestive tract

Neurofibromatosis (von Recklinghausen disease)—peripheral nerve tumors

Other Defects Involving Neural Crest

Albinism

SUMMARY

1. The neural crest arises from neuroepithelial cells along the lateral border of the neural plate. Having left the neural tube, neural crest cells migrate to peripheral locations throughout the body. Some substrates, such as those containing chondroitin sulfate molecules, are not favorable for neural crest cell migration.
2. Neural crest cells differentiate into many types of adult cells, such as sensory and autonomic neurons, Schwann cells, pigment cells, and adrenal medullary cells. Cells from the cranial neural crest also differentiate into bone, cartilage, dentin, dermal fibroblasts, selected smooth muscle, the connective tissue stroma of pharyngeal glands, and several regions of the heart and great vessels.
3. The control of differentiation of neural crest cells seems to be diverse, with some cells being determined before they begin to migrate and others responding to environmental cues along their paths of migration. Trunk neural crest cells cannot differentiate into skeletal elements.
4. Neural crest cells in the trunk follow three main paths of migration: (1) a dorsolateral pathway for pigment cells, (2) a ventral path for cells of the sympathoadrenal lineage, and (3) a pathway leading through the anterior halves of the somites for sensory ganglion–forming cells.
5. Cells of the cranial neural crest form many tissues of the facial region. In the pharyngeal region the pathways of crest cell migration are closely correlated with regions of expression of products of the *Hox-2* gene complex. Cells of the cranial crest are patterned with level-specific instructions, whereas those of the trunk crest are not.
6. A number of genetic diseases and syndromes are associated with disturbances of the neural crest. Neurofibromatosis is often characterized by multiple tumors and pigment disturbances. Disturbances of the cardiac neural crest can result in septation defects in the heart and outflow tract.

REVIEW QUESTIONS

1. How does the segmental distribution of the spinal ganglia occur?
2. A newborn is found to have a septation defect of the heart, hypoparathyroidism, and a slight degree of immunodeficiency. What is the likely common element?
3. What are three major differences between cranial and trunk neural crest?

REFERENCES

Anderson DJ: *Development and plasticity of a neural crest-derived neuroendocrine sublineage*. In Landmesser LT, ed: *The assembly of the nervous system,* New York, 1989, Liss, pp 17-36.

Barald KF: Culture conditions affect the cholinergic development of an isolated subpopulation of chick mesencephalic neural crest cells, *Dev Biol* 135:349-366, 1989.

Bronner-Fraser M, Fraser SE: Cell lineage analysis of the avian neural crest, *Development* 2:(suppl)17-22, 1991.

Erickson CA: *Morphogenesis of the neural crest*. In Browder LW, ed: *Developmental biology,* vol 2, New York, 1986, Plenum.

Erickson CA: Control of pathfinding by the avian trunk neural crest, *Development* 103(suppl):63-80, 1988.

Erickson CA, Loring JF, Lester SM: Migratory pathways of HNK-1-immunoreactive neural crest cells in the rat embryo, *Dev Biol* 134:112-118, 1989.

Fukiishi Y, Morriss-Kay GM: Migration of cranial neural crest cells to the pharyngeal arches and heart in rat embryos, *Cell Tissue Res* 268:1-8, 1992.

Hall BK, Horstadius S: *The neural crest,* London, 1988, Oxford University.

Hunt P, Wilkinson D, Krumlauf R: Patterning the vertebrate head: murine Hox 2 genes mark distinct subpopulations of premigratory and migratory cranial neural crest, *Development* 112:43-50, 1991.

Johnston MC, Vig KWL, Ambrose LJH: *Neurocristopathy as a unifying concept: clinical correlations*. In Riccardi VM, Mulvihill JJ, eds: *Neurofibromatosis (von Recklingshausen disease),* New York, 1981, Raven, pp 97-104.

Jones MC: The neurocristopathies: reinterpretation based upon the mechanism of abnormal morphogenesis, *Cleft Palate J* 27:136-140, 1990.

Kirby ML, Bockman DE: Neural crest and normal development: a new perspective, *Anat Rec* 209:1-6, 1984.

Kuratani SC, Kirby ML: Initial migration and distribution of the cardiac neural crest in the avian embryo: an introduction to the concept of the circumpharyngeal crest, *Am J Anat* 191:215-227, 1991.

Kuratani SC, Kirby ML: Migration and distribution of circumpharyngeal crest cells in the chick embryo, *Anat Rec* 234:263-280, 1992.

Lallier T, Bronner-Fraser M: The role of the extracellular matrix in neural crest migration, *Semin Dev Biol* 1:35-44, 1990.

Le Douarin N: *The neural crest,* Cambridge, England, 1982, Cambridge University.

Lumsden A: Multipotent cells in the avian neural crest, *Trends Neurosci* 12:81-83, 1989.

Lumsden A, Guthrie S: Alternating patterns of cell surface properties and neural crest cell migration during segmentation of the chick hindbrain, *Development* 2(suppl):9-15, 1991.

Morriss-Kay G, Tan S-S: Mapping cranial neural crest cell migration pathways in mammalian embryos, *Trends Genet* 3:257-261, 1987.

Morriss-Kay G, Tucket F: Early events in mammalian craniofacial morphogenesis, *J Craniofac Genet Dev Biol* 11:181-191, 1991.

Newgreen DF, Erickson CA: The migration of neural crest cells, *Int Rev Cytol* 103:89-143, 1986.
Noden DM: The role of the neural crest in patterning of avian cranial skeletal, connective and muscle tissues, *Dev Biol* 96:144-165, 1983.
Noden DM: Origins and patterning of craniofacial mesenchymal tissues, *J Craniofac Genet Dev Biol* 2(suppl):15-31, 1986.
Osumi-Yamashita N, Eto K: Mammalian cranial neural crest cells and facial development, *Dev Growth Differentiation* 32:454-459, 1990.
Patterson PH: Control of cell fate in a vertebrate neurogenic cell lineage, *Cell* 62:1035-1038, 1990.
Quevedo C, Holstein TJ: Molecular genetics and the ontogeny of pigment patterns in mammals, *Pigment Cell Res* 5:328-334, 1992.
Takamura K and others: Association of cephalic neural crest cells with cardiovascular development, particularly that of the semilunar valves, *Anat Embryol* 182:263-272, 1990.
Tan SS, Morriss-Kay G: The development and distribution of the cranial neural crest in the rat embryo, *Cell Tissue Res* 240:403-416, 1985.
Vogel KS, Weston JA: The sympathoadrenal lineage in avian embryos. I. Adrenal chromaffin cells lose neuronal traits during embryogenesis, *Dev Biol* 139:1-12, 1990.
Weston JA: Phenotypic diversification in neural crest-derived cells: the time and stability of commitment during early development, *Curr Top Dev Biol* 20:195-210, 1986.

CHAPTER 14

The Sense Organs

The sense organs arise in large measure from thickened placodes of cells in the ectodermal germ layer (see Fig. 5-9). Forming largely in response to secondary inductions by the central nervous system, the ectodermal placodes can be subdivided into two groups. One group gives rise to a diverse array of sense organs such as the inner ear, the lens of the eye, the olfactory sensory epithelium, and the anterior lobe of the hypophysis. The other group of placodes, which is closely associated with the branchial arches, produces sensory neurons that combine with neural crest–derived neurons to form the sensory ganglia of many of the cranial nerves.

This chapter concentrates on the development of the eyes and ears, the most complex and important sense organs in humans. Discussion of the organs of smell and taste is deferred to Chapter 15, since their development is intimately associated with that of the face and pharynx. The sensory components of the cranial nerves are discussed in Chapter 12.

THE EYE

The eye is a very complex structure that originates from constituents derived from a number of sources, including the wall of the diencephalon, the overlying surface ectoderm, and immigrating cranial mesenchyme. Two basic themes occur throughout the period of ocular development. One is an ongoing series of inductive signals that result in the initial establishment of the major components of the eye. The other is the coordinated differentiation of many of these components.

For normal vision to occur, many complex structures within the eye must properly relate to neighboring structures. For example, the cornea and lens must both become transparent and properly aligned to provide a proper pathway for light to reach the retina. The retina in turn must be configured to both receive concrete visual images and transmit patterned visual signals to the proper parts of the brain through neural processes extending from the retina into the optic nerve.

Early Events in the Establishment of the Eye

Development of the eye is first evident at about 22 days gestation, when the lateral walls of the diencephalon begin to bulge out as **optic grooves** (Fig. 14-1). Within a few days the optic grooves enlarge to form **optic vesicles,** which terminate very close to the overlying surface ectoderm. Apposition of the outer wall of the optic vesicle to the surface ectoderm is essential for the transmission of an important inductive message that stimulates the surface ectodermal cells to thicken and begin forming the lens (Fig. 14-2).

The interaction between optic vesicle and overlying ectoderm was one of the first recognized inductive processes. It was initially characterized by deletion and transplantation experiments conducted on amphibian embryos. When the optic vesicles were removed early, the surface ectoderm remained and differentiated into ordinary ectodermal cells instead of lens fibers. Conversely, when optic vesicles were combined with ectoderm other than eye, the ectoderm was stimulated to form lens fibers (Fig. 14-3). In mammals, an important mechanism underlying the **anophthalmia** (absence of eyes) seen in both the *eyeless* and *fidget* mutants is an interference in the apposition of optic vesicles and surface ectoderm and the attending lack of lens induction. Subsequent research on amphibians has shown that a series of preparatory inductions among future lens ectoderm, adjacent neural plate, and underlying mesoderm condition the ectoderm for its final induction by the optic vesicle.

As the process of lens induction occurs, the outer face

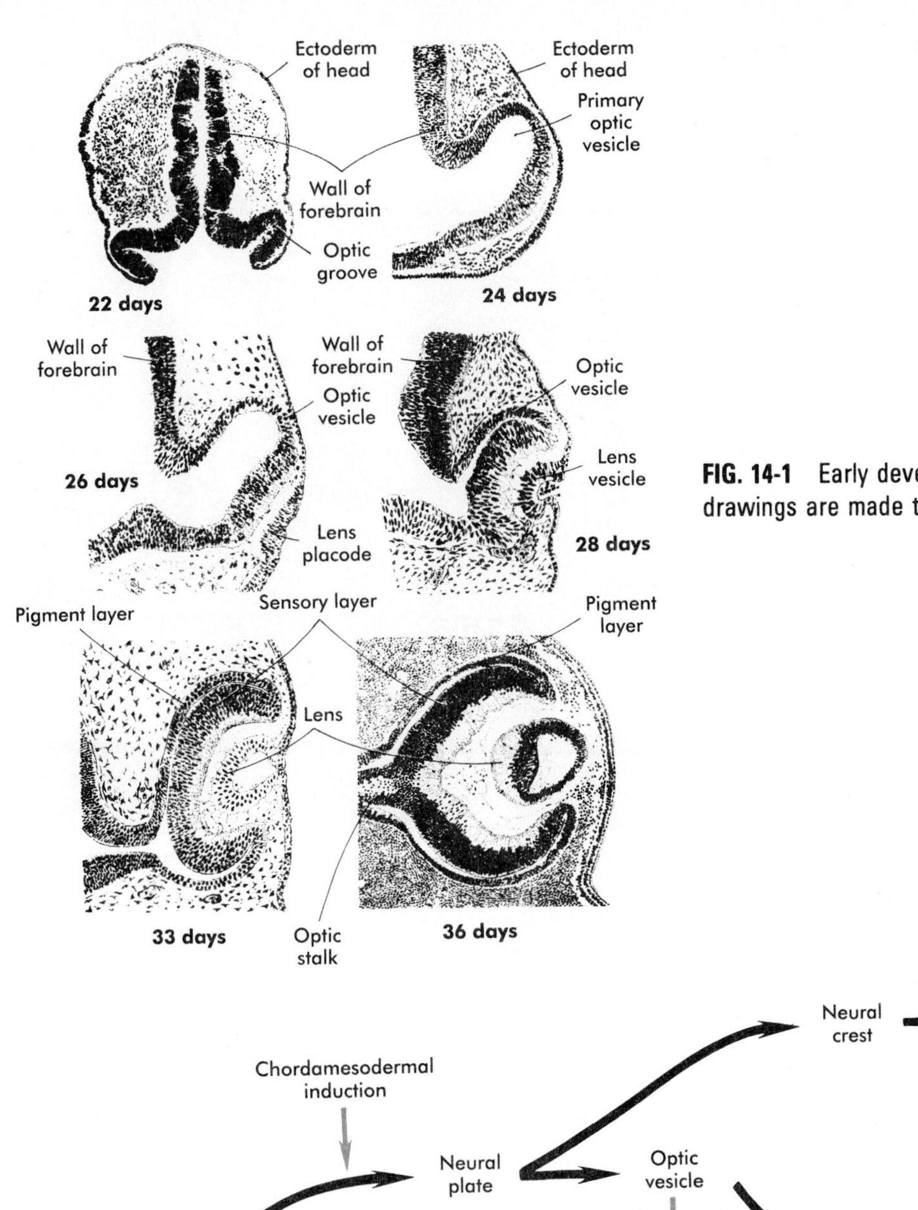

FIG. 14-1 Early development of the human eye. (All drawings are made to the same scale.)

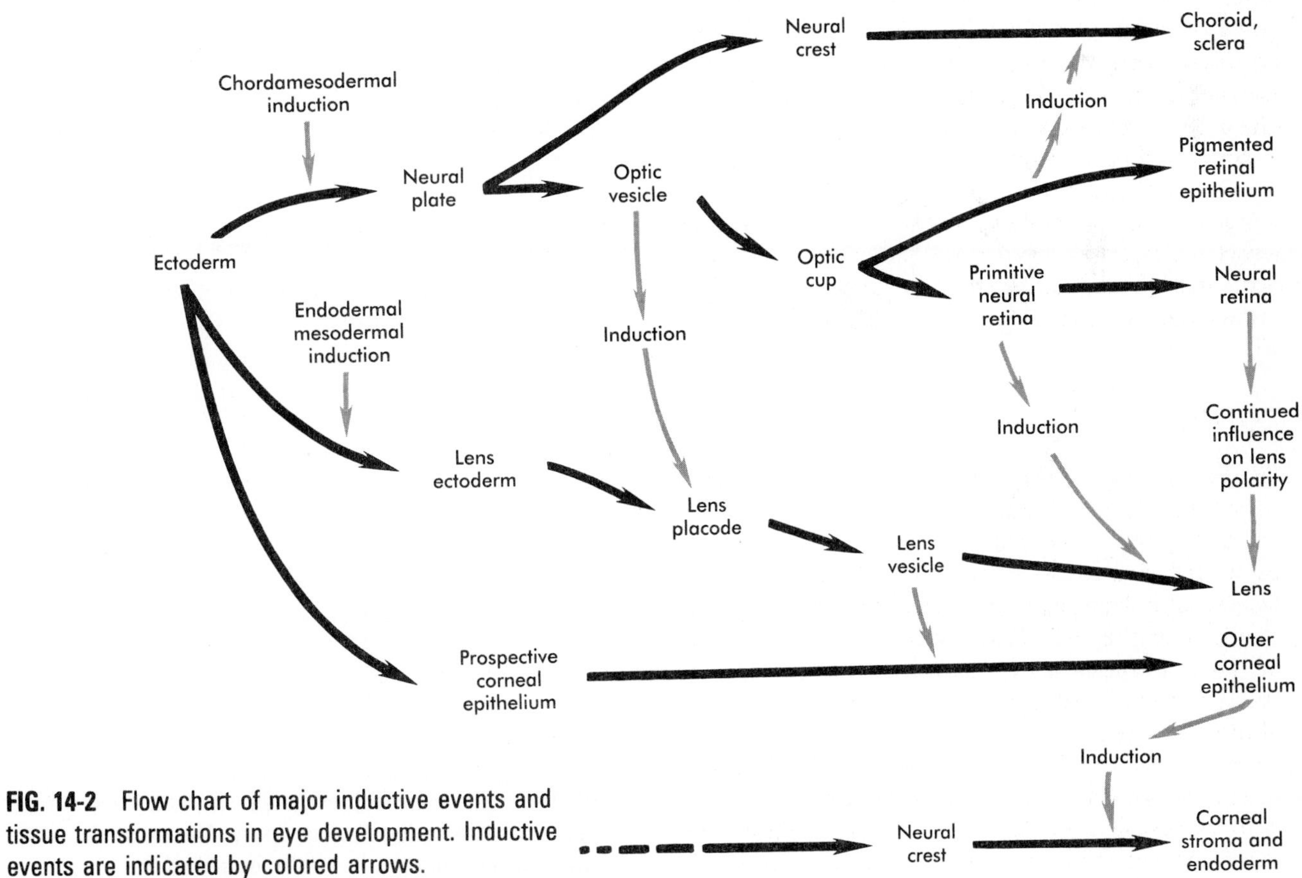

FIG. 14-2 Flow chart of major inductive events and tissue transformations in eye development. Inductive events are indicated by colored arrows.

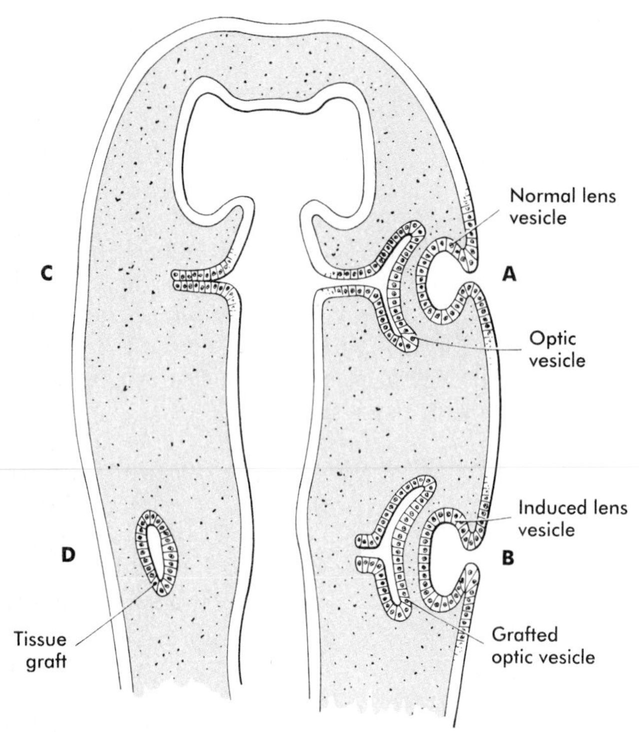

FIG. 14-3 Grafting and deletion experiments related to lens induction. **A,** Normal induction of the lens by the optic vesicle. **B,** Induction of an ectopic lens by a grafted optic vesicle. **C,** No lens is formed after removal of the optic vesicle. **D,** No lens is formed after grafting tissue other than optic vesicle into the prospective lens region.

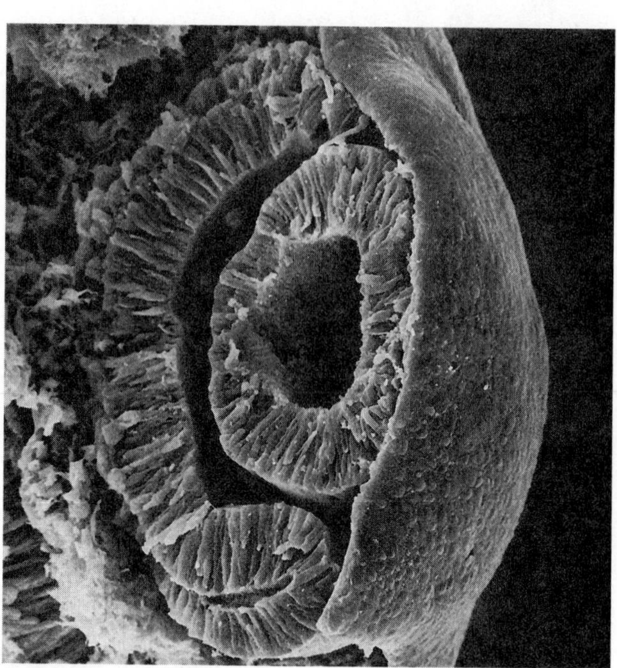

FIG. 14-4 Scanning electron micrograph of the optic cup *(left)* and lens vesicle *(center)* in the chick embryo.
(Courtesy K. Tosney, Ann Arbor, Mich.)

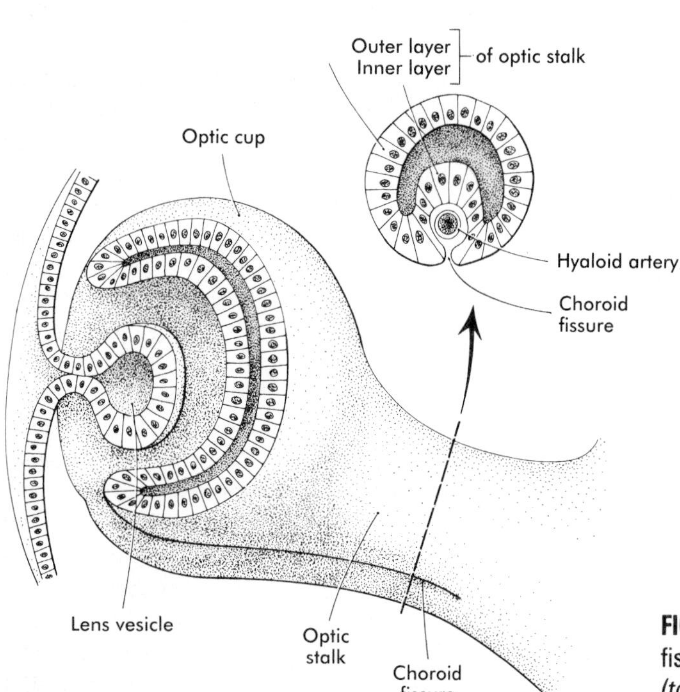

FIG. 14-5 The optic cup and stalk showing the choroid fissure containing the hyaloid artery. The cross section *(top)* is taken from the level of the dashed line.

of the optic vesicle begins to flatten and ultimately becomes concave. This results in the transformation of the optic vesicle to the **optic cup** (see Fig. 14-1). Meanwhile, the induced lens ectoderm thickens and invaginates to form a **lens vesicle,** which detaches from the surface epithelium from which it originated (Figs. 14-1 and 14-4). Then the lens vesicle takes over and becomes the primary agent of a new inductive reaction by acting on the overlying surface ectoderm and causing it to begin corneal development (see Fig. 14-2).

Formation of the optic cup is an asymmetrical process that occurs at the ventral margin of the optic vesicle rather than at its center. This results in the formation of a gap called the **choroid fissure,** which is continuous with a groove in the **optic stalk** (Fig. 14-5). During much of early ocular development the choroid fissure and optic groove form a channel through which the **hyaloid artery** passes into the posterior chamber of the eye. The optic stalk initially represents a narrow neck that connects the optic cup to the diencephalon, but as development progresses, it is invaded by neuronal processes emanating from ganglion cells of the retina. After these processes have made their way to the appropriate regions of the brain, the optic stalk is properly known as the **optic nerve.** Later in development the choroid fissure closes, and no trace of it is seen in the normal iris. Nonclosure of the choroid fissure results in the anomaly of coloboma (see p. 266).

Formation of the Lens

While breaking off from the surface ectoderm, the lens vesicle is roughly spherical, with a large central cavity (see Fig. 14-1). At the end of the sixth week the cells at the inner pole of the lens vesicle begin to elongate in an early step toward their transformation into the long, transparent cells called **lens fibers** (Fig. 14-6).

Differentiation of the lens is a very precise and well-orchestrated process involving several levels of organization. At the cellular level, relatively unspecialized lens epithelial cells undergo a profound transformation into transparent, elongated cells that contain large quantities of specialized **crystallin proteins.** At the tissue level, the entire lens is responsive to signals from the retina and other structures of the eye so that its shape and overall organization are best adapted for the transmission of undistorted light rays from the corneal entrance to the light-receiving cells of the retina.

At the cellular level, cytodifferentiation of the lens consists of the transformation of mitotically active lens epithelial cells into elongated postmitotic lens fiber cells. Up to 90% of the soluble protein in these postmitotic cells consists of crystallin proteins. The mammalian lens contains three major crystallin proteins—**α, β,** and **γ.**

The formation of crystallin-containing lens fibers begins with the elongation of epithelial cells from the inner pole of the lens vesicle (see Fig. 14-1). These cells make up the fibers of the **lens nucleus** (Fig. 14-7). The remainder of the lens fibers arise from transformation of the cuboidal cells of the anterior lens epithelium. During embryonic life, mitotic activity is spread throughout the outer lens epithelial cells. Around the time of birth, mitotic activity ceases in the central region of this epithelium, leaving a germinative ring of mitotically active cells around the central region. Daughter cells from the germinative region move into the equatorial region of cellular elongation, where they cease to divide and take on the cytological characteristics of RNA-producing cells and begin to form crystallin mRNAs. These cells soon elongate tremendously, fill up with crystallins, and transform into secondary lens fibers that form concentric layers around the primary fibers of the lens nucleus. The midline region where secondary lens fibers from opposite points on the equator join is recognized as the anterior and posterior **lens suture** (see Fig. 14-6, *D*). With this arrangement, the lens fibers toward the periphery are successively younger. As long as the lens grows, new secondary fibers move in from the equator onto the outer cortex of the lens.

Patterns of nucleic acid and protein synthesis correlate well with the cytological characteristics of the lens. DNA synthesis is concentrated in the germinative ring in the low outer epithelium (Fig. 14-8, *A*). RNA synthesis is prominent throughout the outer lens epithelium and also in the outer layers of the secondary lens fibers (Fig. 14-8, *B*). Studies with the RNA inhibitor actinomycin D have shown that the mRNAs in the equatorial cells and the secondary lens fibers are relatively long lived, allowing the formation of lens proteins even after the cells have lost their ability to form new RNAs (Table 14-1, p. 259). General patterns of protein synthesis show decreasing activity toward the region of the primary lens nucleus.

The crystallin proteins show a very characteristic pattern and sequence of appearance, with the α-crystallins appearing first in the morphologically undifferentiated epithelial cells. Synthesis of β-crystallins is seen when the lens fibers begin to elongate, whereas the expression of γ-crystallins is restricted to terminally differentiated lens fiber cells. Each of the crystallin protein families contains several members. They show different patterns of activation (some members of a family being coordinately activated) and different patterns of accumulation. There are often pronounced interspecies differences in patterns of crystallin expression. These presumably facilitate the optical clearing of the lens to allow the efficient transmission of light, but many details remain to be elucidated.

Throughout much of its life the lens is under the influence of the retina. Following induction of the lens, chemical secretions of the retina, which accumulate in the vitreous humor behind the lens, appear to stimulate the formation of lens fibers. A striking example of the continued influence of the retina on lens morphology is seen after a

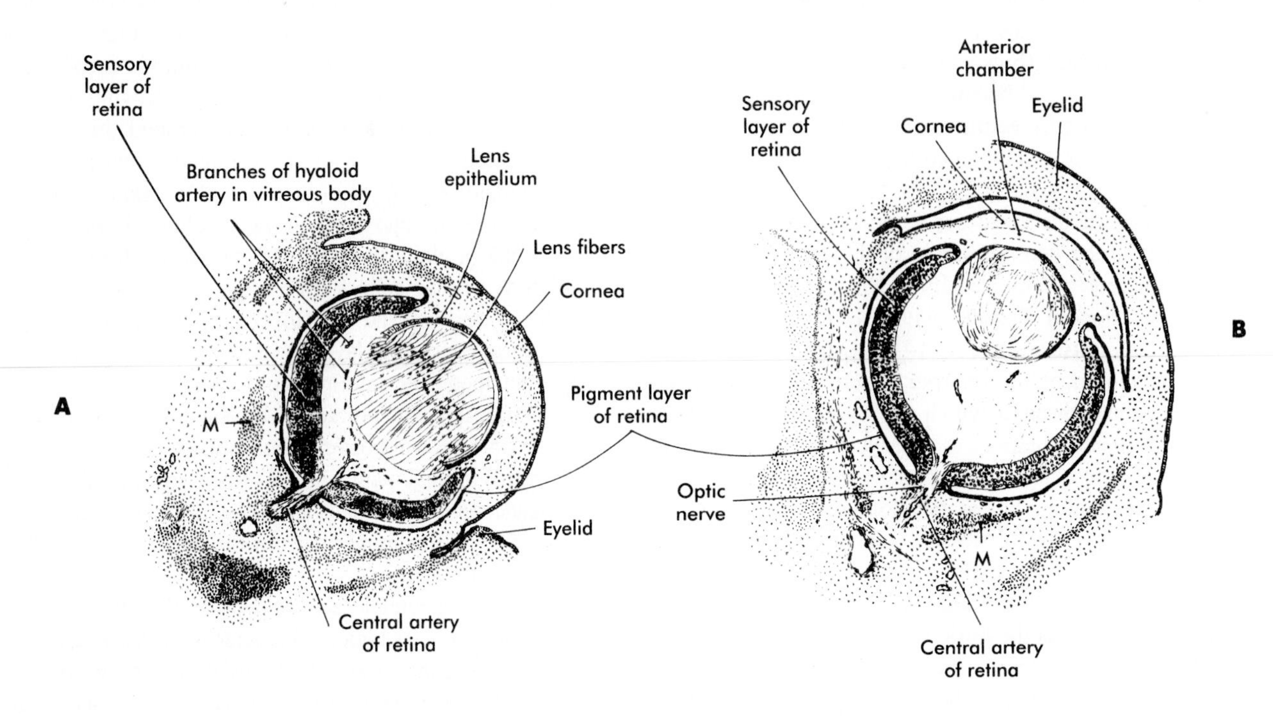

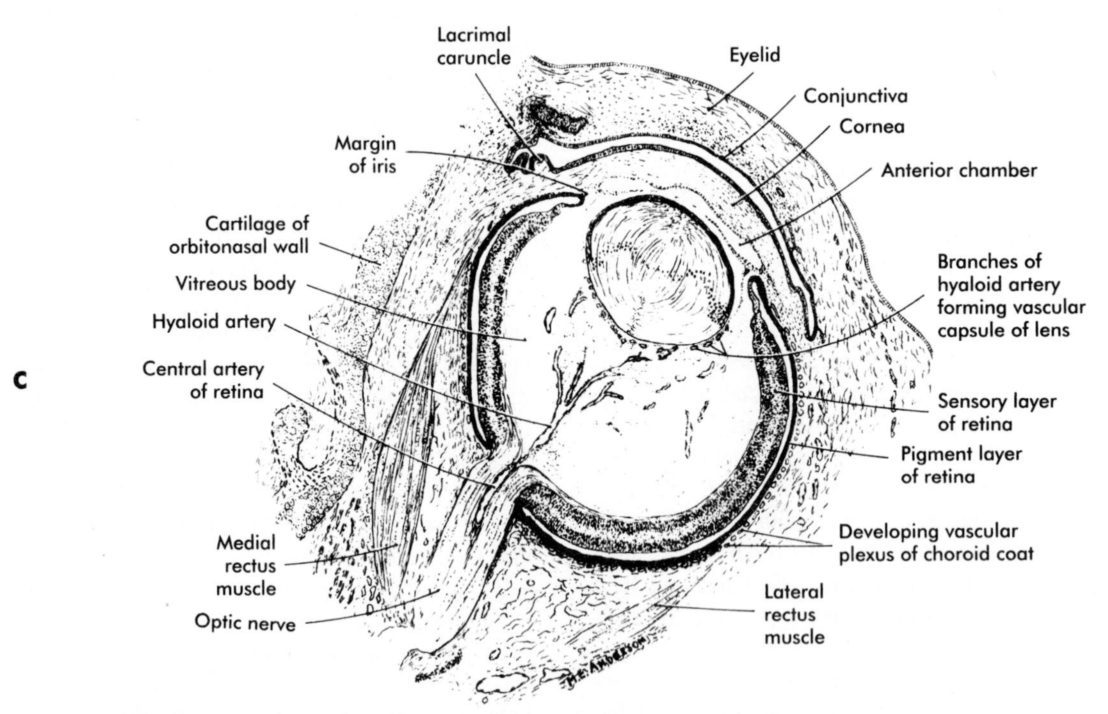

FIG. 14-6 Later stages of eye development drawn from coronal sections through heads of human embryos. **A,** At 7 weeks. **B,** At 9 weeks. **C,** At 10 weeks.

D

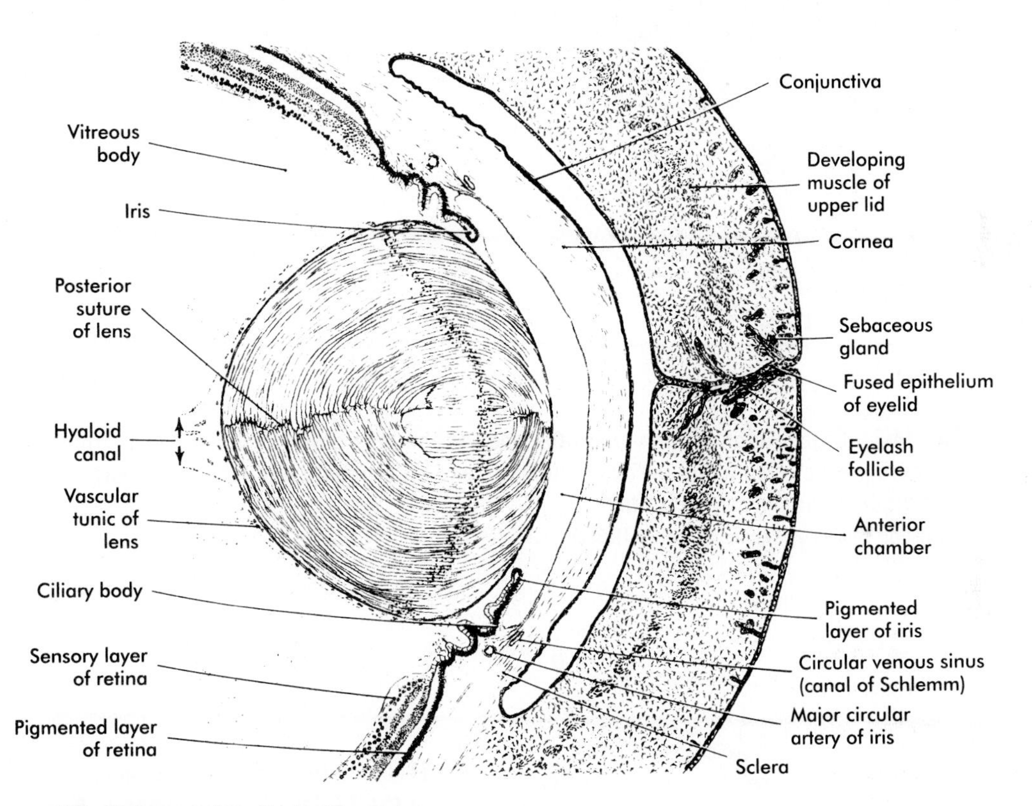

FIG. 14-6—cont'd D, At 19 weeks.
(Modified from Carlson B: *Patten's foundations of embryology,* ed 5, New York, 1988, McGraw-Hill.)

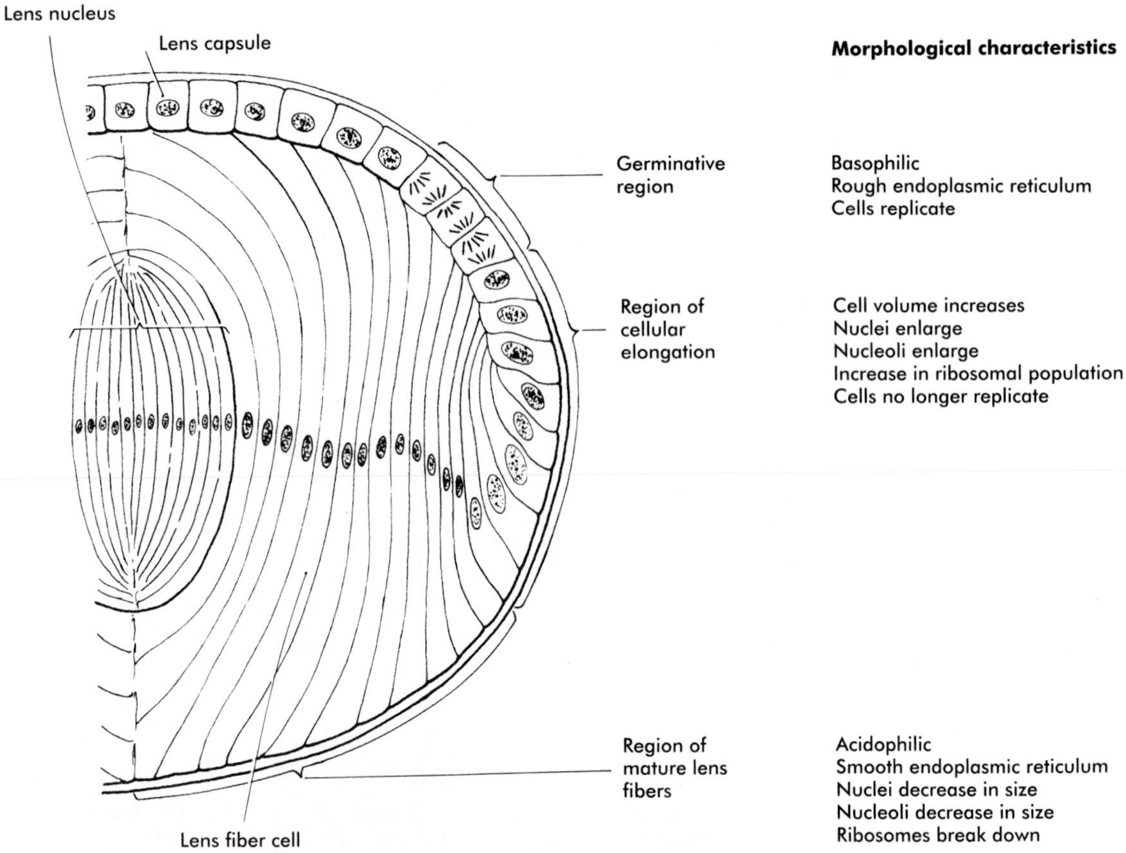

FIG. 14-7 Organization of the vertebrate lens. As the lens grows, epithelial cells from the germinative region stop dividing, elongate, and differentiate into lens fiber cells that produce lens crystallin proteins.

(Based on studies by Papaconstantinou [1967].)

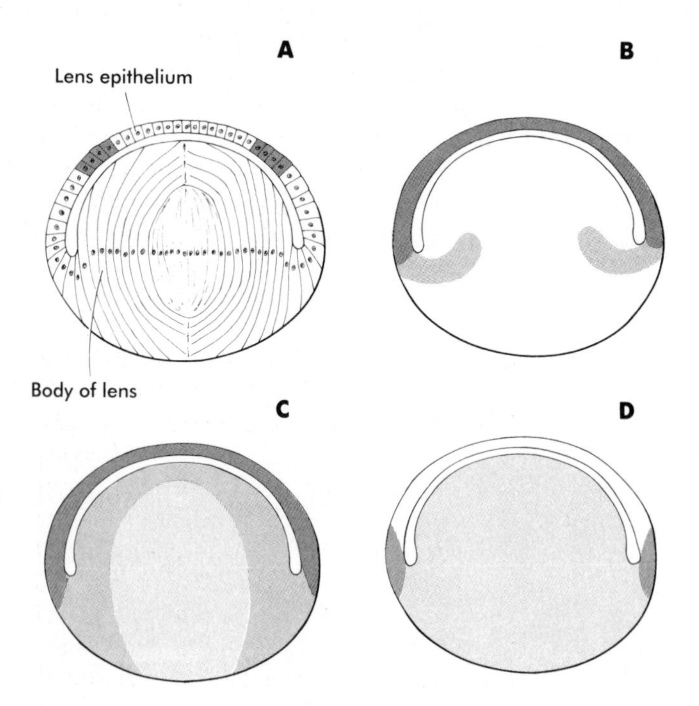

FIG. 14-8 Patterns of macromolecular synthesis in the lens of the 12-day-old chick embryo. **A,** Region of DNA synthesis in the germinative region of the lens epithelium is shaded. **B,** Levels and locations of RNA synthesis in the lens epithelium and lens fibers are indicated by different intensities. **C,** Levels and locations of protein synthesis are indicated by shades. **D,** Approximately 8 hours after the administration of actinomycin D (an inhibitor of RNA synthesis), cells of the lens epithelium no longer synthesize proteins, whereas some protein synthesis continues in the body of the lens itself. This indicates the presence of long-lived mRNA in the lens fiber cells.

(Modified from Reeder R, Bell E: *Science* 150:71-72, 1965.)

TABLE 14-1 Crystallin Protein Expression in Lens Differentiation

Presumptive fibers

Epithelium fibers

Epithelium fibers

DEVELOPMENTAL STAGE	CRYSTALLIN	EPITHELIUM	LENS FIBER
Beginning lens invagination	α		–
	β		–
	γ		–
Lens cup	α		+/–
	β		–
	γ		–
Lens vesicle	α	+/–	++
	β	–	+/–
	γ	–	+/–
Elongation of primary fibers	α	+	++
	β	–	++
	γ	–	+
Embryonic lens	α	++	+++
	β	–	+++
	γ	–	+++

developing lens is rotated so that its outer pole faces the retina. Very rapidly and presumably under the influence of retinal secretions, the low epithelial cells of the former outer pole begin to elongate and form an additional set of lens fibers (Fig. 14-9). A new lens epithelium forms on the corneal side of the rotated lens. Such structural adaptations are striking evidence of a mechanism that ensures correct alignment between the lens and the rest of the visual system throughout development.

Formation of the Cornea

Formation of the cornea is the result of the last of the series of major inductive events in eye formation (see Fig. 14-2), with the lens vesicle acting on the overlying surface ectoderm. This induction results in the transformation of a typical surface ectoderm, consisting of a basal layer of cuboidal cells and a superficial periderm, to a transparent, multilayered structure with a complex extracellular matrix and cellular contributions from several sources.

The inductive influence of the lens stimulates a change in the basal ectodermal cells. They increase in height, largely as a result of the elaboration of secretory organelles (e.g., the Golgi apparatus) on the basal ends of the cells. As these changes are completed, the cells begin to secrete epithelially derived collagen types I, II, and IX to form the **primary stroma** of the cornea (Fig. 14-10).

Using the primary stroma as a basis for migration, neural crest cells around the lip of the optic cup migrate centrally between the primary stroma and the lens capsule. Although mesenchymal in morphology during their migration, these cells become transformed into a cuboidal epithelium called the **corneal endothelium** once their migration is completed. At this point the early cornea consists of (1) an outer epithelium, (2) a still acellular primary stroma, and (3) an inner endothelium.

After the corneal endothelium has formed a continuous layer, its cells synthesize large amounts of **hyaluronic acid** and secrete it into the primary stroma. Because of its pronounced water-binding capacities, hyaluronic acid causes the primary stroma to swell greatly. This provides a proper substrate for the second wave of cellular migration into the developing cornea (Fig. 14-11). These cells, also of neural crest origin, are fibroblastic in nature. They migrate and proliferate in the hyaluronate-rich spaces between layers of collagen in the primary corneal stroma. The migratory phase of cellular seeding of the primary corneal stroma ceases when these cells begin to produce large amounts of

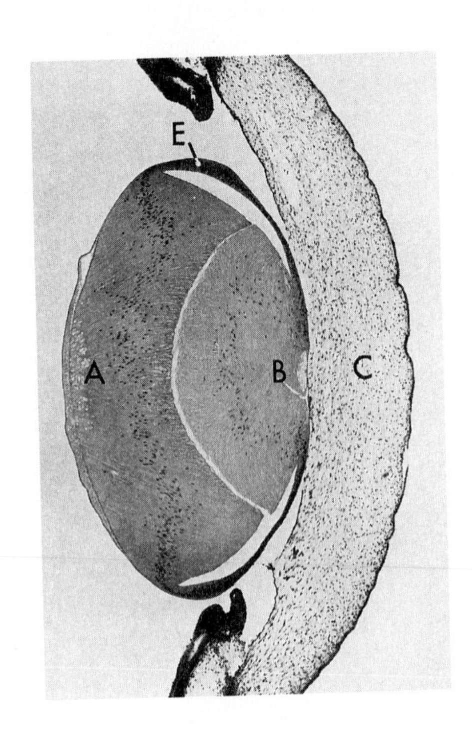

FIG. 14-9 Section through the lens of an 11-day-old chick embryo. At 5 days the lens was surgically reversed so that the anterior epithelial cells *(E)* faced the vitreous body and retina. The formerly low epithelial cells elongated to form new lens fibers *(A)*. Because of the reversal of polarity of the equatorial zone of the lens, new epithelial cells were added over the original mass of lens fibers *(B)* onto the corneal face of the reversed lens. *C*, Cornea.

(From DeHaan RL, Ursprung H, eds: *Organogenesis*, New York, 1965, Holt, Rinehart, Winston.)

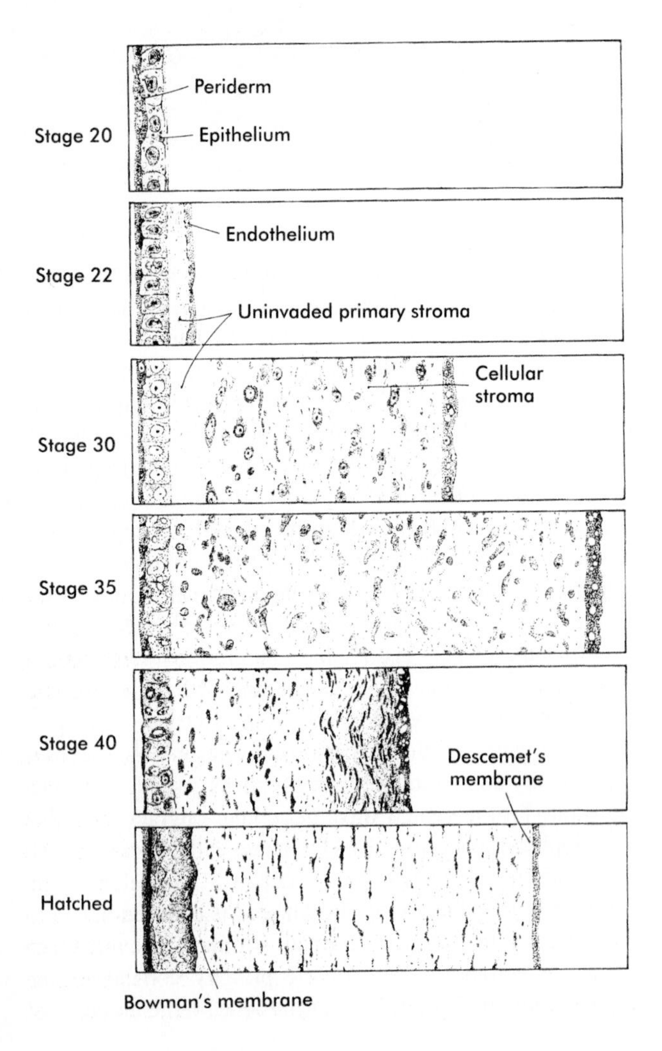

FIG. 14-10 Stages (Hamburger-Hamilton) in the formation of the cornea in the chick embryo.

(Based on studies by Hay and Revel [1969].)

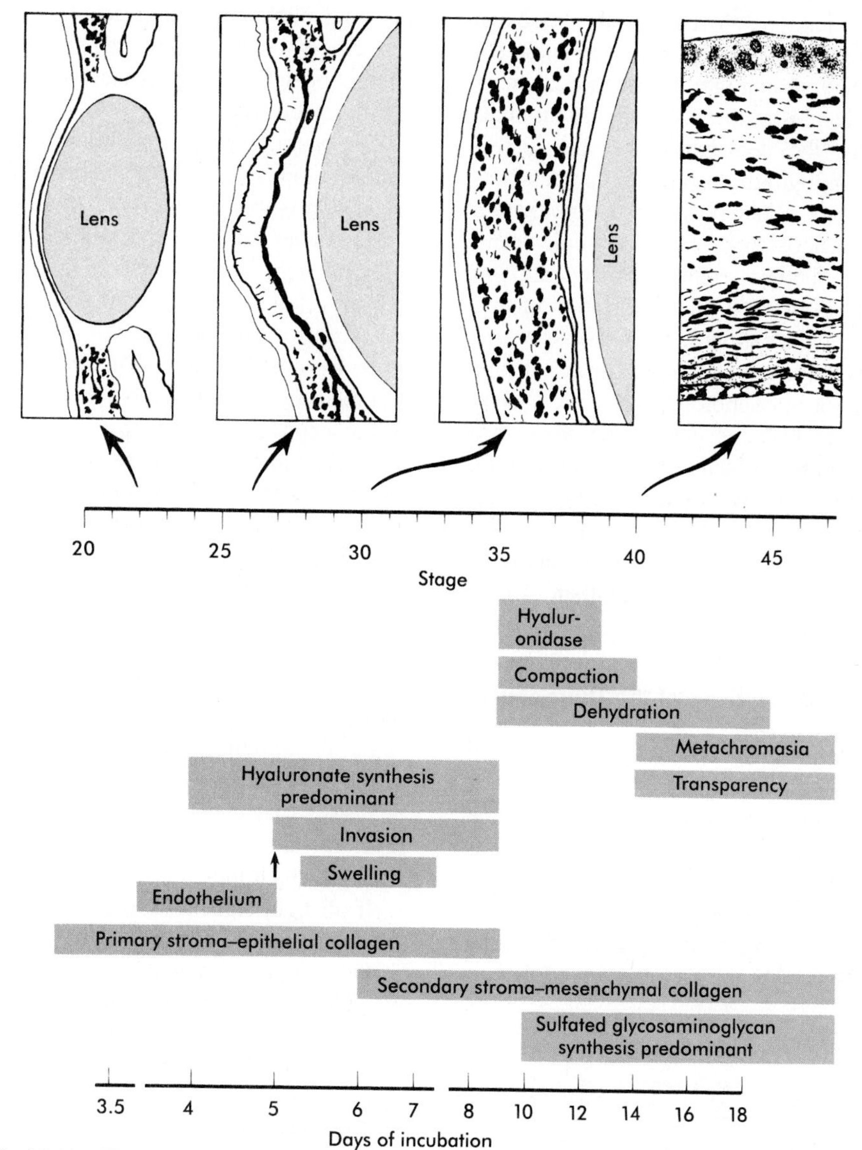

FIG. 14-11 The major events in corneal morphogenesis in the chick embryo. (Based on studies by Toole and Trelstad [1971]; Hay and Revel [1969].)

hyaluronidase, which breaks down much of the hyaluronic acid in the primary stroma. In other parts of the embryo (e.g., the limb bud), there is also a close correlation between high amounts of hyaluronic acid and cellular migration and a cessation of migration with its removal. With removal of hyaluronic acid, the cornea decreases in thickness. Once the migratory fibroblasts have settled, the primary corneal stroma is considered to have been transformed into the **secondary stroma.**

The fibroblasts of the secondary stroma contribute to its organization by secreting coarse collagen fibers to the stromal matrix. Nevertheless, prominent layers of acellular matrix continue to be secreted by both epithelial and endothelial cells of the cornea. These secretions provide the remaining layers that constitute the mature cornea. Listed from outside in, they are (1) the outer epithelium, (2) **Bowman's membrane,** (3) the secondary stroma, (4) **Descemet's membrane,** and (5) the corneal endothelium (see Fig. 14-10).

The final developmental changes in the cornea involve the formation of a transparent pathway free from optical distortion, through which light can enter the eye. A major change is a great increase in transparency, from about 40% to 100% transmission of light. This is accomplished by removing much of the water from the secondary stroma. Initial removal of water occurs with the degradation of much of the water-binding hyaluronic acid. The second phase of dehydration is mediated by **thyroxine,** which is secreted into the blood by the maturing thyroid gland. Thyroxine acts on the corneal endothelium by causing it to pump sodium from the secondary stroma into the anterior chamber of the eye. Water molecules follow the sodium ions, thus effectively completing the dehydration of the corneal stroma. The role of the thyroid gland in this process was demonstrated in two ways. When relatively mature thyroid glands were transplanted onto the extraembryonic membranes of young chick embryos, premature dehydration of the cornea took place. Conversely, the application of thyroid inhibitors retarded the clearing of the cornea.

The other late event in the cornea is a pronounced change in its radius of curvature in relation to that of the eyeball as a whole. This morphogenetic change, which involves a number of mechanical events including intraocular fluid pressure, allows the cornea to work with the lens in bringing light rays into focus on the retina. If irregularities develop in the curvature of the cornea during its final morphogenesis, the individual develops **astigmatism,** which causes distortions in the visual image.

The Retina and Other Derivatives of the Optic Cup

While the lens and cornea are taking shape, profound changes are also occurring in the optic cup (see Fig. 14-1). The inner layer of the optic cup thickens, and the epithelial cells begin a long process of differentiation into neurons and light receptor cells of the **neural retina.** The outer layer of the optic cup remains relatively thin and ultimately becomes transformed into the **pigment layer of the retina** (see Fig. 14-6). At the same time, the outer lips of the optic cup undergo a quite different transformation into the iris and ciliary body, which are involved in controlling the amount of light that enters the eye and the curvature of the lens, respectively.

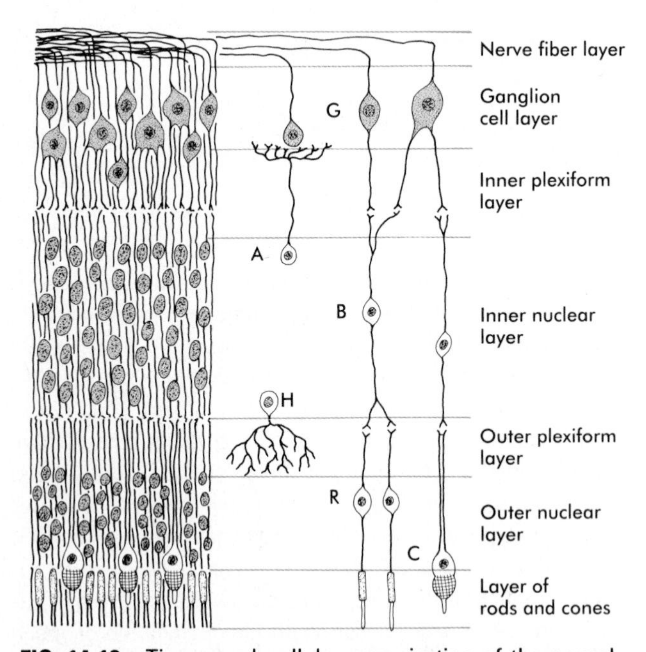

FIG. 14-12 Tissue and cellular organization of the neural retina of a human fetus. *A,* Amacrine cell; *B,* bipolar cell; *C,* cone; *G,* ganglion cell; *H,* horizontal cell; *R,* rod.

The neural retina is a multilayered structure; its embryonic development can be appreciated only after its adult organization is understood (Fig. 14-12). When seen in cross section under a microscope, the neural retina consists of alternating light- and dense-staining strips. These correspond to layers rich in nuclei or cell processes, respectively. The direct sensory pathway in the neural retina is a chain of three neurons that traverse the thickness of the retina. The first element of the chain is the light receptor cell, either a **rod** or a **cone.** A light ray that enters the eye passes through the entire thickness of the neural retina until it impinges on the outer segment of a rod or cone cell (photoreceptor) in the extreme outer layer of the retina. This stimulates that cell, the nucleus of which is located in the **outer nuclear layer.** The sensory cell sends a process toward the **outer plexiform layer,** where it synapses with a process from a bipolar cell located in the **inner nuclear layer.** The other process from the bipolar neuron leads into the **internal plexiform layer** and synapses with the third neuron in the chain, the **ganglion cell.** The bodies of the ganglion cells, which are located in the **ganglion cell layer,** send out long processes that course through the innermost **nerve fiber layer** toward their exit site from the eye, the optic nerve, through which they reach the brain.

If all light signals were processed only through simple three-link series of neurons in the retina, visual acuity would be much less than it actually is. There are many levels of integration by the time a visual pattern is stored in the visual cortex of the brain. The first is in the neural retina. At synapse sites in both the inner and outer plexiform layers of the retina, other cells such as **horizontal** and **amacrine cells** (see Fig. 14-12) are involved in the horizontal redistribution of the simple visual signal. This facilitates integration of components of a visual pattern. Another prominent cell type in the retina is the **Müller glial cell,** which appears to play a role similar to that of other glial cells in the central nervous system.

The Neural Retina. From the original columnar epithelium of the inner sensory layer of the optic cup (see Fig. 14-6), the primordium of the neural retina takes on the form of a mitotically active, thickened pseudostratified columnar epithelium that is organized similar to the early neural tube. During the early stages of development of the retina, its polarity becomes fixed according to the same axial sequence as that seen in the limbs (see Chapter 11). The nasotemporal (anteroposterior) axis is fixed first; this is followed by fixation of the dorsoventral axis. Finally, radial polarity is established.

As the number of cells in the early retina increases, the differentiation of cell types begins. There are two major gradients of differentiation in the retina. The first proceeds roughly linearly from the inner to the outer layers of the retina. The second moves horizontally from the center toward the periphery of the retina.

Differentiation in the first gradient begins with the appearance of ganglion cells and the early definition of the ganglion layer (Fig. 14-13). With the differentiation of the horizontal and amacrine cells, the inner and outer nuclear layers take shape. As the cells within the nuclear layers send out processes, the inner and outer plexiform layers become better defined. The bipolar neurons and the rod and cone cells differentiate last, thus completing the first gradient.

The horizontal gradient of differentiation of the neural retina is based on the outward spread of the first vertical gradient from the center to the periphery of the retina (Fig. 14-14). The retina cannot grow from within. Thus during

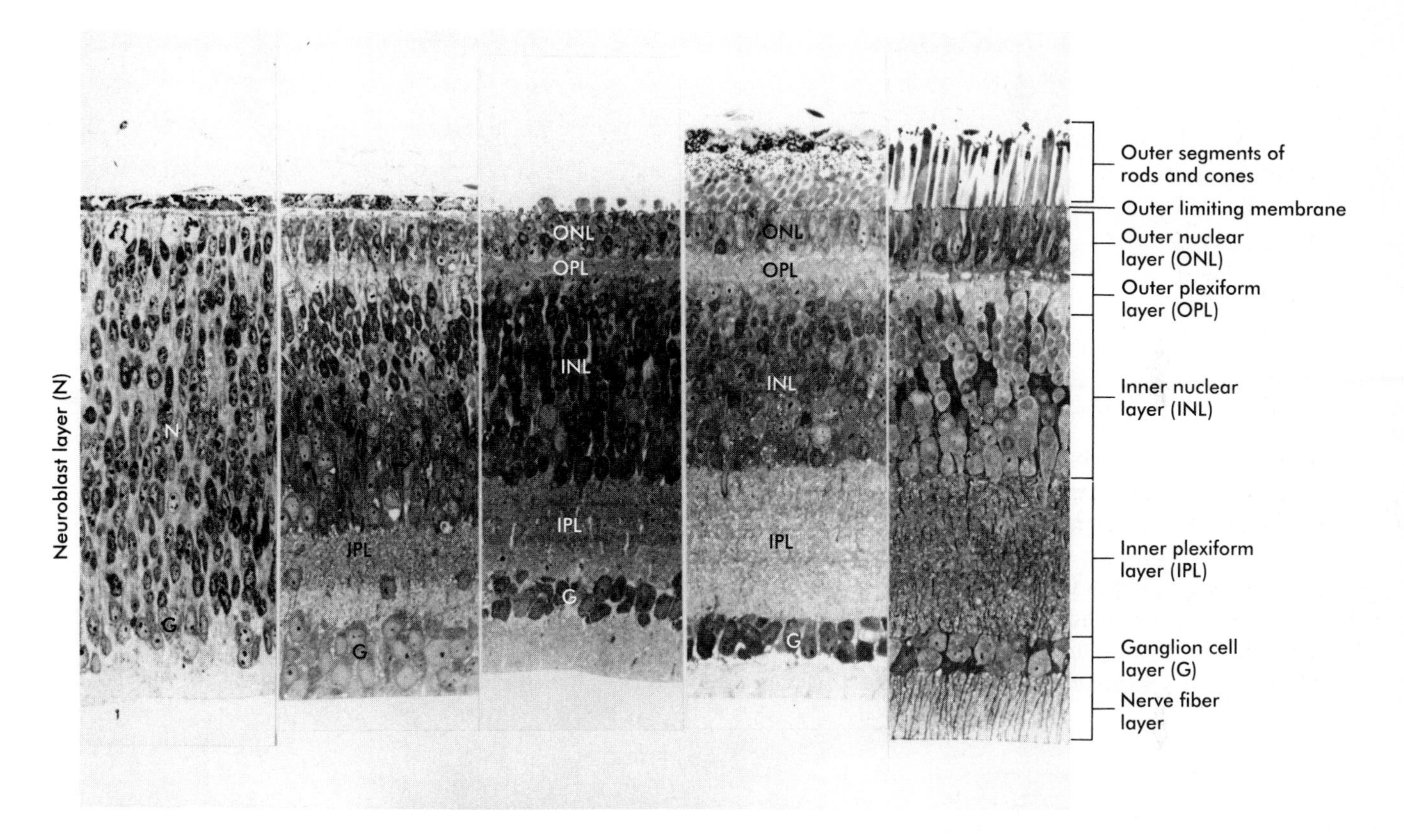

FIG. 14-13 The progressive development of retinal layers in the chick embryo. At the far left, the ganglion cell layer *(G)* begins to take shape from the broad neuroblast layer *(N)*. With time, further layers take shape until all layers of the retina are represented *(far right)*.
(From Sheffield J, Fischman *D: Zeitschr F Zellforsch* 194:407, 1970.)

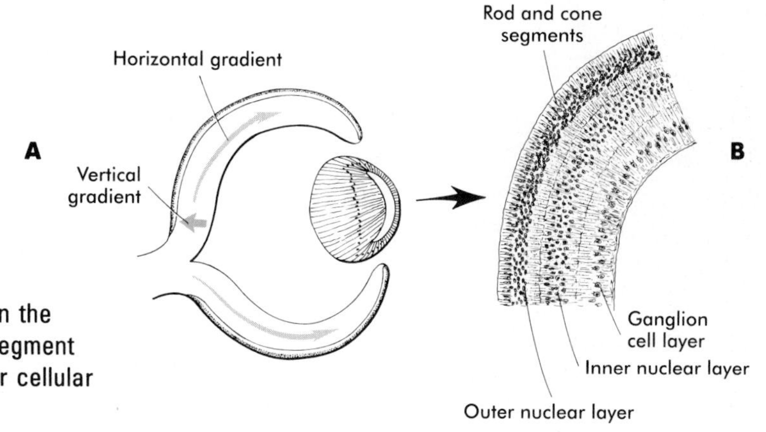

FIG. 14-14 **A,** Horizontal and vertical gradients in the differentiation of layers of the neural retina. **B,** Segment of the embryonic neural retina showing the major cellular layers.

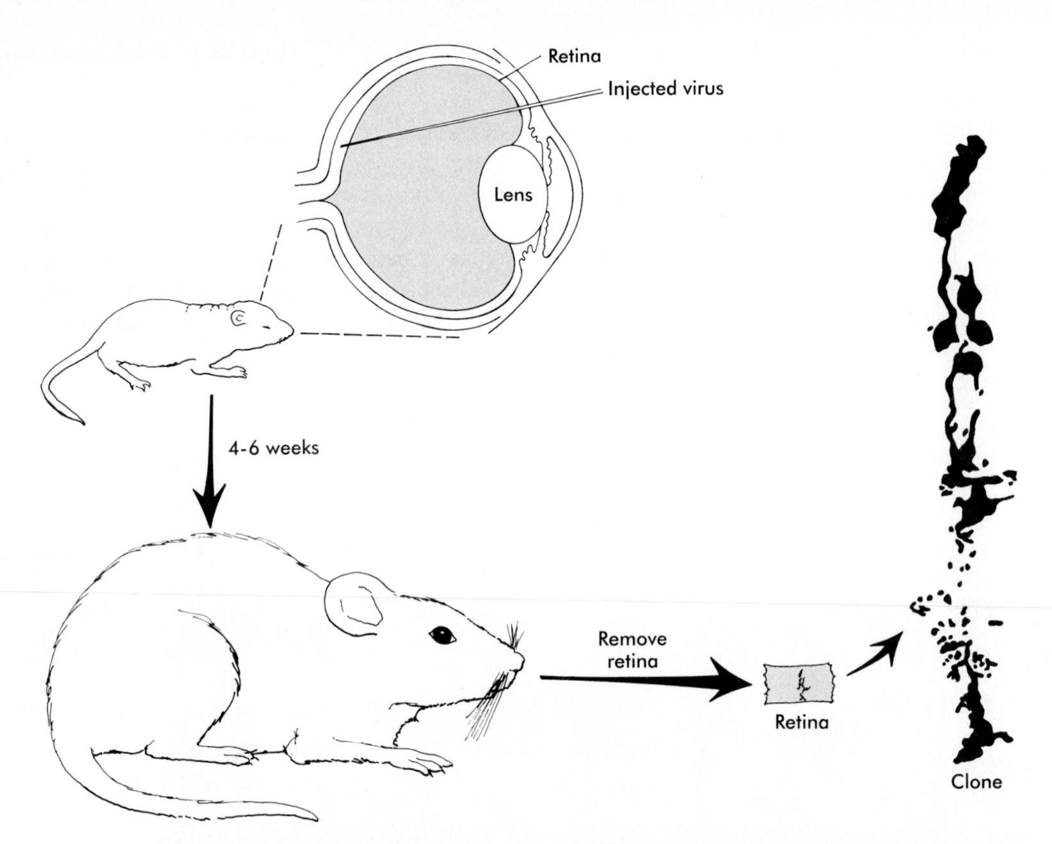

FIG. 14-15 An experiment illustrating origins and lineages of retinal cells in the rat. *Top:* Injection of a retroviral vector that includes the gene for β-galactosidase into the space between the neural and pigmented retinal layers. About 4 to 6 weeks later, the retinas were removed, fixed, and histochemically reacted for β-galactosidase activity. The drawing at the right illustrates a vertical clone of cells derived from a virally infected precursor cell. Several cell types (rods, a bipolar cell, and a Müller glial cell) constituted this clone.

(Modified from Turner DL, Cepko CL: *Nature* 328:131, 1987.)

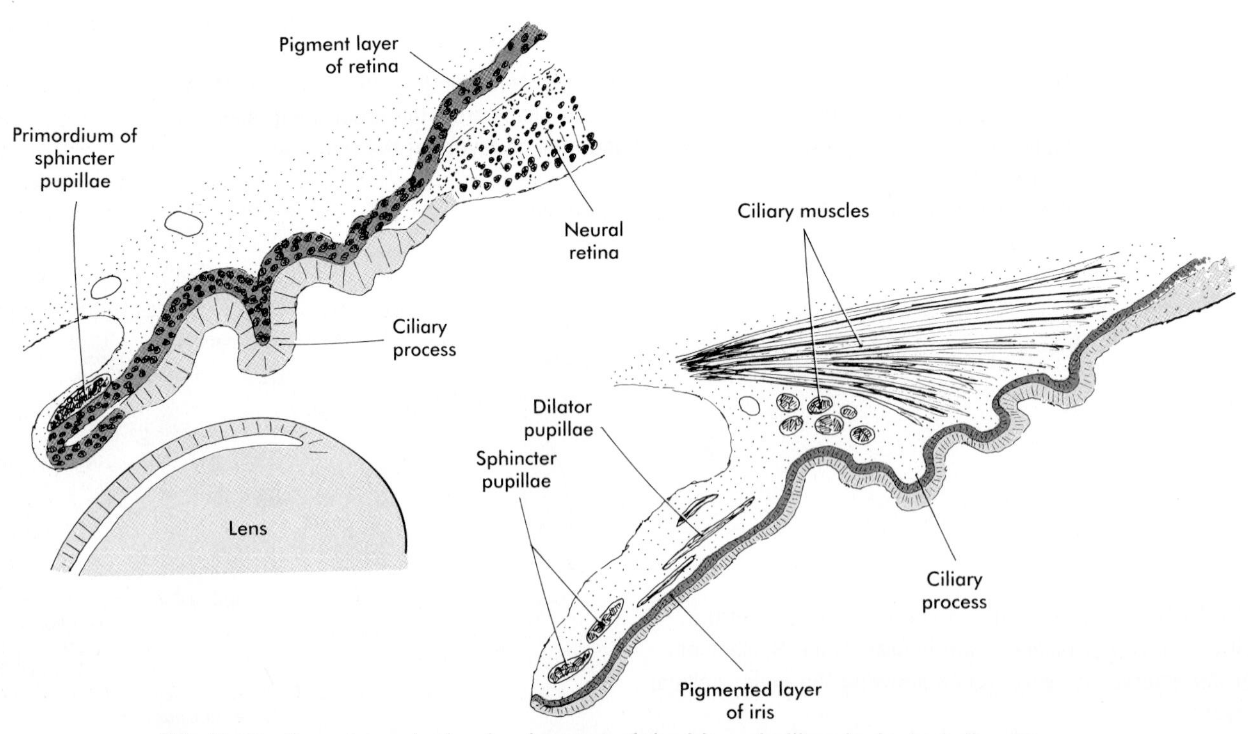

FIG. 14-16 Two stages in the development of the iris and ciliary body, including the sphincter and dilator pupillae muscles.

the phase of growth in the human eye (or throughout life in the case of continuously growing animals such as fish), developmentally immature retinal precursor cells along the edge of the retina undergo mitosis as an ever-expanding concentric ring on the periphery of the retina. Just inside the ring of mitosis, cellular differentiation takes place in a manner corresponding roughly to that of the vertical gradient.

Cell lineage experiments involving the use of retroviral or other tracers (e.g., horseradish peroxidase) introduced into neuronal precursors in the early retina have revealed two significant cellular features of retinal differentiation. First, progeny of a single labeled cell are distributed in a remarkably straight radial pattern following the vertical axis of retinal differentiation. There seems to be little lateral mixing among columns of retinal cells (Fig. 14-15). The second cellular feature of retinal differentiation is that a single labeled precursor cell can give rise to more than one type of differentiated retinal cell.

A later stage in retinal differentiation is the growth of axons from the ganglion cells along the innermost layer of the retina toward the optic stalk. Once the axons reach the optic stalk, they grow into it and make their way toward the visual centers of the brain. During this axonal ingrowth, the axons use a variety of local and positional cues to make very precise connections with the brain. Much of the basic research on the development of retinal connections with the brain has been conducted on fish and amphibian embryos. In general, the lessons learned on these forms are applicable to development of the human visual system.

The Iris and Ciliary Body. At the lip of the optic cup where the developing neural and pigment retinas meet, differentiation of the **iris** and **ciliary body** occurs. Rather than being sensory in function, these structures are involved in modulating the amount and character of light that ultimately impinges on the retina. The iris partially encircles the outer part of the lens, and through contraction or relaxation, it controls the amount of light passing through the lens (Fig. 14-16). The iris contains an inner unpigmented epithelial layer and an outer pigmented layer, which are continuous with the neural and pigmented layers of the retina, respectively. The **stroma of the iris,** which is superficial to the outer pigmented layer of the iris, is of neural crest origin and secondarily migrates into the iris. Within the stroma of the iris lie the primordia of the **sphincter pupillae** and **dilator pupillae** muscles. These muscles are unusual because they are of neurectodermal origin; they seem to arise from the anterior epithelial layer of the iris.

Between the iris and neural retina lies the ciliary body, a muscle-containing structure that is connected to the lens by radial sets of fibers called the **suspensory ligament of the lens.** Through the functions of the ciliary musculature acting through the suspensory ligament, the ciliary body modulates the shape of the lens in focusing light rays on the retina. Normal development of the ciliary body appears to depend on an appropriate amount of intraocular fluid pressure. If some of the fluid of the developing eye is shunted off, a defective ciliary body results.

Eye color is due to levels and distribution of pigmentation in the iris. The bluish color of the iris in most newborns is the result of the intrinsic pigmentation of the outer pigmented layer of the iris. Pigment cells also appear in the iridial stroma in front of the pigmented epithelium. The greater the density of pigment cells in this area, the browner will be the eye color. Definitive pigmentation of the eye gradually develops over the first 6 to 10 months of postnatal life.

The Vitreous Body and Hyaloid Artery System

During early development of the retina, a loose mesenchyme invades the cavity of the optic cup and forms a loose fibrillar mesh along with a gelatinous substance that fills the space between the neural retina and lens. This material is called the **vitreous body.**

During much of embryonic development the vitreous body is supplied by the hyaloid artery and its branches (Fig. 14-17). The hyaloid artery enters the eyeball through the choroid fissure of the optic stalk (see Fig. 14-5), passes through the retina and vitreous body, and terminates in branches to the posterior wall of the lens. As development progresses, the portions of the hyaloid artery in the vitreous body regress, leaving a **hyaloid canal.** The more proximal part of the hyaloid arterial system persists as the **central artery of the retina** and its branches.

The Choroid Coat and Sclera

Outside the optic cup lies a layer of mesenchymal cells, largely of neural crest origin. Reacting to an inductive influence from the pigmented epithelium of the retina, these cells differentiate into structures that provide vascular and mechanical support for the eye. The innermost cells of this layer differentiate into a highly vascular tunic called the **choroid coat** (see Fig. 14-6) and a white, densely collagenous covering known as the **sclera.** The opaque sclera, which serves as a tough outer coating of the eye, is continuous with the cornea. The extraocular muscles, which provide gross movements to the eyeball, attach to the sclera.

The Eyelids and Lacrimal Glands

The eyelids first become apparent during the seventh week as folds of skin that grow over the cornea (Figs. 14-6 and 14-8, *A*). Once their formation has commenced, the eyelids rapidly grow over the eye until they meet and fuse with one another by the end of the ninth week (Fig. 14-18, *B*). The temporary fusion involves only the epithelial layers of the eyelids, resulting in a persisting epithelial lamina between them. Before the eyelids reopen, eyelashes and the

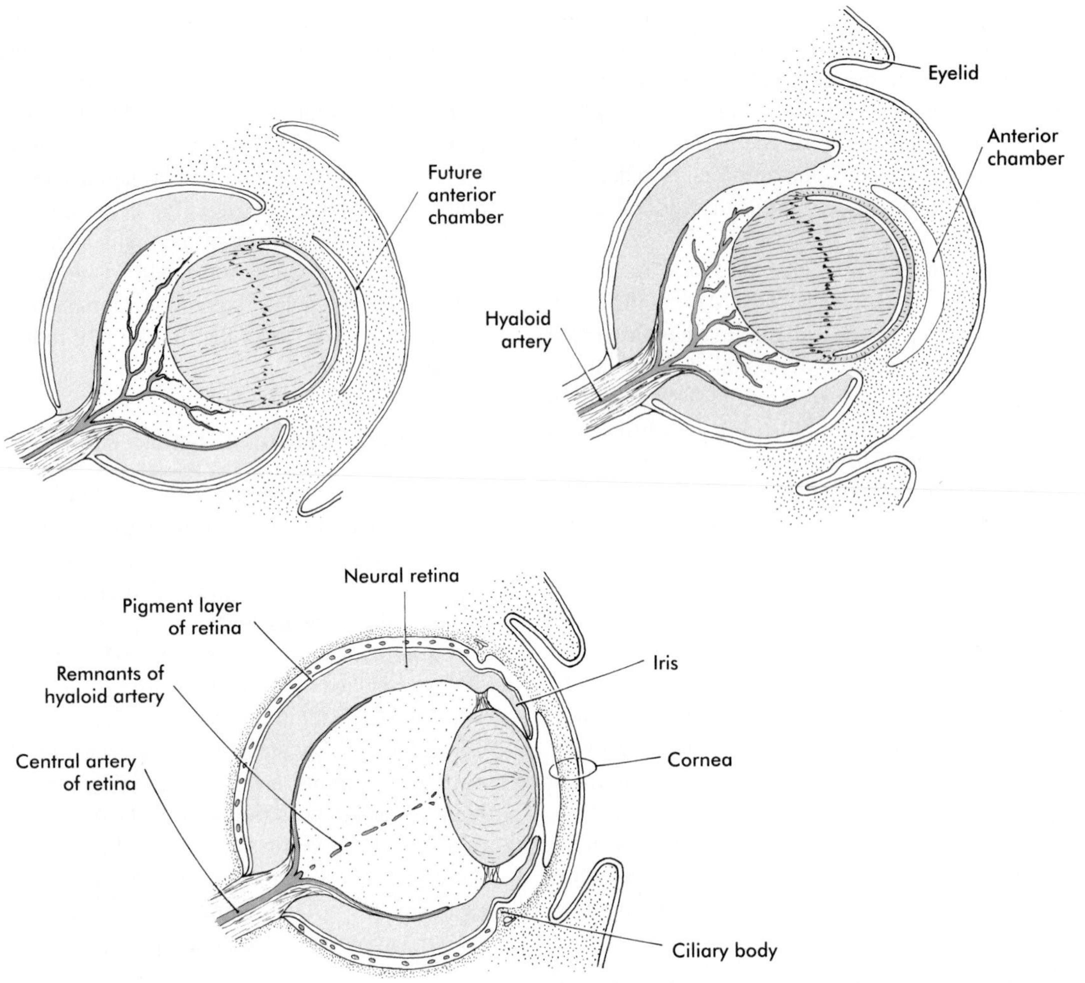

FIG. 14-17 Stages in development and regression of the hyaloid artery in the embryonic eye.

small glands that lie along the margins of the lids begin to differentiate from the common epithelial lamina. Although signs of loosening of the epithelial union of the lids can be seen in the sixth month, reopening of the eyelids normally does not occur until well into the seventh month of pregnancy.

The space between the front of the eyeball and the eyelids is known as the **conjunctival sac.** Multiple epithelial buds grow from the lateral surface ectoderm at about the time when the eyelids fuse. These buds differentiate into the lacrimal glands, which produce a watery secretion that bathes the outer surface of the cornea when mature. This secretion ultimately passes into the nasal chamber by way of the **nasolacrimal duct** (see Chapter 15). The lacrimal glands are not fully mature at birth, and newborns typically do not produce tears when crying. The glands begin to function in lacrimation at about 6 weeks.

Congenital Malformations of the Eyes

Despite the many types of malformations of the eye and visual system, the incidence of most individual types of defects is uncommon. Examples of only a few of the many ocular malformations are given.

Anophthalmos and Microphthalmos. Anophthalmos, the absence of an eye, is very rare and can normally be attributed to lack of formation of the optic vesicle. Since this structure acts as the inductive trigger for much of subsequent eye development, many local inductive interactions involved in the formation of eye structures fail to occur. **Microphthalmos,** which can range from an eyeball that is slightly smaller than normal to one that is almost vestigial, can be associated with a variety of other defects or causes, including intrauterine infections (Fig. 14-19, *A*).

Coloboma of the Iris. Nonclosure of the choroid fissure of the iris during the sixth or seventh week results in its

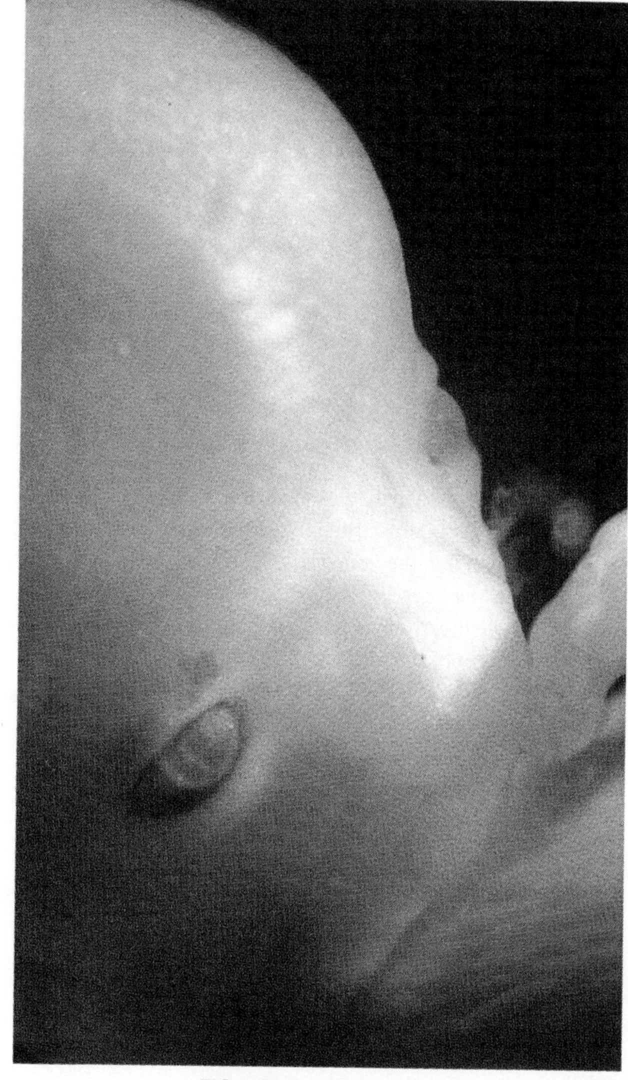

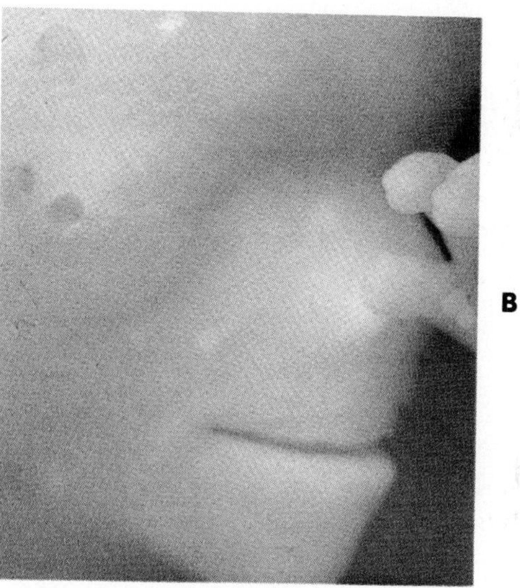

FIG. 14-18 Human fetuses showing **(A)** open (days 44 to 46) and **(B)** closed (week 9) eyelids.

(From England M: *Colour atlas of life before birth,* St Louis, 1983, Mosby.)

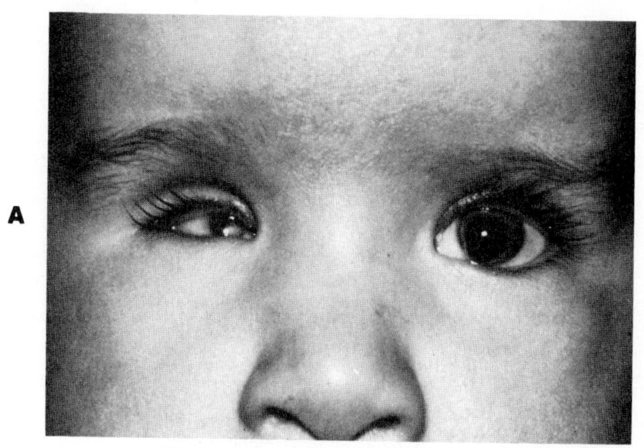

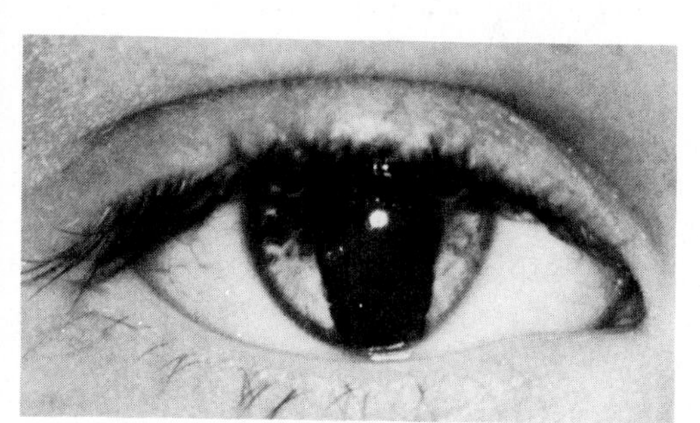

FIG. 14-19 **A**, Microphthalmos of the right eye. **B**, Congenital coloboma of the iris. The fissure is in the region of closure of the choroid fissure.

(**A** from Smith B: *Ophthalmic plastic and reconstructive surgery,* vol 2, St Louis, 1987, Mosby. **B** from Newell F: *Ophthalmology: principles and concepts,* St Louis, 1986, Mosby.)

persistence as a defect called **coloboma iridis** (Fig. 14-19, *B*). The location of colobomas of the iris (typically at the 5 o'clock position in the right eye and the 7 o'clock position in the left) marks the position of the embryonic choroid fissure.

Congenital Cataract. Cataract is a condition characterized by opacity of the lens of the eye. Not so much a structural malformation as a dysplasia, congenital cataracts first came into prominence as one of the triad of defects resulting from exposure of the embryo to the rubella virus.

THE EAR

The ear is a complex structure consisting of three major subdivisions—the external, middle, and internal ear. The **external ear** consists of the **pinna** (auricle), the **external auditory meatus** (external ear canal), and the outer layers of the **tympanic membrane** (eardrum) and functions principally as a sound-collecting apparatus. The middle ear functions as a transmitting device. This function is served by the chain of three middle ear ossicles, which connect the inner side of the tympanic membrane to the oval window of the inner ear. Other components of the middle ear are the middle ear cavity *(tympanic cavity)*, the **auditory tube** (eustachian tube), the middle ear musculature, and the inner layer of the tympanic membrane. The inner ear contains the primary sensory apparatus, which is involved with both hearing and balance. These functions are served by the **cochlea** and **vestibular apparatus,** respectively.

From an embryological standpoint, the ear has a dual origin. The inner ear arises from a thickened ectodermal placode at the level of the rhombencephalon. The structures of the middle and external ear are derivatives of the first and second branchial arches and the intervening first branchial cleft and pharyngeal pouch.

Development of the Inner Ear

Development of the ear begins with preliminary inductions of the surface ectoderm, first by notochord (chordamesoderm) and then by the paraxial mesoderm (Fig. 14-20). These inductions prepare the ectoderm for a third induction, in which the rhombencephalon induces the adjacent surface ectoderm to thicken and form the **otic placode** (Fig.

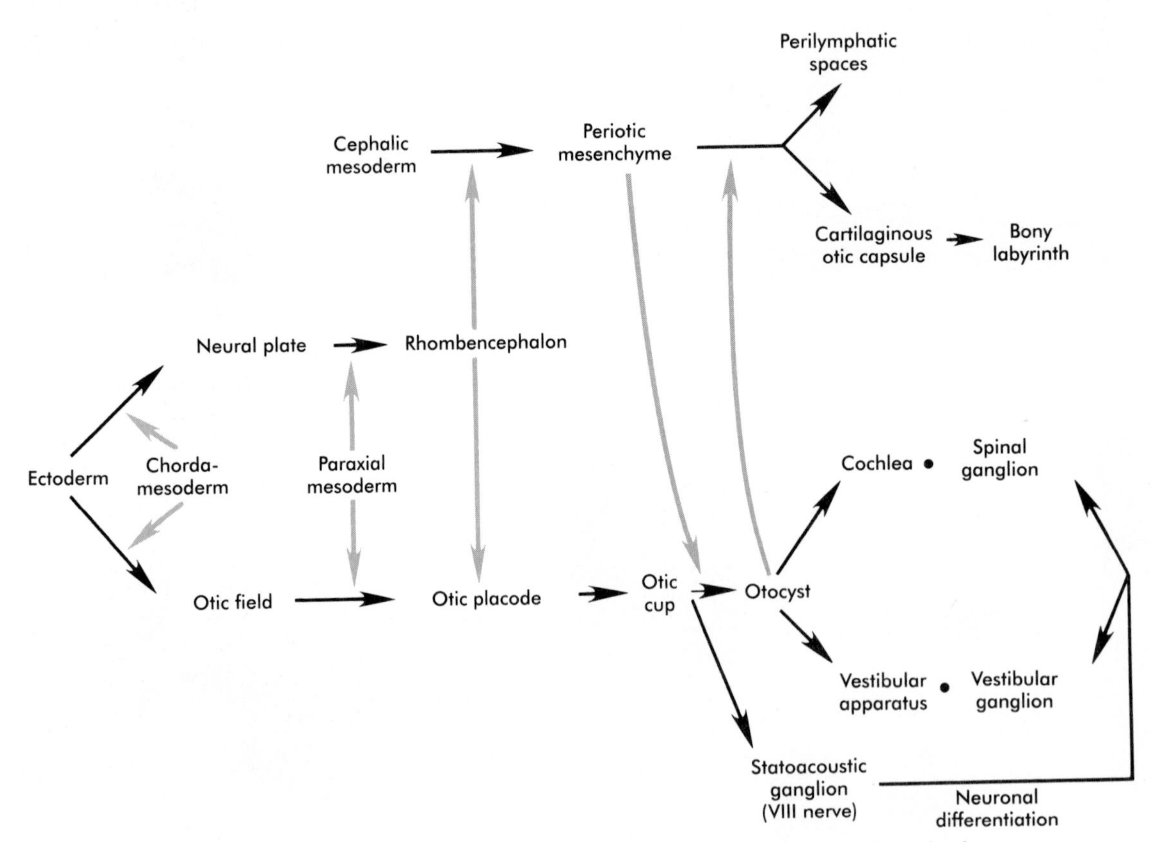

FIG. 14-20 Flow chart of major inductive events and tissue transformations in the developing ear. Colored arrows refer to inductive events.
(Based on studies by McPhee and van de Water [1988].)

14-21). Late in the fourth week the otic placode invaginates and then separates from the surface ectoderm to form the **otic vesicle,** or **otocyst.**

The otic vesicle soon begins to elongate, forming a dorsal vestibular region and a ventral cochlear region (Fig. 14-22, *A*). Quite early, the **endolymphatic duct** arises as a short, fingerlike projection from the dorsomedial surface of the otocyst (Fig. 14-22, *B*). At about 5 weeks, the appearance of two ridges in the vestibular portion of the otocyst foreshadows the formation of two of the **semicircular ducts** (Fig. 14-22, *C*). The cochlear part of the otocyst begins to elongate in a spiral, having attained one complete revolution at 8 weeks and two revolutions by 10 weeks (Fig. 14-22, *C* through *F*). The last half turn of the cochlear spiral (a total of 2½ turns) is not completed until 25 weeks.

The inner ear (membranous labyrinth) is encased in a capsule of skeletal tissue that begins as a condensation of mesodermal mesenchyme around the developing otocyst at 6 weeks gestation. The process of encasement of the otocyst begins with an induction of the surrounding mesenchyme by the epithelium of the otocyst (see Fig. 14-20). This induction stimulates the mesenchymal cells to form a cartilaginous matrix (starting at about 8 weeks). The capsular cartilage then serves as a template for the later formation of the true bony labyrinth. The conversion from the cartilaginous to the bony labyrinth occurs between 16 and 23 weeks gestation.

The sensory neurons that make up the eighth cranial nerve (specifically the **statoacoustic ganglion**) arise from a portion of the medial wall of the otocyst. The cochlear part **(spiral ganglion)** of the eighth nerve fans out in close association with the sensory cells (collectively known as the **organ of Corti**) that develop within the cochlea. Neural crest cells invade the developing statoacoustic ganglion and ultimately form the satellite and supporting cells within it. The sensory cells of the organ of Corti are also derived from the epithelium of the otocyst. They undergo a very complex pattern of differentiation (Fig. 14-23). As in other sensory systems, highly regulated developmental controls ensure precise matching between sensory cells designed to receive sound waves at different frequencies and the neurons that transmit the signals to the brain.

Development of the Middle Ear

Formation of the middle ear is intimately associated with developmental changes in the first and second pharyngeal arches (see Chapter 15). Both the **middle ear cavity** and the **auditory tube** arise from an expansion of the first pharyngeal pouch called the **tubotympanic sulcus** (Fig. 14-24). Such an origin ensures that the entire middle ear cavity and auditory tube are lined with an endodermally derived epithelium. By the end of the second month of pregnancy, the blind end of the tubotympanic sulcus approaches

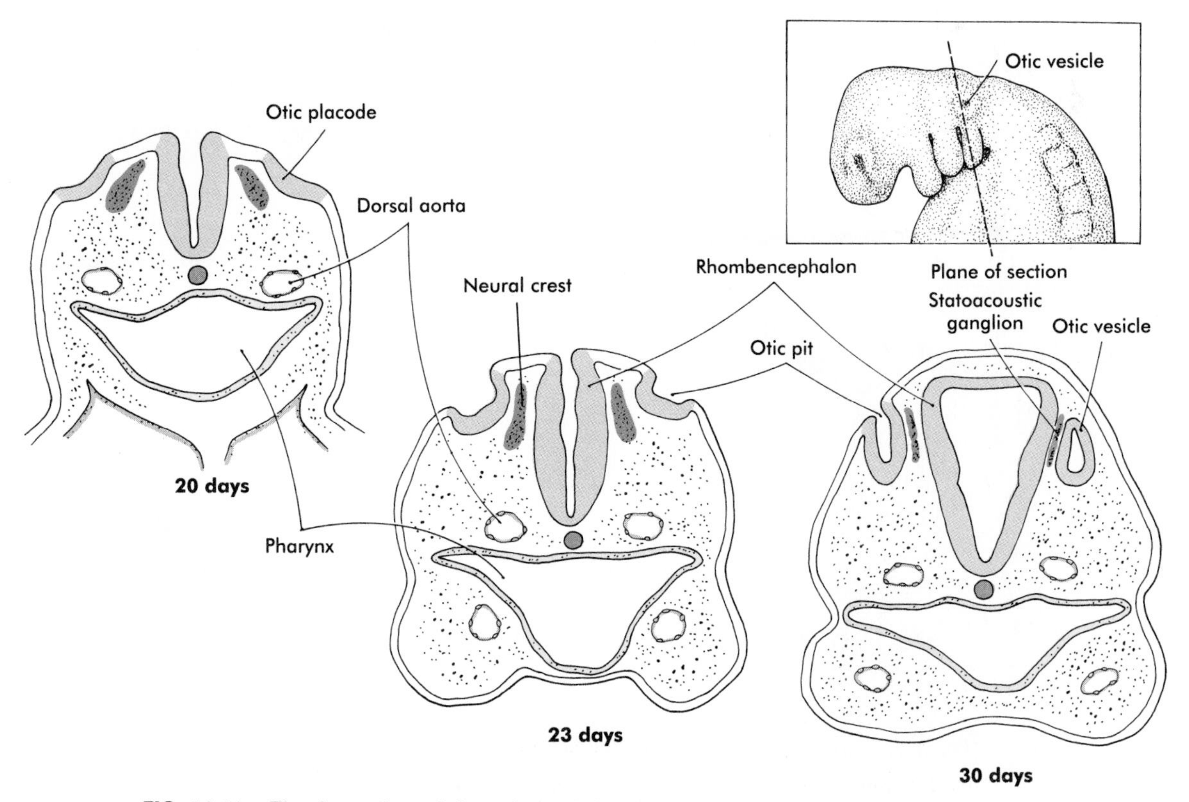

FIG. 14-21 The formation of the otic vesicles from thickened otic placodes.

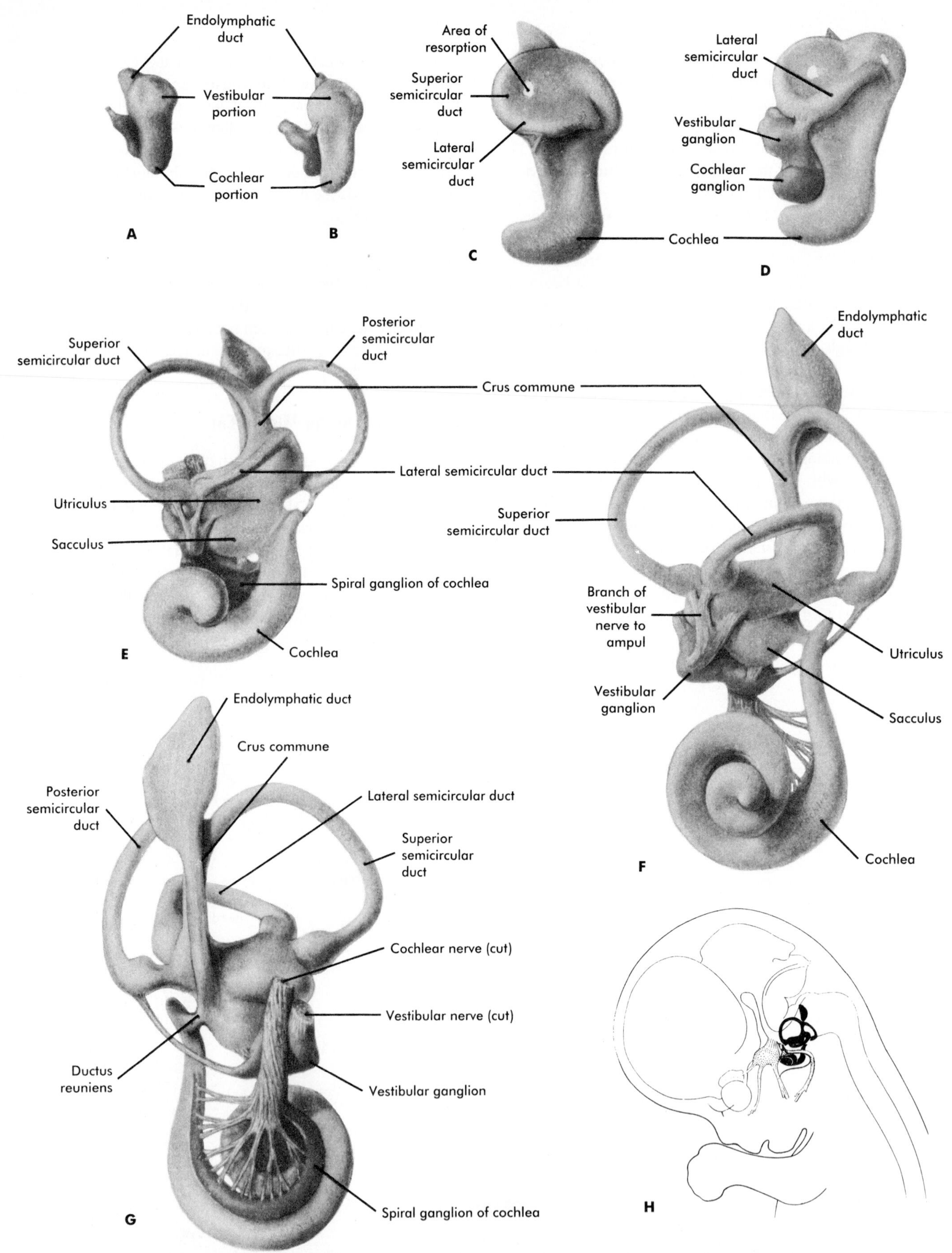

FIG. 14-22 Development of the human inner ear. **A,** At 28 days. **B,** At 33 days. **C,** At 38 days. **D,** At 41 days. **E,** At 50 days. **F,** At 56 days, lateral view. **G,** At 56 days, medial view. **H,** Central reference drawing at 56 days.

(From Carlson B: *Patton's foundations of embryology,* ed 5, New York, 1988, McGraw Hill.)

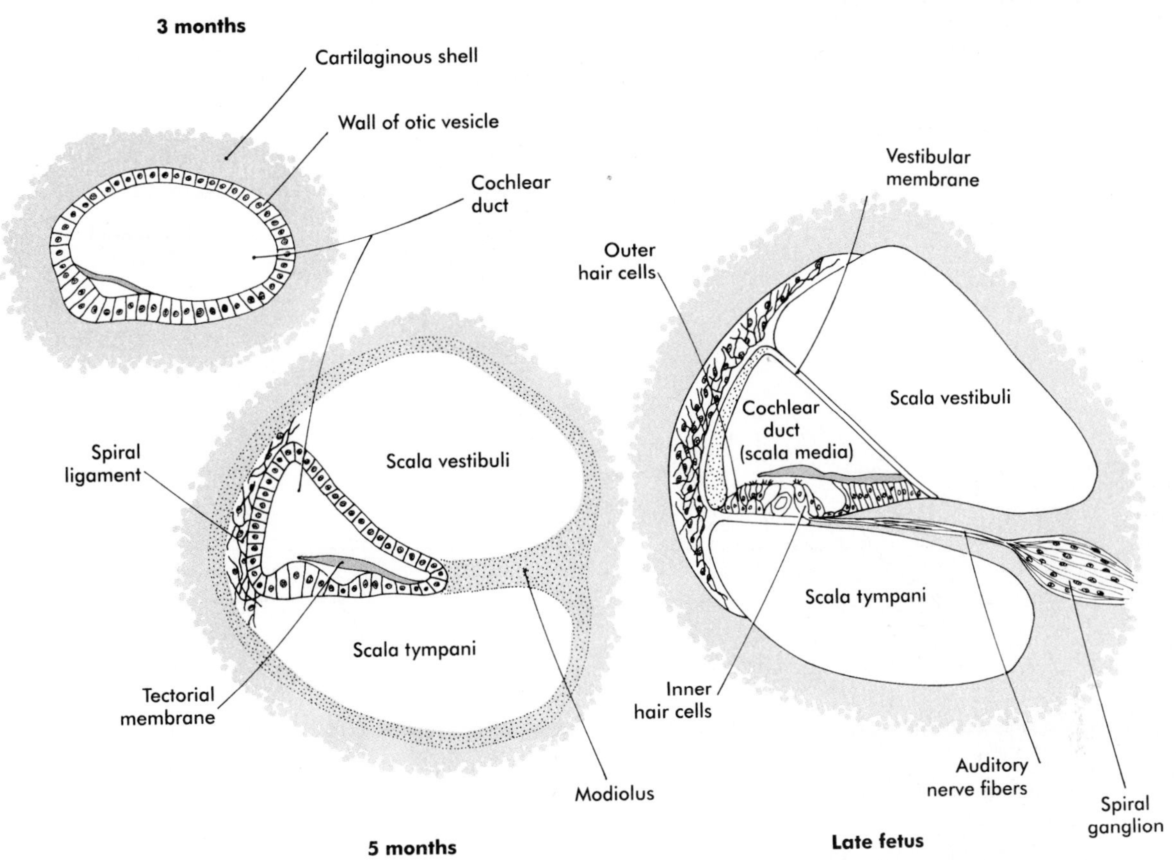

FIG. 14-23 Cross sections through the developing organ of Corti.

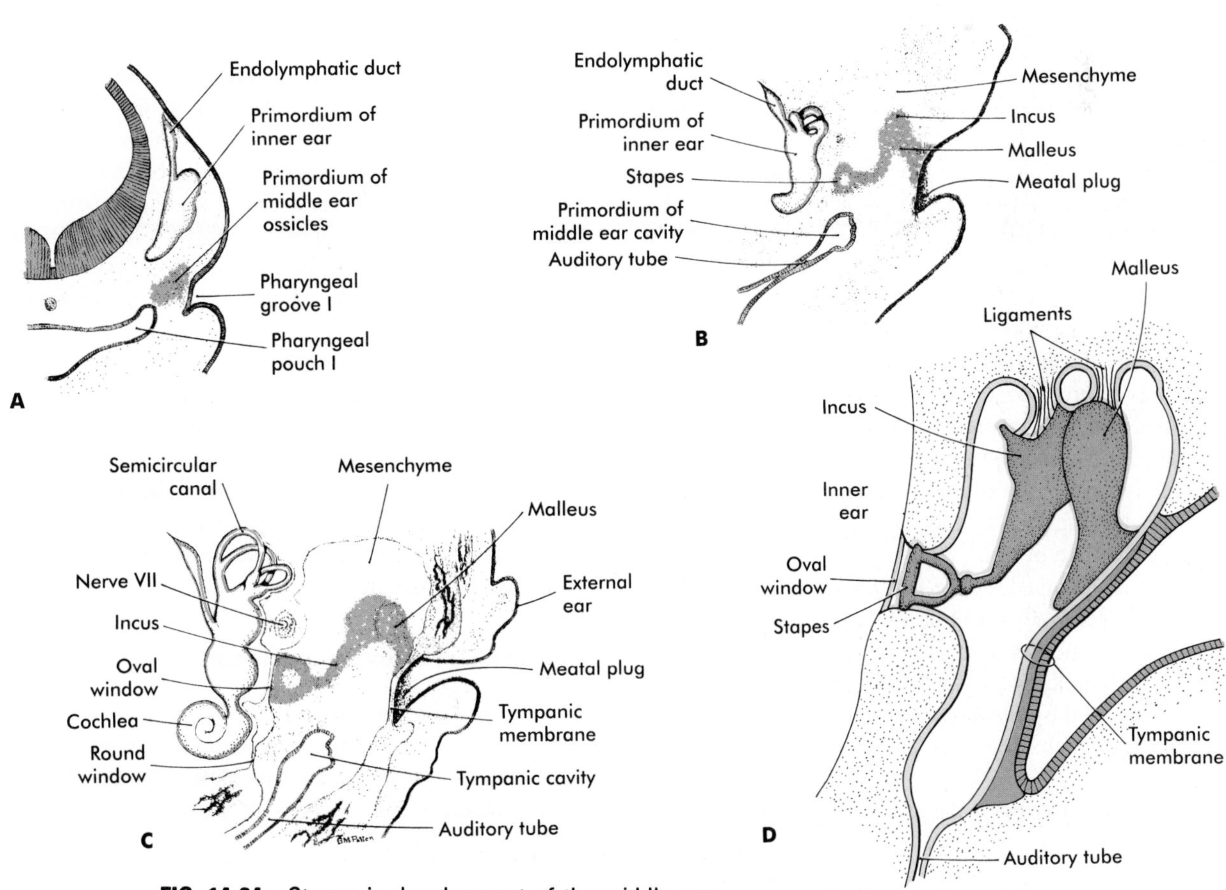

FIG. 14-24 Stages in development of the middle ear.

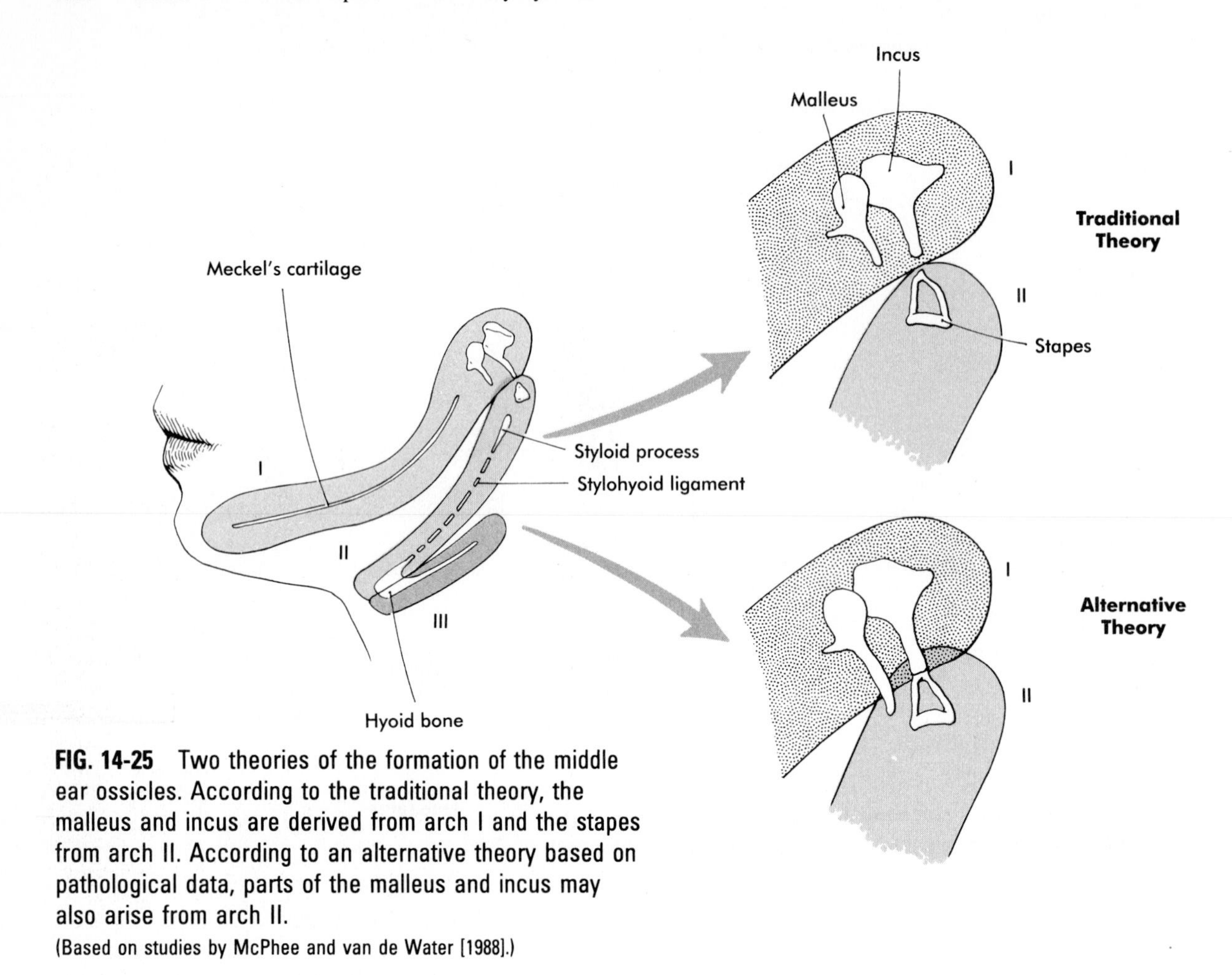

FIG. 14-25 Two theories of the formation of the middle ear ossicles. According to the traditional theory, the malleus and incus are derived from arch I and the stapes from arch II. According to an alternative theory based on pathological data, parts of the malleus and incus may also arise from arch II.
(Based on studies by McPhee and van de Water [1988].)

the innermost portion of the first pharyngeal cleft. Nonetheless, these two structures are still separated by a mass of mesenchyme.

Just dorsal to the end of the tubotympanic sulcus, a conspicuous condensation of mesodermal mesenchyme appears at 6 weeks and gradually takes on the form of the middle ear ossicles. These ossicles, which lie in a bed of very loose embryonic connective tissue, extend from the inner layer of the tympanic membrane to the oval window of the inner ear. Although the middle ear cavity is surrounded by the developing temporal bone, the future middle ear cavity remains filled with loose mesenchyme until late in pregnancy. During the eighth and ninth month, programmed cell death and other resorptive processes gradually clear the middle ear cavity, leaving the auditory ossicles suspended within it. Even at the time of birth, remains of the middle ear connective tissue may dampen the free movement of the auditory ossicles. Free movement of the auditory ossicles is acquired within 2 months after birth. Coincident with the removal of the connective tissue of the middle ear cavity is the expansion of the endodermal epithelium of the tubotympanic sulcus, which ultimately lines the entire middle ear cavity.

The middle ear ossicles themselves have a dual origin. According to comparative anatomical evidence, the **malleus** and **incus** arise from mesoderm of the first branchial arch, whereas the **stapes** originates from second arch mesoderm (Fig. 14-25). However, on the basis of studies of certain human genetic disorders, a second hypothesis has been proposed. According to this, only the head of the malleus and the body of the incus arise from first arch mesoderm, and the remainder of these bones, plus the stapes, are of second arch origin.

Two middle ear muscles help to modulate the transmission of auditory stimuli through the middle ear. The **tensor tympani muscle,** which is attached to the malleus, is derived from first arch mesoderm and is correspondingly innervated by the trigeminal nerve (V). The **stapedius muscle** is associated with the stapes, is of second arch origin, and is innervated by the facial nerve (VII), which supplies derivatives of that arch.

Development of the External Ear

The external ear (pinna) is derived from mesenchymal tissue of the first and second branchial arches that flank the first (hyomandibular) branchial cleft. During the second month, three nodular masses of mesenchyme **(auricular**

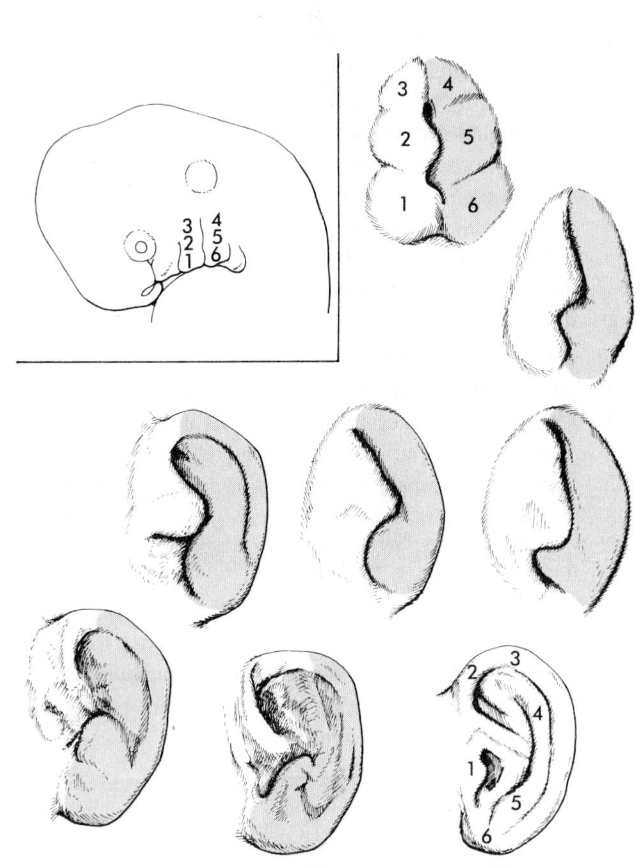

FIG. 14-26 Stages in development of the external ear. Components derived from the mandibular arch (I) are unshaded; those derived from the hyoid arch (II) are shaded.

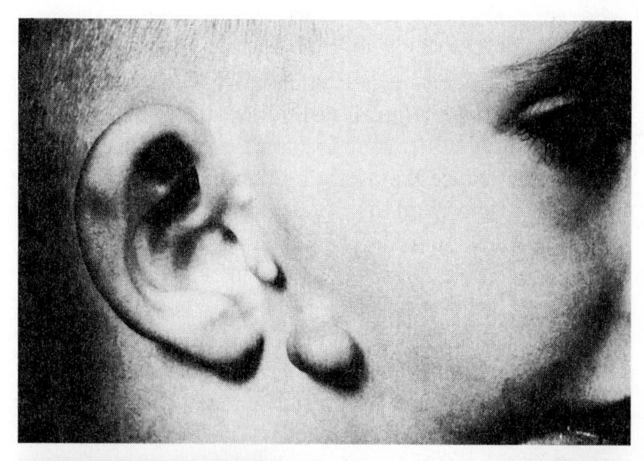

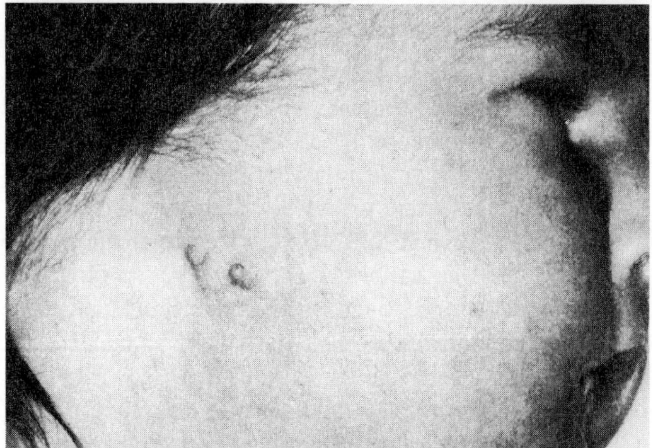

FIG. 14-27 **A,** Auricular anomalies and tags associated with the mandibular arch *(I)* component of the external ear. **B,** Anotia. The external ear is represented only by a couple of small tags.
(Courtesy Mason Barr, Ann Arbor, Mich.)

hillocks) take shape along each side of the first branchial cleft (Fig. 14-26). The auricular hillocks enlarge asymmetrically and ultimately coalesce to form a recognizable external ear. During its formation, the pinna shifts from the base of the neck to its normal location on the side of the head. Because of its intimate association with the pharyngeal arches and its complex origin, the external ear is a sensitive indicator of abnormal development in the pharyngeal region. Other anomalies of the first and second arches are often attended by misshapen or abnormally located external ears.

The external auditory meatus takes shape during the end of the second month by an inward expansion of the first branchial cleft. Early in the third month the ectodermal epithelium of the forming meatus proliferates, forming a solid mass of epithelial cells called the **meatal plug** (see Fig. 14-24). Late during the fetal period (at 28 weeks), a channel within the meatal plate extends the existing external auditory meatus to the level of the tympanic membrane.

The external ear and external auditory meatus are very sensitive to drugs. Exposure to agents such as streptomycin, thalidomide, and salicylates during the first trimester can cause agenesis or atresia of both of these structures.

Congenital Malformations of the Ear

Congenital Deafness. Many disturbances of development of the ear can lead to deafness. Conditions such as **rubella** can lead to maldevelopment of the organ of Corti, with inner ear deafness resulting. Abnormalities of the middle ear ossicles or ligaments, which can be associated with anomalies of the first and second arches, can interfere with transmission, resulting in middle ear deafness. Agenesis or major atresias of the external ear can cause deafness by interfering with the primary collection of sound waves.

Auricular Anomalies. Because of the multiple origins of its components, there is great variety in the normal form of the pinna. Variations include obvious malformations such as **auricular appendages,** or **sinuses** (Fig. 14-27). Many malformations of the external ears are not functionally important but attend other developmental anomalies such as malformations of the kidneys and pharyngeal arches.

SUMMARY

1. The eye begins as a lateral outpocketing (optic groove) of the lateral wall of the diencephalon. The optic grooves enlarge to form the optic vesicles, which induce the overlying ectoderm to form the lens primordium. The optic stalk, which connects the optic cup to the diencephalon, forms a groove containing the hyaloid artery that supplies the developing eye.
2. The lens forms from an ectodermal thickening that invaginates to form a lens vesicle. The cells of the inner wall of the lens vesicle elongate and synthesize lens-specific crystallin proteins. In the growing lens the inner epithelium forms a spherical mass of banana-shaped lens fibers (epithelial cells). The anterior lens epithelium consists of cuboidal epithelial cells. The overall polarity of the lens is under the influence of the retina.
3. The cornea is formed through induction of the surface ectoderm by the lens. After induction, the basal ectodermal cells secrete an extracellular matrix that serves as a substrate for the migration of neural crest cells forming the corneal endothelial layer. The corneal endothelial cells secrete large amounts of hyaluronic acid into the early cornea. This permits the migration of a second wave of neural crest cells into the cornea. These fibroblastlike cells secrete collagen fibers into the coarse corneal stroma matrix. Under the influence of thyroxine, water is removed from the corneal matrix and it becomes transparent.
4. The neural retina differentiates from the inner layer of the optic cup. The outer layer forms the pigment layer of the retina. The neural retina is a complex multilayered structure with three layers of neurons connected by cellular processes. Cellular differentiation in the neural retina follows both vertical and horizontal gradients. Cell processes grow from retinal neurons through the optic stalk to make connections with optic centers within the brain.
5. The iris and ciliary body form from the outer edge of the optic cup. Sphincter and dilator pupillae muscles form within the iris. Eye color is related to levels and distribution of pigmentation within the iris. Outside the optic cup, mesenchyme differentiates into a vascular choroid coat and a tough collagenous sclera. Eyelids begin as folds of skin that grow over the cornea and then fuse, closing off the eyes.
6. The developing eyes are sensitive to a number of teratogens and intrauterine infections. Exposure to these can cause microphthalmia or congenital cataracts. Nonclosure of the choroid fissure results in coloboma.
7. The inner ear arises by an induction of the surface ectoderm by the developing brain. Steps in its formation include ectodermal thickening (placode), invagination to form an otic vesicle, and later growth and morphogenesis into auditory (cochlea) and vestibular (semicircular canals) portions.
8. Development of the middle ear is associated with the first pharyngeal cleft and the arches on either side. Middle ear ossicles and associated muscles take shape within the middle ear cavity.
9. The external ear arises from six modular masses of mesenchyme that take shape in the pharyngeal arch tissue surrounding the first pharyngeal cleft.
10. Congenital deafness can occur after a number of intrauterine disturbances such as a rubella infection. Structural anomalies of the external ear are common.

REVIEW QUESTIONS

1. During a routine physical examination, an infant was found to have a small segment missing from the lower part of one iris. What is the diagnosis, what is the basis for the condition, and why might the infant be sensitive to bright light?
2. Why does a person sometimes get a runny nose while crying?
3. What extracellular matrix molecule is often associated with massive cellular migrations, and where does such an event occur in the developing eye?
4. Why is the hearing of a newborn often not as acute as it is a few months later?
5. Why are malformations or hypoplasia of the lower jaw commonly associated with abnormalities in the shape or position of the ears?

REFERENCES

Anniko M: *Embryonic development of vestibular sense organs and their innervation.* In Romand R, ed: *Development of auditory and vestibular systems,* New York, 1983, Academic Press, pp 375-423.

Anson BJ, Hanson JS, Richany SF: Early embryology of auditory ossicles and associated structures in relation to certain anomalies observed clinically, *Ann Otol Rhinol Laryngol* 69:427-447, 1960.

Bard JBL, Bansal MK, Ross ASA: The extracellular matrix of the developing cornea: diversity, deposition and function, *Development* 103 (suppl):195-205, 1988.

Barishak YR: *Embryology of the eye and its adnexae,* Basel, Switzerland, 1992, Karger.

Coulombre AJ, Coulombre JL: The role of intraocular pressure in the development of the chick eye. III. Ciliary body, *Am J Ophthalmol* 44:85-92, 1957.

Coulombre JL, Coulombre AJ: Lens development: fiber elongation and lens orientation, *Science* 142:1489-1490, 1963.

Hanson JR, Anson BJ, Strickland EM: Branchial sources of auditory ossicles in man, *Arch Otolaryngol* 76:200-215, 1962.

Harris WA, Holt CE: Early events in the embryogenesis of the vertebrate visual system: cellular determination and pathfinding, *Annu Rev Neurosci* 13:155-169, 1990.

Hay ED: Development of the vertebrate cornea, *Int Rev Cytol* 63:263-322, 1980.

Henry JJ: Early tissue interactions leading to embryonic lens formation in *Xenopus laevis, Dev Biol* 141:149-163, 1990.

Hoshino T: Scanning electron microscopy of nerve fibers in human fetal cochlea, *J Electron Microsc Tech* 15:104-114, 1990.

Jakobiec FA, ed: *Ocular anatomy, embryology and teratology,* Philadelphia, 1982, Harper & Row.

Lewis WH: Experimental studies on the development of the eye in amphibia. I. On the origin of the lens, *Am J Anat* 3:505-536, 1904.

Mann I: *The development of the human eye,* ed 3, New York, 1964, Grune & Stratton.

McPhee JR, van de Water TR: *Structural and functional development of the ear.* In Jahn AF, Santos-Sacchi J, eds: *Physiology of the ear,* New York, 1988, Raven, pp 221-242.

Murer-Orlando M and others: Differential regulation of gamma-crystallin genes during mouse lens development, *Dev Biol* 119:260-267, 1987.

Nishimura Y, Kumoi T: The embryologic development of the human external auditory meatus, *Acta Otolaryngol* 112:496-503, 1992.

Noden DM, van de Water TR: *The developing ear: tissue origins and interactions.* In Ruben RW and others, eds: *The biology of change in otolaryngology,* Amsterdam, 1986, Elsevier Science, pp 15-46.

Noden DM, van de Water TR: Genetic analysis of mammalian ear development, *Trends Neurosci* 15:235-237, 1992.

O'Rahilly R: The timing and sequence of events in the development of the human eye and ear during the embryonic period proper, *Anat Embryol* 168:87-99, 1983.

Papaconstantinou J: Molecular aspects of lens differentiation, *Science* 156:338-346, 1967.

Piatigorsky J: Lens differentiation in vertebrates, *Differentiation* 19:134-153, 1981.

Reeder R, Bell E: Short- and long-lived messenger RNA in embryonic chick lens, *Science* 150:71-72, 1965.

Rubel EW: Ontogeny of auditory system function, *Annu Rev Physiol* 46:213-229, 1984.

Spemann H: *Embryonic development and induction,* New Haven, Conn, 1938, Yale University.

Streeter GC: Development of the auricle in the human embryo, *Contr Embryol Carnegie Inst* 14:111-138, 1922.

Toole BP, Trelstad RL: Hyaluronate production and removal during corneal development in the chick, *Dev Biol* 26:28-35, 1971.

Traboulsi EI: Developmental genes and ocular malformation syndromes, *Am J Ophthalmol* 115:105-107, 1993.

van de Water TR: Determinants of neuron-sensory receptor cell interaction during development of the inner ear, *Hear Res* 22:265-277, 1986.

Zwaan J, Hendrix RW: Changes in cell and organ shape during early development of the ocular lens, *Am J Zool* 13:1039-1049, 1973.

CHAPTER 15

Development of the Head and Neck

In previous chapters the development of certain components of the head, (e.g., the nervous system, neural crest, and bones of the skull) has been detailed. The first part of this chapter provides an integrated view of early craniofacial development to show the way the major components are interrelated. The remainder of the chapter concentrates on development of the face, pharynx, and pharyngeal arch system.

EARLY DEVELOPMENT OF THE HEAD AND NECK

Development of the head and neck begins early in embryonic life and continues until the cessation of postnatal growth in the late teens. Cephalization begins with the rapid expansion of the rostral end of the neural plate. Very early the future brain is the dominant component of the craniofacial region. Beneath the brain, the face, which does not take shape until later in embryogenesis, is represented by the **stomodeum.** In the early embryo the stomodeum is sealed off from the primitive gut by the **stomodeal plate,** which breaks down by the end of the first embryonic month (Fig. 15-1). The future cervical region is dominated by the pharyngeal apparatus, consisting of a series of pharyngeal pouches, arches, and clefts. Many components of the face, ears, and glands of the head and neck arise from the pharyngeal region. Also prominent are the paired ectodermal placodes (see Fig. 5-9), which form much of the sensory tissue of the cranial region.

Tissue Components and Segmentation of the Early Craniofacial Region

The early craniofacial region consists of a massive neural tube, beneath which lie the notochord, a ventrally situated digestive tube (pharynx) surrounded by a series of aortic arches, an ectodermal covering, and large masses of neural crest and mesodermally derived mesenchyme filling in the remaining spaces (see Fig. 15-1). Most of these tissue components are organized segmentally. Fig. 15-2 illustrates the registration of the segmentation of the tissue components of the head. As discussed in earlier chapters, distinct patterns of expression of certain homeobox-containing genes are associated with morphological segmentation in some tissues, particularly the central nervous system (Fig. 15-3). The chain of presumed events between segmental patterns of gene expression and the appearance of morphological segmentation in parts of the cranial region still remains obscure.

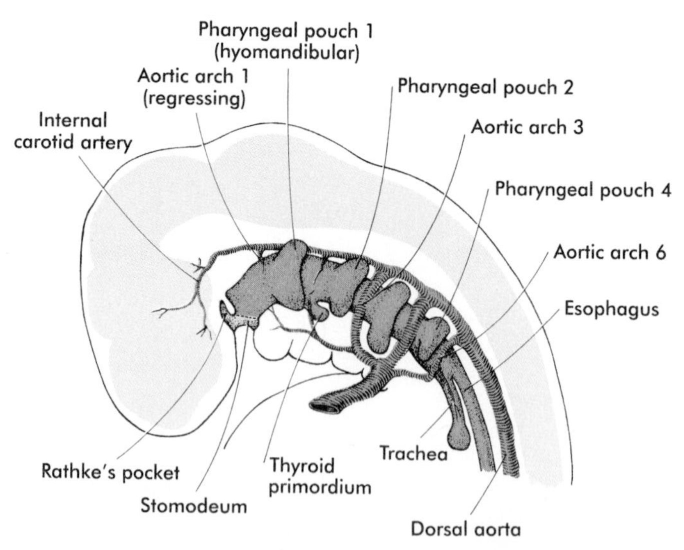

FIG. 15-1 The basic organization of the pharyngeal region of the human embryo at the end of the first month.

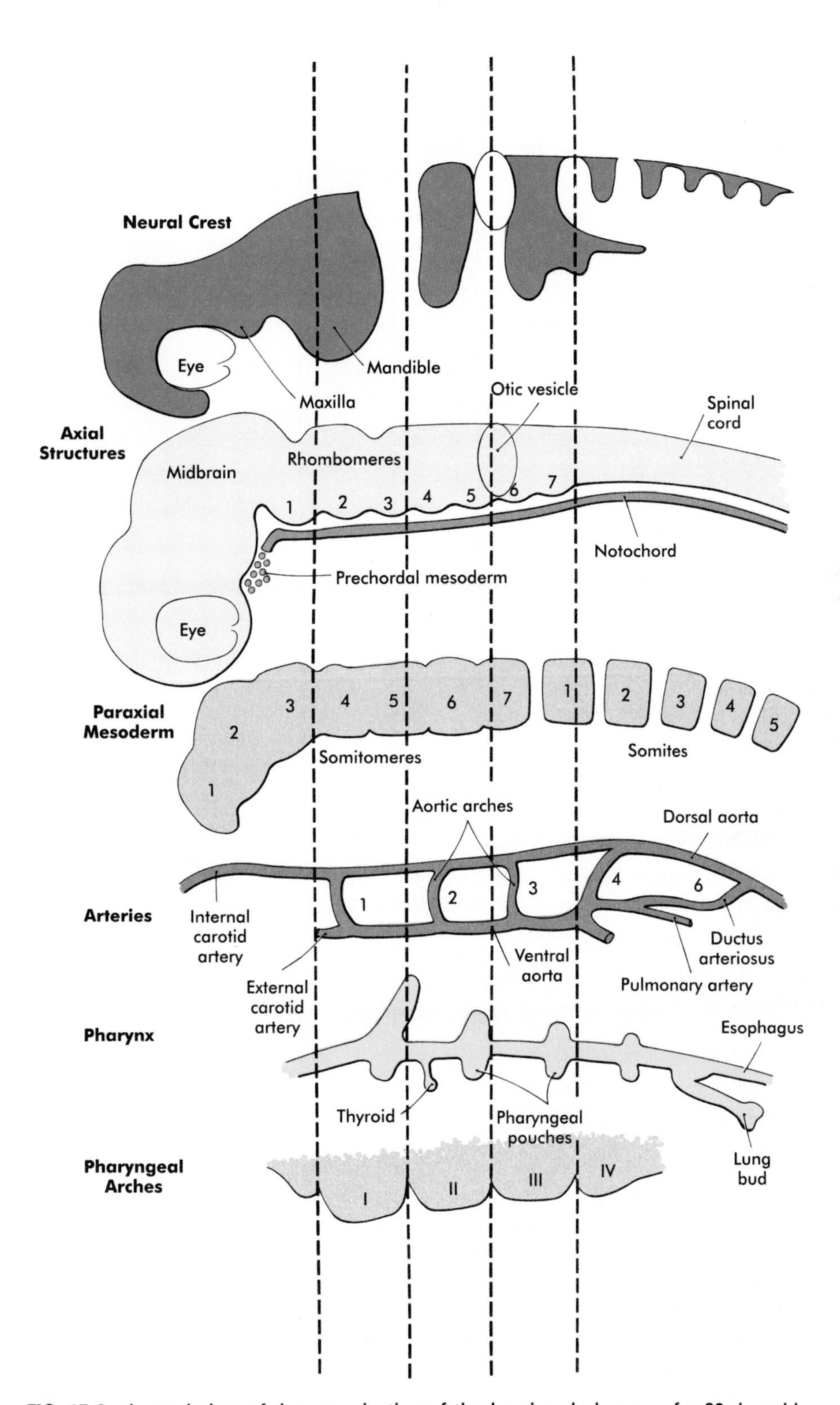

FIG. 15-2 Lateral view of the organization of the head and pharynx of a 30-day-old human embryo, with individual tissue components separated but in register through the dashed lines.

(Based on studies by Noden [1991].)

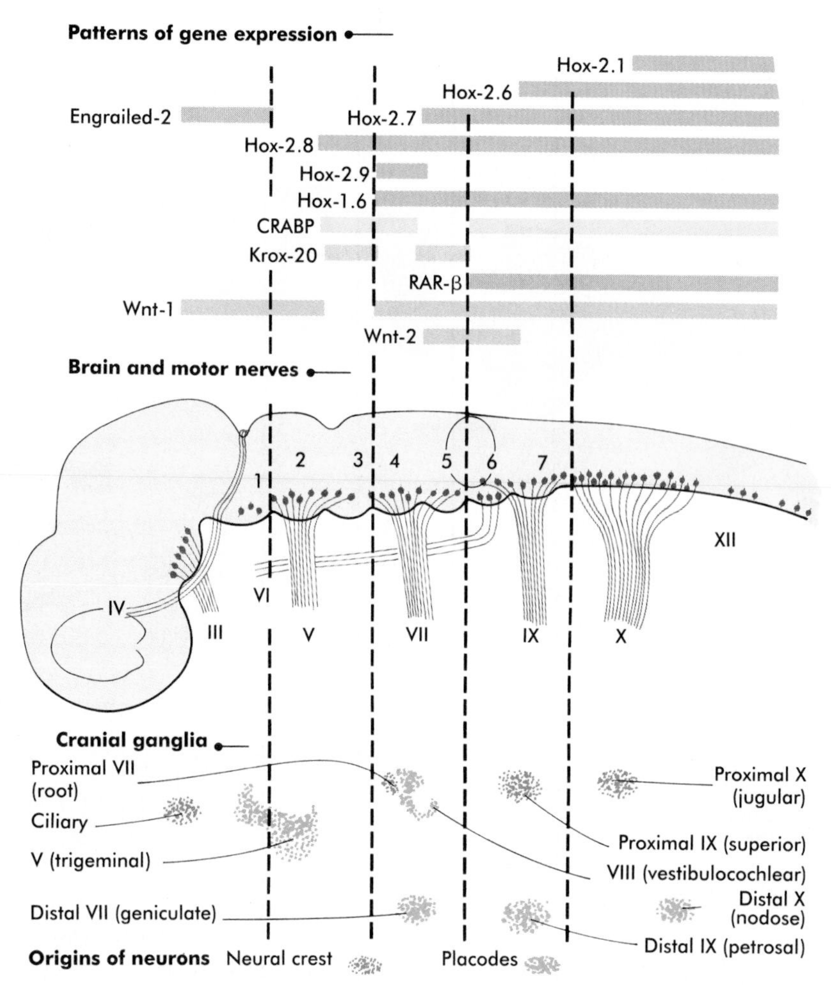

FIG. 15-3 Patterns of homeobox and other gene expression in relation to anatomical landmarks in the early mammalian embryo. The bars refer to craniocaudal levels of expression of a given gene product. *CRABP,* Cytoplasmic retinoic acid–binding protein; *RAR,* retinoic acid receptor. Cranial sensory ganglia derived from neural crest and placodal precursors are laid out in proper register.
(Modified from Noden DM: *J Craniofac Genet Devel Biol* 11:192-213, 1991.)

Early Cellular Migrations and Tissue Displacements in the Craniofacial Region

Early craniofacial development is characterized by a number of massive migrations and displacements of cells and tissues. The neural crest is the first tissue to exhibit massive migratory behavior, with cells emigrating from the nervous system even before closure of the cranial neural tube (see Chapter 13). Initially, segmental groups of neural crest cells are segregated, especially in the pharyngeal region (see Fig. 15-2). However, these populations of cells become confluent during their migrations through the pharyngeal arches. Experiments indicate that the neural crest cells have received some morphogenetic instructions before they commence their migrations. For example, when the presumptive neural crest cells of the second and third branchial arches of avian embryos were replaced with first arch neural crest cells, a duplicate set of first arch structures (essentially the entire lower jaw apparatus) was formed at a level caudal to the of the normal lower jaw of the host embryo.

The early cranial mesoderm consists mainly of the **paraxial** and **prechordal mesoderm** (see Fig. 15-2). Mesenchymal cells originating in the paraxial mesoderm form the connective tissue and skeletal elements of most of the cranium and the dorsal part of the neck. Somitomere-derived myogenic cells undergo extensive migrations to form the bulk of the muscles of the cranial region. Like their counterparts in the trunk and limbs, these myogenic cells integrate with local connective tissue to form muscles. Another similarity with the trunk musculature is that morphogenetic control appears to reside within the connective

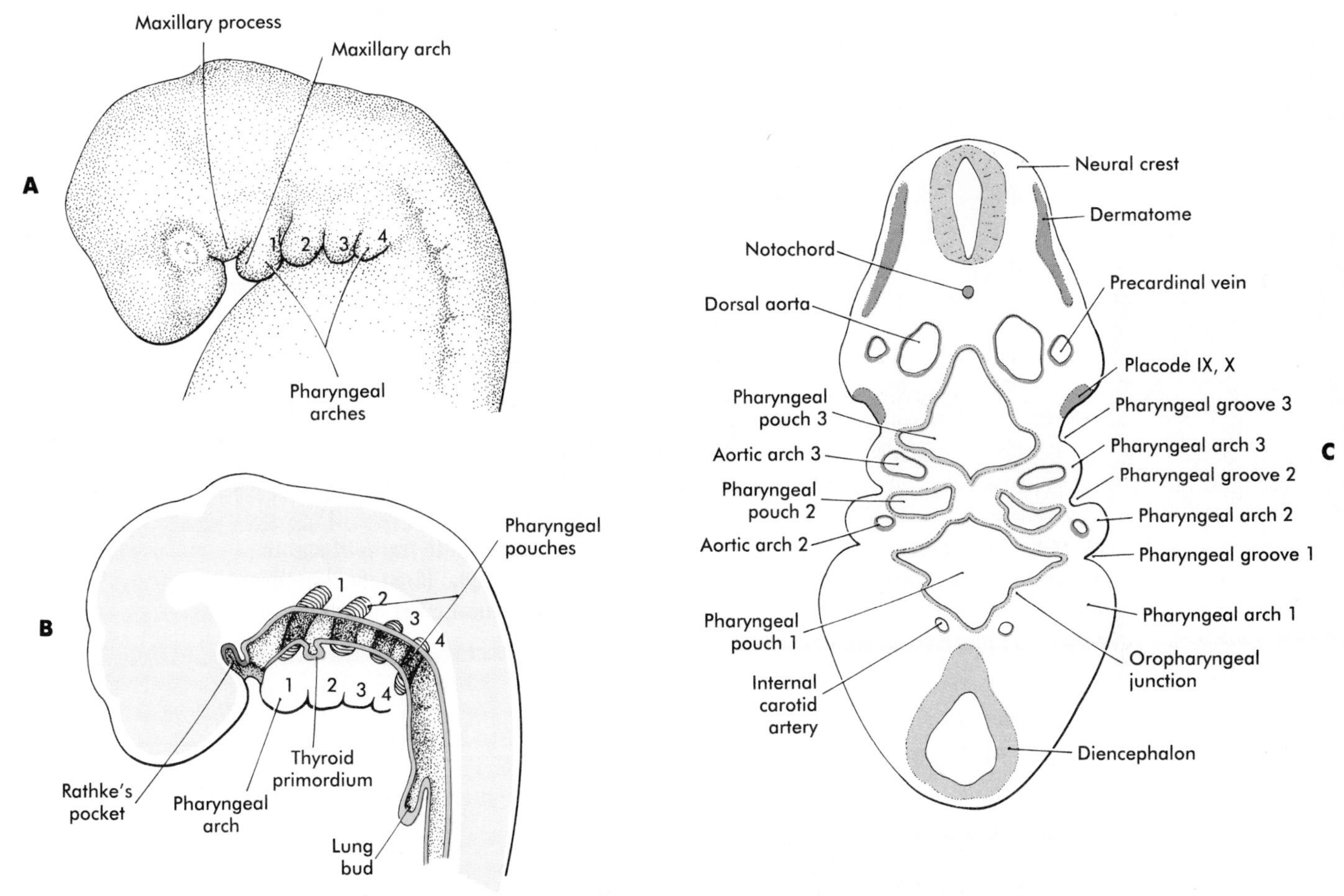

FIG. 15-4 **A** and **B,** Superficial and sagittal views of the head and pharynx of a human embryo during the fifth week. **C,** Cross section through the pharyngeal region of a human embryo of the same age. Because of the strong C curvature of the head and neck of the embryo, a single section passes through the levels of both the forebrain *(bottom)* and hindbrain *(top).*

tissue elements of the muscles rather than in the myogenic cells themselves.

The prechordal mesoderm is a transient mass of cells located in the midline, rostral to the tip of the notochord. Although the fate of these cells is controversial, some investigators feel that the myoblasts contributing to the extraocular muscles take origin from these cells. On their way to the eye, cells of the prechordal mesoderm may pass through the rostral-most somites. Other investigators feel that the extraocular muscles have a purely somitomeric origin.

The **lateral mesoderm** is not well defined in the cranial region. Transplantation experiments have shown that it gives rise to endothelial and smooth muscle cells and, at least in birds, to some portions of the laryngeal cartilages.

Another set of tissue displacements of importance in the cranial region is the joining of cells derived from the ectodermal placodes with those of the neural crest to form parts of sense organs and ganglia of certain cranial nerves (see Fig. 5-9).

FUNDAMENTAL ORGANIZATION OF THE PHARYNGEAL REGION

Because many components of the face are derived from the pharyngeal region, an understanding of the basic organization of this region is important. In the 1-month-old embryo the pharyngeal part of the foregut contains four lateral pairs of endodermally lined outpocketings called **pharyngeal pouches** and an unpaired ventral midline diverticulum, the **thyroid primordium** (Fig. 15-4). If the contours of the ectodermal covering over the pharyngeal region are followed, bilateral pairs of inpocketings called **branchial (pharyngeal) grooves** that almost make contact with the lateral-most extent of the pharyngeal pouches are seen (Fig. 15-4, *C*).

Alternating with the branchial grooves and pharyngeal pouches are paired masses of mesenchyme called **pharyngeal (branchial) arches.** Central to each pharyngeal arch is a prominent artery called an **aortic arch,** which extends

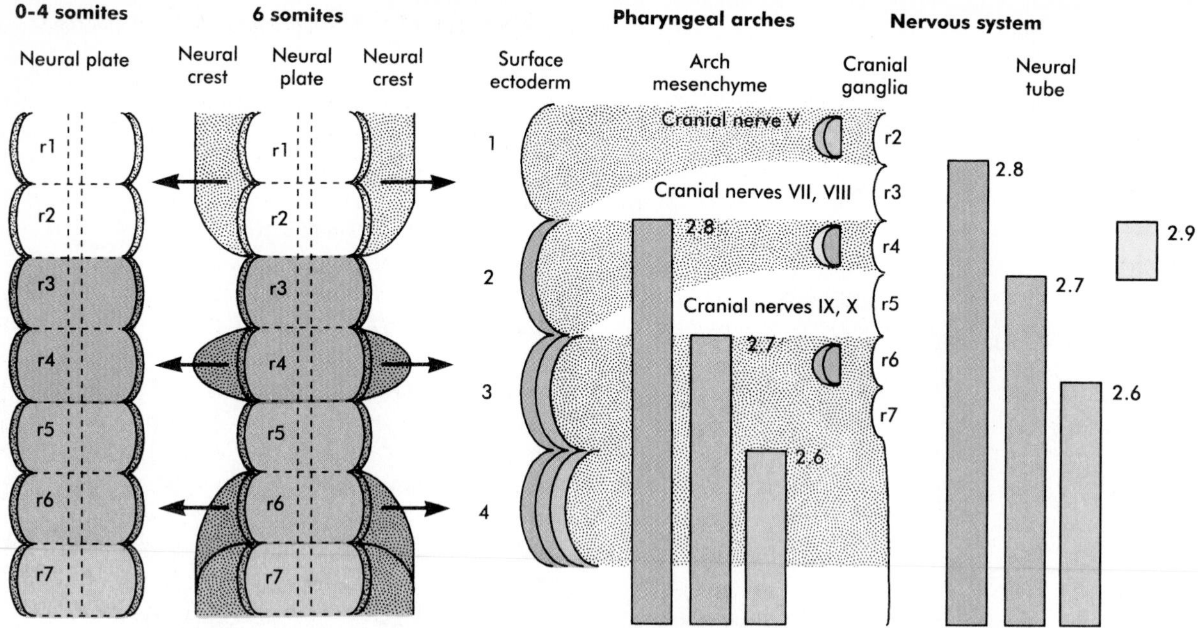

FIG. 15-5 The spread of *Hox* gene expression from the neural plate *(far left)* into the emigrating neural crest *(middle)* and into tissues of the pharyngeal (branchial) arches *(right).* Arrows in the middle diagram indicate directions of neural crest migration. (Modified from Hunt P and others: *Development* 1(suppl):187-196, 1991.)

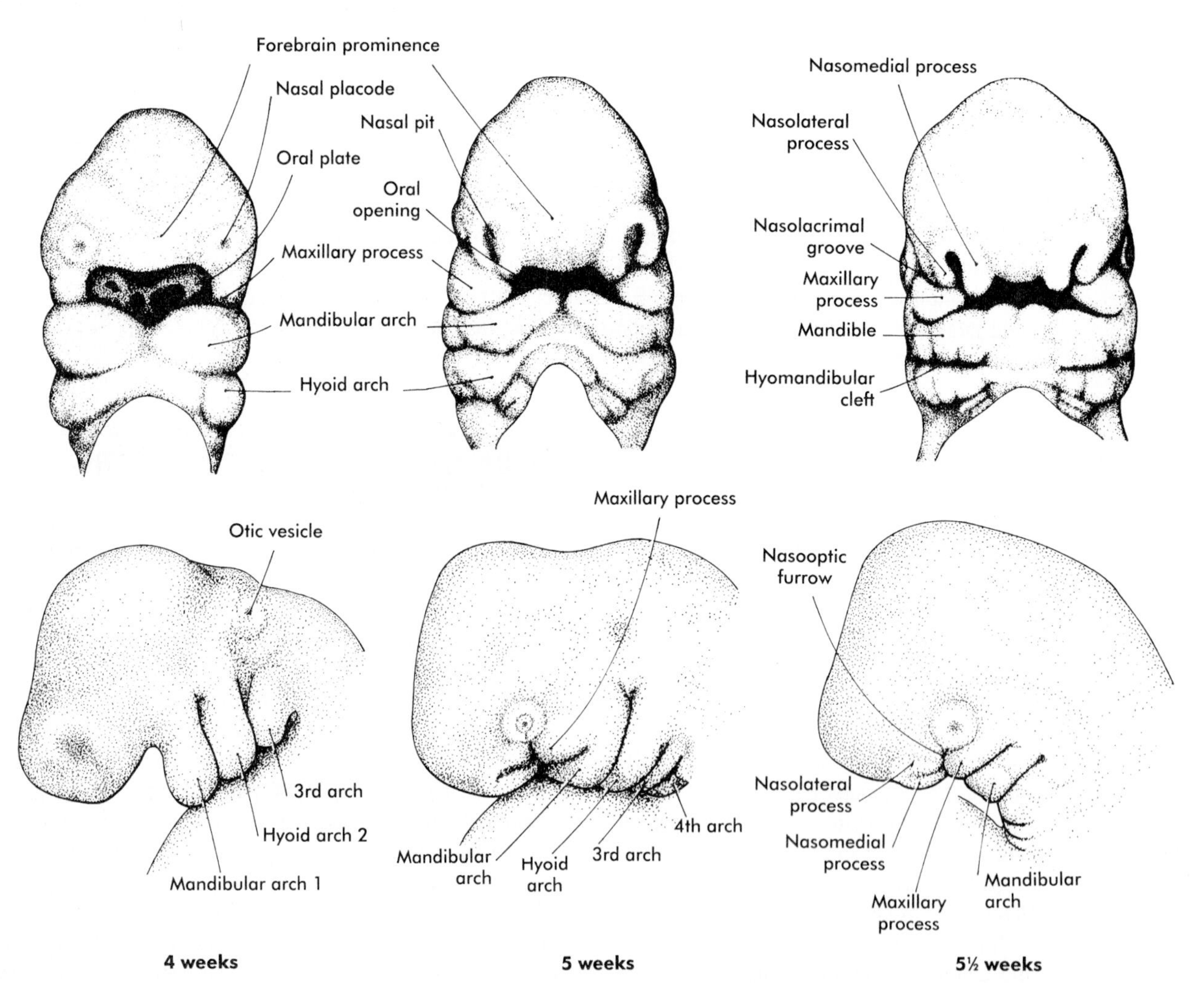

FIG. 15-6 Frontal and lateral views of heads of human embryos from 4 to 8 weeks of age.

from the ventral to the dorsal aorta (see Chapter 18 and Fig. 15-1). The mesenchyme of the pharyngeal arches is of dual origin. The mesenchyme of the incipient musculature originates from mesoderm, specifically the somitomeres. Much of the remaining pharyngeal arch mesenchyme, especially that of the ventral part, is derived from the neural crest, whereas mesoderm makes varying contributions to the dorsal pharyngeal arch mesenchyme. The early formation of the pharyngeal arches is closely associated with the spread of expression of products of the **Hox-2** (B) gene family from the rhombomeres of the neural tube to the pharyngeal arch mesenchyme and ultimately the overlying ectoderm (Fig. 15-5).

DEVELOPMENT OF THE FACIAL REGION

Formation of the Face and Jaws

Structures of the face and jaws originate from several primordia that surround the stomodeal depression of the 4- to 5-week-old human embryo (Fig. 15-6). These primordia consist of an unpaired **frontonasal prominence;** paired **nasomedial processes,** which are components of the horseshoe-shaped olfactory primordia (see p. 285); and paired **maxillary processes** and **mandibular prominences,** both components of the first pharyngeal arches. The primordia that form the face are now recognized to have distinctly different developmental properties. According to present understanding the neural crest cells, which constitute the bulk of the mesenchyme of these primordia, become endowed with specific morphogenetic information before they migrate into the facial primordia. The nature of this information and the way it is used remain poorly understood. Specific patterns of both **retinoic acid receptors** and **retinoic acid–binding proteins** in these structures are likely important in both normal morphogenesis and in retinoic acid–induced anomalies of the facial region.

As in the limb buds, outgrowth of the facial primordia depends on mesenchymal-ectodermal interactions, although the site of the interaction is not marked by an apical ectodermal thickening as it is in the limb bud. Recombination experiments in birds have shown that the mesenchyme from all three facial primordia (frontonasal mass, maxillary, and mandibular primordia) can maintain the thickened apical ectodermal ridge of the limb bud. Although outgrowth of the

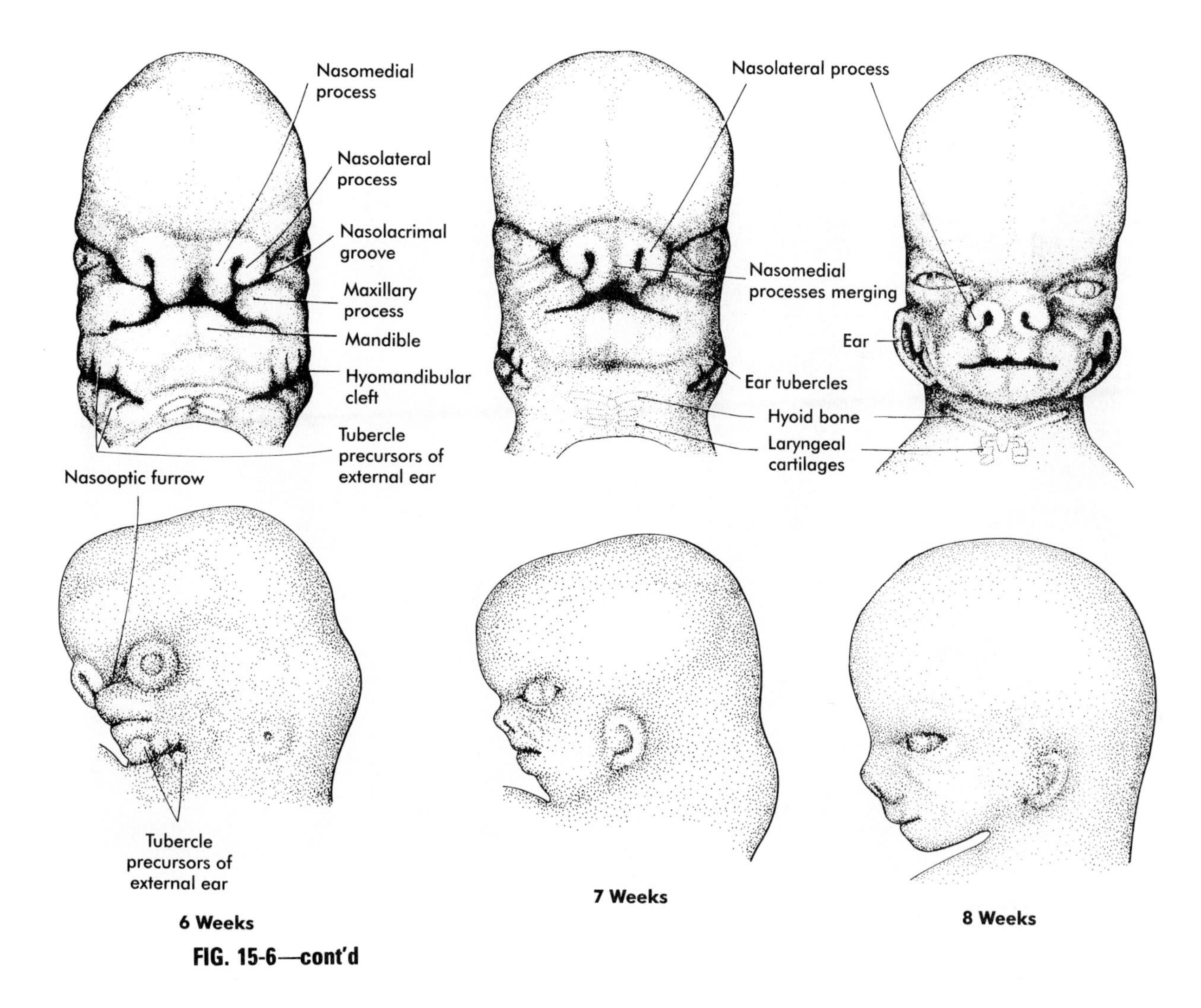

FIG. 15-6—cont'd

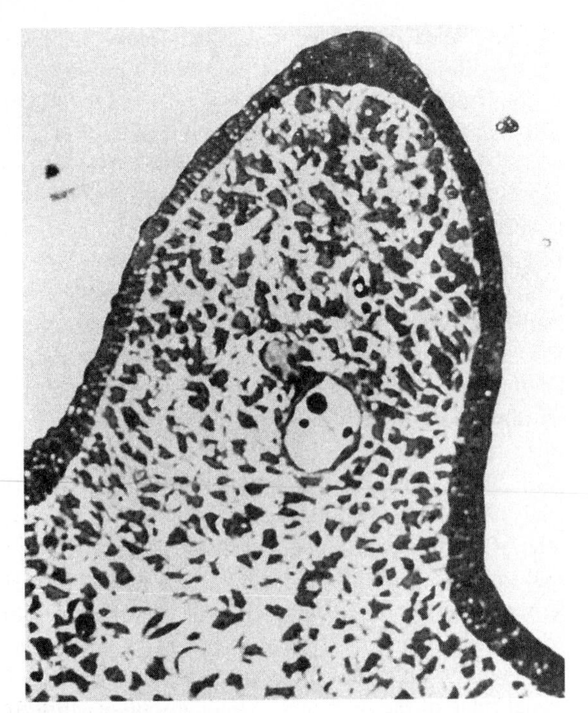

FIG. 15-7 Histological section through a palatal shelf of a mouse embryo showing a similar organization (including apical ectodermal thickening) to that of a limb bud.
(From Carlson B: *Patten's foundations of embryology,* ed 3, New York, 1974, McGraw Hill.)

palatal shelves has received less experimental attention, ectodermal thickenings reminiscent of the apical ectodermal ridge can sometimes be seen at the apical border of the palatal shelves (Fig. 15-7).

Through differential growth during the period between 4 and 8 weeks (see Fig. 15-6), the nasomedial and maxillary processes become relatively more prominent and ultimately fuse to form the upper lip and jaw (Fig. 15-8). As this is occurring, the frontonasal prominence, which was a prominent tissue bordering the stomodeal area in the 4- and 5-week-old embryo, is displaced without contributing significantly to the upper jaw as the two nasomedial processes merge. The merged nasomedial processes form the **intermaxillary segment,** which is a precursor for (1) the **philtrum** of the lip, (2) the **premaxillary component** of the upper jaw, and (3) the **primary palate.**

Between the maxillary process and the nasal primordium is a **nasolacrimal groove** that extends to the developing eye (see Fig. 15-6). The ectoderm of the floor of the nasolacrimal groove thickens to form a solid epithelial cord, which detaches from the groove. The epithelial cord then undergoes canalization and forms the **nasolacrimal duct** and, near the eye, the **lacrimal sac.** The nasolacrimal duct extends from the medial corner of the eye to the nasal cavity (inferior meatus) and in postnatal life acts as a drain for lacrimal fluid. This connection explains why people can get a runny nose when crying. Meanwhile, the expanding na-

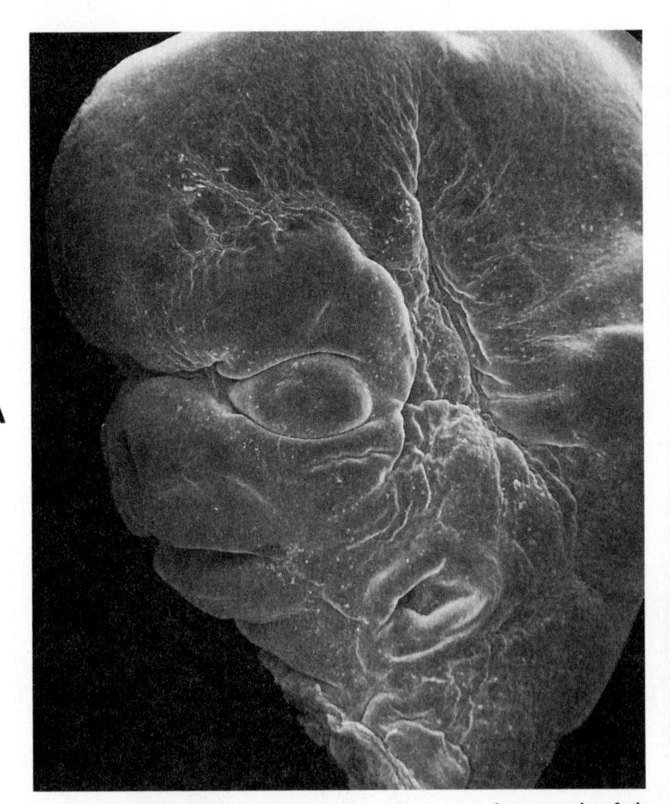

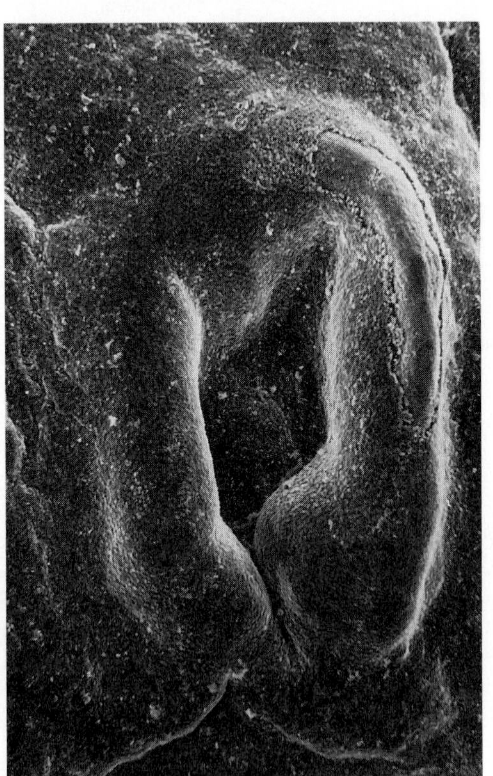

FIG. 15-8 **A,** Scanning electron micrograph of the head of an 8-week-old human embryo. **B,** Higher magnification of the ear, which is located in the neck in **A.**
(From Jirásek JE: *Atlas of human prenatal morphogenesis,* Amsterdam, 1983, Martinus Nijhoff Publishers.)

somedial process fuses with the maxillary process, and over the region of the nasolacrimal groove, the nasolateral process merges with the superficial region of the maxillary process.

The lower jaw is formed in a simpler manner. The bilateral mandibular prominences enlarge, and their medial components merge in the midline, forming the point of the lower jaw. The midline dimple that is seen in the lower jaw of some individuals is a reflection of variation in the degree of merging of the mandibular prominences. A prominent cartilaginous rod called **Meckel's cartilage** differentiates within the lower jaw. Derived from neural crest of the first pharyngeal arch, Meckel's cartilage forms the basis around which the membrane bones (which form the definitive skeleton of the lower jaw) are laid down. Experimental evidence indicates that the rodlike shape of Meckel's cartilage is related to the inhibition of further chondrogenesis by the surrounding ectoderm. If the ectoderm is removed around Meckel's cartilage, large masses of cartilage form instead of a rod. These properties are similar to the inhibitory interactions between ectoderm and chondrogenesis in the limb bud.

Shortly after the basic facial structures take shape, they are invaded by mesodermal cells associated with the first and second pharyngeal arches. These cells form the muscles of mastication (first arch derivatives—innervated by cranial nerve V) and the muscles of facial expression (second arch derivatives—innervated by cranial nerve VII).

Although the basic structure of the face is established between the fourth and eighth week, changes in proportionality of the various regions continue until well after birth. In particular, the midface remains underdeveloped during embryogenesis and even early postnatal life.

Formation of the Palate

The early embryo possesses a common oronasal cavity, but in mammals the **palate** forms between the sixth and tenth week to separate the oral from the nasal cavity. The palate is derived from three primordia—an unpaired **median palatine process** and a pair of **lateral palatine processes** (Fig. 15-9).

The median palatine process is an ingrowth from the newly merged nasomedial processes. As it grows, the me-

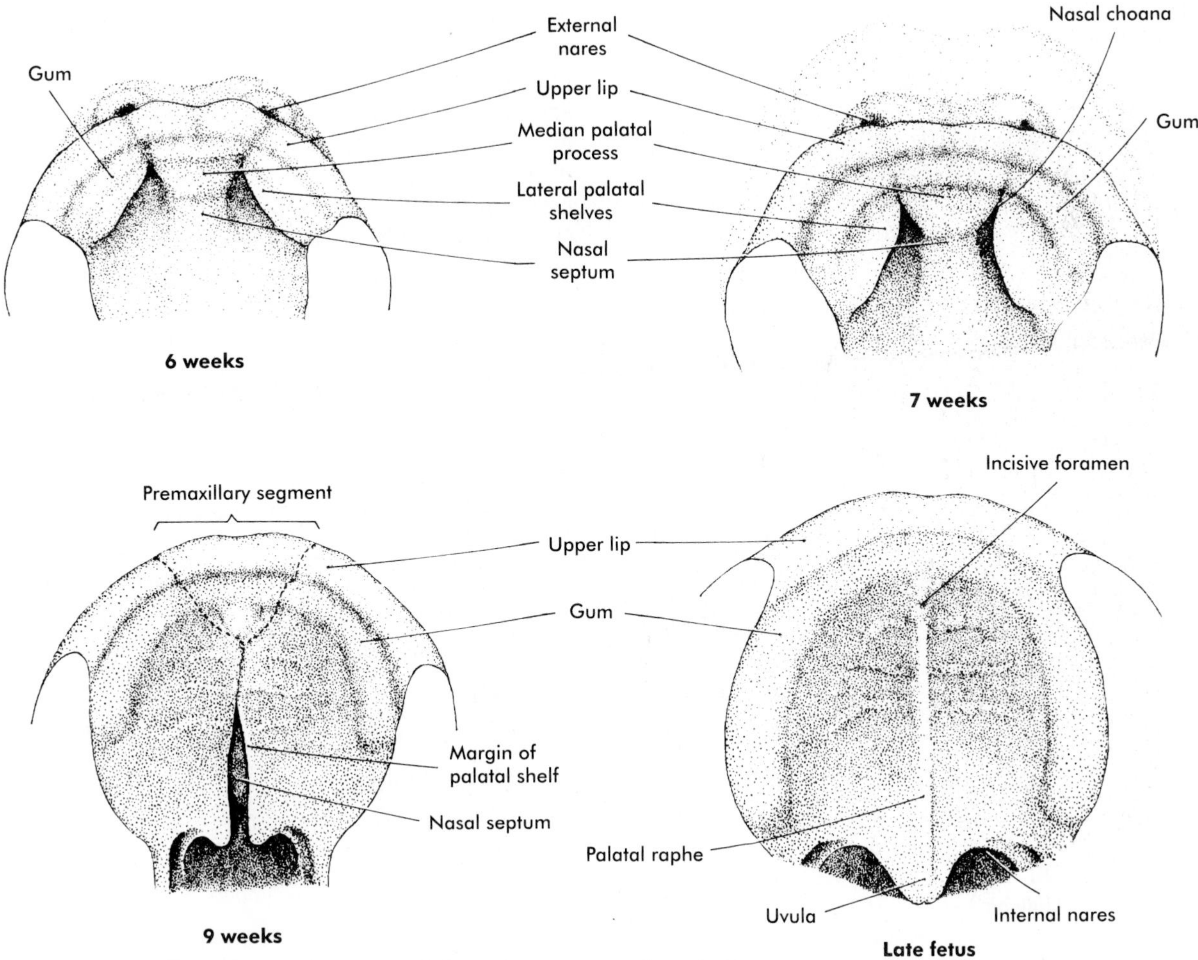

FIG. 15-9 Development of the palate as seen from below.

dian palatine process forms a triangular bony structure called the **primary palate.** In postnatal life the skeletal component of the primary palate is referred to as the **premaxillary component of the maxilla.** The four upper incisor teeth arise from this structure (Fig. 15-10).

The lateral palatine processes, which are the precursors of the **secondary palate,** first appear during the sixth week. At first they grow downward on either side of the tongue (Fig. 15-11). Then during the seventh week the lateral palatine processes **(palatal shelves)** dramatically become dislodged from their positions along the tongue and become oriented perpendicularly to the maxillary processes. The apices of these processes meet in the midline and begin to fuse.

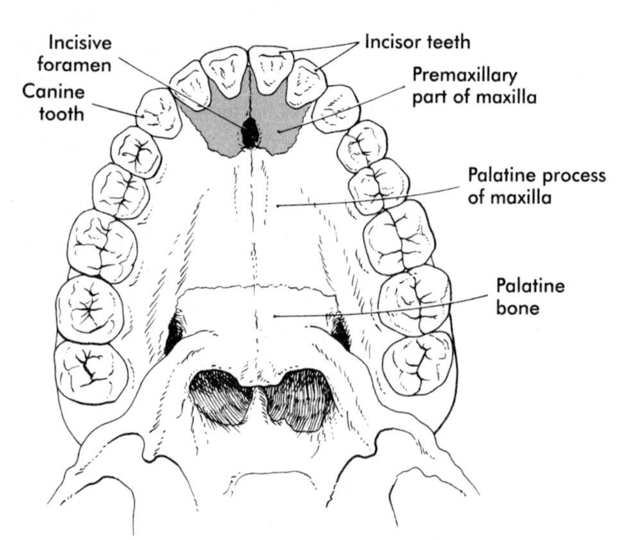

FIG. 15-10 The postnatal bony palate, showing the premaxillary segment.

Despite many years of investigation, the mechanism underlying the elevation of the palatine shelves remains obscure. Swelling of the extracellular matrix of the palatal shelves appears to impart a resiliency that allows them to approximate one another within minutes or hours of their becoming dislodged from along the tongue. There is speculation that the dislodgment, which occurs during a period when the head is growing in height but not width, may be precipated by hiccuping of the embryo. At the time when elevation of the palatal shelves occurs, opening of the jaw, swallowing, and sucking have normally not occurred.

Another structure that is involved in formation of the palate is the **nasal septum** (see Figs. 15-9 and 15-11). This midline structure, which is a downgrowth from the frontonasal prominence, reaches the level of the palatal shelves at the time when the latter fuse to form the definitive secondary palate. Rostrally, the nasal septum is continuous with the primary palate.

At the gross level, the palatal shelves fuse in the midline, but rostrally they also join the primary palate. The midline point of the fusion of the primary palate with the two palatal shelves is marked by the **incisive foramen** (see Fig. 15-10).

Because of its clinical importance, fusion of the palatal shelves has been subject to intensive investigation. When the palatal shelves first make midline contact, each is covered throughout by a homogeneous epithelium. During the process of fusion, however, the midline epithelial seam disappears. The epithelium on the nasal surface of the palate differentiates into a ciliated columnar type, whereas the epithelium takes on a stratified squamous form on the oral surface of the palate. Significant developmental questions are (1) What causes the disappearance of the midline epithelial seam? (2) What signals result in the diverse pathways of differentiation of the epithelium on either side of the palate? Disruption of the midline epithelial seam is known to

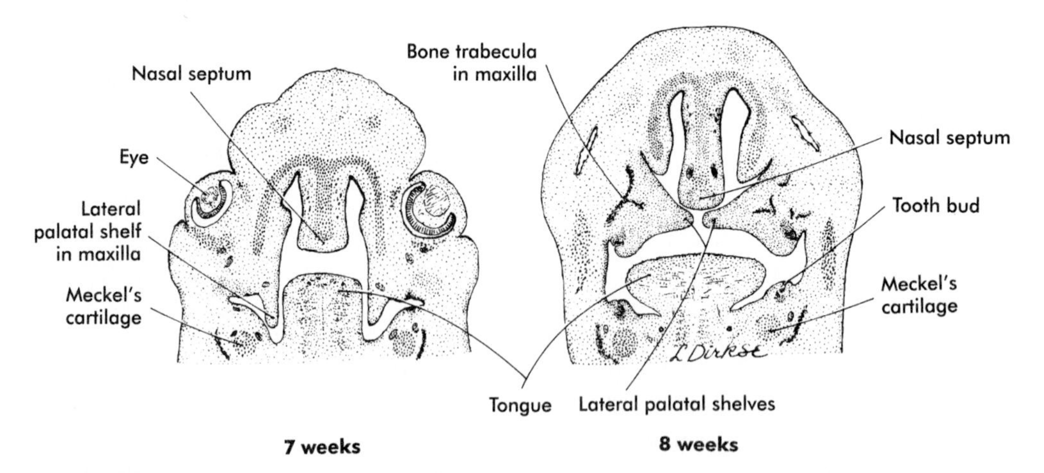

FIG. 15-11 Frontal sections through the human head, showing fusion of the palatal shelves.

(From Patten B: *Human embryology,* ed 3, New York, 1968, McGraw-Hill.)

involve programmed death **(apoptosis)** of many epithelial cells and the migration of some epithelial cells to the nasal and oral surfaces of the palate (Fig. 15-12).

Experiments involving the in vitro culture of single palatal shelves of a number of species have shown clearly that all aspects of epithelial differentiation (cell death in the midline and different pathways of differentiation on the oral and nasal surfaces) can occur in the absence of contact with the opposite palatal shelf. These different pathways of differentiation are not intrinsic to the regional epithelia but are mediated by the underlying neural crest–derived mesenchyme. The mechanism underlying this regional specification of the epithelium remains little understood. According to one model, the underlying mesenchyme produces growth factors that influence the production and regional distribution of extracellular matrix molecules (e.g., type IX collagen). The way these events are received and interpreted by the epithelial cells is unknown.

Formation of the Nose and Olfactory Apparatus

The human olfactory apparatus first becomes visible at the end of the first month as a pair of thickened ectodermal **nasal placodes** located on the frontal aspect of the head (Fig. 15-13, *A*). Although commonly regarded as the end product of an inductive influence by underlying brain tissue, the nasal placodes have recently been considered by some investigators to be patches of ectoderm that detach from the anterolateral edge of the neural plate before its closure.

Soon after their formation the nasal placodes form a surface depression (the **nasal pits**) surrounded by horseshoe-shaped elevations of mesenchymal tissue with the open ends facing the future mouth (see Fig. 15-6). The two limbs of the mesenchymal elevations are the nasomedial and **nasolateral processes.** As the nasal primordia merge toward the midline during the sixth and seventh weeks, the nasomedial processes form the tip and crest of the nose along with part of the nasal septum, and the nasolateral processes form the wings **(alae)** of the nose. The frontonasal process contributes to part of the bridge of the nose.

Meanwhile the nasal pits continue to deepen toward the oral cavity and form substantial cavities themselves (see Fig. 15-13). By the seventh week only a thin **oronasal membrane** separates the oral from the nasal cavities. The oronasal membrane soon breaks down, with the result that the nasal cavities are continuous with the oral cavity through openings behind the primary palate called **nasal choanae.** With the fusion of the lateral palatal shelves, the nasal cavity is considerably lengthened and ultimately communicates with the upper pharynx.

During the third month, shelflike structures called **nasal conchae** form on the lateral wall of the nasal cavity. These structures increase the surface available for conditioning the air within the nasal cavity. Late in fetal life and for several years after birth the paranasal sinuses form as outgrowths

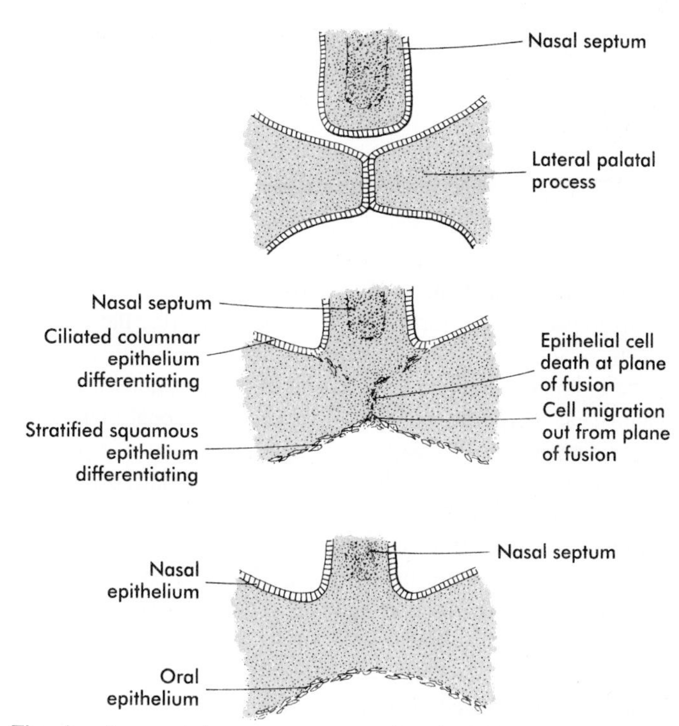

FIG. 15-12 The developmental processes associated with fusion of the palatal shelves and the nasal septum.

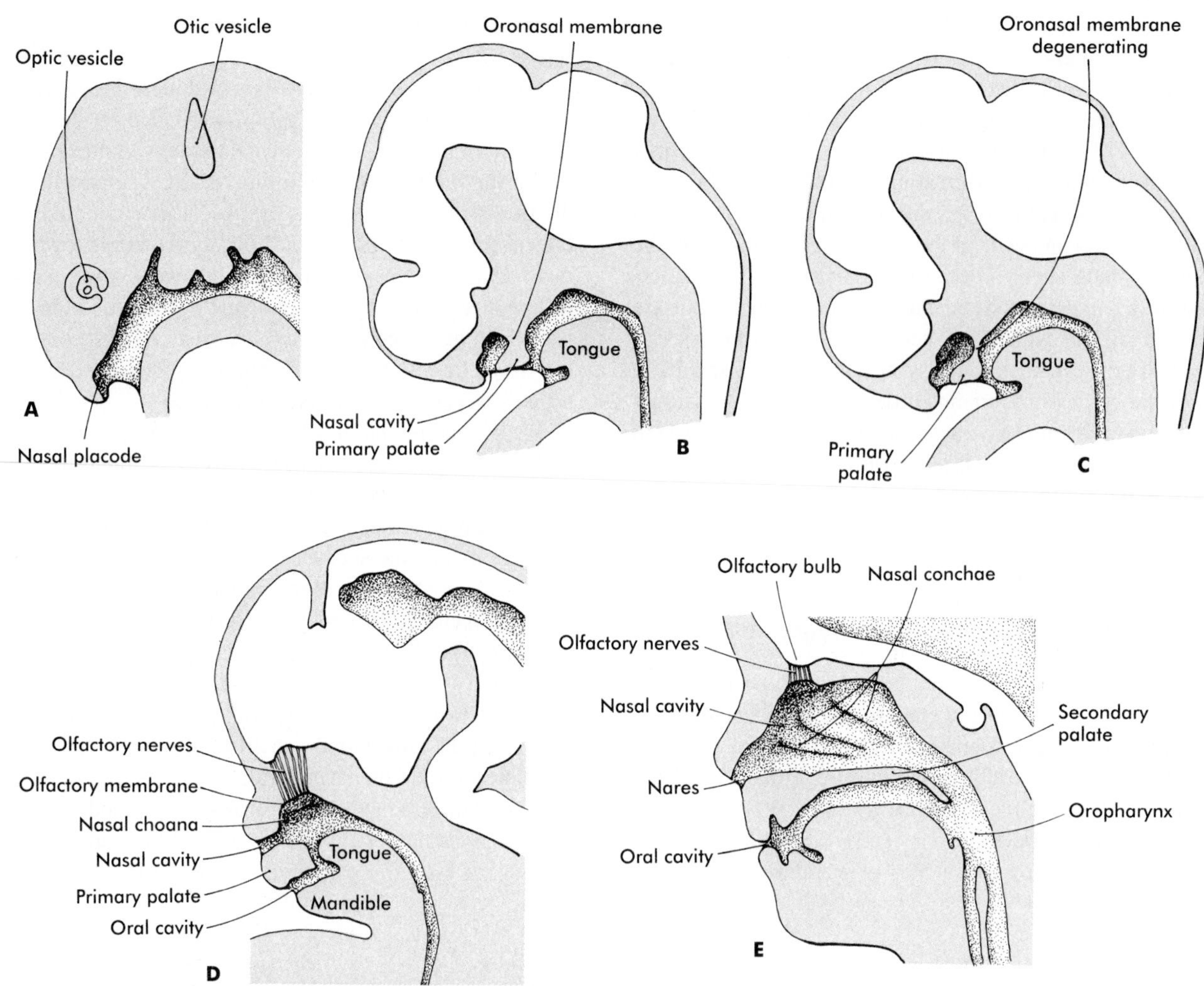

FIG. 15-13 Sagittal sections through embryonic heads with special emphasis on development of the nasal chambers. **A,** At 5 weeks. **B,** At 6 weeks. **C,** At 6½ weeks. **D,** At 7 weeks. **E,** At 12 weeks.

from the walls of the nasal cavities. The size and shape of these structures have a significant impact on the form of the face during its postnatal growth period.

Around midpregnancy a pair of epithelial invaginations into each side of the nasal septum near the palate can be seen. These diverticula, known as **vomeronasal organs,** reach a maximum size of about 6 to 8 mm at around the sixth fetal month and then begin to regress completely or leave small cystic structures. In most mammals and many other vertebrates the vomeronasal organs, which are lined with a modified olfactory epithelium, appear to be involved in the olfaction of food in the mouth or sexual olfactory stimuli. Whether they have any function in humans or are simply atavistic remnants remains to be determined.

The dorsal-most epithelium of the nasal pits undergoes differentiation as a highly specialized olfactory epithelium (see Fig. 15-13). Beginning in the embryonic period and throughout life, the olfactory epithelium has the capacity of forming primitive sensory bipolar neurons, which send axonal projections toward the olfactory bulb of the brain. Preceding axonal ingrowth, some cells break free from this epithelium and migrate toward the brain. Some of these cells may synthesize a substrate for the ingrowth of the olfactory axons. Other cells migrating from the olfactory placode synthesize luteinizing hormone–releasing hormone (LHRH) and become translocated to the hypothalamus (see p. 230). Other cells of the olfactory placode form supporting (sustentacular) cells and glandular cells in the olfactory region of the nose. Physiological evidence demonstrates that the olfactory epithelium is capable of some function in late fetal life but that full olfactory function is not attained until after birth.

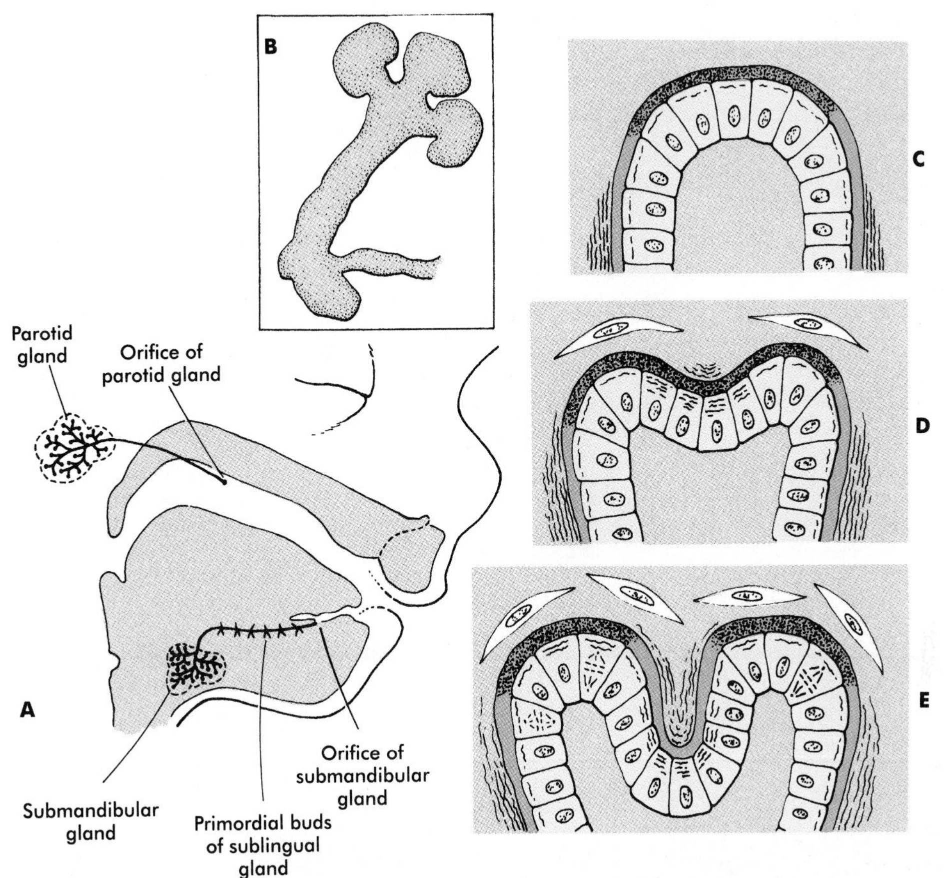

FIG. 15-14 Development of the salivary glands. **A,** Salivary gland development in an 11-week-old human embryo. **B,** Development of salivary gland epithelium in vitro. **C,** Accumulation of newly synthesized glycosaminoglycans *(stipple)* in the basal lamina at the end of a primary lobule. **D,** Early cleft formation is associated with the contraction of bundles of microfilaments in the apices of the epithelial cells lining the cleft. Collagen fibers *(wavy lines)* are lined up lateral to the lobule and in the newly forming cleft. **E,** As the cleft deepens, glycosaminoglycan synthesis is reduced in the cleft, and collagen deposition continues. (**C** through **E** show the relationship between disposition of the extracellular matrix and lobulaton of the glandular primordium.)

Formation of the Salivary Glands

Starting in the sixth week the **salivary glands** originate as solid, ridgelike thickenings of the oral epithelium (Fig. 15-14). Extensive epithelial shifts in the oral cavity make it difficult to determine the germ layer origins of the salivary gland epithelium. The parotid glands are probably derived from ectoderm, whereas the submandibular and sublingual glands are thought to be derived from endoderm.

As with other glandular structures associated with the digestive tract, the development of salivary glands depends on a continuing series of epitheliomesenchymal interactions. Surrounding the early epithelial lobular ingrowths is a basal lamina that differs in composition in areas with different growth potential. Around the stalk and in clefts, the basal lamina contains types I and IV collagen and a basement membrane-1 (BM-1) protoglycan. These components are not found in the regions of the lobules that will undergo further growth. Under the influence of the surrounding mesenchyme, the basal lamina in growing regions loses the collagens and proteoglycans that are associated with stable structures (e.g., the stalks and clefts). In addition to alterations in the basal lamina, branching is associated with the local contraction of ordered microfilaments within the apices of epithelial cells at the branch points. Continued growth at the tips of lobules of the glands is supported by high levels of mitotic activity of the epithelium and the deposition of newly synthesized glycosaminoglycans in the area. The structural and functional differentiation of the epithelium of the salivary gland continues throughout fetal life.

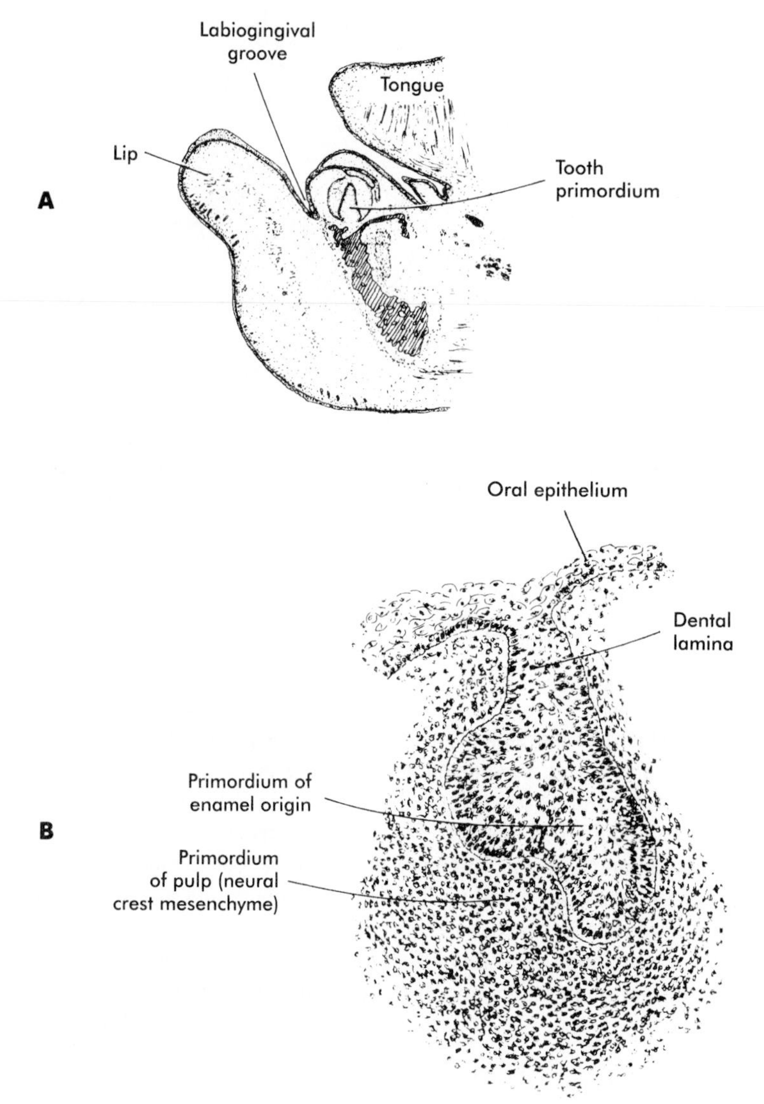

FIG. 15-15 The development of a deciduous tooth. **A,** Parasagittal section through the lower jaw of a 14-week-old human embryo showing the relative location of the tooth primordium. **B,** Tooth primordium in a 9-week-old embryo.

Formation of the Teeth

Teeth are a highly specialized extracellular matrix consisting of two principal components, enamal and dentin, each secreted by a different embryonic epithelium. Tooth development is a highly orchestrated process involving intimate interactions between the epithelia that produce the dentin and enamel. Extending a common theme of development into the macroscopic dimension, teeth undergo an **isoform transition,** with the postnatal replacement of the deciduous teeth by their permanent adult counterparts.

Stages of Tooth Development. Tooth development begins with the migration of neural crest cells into the regions of the upper and lower jaws. Some of these neural crest–derived mesenchymal cells are specified for tooth formation. They act on the overlying oral ectoderm, which thickens into C-shaped bands (**dental laminae**) in the upper and lower jaws. The appearance of the dental laminae during the sixth week is the first manifestation of a continuous series of ectodermal-mesenchymal interactions that continues until tooth formation is virtually completed.

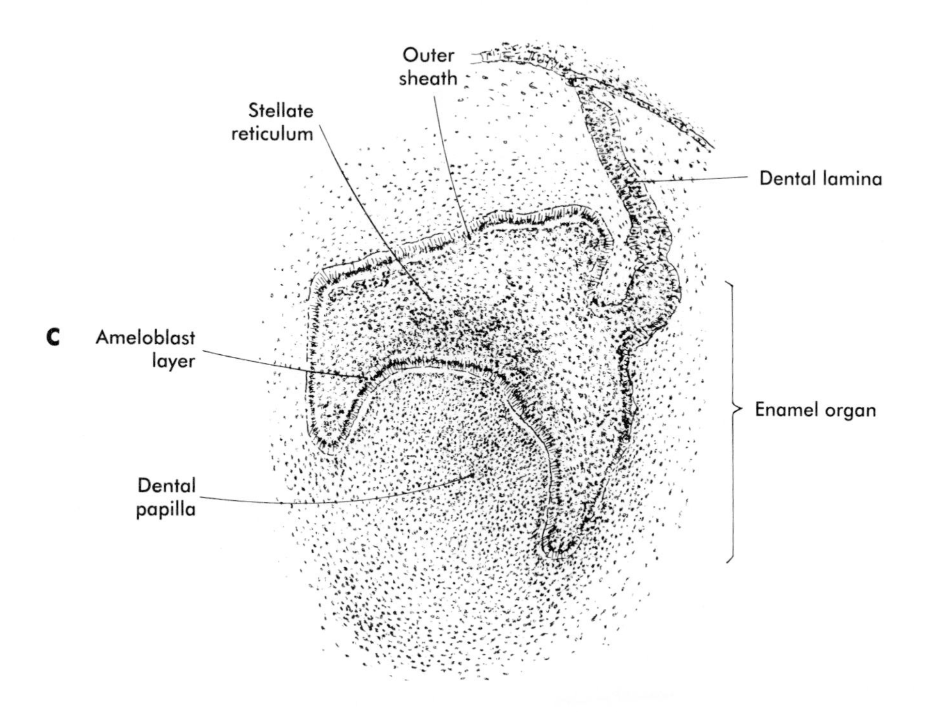

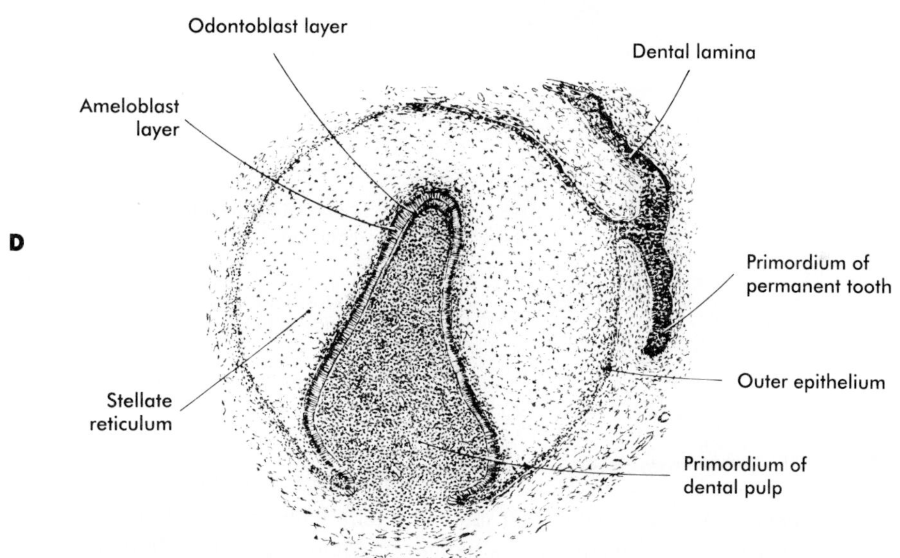

FIG. 15-15—cont'd C, Tooth primordium at the cap stage in an 11-week-old embryo, showing the enamel organ. **D,** Central incisor primordium at the bell stage in a 14-week-old embryo before deposition of enamel or dentin. *Continued.*

Although each of the teeth has a specific time sequence and morphology of development, certain general developmental stages are common to all teeth (Fig. 15-15). As the dental lamina grows into the neural crest mesenchyme, epithelial primordia of the individual teeth begin to take shape as **tooth buds.** In keeping with their interactive mode of development, the tooth buds are associated with condensations of mesenchymal cells. The tooth bud soon expands, passing through a mushroom-shaped **cap stage** before entering the **bell stage.**

By the bell stage the tooth primordium already has a complex structure, even though it has not formed any components of the definitive tooth. The epithelial component, called the **enamel organ,** is still connected to the oral epithelium by an irregular stalk of dental lamina, which soon begins to degenerate. The enamel organ consists of an **outer sheath** of epithelium, a mesenchymelike **stellate reticulum,** and an inner epithelial **ameloblast layer.** Ameloblasts are the cells that begin to secrete the enamel of the tooth. Within the concave surface of the enamel organ is a con-

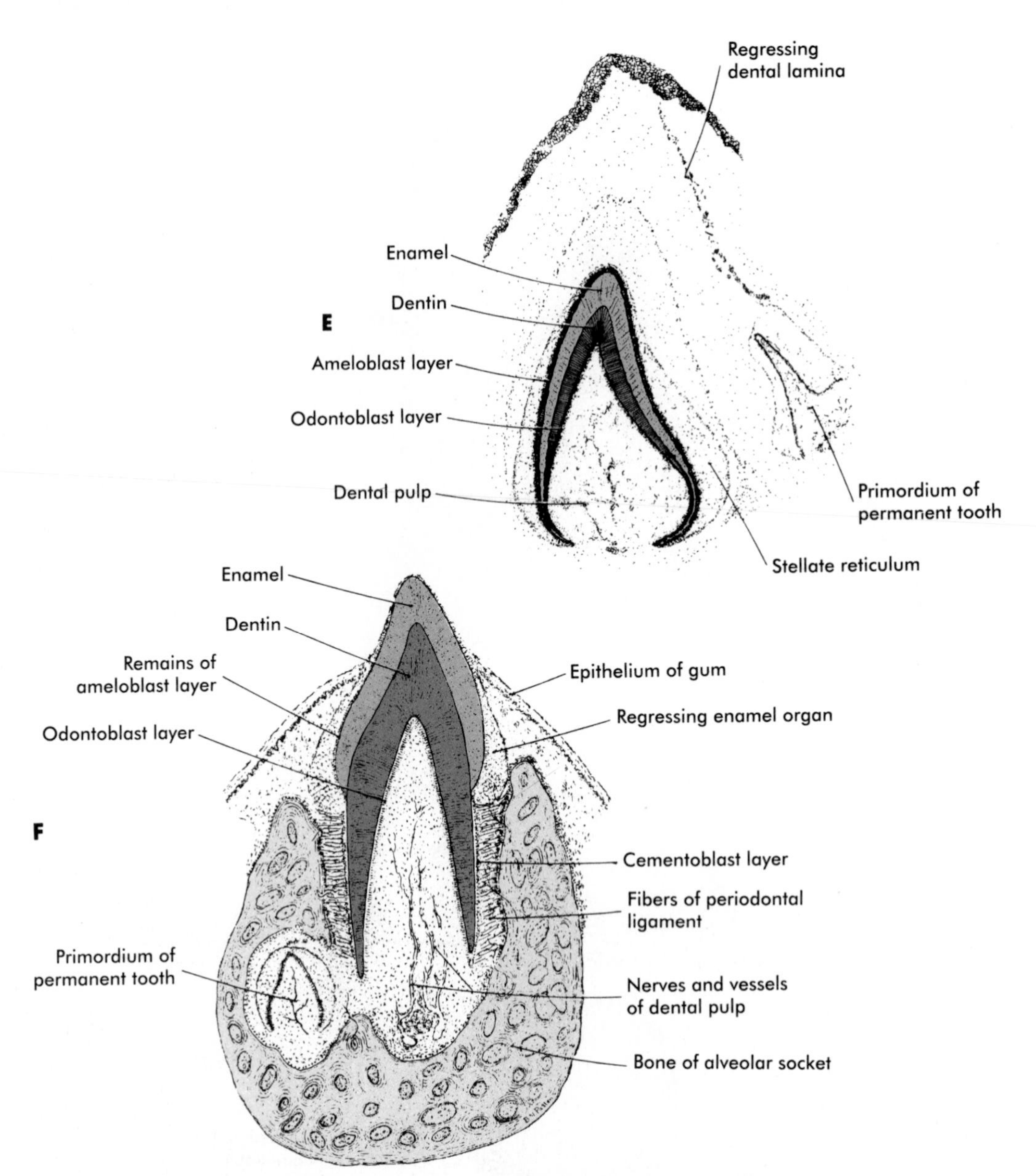

FIG. 15-15—cont'd **E,** Unerupted incisor tooth in a term fetus. **F,** Partially erupted incisor tooth showing the primordium of a permanent tooth near one of its roots. (After Patten.)

densation of neural crest mesenchyme called the **dental papilla.** Cells of the dental papilla opposite the ameloblast layer become transformed into columnar epithelial cells called **odontoblasts** (Fig. 15-16). These cells secrete the dentin of the tooth. Attached to the dental lamina close to the enamel organ is a small bud of the permanent tooth. Although delayed, it goes through the same developmental stages as the deciduous tooth.

Late in the bell stage the odontoblasts and ameloblasts begin to secrete precursors of dentin and enamel, beginning first at the future apex of the tooth. Over a period of several months the definitive form of the tooth takes shape (see Fig. 15-15). Meanwhile a condensation of mesenchymal cells forms around the developing tooth. Cells of this structure, called the **dental sac,** produce specialized extracellular matrix components (**cementum** and the **peridontal ligament**) that provide the tooth a firm attachment to the jaw. While these events are occurring, the tooth elongates and begins to erupt through the gums (**gingiva**).

Tissue Interactions in Tooth Development. Experimental studies involving tissue recombinations have shown that the earliest potential to initiate tooth morphogenesis resides in the dental epithelium. The presumptive dental mesenchyme responds to the epithelial induction by condensing around the forming epithelial bud. With this condensation of dental mesenchyme, the potential to direct tooth for-

A

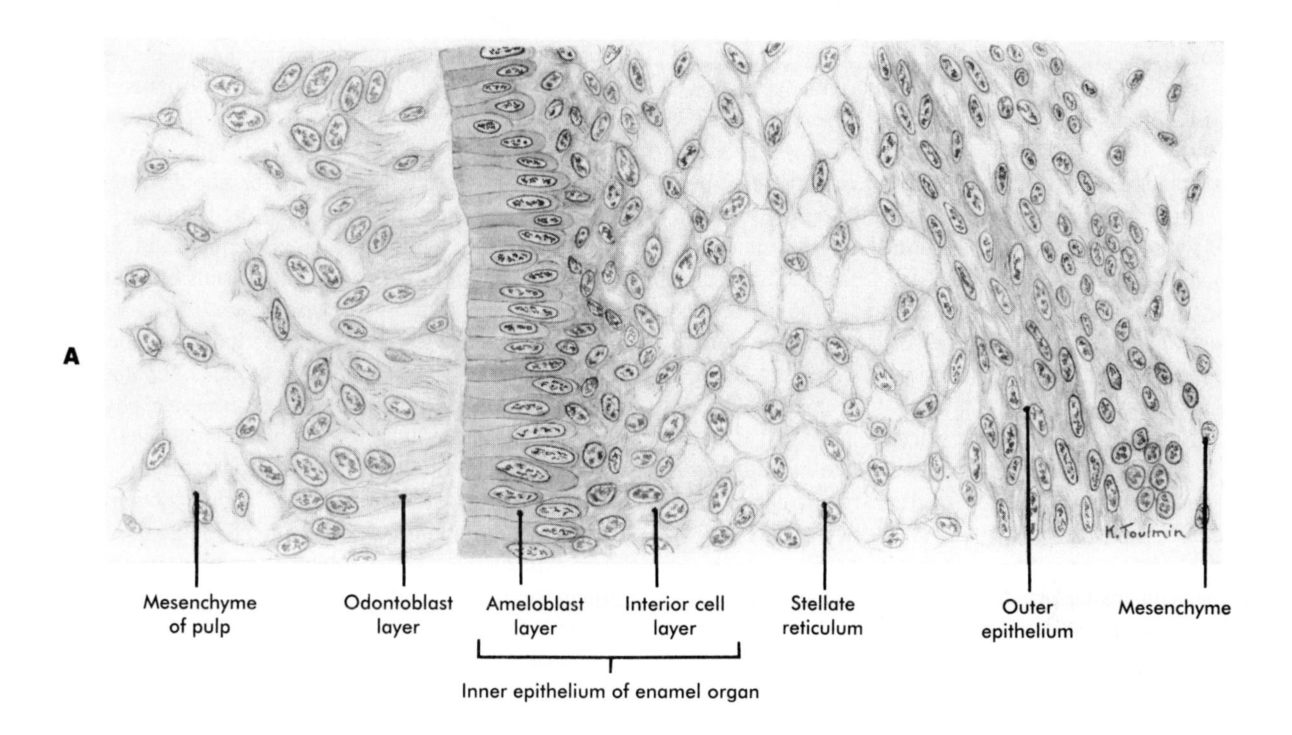

B

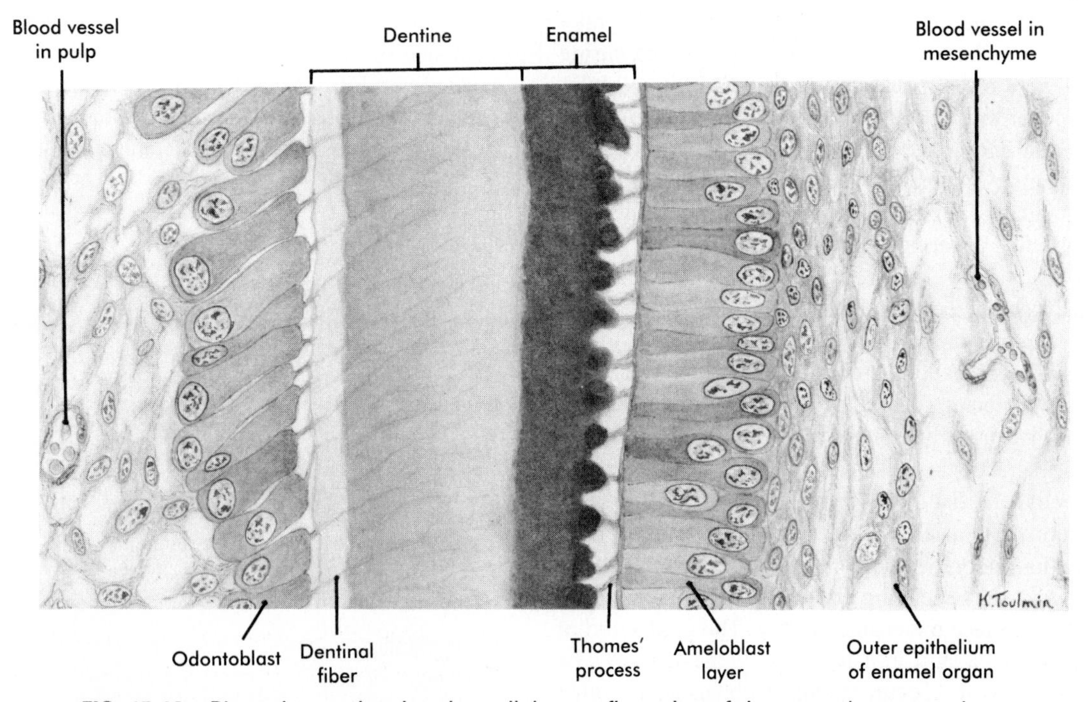

FIG. 15-16 Pig embryos showing the cellular configuration of the enamel organ and adjacent pulp of forming teeth before and after the beginning of deposition of enamel and dentin. **A,** Stage equivalent to a 4-month-old human embryo. **B,** Stage equivalent to a 5-month-old human embryo.

(From Patten B: *Human embryology,* ed 3, New York, 1968, McGraw-Hill.)

mation transfers to the mesenchyme even before the dental cap stage. Information on pattern and specific tooth morphogenesis resides in the neural crest mesenchyme of the dental papilla. Experimental evidence suggests that the neural crest cells are imprinted with some morphogenetic information before or early in their migration from the neural plate.

Recombination experiments have shown that the mesenchymal component determines the specific form of the tooth, as with most other organs formed through epitheliomesenchymal interactions. When molar mesenchyme is combined in vitro with incisor epithelium, a molar tooth takes shape, whereas combining incisor mesenchyme with molar epithelium results in the formation of an incisor. Such experiments do not necessarily mean that the epithelium is without information or influence, since dental papillae are induced in the second arch mesenchyme when mandibular arch epithelium of the mouse is combined with normally non-tooth–forming mesenchyme from the second pharyngeal arch.

The mechanisms underlying the reciprocal inductions in tooth development remain poorly understood, but much of the interaction seems to be mediated through components of the extracellular matrix. For example, the dental epithelium induces the expression of **tenascin** (see Fig. 13-3) and a heparan sulfate–rich proteoglycan called **syndecan** in the mesenchyme underlying the dental papilla. Syndecan appears first; later, tenascin becomes expressed during the early stages of epithelial downgrowth. Both molecules become coexpressed in the condensed mesenchyme of the dental papilla during the bud stage. This is accompanied by proliferation of the condensed mesenchymal cells. Syndecan binds fibroblast growth factor and may be responsible for ensuring that proliferative and growth activities are tightly localized. At a later stage, **fibronectin** may organize the polarization of the odontoblasts through receptors for this molecule located on cells of the dental papilla. Indications are increasing that locally produced growth factors (e.g., **epidermal growth factor**) interact with extracellular matrix molecules to effect the inductive interactions, but more research is needed to verify this hypothesis.

The Formation of Dentin and Enamel. Late in their differentiation, odontoblasts withdraw from the cell cycle, elongate, and start secreting **predentin** from their apical surfaces (which face the enamel organ). The production of predentin signals a shift in patterns of synthesis from type III collagen and fibronectin to type I collagen and other molecules (e.g., **dentin phosphoprotein, dentin osteocalcin**) that characterize the dentin matrix. The first dentin is laid against the inner surface of the enamel organ at the apex of the tooth (see Fig. 15-16, *B*). With the secretion of additional dentin, the accumulated material pushes the odontoblastic epithelium from the odontoblast-ameloblast interface.

Terminal differentiation of the ameloblasts occurs after the odontoblasts begin to secrete predentin. In fact, the presence of predentin may serve as a signal for ameloblast differentiation. Like the odontoblasts, the ameloblasts withdraw from the cell cycle and begin a new pattern of synthesis, producing two classes of proteins—**amelogenins** and **enamelins.** About 5% of enamel consists of organic matrix, and amelogenins account for about 90% of this, with enamelins constituting most of the remainder. Enamelins are secreted before the amelogenins, and they may serve as nuclei for the formation of crystals of **hydroxyapatite,** the dominant inorganic component of enamel.

The amelogenin genes have been cloned and are located on the X and Y chromosomes in humans. Enamel genes have been highly conserved during vertebrate phylogeny. It has been suggested that in early vertebrates, enamel once served as part of an electroreceptor apparatus.

Tooth Eruption and Replacement. Each tooth has a specific time of eruption and replacement (Table 15-1). With the growth of the root, the enamel-covered crown pushes through the oral epithelium. The sequence of eruption begins with the central incisor teeth, usually a few months after birth, and continues generally stepwise until the last of the deciduous molars forms at the end of the second year. A total of 20 deciduous teeth are formed.

Meanwhile, the primordium of the permanent tooth is embedded in a cavity extending into the bone on the lingual side of the alveolar socket in which the tooth is embedded (see Fig. 15-15, *F*). As the permanent tooth develops, its increasing size causes resorption of the root of the deciduous tooth. When a sufficient amount of the root is

TABLE 15-1 The Usual Times of Eruption and Shedding of Deciduous and Permanent Teeth

TEETH	ERUPTION	SHEDDING
DECIDUOUS		
Central incisors	6-8 mo	6-7 yr
Lateral incisors	7-10 mo	7-8 yr
Canines	14-18 mo	10-12 yr
First molars	12-16 mo	9-11 yr
Second molars	20-24 mo	10-12 yr
PERMANENT		
Central incisors	7-8 yr	
Lateral incisors	8-9 yr	
Canines	12-13 yr	
First premolars	10-11 yr	
Second premolars	11-12 yr	
First molars	6-7 yr	
Second molars	12-13 yr	
Third molars	15-25 yr	

destroyed, the deciduous tooth falls out, leaving room for the permanent tooth to take its place. The sequence of eruption of the permanent teeth is the same as that of the deciduous teeth, but an additional 12 permanent teeth (for a total of 32) are formed without deciduous counterparts.

The formation and eruption of teeth are important factors in midfacial growth, much of which occurs after birth. Tooth development and the corresponding growth of the jaw to accommodate them, along with the development of the paranasal sinuses, account for much of the tissue mass of the midface.

Malformations of the Face and Oral Region

Cleft Lip and Palate. Cleft lip and cleft palate are relatively common malformations, with an incidence of approximately 1:1000 and 1:2500 births, respectively. Numerous combinations and degrees of severity exist, ranging from a unilateral cleft lip to a bilateral cleft lip associated with a fully cleft palate.

Structurally, **cleft lip** is the result of the lack of fusion of the maxillary and nasomedial processes. In the most complete form of the defect, the entire premaxillary segment is separated from both maxillae, resulting in bilateral clefts that run through the lip and the upper jaw between the lateral incisors and the canine teeth (Fig. 15-17). The point of convergence of the two clefts is the incisive foramen. The premaxillary segment commonly protrudes past the normal facial contours when viewed from the side. The mechanism frequently underlying cleft lip is hypoplasia of the maxillary process, preventing contact between the maxillary and nasomedial processes from being established.

Cleft palate results from incomplete or absent fusion of the palatal shelves (Fig. 15-18). The defect varies greatly, from involving the entire length of the palate to resulting in a bifid uvula. As with cleft lip, cleft palate is usually multifactorial. Some chromosomal syndromes (e.g., trisomy 13) are characterized by a high incidence of clefts. In other cases, cleft lip and palate can be linked to the action of a chemical teratogen (e.g., anticonvulsant drugs). Experiments on mice have shown that the incidence of cleft palate after exposure to a dose of cortisone is strongly related to the genetic background of the mouse. The higher incidence of cleft palate in females could also be related to the palatal shelves in females fusing about a week later than they do in males, prolonging the susceptible period.

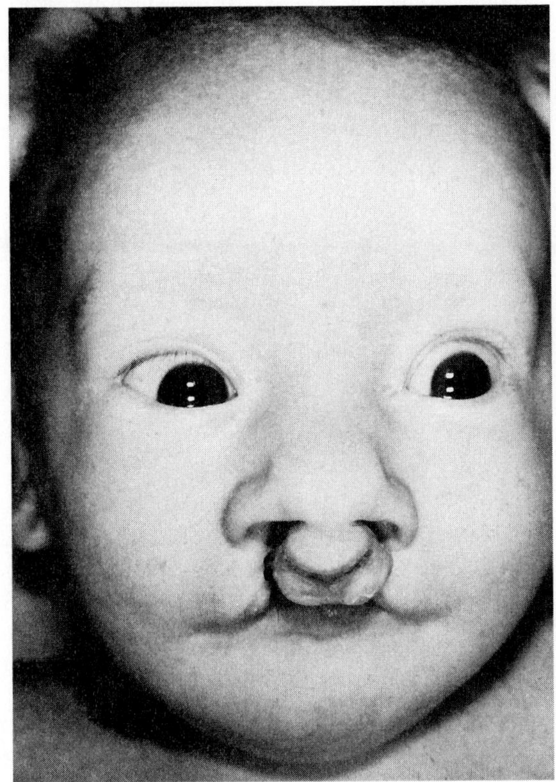

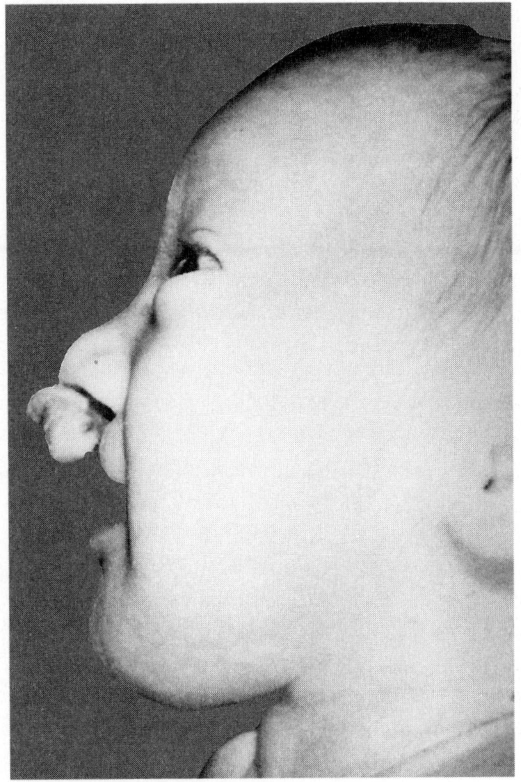

FIG. 15-17 Front and lateral views of an infant with bilateral cleft lip and palate. On the lateral view, note how the premaxillary segment is tipped outward.
(Courtesy A. Burdi, Ann Arbor, Mich.)

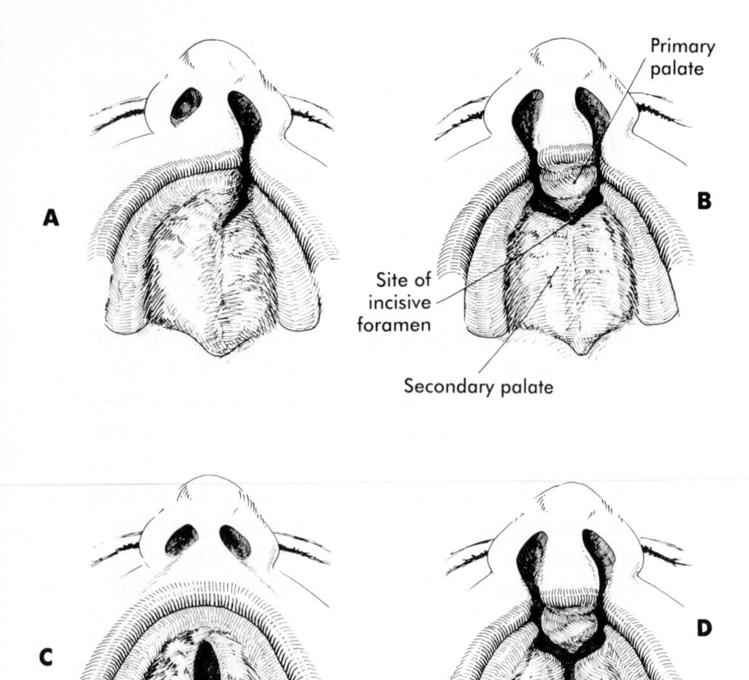

FIG. 15-18 Common varieties of cleft lip and palate. **A,** Unilateral cleft passing through the lip and between the premaxilla (primary palate) and secondary palate. **B,** Bilateral cleft lip and palate similar to that seen in the patient in Fig. 15-17. **C,** Midline palatal cleft. **D,** Bilateral cleft lip and palate continuous with a midline cleft of the secondary palate.

Oblique Facial Cleft. This rare defect results when the nasolateral process fails to fuse with the maxillary process, usually resulting from hypoplasia of one of the tissue masses (Fig. 15-19, *A*). It frequently manifests as an epithelially lined fissure running from the upper lip to the medial corner of the eye.

Macrostomia (Lateral Facial Cleft). An even rarer condition called **macrostomia** (Fig. 15-19, *B*) is the result of hypoplasia or poor merging of the maxillary and mandibular processes. As the name implies, this condition manifests as a very large mouth on one or both sides. In severe cases the cleft can reach almost to the ears.

Median Cleft Lip. Another rare anomaly, **median cleft lip,** results from incomplete merging of the two nasomedial processes (Fig. 15-19, *C*).

Holoprosencephaly. Holoprosencephaly results in a broad spectrum of defects, all based on defective formation of the forebrain (prosencephalon) and structures whose normal formation depends on influences from the forebrain. The defect arises in early pregnancy when the forebrain is taking shape, and the brain defects usually involve archencephalic structures (e.g., the olfactory system). Because of the influence of the brain on surrounding structures, primary defects of the forebrain often manifest externally as facial malformations, typically a reduction in tissue of the frontonasal process.

In extreme cases, holoprosencephaly can take the form of cyclopia (see Fig. 8-16), in which the near absence of upper and midfacial tissue results in a convergence and fusion of the optic primordia. Reduction defects of the nose can also be components of this condition. The nose can be either absent or represented by a tubular **proboscis** (or two

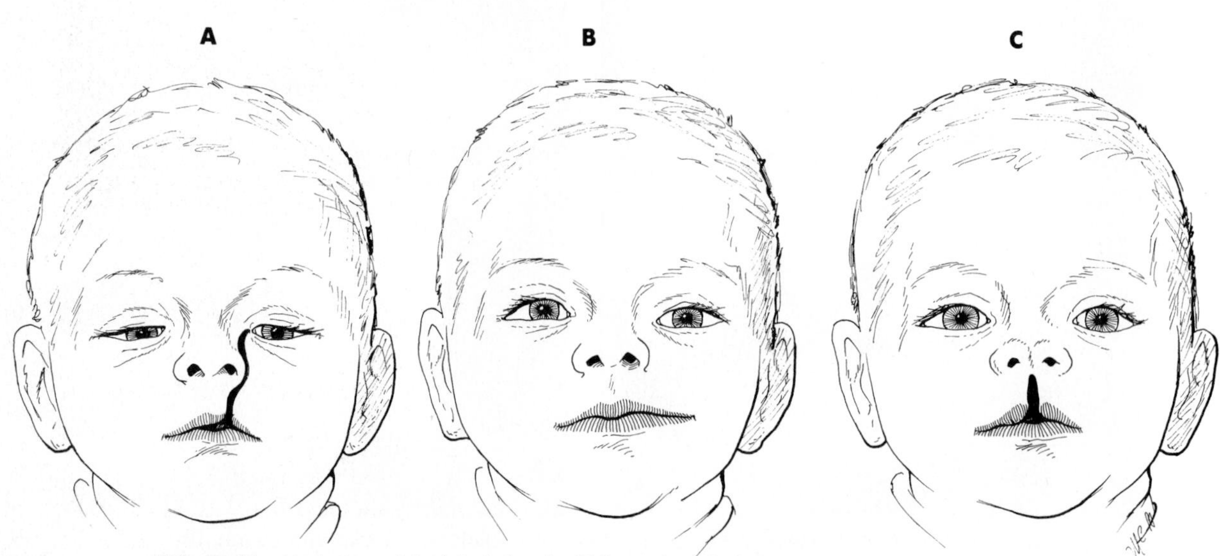

FIG. 15-19 Varieties of facial clefts. **A,** Oblique facial cleft combined with a cleft lip. **B,** Macrostomia. **C,** Medial cleft lip with a partial nasal cleft.

such structures), sometimes even located above the eye. Midline defects of the upper lip can also be attributed to holoprosencephaly (see Fig. 15-19, *C*).

Some cases of holoprosencephaly (e.g., **Meckel's syndrome,** which includes midline cleft lip, olfactory bulb absence or hypoplasia, and nasal abnormalities) can be attributed to genetic causes. Meckel's syndrome is an autosomal recessive condition. Most cases of holoprosencephaly appear to be multifactorial, although maternal consumption of alcohol during pregnancy is suspected to be a leading cause of this condition. Trisomies of chromosomes 13 and 18 are commonly associated with holoprosencephaly.

Frontonasal Dysplasia. Frontonasal dysplasia encompasses a variety of forms of nasal malformations that result from an excess of tissue in the frontonasal process. The spectrum of anomalies usually includes a broad nasal bridge and **hypertelorism** (an excessive distance between the eyes). In very severe cases the two external nares are separated, often by several centimeters, and a median cleft lip can also occur (Fig. 15-20).

Abnormalities of the Teeth. A wide variety of conditions affect the teeth. Some are abnormalities of pattern, which could be a reflection of abnormal morphogenetic instructions of the early neural crest or secondary to growth defects of the jaw. Abnormal morphogenesis of individual teeth is common, with frequent variations including extra or distorted roots, enamel "pearls" (small masses of enamel) along the tooth, and abnormally shaped crowns. At a different level are **amelogenesis imperfecta** and **dentinogenesis imperfecta.** These rare genetic conditions are characterized by a defect in expression of one or more matrix proteins of the enamel or dentin. Progress is being made in identifying the specific genes that are defective in these conditions.

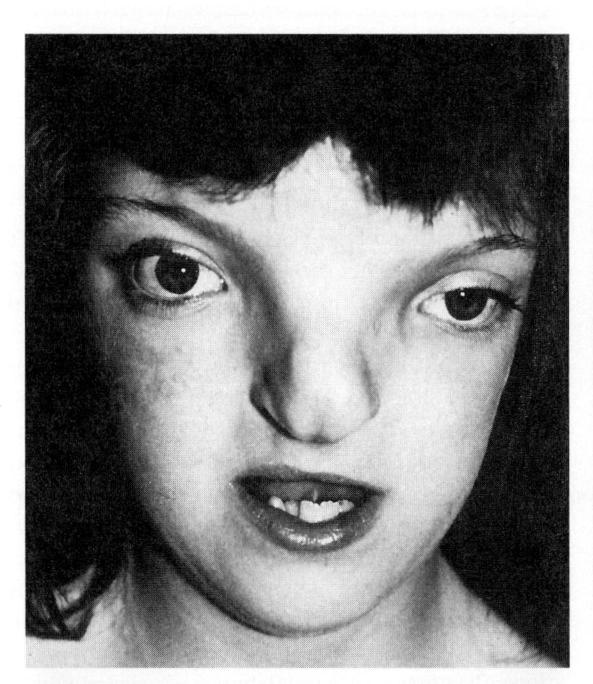

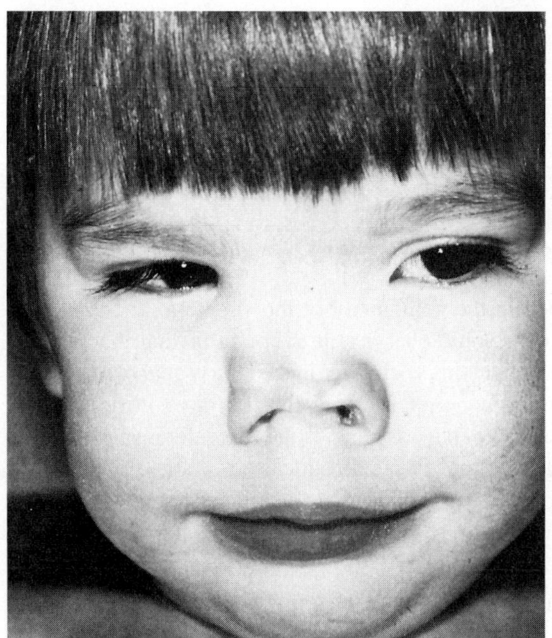

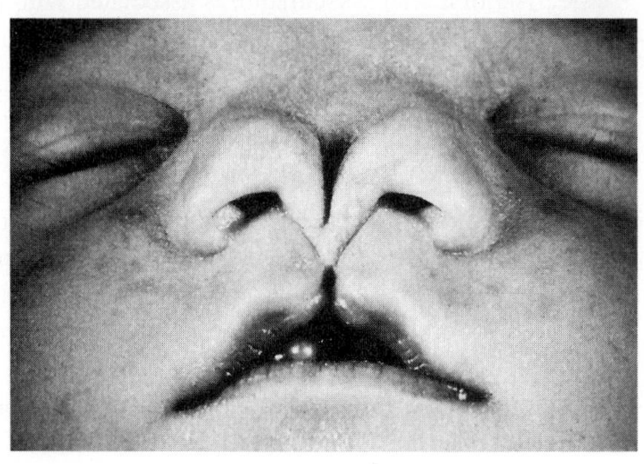

FIG. 15-20 Varying degrees of frontonasal dysplasia.
(Courtesy A. Burdi, Ann Arbor, Mich.)

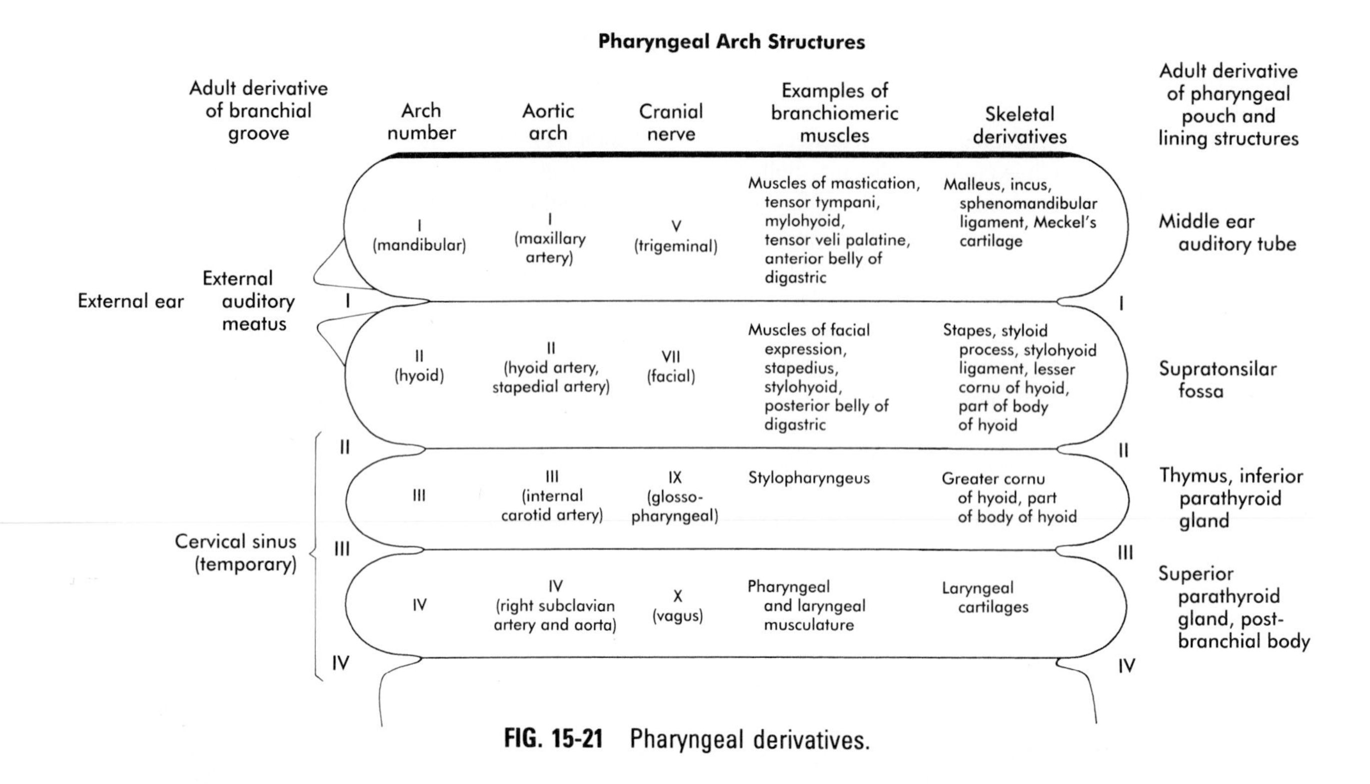

FIG. 15-21 Pharyngeal derivatives.

DEVELOPMENT OF THE PHARYNX AND ITS DERIVATIVES

Considering the complexity of the structural arrangements of the embryonic pharynx, it is not surprising that a wide variety of structures originate in the pharyngeal region. This provides many opportunities for defective development. This section details aspects of later development that lead to the formation of specific structures. Adult derivatives of the regions of the pharynx and pharyngeal arches are summarized in Fig. 15-21.

External Development of the Pharyngeal Region

Externally, the pharyngeal (branchial) region is characterized by four pharyngeal arches and branchial grooves interposed between the arches (see Fig. 15-4). These structures give rise to a diverse array of derivatives.

The Pharyngeal Arches. In addition to being packed with mesenchyme (mainly of neural crest origin except for the premuscle mesoderm, which migrates from the somitomeres), each pharyngeal arch is associated with a major artery (aortic arch) and a cranial nerve (see Fig. 15-21). Each also contains a central rod of precartilaginous mesenchyme, which is transformed into characteristic adult skeletal derivatives. Understanding the relationship between the pharyngeal arches and their innervation and vascular supply is very important, since tissues often maintain their relationship with their original nerve as they migrate out or become displaced from their original location in the pharyngeal arch system.

The **first pharyngeal arch** (mandibular) contributes mainly to structures of the face (both mandibular and maxillary portions) and ear (see Fig. 15-21). Its central cartilaginous rod, Meckel's cartilage, is a prominent component of the embryonic lower jaw until it is surrounded by locally formed intramembranous bone, which forms the definitive jaw. More dorsally, Meckel's cartilage forms the **sphenomandibular ligament,** the **anterior ligament of the malleus,** and the **malleus** (Fig. 15-22). Additionally, the **incus** arises from a primordium of the **quadrate cartilage.** The first arch musculature is associated with the masticatory apparatus, the pharynx, and the middle ear. A common feature of these muscles is their innervation by the trigeminal nerve (V).

The **second pharyngeal arch** (hyoid) also forms a string of skeletal structures from the body of the hyoid bone to the **stapes** of the middle ear. Although much of the second arch mesoderm migrates to the face to form the **muscles of facial expression,** additional muscles form with other second arch skeletal derivatives. These muscles are innervated by the facial nerve (VII).

The **third** and **fourth pharyngeal arches** are otherwise unnamed. The third arch gives rise to muscle and skeletal

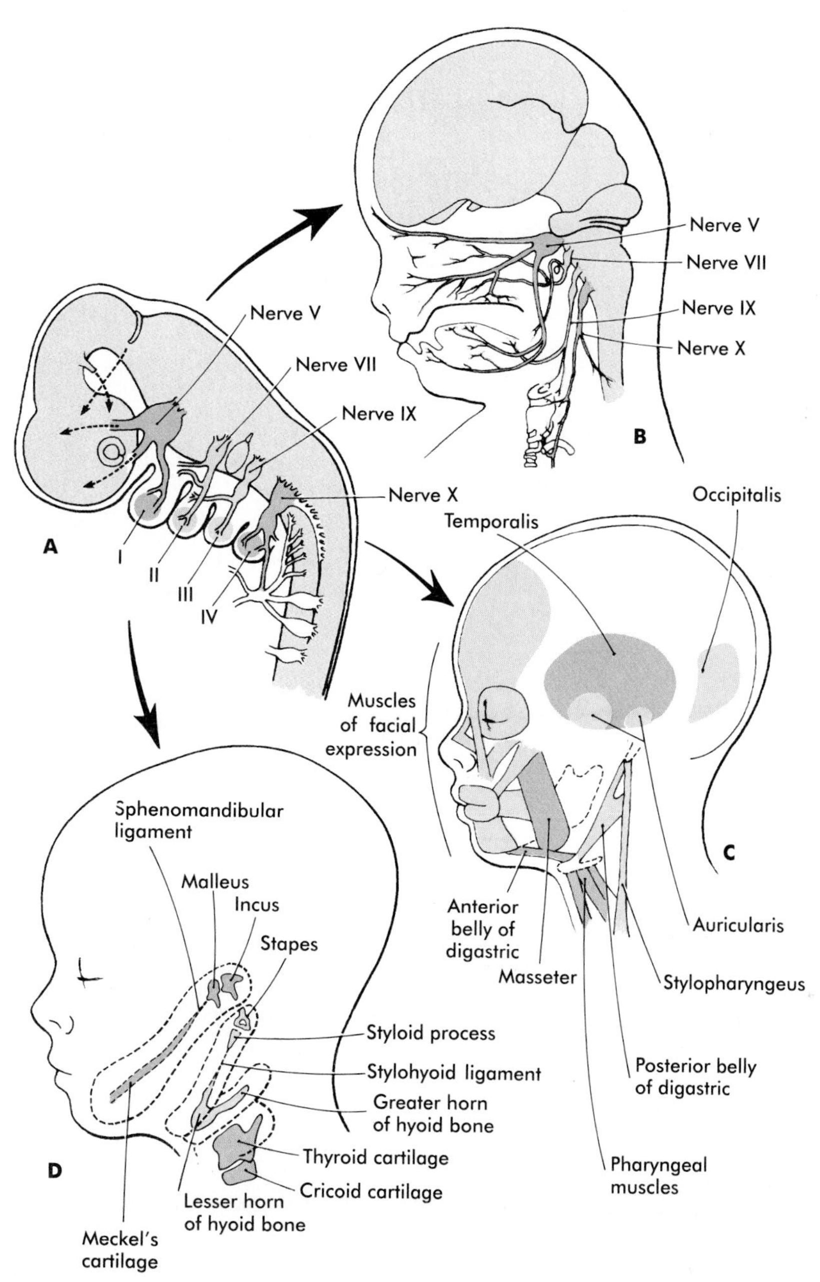

FIG. 15-22 The pharyngeal arch system **(A)** and the adult derivatives of the neural **(B)**, muscular **(C)**, and skeletal **(D)** components of the arches.

derivatives related to the hyoid bone and upper pharynx. Its one muscular derivative (stylopharyngeus) is innervated by the glossopharyngeal nerve (IX). The fourth arch gives rise to certain muscles and cartilages of the larynx and lower pharynx. The muscles are innervated by the vagus nerve (X), which also grows into the thoracic and abdominal cavities.

A sixth pharyngeal arch is presumed to be the equivalent of the sixth gill (branchial) arch of lower vertebrates. The fifth arch is not represented in the human embryo.

The Branchial Grooves. The first branchial groove is the only one that persists as a recognizable adult structure, the **external auditory meatus.** The remaining grooves (II to IV) are partially covered by the enlarged external portion of the second arch (a phylogenetic homologue of the operculum [gill cover] of fish). During the period of their overshadowing by the hyoid arch, grooves II to IV are collectively known as the **cervical sinus** (Fig. 15-23). As development progresses the cervical sinus disappears and the external contours of the neck become smooth.

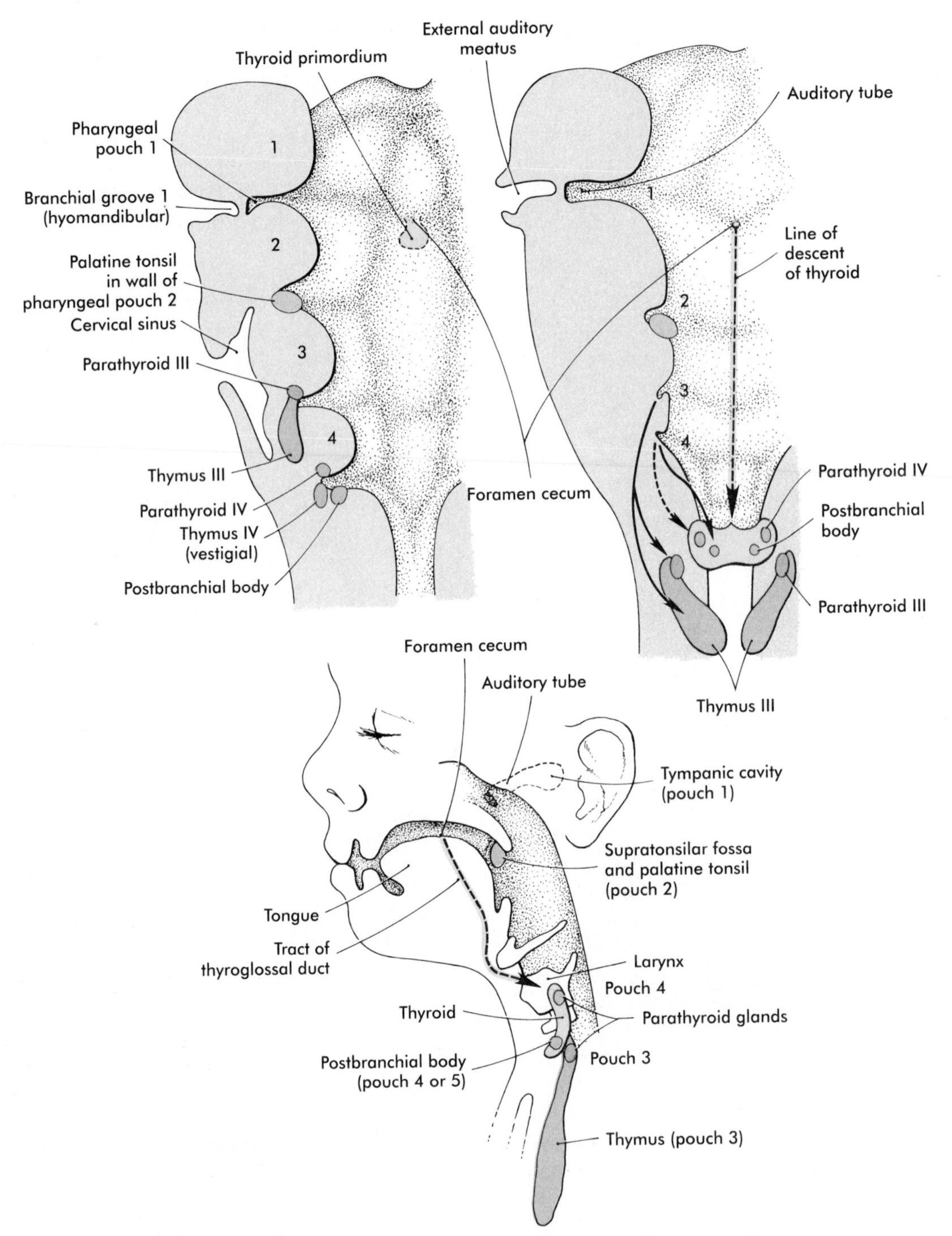

FIG. 15-23 The embryonic origins and pathways of primordia of glands derived from the pharyngeal pouches and the floor of the pharynx.

The Pharynx and Pharyngeal Pouches. The embryonic pharynx is directly converted to the smooth-walled pharynx of the adult. Of greater developmental interest is the fate of the pharyngeal pouches and their lining epithelium.

As with their corresponding first branchial clefts, the **first pharyngeal pouches** become intimately involved with formation of the ear. The ends expand to become the **tympanic cavity** of the middle ear, and the remainder becomes the **auditory (eustachian) tube,** which connects the middle ear with the pharynx (see Fig. 14-24).

The **second pharyngeal pouches** become shallower and less conspicuous as development progresses. Late in the fetal period, patches of lymphoid tissue form aggregations in the walls to form the **palatine (faucial) tonsils.** The pouches are represented only as the **supratonsillar fossae.**

The **third pharyngeal pouch** is a more complex structure, consisting of a solid, dorsal epithelial mass and a hollow, elongated ventral portion (see Fig. 15-23). By the fifth week of gestation, cells identifiable as **parathyroid** tissue can be recognized in the endoderm of the solid dorsal masses. The ventral elongations of the third pouches differentiate into the epithelial portion of the **thymus gland.** The primordia of both the thymus and the parathyroid glands lose their connection with the third pharyngeal pouch and migrate caudally to their site of origin. Although the parathyroid III primordia initially comigrate with the thymic primordia, they ultimately continue to migrate toward the midline. There they join up with the thyroid gland, passing the parathyroid primordia of the fourth pouch to form the **inferior parathyroid glands.** The third pharyngeal pouch disappears.

The **fourth pharyngeal pouch** is organized somewhat similar to the third, with a solid, bulbous, dorsal parathyroid IV primordium. It also contains a small ventral epithelial outpocketing, which contributes a minor component to the thymus in some species. In humans the thymic component of the fourth pouch is vestigial. At the ventral-most part of each fourth pouch is another structure called the **postbranchial (ultimobranchial) body** (see Fig. 15-23). These structures are sometimes referred to as rudimentary fifth pharyngeal pouches.

Uncertainty surrounds the cellular origins and composition of the postbranchial bodies. According to one view the postbranchial bodies arise purely from pharyngeal endoderm. Another theory suggests that neural crest cells migrate into the postbranchial bodies and ultimately become the secretory component of these structures. An ectodermal placodal source has also been proposed for the postbranchial bodies.

As with their counterparts from the third pouch, the parathyroid IV primordia lose their connection with the fourth pouch and migrate toward the thyroid gland as the **superior parathyroid glands.** The postbranchial bodies also migrate toward the thyroid, where they become incorporated as **parafollicular** or **C cells.** The parafollicular cells, which are of neural crest origin, produce the polypeptide hormone **calcitonin,** which acts to reduce the concentration of calcium in the blood. The parathyroid glands produce **parathyroid hormone,** which causes an increase in blood calcium.

Midline Structures Arising from the Pharynx. An unpaired primordium of the **thyroid gland** appears in the ventral midline of the pharynx between the first and second pouches (see Fig. 15-23). Starting during the fourth week as an endodermal thickening just caudal to the median tongue bud (tuberculum impar), the thyroid primordium soon elongates to form a prominent downgrowth called the **thyroid diverticulum.** Caudal extension of the thyroid diverticulum continues during development of the pharynx. During its caudal migration the tip of the thyroid diverticulum expands and bifurcates to form the thyroid gland itself, which consists of two main lobes connected by an isthmus. For some time the gland remains connected to its original site of origin by a narrow **thyroglossal duct.** By about the seventh week when the thyroid has reached its final location at the level of the second and third tracheal cartilages, the thyroglossal duct has largely regressed. Nevertheless, in almost half the population the distal portion of the thyroglossal duct persists as the **pyramidal lobe of the thyroid.** The original site of the thyroid primordium persists as the **foramen cecum,** a small blind pit at the base of the tongue.

The thyroid gland undergoes histodifferentiation and begins functioning relatively early in embryonic development. By the tenth week of gestation, follicles containing some colloid material are evident, and a few weeks thereafter the gland begins to synthesize noniodinated **thyroglobulin.** Secretion of **triiodothyronine,** one of the forms of thyroid hormone, is detectable by late in the fourth month.

The Thymus and Lymphoid Organs. The paired endodermal thymic primordia begin to migrate from their pharyngeal pouch origins during the sixth week. Their path of migration takes them through a substrate of mesenchymal cells until they reach the area of the future mediastinum behind the sternum. By the end of their migration, the two closely apposed thymic lobes are still epithelial structures. Soon, however, they become invested with a capsule of neural crest–derived connective tissue, which also forms septa among the endodermal epithelial cords. In the absence of neural crest the thymus fails to develop. Thus an interaction between the neural crest and endodermal components of the thymic primordia conditions the latter for subsequent differentiation of thymic structure and function.

At about 9 to 10 weeks gestation, blood-borne thymocyte precursors **(prothymocytes),** which originate in the hematopoietic tissue, begin to invade the epithelial thymus, presumably in response to the secretion of peptides by the thymus. Within the thymus the prothymocytes force apart the epithelial cells, causing them to form a spongy **epithelial reticulum.** Under the influence of the thymic epithelium the prothymocytes proliferate and redistribute, forming the cortical and medullary regions of the thymus. By 14 to 15 weeks, blood vessels grow into the thymus, and a week later, some epithelial cells aggregate into small, spherical **Hassal's corpuscles.** At this point the overall organization of the thymus is the same as in adults. Functionally, the action of various **thymic hormones** causes the thymus to condition or instruct the prothymocytes immigrating into it to become competent members of the **T-lymphocyte** family. The T lymphocytes leave the thymus and populate other lymphoid organs (e.g., lymph nodes, spleen) as fully functional immune cells.

The T lymphocytes are principally involved in **cellular immune responses.** Another population of lymphocytes that also originates in the bone marrow is instructed to become **B lymphocytes,** which are the mediators of **humoral immune responses.** B-lymphocyte precursors **(pro B cells)** must also undergo conditioning to become fully functional, but their conditioning does not occur in the thymus. In birds, the pro B cells pass through a cloacal lymphoid organ known as the **bursa of Fabricius,** where conditioning occurs. Humans do not possess a bursa, but its functional equivalent is assumed to exist, although it has not been discovered. B-lymphocyte conditioning is thought to occur in the bone marrow; in early embryos, conditioning possibly occurs in the liver.

The thymus and the bursa or mammalian equivalent are commonly referred to as **central lymphoid organs.** The lymphoid structures that are seeded by B and T lymphocytes are called **peripheral lymphoid organs.** (Fig. 15-24 shows the development and function of the lymphoid system.)

Formation of the Tongue. The tongue begins to take shape from a series of ventral swellings in the floor of the pharynx about the same time as the palate forms in the mouth. Major shifts in positions of tissues of the tongue occur, making the characteristics of the adult form difficult to comprehend without knowledge of the basic elements of its embryonic development.

In 5-week-old embryos the tongue is represented by a

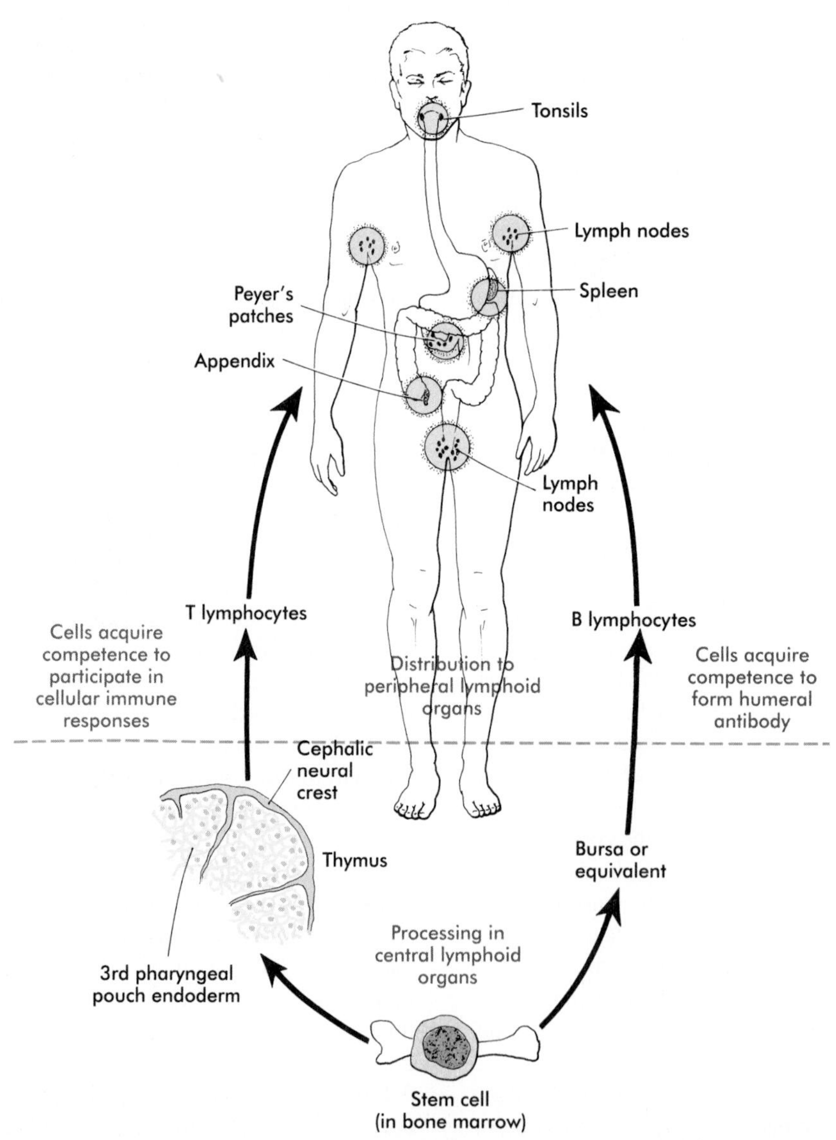

FIG. 15-24 The embryonic development of the lymphoid system.

pair of **lateral lingual swellings** in the ventral regions of the first pharyngeal arches and two median unpaired swellings. The **tuberculum impar** is located between the first and second arches, and the **copula** (yoke) unites the second and third arches (Fig. 15-25, *A* and *B*). The foramen cecum, which marks the original location of the thyroid primordium, serves as a convenient landmark delineating the border between the original tuberculum impar and the copula. Caudal to the copula is another swelling that represents the **epiglottis.**

Growth of the body of the tongue is accomplished by a great expansion of the lateral lingual swellings, with a minor contribution by the tuberculum impar (Fig. 15-25, *C* and *D*). The root of the tongue is derived from the copula along with additional ventromedial tissue between the third and fourth branchial arches.

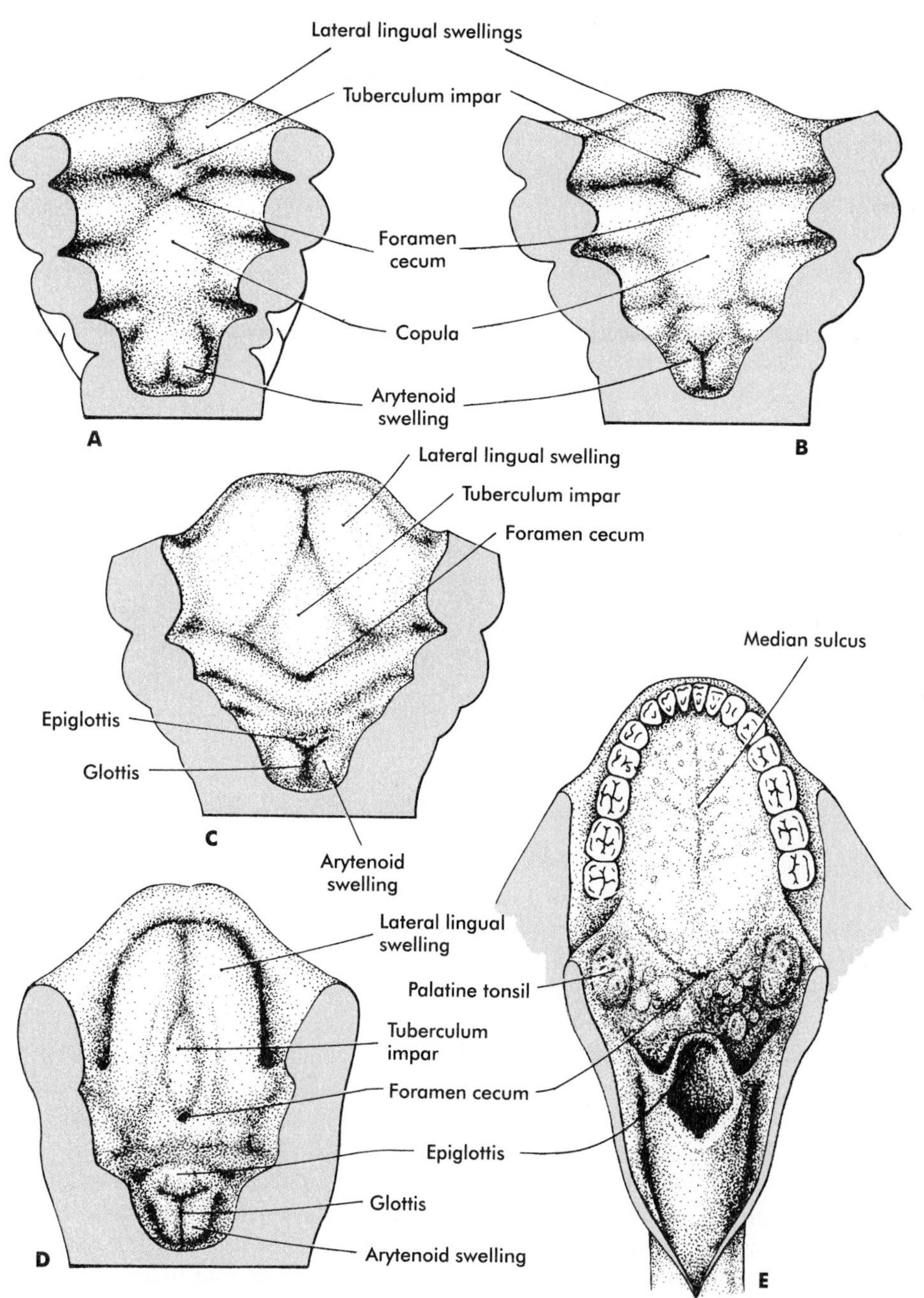

FIG. 15-25 Development of the tongue as seen from above. **A,** At 4 weeks. **B,** Late in the fifth week. **C,** Early in the sixth week. **D,** Middle of the seventh week. **E,** The adult form.

Because of its innervation by the hypoglossal nerve (XII), the musculature of the tongue is assumed to arise from the **occipital (postotic) myotomes** and to have migrated a considerable distance into the lingual swellings, retaining its original nerve supply. The general sensory innervation of the tongue accurately reflects the pharyngeal arch origins of the epithelium. Thus the lingual epithelium over the body of the tongue is innervated by the trigeminal nerve (V), in keeping with the first arch origins of the lateral lingual swellings. Correspondingly, the root of the tongue is innervated by the glossopharyngeal nerve (IX—third arch) and vagus nerve (X—fourth arch). The epithelium of the second arch is overgrown by that of the third arch; therefore there is no general sensory innervation of the tongue by the seventh nerve.

The seventh (facial) and ninth nerves innervate the taste buds. The contribution of the seventh nerve is facilitated by its **chorda tympani** branch, which joins with the trigeminal nerve and thus has access to the body of the tongue. The taste buds, which appear during the seventh week of gestation, are the result of an interaction between the lingual epithelium and special visceral afferent fibers of nerves VII and IX. Considerable evidence indicates that the fetus is able to taste, and it has been postulated that the fetus uses the taste function to monitor its intraamniotic environment.

Anomalies and Syndromes Involving the Pharynx and Pharyngeal Arches

Syndromes Involving the First Pharyngeal Arch. Several syndromes involve hypoplasia of the mandible and other structures arising from the first pharyngeal arch. However, the mechanisms underlying the hypoplasia are not clear. Recent research with transgenic mice involving ectopic expression of the homeobox gene *Hox-1.1* (A7) has produced mice with abnormalities of first arch derivatives and other anomalies of the cranial region.

The **Pierre Robin syndrome** involves extreme **micrognathia** (small mandible), cleft palate, and associated defects of the ear. An imbalance often exists between the size of the tongue and the very hypoplastic jaw, which can lead to respiratory distress due to mechanical interference of the pharyngeal airway by the relatively large tongue. Although many cases of Pierre Robin syndrome are sporadic, others appear to have a genetic basis.

Treacher Collins syndrome (mandibulofacial dysostosis) is typically inherited as an autosomal dominant condition. It includes a wide variety of anomalies, not all of which are found in the same patient. Common components of the syndrome include hypoplasia of the mandible and facial bones, malformations of the external and middle ears, high or cleft palate, faulty dentition, and coloboma-type defects of the lower eyelid (Fig. 15-26).

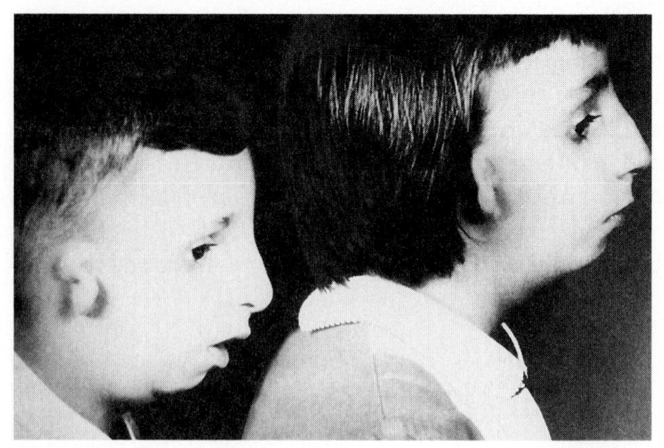

FIG. 15-26 Two siblings with Treacher Collins' syndrome. (Courtesy A. Burdi, Ann Arbor, Mich.)

The most extreme form of first arch hypoplasia is **agnathia,** in which the lower jaw basically fails to form (Fig. 15-27). In severe agnathia the external ears remain in the ventral cervical region and may even join in the ventral midline.

Lateral Cysts, Sinuses, and Fistulas. This class of structural malformations can be related directly to the abnormal persistence of branchial grooves, pharyngeal pouches, or both. A **cyst** is a completely enclosed, epithelially lined cavity that may be derived from persistence of part of a pharyngeal pouch, branchial groove, or cervical sinus. A **sinus** is closed on one end and open to the outside or to the pharynx. A **fistula** (Latin for "pipe") is an epithelially lined tube that is open at both ends—in this case to the outside and to the pharynx.

The postnatal location of these structures accurately marks the location of their embryonic precursors. External openings of cysts are typically found anterior to the sternocleidomastoid muscle in the neck (Fig. 15-28). Although present from birth, cervical cysts are often not manifest until after puberty. At that time they expand because of increased amounts of secretions by the epithelium that lines the inner surface of the cyst, corresponding to maturational changes in the normal epidermis.

Preauricular sinuses or **fistulas,** found in a triangular region in front of the ear, are also common, but they are not homologous to the corresponding cervical structures. These structures are assumed to represent persistent clefts between preauricular hillocks on the first and second arches. True fistulas **(cervicoaural fistulas)** represent persisting ventral portions of the first branchial groove. These extend from a pharyngeal opening to somewhere along the auditory tube or even the external auditory meatus.

Thyroglossal Duct Remnants. A variety of abnormal structures can persist along the pathway of the thyroglossal duct. **Ectopic thyroid tissue** can be found anywhere along the pathway of migration of the thyroid primordium from the foramen cecum in the tongue to the isthmus of the normal thyroid gland (Fig. 15-29). This fact must be consid-

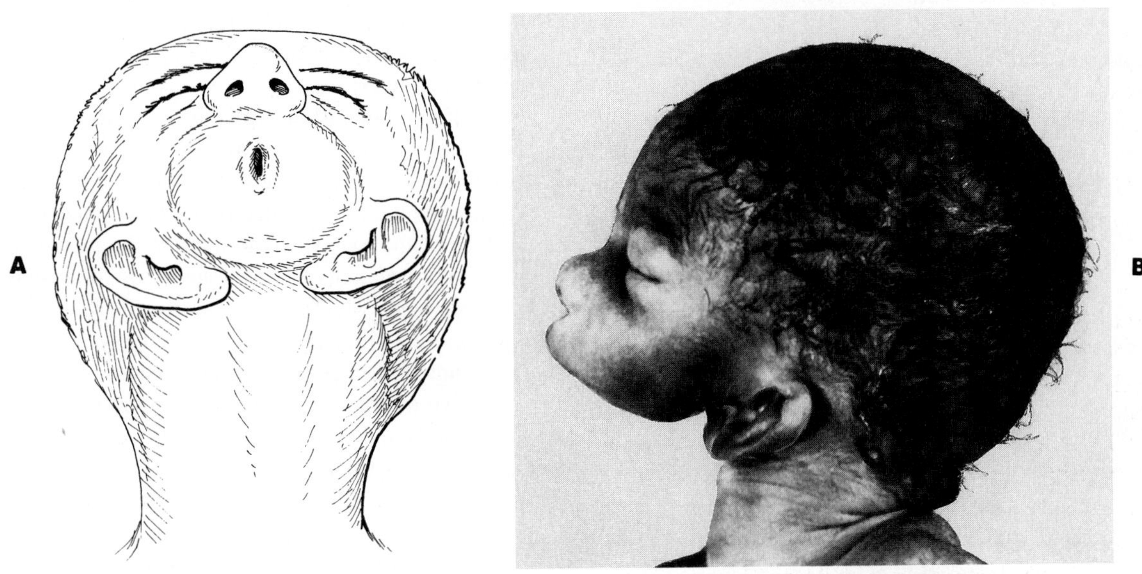

FIG. 15-27 **A,** Ventral view of the upturned face of an infant with agnathia. **B,** Lateral view of a fetus with agnathia. Note the cervical location of the external ears.
(Courtesy Mason Barr, Ann Arbor, Mich.)

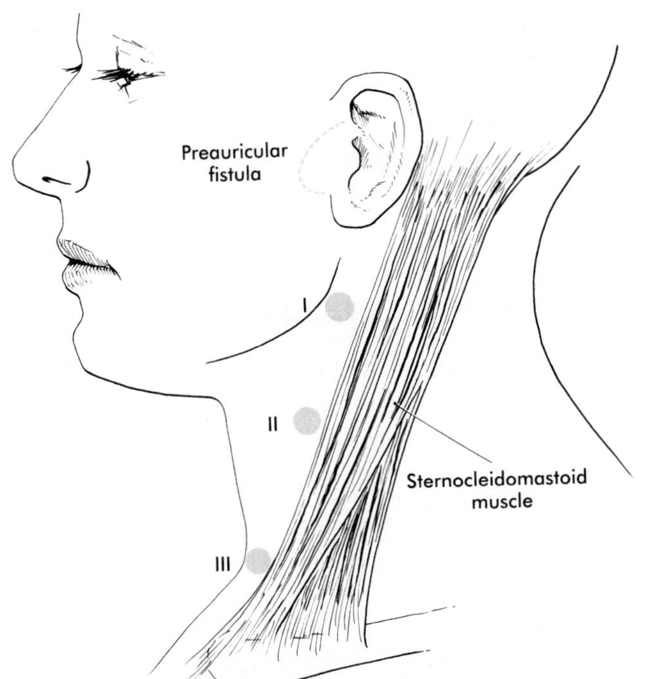

FIG. 15-28 Common locations of lateral cervical (branchial) cysts and sinuses and preauricular fistulas. The Roman numerals refer to the cervical cleft origin of the cysts.

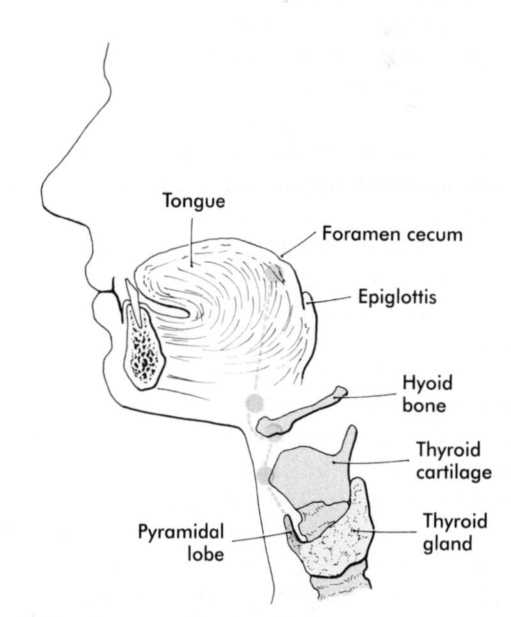

FIG. 15-29 Common locations *(red circles)* of thyroglossal duct remnants.

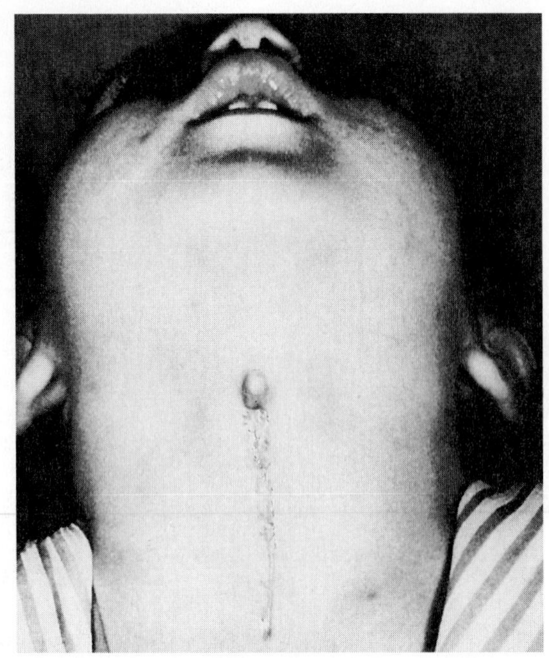

FIG. 15-30 Individual with a thyroglossal duct sinus in the ventral midline of the neck.
(Courtesy A. Burdi, Ann Arbor, Mich.)

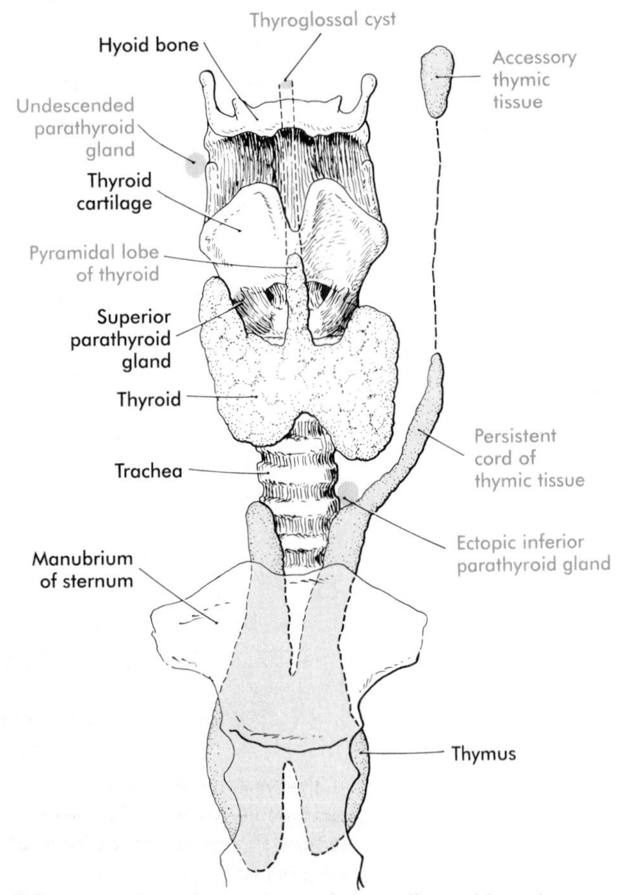

FIG. 15-31 Locations where abnormally positioned pharyngeal glands or portions of glands can be found.

ered in the clinical diagnosis or surgical treatment of carcinomas and other conditions affecting thyroid tissue. Less common are midline cysts or sinuses involving the former thyroglossal duct (Fig. 15-30). Because of their location, they can usually be easily distinguished from their lateral cervical counterparts.

Malformations of the Tongue. The most common malformation of the tongue is **ankyloglossia** (tongue-tie). This condition is caused by less-than-normal regression of the **frenulum,** the thin midline tissue that connects the ventral surface of the tongue to the floor of the mouth. Less common malformations of the tongue are **macroglossia** and **microglossia,** which are characterized by hyperplasia and hypoplasia of lingual tissue, respectively.

Ectopic Parathyroid or Thymic Tissue. Because of their extensive migrations during early embryogenesis, parathyroid glands and components of the thymus gland are often found in abnormal sites (Fig. 15-31). Typically, this displacement is not accompanied by functional abnormalities, but awareness of the possibility of ectopic tissue or even supernumerary parathyroid glands is important for the surgeon.

DiGeorge Syndrome. DiGeorge syndrome is a cranial neural crest deficiency and is manifest by immunological defects and hypoparathyroidism (see box on p. 249). The underlying pathology is failure of differentiation of the thymus and parathyroid glands. Associated anomalies are malformations of first arch structures and defects of the outflow tract of the heart, which contains an important cranial neural crest contribution as well.

SUMMARY

1. The early craniofacial region arises from the rostral portions of the neural tube, the notochord, and the pharynx, which is surrounded by a series of aortic arches. Between these structures and the overlying ectoderm are large masses of neural crest and mesodermally derived mesenchyme. A number of these components show evidence of anatomical segmentation or segmental patterns of gene expression.
2. Massive migrations of segmental groups of neural crest cells provide the mesenchyme for much of the facial region. The musculature of the craniofacial region is derived from somitomeric mesoderm or the occipital somites. The connective tissue component of the facial musculature is of neural crest origin.
3. The pharyngeal (branchial) region is organized around paired mesenchymal pharyngeal arches, which alternate with endodermally lined pharyngeal pouches and ectodermally lined branchial grooves.
4. The face and lower jaw arise from an unpaired frontonasal prominence and paired nasomedial, maxillary, and mandibular processes. Through differential growth and fusion, the nasomedial

processes form the upper jaw and lip, and the frontonasal prominence forms the upper part of the face. The expanding mandibular processes merge to form the lower jaw and lip. A nasolacrimal groove between the nasomedial and maxillary processes ultimately becomes canalized to form the nasolacrimal duct, which connects the orbit to the nasal cavity.

5. The palate arises from the fusion of an unpaired median palatine process and paired lateral palatine processes. The former forms the primary palate and the latter, the secondary palate.
6. The olfactory apparatus begins as a pair of thickened ectodermal nasal placodes. As these sink to form nasal pits, they are surrounded by horseshoe-shaped nasomedial and nasolateral processes. The former form the bridge and septum of the nose, and the nasolateral processes form the alae of the nose. The deepening nasal pits break into the oral cavity, and only later are the nasal cavities separated from the oral cavity by the palate.
7. The salivary glands arise as epithelial outgrowths of the oral epithelium. Through a series of continuing interactions with the surrounding mesenchyme, the expanding glandular epithelium branches and differentiates.
8. Teeth form from interactions between oral ectoderm (dental lamina) and neural crest mesenchyme. A developing tooth is first a tooth bud, which then passes through a cap and bell stage. Late in the bell stage, ectodermal cells (ameloblasts) of the epithelial enamel organ begin to form enamel, and the neural crest–derived epithelium (odontoblasts) begins to secrete dentin. Precursors of the permanent teeth form dental primordia along the more advanced primary teeth.
9. Malformations of the face are relatively common. Many, such as cleft lip and cleft palate, represent the persistence of the structural arrangements that are normal for earlier embryonic stages. Others, such as hypertelorism, result from growth disturbances in the frontonasal process. Most facial malformations appear to be multifactorial in origin, involving both genetic susceptibility and environmental causes.
10. Components of the pharynx (branchial grooves, pharyngeal arches, and pharyngeal pouches) give rise to a wide variety of structures. The first arch gives rise to the upper and lower jaws and associated structures. The first groove and pouch, along with associated mesenchyme from the first and second arches, form the many structures of the external and middle ear. The second, third, and fourth branchial grooves become obliterated and form the outer surface of the neck, and the components of the second through fourth arches form the pharyngeal skeleton and much of the musculature and connective tissue of the pharyngeal part of the neck. The endoderm of the third and fourth pouches forms the thymus and parathyroid glands. The thyroid gland arises from an unpaired ventral endodermal outgrowth of the upper pharynx.
11. The tongue originates from multiple ventral swellings in the floor of the pharynx. The bulk of the tongue comes from the paired lateral lingual swellings in the region of the first pharyngeal arches. The unpaired tuberculum impar and copula also contribute to the formation of the tongue. The tongue musculature arises from the occipital somites, along with the hypoglossal nerve (cranial nerve XII), which supplies the muscles. General sensory innervation of the tongue (from cranial nerves V, IX, and X) corresponds with the embryological origin of the innervated part of the tongue. Cranial nerves VII and IX innervate the taste buds.
12. Many malformations of the lower face and jaw are related to hypoplasia of the first pharyngeal arches. Cysts, sinuses, and fistulas of the neck are commonly caused by abnormal persistence of branchial grooves or pharyngeal pouches. Ectopic glandular tissue (thyroid, thymus, or parathyroid) is explained by the persistence of tissue rests along the pathway of migration of the glands. Certain syndromes (e.g., DiGeorge syndrome), which seemingly affect disparate organs, can be attributed to neural crest defects.

REVIEW QUESTIONS

1. A 15-year-old boy with mild acne developed a tender boil along the anterior border of the sternocleidomastoid muscle. What would be included in a differential diagnosis of this condition?
2. The physician of the 15-year-old boy determined that the boy had a congenital cyst that needed to be removed surgically. What should the surgeon consider during removal of the cyst?
3. Certain types of thyroid tumors are common in young women. In planning therapy, whether radiation or surgery, what embryological consideration must the clinician keep in mind?
4. A woman who took an experimental anticonvulsant drug during the tenth week of pregnancy gave birth to an infant with bilateral cleft lip and cleft palate. She sued the physician, and you are called in as an expert witness for the defense. What would be the basis for your case?
5. A woman who averaged two mixed drinks a day during pregnancy gave birth to an infant who was mildly retarded and who had a small notch in an upturned upper lip and a reduced olfactory sensitivity. What is the basis for this constellation of defects?

REFERENCES

Balling R and others: Craniofacial abnormalities induced by ectopic expression of the homeobox gene *Hox-1.1* in transgenic mice, *Cell* 58:337-347, 1989.

Bernfield M and others: *Remodelling of the basement membrane as a mechanism of morphogenetic tissue interaction.* In Trelstad RL, ed: *The role of extracellular matrix in development,* New York, 1984, Liss, pp 545-572.

Bockman DE, Kirby ML: Dependence of thymus development on derivatives of the neural crest, *Science* 223:498-500, 1984.

Bradley RB, Mistretta CM: Fetal sensory receptors, *Physiol Rev* 55:352-382, 1975.

Burdi AR: Sexual differences in closure of the human palatal shelves, *Cleft Palate J* 6:1-7, 1969.

Chenevix-Trench G and others: Cleft lip with or without cleft palate: associations with transforming growth factor alpha and retinoic acid receptor loci, *Am J Hum Genet* 51:1377-1385, 1992.

Farbman AI: *Developmental neurobiology of the olfactory system.* In Getchell TV and others, eds: *Smell and taste in health and disease,* New York, 1991, Raven, pp 19-33.

Ferguson MJ, Honig LS: Epithelial-mesenchymal interactions during vertebrate palatogenesis, *Curr Top Dev Biol* 19:137-163, 1984.

Gans C, Northcutt RG: Neural crest and the origin of vertebrates: a new head, *Science* 220:268-274, 1983.

Grobstein C: Epithelio-mesenchymal specificity in the morphogenesis of mouse submandibular rudiments in vitro, *J Exp Zool* 124:383-413, 1953.

Hunt P and others: Homeobox genes and models for patterning the hindbrain and branchial arches, *Development* 1(suppl):187-196, 1991.

Karavanova I, Vainio S, Thesleff I: Transient and recurrent expression of the Egr-1 gene in epithelial and mesenchymal cells during tooth morphogenesis suggests involvement in tissue interactions and in determination of cell fate, *Mech Dev* 39:41-50, 1992.

Kollar EJ: *Tooth development and dental patterning.* In Connelly T, Brinkley L, Carlson B, eds: *Morphogenesis and pattern formation,* New York, 1981, Raven, pp 87-102.

Kraus BS, Kitimura H, Latham RA: *Atlas of developmental anatomy of the face,* New York, 1966, Hoeber Harper.

Linask KK and others: Transforming growth factor-beta receptor profiles of human and murine embryonic palate mesenchymal cells, *Exp Cell Res* 192:1-9, 1991.

Lumsden AGS: Spatial organization of the epithelium and the role of neural crest cells in the initiation of the mammalian tooth germ, *Development* 103:155-169, 1988.

Maden M and others: Domains of cellular retinoic acid-binding protein I (CRABP I) expression in the hindbrain and neural crest of the mouse embryo, *Mech Dev* 37:13-23, 1992.

Mazzola RF: Congenital malformations in the frontonasal area: their pathogenesis and classification, *Clin Plast Surg* 3:573-609, 1976.

Mina M, Kollar EJ: The induction of odontogenesis in non-dental mesenchyme combined with early murine mandibular arch epithelium, *Arch Oral Biol* 32:123-127, 1987.

Noden DM: Cell movements and control of patterned tissue assembly during craniofacial development, *J Craniofac Genet Dev Biol* 11:192-213, 1991.

Noden DM: Vertebrate craniofacial development: the relation between ontogenetic process and morphological outcome, *Brain Behav Evol* 38:190-225, 1991.

Patten BM: *The normal development of the facial region.* In Pruznznsky S, ed: *Congenital anomalies of the face and associated structures,* Springfield, Ill, 1961, Charles C. Thomas, pp 11-45.

Poswillo DE: Neural crest and craniofacial disorders: summary and conclusions, *Am J Med Genet* 4:1-5, 1988.

Richman JM, Tickle C: Epithelial-mesenchymal interactions in the outgrowth of limb buds and facial primordia in chick embryos, *Dev Biol* 154:299-308, 1992.

Salzer GM, Zenker W: Das juxtaorale Organ, *Bibl Anat* 3:1-113, 1962.

Slavkin HC: Molecular determinants of tooth development: a review, *Crit Rev Oral Biol Med* 1:1-16, 1990.

Sperber GH: *Craniofacial embryology,* ed 4, London, 1989, Butterworth.

Stricker M and others, eds: *Craniofacial malformations,* Edinburgh, 1990, Churchill Livingstone.

Tan SS, Morriss-Kay G: The development and distribution of the cranial neural crest in the rat embryo, *Cell Tissue Res* 240:403-416, 1985.

Thesleff I, Hurmerinta K: Tissue interactions in tooth development, *Differentiation* 18:75-88, 1981.

Thesleff I, Partanen A-M, Vainio S: Epithelial-mesenchymal interactions in tooth morphogenesis: the roles of extracellular matrix, growth factors, and cell surface receptors, *J Craniofac Genet Dev Biol* 11:229-237, 1991.

Thorogood P, Tickle C, eds: Craniofacial development, *Development* 103(suppl):1-257, 1988.

Vainio S, Thesleff I: Sequential induction of syndecan, tenascin and cell proliferation associated with mesenchymal cell condensation during early tooth development, *Differentiation* 50:97-105, 1992.

Wedden SE: Epithelial-mesenchymal interactions in the development of chick facial primordia and the target of retinoid action, *Development* 99:341-351, 1987.

Zimmerman EF, ed: Palate development: normal and abnormal, cellular and molecular aspects, *Curr Top Dev Biol* 19:1-243, 1984.

CHAPTER 16

The Digestive and Respiratory Systems and Body Cavities

The initial formation of the digestive system by the lateral folding of the endodermal germ layer into a tube is described in Chapter 5. From the simple tubular gut, development of the digestive system can be viewed at several levels, ranging from the elongation and gross twistings and foldings of the digestive tube itself, to the series of inductions and tissue interactions that provide the basis for development of the digestive glands, and to the biochemical maturation of the secretory and absorptive epithelia associated with the digestive tract.

Formation of the respiratory system begins with a very unimposing ventral outpocketing of the foregut. Soon, however, this outpocketing embarks on a unique course of development while still following some of the basic patterns of epithelial-mesenchymal interactions characteristic of other gut-associated glands. Initially, both the digestive and respiratory systems form in a common body cavity, but functional considerations later necessitate the division of this primitive body cavity.

THE DIGESTIVE SYSTEM

Chapter 5 describes the formation of the primitive endodermal digestive tube, which is bounded at its cephalic end by the **stomodeal plate** and at its caudal end by the **cloacal plate** (see Fig. 5-19). Because of its intimate relationship with the yolk sac through the **yolk stalk,** the gut can be divided into a **foregut,** an open-bottomed **midgut,** and a **hindgut.** As early as the end of the first month, small endodermal diverticula, which represent primordia of the major digestive glands, can be identified (Fig. 16-1). (Development of the pharynx and its glandular derivatives is discussed in Chapter 15.)

Formation of the gut tube proper involves continuous elongation, herniation past the body wall, rotation and folding for efficient packing into the body cavity, and histogenesis and functional maturation. While these processes are occurring, the primordial digestive glands and respiratory structures are growing in complex branching patterns as a result of continuous epithelial-mesenchymal interactions. These interactions also occur in the developing digestive tube itself, with specific regional mesenchymal influences determining the character of the epithelium lining that part of the digestive tract.

Formation of the Esophagus

Just caudal to the most posterior pharyngeal pouches of the 4-week-old embryo the pharynx becomes abruptly narrowed, and a small ventral outgrowth (lung bud) appears (see Fig. 5-19). The region of foregut just caudal to the lung bud is the **esophagus.** This segment is initially very short, with the stomach seeming to reach almost to the pharynx. During the second month of development, during which the gut elongates considerably, the esophagus assumes nearly postnatal proportions in relation to the location of the stomach.

Although the esophagus grossly resembles a simple tube, it undergoes a series of striking differentiative changes at the tissue level. In its earliest stages the endodermal lining epithelium of the esophagus is stratified columnar. By the eighth week the epithelium has partially occluded the lumen of the esophagus, and large vacuoles appear (Fig. 16-2). In succeeding weeks the vacuoles coalesce and the esophageal lumen recanalizes but with a multilayered ciliated epithelium. During the fourth month this epithelium finally becomes replaced with the stratified squamous epithelium that characterizes the mature esophagus.

Deeper in the esophageal wall, layers of muscle also differentiate. Very early (5 weeks gestation) the primordium

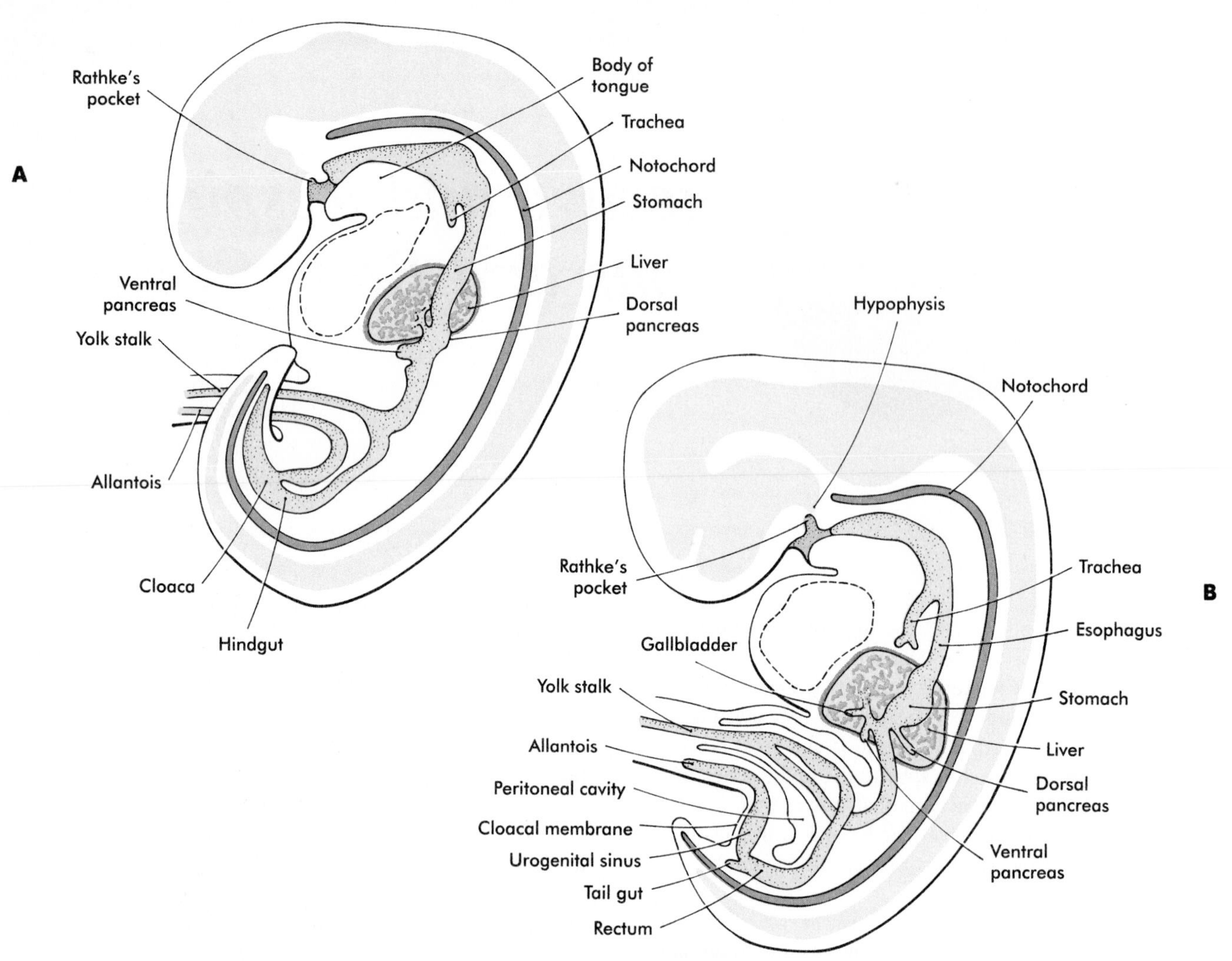

FIG. 16-1 Early stages in the formation of the digestive tract as seen in sagittal section. **A,** Early in the fifth week. **B,** Early in the sixth week.

of the inner circular muscular layer of esophagus is recognizable, and by 8 weeks the outer longitudinal layer of muscle is beginning to take shape. The esophageal wall contains both smooth and skeletal muscle. The skeletal muscle is derived from posterior pharyngeal arch mesoderm, whereas the smooth muscle cells differentiate from the local splanchnic mesoderm associated with the gut. All esophageal musculature is innervated by the vagus nerve (X).

Malformations of the Esophagus

The most common anomalies of the esophagus are associated with abnormalities of the developing respiratory tract (see p. 329). Other rather rare anomalies are **stenosis** and **atresia** of the esophagus. Stenosis is normally attributed to abnormal recanalization of the esophagus after epithelial occlusion of its lumen. Atresia of the esophagus is most commonly associated with abnormal development of the respiratory tract. In both of these conditions, impaired swallowing by the fetus can lead to an excessive accumulation of amniotic fluid **(polyhydramnios).** Just after birth a newborn with these anomalies commonly has difficulty swallowing milk, and regurgitation and choking while drinking are indications for examination of the patency of the esophagus.

Formation of the Stomach

Very early in the formation of the digestive tract the **stomach** is recognizable as a dilated region with a shape remarkably similar to that of the adult stomach (see Fig. 16-1). The early stomach is suspended from the dorsal body wall by a portion of the dorsal mesentery called the **dorsal mesogastrium.** It is connected to the ventral body wall by a ventral mesentery that also encloses the developing liver (Fig. 16-3).

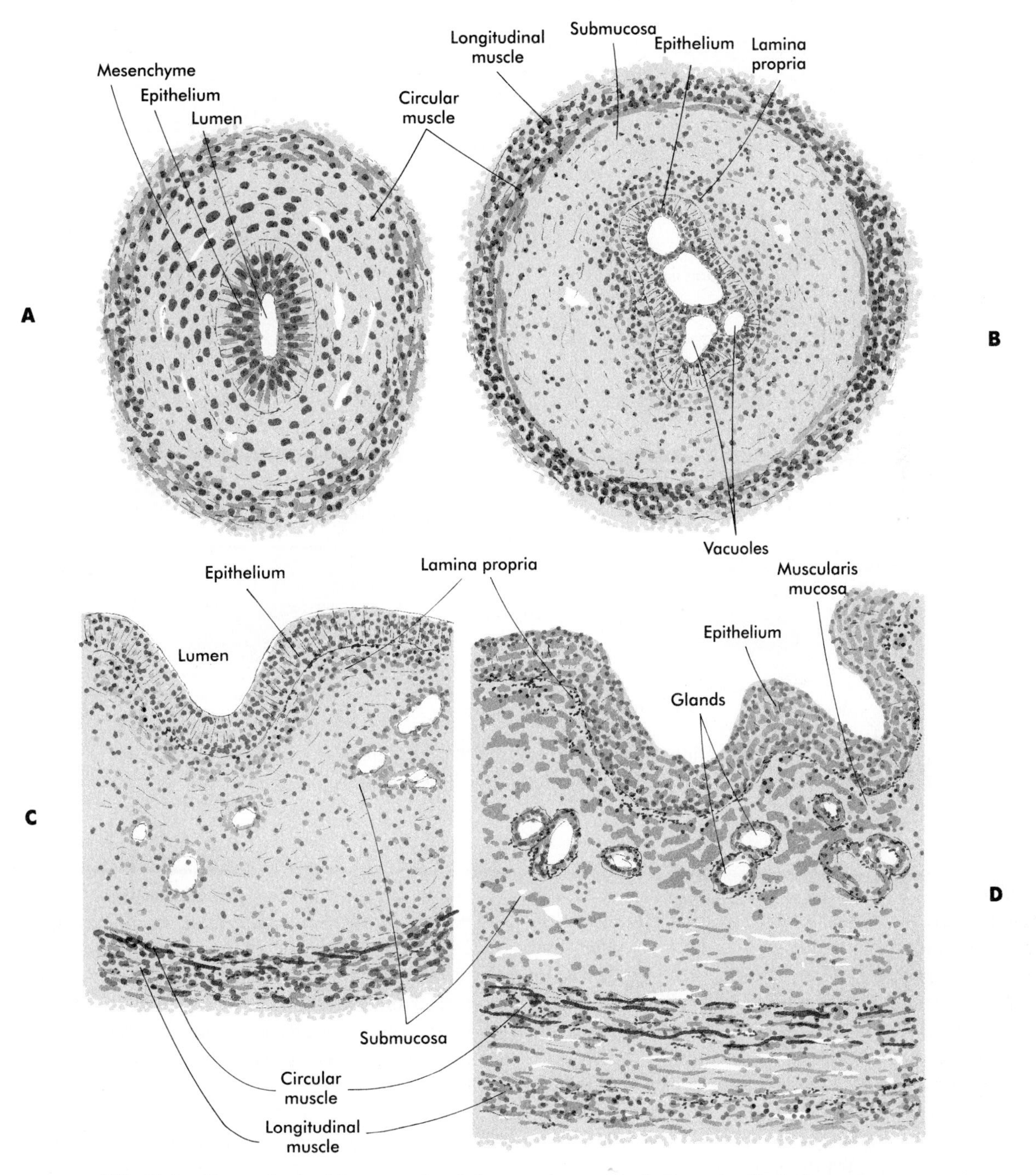

FIG. 16-2 Stages in the histogenesis of the esophagus. **A,** At 7 weeks. **B,** At 8 weeks. **C,** At 12 weeks. **D,** At 34 weeks.

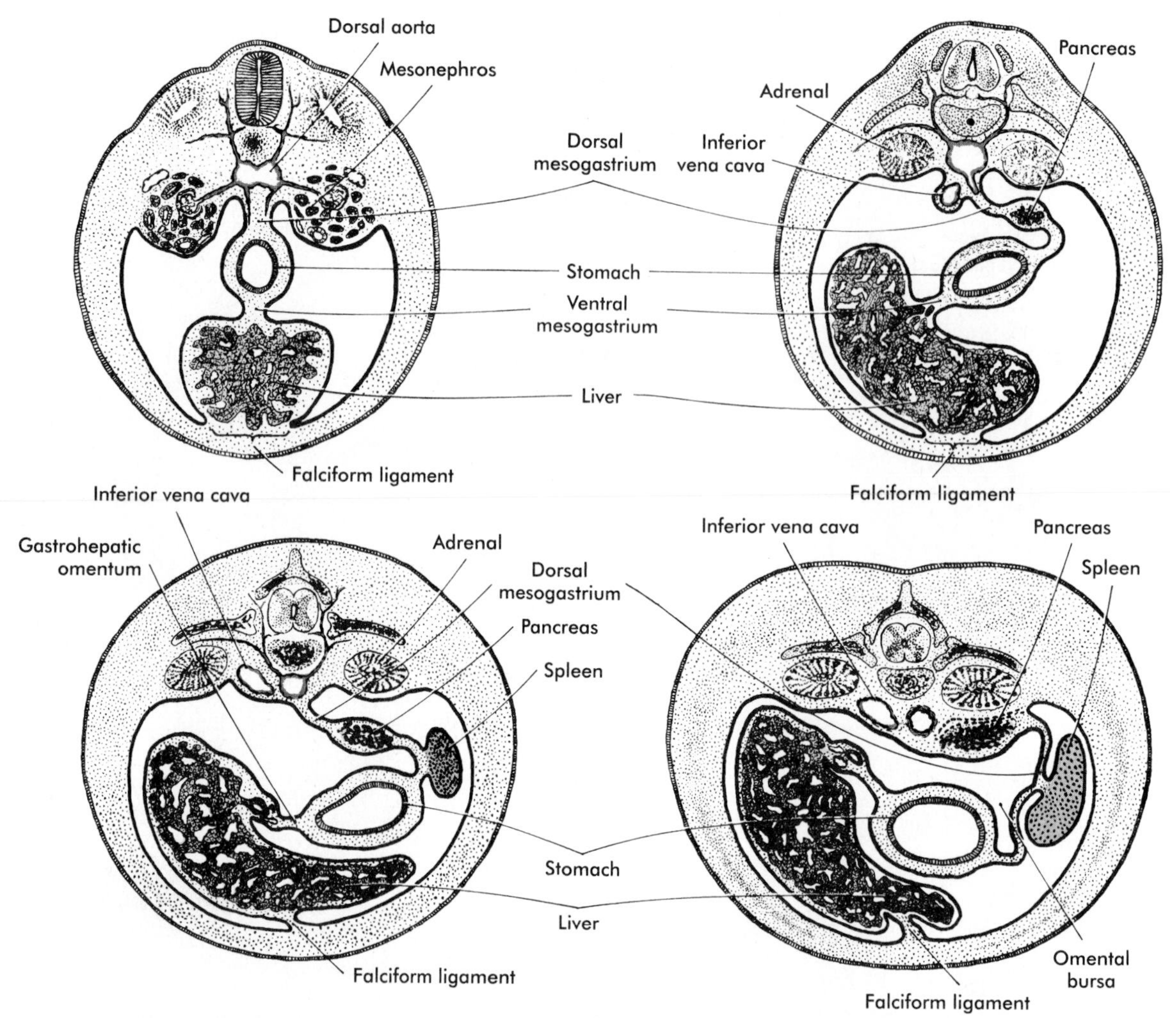

FIG. 16-3 Cross sections through the level of the developing stomach, showing changes in the relations of the mesenteries as the stomach rotates.

When the stomach first appears, its concave border faces ventrally and its convex border faces dorsally. Two concomitant positional shifts bring the stomach to its adult configuration. The first is an approximately 90-degree rotation about its craniocaudal axis, so its originally dorsal convex border faces left and its concave border faces right. The other positional shift consists of a minor tipping of the caudal (pyloric) end of the stomach in a cranial direction, so the long axis of the stomach is positioned somewhat diagonally across the body (Fig. 16-4).

During rotation of the stomach, the dorsal mesogastrium is carried with it, leading to the formation of a pouchlike structure called the **omental bursa** (bursa comes from a Latin word meaning "sac" or "pouch"). Both the spleen and tail of the pancreas are embedded in the dorsal mesogastrium (see Fig. 16-3). As the stomach rotates, the dorsal mesogastrium and the omental bursa that it encloses enlarge dramatically. Soon part of the dorsal mesogastrium, which becomes the **greater omentum,** overhangs the transverse colon and portions of the small intestines as a large, double flap of fatty tissue (Fig. 16-5). The two sides of the greater omentum ultimately fuse, obliterating the omental bursa within the greater omentum. The rapidly enlarging liver occupies an increasingly large portion of the ventral mesentery.

At the histological level the **gastric mucosa** begins to take shape late in the second month with the appearance of folds **(rugae)** and the first **gastric pits.** During the early fetal period the individual cell types that characterize the gastric mucosa begin to differentiate. Biochemical and cytochemical studies have shown the gradual functional differentiation of specific cell types during the late fetal period. In most mammals, including humans, cells of the gastric mucosa begin to secrete hydrochloric acid shortly before birth.

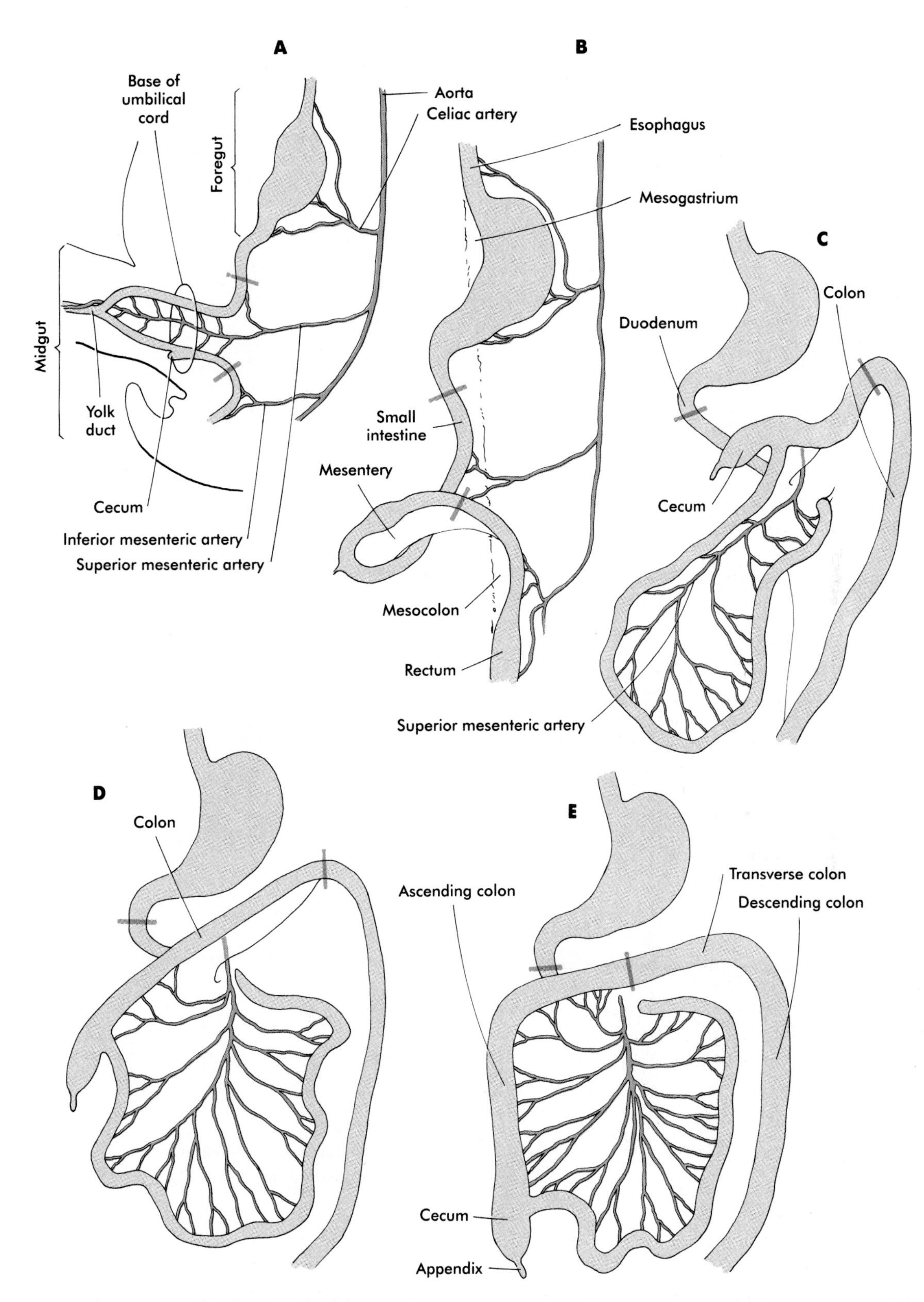

FIG. 16-4 Stages in the development and rotation of the gut. **A,** At 5 weeks. **B,** At 6 weeks. **C,** At 11 weeks. **D,** At 12 weeks. **E,** Fetal period. Areas between the green lines represent the midgut, which is supplied by the superior mesenteric artery.

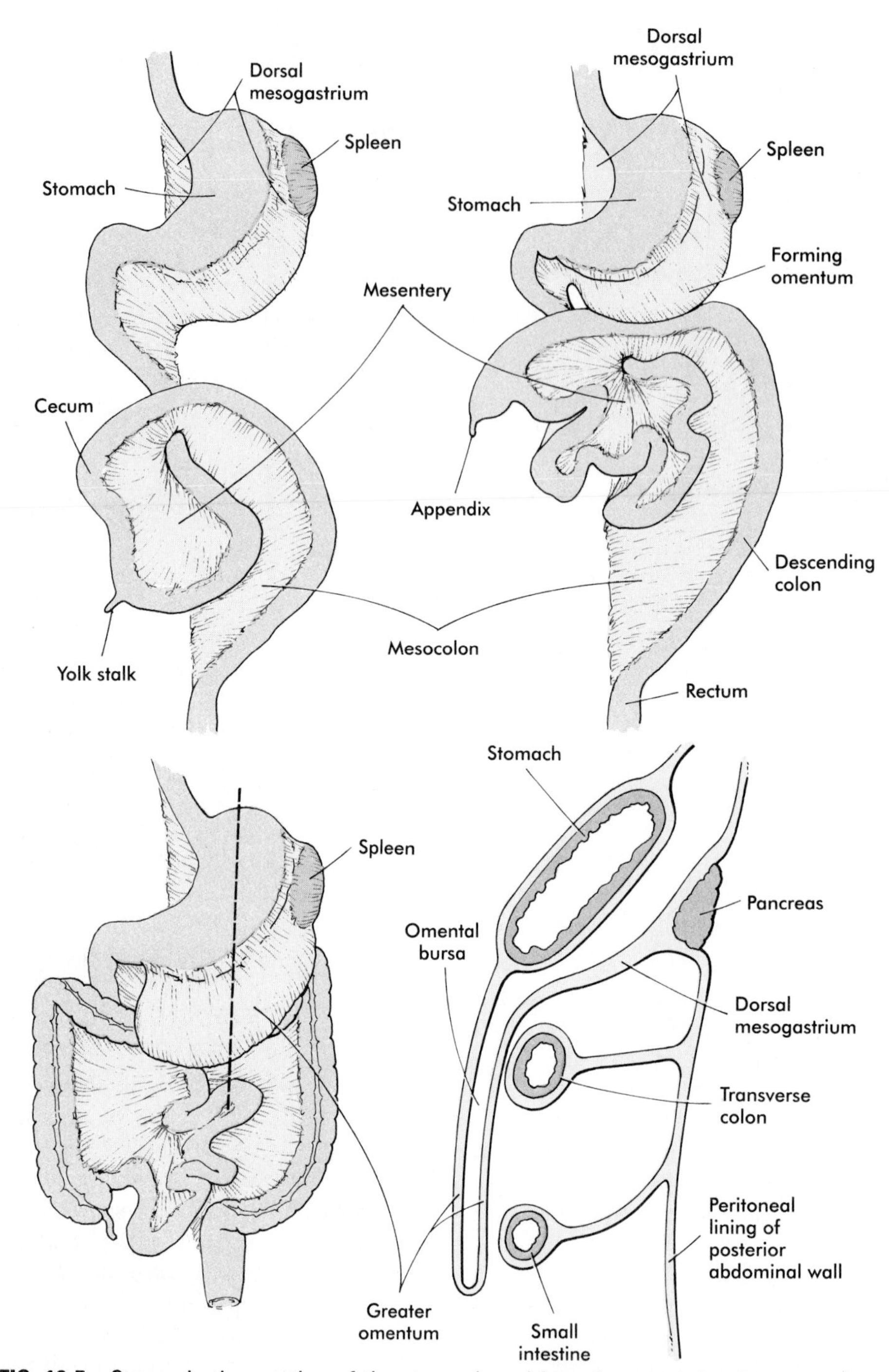

FIG. 16-5 Stages in the rotation of the stomach and intestines and development of the greater omentum. A section through the level of the dashed line *(lower left)* is shown *(lower right)*.

Malformations of the Stomach

Pyloric Stenosis. Pyloric stenosis, which appears to be more physiological than anatomical, consists of hypertrophy of the circular layer of smooth muscle that surrounds the pyloric (outlet) end of the stomach. The hypertrophy causes a narrowing (stenosis) of the pyloric opening and impedes the passage of food. Several hours after a meal the infant violently vomits (projectile vomiting) the contents of that meal. The enlarged pyloric end of the stomach can often be palpated on physical examination. Although pyloric stenosis is commonly treated by a simple surgical inci-

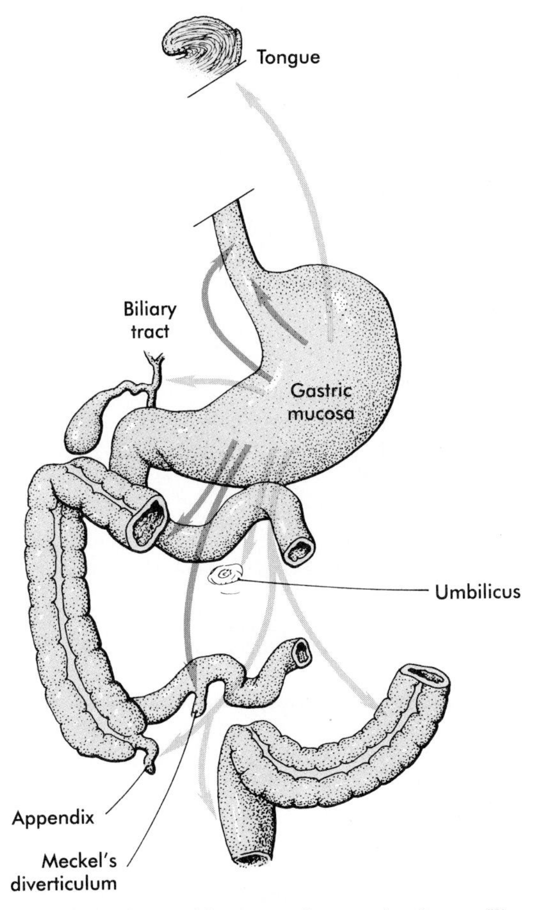

FIG. 16-6 Locations of heterotopic gastric tissue. The red arrows point to the most frequently occurring sites. The pink arrows indicate less common sites of occurrence.
(Based on Gray and Skandalakis [1972].)

TABLE 16-1 Derivatives of Regions of the Primitive Gut

BLOOD SUPPLY	ADULT DERIVATIVES
FOREGUT	
Celiac artery (lower esophagus to duodenum)	Pharynx Esophagus Stomach Upper duodenum Glands of pharyngeal pouches, respiratory tract, liver and gallbladder, pancreas
MIDGUT	
Superior mesenteric artery	Lower duodenum Jejunum and ileum Cecum and vermiform appendix Ascending colon Cranial half of transverse colon
HINDGUT	
Inferior mesenteric artery	Caudal half of transverse colon Descending colon Rectum Superior part of anal canal

sion through the layer of circular smooth muscle of the pylorus, the hypertrophy sometimes diminishes untreated several weeks after birth. The pathogenesis of this defect remains unknown, but it seems to have a genetic basis. Pyloric stenosis is much more common in males than in females, and the incidence has been reported as from 1:200 to 1:1000 infants.

Heterotopic Gastric Mucosa. Heterotopic gastric mucosa has been found in a variety of otherwise normal organs (Fig. 16-6). This condition is often of clinical significance because if the heterotopic mucosa secretes hydrochloric acid, ulcers can form in unexpected locations.

Formation of the Intestines

The intestines are formed from the posterior part of the foregut, the midgut, and the hindgut (Table 16-1). Two points of reference are useful in understanding the gross transformation of the primitive gut tube from a relatively straight cylinder to the complex folded arrangement characteristic of the adult intestinal tract. The first is the yolk stalk, which extends from the floor of the midgut to the yolk sac. In the adult the site of attachment of the yolk stalk is on the small intestine about 2 feet cranial to the junction between small and large intestine **(ileocecal junction).** On the dorsal side of the primitive gut an unpaired ventral branch of the aorta, the **superior mesenteric artery,** and its branches feed the midgut (see Fig. 16-4). The superior mesenteric artery itself serves as a pivot point about which later rotation of the gut occurs.

As early as the fifth week, rapid growth of the gut tube causes it to buckle out in a hairpin-like loop. The major change that causes the intestines to assume their adult positions is a counterclockwise rotation of the caudal limb of the intestinal loop (with the yolk stalk attachment and superior mesenteric artery as reference points) around the cephalic limb from its ventral aspect. The main consequence of this rotation is to bring the future colon across the small intestine, so it can readily assume its C-shaped position along the ventral abdominal wall (see Fig. 16-4). Behind the colon the small intestine undergoes a great elongation and becomes packed in its characteristic position in the abdominal cavity.

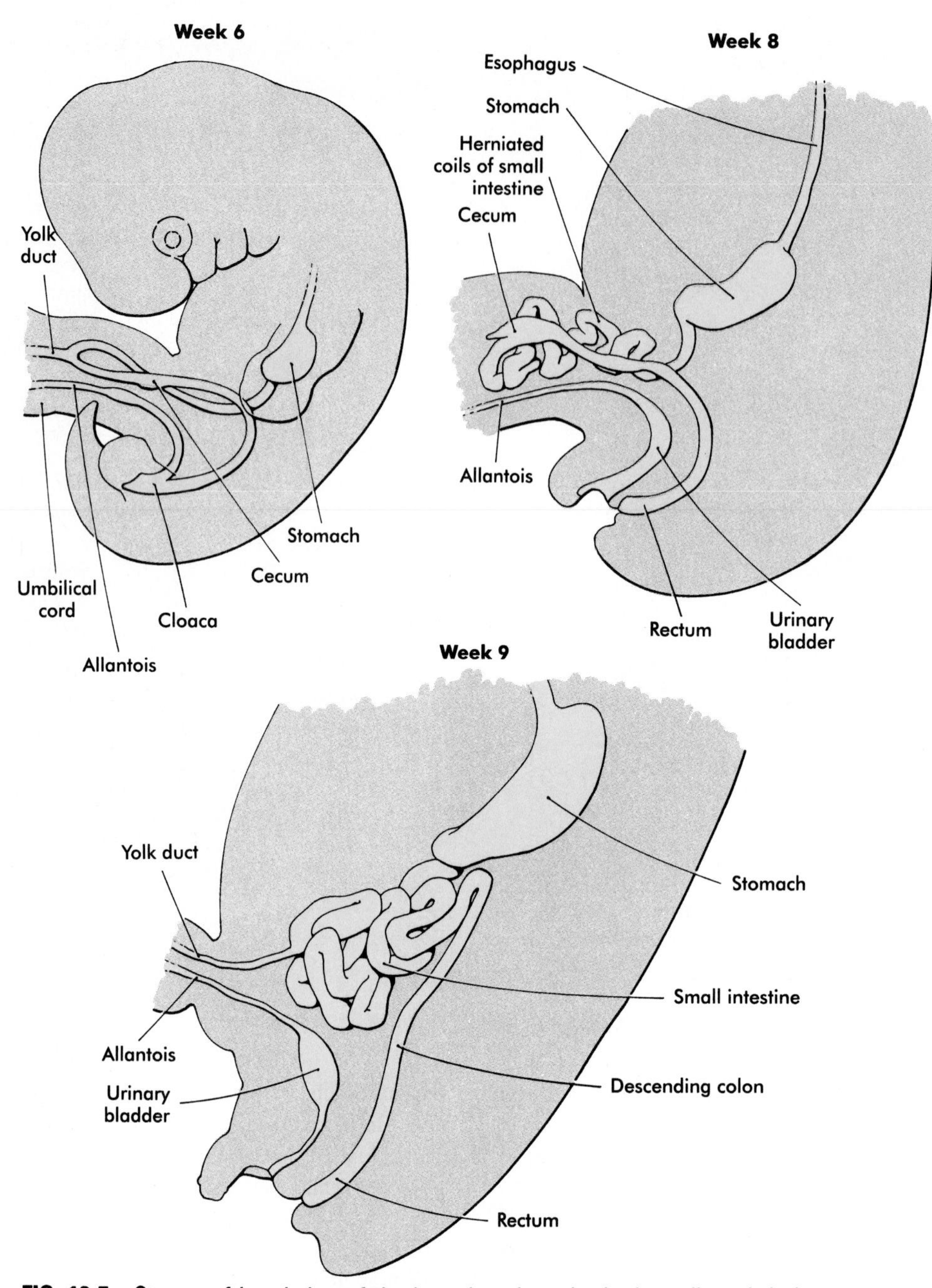

FIG. 16-7 Stages of herniation of the intestines into the body stalk and their return.

The rotation and other positional changes of the gut partly occur because the length of the gut increases much more than the length of the embryo. From almost the first stages the volume of the expanded gut tract is greater than the body cavity can accommodate. Consequently, the developing intestines herniate into the body stalk (the umbilical cord after further development) (Fig. 16-7). Intestinal herniation begins as early as the sixth or seventh week of embryogenesis. By the tenth week the abdominal cavity has enlarged sufficiently to accommodate the intestinal tract, and the herniated intestinal loops begin to move through the intestinal ring into the abdominal cavity. Coils of small intestine return first. As they do, they force the distal part of the colon, which was never herniated, to the left side of the peritoneal cavity, thus establishing the definitive position of the descending colon. After the small intestine has taken its intraabdominal position, the herniated proximal part of the colon also returns, with its cecal end swinging to the right and downward (see Fig. 16-4).

During these coilings, herniations, and return move-

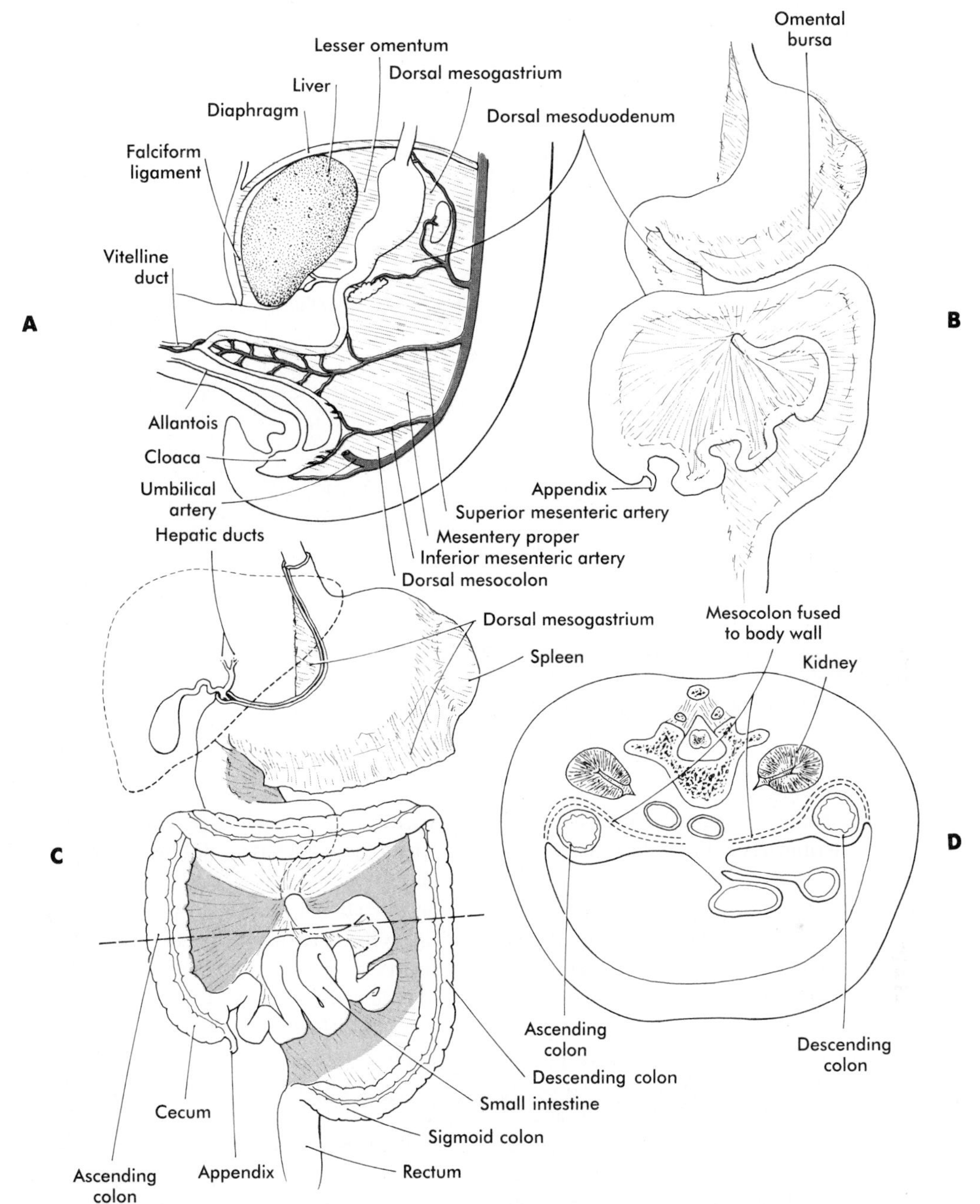

FIG. 16-8 Stages in the development of the mesenteries. **A,** At 5 weeks. **B,** In the third month. **C,** During late fetal period. **D,** Cross section through the dashed line in **C.** In **C,** the shaded areas represent regions where the mesentery is fused to the dorsal body wall.

ments, the intestines are suspended from the dorsal body wall by a mesentery (Fig. 16-8). As the intestines assume their definitive positions within the body cavity, their mesenteries follow. Parts of the mesentery associated with the duodenum and colon (mesoduodenum and mesocolon) fuse with the peritoneal lining of the dorsal body wall.

Starting in the sixth week the primordium of the **cecum** becomes apparent as a swelling in the caudal limb of the midgut (see Fig. 16-4). In succeeding weeks the cecal enlargement becomes so prominent that the distal small intestine enters the colon at a right angle.

The tip of the cecum elongates, but its diameter does not increase in proportion to the rest of the cecum. This wormlike appendage is aptly called the **vermiform appendix.**

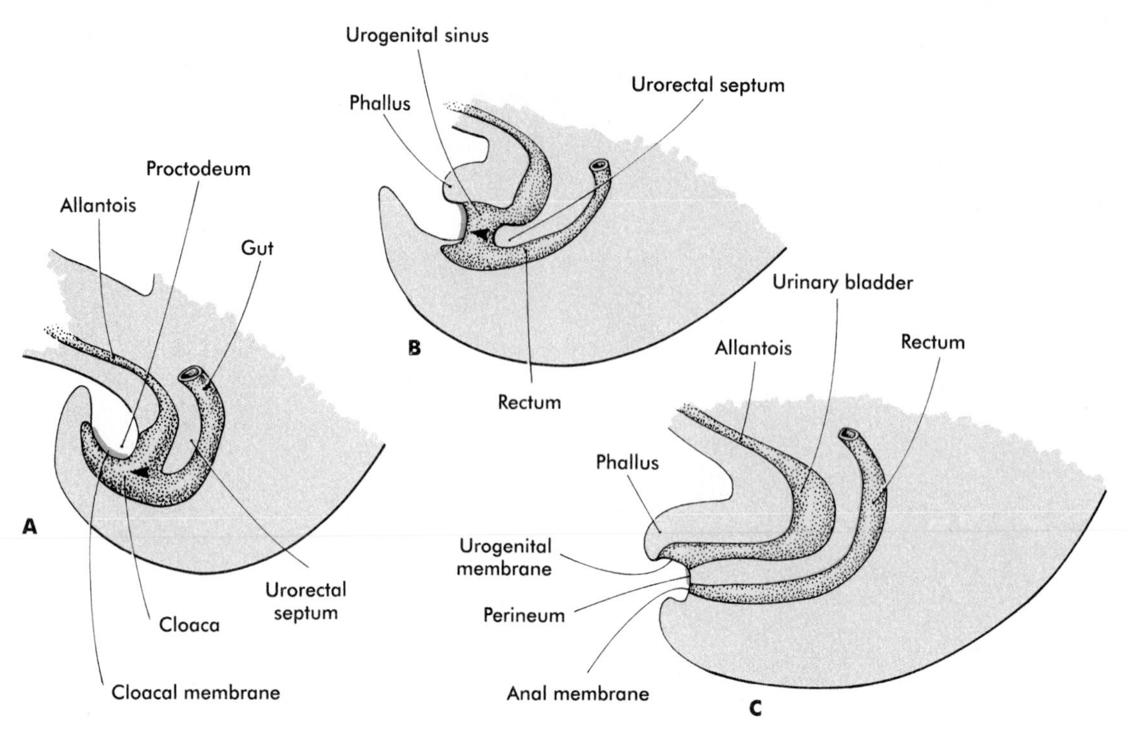

FIG. 16-9 Stages in the subdivision of the common cloaca by the urorectal septum. **A,** In the fifth week. **B,** In the sixth week. **C,** In the eighth week.

Partitioning of the Cloaca. In the early embryo the caudal end of the hindgut terminates in the endodermally lined **cloaca,** which in lower vertebrates serves as a common outlet for digestive and urinary wastes and gametes. The cloaca also includes the base of the allantois, which later expands as a common **urogenital sinus** (see Chapter 17). A **cloacal (proctodeal) membrane** consisting of apposed layers of ectoderm and endoderm acts as a barrier between the cloaca and an ectodermal depression known as the **proctodeum** (Fig. 16-9). A shelf of mesodermal tissue called the **urorectal septum** is situated between the hindgut and the base of the allantois. During the sixth and seventh weeks the urorectal septum grows toward and ultimately fuses with the cloacal membrane, dividing it into an **anal membrane,** which blocks the end of the hindgut, and a **urogenital membrane,** which seals the urogenital sinus from the exterior. By the end of the eighth week the anal membrane ruptures and affords free access between the hindgut and the exterior of the body. The area where the urorectal septum fuses with the cloacal membrane becomes the **perineal body,** which represents the partition between the digestive and urogenital systems.

Histogenesis of the Intestinal Tract. Shortly after initial formation the intestinal tract consists of a simple layer of columnar endodermal epithelium surrounded by a layer of splanchnopleural mesoderm. Three major phases are involved in the histogenesis of the intestinal epithelium: (1) an early phase of epithelial proliferation and morphogenesis, (2) an intermediate period of cellular differentiation in which the distinctive cell types characteristic of the intestinal epithelium appear, and (3) a final phase of biochemical and functional maturation of the different types of epithelial cells. The mesenchymal wall of the intestine also differentiates into several layers of highly innervated smooth muscle and connective tissue. An overall craniocaudal gradient of differentiation is present within the developing intestine.

Early in the second month the epithelium of the small intestine begins a phase of rapid proliferation that results in the epithelium temporarily occluding the lumen by 6 to 7 weeks gestation. Within a couple of weeks, recanalization of the intestinal lumen has occurred. At about this time, small, cracklike secondary lumina appear beneath the surface of the multilayered epithelium, and aggregates of mesoderm push into the epithelium. A combination of coalescence of the secondary lumina with continued mesenchymal upgrowth beneath the epithelium results in the formation of numerous, fingerlike **intestinal villi,** which greatly increase the absorptive surface of the intestinal surface. By this time the epithelium has transformed from a stratified into a simple columnar type.

With the formation of villi, pitlike **intestinal crypts** also form at the base of the villi. The crypts contain **epithelial stem cells,** which have a high rate of mitosis and serve as the source of epithelial cells for the entire intestinal surface (Fig. 16-10). Autoradiographic studies have shown that

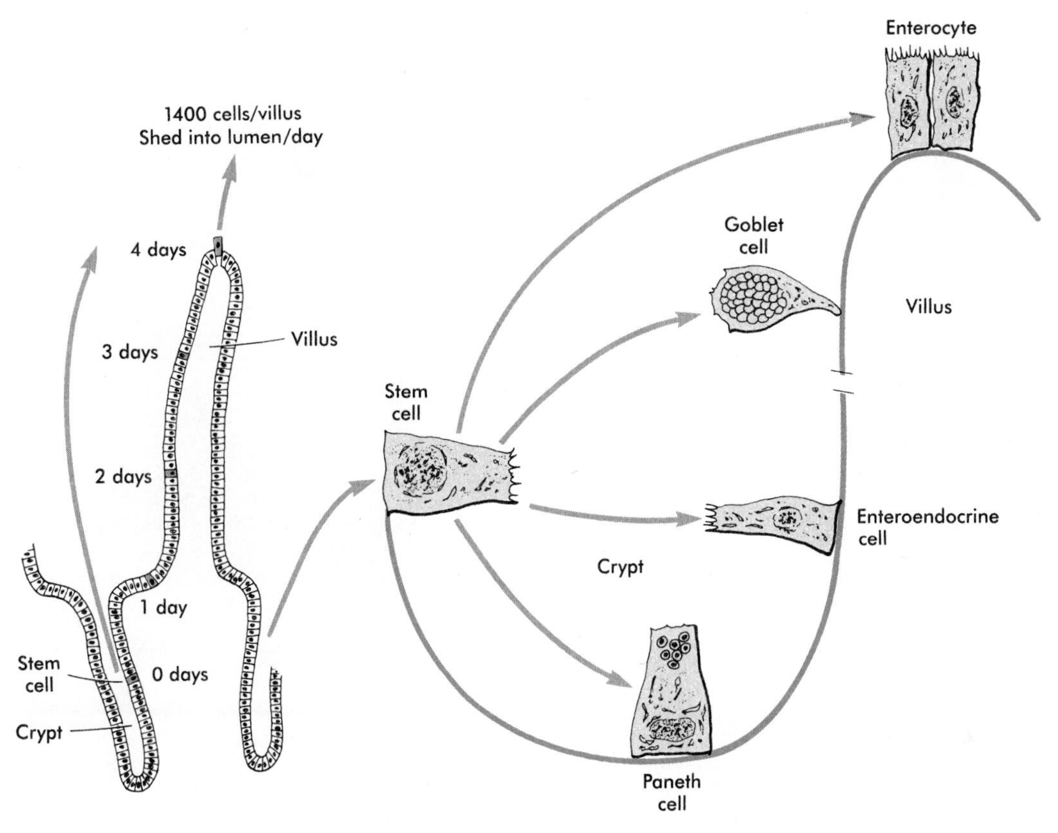

FIG. 16-10 Differentiation of intestinal epithelial cells from stem cells located in the crypts. The time scale shows the typical course of migration of daughter cells from their generation from the stem cell population to their being shed from the villus into the intestinal lumen.

over 3 to 4 days, epithelial cells originating in the crypts migrate up the villus and are ultimately shed from the tips of the villi. The intestinal epithelium is continuously renewed by this mechanism.

Shortly after the formation of the crypts, individual epithelial stem cells within them begin to form each of the four types of mature epithelial cells found in the intestinal epithelial lining.* By the end of the second trimester of pregnancy, all cell types found in the adult intestinal lining have differentiated, but many of these cells do not possess adult functional patterns. A number of specific biochemical patterns of differentiation are present as early as 12 weeks gestation and mature during the fetal period. For example, **lactase,** an enzyme that breaks down the disaccharide **lactose** (milk sugar), is one of the digestive enzymes synthesized in the fetal period. Further biochemical differentiation of the intestine occurs after birth, often in response to specific dietary patterns.

Histodifferentiation of the intestinal tract is not an isolated property of the individual tissue components of the intestinal wall. During the early embryonic period and sometimes into postnatal life the epithelial and mesodermal components of the intestinal wall communicate by inductive interactions. Interspecies recombination experiments show that the gut mesoderm exerts a regional influence on intestinal epithelial differentiation (e.g., whether the epithelium differentiates into a duodenal or colonic phenotype). Once regional determination is set, however, the controls for biochemical differentiation of the epithelium are inherent. This pattern of inductive influence and the epithelial reaction are very similar to those outlined earlier for dermal-epidermal interactions in the developing skin (see Chapter 10).

Final enzymatic differentiation of intestinal absorptive cells is strongly influenced by glucocorticoids, and the underlying mesoderm appears to mediate this hormonal effect. In a converse inductive influence the intestinal endoderm induces the differentiation of smooth muscle in the wall of the intestine. The mechanism of the interactions between endoderm and mesenchyme are not completely understood. The mesoderm may exert some effects directly on the

*For a number of years, it was thought that the peptide hormone–secreting cells in the intestinal tract were derived from neural crest. This assertion has not been supported by transplantation and marking experiments.

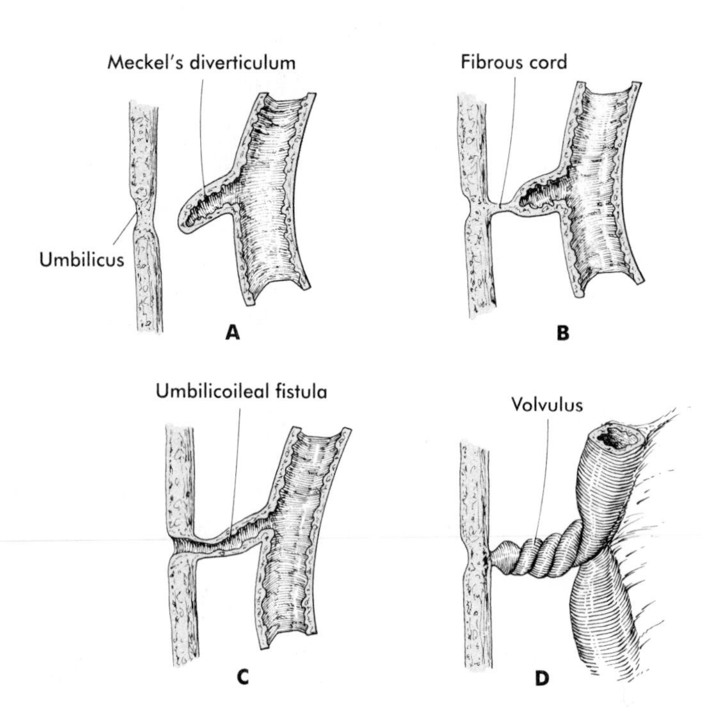

FIG. 16-11 Varieties of vitelline duct remnants. **A,** Meckel's diverticulum. **B,** Fibrous cord connecting a Meckel's diverticulum to the umbilicus. **C,** Umbilicoileal (vitelline) fistule. **D,** Volvulus caused by rotation of the intestine around a vitelline duct remnant.

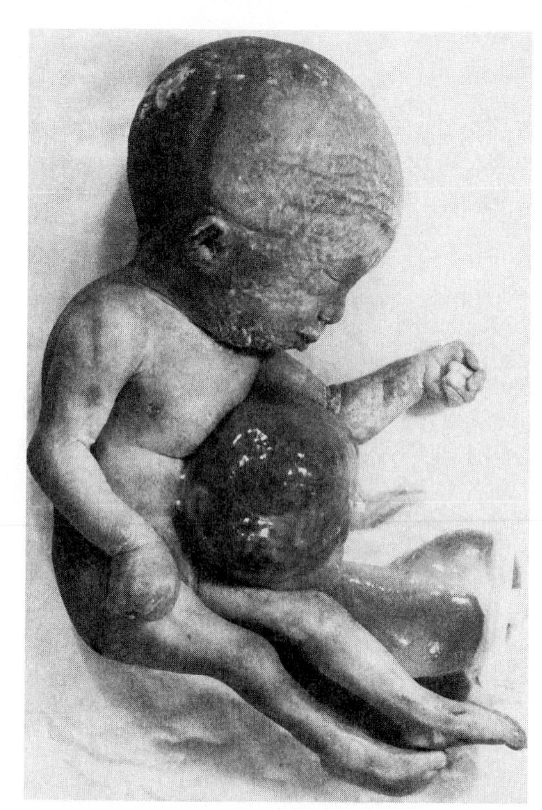

FIG. 16-12 Omphalocele in a stillborn. Loops of small intestine can be clearly seen through the nearly transparent amniotic membrane that covers the omphalocele.
(Courtesy Mason Barr, Ann Arbor, Mich.)

endoderm, whereas extracellular matrix molecules secreted by the mesodermal cells may mediate other effects.

Although the intestine develops many functional capabilities during the fetal period, no major digestive function occurs until feeding begins after birth. The intestines of the fetus contain a greenish material called **meconium,** which is a mixture of lanugo hairs and vernix caseosa sloughed from the skin, desquamated cells from the gut, bile secretion, and other materials swallowed with the amniotic fluid.

Malformations of the Intestinal Tract

Duodenal stenosis and atresia. Duodenal stenosis and atresia typically result from absent or incomplete recanalization of the duodenal lumen after it is plugged by endothelium. These malformations are rare.

Vitelline duct remnants. The most common family of anomalies of the intestinal tract is some form of persistence of the **vitelline (yolk) duct.** The most common member of this family is **Meckel's diverticulum,** which is present in 2% to 4% of the population. A typical Meckel's diverticulum is a blind pouch a few centimeters long located on the antimesenteric border of the ileum about 50 cm cranial from the ileocecal junction (Fig. 16-11, *A*). This structure represents the persistent proximal portion of the yolk stalk. Simple Meckel's diverticula are often asymptomatic, but they occasionally become inflamed or contain ectopic tissue (e.g., gastric, pancreatic), which can cause ulceration.

In some cases a ligament connects a Meckel's diverticulum to the umbilicus (Fig. 16-11, *B*), or a simple **vitelline ligament** that may have an associated persisting **vitelline artery** can connect the intestine to the umbilicus. Occasionally the intestine rotates about such a ligament, causing a condition known as **volvulus** (Fig. 16-11, *D*). This can lead to strangulation of the bowel.

A persistent vitelline duct can take the form of a **vitelline fistula** (Fig. 16-11, *C*), which constitutes a direct connection between the intestinal lumen and the outside of the body via the umbilicus. Rarely, a **vitelline duct cyst** is present along the length of a vitelline ligament.

Omphalocele. **Omphalocele** represents the failure of return of the intestinal loops into the body cavity during the tenth week. After birth, the intestinal loops can be easily seen within an almost transparent sac consisting of amnion on the outside and peritoneal membrane on the inside (Fig. 16-12). The incidence of omphalocele is approximately 1:3500 births, but half of the infants with this condition are stillborn.

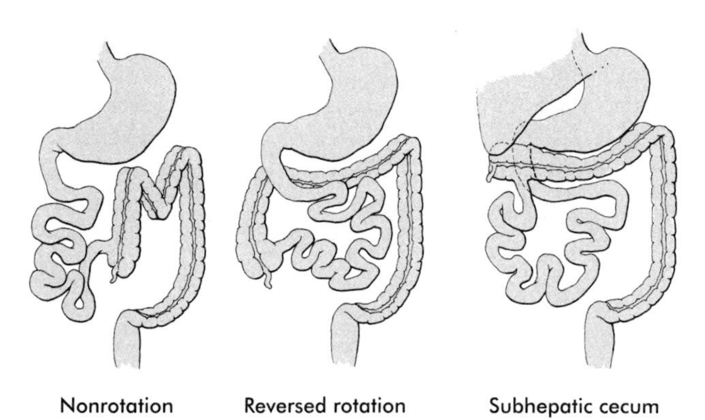

FIG. 16-13 Types of abnormal rotations of the gut.

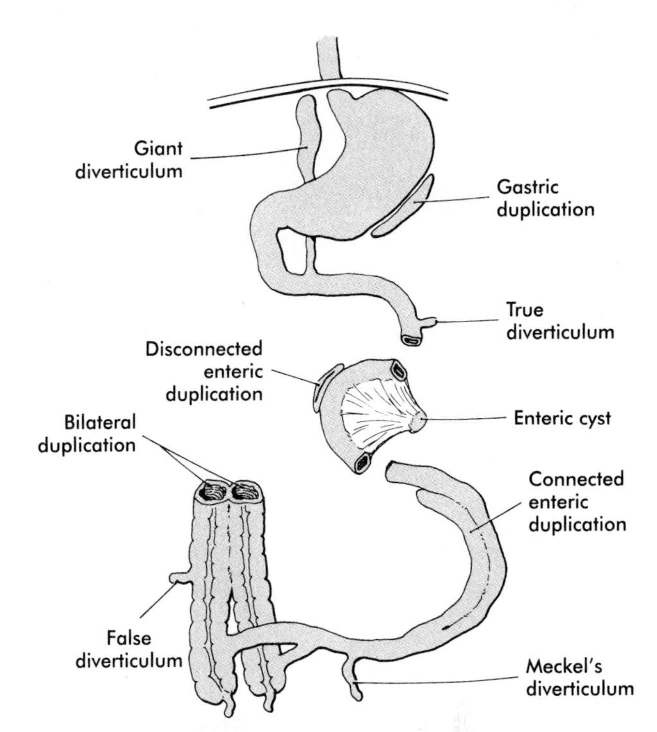

FIG. 16-14 Types of diverticula and duplications that can occur in the digestive tract.
(Based on Gray and Skandalakis [1972].)

Congenital umbilical hernia. In this condition, which is especially common in premature births, the intestines return normally into the body cavity, but the musculature (rectus abdominis) of the ventral abdominal wall fails to close the umbilical ring, allowing a varying amount of omentum or bowel to protrude through the umbilicus. In contrast to omphalocele the protruding tissue in an umbilical hernia is covered by skin rather than amniotic membrane.

Both omphalocele and congenital umbilical hernia are associated with closure defects in the ventral abdominal wall. If these defects are large, they may be accompanied by massive protrusion of abdominal contents or with other closure defects such as exstrophy of the bladder (see Chapter 17).

Abnormal rotation of the gut. Sometimes the intestines undergo no or abnormal rotation as they return to the abdominal cavity. This can result in a wide spectrum of anatomical anomalies (Fig. 16-13). In most cases, these are asymptomatic, but occasionally they can lead to volvulus or another form of strangulation of the gut.

Intestinal duplications, diverticula, and atresia. As with the esophagus and duodenum, the remainder of the intestinal tract is susceptible to various anomalies that seem to be based on incomplete recanalization of the lumen after the stage of temporary blocking of the lumen by epithelium during the first trimester. Some of the variants of these conditions are shown in Fig. 16-14.

Aganglionic megacolon (Hirschsprung's disease). The basis of this condition, which is manifest by great dilatation of certain segments of the colon, is the absence of parasympathetic ganglia in the affected walls of the colon. This has been attributed to defective migration of neural crest cells into that portion of the hindgut early in the second month of pregnancy. Evidence from mutant mice developing aganglionic segments of the bowel strongly suggests that the environment of the gut wall inhibits the migration of neural crest cells into the affected segment of gut. This was shown by experiments in which crest cells from mutant mice were capable of colonizing normal gut, but normal crest cells could not migrate into gut segments of mutant mice. Some evidence indicates that accumulations of laminin in the gut wall may serve as a stop signal for neural crest migration. The distal colon is the most affected region for aganglionosis, but in a small number of cases, aganglionic segments extend as far cranially as the ascending colon. Estimates of the frequency of megacolon vary widely—from 1:1000 to 1:30,000 births.

Imperforate anus. Imperforate anus includes a spectrum of anal defects that can range from a simple membrane covering the anal opening (persistence of the anal membrane) to atresia of varying lengths of the anal canal and/or rectum. Grossly, all are characterized by the absence of an anal opening (Fig. 16-15). Any examination of a newborn must include a determination of the presence of an anal opening. Of considerable importance when considering the surgical treatment of imperforate anus is the extent of the atretic segment. Treatment of a persistent anal membrane can be trivial, whereas more extensive defects, especially those involving the anal musculature, are often very challenging surgical problems.

In the early days of anal reconstructive surgery a clever method of determining the extent of the anal defect was devised. The infant was tipped upside down. A coin was placed over the skin covering the anal opening and an x-ray film was taken. An air bubble collected at the blind end of the hindgut. By determining the distance between the air

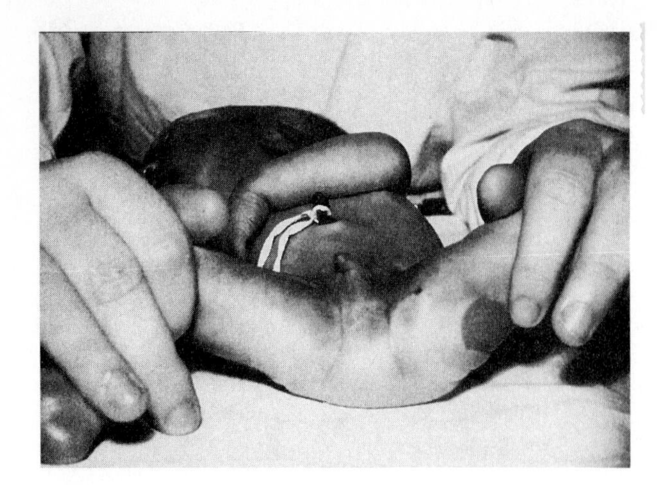

FIG. 16-15 Anal atresia in a newborn. No trace of an anal opening is seen.
(Courtesy Mason Barr.)

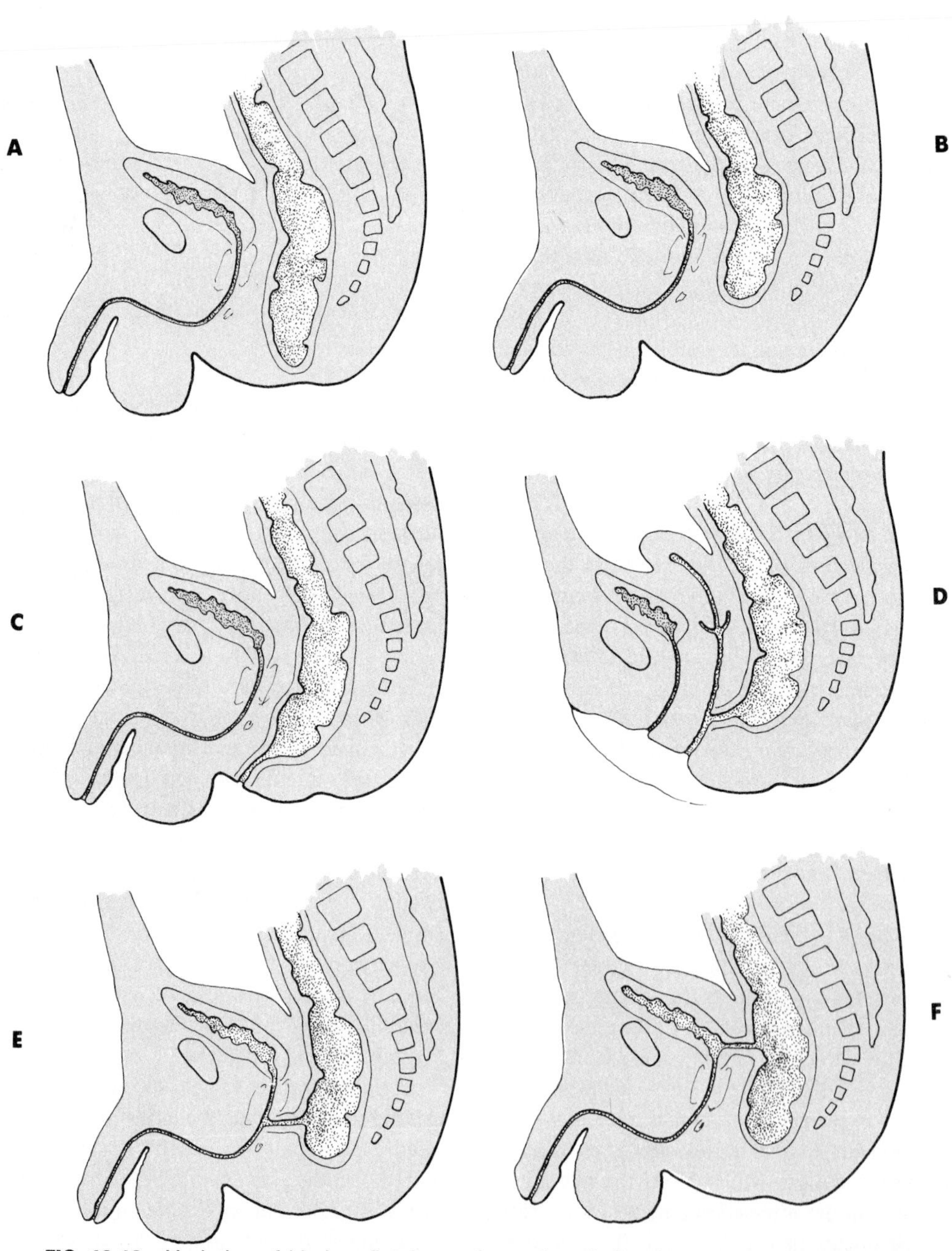

FIG. 16-16 Varieties of hindgut fistulas and atresias. **A,** Persistent anal membrane. **B,** Anal atresia. **C,** Anoperineal fistula. **D,** Rectovaginal fistula. **E,** Rectourethral fistula. **F,** Rectovesical fistula.

bubble and the coin, the surgeon could determine the amount of tissue that had to be opened to reconstruct the atretic or imperforate anus.

Hindgut fistulas. In many cases, anal atresia is accompanied by a fistula linking the patent portion of the hindgut to another structure in the region of the original urogenital sinus region. Common types of fistulas connect the hindgut with the vagina, the urethra, or bladder, and others may lead to the surface in the perineal area (Fig. 16-16).

Glands of the Digestive System

The glands of the digestive system arise through inductive processes between the early epithelial outgrowths and the surrounding mesenchyme. The various glandular epithelia have considerably different requirements in the types of mesenchyme that can support their development. For example, in tissue recombination experiments, pancreatic epithelium undergoes typical development when juxtaposed with mesenchyme from almost any source. The development of salivary gland epithelium, on the other hand, is supported by mesenchyme from lung or accessory sexual glands but not by many other types of mesenchyme. Inductive support of hepatic (liver) epithelium follows a distinctive pattern. Normal epithelial development is supported by mesenchyme derived from lateral plate or intermediate mesoderm, but axial mesenchyme (either somitic or neural crest) fails to support hepatic differentiation. The inductive properties of certain glandular mesenchymes may be correlated with different modes of vascularization of these mesenchymes (see Chapter 18).

Formation of the Liver. Early in the third week an endodermal hepatic diverticulum arises from the floor of the foregut and grows into the mesenchyme of the **septum transversum** (see Fig. 5-19). The hepatic diverticulum is the manifestation of a series of inductive processes already beginning to take place (Fig. 16-17). The original hepatic diverticulum branches into many hepatic cords, which are closely associated with splanchnic mesoderm of the septum transversum. The mesoderm supports continued growth of the hepatic endoderm. A number of experimental studies have shown that mesoderm from either the splanchnopleural or somatopleural components of the lateral plate mesoderm can support further hepatic growth and differentiation, whereas paraxial mesoderm has only a limited capacity to support hepatic development.

In addition to cords of hepatic endoderm, a system of bile drainage ducts forms in the developing liver. Near the area where the hepatic ducts become confluent, a dilatation foreshadows the further development of the **gallbladder** (Fig. 16-18). The hepatic cords form a series of loosely packed and highly irregular sheets that alternate with mesodermally lined **sinusoids,** through which blood percolates and exchanges nutrients with the **hepatocytes.** The devel-

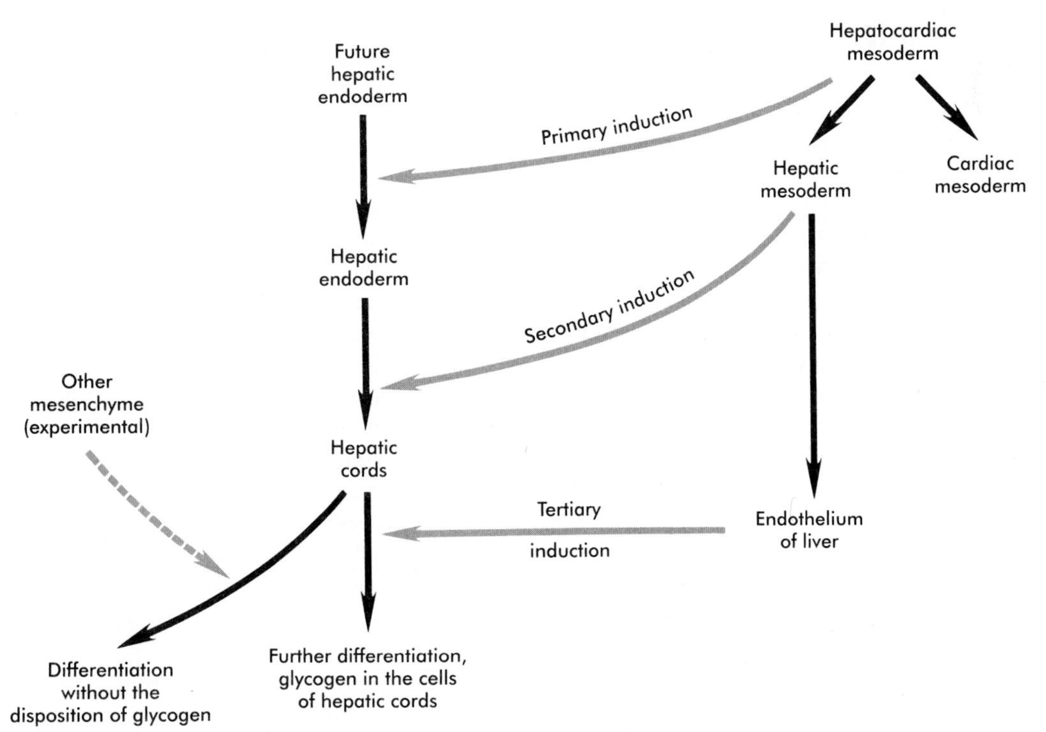

FIG. 16-17 Tissue interactions in the morphogenesis of the endodermal component of the liver.

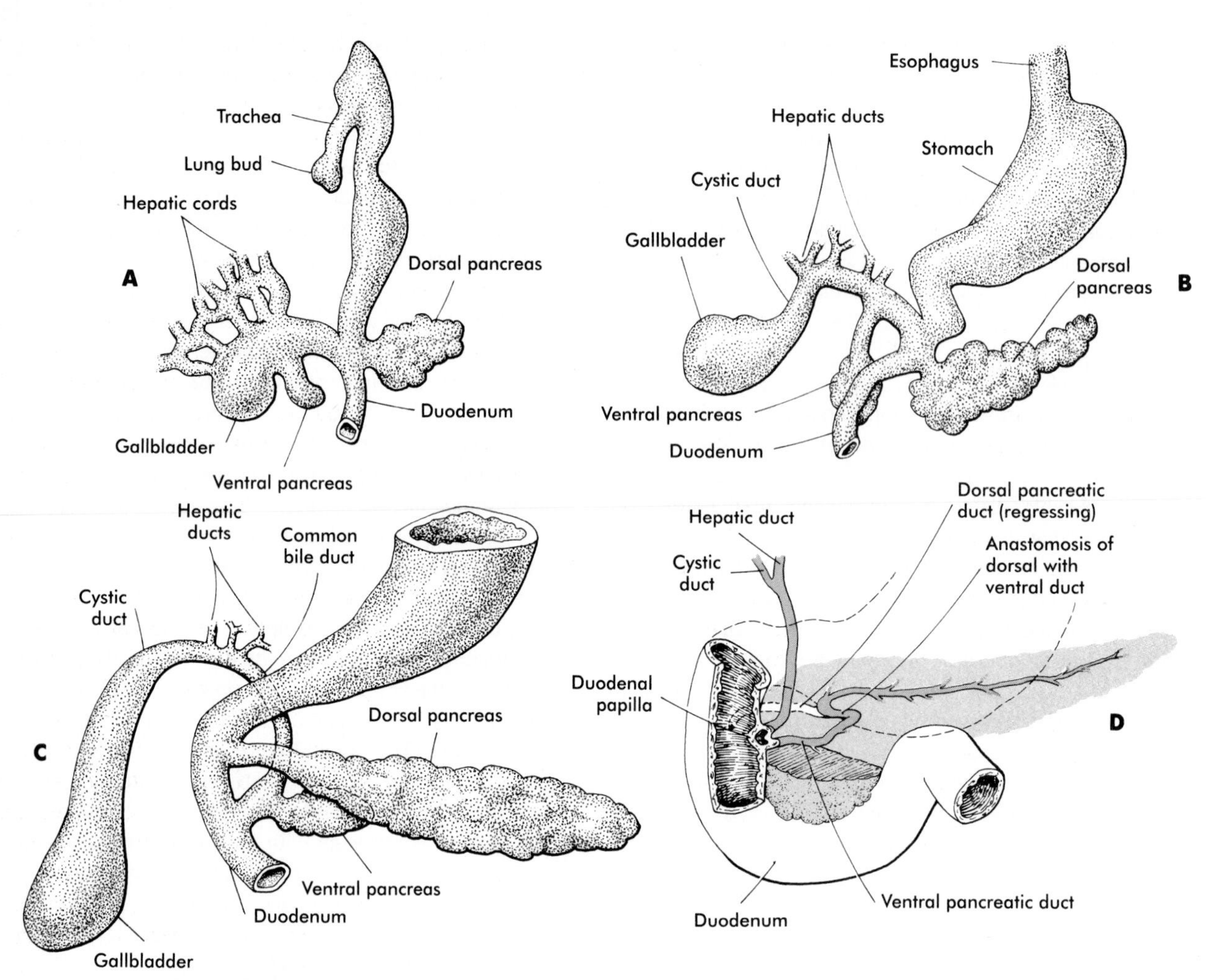

FIG. 16-18 Development of the hepatic and pancreatic primordia from the ventral aspect. **A,** In the fifth week. **B,** In the sixth week. **C,** In the seventh week. **D,** In the late fetus, showing fusion of the dorsal and ventral pancreatic ducts and regression of the distal portion of the dorsal duct.

oping liver is richly vascularized, with many major vessels passing through it in the embryonic period (see Chapter 18).

The entire liver soon becomes too large to be contained in the septum transversum, and it protrudes into the ventral mesentery within the abdominal cavity. As it continues to expand, the rapidly growing liver remains covered by a glistening, translucent layer of mesenteric tissue that now serves as the connective tissue capsule of the liver. Between the liver and the ventral body wall is a thin, sickle-shaped piece of ventral mesentery—the **falciform ligament.** The ventral mesentery between the liver and the stomach is the **lesser omentum** (see Fig. 16-3).

Development of hepatic function. Development of the liver is not only a matter of increasing its mass and structural complexity. As the liver develops, its cells gradually acquire the capacity to perform the many biochemical functions characterizing the mature, functioning liver. One characteristic major function of the liver is to produce the plasma protein **serum albumin.** The mRNA for this protein has been detected in mammalian hepatocytes during the earliest stages of their ingrowth into the hepatic mesoderm. Whether the capacity to produce serum albumin is acquired after the first or second hepatic induction is not certain.

A major function of the adult liver is the synthesis and storage of **glycogen,** which serves as a carbohydrate reserve. As the fetal period progresses, the liver actively stores glycogen. This function is strongly influenced by adrenocortical hormones and indirectly by the anterior pituitary. Similarly, the fetal period includes the functional development of the system of enzymes involved in the synthesis of urea from nitrogenous metabolites. By birth, these have attained full functional capacity.

A major function of the embryonic liver is the production of blood cells. After yolk sac hematopoiesis the liver is one of the chief sites of intraembryonic blood formation. Probably arising from the mesenchyme of the septum transversum, hematopoietic cells appear in small clusters among the hepatic parenchymal cells.

At approximately 12 weeks gestation the hepatocytes begin to produce **bile,** largely through the breakdown of hemoglobin. The bile drains down the newly formed bile duct system and is stored in the gallbladder. As bile is released into the intestines, it stains the other intestinal contents a dark green, which is one of the characteristics of meconium.

Formation of the Pancreas. Shortly after the hepatic primordium first appears, two pancreatic buds begin to grow out of the dorsal and ventral walls of the foregut (see Fig. 5-19). The dorsal bud is a direct outgrowth from the duodenal endoderm; the ventral bud arises from the endoderm of the hepatic diverticulum (see Fig. 16-18). During the phase of early growth the dorsal pancreas becomes considerably larger than the ventral pancreas. At about the same time the duodenum rotates to the right and forms a C-shaped loop, carrying the ventral pancreas and common bile duct behind it and into the dorsal mesentery. The ventral pancreas soon makes contact and fuses with the dorsal pancreas.

Both dorsal and ventral pancreas possess a large duct. After fusion of the two pancreatic primordia, the main duct of the ventral pancreas makes an anastomotic connection with the duct of the dorsal pancreas. The portion of the dorsal pancreatic duct between the anastomotic connection and the duodenum normally regresses, leaving the main duct of the ventral pancreas **(duct of Wirsung)** the definitive outlet from the pancreas into the duodenum.

The pancreatic primordia grow through several epithelial-mesenchymal interactions that are normally seen in the development of glands associated with the gut. The events leading to the establishment of a population of pancreatic precursor cells are poorly understood. Once the phase of outgrowth begins, the glandular epithelium of the pancreas takes shape by the sequential budding of cords of cells derived from the population of pancreatic founder cells (Fig. 16-19). Although the pancreatic epithelium must interact with mesenchyme for continued growth and branching in vivo, results of in vitro experiments show that the epithelium can interact with several types of mesenchyme or a noncellular embryo extract and still grow and branch properly. As the pancreas grows, it develops a prominent system of secretory ducts that ramify throughout the parenchyma of the gland.

The pancreas is a dual organ with both endocrine and exocrine functions. The exocrine portion consists of large numbers of **acini,** which are connected with the secretory duct system. The endocrine component consists of roughly a million richly vascularized **islets of Langerhans,** which are scattered among the acini.

Differentiation of the acini is divided into three phases (see Fig. 16-19). The first, called the **predifferentiated state,** occurs while the pancreatic primordia are first taking shape. A population of pancreatic founder cells that exhibits virtually undetectable levels of digestive enzyme activity is established. As the pancreatic buds begin to grow outward, the epithelium undergoes a transition into a second, **protodifferentiated state.** During this phase the exocrine cells synthesize low levels of many hydrolytic enzymes that they will ultimately produce. After the main period of outgrowth the pancreatic acinar cells pass through another transition before attaining a third, **differentiated state.** By this time, they have acquired an elaborate protein-synthesizing apparatus, and the inactive forms of the polypeptide digestive enzymes are stored in the cytoplasm as **zymogen granules.** Glucocorticoid hormones from the fetal adrenal cortex stimulate increased production of a number of digestive enzymes.

Development of the islets of Langerhans follows a somewhat different course from that of the acini. They are formed from groups of epithelial cells that bud from the pancreatic duct system. The sequence of appearance of the various types of islet cells is well defined. The first to differentiate are the **α cells,** which secrete **glucagon.** This is followed by the appearance of **β cells,** which secrete **insulin.** During the second phase of pancreatic differentiation (protodifferentiated state), the levels of glucagon synthesis considerably exceed those of insulin. Even later, a third population of islet cells begins to secrete **somatostatin.** By the third phase of pancreatic development, secretory granules are evident in the cytoplasm of most islet cells. Insulin and glucagon are present in the fetal circulation by the end of the fifth month of gestation.

Anomalies of the Liver and Pancreas. Many minor variations in the shape of the liver or bile ducts occur, but these normally have no functional significance. One of the most serious malformations involving the liver is **biliary atresia.** This can involve any level, from the tiny bile canaliculi to the major bile-carrying ducts. Newborns with this condition typically develop severe **jaundice** shortly after birth. Some cases can be treated surgically; for others, a liver transplant is necessary.

On rare occasions a ring of pancreatic tissue completely encircles the duodenum, forming an **annular pancreas** (Fig. 16-20). This can sometimes cause obstruction of the duodenum after birth. The cause of annular pancreas is not established, but the most commonly accepted explanation is that outgrowths from a bifid ventral pancreas may encircle the duodenum from both sides.

Heterotopic pancreatic tissue can occasionally be found along the digestive tract, and it occurs most frequently in the duodenum or mucosa of the stomach (Fig. 16-21). About 6% of Meckel's diverticula contain heterotopic pancreatic tissue.

FIG. 16-19 Stages of structural and functional differentiation of the pancreas. The green areas represent primitive islets.

(Modified from Rutter W: *Handbook of physiology, endocrinology I*, Washington, DC, 1972, American Physiological Society, p 32.)

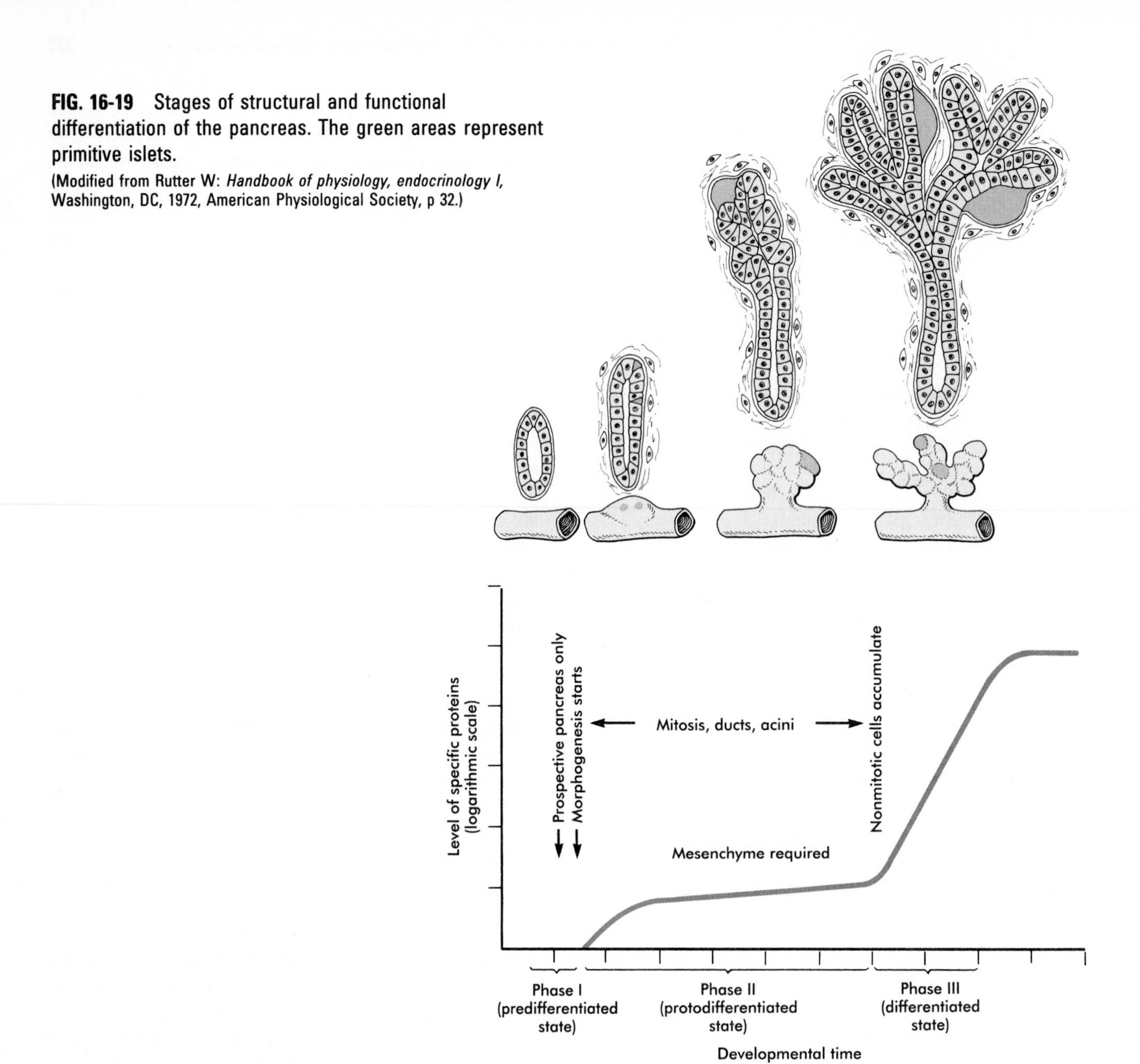

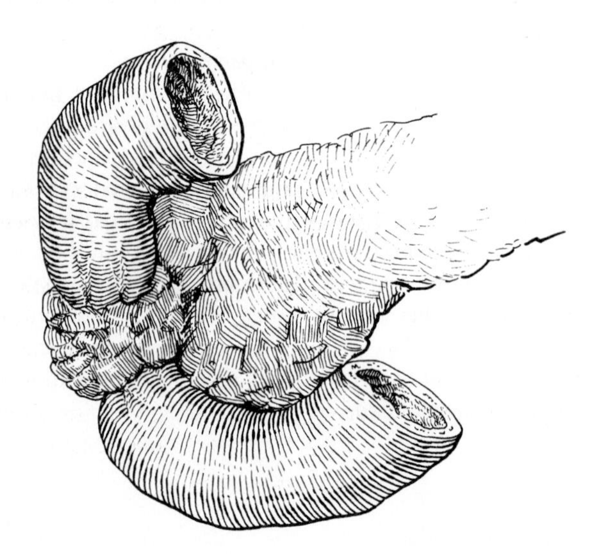

FIG. 16-20 Annular pancreas encircling the duodenum.

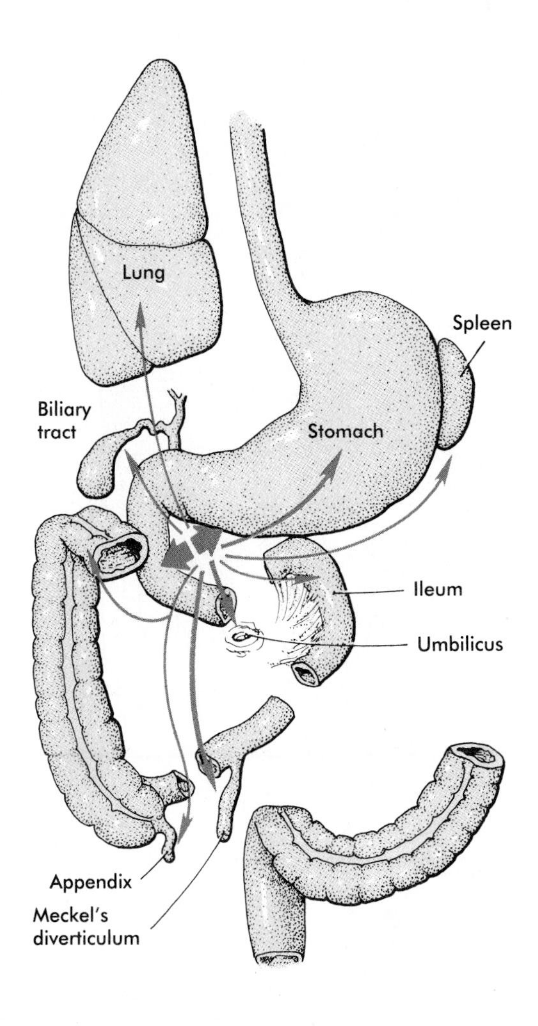

FIG. 16-21 Most common locations in which heterotopic pancreatic tissue can be found. The thickness of the arrows corresponds to the frequency of heterotopic tissue in that location.
(Based on Gray and Skandalakis [1972].)

THE RESPIRATORY SYSTEM

The respiratory system is first seen during the fourth week as an inconspicuous midline **laryngotracheal groove** in the ventral midline at the posterior limit of the pharyngeal region. By the fifth week of gestation, further outgrowth has transformed the laryngotracheal groove into a well-defined **respiratory diverticulum** (see Fig. 5-19) that grows into the splanchnic mesoderm almost parallel to the esophagus. Through a series of interactions with the surrounding mesoderm, the respiratory diverticulum elongates into a tracheal portion and begins to form the first of 23 sets of bifurcations **(lung buds)** that continue into postnatal life.

Formation of the Larynx

During the fourth and fifth weeks of gestation a rapid proliferation of the fourth and sixth branchial arch mesenchyme around the site of origin of the respiratory bud converts the opening slit from the esophagus into a T-shaped **glottis** that is bounded by two lateral **arytenoid swellings** and a cranial **epiglottis**. The fourth and sixth arch mesenchyme surrounding the laryngeal orifice ultimately differentiates into the **thyroid, cricoid,** and **arytenoid cartilages,** which form the skeletal supports of the **larynx.** Like the esophagus, the lumen of the larynx undergoes a temporary epithelial occlusion. In the process of recanalization during the ninth and tenth week, a pair of lateral folds and recesses form the structural basis for the **vocal cords** and adjacent **laryngeal ventricles.** The somitomere-derived musculature of the larynx is innervated by branches of the vagus nerve (X), with that associated with the fourth arch innervated by the **superior laryngeal nerve** and that of the sixth arch innervated by the **recurrent laryngeal nerve.**

Formation of the Trachea and Bronchial Tree

After its initial appearance the respiratory diverticulum undergoes considerable elongation before a pair of bronchial buds appears at the end. The straight portion of the respiratory diverticulum is the primordium of the **trachea.** The **bronchial buds,** which ultimately become the primary bronchi, give rise to additional buds—three on the right and two on the left. These buds become the **secondary,** or **stem, bronchi,** and their numbers presage the formation of the three lobes of the right lung and the two lobes of the

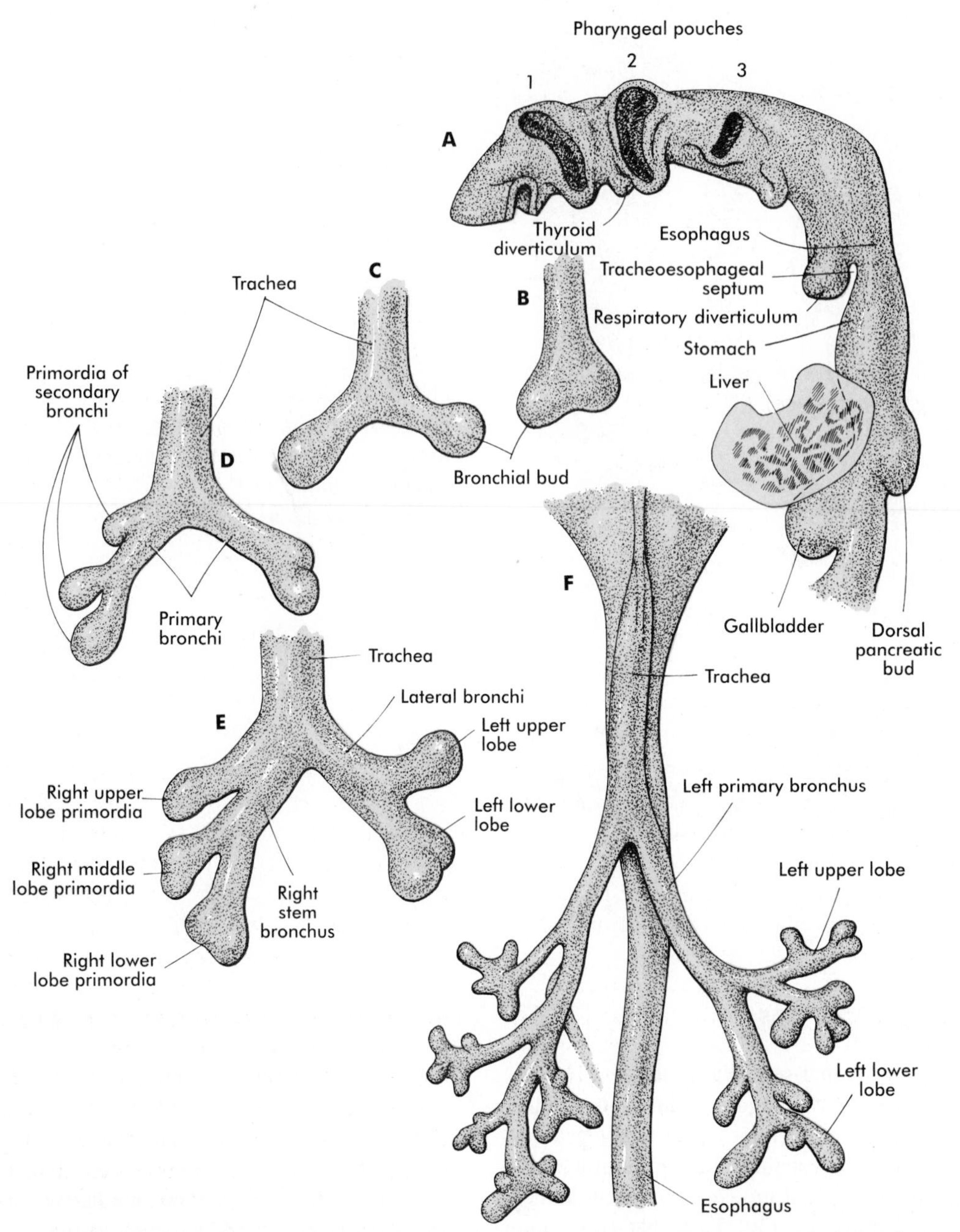

FIG. 16-22 The development of the major branching patterns of the lungs. **A,** Lateral view of the pharynx, showing the respiratory diverticulum in a 4-week-old embryo. **B,** At 4 weeks. **C,** At 32 days. **D,** At 33 days. **E,** At the end of the fifth week. **F,** Early in the seventh week.

left (Fig. 16-22). From this point, each secondary bronchial bud undergoes a long series of branchings until a maximum 23 successive orders of branching have occurred. Morphogenesis of the lung continues after birth, and stabilization of the morphological pattern of the lungs does not occur until several years later.

The mesoderm surrounding the respiratory endoderm controls the nonbudding of the tracheal primordium and the frequent budding of the bronchial tree. Numerous tissue recombination experiments have shown that the mesoderm surrounding the trachea inhibits branching, whereas that surrounding the bronchial buds promotes branching. If tracheal endoderm is combined with bronchial mesoderm, abnormal budding is induced. Conversely, tracheal mesoderm inhibits bronchial budding. Mesoderm of certain other organs such as salivary glands can promote budding of the bronchial endoderm, but a pattern of branching characteristic of the mesoderm is induced. A mesoderm capable of promoting or sustaining budding must maintain a high rate of proliferation of the epithelial cells. Generally, the pattern of the epithelial organ is largely determined by the mesoderm. Structural and functional differentiation of the epithelium is a specific property of the epithelial cells, but the epithelial phenotype corresponds to the region dictated by the mesoderm.

The specific factors that control pulmonary branching are being investigated, but a definitive explanation is not known. Epidermal growth factor and the extracellular form of transforming growth factor-β (TGF-β) are associated with the growth and branching of lung buds. TGF-β is colocalized with types I and III collagen, fibronectin, and proteoglycans, and these molecules are concentrated along ducts and in crotches of epithelial branches, suggesting that they are involved in the stabilization of already-formed structures. In contrast, these substances are not found in significant concentrations around the rapidly expanding portions of the epithelial buds.

As with branching morphogenesis, the formation and maintenance of epithelially lined ducts involves its own set of molecular components. For example, the recently discovered protein **epimorphin** is important in the formation of epithelial tubes. Epimorphin is located in the mesenchyme and appears to provide a signal that allows overlying epithelial cells to establish proper polarity or cell arrangements. In the developing lung the developing epithelial ducts become disorganized and do not form lumens if epimorphin is blocked by specific antibodies.

The protcoglycan **syndecan** is important for maintaining the stability of epithelial sheets along tubules or ducts. Syndecan seems to play a similar role in lung development when it is found along ducts but not in areas where branching is occurring in terminal saccular regions of the developing airway.

Stages in Lung Development

Lung development has been divided into several structural and functional stages.

Embryonic Stage (Weeks 4 to 7). The embryonic stage includes the initial formation of the respiratory diverticulum up to the formation of all major bronchopulmonary segments. During this period the developing lungs grow into and begin to fill the bilateral **pleural cavities.** These represent the major components of the thoracic body cavity above the pericardium (Fig. 16-23).

Pseudoglandular Stage (Weeks 8 to 16). The pseudoglandular stage is the period of major formation and growth of the duct systems within the bronchopulmonary segments before their terminal portions form respiratory components. The histological structure of the lung resembles that of a gland (Fig. 16-24), thus providing the basis for the designation of this stage.

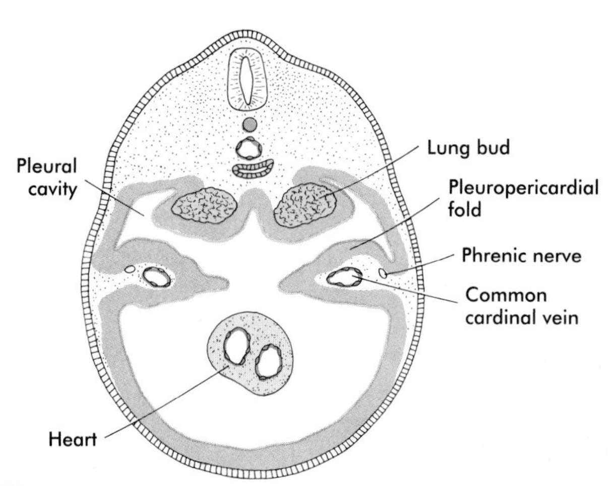

FIG. 16-23 Cross section through the thorax showing the lung buds growing into the pleural cavities. The pleuropericardial folds separate the future pleural from the pericardial cavities.

Canalicular Stage (Weeks 17 to 26). The canalicular stage is characterized by the formation of **respiratory bronchioles** as the result of budding of the terminal components of the system of bronchioles that formed during the pseudoglandular stage. The other major event during this stage is the intense ingrowth of blood vessels into the developing lungs and the close association of capillaries with the walls of the respiratory bronchioles (see Fig. 16-24). Occasionally a fetus born toward the end of this period can survive with intensive care, but respiratory immaturity is the principal reason for poor viability.

Terminal Sac Stage (Weeks 26 to Birth). During the terminal sac stage the terminal air sacs **(alveoli)** bud off the respiratory bronchioles that largely formed during the canalicular stage. The epithelium lining the alveoli differentiates into two types of cells—**type I alveolar cells** (pneumocytes), across which gas exchange occurs, and **type II secretory epithelial cells.** These latter cells form **pulmonary surfactant,** the material that spreads over the surface of the alveoli to reduce surface tension and facilitate expansion of the alveoli during breathing. Research involving specific markers of the epithelial cells has shown that the type II cells form first in the alveolar lining. After proliferation, some type II cells flatten, lose their characteristic secretory function, and undergo terminal differentiation into type I pneumocytes. Other type I cells may differentiate directly from a pool of epithelial precursor cells in the early alveolar lining. With increasing amounts of pulmonary surfactant being formed, the fetus has a correspondingly greater chance of survival if born prematurely. In the fetus the respiratory passageways in the lungs are filled with fluid (see Chapter 19).

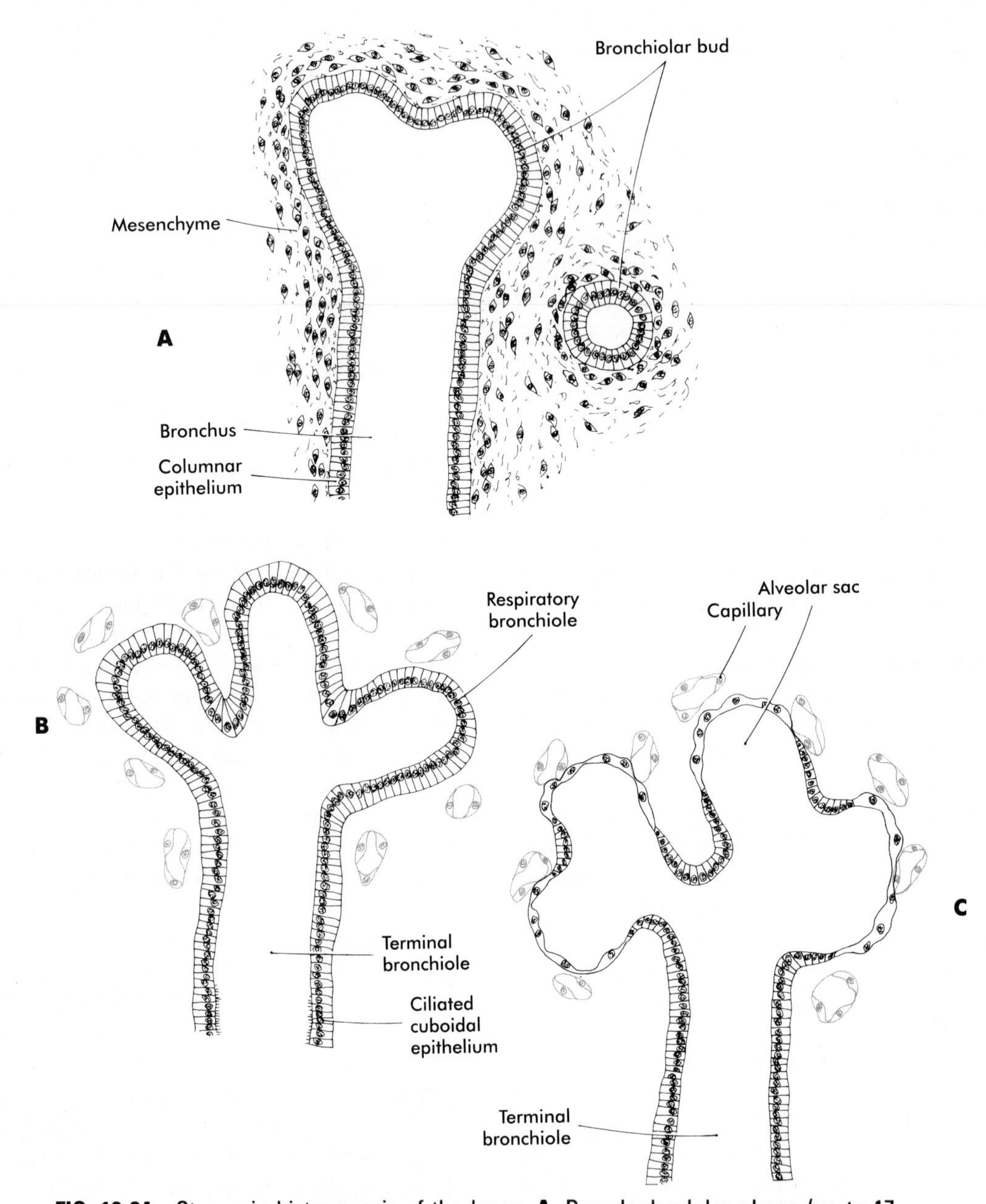

FIG. 16-24 Stages in histogenesis of the lungs. **A,** Pseudoglandular phase (up to 17 weeks). **B,** Canalicular phase (17 to 26 weeks). **C,** Terminal sac phase (26 weeks to birth).

Postnatal Stage. At birth the mammalian lung is far from mature. An estimated 90% or more of the roughly 300 million alveoli found in the mature human lung are formed after birth. The major mechanism for this increase is the formation of secondary connective tissue septa that divide existing alveolar sacs. When they first appear, the secondary septa are relatively thick. In time they transform into thinner mature septa capable of full respiratory exchange function.

Malformations of the Respiratory System

Tracheoesophageal Fistulas. The most common family of malformations of the respiratory tract is related to abnormal separation of the tracheal bud from the esophagus during early development of the respiratory system. Many common anatomical varieties of tracheoesophageal fistulas exist (Fig. 16-25), but virtually all involve the stenosis or atresia of a segment of trachea or esophagus and an abnormal connection between them. These are manifest early after birth by the newborn's choking or regurgitation of milk when feeding.

Tracheal or Pulmonary Atresia. These rare malformations are incompatible with life. They are probably caused by fundamental defects in the epithelial-mesenchymal interactions on which formation of the respiratory system depends.

Gross Malformations of the Lungs. Because of their structural complexity, the lungs are subject to a variety of structural variations or malformations (e.g., abnormal lobation). These are usually asymptomatic, but recognition of the possibility of these variations from normal is important for the pulmonary surgeon.

Respiratory Distress Syndrome (Hyaline Membrane Disease). This condition is often manifest in infants born prematurely and is characterized by labored breathing. In infants who die of this condition, the lungs are underinflated and the alveoli are partially filled with a proteinaceous fluid that seems to form a membrane over the respiratory surfaces. This is related to insufficiencies in the formation of surfactant by the type II alveolar cells.

Congenital Cysts in the Lung. Abnormal cystic structures can form in the lung or other parts of the respiratory tract. These can range from large single cysts to numerous small cysts located throughout the parenchyma of the lung. They may be associated with polycystic kidneys. If the cysts are numerous, they can cause respiratory distress.

THE BODY CAVITIES

Formation of the Common Coelom and Mesentery

As the lateral mesoderm of the early embryo splits and then folds laterally, the space between the somatic and splanchnic layers of mesoderm becomes the common **intraembryonic coelom** (Fig. 16-26) (see Chapter 5). The same folding process that results in the completion of the ventral body wall and the separation of the intraembryonic from the extraembryonic coelom also brings the two layers of

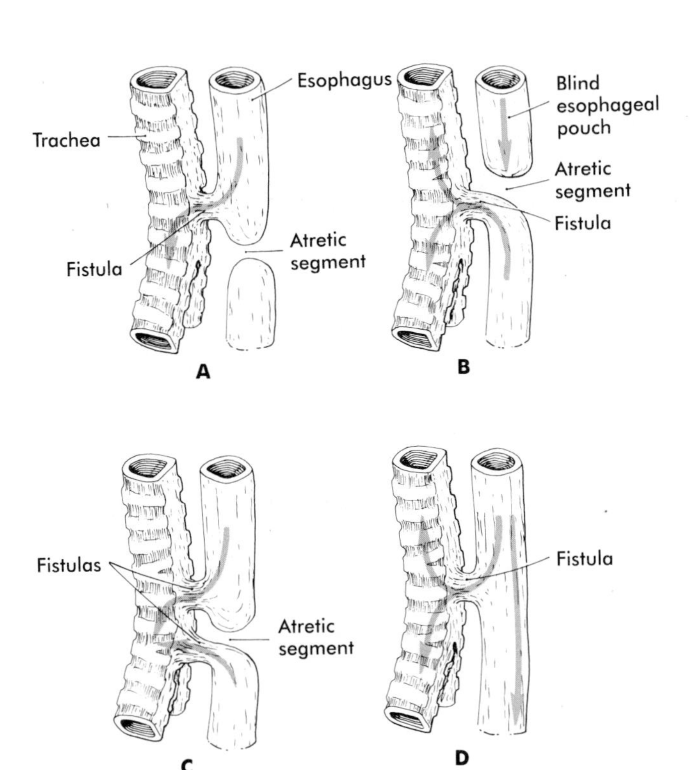

FIG. 16-25 Varieties of tracheoesophageal fistulas. **A,** Fistula above atretic esophageal segment. **B,** Fistula below atretic esophageal segment. **C,** Fistulas above and below atretic esophageal segment. **D,** Fistula between patent esophagus and trachea.

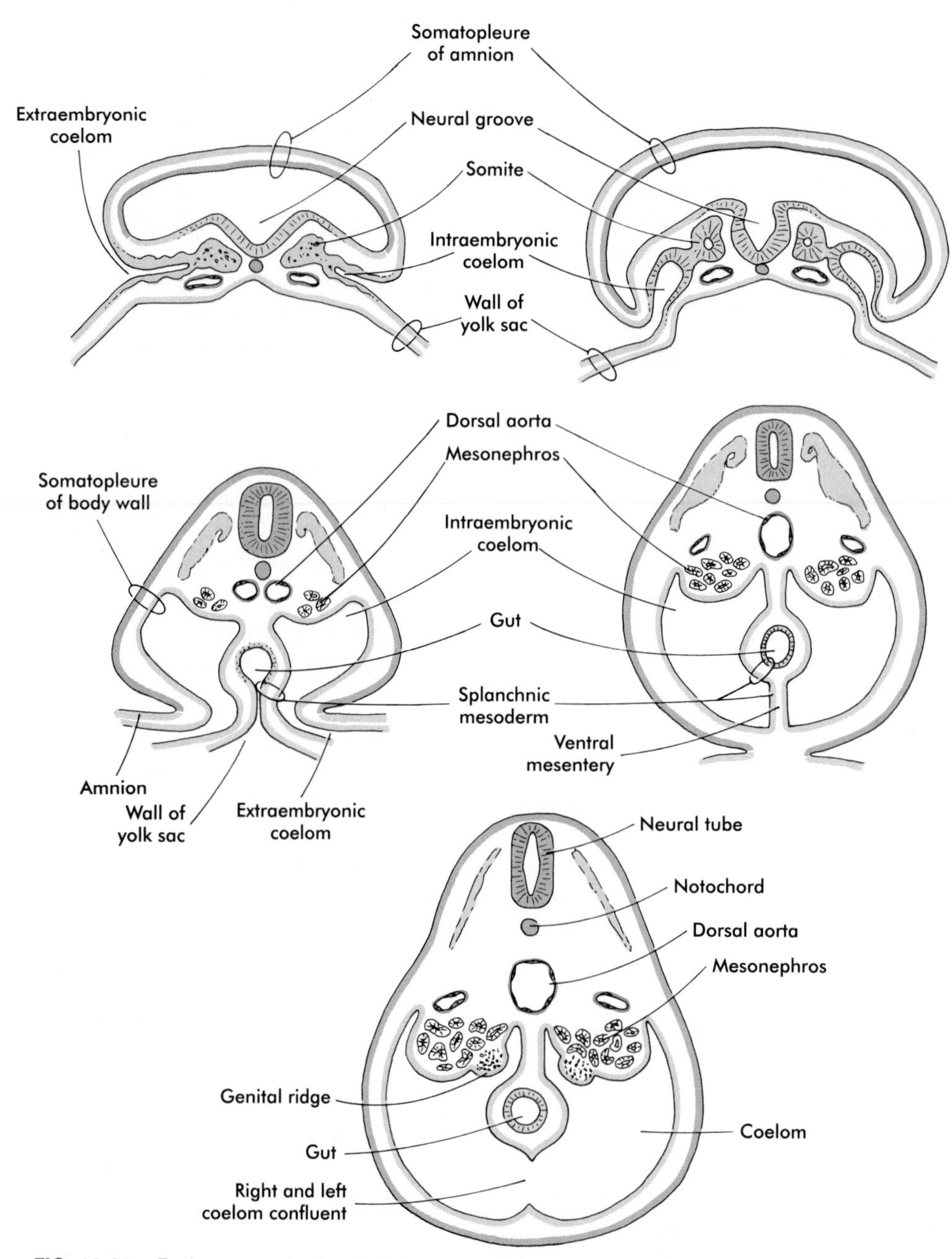

FIG. 16-26 Early stages in the development of the coelom and mesenteries. (Adapted from Carlson B: *Patten's foundations of embryology,* ed 5, New York, 1988, McGraw-Hill.)

splanchnic mesoderm around the newly formed gut as the **primary,** or **common, mesentery.** The primary mesentery suspends the gut from the dorsal body wall as the **dorsal mesentery** and attaches it to the ventral body wall as the **ventral mesentery.** This placement effectively divides the coelom into right and left components. Soon, however, most of the ventral mesentery breaks down and causes a confluence of the right and left halves of the coelom. In the region of the developing stomach and liver the ventral mesentery persists, forming the **ventral mesogastrium** and the falciform ligament of the liver (see Fig. 16-3). Farther cranially the tubular primordium of the heart is similarly supported by a **dorsal mesocardium** and briefly by a **ventral mesocardium,** which soon breaks down.

Formation of the Septum Transversum and Pleural Canals

A major factor in division of the common coelom into thoracic and abdominal components is the **septum transver-**

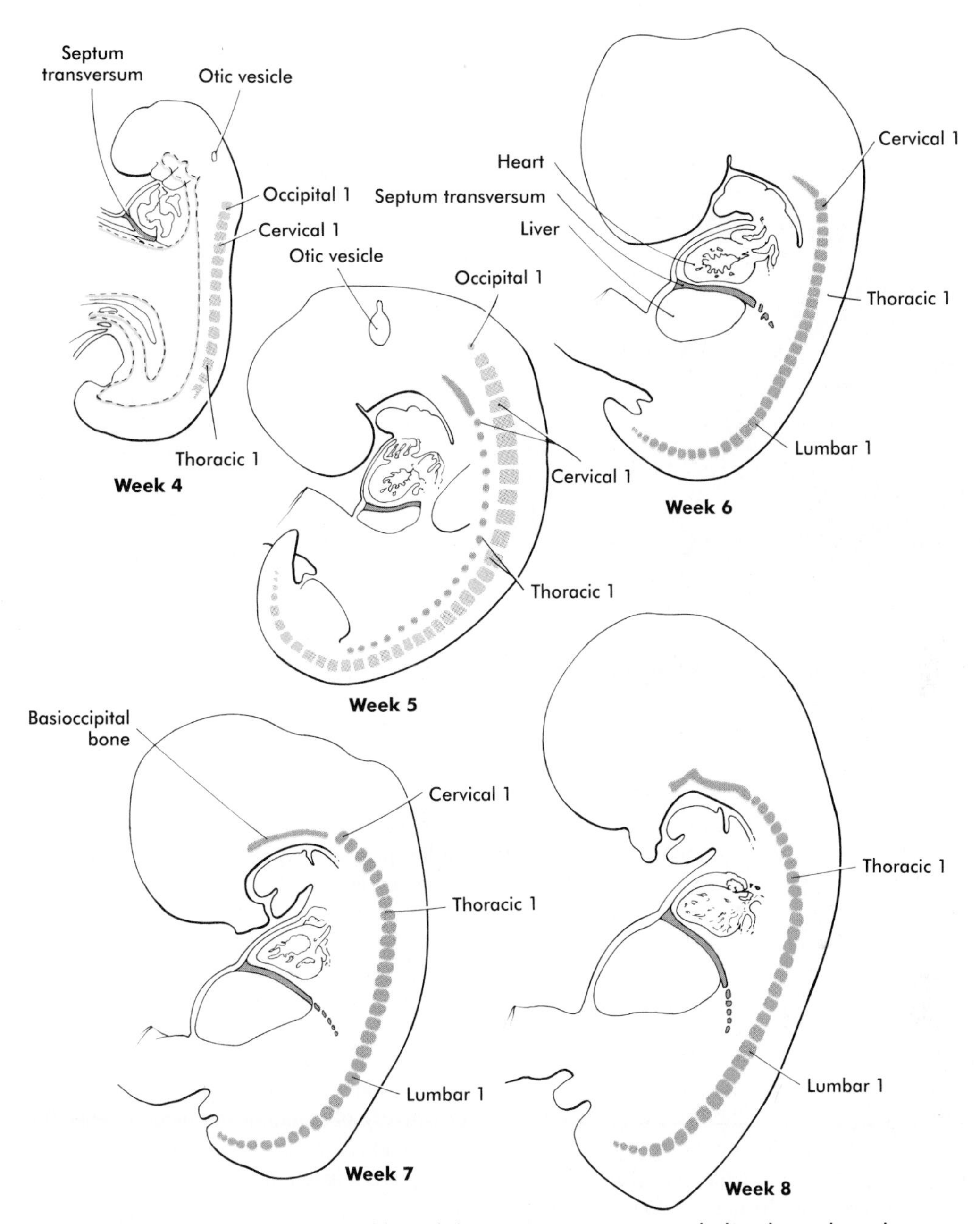

FIG. 16-27 Changes in the position of the septum transversum during the embryonic period. The gray repeating structures are somites. The olive repeating structures are elements of the axial skeleton.

sum. This septum grows from the ventral body wall as a semicircular shelf, separating the heart from the developing liver (Fig. 16-27). During the liver's early development a major portion is embedded in the septum transversum. Ultimately, the septum transversum constitutes a significant component of the diaphragm (see p. 332).

The expanding septum transversum serves as a partial partition between the pericardial and peritoneal portions of the coelom. By the time the expanding edge of the septum transversum reaches the floor of the foregut, it has almost cut the common coelom into two parts. Two short channels located on either side of the foregut, however, connect the two major parts (Fig. 16-28). Initially known as the **pleural (pericardioperitoneal) canals,** these channels represent the spaces into which the developing lungs grow. The pleural canals enlarge greatly as the lungs increase in size and ultimately form the **pleural cavities.**

The pleural canals are partially delimited by two paired folds of tissue, the pleuropericardial and pleuroperitoneal folds. The **pleuropericardial folds** (see Fig. 16-23) are

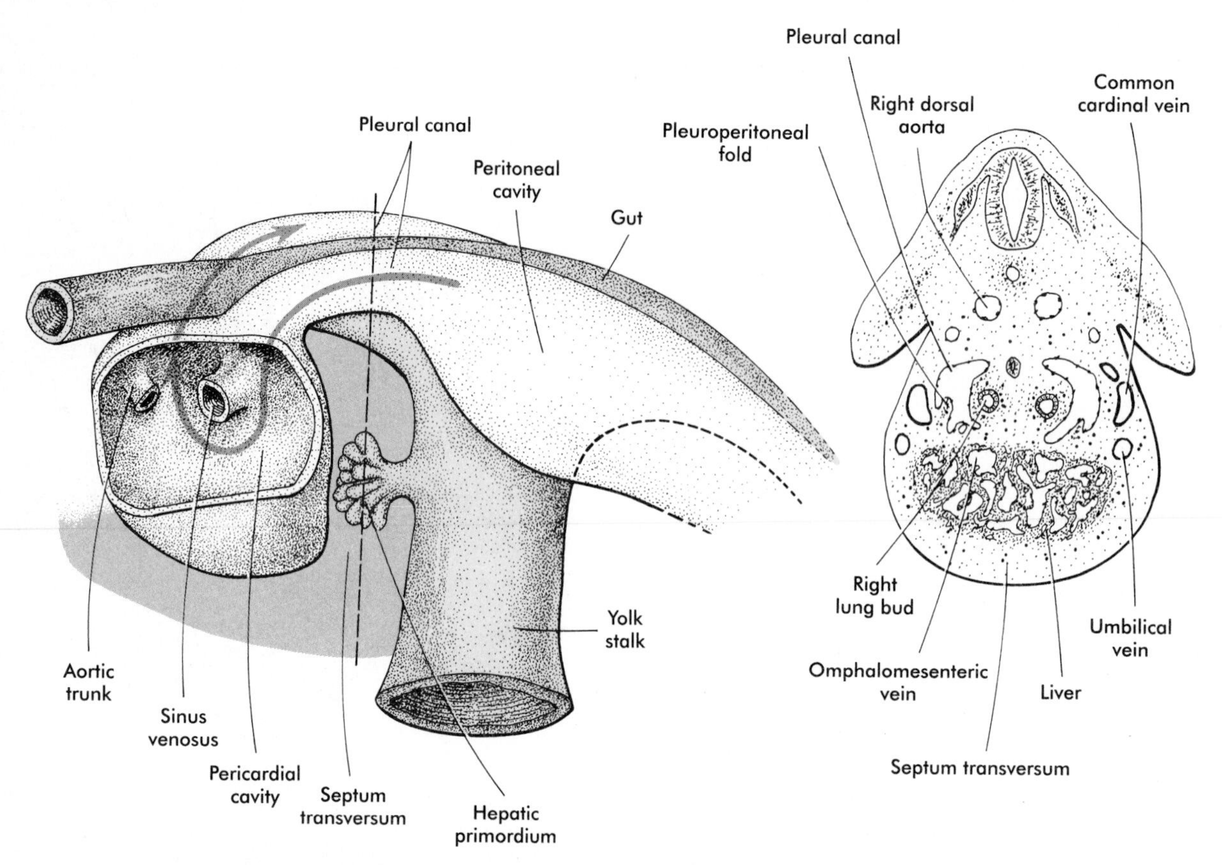

FIG. 16-28 The relationships among the pericardial cavity, the pleural canals, and the peritoneal cavity. The red arrow passes from the left pleural cavity into the pericardial cavity and then into the right pleural canal. The dashed line on the left represents the level of the cross section on the right.

ridges of tissue associated with the common cardinal veins, which bulge into the dorsolateral wall of the coelom as they arch toward the midline of the thoracic portion of the coelom and enter the sinus venosus of the heart (Fig. 16-29). Initially, the pleuropericardial folds are not large and cause only a narrowing at the junction of the pericardial cavity and pleural canals. However, as the lungs expand, the folds form prominent shelves that meet at the midline, forming the fibrous (parietal) layer of the pericardium.

Associated with the pleuropericardial folds are the paired **phrenic nerves.** These arise from joined branches of cervical roots 3, 4, and 5 and supply the muscle fibers of the diaphragm. With the shifts in positions of various components of the body during growth, the diaphragm ultimately descends to the level of the lower thoracic vertebrae. As it does, it carries the phrenic nerves with it. Even in adults the pathway of the phrenic nerves through the fibrous pericardium is a reminder of their early association with the pleuropericardial folds.

At the caudal ends of the pleural canals, another pair of folds, the **pleuroperitoneal folds,** becomes prominent as the expanding lungs push into the mesoderm of the body wall. The pleuroperitoneal folds occupy successively greater portions of the pleural canal until they fuse with the septum transversum and the mesentery of the esophagus, effectively obliterating the pleural canal (Fig. 16-30). All connections between the abdominal cavity and the thoracic cavity are thus eliminated.

Formation of the Diaphragm

The **diaphragm,** which separates the thoracic from the abdominal cavity in adults, is a composite structure derived from several embryonic components (see Fig. 16-30). The large ventral component of the diaphragm arises from the septum transversum, which fuses with the ventral part of the esophageal mesentery. Converging on the esophageal mesentery from the dorsolateral sides are the pleuroperitoneal folds. These components form the bulk of the diaphragm. As the lungs continue to grow, their caudal tips excavate additional space in the body wall. The body wall mesenchyme separated from the body wall proper becomes a third component of the diaphragm by forming a thin rim of tissue along its dorsolateral borders.

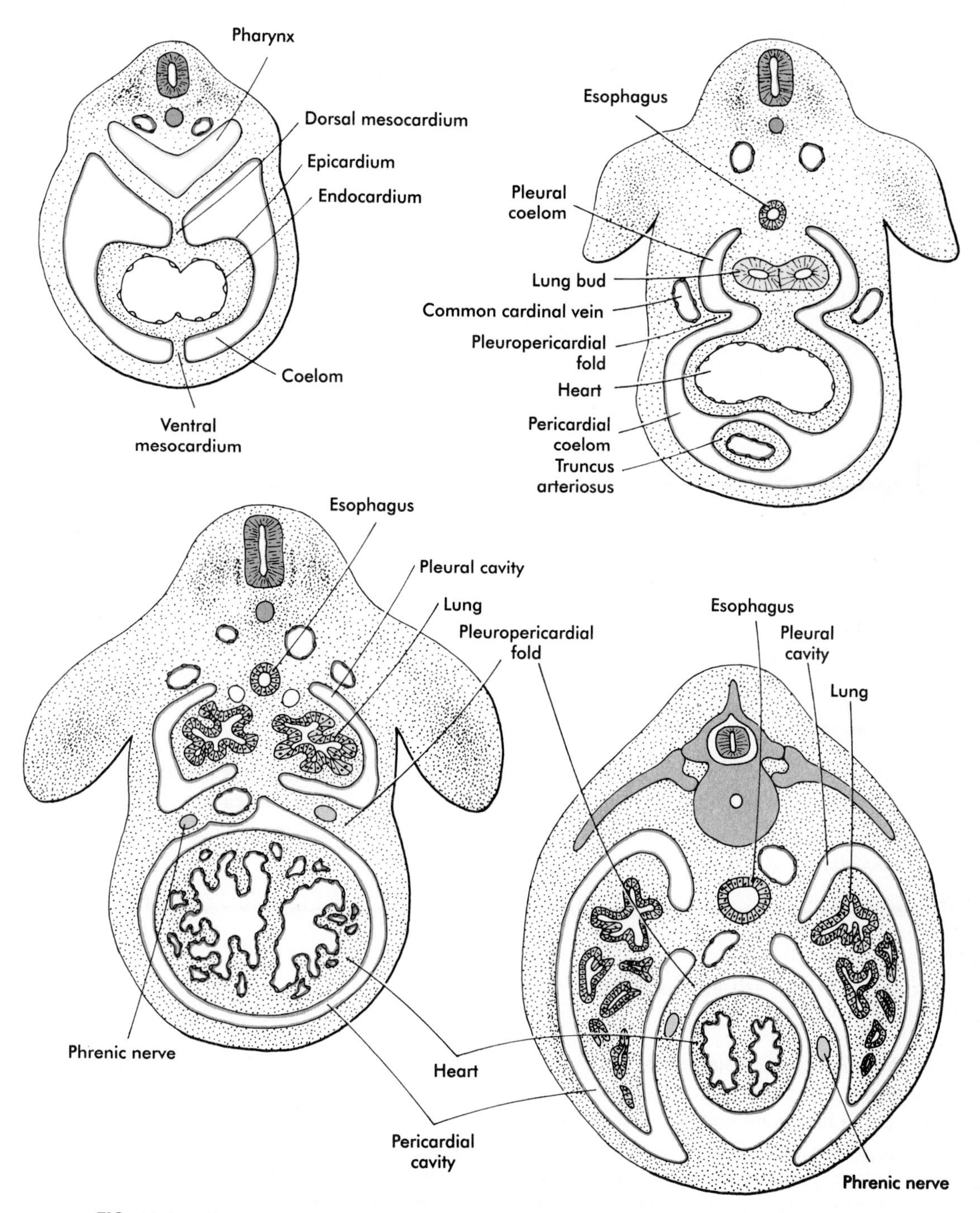

FIG. 16-29 The development of the pleuropericardial folds.
(Adapted from Carlson B: *Patten's foundations of embryology,* ed 5, New York, 1988, Mc-Graw-Hill.)

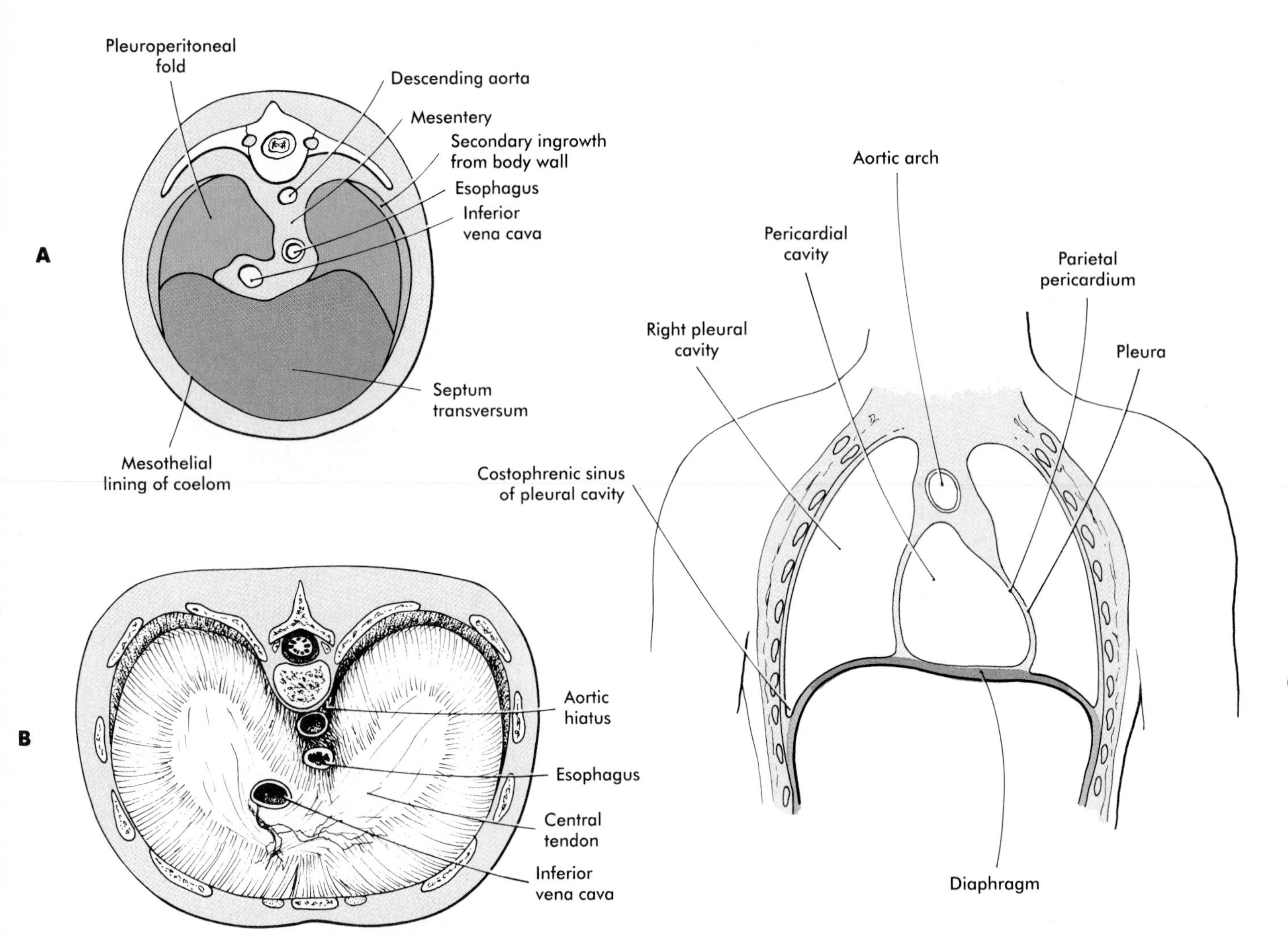

FIG. 16-30 Stages in the formation of the diaphragm. **A,** The components making up the embryonic diaphragm. **B,** The adult diaphragm as seen from the thoracic side. **C,** Frontal section showing relations of the diaphragm to the pleural and pericardial cavities.

Malformations of the Body Cavities, Diaphragm, and Body Wall

Ventral Body Wall Defects, Ectopia Cordis, Gastroschisis, and Omphalocele. The opposing sides of the body wall occasionally fail to fuse as the embryo assumes its cylindrical shape late in the first month. Several defective mechanisms such as hypoplasia of the tissues can account for these defects. A quantitatively minor defect in closure of the thoracic wall is manifest as **failure of sternal fusion** (Fig. 16-31). If growth of the two sides of the thoracic wall is severely defective, the heart can form outside the thoracic cavity, resulting in **ectopia cordis** (Fig. 16-32).

Closure defects of the ventral abdominal wall can lead to similar gross malformations. In many cases of omphalocele (see Fig. 16-12), hypoplasia of the abdominal wall itself or deficiencies of abdominal musculature are evident. More serious cases involve evisceration of the abdominal contents through a fissure between the umbilicus and sternum **(gastroschisis)** (Fig. 16-33). Caudal to the umbilicus an associated closure defect of the urinary bladder (exstrophy of the bladder [see Chapter 17]) is common.

Diaphragmatic Hernias. Incomplete fusion or hypoplasia of one or more of the components of the diaphragm can lead to an open connection between the abdominal and thoracic cavities. If the defect is large enough, a variety of structures in the abdominal cavity (usually part of the stomach or intestines) can herniate into the thoracic cavity, or more rarely, thoracic structures can penetrate into the abdominal cavity. Minor cases of herniation can cause digestive symptoms. In the case of major defects the herniation

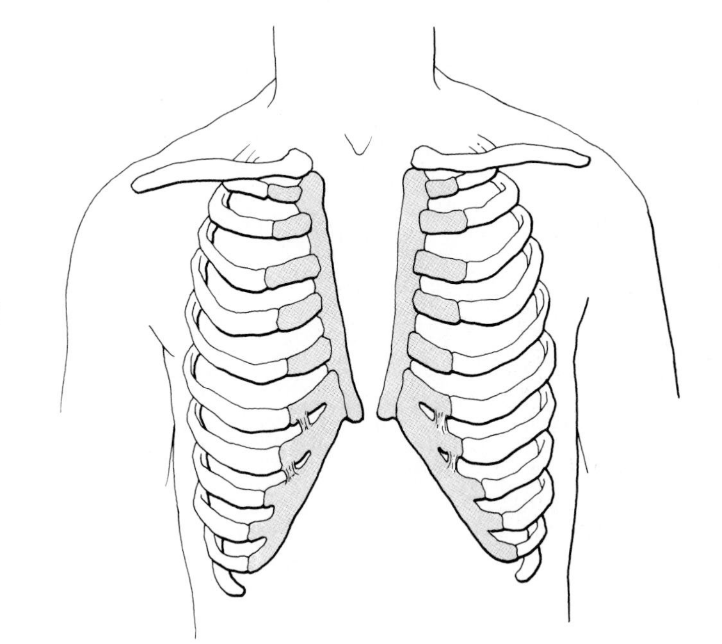

FIG. 16-31 Failure of fusion of the paired components of the embryonic sternum.

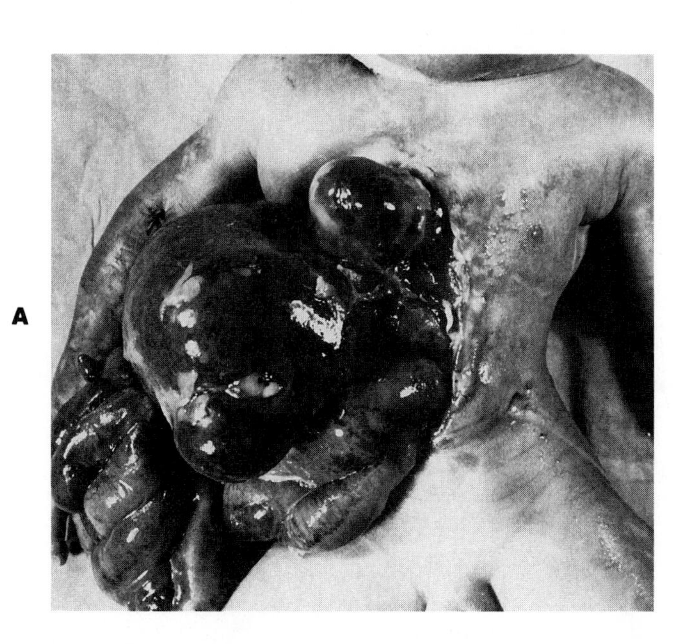

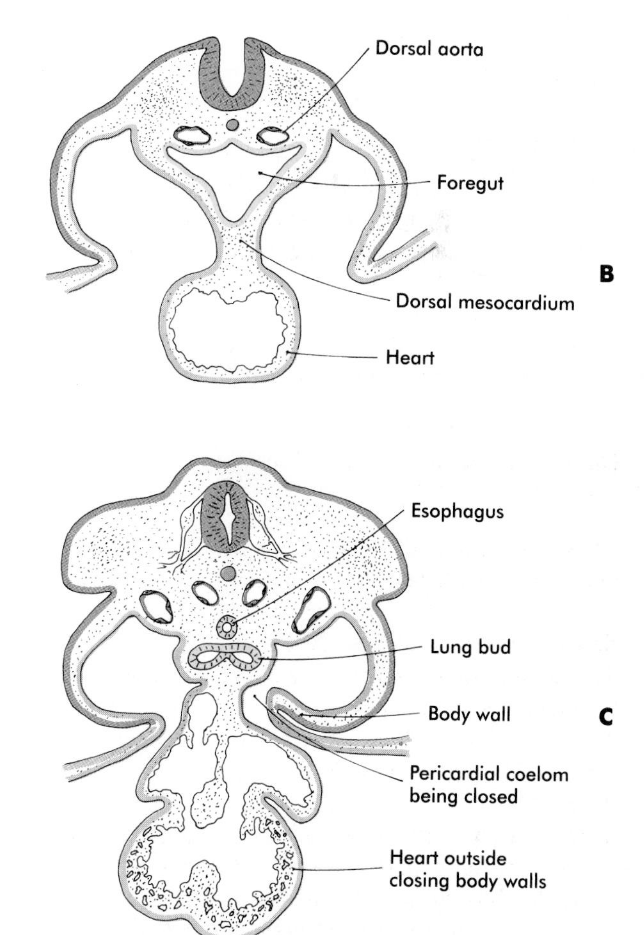

FIG. 16-32 Ectopia cordis. **A,** Fetus with a major ventral abdominal wall defect that combines gastroschisis and ectopia cordis. **B** and **C,** Cross sections illustrating the inability of the folding sides of the body wall to encompass the developing heart, resulting in ectopia cordis.

(Courtesy Mason Barr, Ann Arbor, Mich.)

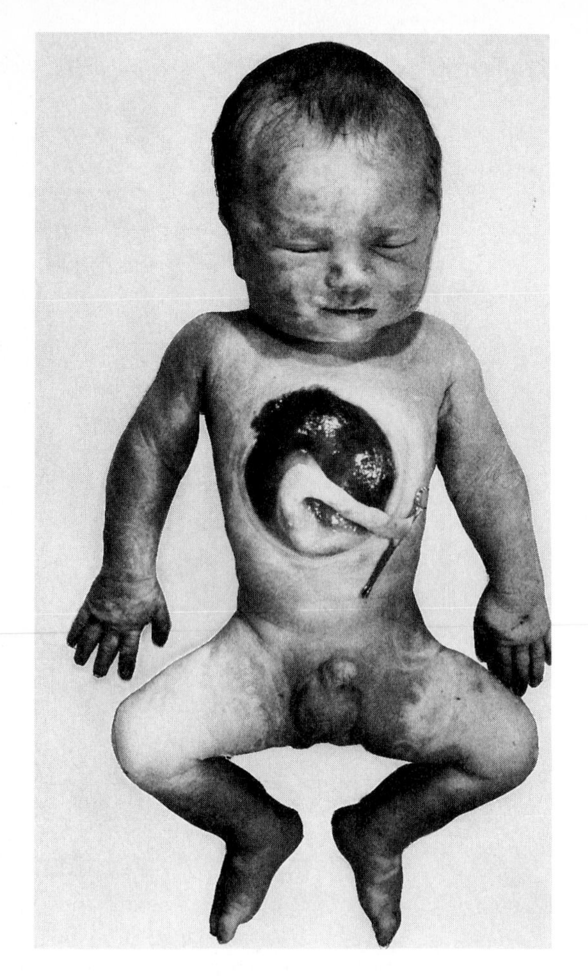

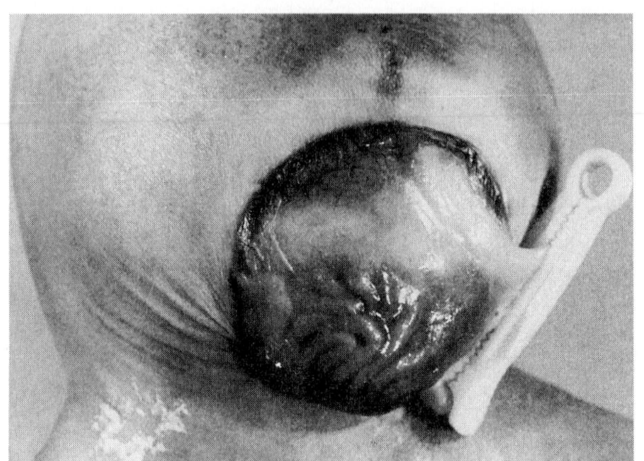

FIG. 16-33 Defective closure of the ventral abdominal wall cranial **(A)** and caudal **(B)** to the umbilicus.
(Courtesy Mason Barr, Ann Arbor, Mich.)

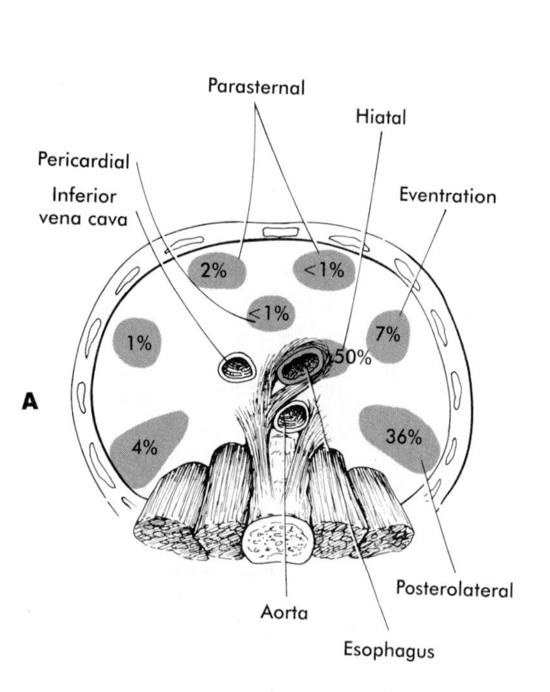

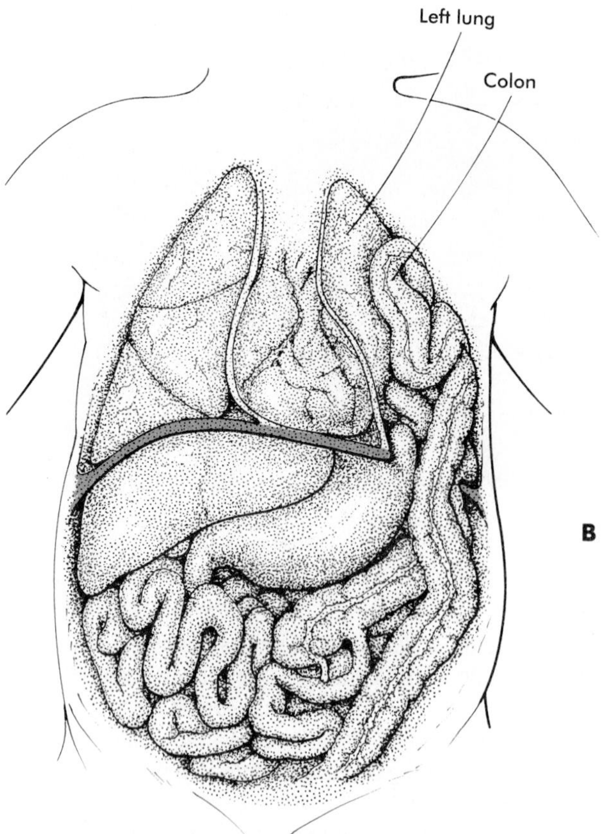

FIG. 16-34 **A,** Common sites of diaphragmatic hernias. Percentages of occurrence are indicated. **B,** A diaphragmatic hernia with intestines entering the left pleural cavity and compressing the left lung.

of massive portions of the intestines can press against the heart or lungs and interfere with their function. Some common sites of defects in the diaphragm are shown in Fig. 16-34.

SUMMARY

1. The digestive system arises from the primitive endodermally lined gut tube, which is bounded cranially by the stomodeal plate and caudally by the cloacal plate. The gut is divided into foregut, midgut, and hindgut segments, with the midgut opening into the yolk sac. Development of virtually all parts of the gut depends on epithelial-mesenchymal interactions. Responding to such interactions, primordia of the respiratory system, the liver, the pancreas, and other digestive glands bud out from the original gut tube.
2. The esophagus takes shape as a simple tubular structure between the pharynx and stomach. At one stage the epithelium occludes the lumen of the esophagus; the lumen later recanalizes. The developing stomach is suspended from a dorsal and ventral mesogastrium. Through two types of rotation, the stomach attains its adult position. Common malformations of the stomach are pyloric stenosis, which interferes with emptying of the stomach, and ectopic gastric mucosa, which can produce ulcers in unexpected locations.
3. As they grow, the intestines form a hairpin loop that herniates into the body stalk. Further growth of the small intestine causes small intestinal loops to accumulate in the body stalk. While the intestines retract into the body cavity, they rotate around the superior mesenteric artery. This results in the characteristic positioning of the colon around the small intestine in the abdominal cavity. During these changes in position, parts of the dorsal mesentery fuse with the peritoneal lining of the dorsal body wall. In the posterior part of the gut the urorectal septum partitions the cloaca into the rectum and urogenital sinus.
4. During its differentiation the lining of the intestinal tract passes through phases of (1) epithelial proliferation, (2) cellular differentiation, and (3) biochemical and functional maturation. Like the esophagus, the small intestine goes through a period of occlusion of the lumen by the epithelium. At later stages, intestinal crypts located at the base of villi contain epithelial stem cells, which supply the entire intestinal epithelial surface with a variety of epithelial cells.
5. The intestinal tract is subject to a variety of malformations, including local stenosis, atresia, duplications, diverticula, and abnormal rotation. Incomplete resorption of the vitelline duct can give rise to Meckel's diverticulum, vitelline duct ligaments, cysts, or fistulas. Omphalocele is the failure of the intestines to return to the body cavity from the body stalk. Aganglionic megacolon is caused by the failure of parasympathetic neurons to populate the distal part of the colon. Failure of the anal membrane to break down (imperforate anus) may be associated with fistulas connecting the digestive tract to various regions of the urogenital system.
6. Digestive glands arise as epithelial diverticula from the gut. Their formation and further outgrowth is based on interactions with the surrounding mesenchyme. The primordium of the liver arises in the septum transversum, but as it expands, it protrudes into the ventral mesentery. As it develops, the liver acquires the capacity to synthesize and secrete serum albumin and to store glycogen among other biochemical functions. The pancreas grows out as dorsal and ventral pancreatic buds that ultimately fuse to form a single pancreas. Within the pancreas the epithelium forms exocrine components, which secrete digestive enzymes, and endocrine components (islets of Langerhans), which secrete insulin and glucagon.
7. The respiratory system arises as a ventral outgrowth from the gut just caudal to the pharynx. Through epithelial-mesenchymal interactions, the tip of the respiratory diverticulum undergoes up to 23 sets of dichotomous branchings. Other interactions with the surrounding mesenchyme stabilize the tubular parts of the respiratory tract by inhibiting branching. Lung development goes through several stages: (1) the embryonic stage, (2) the pseudoglandular stage, (3) the canalicular stage, (4) the terminal sac stage, and (5) the postnatal stage.
8. Important malformations of the respiratory tract include tracheoesophageal fistulas, which result in abnormal connections between the trachea and esophagus. Atresia of components of the respiratory system is rare, but anatomical variations in the morphology of the lungs are common. Respiratory distress syndrome, commonly seen in premature infants, is related to insufficiencies in the formation of pulmonary surfactant by type II alveolar cells.
9. In its most basic condition the intraembryonic coelom is separated into right and left components by the dorsal and ventral mesenteries, which suspend the gut. Except for the region of the stomach and liver, the ventral mesentery disappears. In the region of the heart the dorsal mesocardium persists and the ventral mesocardium disappears.
10. The septum transversum divides the coelom into thoracic and abdominal regions, which are connected

by pleural canals. The developing lungs grow into the pleural canals, which are partially delimited by paired pleuropericardial and pleuroperitoneal folds. The definitive diaphragm is formed from (1) the septum transversum, (2) pleuroperitoneal folds, and (3) ingrowths from body wall mesenchyme.

11. Quantitative deficiencies in ventral body wall tissue can result in abnormalities ranging from failure of sternal fusion to ectopia cordis in the thorax and omphalocele to gastroschisis and/or exstrophy of the bladder in the abdomen. Defects in the diaphragm are diaphragmatic hernias and can result in intestines herniating into the thoracic cavity.

REVIEW QUESTIONS

1. A young girl was troubled for several years with moderately severe abdominal pain that appeared after eating. After passing through several pediatric and medical clinics without obtaining relief, she was sent to a psychiatrist, who also could not resolve her symptoms. Finally, an astute physician suspected that her symptoms might be caused by a congenital anomaly, and further testing proved this suspicion to be correct. What was the diagnosis?
2. During its first feeding, a newborn begins to choke. What congenital anomalies should be included in the differential diagnosis?
3. A newborn took its first feeding of milk without incident but an hour later was crying in pain and vomited the milk with considerable force. Examination of the infant revealed a hard mass near the midline in the upper region of the abdomen. What was the diagnosis?
4. An infant was noted to extrude a small amount of mucus and fluid from the umbilicus when crying or straining. This should make the physician think of what congenital anomaly in the differential diagnosis?
5. A newborn was given a cursory physical examination and was taken home by the mother 1 day after delivery. Several days later the mother brought the child to the clinic. The newborn was in obvious severe discomfort with a swollen abdomen. Physical examination revealed that an important congenital anomaly had been overlooked at the original examination. What was that anomaly?

REFERENCES

Alescio T, Cassini A: Induction in vitro of tracheal buds by pulmonary mesenchyme grafted on tracheal epithelium, *J Exp Zool* 150:83-94, 1962.

Avery ME, Wang N-S, Taeusch HW: The lung of the newborn infant, *Sci Am* 228:74-85, 1973.

Boyden EA: Development of the human lung, *Pract Pediatr* IV:1-17, 1975.

Brauker JH, Trautman MS, Bernfield M: Syndecan, a cell surface proteoglycan, exhibits a molecular polymorphism during lung development, *Dev Biol* 111:213-220, 1991.

Burri PH: Fetal and postnatal development of the lung, *Ann Rev Physiol* 46:617-628, 1984.

Cascio S, Zaret KS: Hepatocyte differentiation initiates during endodermal-mesenchymal interactions prior to liver formation, *Development* 113:217-225, 1991.

Chen J-M, Little CD: Cellular events associated with lung branching morphogenesis including the deposition of collagen type IV, *Dev Biol* 120:311-321, 1987.

Colony PC: *Successive phases of human fetal development.* In Kretchmer N, Minkowski A, eds: *Nutritional adaptation of the gastrointestinal tract of the newborn,* New York, 1983, Raven Press, pp 3-28.

Cullen TS: *Embryology, anatomy, and diseases of the umbilicus, together with diseases of the urachus,* Philadelphia, 1916, Saunders.

Elliott R: A contribution to the development of the pericardium, *Am J Anat* 48:355-390, 1933.

Githens S: *Differentiation and development of the exocrine pancreas in animals.* In Go VL and others, eds: *The exocrine pancreas: biology, pathobiology and diseases,* New York, 1986, Raven Press, pp 21-32.

Gray SW, Skandalakis JE: *Embryology for surgeons,* Philadelphia, 1972, Saunders.

Heine UI and others: Colocalization of TGF-beta 1 and collagen I and III, fibronectin and glycosaminoglycans during lung branching morphogenesis, *Development* 109:29-36, 1990.

Hilfer SR, Rayner RM, Brown JW: Mesenchymal control of branching pattern in the fetal mouse lung, *Tissue Cell* 17:523-538, 1985.

Hirai Y and others: Epimorphin: a mesenchymal protein essential for epithelial morphogenesis, *Cell* 69:471-481, 1992.

Hollinshead WH: Embryology and anatomy of the anal canal and rectum, *Dis Colon Rectum* 5:18-22, 1962.

Hollinshead WH: Embryology and surgical anatomy of the colon, *Dis Colon Rectum* 5:23-27, 1962.

Jackson CM: On the developmental topography of the thoracic and abdominal viscera, *Anat Rec* 3:361-396, 1909.

Jacobs-Cohen RJ and others: Inability of neural crest cells to colonize the presumptive aganglionic bowel of *ls/ls* mutant mice: requirement for a permissive environment, *J Comp Neurol* 255:425-438, 1987.

Joyce-Bradley MF, Brody JS: Ontogeny of pulmonary alveolar epithelial markers of differentiation, *Dev Biol* 137:331-348, 1990.

Kedinger M and others: Epithelial-mesenchymal interactions in intestinal epithelial differentiation, *Scand J Gastroenterol* 23(suppl 151):62-69, 1988.

Le Douarin NM: An experimental analysis of liver development, *Med Biol* 53:427-455, 1975.

Le Douarin NM: On the origin of pancreatic endocrine cells, *Cell* 53:169-171, 1988.

Lewis FT: The form of the stomach in human embryos with notes upon the nomenclature of the stomach, *Am J Anat* 13:477-503, 1912.

Mall FP: Development of the human coelom, *J Morphol* 12:395-453, 1897.

Mathan M, Moxey PC, Trier JS: Morphogenesis of fetal rat duodenal villi, *Am J Anat* 146:73-92, 1976.

McGowan SE: Extracellular matrix and the regulation of lung development and repair, *FASEB J* 6:2895-2904, 1992.

Moens CB and others: A targeted mutation reveals a role for N-*myc* in branching morphogenesis in the embryonic mouse lung, *Genes Dev* 6:691-704, 1992.

O'Rahilly R: The timing and sequence of events in the development of the human digestive system and associated structures during the embryonic period proper, *Anat Embryol* 153:123-136, 1978.

O'Rahilly R, Boyden EA: The timing and sequence of events in the development of the human respiratory system during the embryonic period proper, *Z Anat Entwickl-Gesch* 141:237-250, 1973.

O'Rahilly R, Tucker JA: The early development of the larynx in staged human embryos, *Ann Otolaryngol Rhinol Laryngol* 82(suppl 7):1-27, 1973.

Pictet R, Rutter WJ: *Development of the embryonic endocrine pancreas.* In *Handbook of physiology, section 7: endocrinology,* vol 1, Washington, DC, 1972, American Physiological Society, pp 25-66.

Ponder BAJ and others: Derivations of mouse intestinal crypts from single progenitor cells, *Nature* 313:689-691, 1985.

Potten CS, Loeffler M: Stem cells: attributes, cycles, spirals, pitfalls and uncertainties: lessons for and from the crypt, *Development* 110:1001-1020, 1990.

Rudnick D: Development of the digestive tube and its derivatives, *Ann NY Acad Sci* 55:109-116, 1952.

Teitleman G, Lee JK: Cell lineage analysis of pancreatic islet cell development: glucagon and insulin precursors arise from catecholaminergic precursors present in the pancreatic duct, *Dev Biol* 121:454-466, 1987.

Thompson ABR, Keelan M: The development of the small intestine, *Can J Physiol Pharmacol* 64:13-29, 1986.

Wells LJ: Development of the human diaphragm and pleural sacs, *Carnegie Contr Embryol* 35:107-134, 1954.

Wells LJ, Boyden EA: The development of the human bronchopulmonary segments in human embryos of horizons XVII to XIX, *Am J Anat* 95:163-201, 1954.

Wessels NK: *Tissue interactions and development,* Menlo Park, Calif, 1977, WA Benjamin.

Yasugi S, Mizuno T: Mesenchymal-epithelial interactions in the organogenesis of the digestive tract, *Zool Sci* 7:159-170, 1990.

CHAPTER 17

The Urogenital System

The urogenital system arises from the intermediate mesoderm of the early embryo (see Fig. 5-10). Several major themes underlie the development of urinary and genital structures from this common precursor. The first is the interconnectedness of urinary and genital development, where early components of one system are taken over by another during its later development. A second is the recapitulation during human ontogeny of kidney types (the equivalent of organ isoforms) that are terminal forms of the kidney in lower vertebrates. A third theme is the dependence of differentiation and maintenance of many structures in the urogenital system on epithelial-mesenchymal interactions. Finally, the sexual differentiation of many structures passes from an indifferent stage, in which male and female differences are not readily apparent, to a male or female pathway depending on the presence of specific promoting or inhibiting factors acting on the structure. Although phenotypic sex is genetically determined, genetic sex can be overridden by environmental factors, leading to a discordance between the two.

THE URINARY SYSTEM

The urinary system begins to take shape before any gonadal development is evident. Embryogenesis of the kidney begins with the formation of an elongated pair of excretory organs similar in structure and function to the kidneys of lower vertebrates. These early forms of the kidney are later supplanted by the definitive metanephric kidneys, but as they regress, certain components are retained to be reused by other components of the urogenital system. At all stages of kidney development the renal tubules and ducts exhibit the tissue-specific expression of **Pax-2,** a paired box-containing gene.

Early Forms of the Kidney

The common representation of mammalian kidney development includes three successive phases beginning with the appearance of the **pronephros,** the developmental homologue of the type of kidney found in only the lowest vertebrates. In human embryos the first evidence of a urinary system consists of the appearance of a few segmentally arranged sets of epithelial cords that differentiate from the anterior intermediate mesoderm at about 22 days gestation. These structures are more appropriately called **nephrotomes.** The nephrotomes connect laterally, with a pair of **primary nephric (pronephric) ducts,** which grow toward the cloaca (Fig. 17-1).

As the primary nephric ducts extend caudally, they appear to stimulate the intermediate mesoderm to form additional segmental sets of tubules. These tubules are structurally equivalent to the **mesonephric tubules** of fishes and amphibians. A typical mesonephric unit consists of a vascular **glomerulus,** which is partially surrounded by an epithelial glomerular capsule. The glomerular capsule is continuous with a contorted mesonephric tubule, which is surrounded by a mesh of capillaries. Each mesonephric tubule empties separately into the continuation of the primary nephric duct, which becomes known as the **mesonephric (wolffian) duct.**

The formation of pairs of mesonephric tubules occurs along a craniocaudal gradient. Mesonephric tubules take shape slightly behind the caudal extension of the mesonephric ducts. By the end of the fourth week of gestation the mesonephric ducts attach to the cloaca, and a continuous lumen is present throughout each. Very near its attachment site to the cloaca, the mesonephric duct develops an epithelial outgrowth called the **ureteric bud** (see Fig. 17-1, *A*). Early in the fifth week the ureteric bud begins to grow

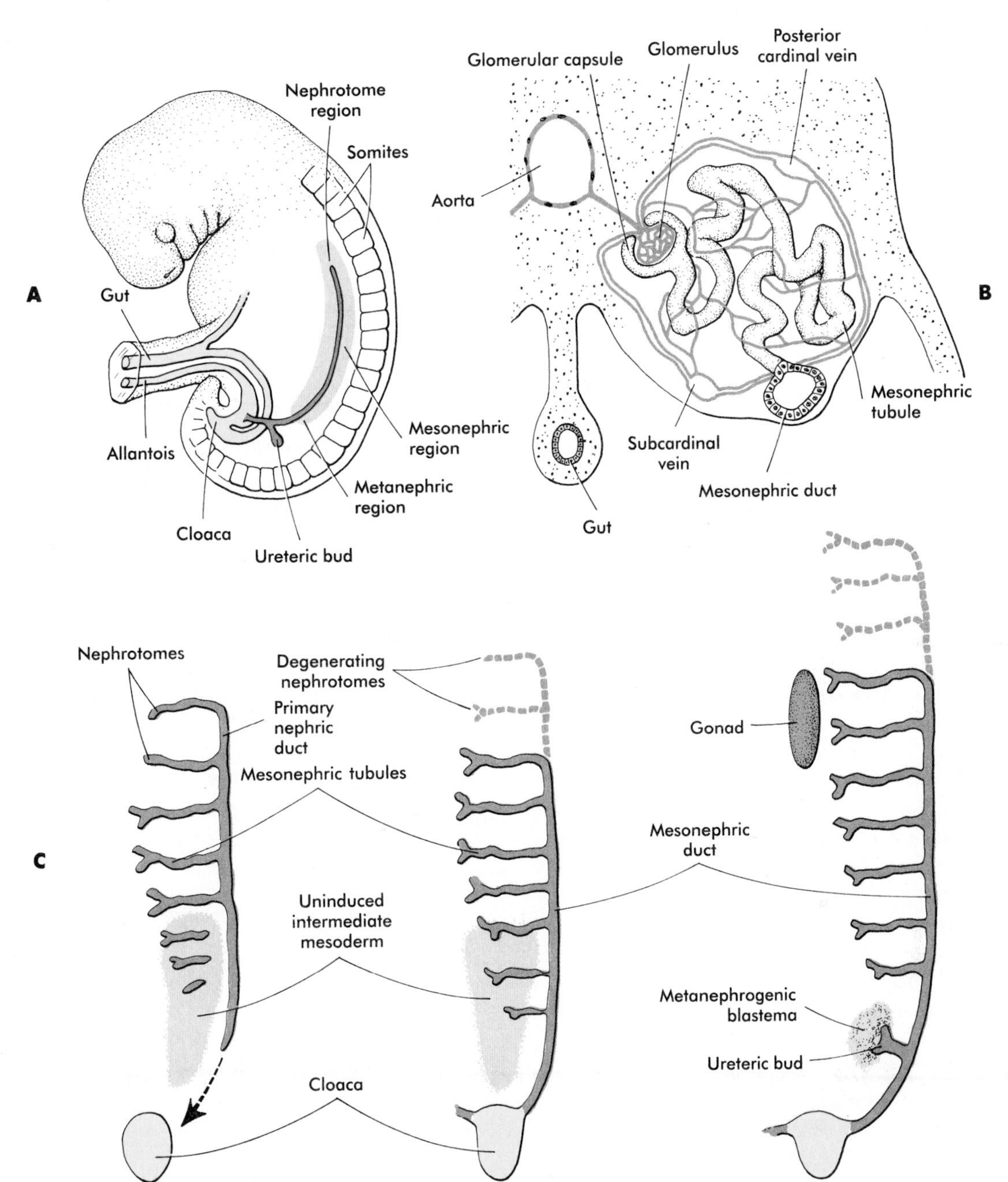

FIG. 17-1 Early stages in establishment of the urinary system. **A,** Subdivision of the intermediate mesoderm into areas that will form nephrotomes, mesonephros, and metanephros. **B,** Cross section through mesonephros showing a well-developed mesonephric tubule and its associated vasculature. **C,** The caudal progression of formation of the mesonephros and degeneration of the most cranial segments of the primitive kidney.

into the most posterior region of the intermediate mesoderm. It then sets up a series of continuous inductive interactions leading to the formation of the definitive kidney, the **metanephros.**

Although there is evidence of function in the mammalian mesonephric kidney, the physiology of the mesonephros has not been extensively investigated. Urine formation in the mesonephros begins with a filtrate of blood from the glomerulus into the glomerular capsule. This filtrate then flows into the tubular portion of the mesonephros, where the selective resorption of ions and other substances occurs. The return of resorbed materials to the blood is facilitated by the presence of a dense plexus of capillaries around the mesonephric tubules.

The structure of the human mesonephros is very similar to that of fishes and aquatic amphibians, and it functions

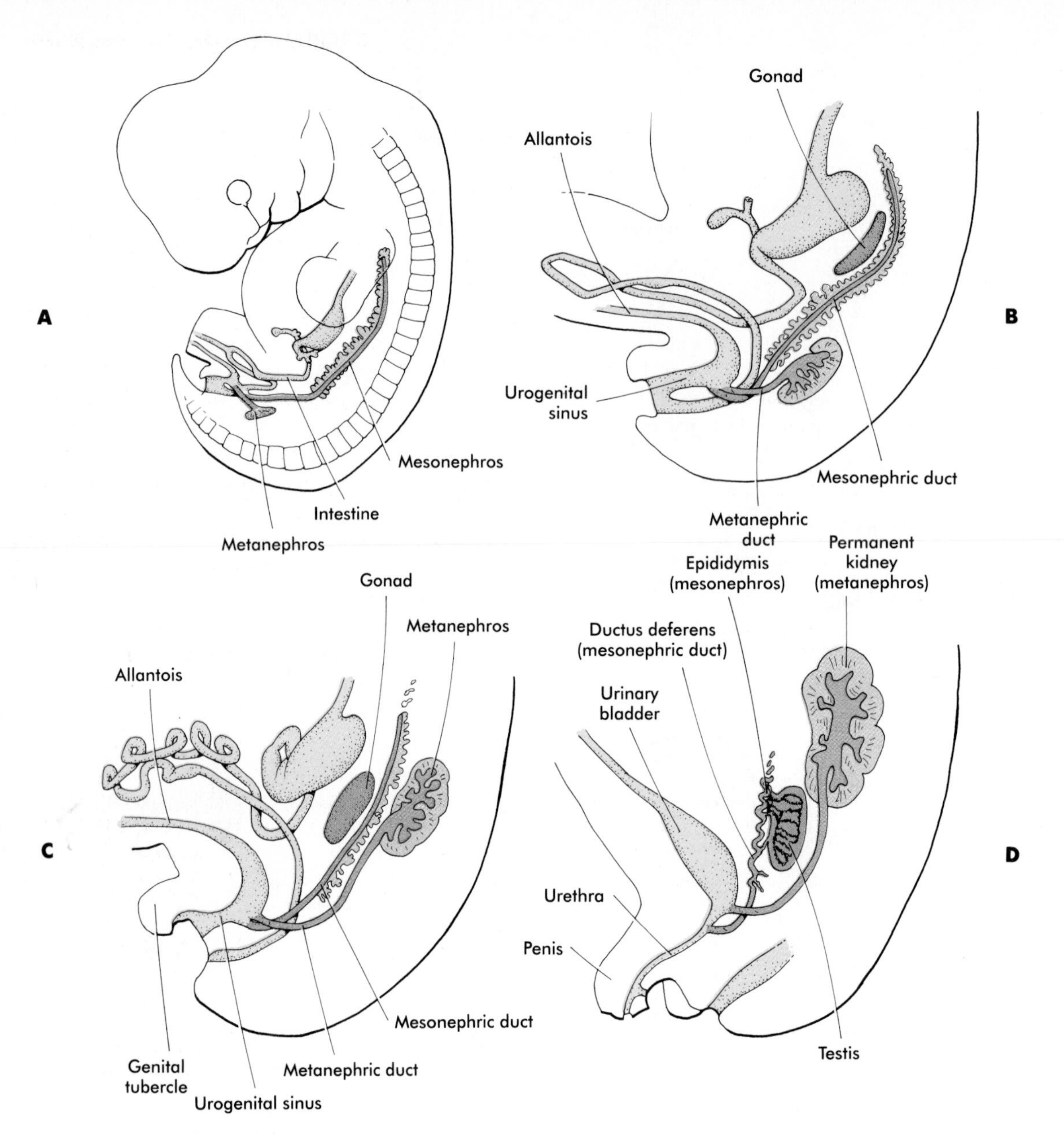

FIG. 17-2 Stages in the formation of the metanephros. **A,** At 6 weeks. **B,** At 7 weeks. **C,** At 8 weeks. **D,** At 3 months (male).

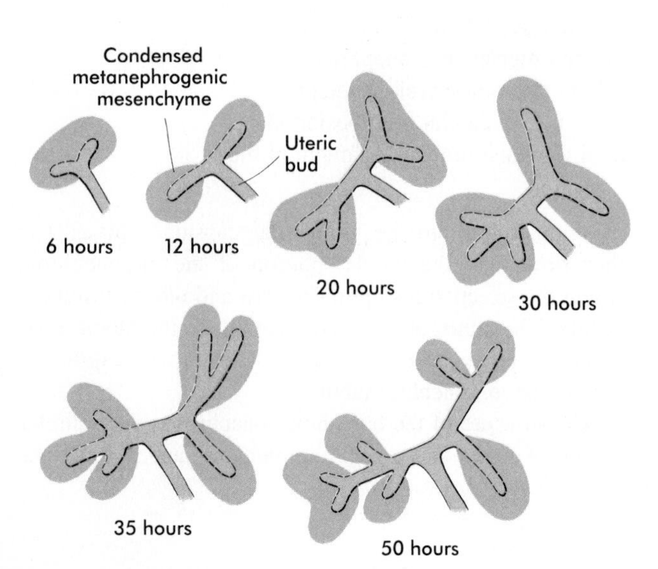

FIG. 17-3 Early development of the metanephric primordium in the mouse in culture. Over a 50 hour interval the ureteric bud *(orange)* and metanephrogenic blastema *(green)* become subdivided into separate lobes. (Based on studies by Saxén.)

principally to filter and remove body wastes. Because these species and the amniote embryo exist in an aquatic environment, there is little need to conserve water. Therefore the mesonephros does not develop a medullary region and an elaborate system for concentrating urine as the adult human kidney must.

The mesonephros is most prominent while the definitive metanephros is taking shape. Although it rapidly regresses as a urinary unit after the metanephric kidneys become functional, the mesonephric ducts and some of the mesonephric tubules persist in the male and become incorporated as integral components of the genital duct system.

The Metanephros

Development of the metanephros begins early in the fifth week when the ureteric bud **(metanephric diverticulum)** grows into the posterior portion of the intermediate mesoderm (Fig. 17-2). Mesenchymal cells of the intermediate mesoderm condense around the metanephric diverticulum to form the **metanephrogenic blastema.**

The morphological basis for the development of the metanephric kidney is the elongation and branching (up to 14 or 15 times) of the metanephrogenic diverticulum, which becomes the **collecting (metanephric) duct** system of the metanephros, and the formation of renal tubules from mesenchymal condensations (metanephrogenic blastema) located around the tips of the branches. The mechanism underlying these events is a series of reciprocal inductive interactions between the tips of the branches of the metanephric ducts and the surrounding metanephrogenic blastemal cells. Without the metanephric duct system, tubules do not form; conversely, the metanephrogenic mesoderm acts on the metanephric duct system, inducing its characteristic branching.

The formation of individual functional units **(nephrons)** in the developing metanephros involves three mesodermal cell lineages—epithelial cells derived from the metanephric diverticulum, mesenchymal cells of the metanephrogenic blastema, and ingrowing vascular endothelial cells. The earliest stage is the condensation of mesenchymal blastemal cells around the terminal bud of the ureteric (metanephric) duct (Fig. 17-3). The preinduced mesenchyme contains several interstitial proteins, such as types I and III collagen and fibronectin. As the mesenchymal cells condense after an inductive event, these proteins are lost and replaced with epithelial type proteins (type IV collagen, laminin, and heparin sulfate proteoglycan), which are ultimately localized to the basement membranes (Fig. 17-4).

As the terminal bud of the metanephric duct branches, the surrounding mesenchyme splits into two parts. A single condensation of mesenchymal cells goes through a defined series of stages to form a renal tubule. After a growth phase, mitotic activity within the rounded blastemal mesenchyme decreases, and the primordium of the tubule assumes a comma shape (Fig. 17-5, *A*). Within the comma a

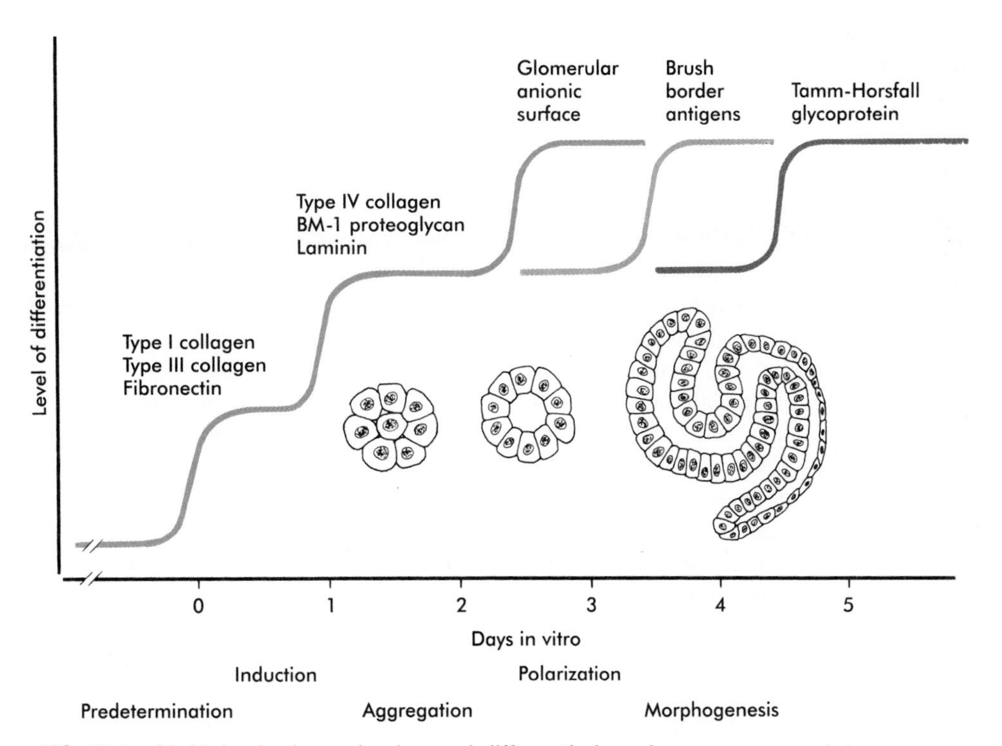

FIG. 17-4 Multiphasic determination and differentiation of mouse metanephric mesoderm in vitro.

(Modified from Saxén L and others: *Biology of human growth,* New York, 1981, Raven Press.)

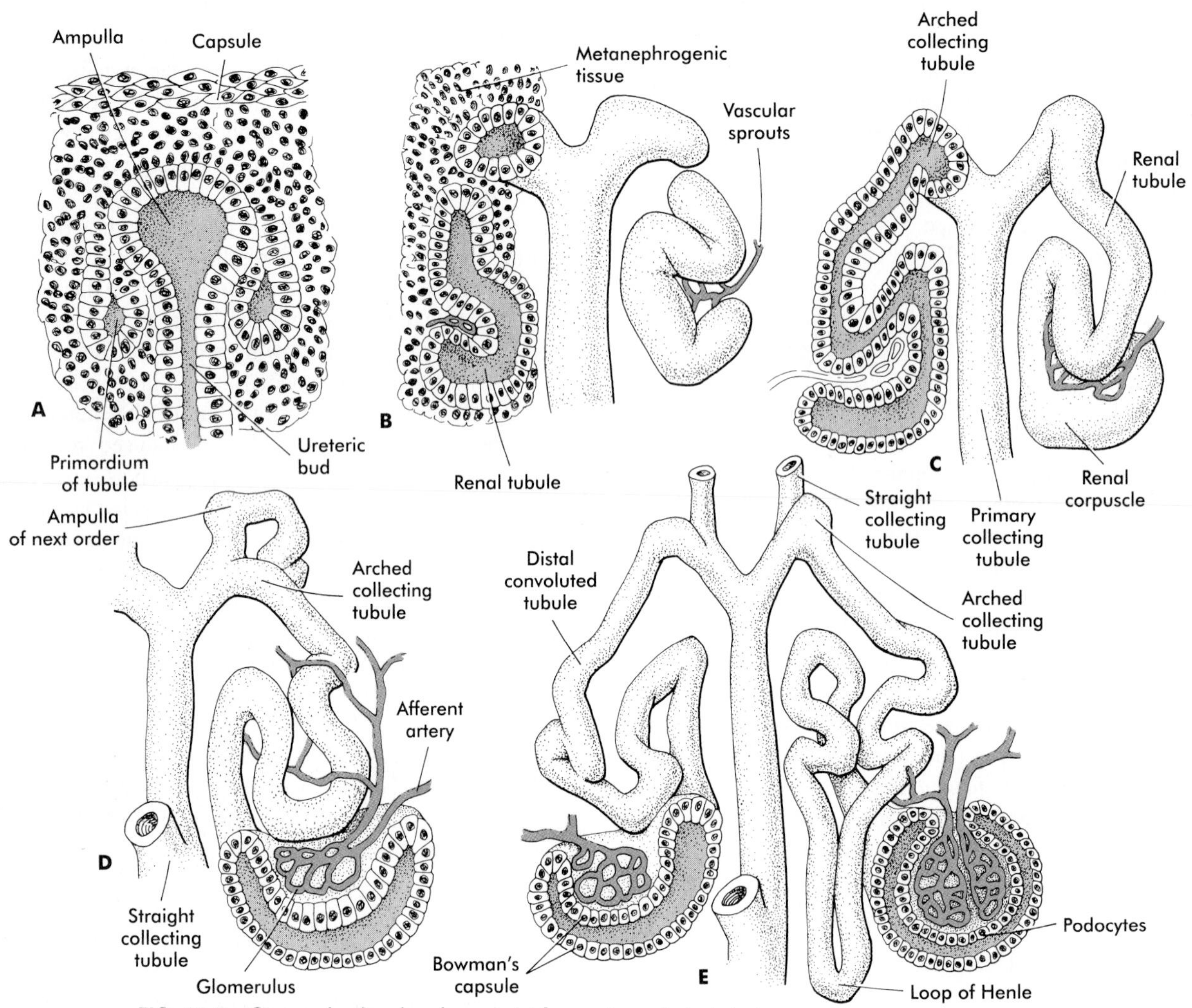

FIG. 17-5 Stages in the development of a metanephric tubule.

group of cells farthest from the end of the metanephric duct becomes polarized, forming a central lumen and a basal lamina on the outer surface. This marks the transformation of the induced mesenchymal cells into an epithelium, specifically the specialized **podocytes,** which ultimately surround the vascular endothelium of the glomerulus.

A consequence of this epithelial transformation is the formation of a slit just beneath the transforming podocyte precursors in the tubular primordium (Fig. 17-5, *B*). Precursors of vascular endothelial cells grow into this slit, which will ultimately form the glomerulus. Research using interspecific hybrids has shown that the vascular endothelium migrates into the developing metanephros from outside the region of the intermediate mesoderm. Induced metanephric mesenchyme stimulates the ingrowth of endothelial cells, possibly by the release of a factor similar to fibroblast growth factor (FGF). Uninduced mesenchyme does not possess this capability. The endothelial cells are connected with branches from the dorsal aorta, and they form a complex looping structure that ultimately becomes the renal glomerulus. Cells of the glomerular endothelium and the adjoining podocyte epithelium form a thick basement membrane between them. This basement membrane later serves as an important component of the renal filtration apparatus.

As the glomerular apparatus of the nephron takes shape, another slit forms in the comma-shaped tubular primordium, transforming it into an S-shaped structure (Fig. 17-5, *C*). Cells in the rest of the tubule primordium also undergo an epithelial transformation to form the remainder of the renal tubule. This transformation involves the acquisition of polarity by the differentiating epithelial cells. It is correlated with the deposition of laminin in the extracellular matrix along the basal surface of the cells and the concentration of the integral membrane glycoprotein **uvomorulin,** which seals the lateral borders of the cells (Fig. 17-6).

Differentiation of the renal tubule progresses from the glomerulus to the proximal and then distal convoluted tubule. During differentiation of the nephron a portion of the tubule develops into an elongated hairpin loop that extends into the medulla of the kidney as the **loop of Henle.** During differentiation the tubular epithelial cells develop mo-

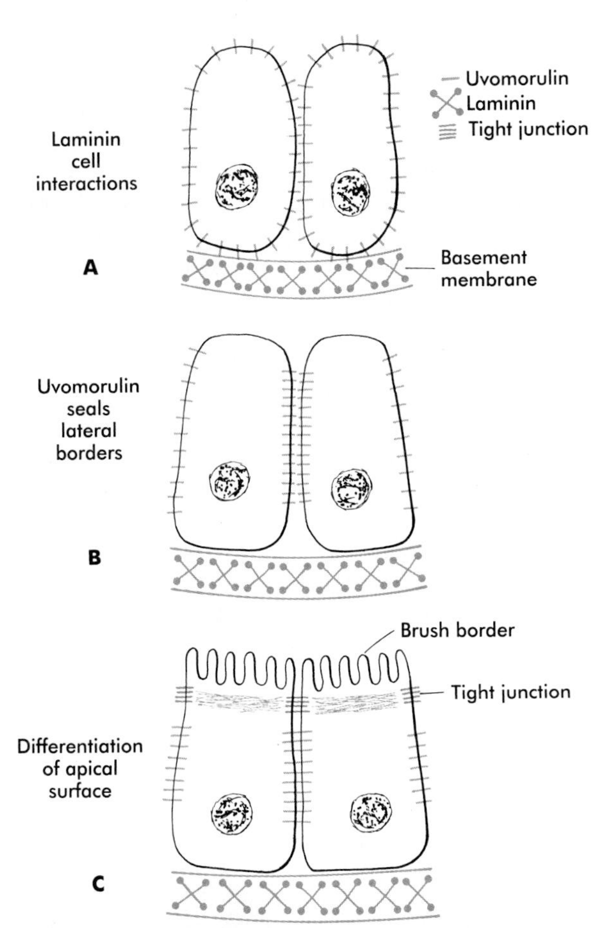

FIG. 17-6 Stages in the transformation of renal mesenchyme into epithelium, with emphasis on the role of laminin and uvomorulin. **A,** Development of polarity is triggered by interactions between laminin and the cell surface, but uvomorulin is still distributed in a nonpolar manner. **B,** Uvomorulin redistribution occurs, and uvomorulin interactions seal the lateral borders of the cells. **C,** The apical border of epithelial cells differentiates, as seen by formation of a brush border. (Based on studies by Ekblom [1989].)

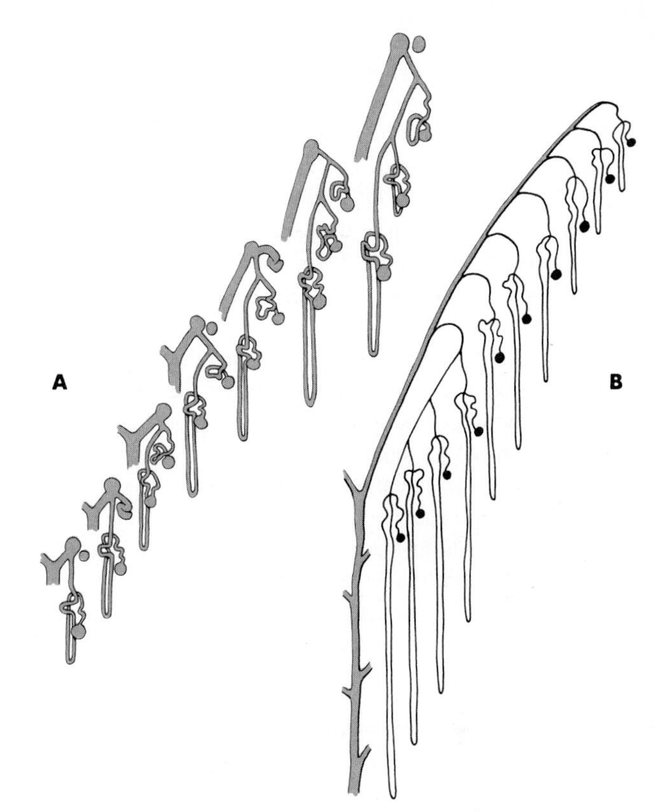

FIG. 17-7 Formation of arcades of nephrons in the developing human metanephros. **A,** Early stages. **B,** The arrangement of nephrons at the time of birth. (Based on studies by Osathanondh and Potter [1963].)

lecular features characteristic of the mature kidney (e.g., brush border antigens, the Tamm-Horsfall glycoprotein).

The mechanisms underlying induction of the nephron remain obscure. Even though induction can occur in vitro, with the inductor and metanephrogenic mesenchyme separated by a porous filter, the inductive interaction appears to be mediated by contact between cellular processes of inductor and responding tissue within pores of the filter. The metanephrogenic tissue responds to the induction by undergoing (1) a burst of cellular proliferation, (2) degradation of interstitial-type proteins from the extracellular matrix, and (3) synthesis of epithelial-type proteins of the extracellular matrix and cytoskeleton. Recent research has shown that both tubule formation from metanephrogenic mesenchyme and branching of the ureteric duct can occur in vitro in the absence of the inducing tissue. The formation of tubules can occur in the presence of epidermal growth factor and a pituitary gland extract, along with a type IV collagen–containing matrix. Branching of the ureteric bud is supported by a matrix consisting of gelled type I collagen.

Polypeptide growth factors originating in the renal primordium have important effects on growth and differentiation of the nephrons. Several of these (e.g., insulinlike growth factor [IGF], epidermal growth factor [EGF], transforming growth factor [TGF]-α, and nerve growth factor [NGF]) exert positive effects on tubular development. Another growth factor, TGF-β, appears to act by retarding the differentiation of metanephric tubules. The presence of **transferrin,** an iron-binding protein produced by the liver, is required for the proliferation of epithelial cells in the developing tubule. As the epithelial cells differentiate from renal mesenchyme, transferrin receptors appear on their surfaces. Because information is still insufficient, an integrated scenario about the role of growth factors in kidney development cannot be constructed.

Growth of the kidney involves the formation of approximately 15 successive generations of nephrons in its peripheral zone, with the outermost nephrons less mature than those farther inward. Development of the internal architecture of the kidney is very complex, involving the formation of highly ordered arcades of nephrons (Fig. 17-7). Details are beyond the scope of this text.

Later Changes in Kidney Development

While the many sets of nephrons are differentiating, the kidney becomes progressively larger. The branched system of ducts also becomes much larger and more complex, forming the pelvis and system of **calyces** of the kidney (Fig. 17-8). These structures collect the urine and funnel it into the ureters. During much of the fetal period the kidneys are divided into grossly visible lobes. By birth the lobation is already much less evident, and it disappears during the neonatal period.

When they first take shape, the metanephric kidneys are located deep in the pelvic region. During the late embryonic and early fetal period, they undergo a pronounced shift in position that moves them into the abdominal region. This shift is partly due to actual migration and partly to a marked expansion of the caudal region of the embryo. Two concurrent components to the migration occur. One is a caudocranial shift from the level of the fourth lumbar to the first lumbar or even twelfth thoracic vertebra (Fig. 17-9). The other is a lateral displacement. These changes bring the kidneys into contact with the adrenal glands, which form a cap of glandular tissue on the cranial pole of each kidney. During their migration, the kidneys also undergo a 90-degree rotation. As they are migrating out of the pelvic cavity, the kidneys slide over the large umbilical arteries, which branch from the caudal end of the aorta. All these changes take place behind the peritoneum, since the kidneys are retroperitoneal organs. During the early phases of migration of the metanephric kidneys, the mesonephric kidneys regress. The mesonephric ducts, however, are retained as they become closely associated with the developing gonads.

Although normally supplied by one large renal artery branching directly from the aorta, the adult kidney consists of five vascular lobes. The arteries feeding each of these lobes were originally segmental vessels that supplied the mesonephros and were taken over by the migrating metanephros. Their aortic origins are typically reduced to the single pair of renal arteries, but anatomical variations are common.

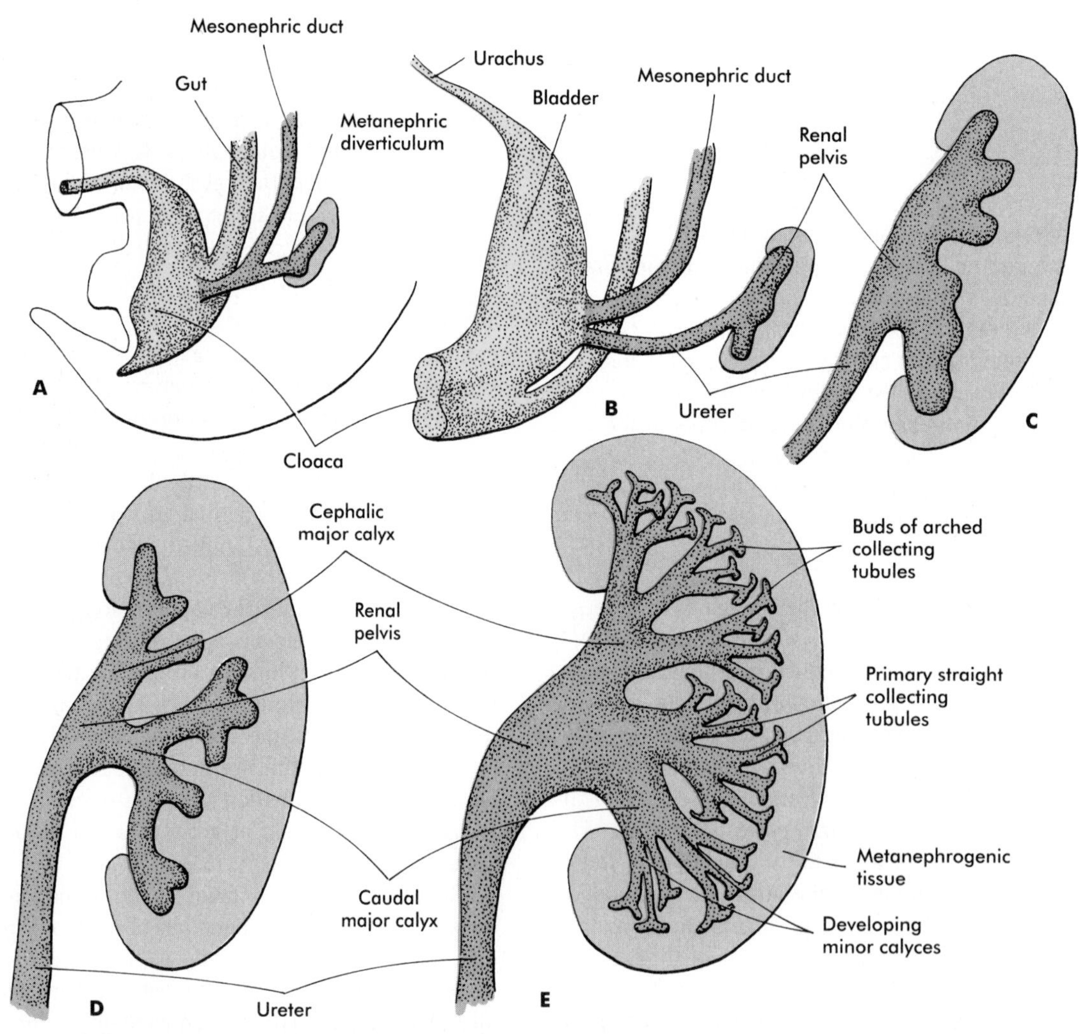

FIG. 17-8 Later changes in the development of the metanephros.

Formation of the Urinary Bladder

The division of the cloaca into the rectum and urogenital sinus region was already introduced in Chapter 16 (see Fig. 16-9). The urogenital sinus is continuous with the allantois, which has an expanded base continuous with the urogenital sinus and an attenuated tubular process that extends into the body stalk on the other end. The dilated base of the allantois continues to expand to form the **urinary bladder,** and its attenuated distal end solidifies into the cordlike **urachus,** which ultimately forms a ligament leading from the bladder to the umbilical region (see Fig. 17-18).

As the bladder grows, its expanding wall incorporates the mesonephric ducts and the ureteric buds (Fig. 17-10). The result is that these structures open separately into the posterior wall of the bladder. Through a poorly defined mechanism possibly involving mechanical tension exerted by the migrating kidneys, the ends of the ureters open into the bladder laterally and cephalically to the mesonephric ducts. The region bounded by these structures is called the **trigone** of the bladder. At the entrance of the mesonephric ducts, the bladder becomes sharply attenuated. This region, originally part of the urogenital sinus, forms the **urethra,** which serves as the outlet of the bladder (see p. 364).

Congenital Anomalies of the Urinary System

Anomalies of the urinary system are relatively common (3% to 4% of live births). Many are asymptomatic, and others manifest only later in life. Fig. 17-11 summarizes the locations of many of the more frequently encountered malformations of the urinary system.

Renal Agenesis. **Renal agenesis** is the unilateral or bilateral absence of any trace of kidney tissue (Fig. 17-12). Unilateral renal agenesis is seen in roughly 0.1% of adults,

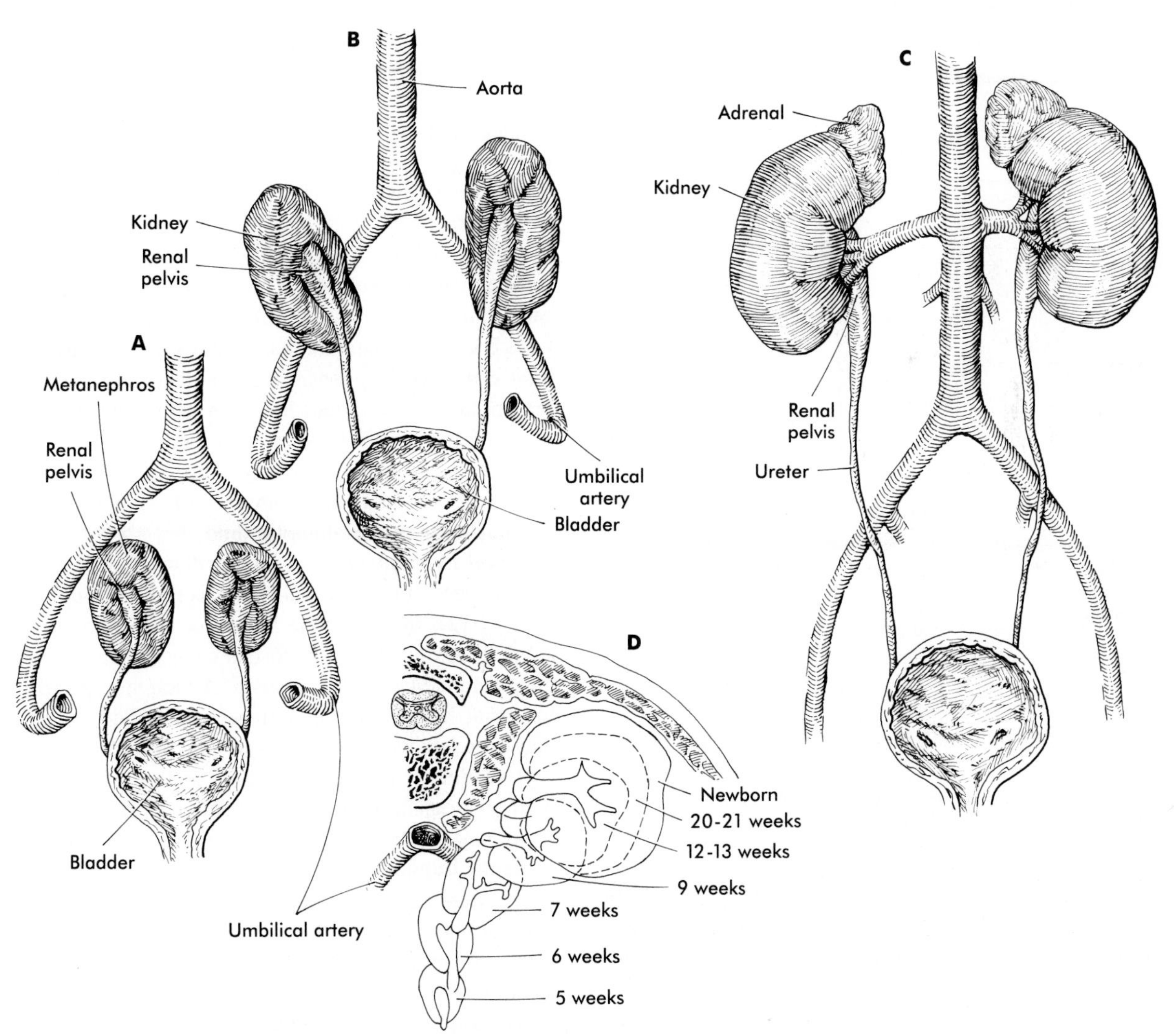

FIG. 17-9 **A-C,** Migration of the kidneys from the pelvis to their definitive adult level. **D,** Cross section of the pathway of migration of the kidneys out of the pelvis.

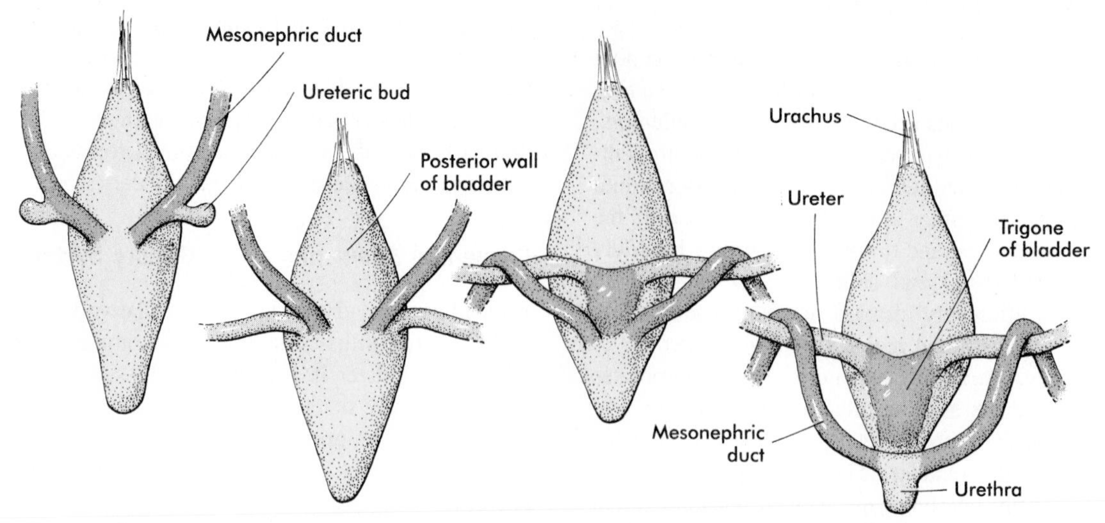

FIG. 17-10 Dorsal views of the developing urinary bladder showing changing relationships of the mesonephric ducts and the ureters as they approach and become incorporated into the bladder. In the two on the right, note the incorporation of portions of the walls of the mesonephric ducts into the trigone of the bladder.
(Modified from Sadler T: *Langman's medical embryology*, ed 6, Baltimore, 1990, Williams & Wilkins.)

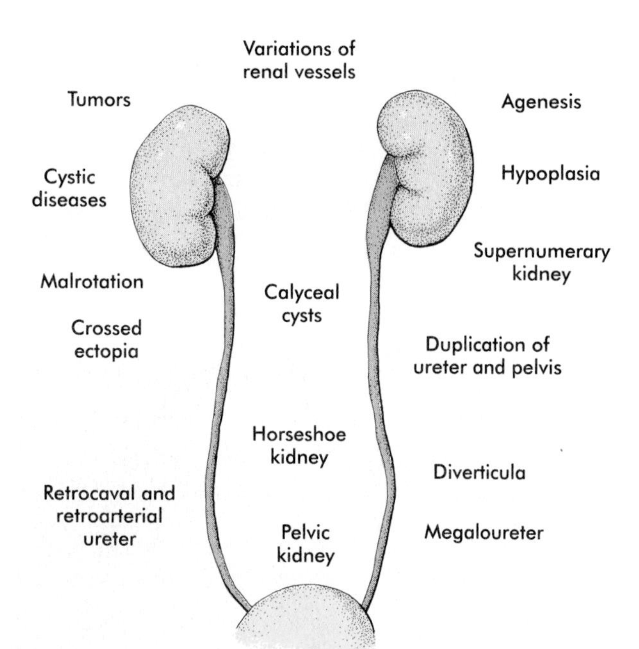

FIG. 17-11 Types and sites of anomalies of the kidneys and ureters.
(Modified from Gray, Skandalakis: *Embryology for surgeons*, New York, 1972, Saunders.)

whereas bilateral renal agenesis occurs in 1:3000 to 1:4000 newborns. The ureter may be present. This anomaly is usually ascribed to a faulty inductive interaction between the ureteric bud and the metanephrogenic mesenchyme. It could also be the result of a faulty connection between the two or a defect or defects in the inductive process itself. Individuals with unilateral renal agenesis are often asymptomatic, but typically the single kidney undergoes **compensatory hypertrophy** to maintain a normal functional mass of renal tissue.

An infant born with bilateral renal agenesis dies within a few days after birth. Because of the lack of urine output, reduction in the volume of amniotic fluid **(oligohydramnios)** during pregnancy is often an associated feature. Infants born with bilateral renal agenesis characteristically exhibit **Potter facies,** consisting of a flattened nose, wide interpupillary space, a receding chin, and large, low-set ears (Fig. 17-13). Potter facies is classified as a secondary disruption resulting from mechanical pressure of the uterus on the face of the fetus in the absence of the normal mechanical buffering of the amniotic fluid.

Renal Hypoplasia. An intermediate condition between renal agenesis and a normal kidney is **renal hypoplasia** (see Fig. 17-12), in which one kidney or more rarely both are substantially smaller than normal even though a certain degree of function may be retained. Although a specific cause for renal hypoplasia has not been identified, some cases may be related to deficiencies in growth factors or their receptors during critical phases of metanephrogenesis. As with renal agenesis, the normal counterpart of a hypoplastic kidney is likely to undergo compensatory hypertrophy.

Renal Duplications. Renal duplications range from a simple duplication of the renal pelvis to a completely separate **supernumerary kidney.** Like hypoplastic kidneys, renal duplications may be asymptomatic, although the incidence of renal infections may be increased. Many variants of **duplications of the ureter** have also been described (see Fig. 17-12). Duplication anomalies are commonly attributed to splitting or wide separation of branches of the ureteric bud.

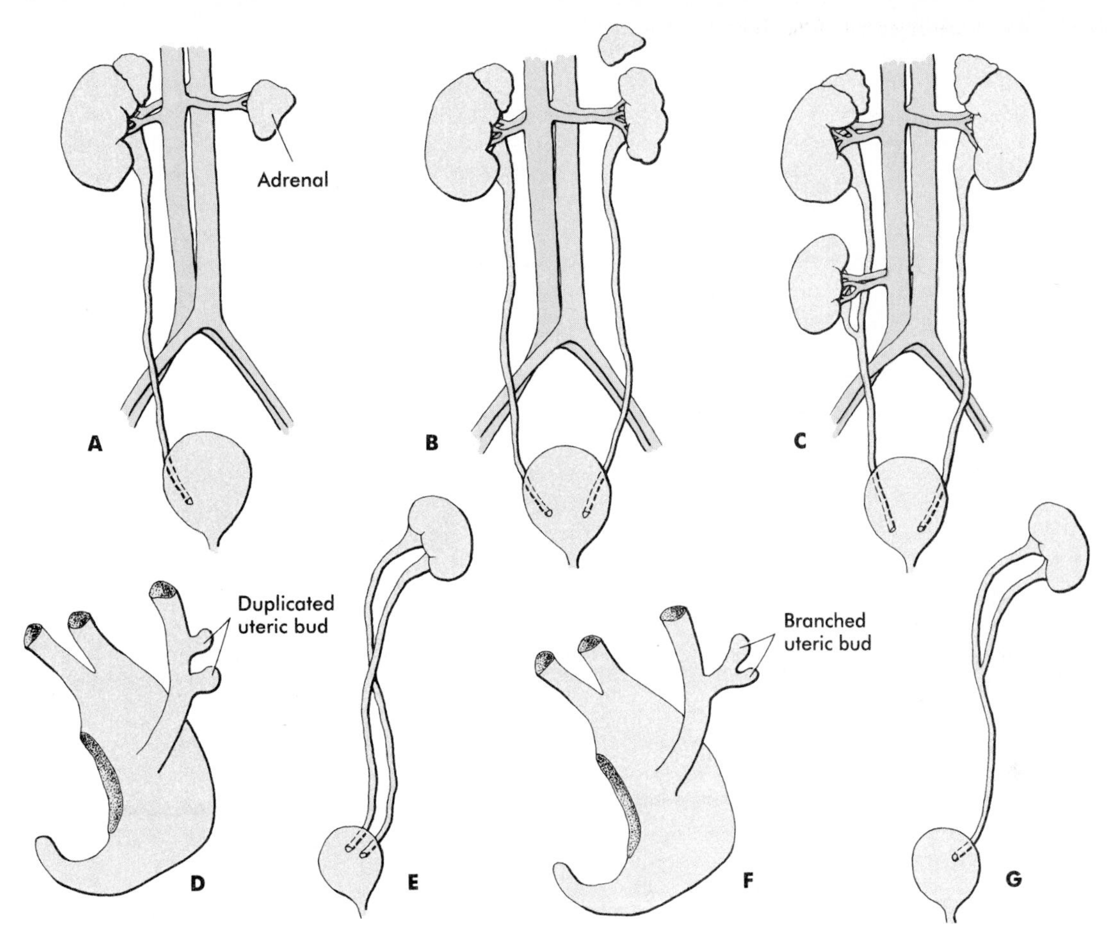

FIG. 17-12 Common renal anomalies. **A,** Unilateral renal agenesis. The ureter is also missing. **B,** Unilateral renal hypoplasia. **C,** Supernumerary kidney. **D** and **E,** Complete duplication of ureter, presumably arising from two separate ureteric buds. **F** and **G,** Partial duplication of ureter, presumably arising from a bifurcated ureteric bud.

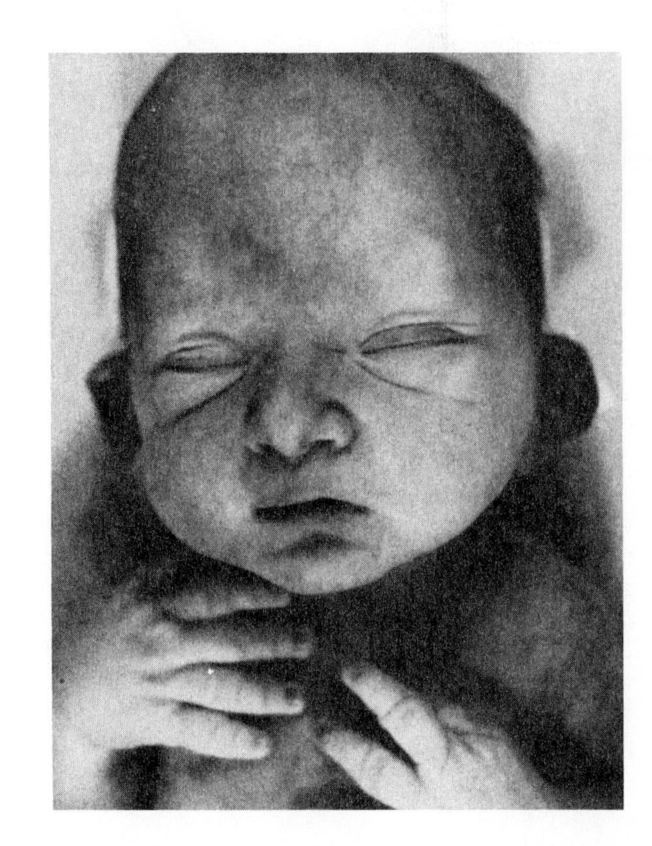

FIG. 17-13 The Potter facies, which is characteristic of a fetus who has been exposed to oligohydramnios. Note the flattened nose, low-set ears, wide interpupillary space, and large hands and fingers.

(From Wigglesworth, Singer: *Textbook of fetal and perinatal pathology,* London, 1991, Blackwell Scientific.)

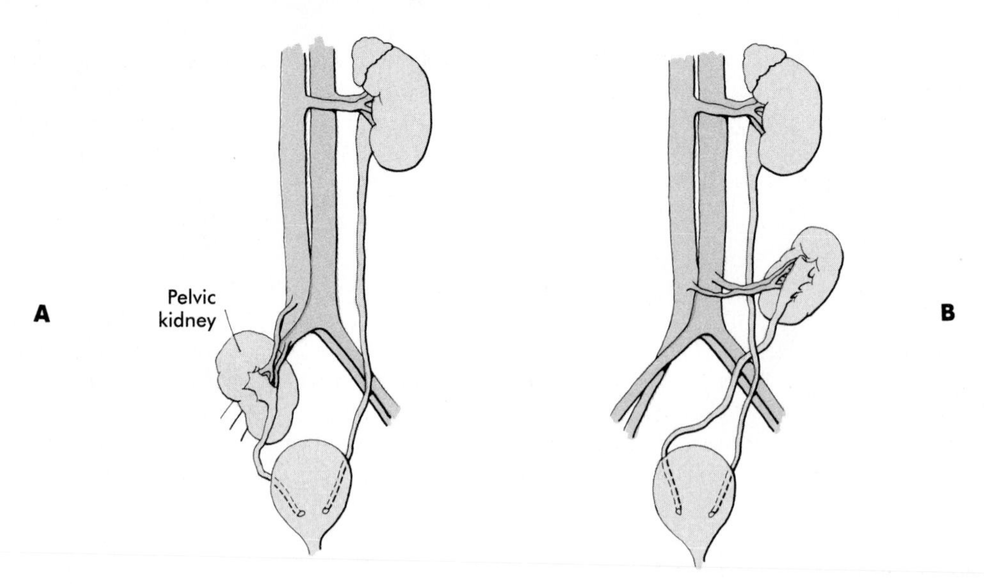

FIG. 17-14 Migration defects of the kidney. **A,** Pelvic kidney. **B,** Crossed ectopia. The right kidney has crossed the left ureter and has migrated only part of the normal distance.

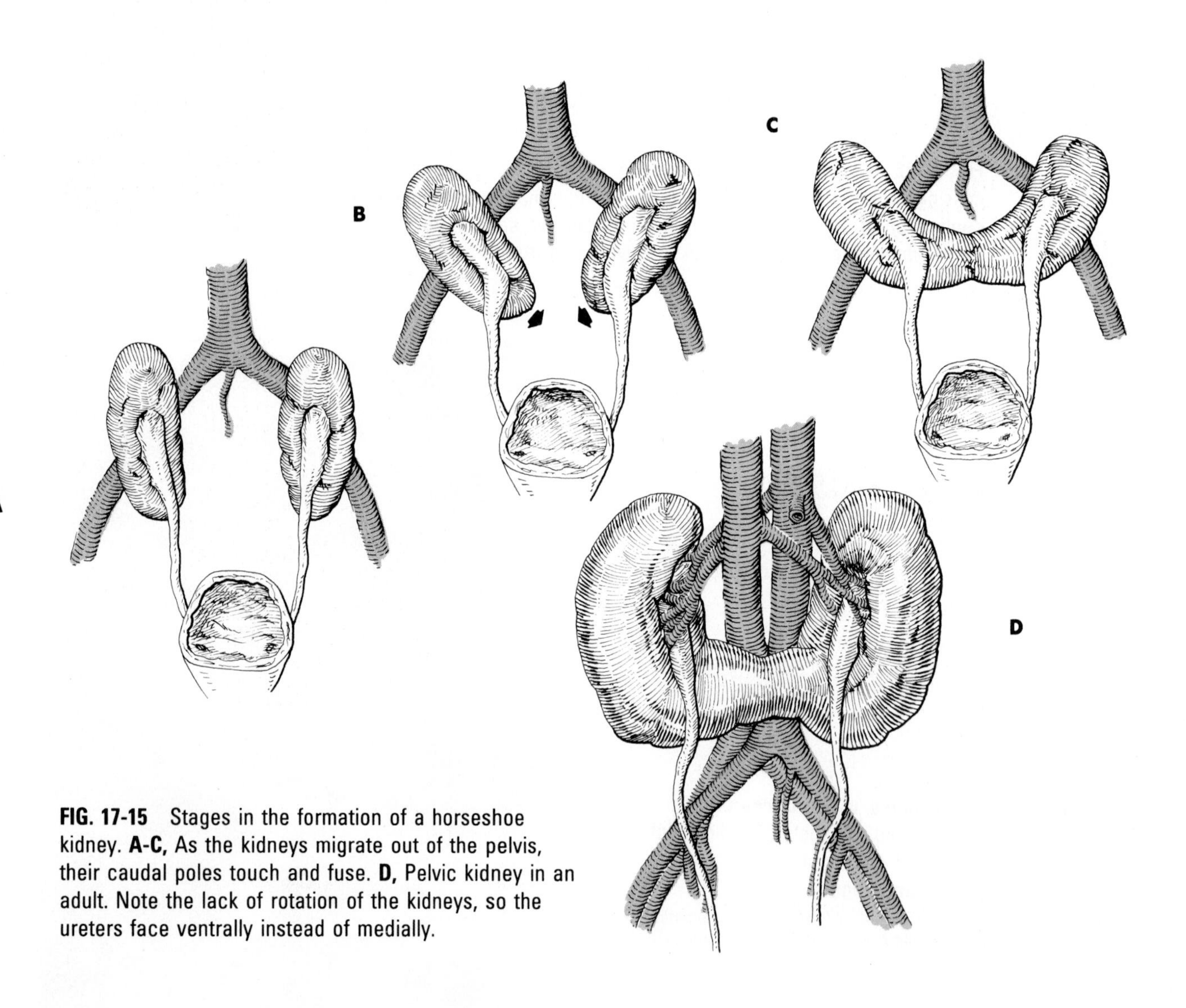

FIG. 17-15 Stages in the formation of a horseshoe kidney. **A-C,** As the kidneys migrate out of the pelvis, their caudal poles touch and fuse. **D,** Pelvic kidney in an adult. Note the lack of rotation of the kidneys, so the ureters face ventrally instead of medially.

Anomalies of Renal Migration and Rotation. The most common disturbance of renal migration leaves a kidney in the pelvic cavity (Fig. 17-14). This is usually associated with malrotation of the kidney as well, so the hilus of the **pelvic kidney** faces anteriorly instead of toward the midline. Another category of migratory malformation is **crossed ectopia,** in which one kidney and its associated ureter are found on the same side of the body as the other kidney (Fig. 17-14). In this condition the ectopic kidney may be fused with the normal kidney.

In the condition of **horseshoe kidney,** occurring in as many as 1:400 individuals, the kidneys are typically fused at their inferior poles (Fig. 17-15). Horseshoe kidneys cannot migrate out of the pelvic cavity because the inferior mesenteric artery, coming off the aorta, blocks them. In most cases, horseshoe kidneys are asymptomatic, but occasionally pain or obstruction of the ureters may occur. This condition may be associated with anomalies of other internal organs. Pelvic kidneys are subject to an increased incidence of infections and obstructions of the ureters.

Anomalies of the Renal Arteries. Instead of a single renal artery branching off each side of the aorta, duplications or major extrarenal branches of the renal artery are common. Because of the appropriation of segmental arterial branches to the mesonephros by the metanephros, consolidation of the major external arterial supply to the kidney occasionally does not occur.

Polycystic Disease of the Kidney. Congenital polycystic disease of the kidney, an autosomal recessive condition in infants, is typically manifest by the presence of large numbers of cysts of different sizes within the parenchyma of the kidney (Fig. 17-16). Neither the origin nor the pathogenesis of this condition has been definitively ascertained. According to one theory, it is caused by the lack of connection between the tubular part of the nephron and the collecting duct system. The cysts have also been attributed to dilatations of the collecting duct system. Recent studies on mice with genetic polycystic disease have demonstrated defective expression of EGF and high levels of certain oncogene products in the tubules of polycystic kidneys. Cysts of other organs, especially the liver and pancreas, are frequently associated with polycystic kidneys.

Ectopic Ureteral Orifices. Ureters may open into a variety of ectopic sites (Fig. 17-17). Because of the continuous supply of urine flowing through them, these sites are symptomatic and usually relatively easy to diagnose. Their embryogenesis is commonly attributed to abnormal origins of the ureteric buds in the early embryo.

FIG. 17-16 Polycystic kidneys.
(Courtesy Mason Barr, Ann Arbor, Mich.)

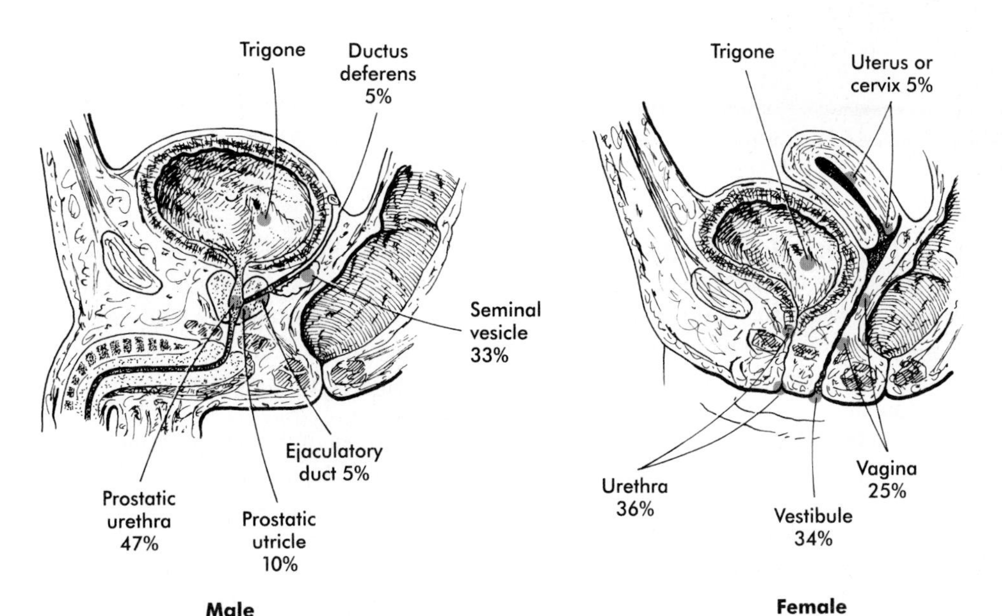

FIG. 17-17 Common sites of ectopic ureteral orifices.
(Modified from Gray and Skandalakis [1972].)

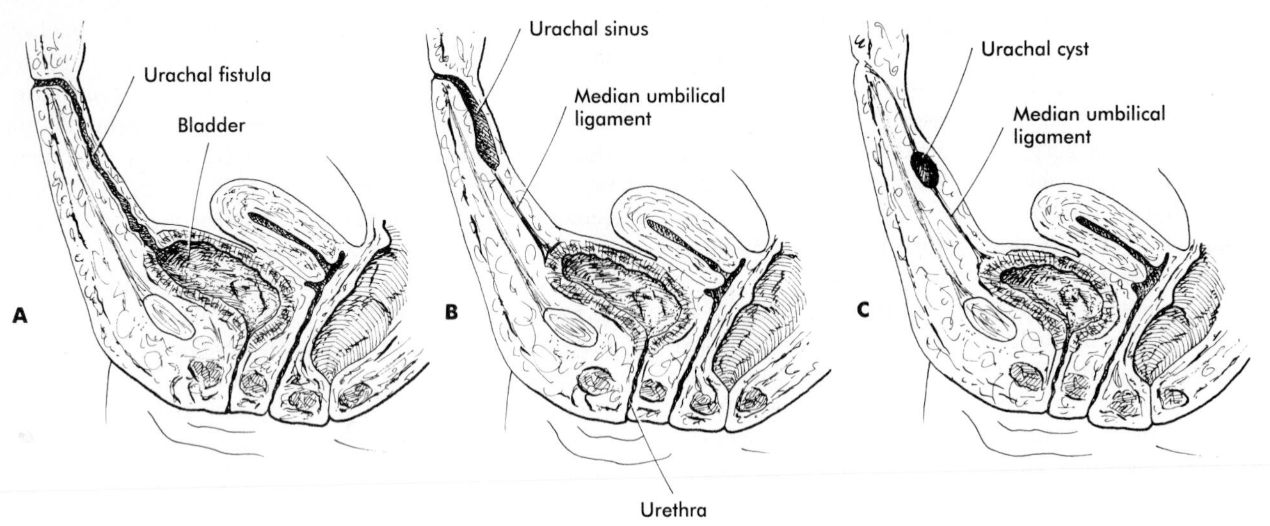

FIG. 17-18 Anomalies of the urachus. **A,** Urachal fistula. **B,** Urachal sinus. **C,** Urachal cyst.

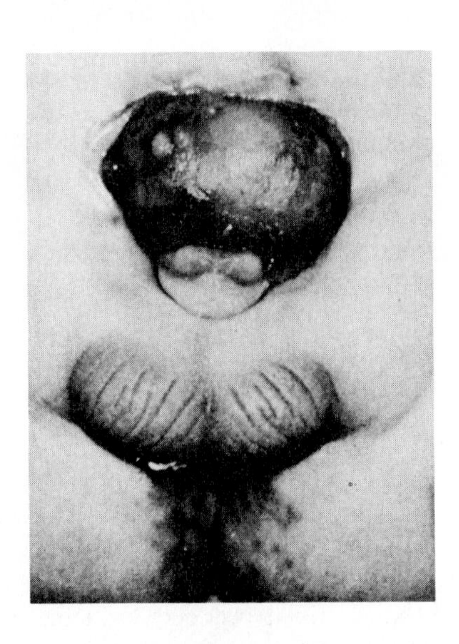

FIG. 17-19 Exstrophy of the bladder in a male infant, showing protrusion of the posterior wall of the bladder through a defect in the lower abdominal wall. At the base of the open bladder is an abnormal, partially bifid penis with an open urethra (not seen) on its dorsal surface. A wide, shallow scrotum is separated from the penis. (From Crowley: *An introduction to clinical embryology,* Chicago, 1974, Mosby.)

Cysts, Sinuses, and Fistulas of the Urachus. If parts of the lumen of the allantois fail to become obliterated, **urachal cysts, sinuses,** or **fistulas** can form (Fig. 17-18). In the case of a urachal fistula, urine seeps from the umbilicus. Urachal sinuses or cysts may swell in later life if they are not evident in the infant.

Exstrophy of the Bladder. Exstrophy of the bladder is a major defect in which the urinary bladder opens broadly onto the abdominal wall (Fig. 17-19). Rather than being a primary defect of the urinary system, it is most commonly attributed to an insufficiency of mesodermal tissue of the ventral abdominal wall. Although the ventral body wall may be closed with ectoderm initially, in the absence of mesoderm it breaks down, and degeneration of the anterior wall of the bladder typically follows. In males, exstrophy of the bladder commonly involves the phallus, and a condition called **epispadias** results (see p. 369).

THE GENITAL SYSTEM

Development of the genital system is one phase in the overall sexual differentiation of an individual (Fig. 17-20). Sexual determination begins at the time of fertilization, when a Y chromosome or an additional X chromosome is joined to the X chromosome already in the egg. This phase represents the genetic determination of gender. Although the genetic sex of the embryo is fixed at fertilization, the gross phenotypic sex of the embryo is not manifest until the sev-

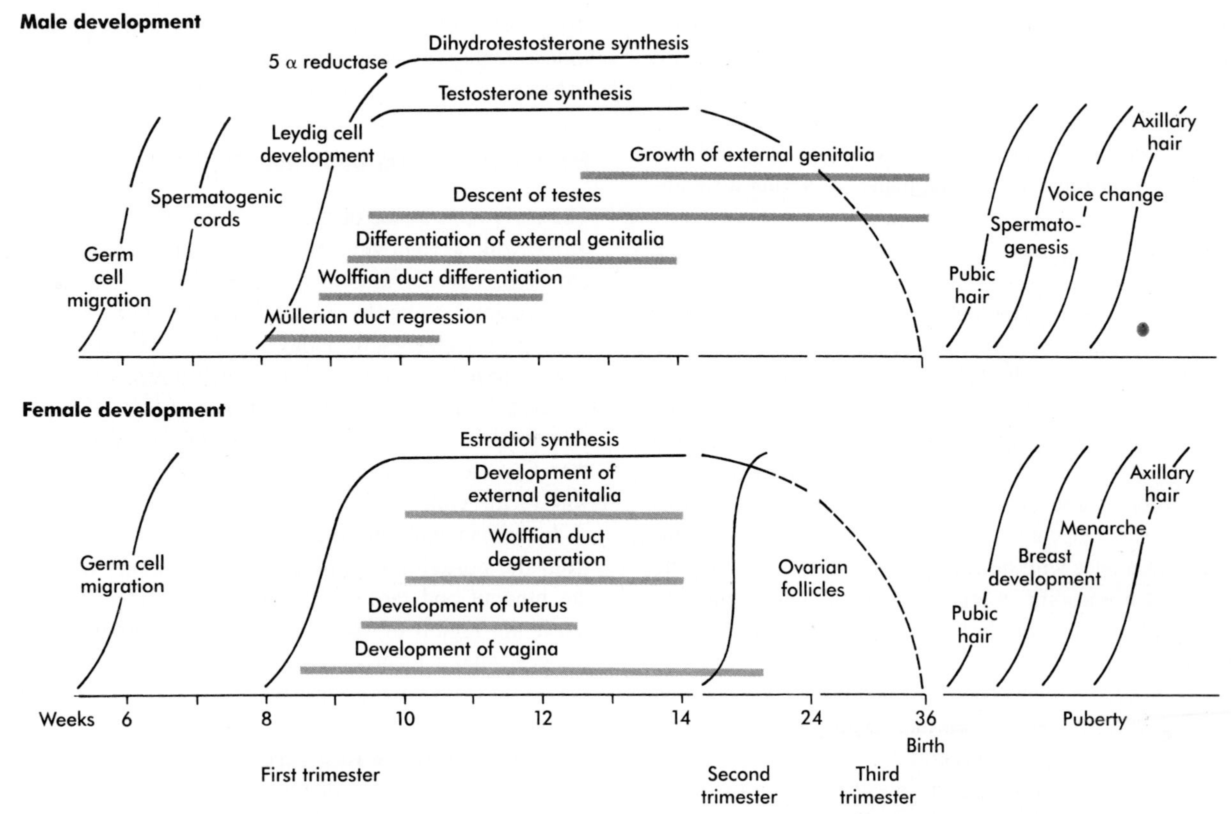

FIG. 17-20 Major events in the sexual differentiation of male and female human embryos.

enth week of development. Before that time the principal morphological indicator of the embryo's sex is the presence or absence of the sex chromatin (Barr body) in the female. The Barr body is the result of inactivation of one of the X chromosomes. During this morphologically **indifferent stage** of sexual development, the gametes migrate into the gonadal primordia from the yolk sac.

The phenotypic differentiation of sex begins with the gonads and progresses with gonadal influences on the sexual duct systems. Similar influences on differentiation of the external genitalia and finally on the development of the secondary sexual characteristics (e.g., body configuration, breasts and hair patterns) complete the events that constitute the overall process of sexual differentiation. Sexual differentiation of the brain, which has an influence on behavior, also occurs.

Under certain circumstances an individual's genetic sex can be overridden by environmental factors, so the genotypic and phenotypic sex do not correspond. An important general principle is that the development of phenotypic maleness requires the action of substances produced by the testis. In the absence of specific testicular influences or the ability to respond to them, a female phenotype results. The female phenotype is the baseline, or default, condition, which must be acted on by male influences to produce a male phenotype.

The Genetic Determination of Sex

Since 1923, scientists have recognized that the XX and XY chromosomal pairings represent the genetic basis for human femaleness and maleness. For many decades, they believed that the presence of two X chromosomes was the sex-determining factor, but in 1959, it was established that the differentiation between maleness and femaleness depended on the presence of a Y chromosome. Nevertheless, the link between the Y chromosome and determination of the testis remained obscure. During recent decades, three candidates for the testis-determining factor have been proposed.

The first was the **H-Y antigen,** a minor histocompatibility antigen present on the cells of males but not females. The H-Y antigen has been mapped to the long arm of the human Y chromosome. It has been considered to be the product of the mammalian testis-determining gene. However, a strain of mice *(Sxr)* was found to produce males in the absence of the H-Y antigen. *Sxr* mice were found to have a transposition of a region of the Y chromosome onto the X chromosome, but the locus coding for the H-Y antigen was not included. In addition, certain phenotypic human males with an XX genotype were shown to be missing the genetic material for the H-Y antigen.

The next candidate for the sex-determining gene was a locus on the short arm of the Y chromosome called the *zinc-finger Y (ZFY)* gene. With DNA hybridization techniques this gene has been found in both XX male humans and mice in which small pieces of the X and Y chromosomes were swapped during crossing-over in meiosis. Conversely, this gene was missing in some rare XY human females. However, certain XX males were found to lack the gene, and other rare cases of anomalies of sexual differentiation did not show a correspondence between sexual phenotype and the expected presence or absence of the *ZFY* gene.

The most recent candidate for the testis-determining gene is one called *Sry,* which is also located within a 35-kilobase region on the short arm of the Y chromosome (Fig. 17-21). The *Sry* gene encodes a 220-amino acid nonhistone protein belonging to a family of high-mobility group proteins that seems to be important in maintaining chromosomal structure and gene activation. After the gene was cloned, it was detected in many cases of sex reversal, including XX males with no *ZFY* genes. The *Sry* gene on the human Y chromosome is located near the homologous region, making it susceptible to translocation to the X chromosome.

The *Sry* gene is also absent in a strain of XY mice that are phenotypically female. Further experimental evidence consisted of producing transgenic mice with the insertion of a 14-kilobase fragment of the Y chromosome that contained the *Sry* gene. Many of the transgenic XX mice developed into phenotypic males with normal testes and male behavior. In situ hybridization studies in mice have shown expression of the *Sry* gene product occurs in male gonadal tissue at the time of sex determination. Both *Sry* and *Zfy* genes are expressed as early as the two-cell stage in mouse embryos, suggesting that mammalian sex determination may begin before gonadal differentiation.

Migration of Germ Cells into the Gonads

The early appearance of primordial germ cells in the lining of the yolk sac and their migration to the gonads in human embryos is briefly described in Chapter 1. Descriptive and experimental studies have shown that primordial germ cells in the mouse can first be demonstrated in the epiblast. These cells appear to pass through the early primitive streak and can next be located as a small cluster of cells in the extraembryonic mesoderm near the base of the allantois. They then become associated with the endoderm of the posterior wall of the yolk sac (see Fig. 1-1).

In human embryos the primordial germ cells migrate from the posterior wall of the yolk sac along the wall of the hindgut and through the dorsal mesentery until they reach the region of the newly appearing **genital ridges.** Experimental evidence suggests that the initial stages of migration of primordial germ cells at some distance from the gonads are accomplished by active ameboid movement of the cells in response to a permissive extracellular matrix substrate. Tissue displacements through differential growth of the posterior region of the embryo may also contribute. During their migration the germ cells proliferate in response to mitogenic factors such as **leukemia inhibitory factor** and **mastocyte growth factor** (stem-cell factor or Steel factor).

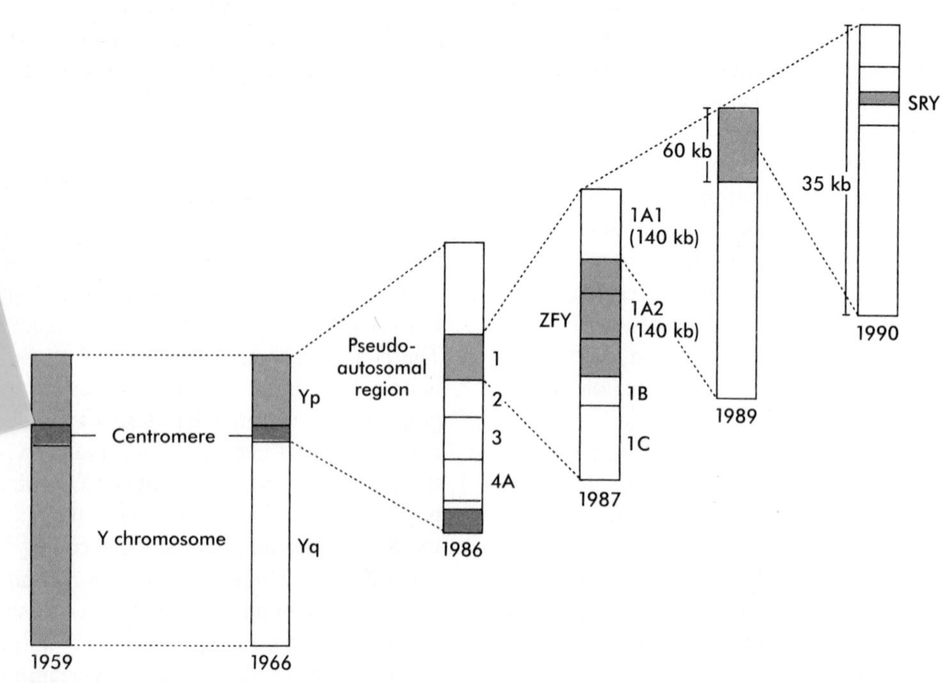

FIG. 17-21 A history in the progress of localization of the sex-determining gene on the Y-chromosome.
(Based on studies by Sultan and others [1991].)

As the germ cells approach the genital ridges late in the fifth week of development, they are influenced by chemotactic factors secreted by the newly forming gonads. Such influences have been demonstrated by grafting embryonic tissues (e.g., hindgut, which contains dispersed germ cells) into the body cavity of a host embryo. The primordial germ cells of the graft typically concentrate on the side of the graft nearest the genital ridges of the host or sometimes migrate into the genital ridges from the graft. Approximately 1000 to 2000 primordial germ cells enter the genital ridges. Once the primordial germ cells have penetrated the genital ridges, their migratory behavior ceases. The reason for this is not known, but it may occur because the area of the genital ridges that the germ cells enter lacks a basement membrane, whereas a well-defined basement membrane underlies the epithelium of the dorsal mesentery along which primordial germ cells migrate. The basement membrane may represent the substrate that permits migration of the primordial germ cells. Some primordial germ cells follow inappropriate migratory pathways, leading them to settle into extragonadal sites. These cells normally start to develop as oogonia, regardless of genotype; they then degenerate. In rare instances, however, they persist in ectopic sites such as the mediastinum and ultimately give rise to **teratomas** (see Chapter 1).

The Establishment of Gonadal Sex

Origin of the Gonads. The gonads arise from an elongated region of steroidogenic mesoderm along the ventromedial border of the mesonephros. Cells in the cranial part of this region condense to form the **adrenocortical primordia,** and those of the caudal part become the genital ridges, which are identifiable midway through the fifth week. The early genital ridges consist of two major populations of cells: one derived from the **coelomic epithelium** and the other arising from the **mesonephric ridge.**

Differentiation of the Testes. When the genital ridges first appear, those of males and females are morphologically indistinguishable (indifferent stage). The general principle underlying gonadal differentiation is that under the influence of the *Sry* gene on the Y chromosome, the indifferent gonad differentiates into a testis (Fig. 17-22). In the absence of expression of products of this gene, the gonad later differentiates into an ovary.

In males, transcripts of the *Sry* gene are detected only in the genital ridge just at the onset of differentiation of the testis. Neither expression of the *Sry* gene nor later differentiation of the testis depends on the presence of germ cells. The sex-determining genes act on the somatic portion of the testis and not on the germ cells or in conjunction with them. A great deal remains to be learned about the events leading from expression of the *Sry* gene product to the overt differentiation of the testis. Current evidence strongly suggests that the *Sry* gene product acts principally as a switch and that other genes necessary for guiding the differentiation are present on other areas of the Y chromosome and even on the X chromosome. The way the actions of specific genes relate to specific morphological events is unknown.

Timing is important in differentiation of the testis. The testis develops more rapidly than the ovary. Precursors of the Sertoli cells must be prepared to receive the genetic signals for testicular differentiation by a certain time. If not, the primordial germ cells begin to undergo meiosis and the gonad differentiates into an ovary.

The morphology of early gonadal differentiation has been controversial, with several scenarios of cell lineage

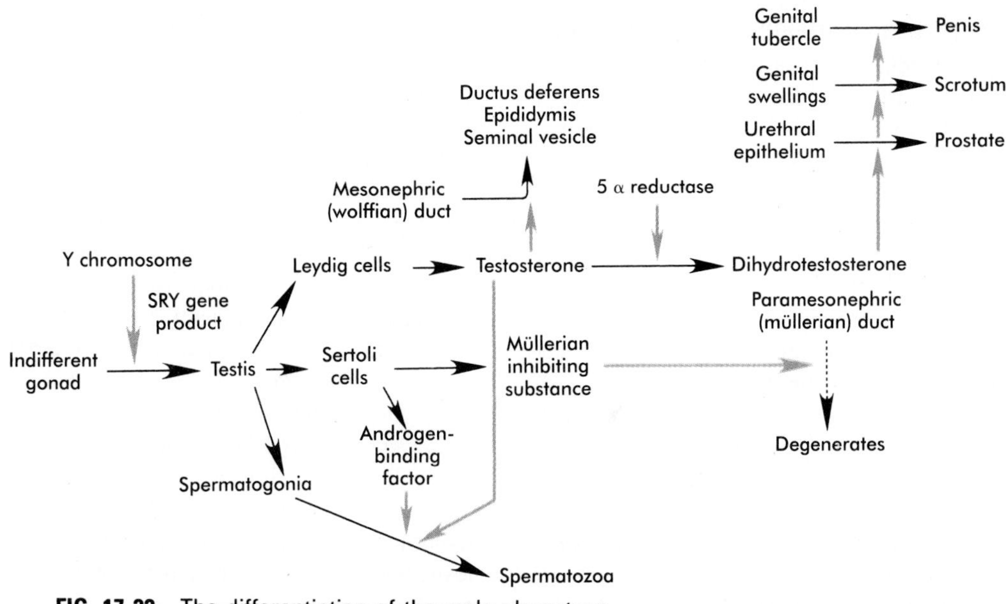

FIG. 17-22 The differentiation of the male phenotype.

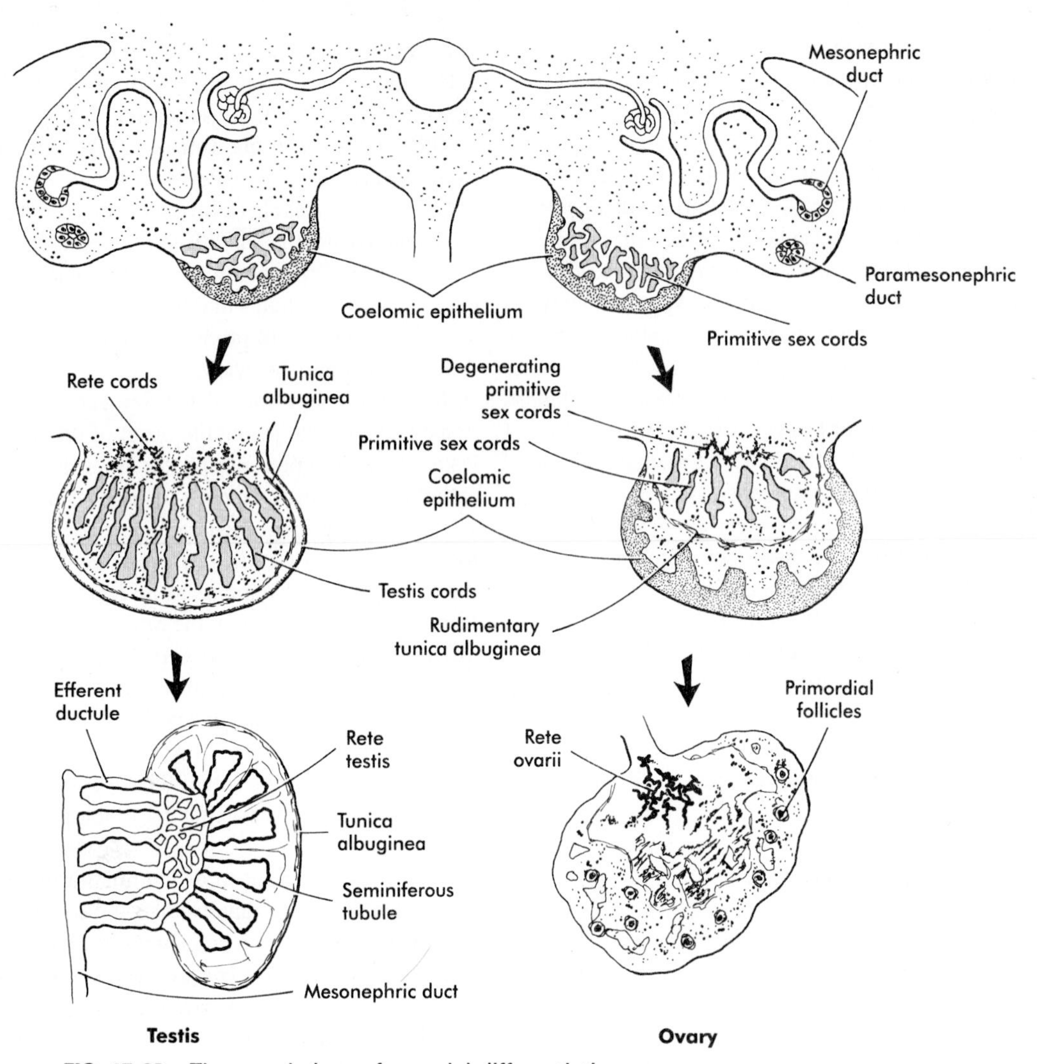

FIG. 17-23 The morphology of gonadal differentiation.

and interactions being proposed. According to recent morphological evidence, the genital ridges first appear midway in the fifth week through proliferation of **coelomic epithelial cells** along the medial border of the mesonephros (Fig. 17-23). Later in the fifth week the primordial germ cells enter the early genital ridge, and the coelomic epithelium sends short epithelial pillars toward the interior of the gonad. Early in the sixth week a set of **primitive sex cords** begins to grow into the genital ridge from the mesonephros, and the primordial germ cells migrate into the primitive sex cords.

Late in the sixth week the testis shows evidence of differentiation. The primitive sex cords enlarge and are better defined, and their cells are thought to represent the precursors of the Sertoli cells. As the sex cords differentiate, they are separated from the surface epithelium (germinal epithelium) by a dense layer of connective tissue called the **tunica albuginea.** The deepest portions of the testicular sex cords are in contact with the fifth to twelfth sets of mesonephric nephrons. The outer portions of the testicular sex cords form the **seminiferous tubules,** and the inner portions become meshlike and ultimately form the **rete testis.** The rete testis ultimately joins the **efferent ductules,** which are derived from the mesonephric ducts.

For the first 2 months of development, **Leydig cells** are not present in the embryonic testis. These appear during the eighth week and soon begin to synthesize androgenic hormones (testosterone and androstenedione). This endocrine activity is important because differentiation of the male sexual duct system and the external genitalia depends on the sex hormones secreted by the fetal testis. Fetal Leydig cells secrete their hormonal products at just the period when differentiation of the hormonally sensitive genital ducts takes place (9 to 14 weeks). After weeks 17 and 18 the Leydig cells gradually involute and do not reappear until puberty, when they stimulate spermatogenesis. The fetal Leydig cells can be viewed as a cellular isoform that is later replaced by the definitive adult form of the cells.

During late embryonic and fetal periods and after birth the primordial germ cells in the testis divide slowly by mitosis, but the fetal Sertoli cells are insensitive to androgens and fail to mature. The embryonic testis may secrete a

TABLE 17-1 Homologies in the Male and Female Urogenital Systems

INDIFFERENT STRUCTURE	MALE DERIVATIVE	FEMALE DERIVATIVE
Genital ridge	Testis	Ovary
Primordial germ cells	Spermatozoa	Ova
Sex cords	Seminiferous tubules (Sertoli cells)	Follicular cells
Mesonephric tubules	Efferent ductules	Eoophoron
	Paradidymis	Paroophoron
Mesonephric (wolffian) ducts	Appendix of epididymis	Appendix of ovary
	Epididymal duct	Gartner's duct
	Ductus deferens	
	Ejaculatory duct	
Paramesonephric (müllerian) ducts	Appendix of testis	Uterine tubes
	Prostate utricle	Uterus
		Upper vagina
Definitive urogenital sinus (lower part)	Penile urethra	Lower vagina
		Vaginal vestibule
Early urogenital sinus (upper part)	Urinary bladder	Urinary bladder
	Prostatic urethra	Urethra
Genital tubercle	Penis	Clitoris
Genital folds	Floor of penile urethra	Labia minora
Genital swellings	Scrotum	Labia majora

meiosis-inhibiting factor, but such a factor has not been isolated or characterized. The environment of the testis does not become favorable for meiosis and spermatogenesis until puberty. Embryonic Sertoli cells produce müllerian inhibitory substance (see p. 358), which is involved in shaping the sexual duct system.

Differentiation of the Ovaries. In the absence of specific testis-differentiation signals, the gonads ultimately differentiate into ovaries. In contrast to the testes, the presence of viable germ cells is essential for ovarian differentiation. If primordial germ cells fail to reach the genital ridges or if they are abnormal (e.g., XO) and degenerate, the gonad regresses and **streak ovaries** (vestigeal ovaries) result.

After the primordial germ cells have entered the future ovary, they remain concentrated in the outer cortical region or near the corticomedullary border. Like the testis, the ovary contains primitive sex cords in the medullary region, but these are not as well developed as those in the testis. The origin of the cells that form the ovarian follicles has not been established. Three sites of origin have been proposed for the follicular epithelial cells: (1) the coelomic epithelium (secondary sex cords), (2) the primitive sex cords of mesonephric origin, and (3) a combination of the two. The last possibility is in accordance with the presence of two distinct cell types—light and dark—within the early follicular epithelium.

The primary germ cells, now properly called **oogonia,** proliferate by mitosis from the time of their entry into the gonad until the start of the fourth month of gestation. At that time, some oogonia in the inner medullary region of the ovary enter prophase of the first meiotic division, possibly under the influence of a meiosis-stimulating factor emanating from the mesonephros. This influence may be associated with clumps of mesonephrically derived epithelia called the medullary **rete ovarii.** The meiotic oogonia, now called **oocytes,** become associated with follicular cells and form **primordial follicles** (see Fig. 1-5). Meanwhile, the oogonia in the cortical region of the ovary continue to divide mitotically. The oogonia and early oocytes are connected by intercellular cytoplasmic bridges that may play a role in synchronization of their development. By week 22 follicular development is well underway throughout the ovary. The oocytes continue in meiosis until they reach the diplotene stage of prophase of the first meiotic division. Meiosis is then arrested, and the oocytes remain in this stage until the block is removed. This occurs days before ovulation in the adult, sometimes as late as 50 years after they entered the block.

In the fetal ovary an inconspicuous tunica albuginea forms at the corticomedullary junction. The cortex of the ovary is the dominant component, and it contains the vast majority of oocytes. The medulla fills with connective tissue and blood vessels that are derived from the mesonephros. The testis, on the other hand, is characterized by a dominance of the medullary component located inside a prominent tunica albuginea.

The developing ovary does not maintain a relationship with the mesonephros. Normally, the mesonephric tubules that are equivalent to those that form the efferent ductules in the male degenerate in the female embryo, leaving only a few remnants (Table 17-1).

THE SEXUAL DUCT SYSTEM

Like the gonads, the sexual ducts pass through an early indifferent stage. As the fetal testes begin to function in the male, their secretion products act on the indifferent ducts, causing some components of the duct system to develop further and others to regress. In females the absence of testicular secretory products results in the preservation of structures that regress and the regression of structures that persist in the male.

The Indifferent Sexual Duct System

The indifferent sexual duct system consists of the mesonephric (wolffian) ducts and the **paramesonephric (müllerian) ducts** (Fig. 17-24). The paramesonephric ducts appear between 44 and 48 days of gestation as longitudinal invaginations of the surface epithelium along the mesonephric ridge lateral to the mesonephric ducts. The invaginations soon become epithelial cords, which grow caudally and terminate on the urogenital sinus between the ends of the mesonephric ducts without breaking into the sinus. These cords then develop a lumen in a craniocaudal direction. The cranial end of the paramesonephric duct opens into the coelomic cavity as a funnel-shaped structure. The fate of the indifferent genital ducts depends on the sex of the gonad.

The Sexual Duct System of Males

Development of the sexual duct system in the male depends on secretions from the testis. Under the influence of **müllerian inhibiting substance,** a glycoprotein secreted by the Sertoli cells of the testes at 8 weeks gestation, the paramesonephric ducts degenerate, leaving only remnants at their cranial and caudal ends (Figs. 17-25 and 17-26 and Table 17-1). The molecular structure of müllerian inhibitory substance closely resembles that of TGF-β.

Under the influence of testosterone, one of the male hormones secreted by the Leydig cells of the testes, the mesonephric ducts continue to develop even though the mesonephric kidneys are degenerating. The mesonephric ducts differentiate into the paired **ductus deferens,** which con-

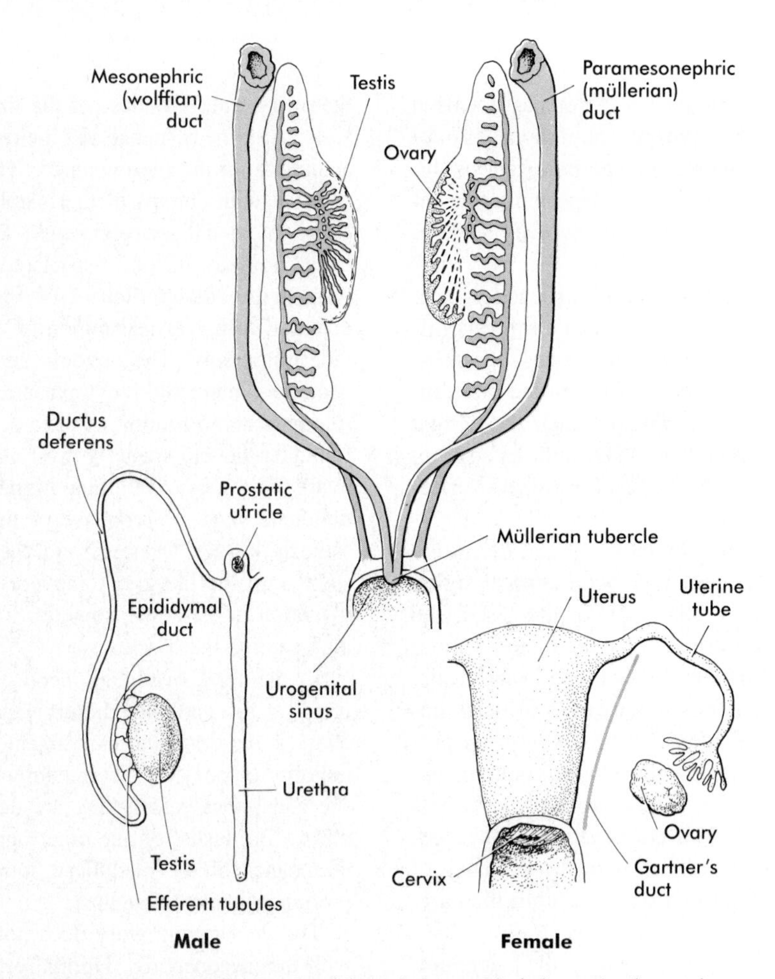

FIG. 17-24 The indifferent condition of the genital ducts in the embryo at approximately 6 weeks.

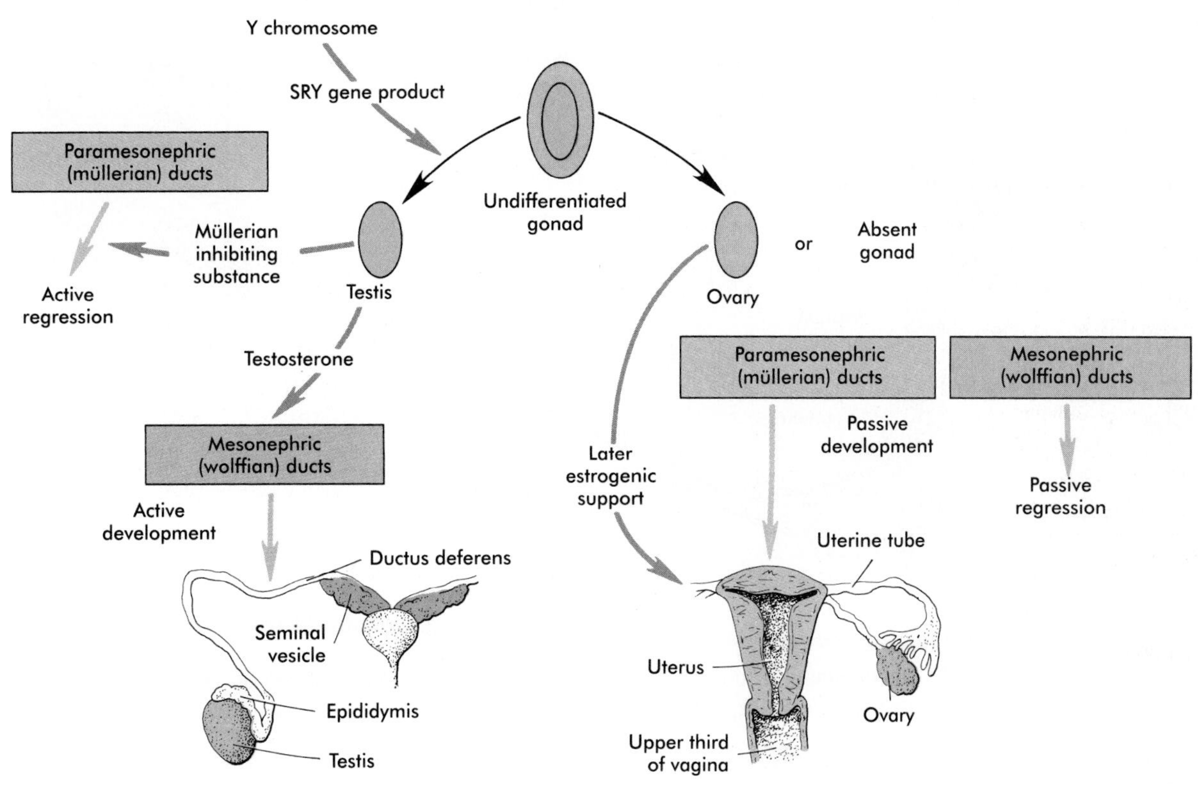

FIG. 17-25 Factors involved in sexual differentiation of the genital tract.
(Based on studies by Hutson and others, [1989].)

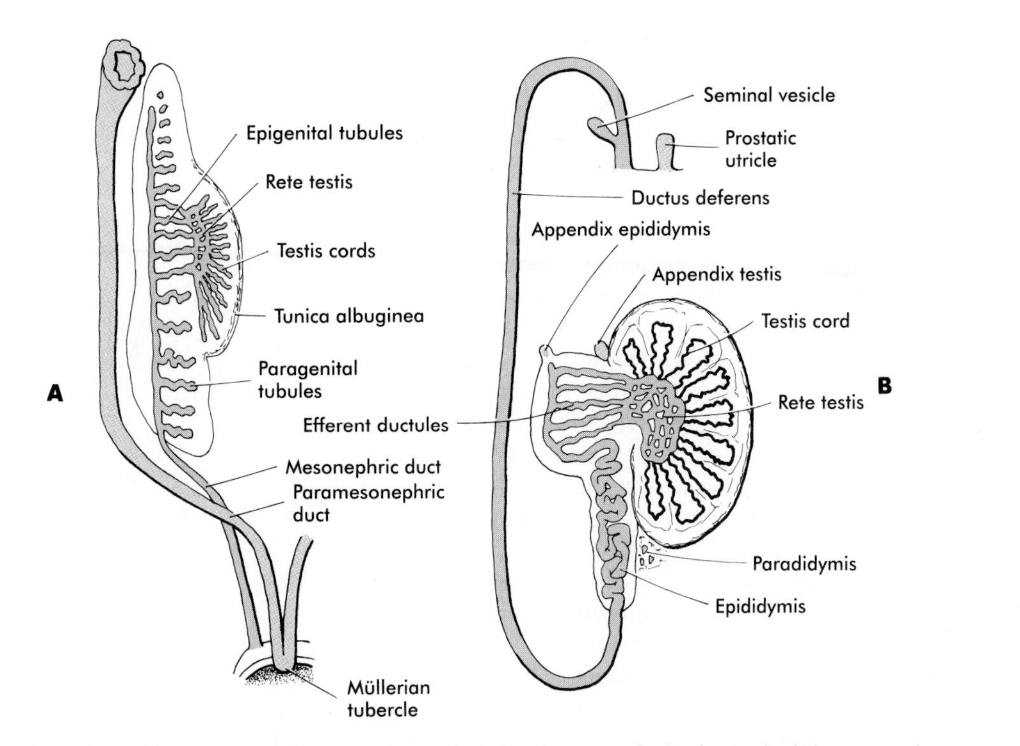

FIG. 17-26 Development of the male genital duct system. **A,** At the end of the second month. **B,** In the late fetus.
(Modified from Sadler T: *Langman's medical embryology,* ed 6, Baltimore, 1990, Williams & Wilkins.)

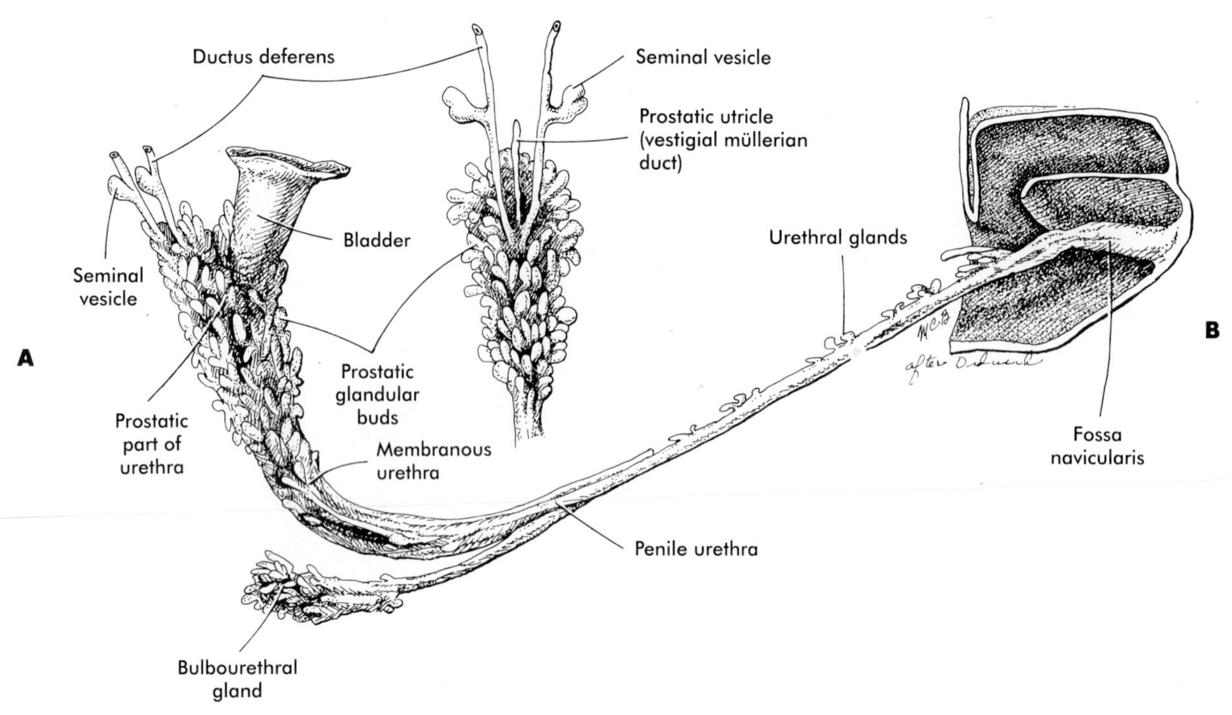

FIG. 17-27 Development of the male urethra and accessory sex glands in an embryo of approximately 16 weeks. **A,** Lateral view. **B,** Dorsal view of prostatic region.
(After Didusch.)

stitutes the path of sperm transport from the testis to the urethra. Portions of degenerating mesonephric tubules may persist near the testis as the **paradidymis.**

Associated with development of the male genital duct system (both the ductus deferens and the urethra) is the formation of the male accessory sex glands, the **seminal vesicles,** the **prostate,** and the **bulbourethral glands** (Fig. 17-27). These glands arise as epithelial outgrowths from their associated duct systems (seminal vesicles from the ductus deferens and the others from the urethra), and their formation involves epithelial-mesenchymal interactions similar to those of other glands. In addition, these glands depend on androgenic stimulation for their development. Specifically, the mesenchymal cells develop androgen receptors and appear to be the primary targets of the circulating androgenic hormones. After stimulation by the androgens the mesenchymal cells act on the associated epithelium, causing it to differentiate with gland-specific characteristics.

Tissue recombination experiments in which glandular mesoderm from mice with **testicular feminization syndrome** (lack of testosterone receptors resulting in no response to testosterone) was combined with normal epithelium demonstrated that the mesodermal component of the glandular primordia is the hormonal target. Differentiation of the epithelium did not occur. In contrast, when normal glandular mesoderm was combined with epithelium from animals with testicular feminization syndrome, normal development took place.

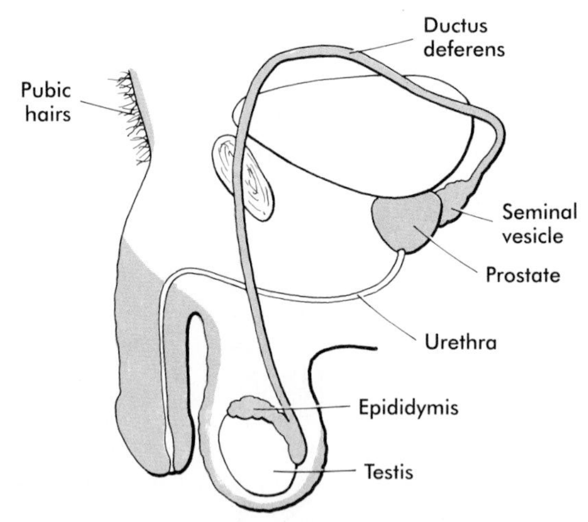

FIG. 17-28 Regions of the male reproductive tract sensitive to testosterone (gray) and dihydrotestosterone (blue).
(Based on studies by Imperato-McGinley and others [1974].)

In the embryo the tissues around the urogenital sinus synthesize an enzyme **(5-α-reductase)** that converts testosterone to dihydrotestosterone. Through the action of appropriate receptors of either form of testosterone, critical tissues of the male reproductive tract are maintained and grow (Fig. 17-28).

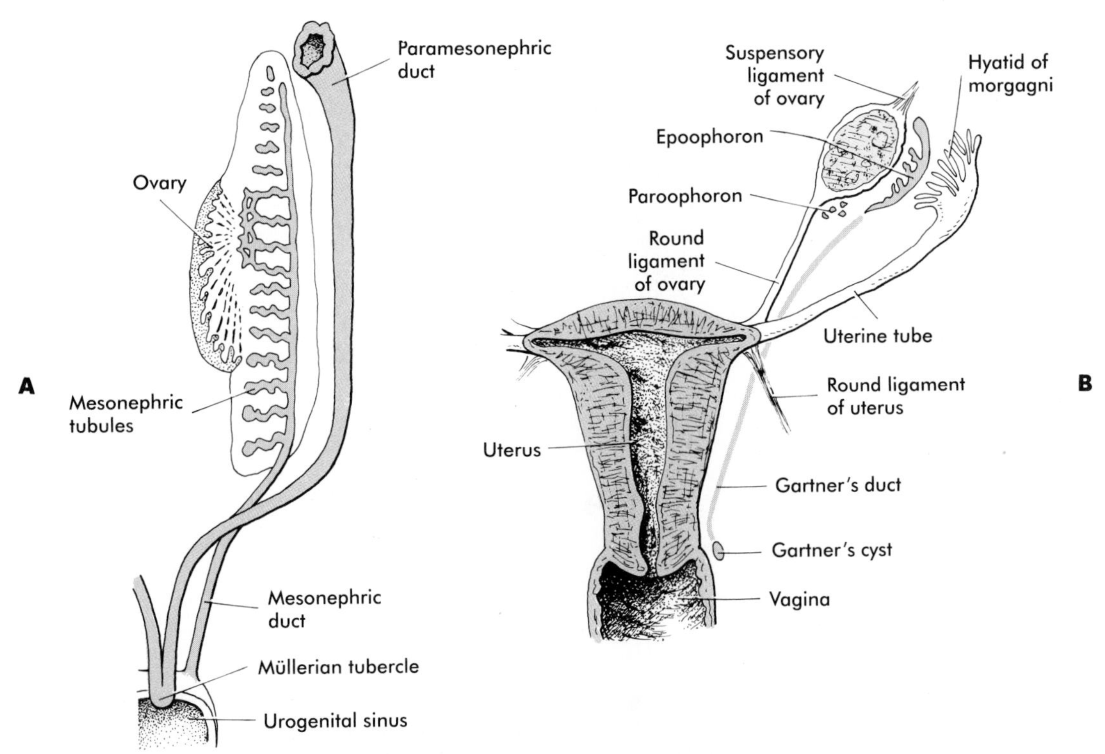

FIG. 17-29 Development of the female genital duct system. **A,** At the end of the second month. **B,** Mature condition.
(Modified from Sadler T: *Langman's medical embryology,* ed 6, Baltimore, Williams & Wilkins.)

The Sexual Duct System of Females

If ovaries are present or if the gonads are absent or dysgenic, the sexual duct system differentiates into a female phenotype. In the absence of testosterone secreted by the testes, the mesonephric ducts regress, leaving only rudimentary structures (see Table 17-1). In contrast, the absence of müllerian inhibitory substance allows the paramesonephric ducts to continue to develop into the major structures of the female genital tract (Fig. 17-29).

The cranial portions of the paramesonephric ducts become the **uterine tubes,** with the cranial openings into the coelomic cavity persisting as the fimbriated ends. Toward their caudal ends the paramesonephric ducts begin to approach the midline and cross the mesonephric ducts ventrally. This crossing and ultimate meeting in the midline are caused by the medial swinging of the entire urogenital ridge (Fig. 17-30). The region of midline fusion of the paramesonephric ducts ultimately becomes the uterus, and the ridge tissue that is carried along with the paramesonephric ducts forms the **broad ligament of the uterus.**

The formation of the **vagina** remains poorly understood, and several explanations for its origin have been posited. According to one commonly held hypothesis the fused paramesonephric ducts form the upper part of the vagina and epithelial tissue from the müllerian tubercle **(vaginal plate)** hollows out to form the lower part. According to this model the hymen is the result of the penetration of mesenchyme between the fused ends of the paramesonephric ducts and the epithelium of the vaginal plate. More recently, several investigators have suggested that the most caudal portions of the mesonephric ducts participate in formation of the vagina by either directly contributing cells to its wall or acting on the paramesonephric tissue inductively. Full development of the female reproductive tract depends on estrogenic hormones secreted by the fetal ovaries.

Descent of the Gonads

Descent of the Testes. The testes do not remain in their original site of development; they migrate from their intraabdominal location into the scrotum (Fig. 17-31). As with the kidneys, the testes are retroperitoneal structures, and their descent occurs behind the peritoneal epithelium.

Control of testicular descent has been divided into three phases: The first is associated with the enlargement of the testes and the concomitant regression of the mesonephric kidneys. This causes some caudal displacement of the testes. The second phase, commonly called **transabdominal descent,** brings the testes down to the level of the inguinal ring but not into the scrotum. Control of this phase has been attributed to müllerian inhibitory substance and the regression of the paramesonephric ducts. The third phase, called

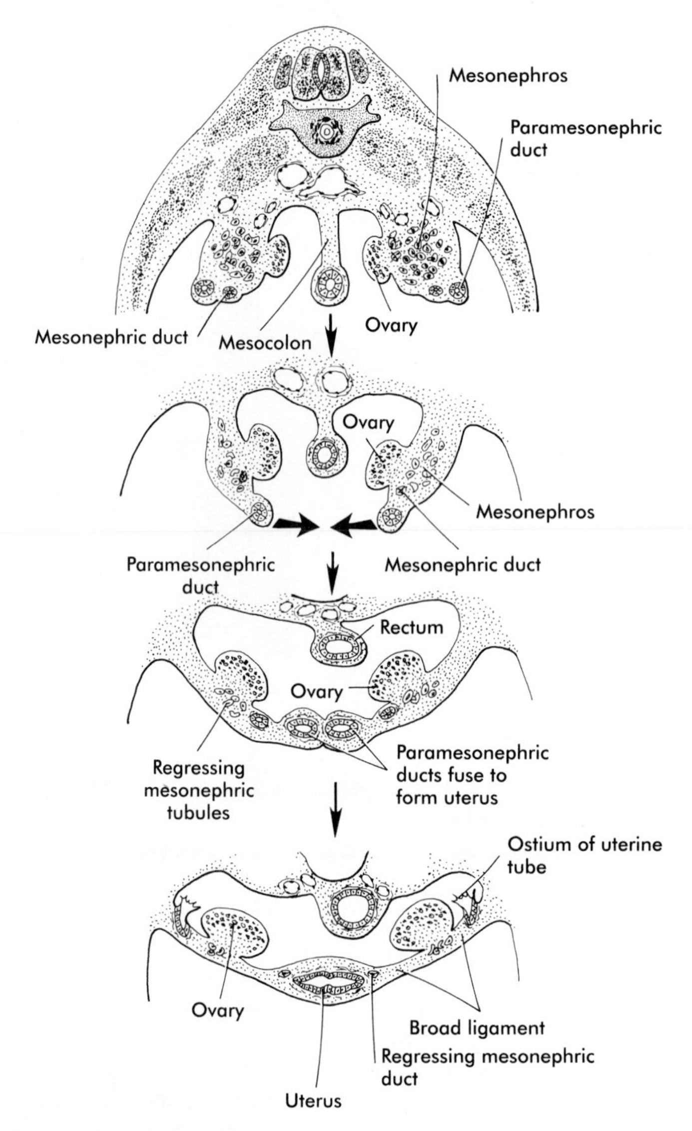

FIG. 17-30 Formation of the broad ligament in the female embryo.

transinguinal descent, brings the testes into the scrotum. This phase involves both the action of testosterone and the guidance of the **inguinal ligament of the mesonephros,** which in later development is called the **gubernaculum.** Whether the gubernaculum actively pulls the testis into the scrotum or just acts as a fixed point while the other tissues grow has not been resolved. Testicular descent begins during the seventh month and may not be completed until birth. As it descends into the scrotum, the testis slides behind an extension of the peritoneal cavity—the vaginal process (see Fig. 17-31, *C*). Although this cavity largely closes off with maturation of the testis, it remains as a potential mechanical weak point. With straining, it can open and permit the herniation of intestine into the scrotum.

Descent of the Ovaries. Although not as dramatically as the testes, the ovaries also undergo a distinct caudal shift in position. In conjunction with their growth and the crossing over of the paramesonephric ducts, the ovaries move caudally and laterally. Their position is stabilized by two ligaments, both of which are remnants of structures associated with the mesonephros. Cranially, the **diaphragmatic ligament of the mesonephros** becomes the **suspensory ligament of the ovary.** Caudally, the inguinal ligament of the mesonephros develops into the **round ligament of the ovary** and, even more caudally, the **round ligament of the uterus** (see Fig. 17-29).

THE EXTERNAL GENITALIA

The Indifferent Stage

The external genitalia are derived from a complex of mesodermal tissue located around the cloaca. A very early midline elevation called the **genital eminence** is situated just cephalic to the proctodeal depression. This structure soon develops into a prominent **genital tubercle** (Fig. 17-32), which is flanked by a pair of **genital folds** extending

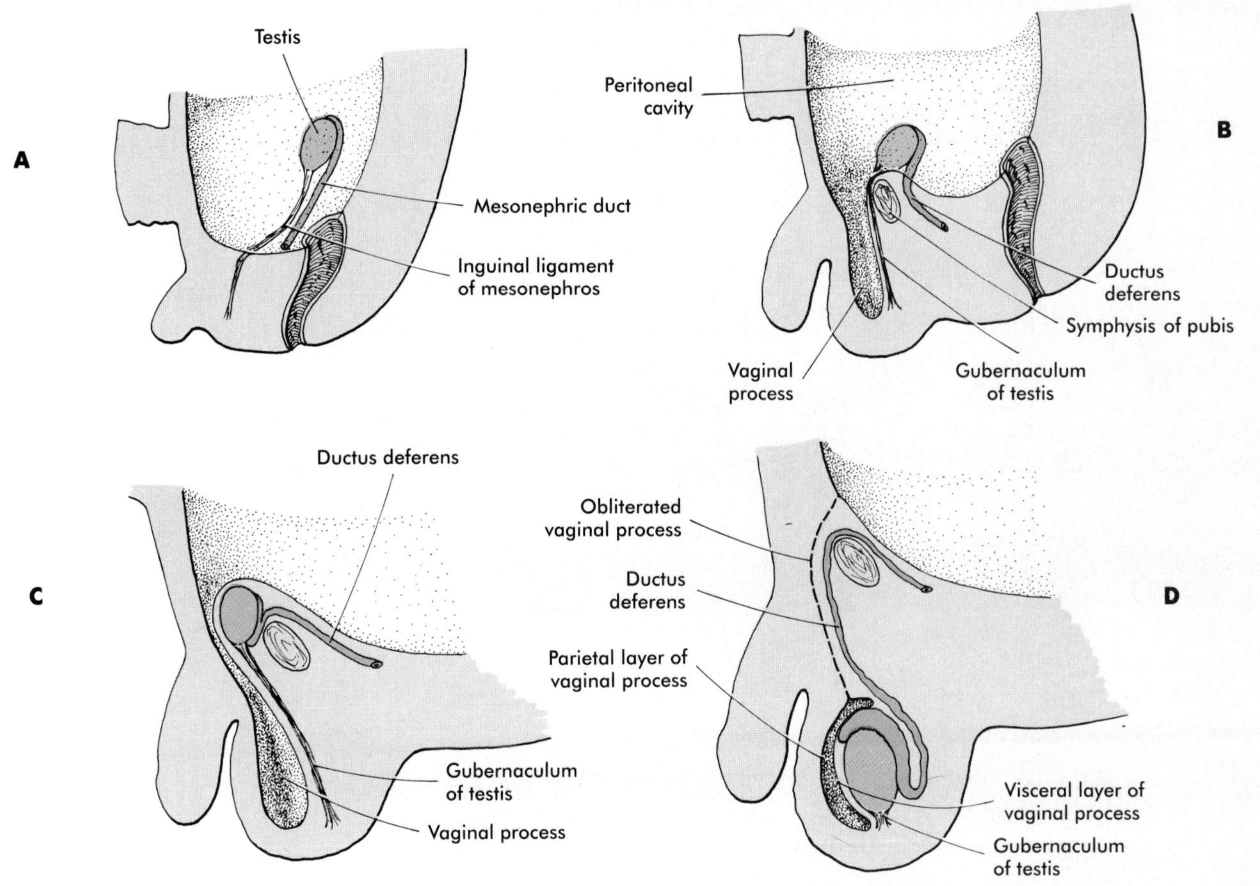

FIG. 17-31 Descent of the testis in the male fetus. **A,** In the second month. **B,** In the third month. **C,** In the seventh month. **D,** At term.

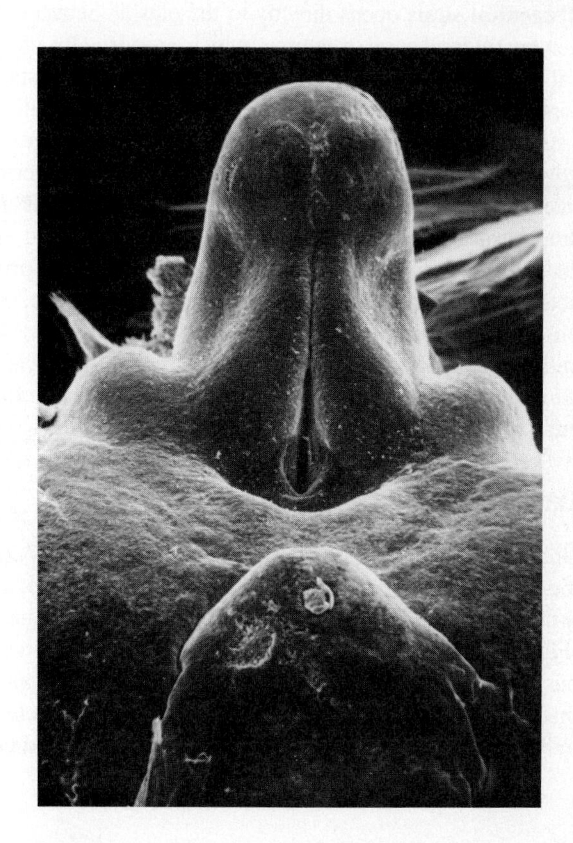

FIG. 17-32 Scanning electron micrograph of the inferior aspect of the indifferent external genitalia of a human embryo at the end of the eighth week of development.
(From Jirásek JE: *Atlas of human prenatal morphogenesis,* Amsterdam, 1983, Martinus Nijhoff Publishers.)

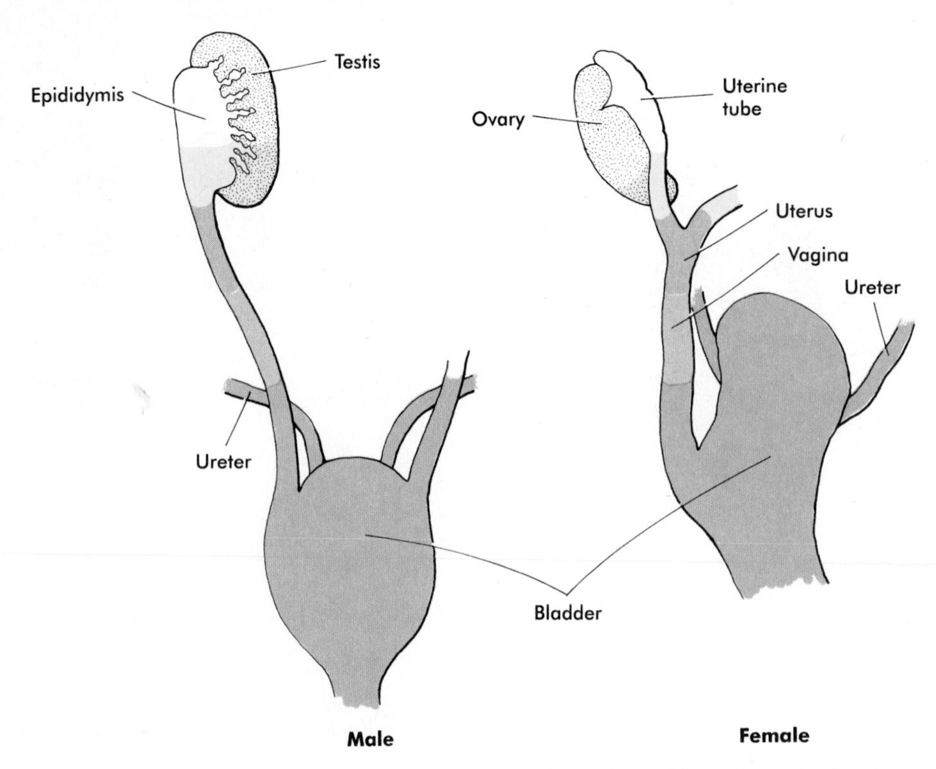

FIG. 17-33 Gradients of *Hox-4* gene expression in the male and female genitalia of mouse embryos.
(Based on studies by Dollé et al [1991].)

toward the proctodeum. Somewhat lateral to these are paired **genital swellings** (see Fig. 17-34). When the genital membrane breaks down during the eighth week, the **urogenital sinus** opens directly to the outside between the genital folds. These structures, which are virtually identical in male and female embryos during the indifferent stage, form the basis for development of the external genitalia.

As with the developing limb, outgrowth of the genital tubercle depends on a continuing ectodermal-mesodermal interaction, although an apical ectodermal ridge is not present on the genital tubercle. In another similarity with the limb bud, there is a gradient of *Hox-4* (D) gene products in the genital tubercle and genital ducts in mice, and this is presumably true in humans as well (Fig. 17-33). Finally, pieces of mesenchyme from the genital tubercle of mice demonstrate polarizing activity when grafted into limb buds of chick embryos (see Chapter 10).

External Genitalia of Males

Under the influence of dihydrotestosterone (see Fig. 17-28), the genital tubercle greatly elongates to form the penis, and the genital swellings enlarge to form the scrotal pouches (Fig. 17-34). As this growth is occurring, the urogenital sinus becomes continuous with a groove that develops along the caudal face of the genital tubercle. This groove closes to become the penile part of the urethra, and the closed urogenital sinus becomes the prostatic portion of the urethra. In the distal-most part of the penis a solid cord of epithelial cells grows from the glans to meet the penile urethra. When this cord canalizes, the formation of the urethra in the male is complete. The line of fusion along the urethra and passing through the scrotal swellings is the **raphe**.

External Genitalia of Females

In females the pattern of external genitalia follows that of the indifferent stage closely (Fig. 17-34). The genital tubercle becomes the **clitoris,** the genital folds become the **labia minora,** and the genital swellings develop into the **labia majora.** The urogenital sinus remains open as the vestibule, into which both the urethra and vagina open. The female **urethra,** developing from the more cranial part of the urogenital sinus, is equivalent to the prostatic urethra of the male.

MALFORMATIONS OF THE GENITAL SYSTEM

Abnormalities of Sexual Differentiation

Turner Syndrome (Gonadal Dysgenesis). Turner syndrome results from a chromosomal anomaly (45,XO) (see Chapter 8). Individuals with Turner syndrome possess primordial germ cells that degenerate shortly after reaching the gonads. Differentiation of the gonad fails to occur, leading

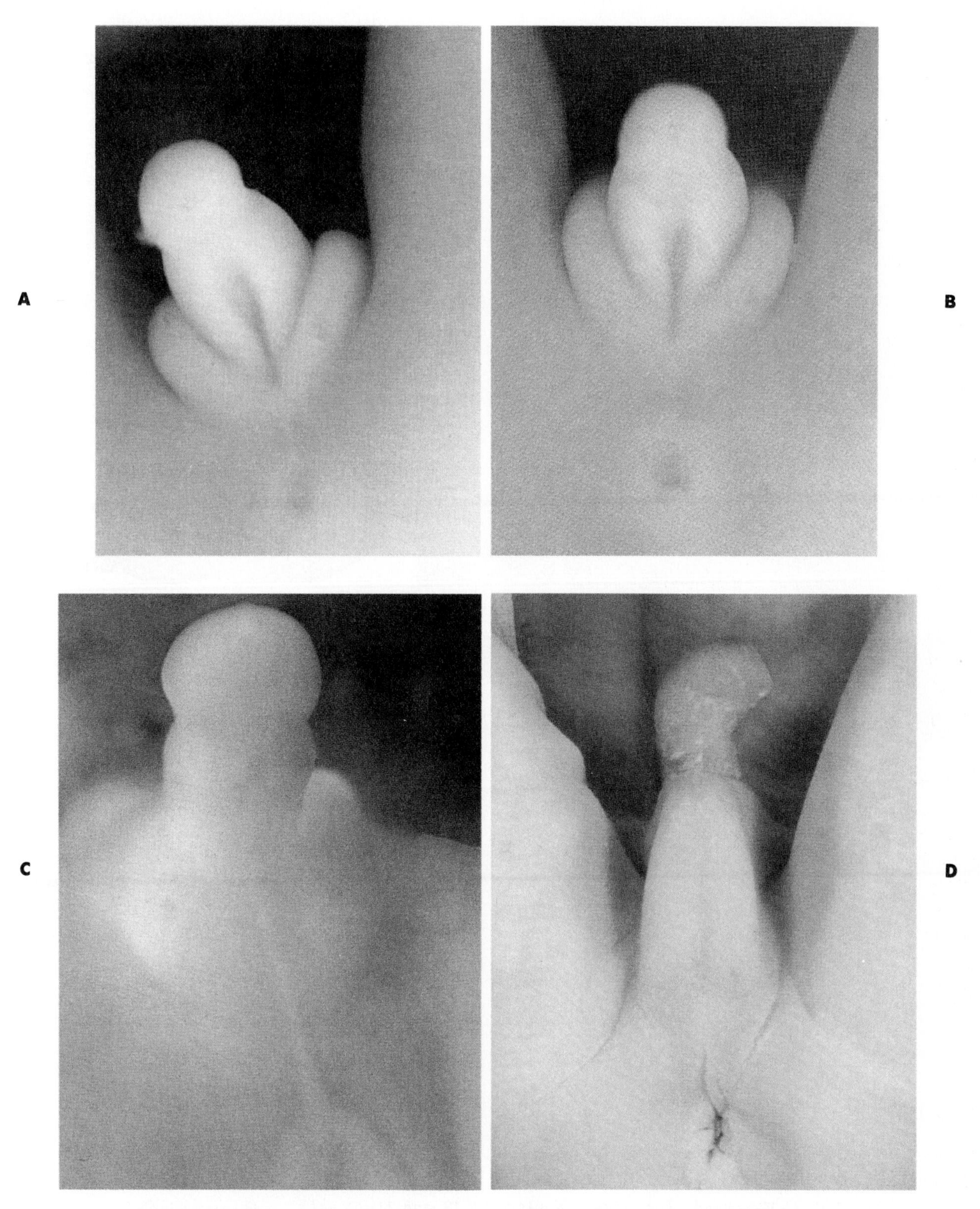

FIG. 17-34 Differentiation of the external genitalia of male embryos. **A,** Indifferent stage, early week 9. **B-D,** Male, weeks 9, 10, and 18.
(From England M: *Color atlas of life before birth,* Chicago, 1983, Mosby.)

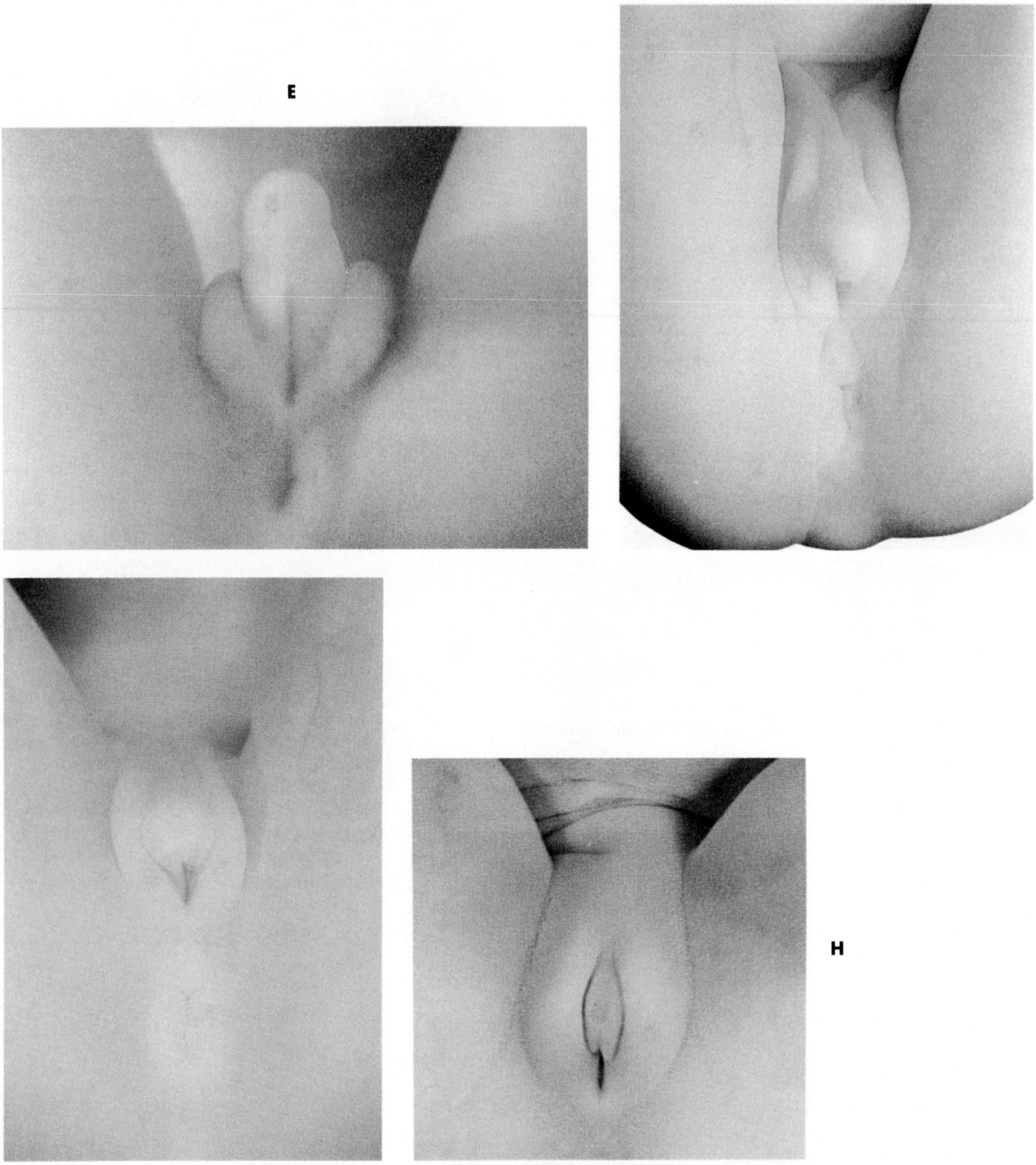

E

F

G

H

FIG. 17-34—cont'd E-H, Female, weeks 9, 12, 13, and 20.

to the formation of a **streak gonad.** In the absence of gonadal hormones the genitalia develop along female lines but remain infantile. The mesonephric duct system regresses for lack of androgenic hormonal stimulation.

True Hermaphroditism. Individuals with true hermaphroditism, which is an extremely rare condition, possess both testicular and ovarian tissue. In cases of genetic mosaicism, both an ovary and a testis may be present; in other cases, ovarian and testicular tissue are present in the same gonad **(ovotestis).** Most true hermaphrodites have a 46,XX chromosome constitution, and the external genitalia are basically female, although typically the clitoris is hypertrophied. Such individuals are usually reared as females.

Female Pseudohermaphroditism. Female pseudohermaphrodites are genetically female (46,XX) and are sex chromatin positive. The internal genitalia are typically female, but the external genitalia are masculinized, either from excessive production of androgenic hormones by the adrenal cortex **(congenital virilizing adrenal hyperplasia)** or from inappropriate hormonal treatment of pregnant women. The degree of external masculinization can vary from simple clitoral enlargement to partial fusion of the labia majora into a scrotumlike structure (see Fig. 8-12).

Male Pseudohermaphroditism. Male pseudohermaphrodites are sex chromatin negative (46,XY). Because this condition is commonly caused by inadequate hormone production by the fetal testes, the phenotype can vary. It is commonly associated with hypoplasia of the phallus, and there may be varying degrees of persistence of paramesonephric duct structures.

Testicular Feminization Syndrome. Individuals with this condition are genetic males (46,XY) and possess internal testes, but they typically have a normal female external phenotype and are raised as females (see Fig. 10-12). Often testicular feminization is not discovered until the person seeks treatment for amenorrhea or is tested for sex chromatin before athletic events. The testes typically produce testosterone, but because of a deficiency in receptors caused by a mutation on the X chromosome, the testosterone is unable to act on the appropriate tissues. Because müllerian inhibitory substance is produced by the testes, the uterus and upper part of the vagina are absent.

Vestigial Structures from the Embryonic Genital Ducts

Vestigial structures are remnants from the regression of embryonic genital ducts, which is rarely complete. They are so common that they are not necessarily malformations, although they can become cystic.

Mesonephric Duct Remnants. In males a persisting blind cranial end of the mesonephric duct can appear as the **appendix of the epididymis** (see Fig. 17-26). Remnants of a few mesonephric tubules caudal to the efferent ductules occasionally appear as the **paradidymis.**

In females, remains of the cranial parts of the mesonephros may persist as the **epoophoron,** or **paroophoron** (see Fig. 17-29). The caudal part of the mesonephric ducts is often seen in histological sections along the uterus or upper vagina as **Gartner's ducts.** Portions of these duct remnants sometimes enlarge to form cysts.

Paramesonephric Duct Remnants. The cranial tip of the paramesonephric duct may remain as the small **appendix of the testis** (see Fig. 17-26). The fused caudal ends of the paramesonephric ducts are commonly seen in the prostate gland as a small midline **prostatic utricle,** which represents the rudimentary uterine primordium. In newborn males the prostatic utricle is typically slightly enlarged because of the influence of maternal estrogenic hormones during pregnancy, but it regresses soon after birth. This structure can enlarge to form a uteruslike structure in some cases of male pseudohermaphroditism.

In females a small part of the cranial tip of the paramesonephric duct may persist at the fimbriated end of the uterine tube as the **hydatid of Morgagni** (see Fig. 17-29).

Other Abnormalities of the Genital Duct System

Males. Abnormalities of the mesonephric duct system are relatively rare, but duplications or diverticulae of the ductus deferens or urethra can occur. There is an interesting correlation of absent or rudimentary ductus deferens in males with cystic fibrosis. This may be the result of defect in the gene situated along the gene causing the cystic fibrosis. If the Sertoli cells do not produce müllerian inhibitory substance, the paramesonephric ducts form a uterus and uterine tubes in an otherwise normal male, resulting in the **persistent müllerian duct syndrome.** Because the testes adhere to the uterine tubes, this condition is often associated with cryptorchidism. Any of the accessory sex glands in the male may be absent or hypoplastic. This usually results in infertility.

Females. Malformations of the uterus or vagina are attributed to abnormalities of fusion or regression of the caudal ends of the paramesonephric ducts (Fig. 17-35). Uterine anomalies range from a small septum extending from the dorsal wall of the uterus to complete duplication of the uterus and cervix. Numerous successful pregnancies have been recorded in women with uterine malformations. **Agenesis of the vagina** has been attributed to a failure of formation of the epithelial vaginal plate from the site of joining of the müllerian tubercle with the urogenital sinus.

Abnormalities of Testicular Descent

Cryptorchidism. Undescended testes are common in premature males and are seen in about 3% of term males. Normally, the testes of these individuals descend into the scrotum within the first few months after birth. If they do not, the condition of **cryptorchidism** results. Cryptorchid-

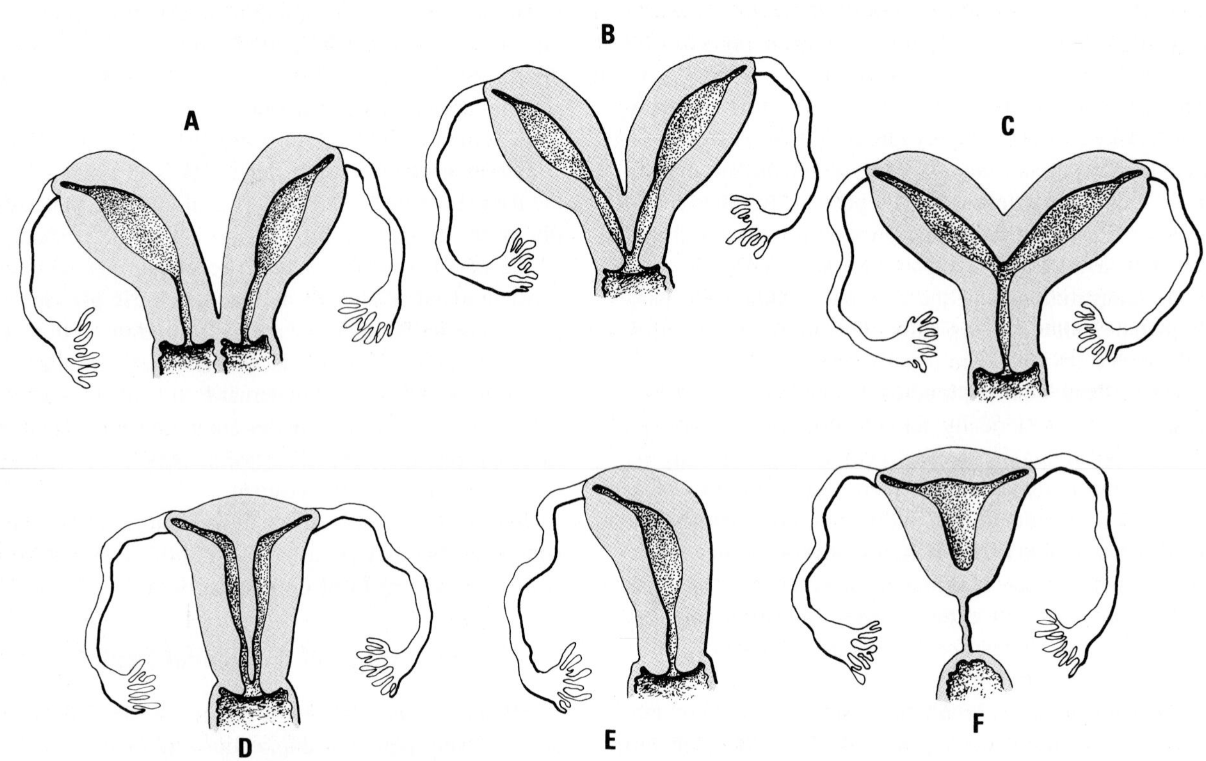

FIG. 17-35 Abnormalities of the uterus and vagina. **A,** Double uterus and double vagina. **B,** Double uterus and single vagina. **C,** Bicornuate uterus. **D,** Septate uterus. **E,** Unicornuate uterus. **F,** Atresia of the cervix.

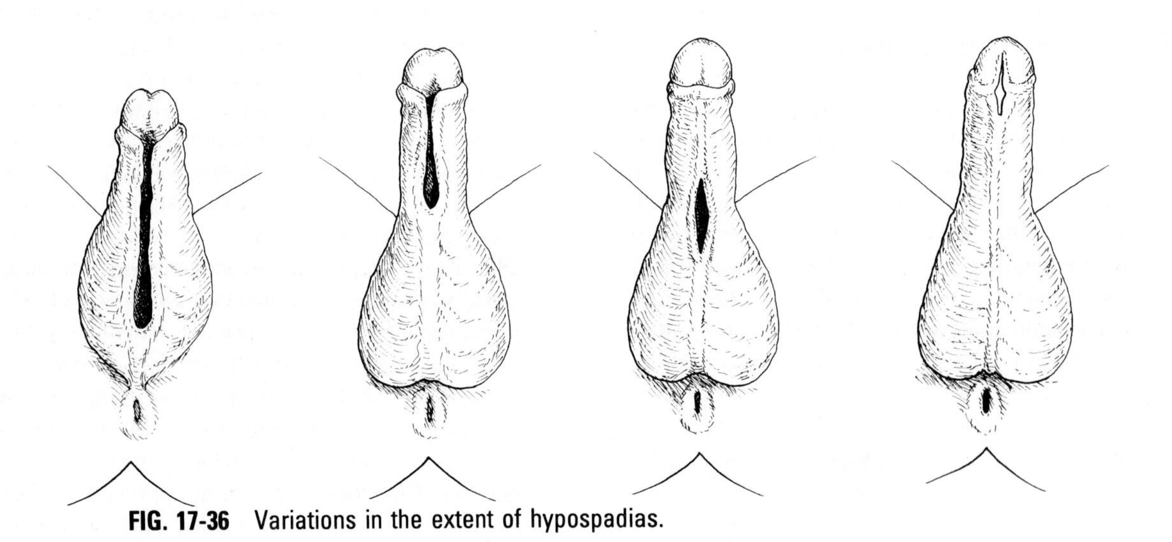

FIG. 17-36 Variations in the extent of hypospadias.

ism results in sterility because spermatogenesis does not normally occur at the temperature of the body cavity. There is also a fiftyfold greater incidence of malignancy in undescended testes.

Ectopic Testes. A testis occasionally migrates to some site other than the scrotum, including the thigh, the perineum, and the ventral abdominal wall. Because of the elevated temperature of the surrounding tissues, ectopic testes produce reduced numbers of viable spermatozoa.

Congenital Inguinal Hernia. If the peritoneal canal that leads into the fetal scrotum fails to close, a condition called **persistent processus vaginalis** occurs. This space may become occupied by loops of bowel that herniate into the scrotum.

Malformations of the External Genitalia

Males. The most common malformation of the penis is **hypospadias,** in which the urethra opens onto the ventral surface of the penis rather than at the end of the glans (Fig. 17-36). The degree of hypospadias can range from a mild ventral deviation of the urethral opening to an elongated opening representing an unfused portion of the urogenital sinus. In the more severe varieties the penis is often bowed ventrally **(chordee).**

Isolated **epispadias,** with the urethra opening on the dorsal surface of the penis, is very rare. A dorsal groove on the penis is commonly associated with exstrophy of the bladder (see Fig. 17-19).

Duplication of the penis occurs most commonly in association with exstrophy of the bladder and appears to result from the early separation of the tissues destined to form the genital tubercle. Duplication of the penis very rarely occurs in the absence of exstrophy of the bladder.

Congenital absence of the penis (or clitoris in the female) is rare. With recent knowledge derived from experimental embryology, this condition may be explained on the basis of inadequate ectodermal-mesodermal interactions in the development of the genital tubercle.

Females. Anomalies of the external genitalia in females can range from hormonally induced enlargement of the clitoris to duplications. Exposure to androgens may also masculinize the genital swellings, resulting in scrotalization of the labia majora. Depending on the degree of severity, wrinkling of the skin and partial fusion may occur.

SUMMARY

1. The urogenital system arises from the intermediate mesoderm. The urinary system arises before gonadal development begins.
2. Kidney development begins with the formation of pairs of nephrotomes that connect with a pair of primary nephric ducts. Farther caudally to the nephrotomes, pairs of mesonephric tubules form in a craniocaudal sequence and connect to the primary nephric ducts, which become known as mesonephric ducts. In the caudal part of each mesonephric duct a ureteric bud grows out and induces the surrounding mesoderm to form the metanephros.
3. Within the developing metanephros, nephrons (functional units of the kidney) form from three sources: the metanephrogenic blastema, the metanephrogenic diverticulum, and ingrowing vascular endothelial cells. Nephrons continue to form throughout fetal life. Several growth factors and transferrin are involved in the induction and differentiation of the nephron.
4. The kidneys arise in the pelvic basin, and during the late embryonic and early fetal period, they shift into the abdominal region where they become associated with the adrenal glands. The urinary bladder arises from the base of the allantois.
5. The urinary system is subject to a variety of malformations. Most severe is renal agenesis, which is probably caused by faulty induction in the early embryo. Abnormal migration can result in pelvic or other ectopic kidneys or horseshoe kidney. Polycystic disease of the kidney is associated with cysts in other internal organs. Faulty closure of the allantois results in urachal cysts, sinuses, or fistulas.
6. Sex determination begins at fertilization by the contribution of an X or a Y chromosome to the egg by the sperm. The early embryo is sexually indifferent. Through the action of the *Sry* gene the indifferent gonad in the male develops into a testis. In the absence of this gene the gonad becomes an ovary.
7. Gonadal differentiation begins after migration of the primordial germ cells into the indifferent gonads. Under the influence of *Sry* gene products the testis begins to differentiate. In the embryonic testis, Leydig cells secrete testosterone and Sertoli cells produce müllerian inhibitory substance. In the absence of *Sry* expression, the gonad differentiates into an ovary and contains follicles.
8. The sexual duct system consists of the mesonephric (wolffian) and paramesonephric (müllerian) ducts. The duct system is originally indifferent. In the male, müllerian inhibitory substance causes regression of the paramesonephric duct system, and testosterone causes further development of the mesonephric duct system. In the female the mesonephric ducts regress in the absence of testosterone, and the paramesonephric ducts persist in the absence of müllerian inhibitory substance.
9. In males the mesonephric ducts form the ductus deferens and give rise to the male accessory sex glands. In females the paramesonephric ducts form the uterine tubes, the uterus, and part of the vagina.
10. The testes descend from the abdominal cavity into the scrotum later in development. The ovaries also shift to a more caudal position. Faulty descent of the testes is cryptorchidism and is associated with sterility and testicular tumors.
11. The external genitalia also begin in an indifferent condition. Basic components of the external genitalia are the genital tubercle, genital folds, and genital swellings. Under the influence of testosterone the genital tubercle elongates into a phallus, and the genital folds fuse to form the penile urethra. The genital swellings form the scrotum. In the female the genital tubercle forms the clitoris, the genital folds form the labia minora, and the genital swellings form the labia majora.

12. If an individual possesses only one X chromosome (XO), Turner syndrome results. Such individuals have a female phenotype with streak gonads. True hermaphroditism or pseudohermaphroditism can result from a variety of causes. Testicular feminization is found in genetic males lacking testosterone receptors. Such individuals are phenotypic females. Major abnormalities of the sexual ducts are rare, but they can lead to duplications or the absence of the uterus in females.

REVIEW QUESTIONS

1. A female Olympic athlete is subjected to a sex chromatin test and is told that she cannot compete because she is a male. What is the likely basis for the confusion, and what would be the status of her gonads and genital duct systems?
2. Drops of a yellowish fluid were observed around the umbilicus of a young infant. What is a likely diagnosis and what is the embryological basis?
3. When living donors are used for kidney transplantation procedures, one of the first rules is to be sure that the donor has two functional kidneys. What is the embryological basis for this rule?
4. A woman who gained relatively little weight during pregnancy gives birth to an infant with large, low-set ears; a flattened nose; and a wide interpupillary space. Within hours after birth the infant is obviously in great distress, and it dies after 2 days. What is the diagnosis?
5. A seemingly normal woman experiences pelvic pain during the later stages of pregnancy. An ultrasound examination reveals that she has a bicornuate uterus. What is the embryological basis for this condition?

REFERENCES

Acien P: Embryological observations on the female genital tract, *Hum Reprod* 7:437-445, 1992.

Anguiano A and others: Congenital absence of the vas deferens: a primarily genital form of cystic fibrosis, *JAMA* 267:1794-1797, 1992.

Byskov AG: Differentiation of mammalian embryonic gonad, *Physiol Rev* 66:71-117, 1986.

Carre-Eusebe D and others: Variants of the anti-müllerian hormone gene in a compound heterozygote with the persistent müllerian duct syndrome and his family, *Hum Genet* 90:389-394, 1992.

Cunha GR: *Androgenic effects upon prostatic epithelium are mediated via trophic influences from stroma.* In *New approaches to the study of benign hyperplasia,* New York, 1984, Alan R Liss, pp 81-102.

Cunha GR and others: Hormone-induced morphogenesis and growth: role of mesenchymal-epithelial interactions, *Recent Prog Horm Res* 39:559-598, 1983.

De Felici MS, Dolci S, Pesce M: Cellular and molecular aspects of mouse primordial germ cell migration and proliferation in culture, *Int J Dev Biol* 36:205-213, 1992.

Dollé P and others: *Hox-4* genes and the morphogenesis of mammalian genitalia, *Genes Dev* 5:1767-1776, 1991.

Dressler GR and others: *Pax2,* a new murine paired-box-containing gene and its expression in the developing excretory system, *Development* 109:787-795, 1990.

Ekblom P: Developmentally regulated conversion of mesenchyme to epithelium, *FASEB J* 3:2141-2150, 1989.

Gattone VH and others: Defective epidermal growth factor gene expression in mice with polycystic kidney disease, *Dev Biol* 138:225-230, 1990.

Ginsburg M, Snow MHL, McLaren A: Primordial germ cells in the mouse embryo during gastrulation, *Development* 110:521-528, 1990.

Godin I, Wylie C, Heasman J: Genital ridges exert long-range effects on mouse primordial germ cell numbers and direction of migration in culture, *Development* 108:357-363, 1990.

Gubbay J and others: A gene mapping to the sex-determining region of the mouse Y chromosome is a member of a novel family of embryonically expressed genes, *Nature* 346:245-250, 1990.

Hammerman MR, Rogers SA, Ryan G: Growth factors and metanephrogenesis, *Am J Physiol* 26231:F523-F532, 1992.

Harley VR and others: *Science* 255:453-456, 1992.

Huhtaniemi I, Pelliniemi LJ: Fetal Leydig cells: cellular origin, morphology, life span, and special functional features, *Proc Soc Exp Biol Med* 201:125-140, 1992.

Hutson JM and others: *Müllerian inhibiting substance.* In Burger H, de Krester, D, eds: *The testis,* ed 2, Raven Press, New York, 1989, pp 143-179.

Jirásek JE: *Normal sex differentiation.* In Droegemueller W, Sciarra JJ, eds: *Gynecology and obstetrics, revised edition,* Philadelphia, 1991, JB Lippincott, pp 1-16.

Josso N: Anti-müllerian hormone and Sertoli cell function, *Horm Res* 38(suppl, 2):72-76, 1992.

Josso N, Picard J-Y: Anti-müllerian hormone, *Physiol Rev* 66:1038-1090, 1986.

Koopman P and others: Expression of a candidate sex-determining gene during mouse testis differentiation, *Nature* 348:450-452, 1991.

Lobaccaro J-M, Sultan C: La differenciation sexuelle normale: genetique et endocrinologie moleculaires, *CR Soc Biol* 186:314-331, 1992.

Makabe S and others: *Migration of germ cells, development of the ovary, and folliculogenesis.* In Familiari G, Makabe S, Motta PM, eds: *Ultrastructure of the ovary,* Norwell, Mass, 1991, Kluwer Academic Publishers, pp 1-27.

Merchant-Larios H: *Germ and somatic cell interactions during gonadal morphogenesis.* In Van Blerkom J, Motta PM, eds: *Ultrastructure of reproduction,* Boston, 1984, Martinus Nijhoff Publishers, pp 19-30.

Mittowch U: Sex determination and sex reversal: genotype, phenotype, dogma and semantics, *Hum Genet* 89:467-479, 1992.

Muller J, Shakkebaek NE: The prenatal and postnatal development of the testis, *Baillieres Clin Endocrinol Metab* 6:251-271, 1992.

O'Rahilly RO, Muecke EC: The timing and sequence of events in the development of the human urinary system during the embryonic period proper, *Z Anat Entwickl-Gesch* 138:99-109, 1972.

Osathanondh V, Potter EL: Development of the human kidney as shown by microdissection. I, II, III, *Arch Pathol* 76:271-302, 1963.

Reyes FI, Winter JSD, Faiman C: *Endocrinology of the fetal testis*. In Burger H, de Kretser D, eds: *The testis,* ed 2, New York, 1989, Raven Press, pp 119-142.

Sariola H: *Mechanisms and regulation of the vascular growth during kidney differentiation*. In Feinberg RN, Sherer GK, Auerbach R, eds: *The development of the vascular system,* Basel, Switzerland, 1991, S Karger, pp 69-80.

Sariola H and others: Dependence of kidney morphogenesis on the expression of nerve growth factor, *Science* 254:571-573, 1991.

Satoh M: Histogenesis and organogenesis of the gonad in human embryos, *J Anat* 177:85-107, 1991.

Saxén L: *Organogenesis of the kidney,* Cambridge, England, 1987, Cambridge University Press.

Sinclair AH and others: A gene from the human sex-determining region encodes a protein with homology to a conserved DNA binding motif, *Nature* 346:240-244, 1990.

Sultan C and others: Sry and male sex determination, *Horm Res* 36:1-3, 1991.

Sultan C and others: Ambiguite sexuelle, *Arch Fr Pediatr* 50:69-80, 1993.

Wartenberg H: *Differentiation and development of the testes*. In Burger H, de Kretser D, eds: *The testis,* New York, 1989, Raven Press, pp 67-118.

Wartenberg H: *Ultrastructure of fetal ovary including oogenesis*. In Van Blerkom J, Motta PM, eds: *Ultrastructure of human gametogenesis and early embryogenesis,* Norwell, Mass, 1989, Kluwer Academic Publishers, pp 61-84.

Zwingman T and others: Transcription of the sex-determining region genes *Sry* and *Zfy* in the preimplantation mouse embryo, *Proc Natl Acad Sci USA* 90:814-817, 1993.

CHAPTER 18

The Cardiovascular System

This chapter follows the development of the heart from a simple tubular structure to the four-chambered organ that can assume the full burden of maintaining an independent circulation at birth. Similarly, the pattern of blood vessels is traced from their first appearance to an integrated system that carries blood to all parts of the embryo and the placenta. (The early stages in the establishment of the heart and blood vessels are described in Chapter 5 [see Figs. 5-15 through 5-19], and the general plan of the embryonic circulation is summarized in Figure 5-25. Hematopoiesis and the molecular control of the differentiation of erythrocytes are covered in Chapter 9.)

Functionally, the embryonic heart needs only to act like a simple pump that maintains the flow of blood through the body of the embryo and into the placenta, where fetal wastes are exchanged for oxygen and nutrients. An equally important function, however, is to anticipate the radical changes in the circulation that take place at birth as a consequence of the abrupt cutting off of the placental circulation and the initiation of breathing. To meet the complex requirements of the postnatal circulatory system, the embryonic heart must develop four chambers that can receive or pump the full flow of blood circulating throughout the body. The heart must also adapt to the condition of the fetal lungs, which are poorly developed and for much of the fetal period do not possess a vasculature that can accommodate a large flow of blood. This physiological dilemma is resolved by the presence of two shunts that allow each chamber of the heart to handle large amounts of blood while sparing the underdeveloped pulmonary vascular channels.

Cardiac morphogenesis involves intrinsic cellular and molecular interactions, but these must occur against a background of ongoing mechanical function. Some of these mechanisms remain elusive, but others are becoming better defined through research on normal and abnormal cardiac development.

Development of the vasculature at the level of gross patterns of arteries and veins has been well understood for many years. Only recently have new cellular and molecular markers enabled investigators to question the cellular origins of the arteries and veins in specific organs or regions of the body. Studies on mechanisms of vascular differentiation are for the most part still in their infancy.

EARLY DEVELOPMENT OF THE HEART

Cellular Origins

The early tubular heart (see Fig. 5-17) is derived from several tissue sources. The atrial and ventricular myocardial tissue arises from a horseshoe-shaped bilateral primordium of splanchnic mesoderm (see Fig. 5-15), which coalesces in the ventral midline to form the bulk of the early heart tube. The outflow tract has a quite different cellular origin, with endothelial components arising from cephalic paraxial and lateral mesoderm in the region of the otic placode. A major cellular component of the wall of the outflow tract of the heart is derived from cranial neural crest, specifically from the level between the midotic placode to the caudal end of the third somite (Fig. 18-1). These cranial components integrate with the bilateral cardiac primordia while in the cervical region. As the heart descends into the thoracic cavity, the cranially derived cells of the outflow tract accompany it.

Looping of the Heart

When the cardiac tube first forms late in the third week, it is bilaterally symmetrical (see Fig. 5-17). Soon it under-

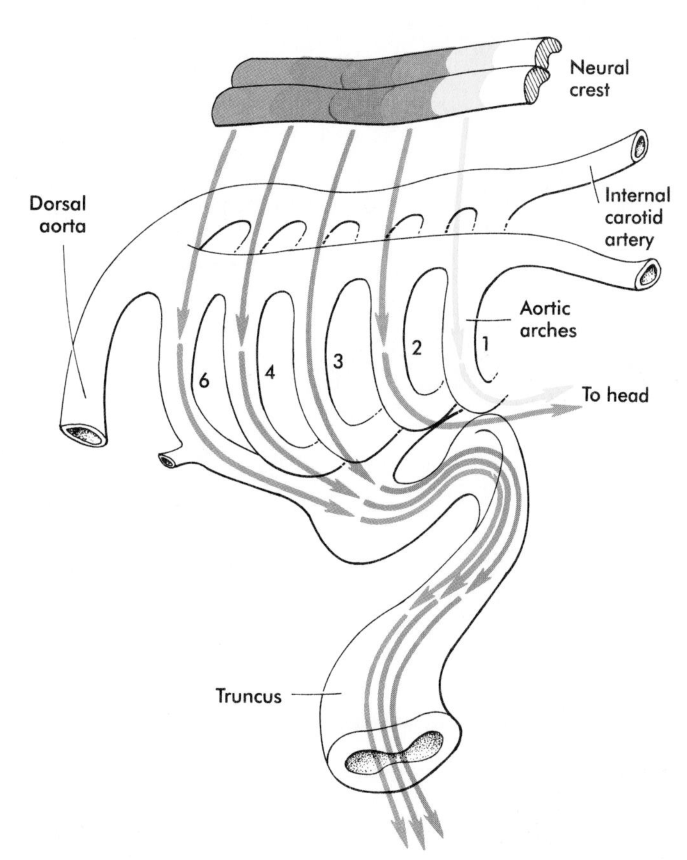

FIG. 18-1 The migration of cranial neural crest cells through the pharyngeal region and into the outflow tract of the heart, where they participate in formation of the truncoconal ridges.
(Modified from Kirby ML, Waldo KL: *Circulation* 82:332-340, 1990.)

goes a characteristic dextral looping, making it the first asymmetrical structure to appear in the embryonic body. The underlying mechanisms of the morphological changes in the looping process remain poorly defined. Most investigators agree that cardiac looping is likely to have a multifactorial basis, with factors as diverse as microtubules, asymmetrically located bundles of actin, pressure of the cardiac jelly, and changes in the shape of individual myocardial cells proposed as bases for looping. More recently the existence of diffusible morphogens that may promote and inhibit looping has been postulated. Local mechanical conditions are also widely believed to participate in the looping process.

The result of cardiac looping is an S-shaped heart in which the originally caudal inflow part of the heart **(atrium)** becomes positioned dorsal to the outflow tract. In the early heart the outflow tract is commonly called the **bulbus cordis** (Fig. 18-2). In early looping, the cranial limb of the S represents the bulbus cordis, the middle limb represents the ventricular part of the heart, and the caudal limb represents a common atrium. Somewhat later the common atrium bulges on either side (Fig. 18-3), and an internal septum begins to divide the ventricle into right and left chambers. The outflow tract (bulbus cordis of the early heart) retains its gross tubular appearance. Its distal part, which leads directly into the aortic arch system, is called the **truncus arteriosus.** The shorter transitional segment between the truncus and the ventricle is called the **conus arteriosus.** The conus is separated from the ventricles and truncus by faint grooves.

Early Partitioning of the Heart

Early in heart development the atrium becomes partially separated from the ventricle by the formation of thickened **atrioventricular endocardial cushions.** A similar but less pronounced thickening forms at the junction between the ventricle and the outflow tract (Fig. 18-4). In these areas, the cardiac jelly, which is organized like a thick basement membrane, protrudes into the endocardial canal. The endocardial cushions function as primitive valves that assist in the forward propulsion of blood.

Acting in response to an inductive action by the underlying myocardium, cells from the endocardium in the two areas lose their epithelial character and transform into mesenchymal cells, which migrate into the cardiac jelly. Only the endocardial cells in the atrioventricular region and the proximal outflow tract can respond to the induction and

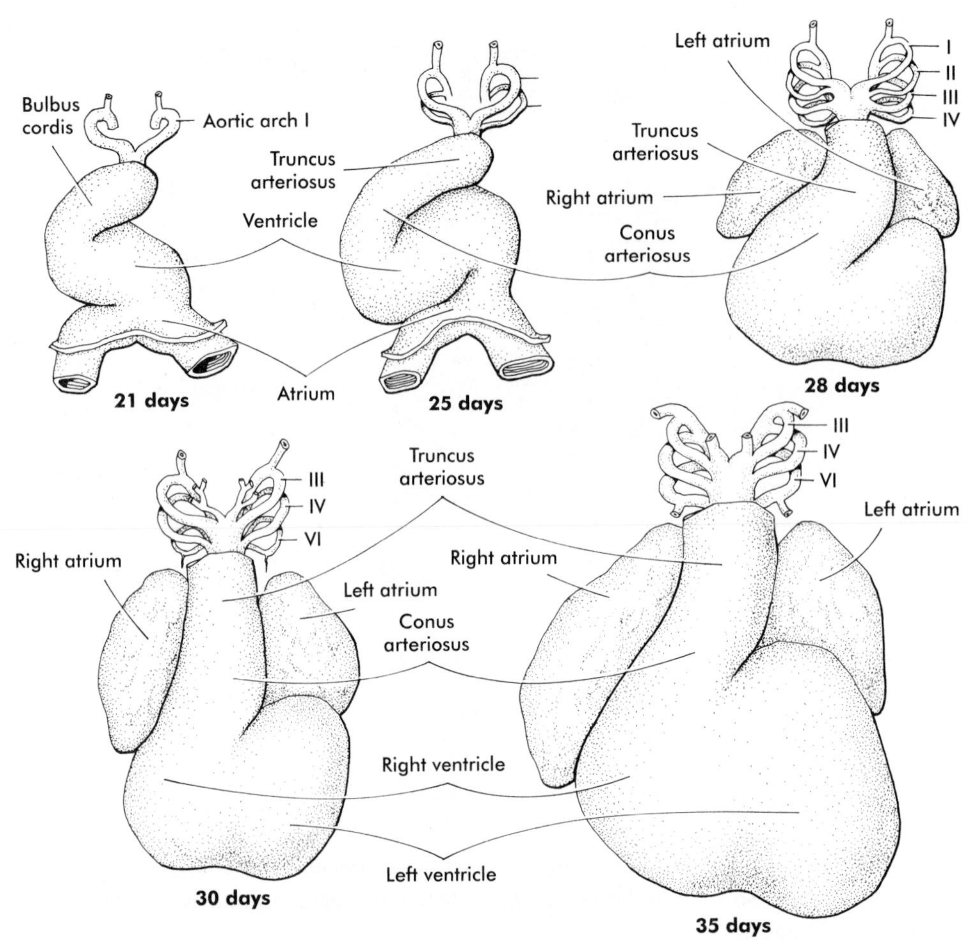

FIG. 18-2 Ventral views of human embryonic hearts, illustrating bending of the cardiac tube and the establishment of its regional divisions.

transform into mesenchyme both in vivo and in vitro. Atrial and ventricular endocardial cells do not lose their epithelial character. The early heart is segmented molecularly, and the segments of endocardium that can undergo a mesenchymal transformation express the *Hox-7* gene, whereas the atrial and ventricular endocardium do not express it. Endocardial cells contain the cell adhesion molecule N-CAM on their surfaces. Those cells that transform into mesenchyme downregulate the production of N-CAM. This presumably facilitates their transformation into mobile cells.

In vivo studies have shown that just before transformation of the endocardial cells into mesenchyme, 20 to 50 nm particles appear in the extracellular matrix (cardiac jelly) beneath these cells but not beneath other areas of endocardium. Produced by the underlying myocardial cells, these particles, called **adherons,** contain a molecular complex of proteoglycan, fibronectin, and a number of proteins (see Fig. 18-4). These particles can induce atrioventricular endothelial cells to transform into mesenchyme in vitro. The induction of invasion of the cardiac jelly by atrioventricular endothelial cells is also accompanied by the activity of **transforming growth factors-β 3.** If this growth factor is inactivated, transformation fails to occur. Some evidence indicates that the transformed mesenchymal cells secrete **proteases,** which could destroy the inductively active adherons and restore morphogenetic stability to the endocardial cushion regions.

These cellular and molecular events form the basis for the early formation of the major heart valves. Disturbances in these processes could account for the genesis of many malformations of the heart.

DEVELOPMENT OF THE VASCULAR SYSTEM

Formation of Embryonic Blood Vessels

The early embryo is devoid of blood vessels. Although blood islands appear in the wall of the yolk sac and extraembryonic vascular channels form in association with them (see Fig. 5-18), much of the vasculature of the embryonic body is derived from intraembryonic sources. During the early period of somite formation, networks of small vessels rapidly appear in many regions of the embryonic body.

Some vascular channels coalesce to form larger vessels, and others remain similar to capillaries or disappear. A fundamental principle of early vasculogenesis is dynamic

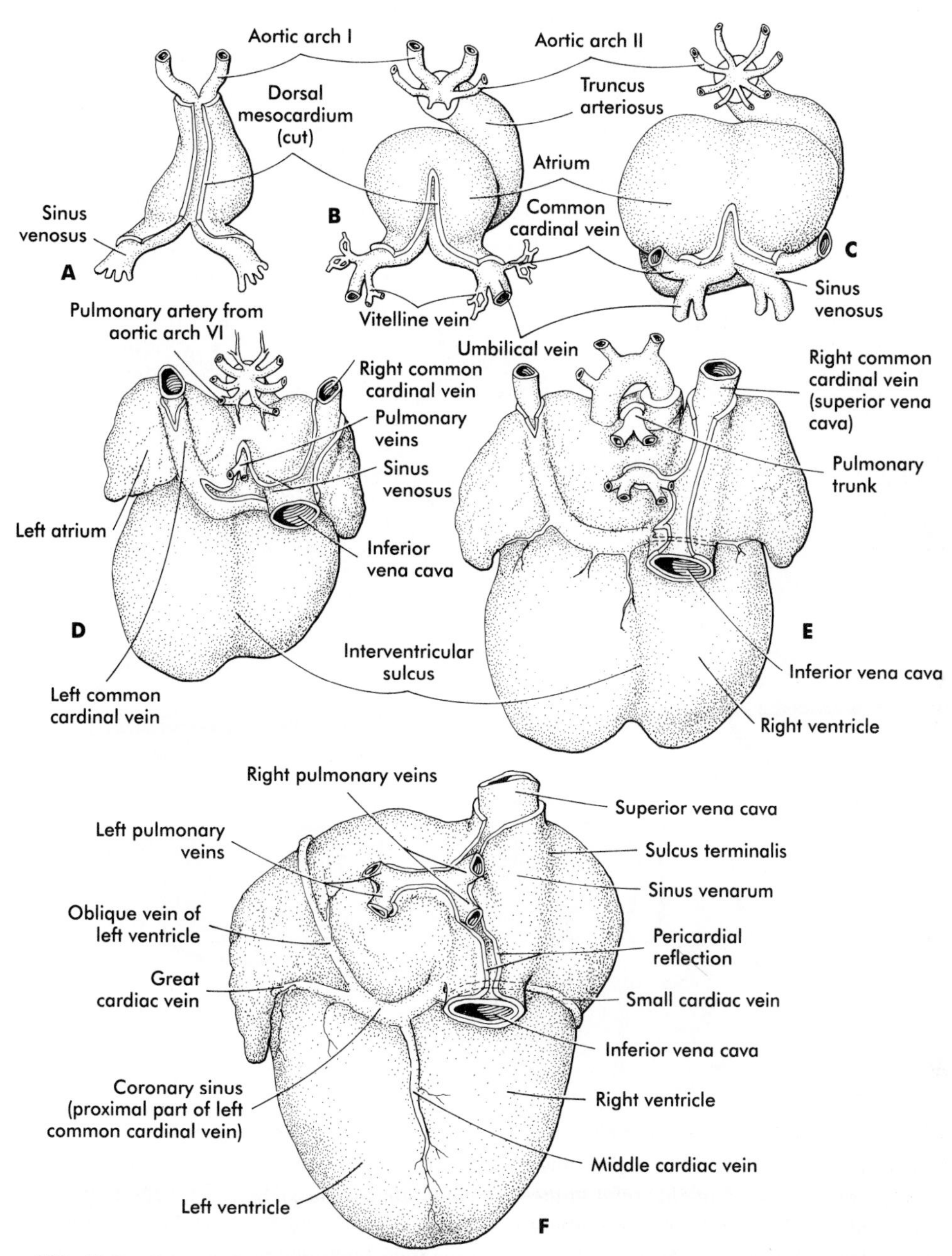

FIG. 18-3 Dorsal views of the development of the human heart, showing changes in the venous input channels into the heart.

change associated with growth of the structures with which the vascular channels are associated. The designation of vessels as arteries and veins is not fixed in the early embryo, since the direction of flow of blood in a given channel can easily be reversed. Although the major patterns of vascular channels are recognizable for a given species, the seemingly haphazard manner of recruiting new capillaries to become components of larger vessels largely accounts for the frequent minor variations in the vascular pattern of an individual. The frequency of anatomical variations is particularly pronounced in the venous system.

Detailed descriptive studies and transplantation experiments involving intrinsic cellular labels or particularly graft-specific monoclonal antibody labels have shown that **angioblasts** (endothelial cell precursors) arise from most mesodermal tissues of the body except for the prechordal mesoderm (Table 18-1). Embryonic blood vessels form from angioblasts by three main mechanisms. Many of the larger blood vessels, such as the dorsal aortae, are formed by the coalescence of angioblasts in situ. Other equally large channels, such as the endocardium, are formed by angioblasts migrating into the region for other sites. Other vessels, especially the intersegmental vessels of the main body axis, arise as vascular sprouts from existing larger ves-

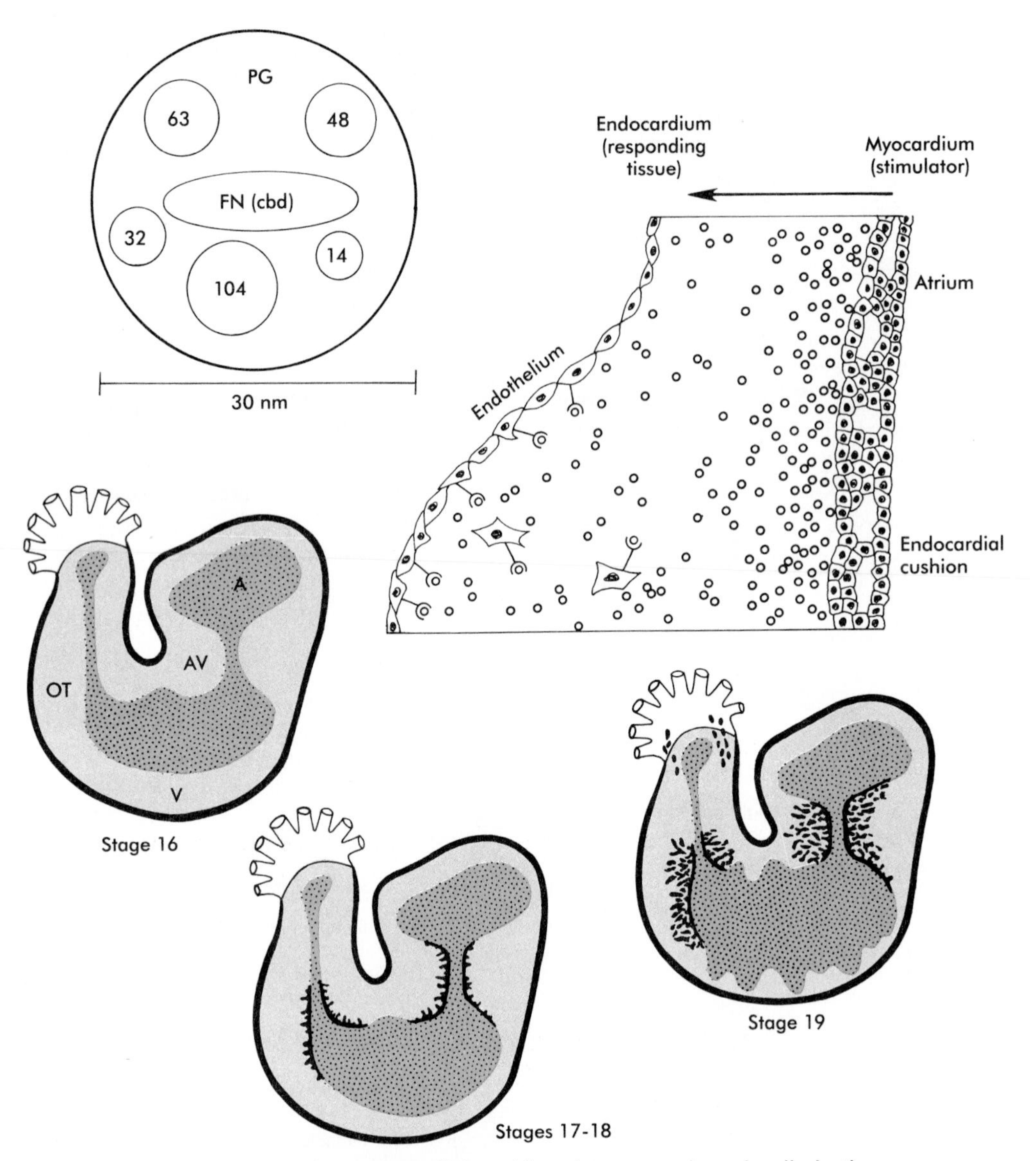

FIG. 18-4 Seeding of the endocardial cushions by mesenchymal cells in the developing avian heart. The higher power drawing in the upper right shows details of the transformation of endocardial to mesenchymal cells. Also shown is a hypothetical model of the 30-nm adheronlike particle that appears to induce this transformation in the endocardial cells. *A,* Atrium; *FN, (cbd),* fibronectin (cell-binding domain); *PG,* proteoglycan; numbered circles indicate other proteins.
(Based on studies by Bolender and Markwald [1991].)

TABLE 18-1 Distribution of Endogenous Angioblasts in Embryonic Tissues

CEPHALIC TISSUES	ANGIOBLASTS	TRUNK TISSUES	ANGIOBLASTS
Paraxial mesoderm	+	Whole somites	+
Lateral mesoderm	+	Dorsal half somites	+
Prechordal mesoderm	–	Segmental plate mesoderm	+
Notochord	–	Lateral somatic mesoderm	+
Brain	–	Lateral splanchnic mesoderm	+
Neural crest	–	Spinal cord	–

From Noden DM: *Ann NY Acad Sci* 588:236-249, 1990.

sels. Many of the angioblasts of the trunk are originally associated with the splanchnic mesoderm.

The walls of blood vessels in most of the trunk and the extremities are derived from local mesoderm that becomes associated with the endothelium of the larger vessels. In the head and many areas of the aortic arch system, mesenchyme derived from neural crest ectoderm constitutes significant portions of the vascular walls (e.g., the smooth muscle cells).

As with myoblasts, angioblasts appear to react to local environmental cues that determine the specific morphological pattern of a blood vessel. Tracing studies of transplanted angioblasts have shown that some can migrate long distances. Angioblasts taken from sites far away from their place of grafting become integrated into morphologically normal blood vessels in the areas in which they settle.

Local factors also influence the initiation of vasculogenesis. In some organs (e.g., the liver) or parts of organs (e.g., the bronchi of the respiratory system), the blood vessels supplying the regions arise from local mesoderm, whereas other organs (e.g., metanephric kidneys) or parts of organs (e.g., the alveolar parts of the lungs) are supplied by blood vessels that grow into the mesenchyme from other tissues. In the latter type of vascularization mechanism, evidence is increasing that the organ primordia produce **angiogenesis factors** that stimulate the growth of vascular sprouts into the glandular mesenchyme. Many putative angiogenesis factors have not been molecularly defined, but **α-fibroblast growth factor** or molecules with regions homologous to regions of this molecule have been shown to have angiogenic activity in a number of developing structures.

Development of the Arteries

The Aortic Arches and Their Derivatives. The system of aortic arches in early human embryos is organized along the same principles as the system of arteries supplying blood to the gills of many aquatic lower vertebrates. Blood exits from a common ventricle in the heart into a ventral aortic root, from which it is distributed through the branchial arches by pairs of aortic arches (Fig. 18-5). In gilled vertebrates the aortic arch arteries branch into a capillary bed where the blood becomes reoxygenated as it passes through the gills. In mammalian embryos the aortic arches remain continuous vessels because gas exchange occurs in the placenta and not in the branchial arches. The aortic arches empty into paired dorsal aortae where the blood enters the regular systemic circulation. In human embryos, all aortic arches are never present at the same time. Their formation and remodeling show a pronounced craniocaudal gradient.

The developmental anatomy of the aortic arch system well illustrates the principle of morphological adaptation of the vascular bed during different stages of embryogenesis (Table 18-2). Continued development of the cranial and cer-

TABLE 18-2 Adult Derivatives of the Aortic Arch System

	RIGHT SIDE	LEFT SIDE
AORTIC ARCHES		
1	Disappearance of most of structure Part of maxillary artery	Disappearance of most of structure Part of maxillary artery
2	Disappearance of most of structure Hyoid and stapedial arteries	Disappearance of most of structure Hyoid and stapedial arteries
3	Ventral part—common carotid artery Dorsal part—internal carotid artery	Ventral part—common carotid artery Dorsal part—internal carotid artery
4	Proximal part of right subclavian artery	Part of arch of aorta
5	Rarely recognizable, even in early embryo	Rarely recognizable, even in early embryo
6 (pulmonary)	Part of right pulmonary artery	Ductus arteriosus Part of left pulmonary artery
VENTRAL AORTIC ROOTS		
Cranial to 3rd arch	External carotid artery	External carotid artery
Between 3rd and 4th arches	Common carotid artery	Common carotid artery
Between 4th and 6th arches	Brachiocephalic artery	Ascending part of aorta
DORSAL AORTIC ROOTS		
Cranial to 3rd arch	Internal carotid artery	Internal carotid artery
Between 3rd and 4th arches	Disappearance of structure	Disappearance of structure
Between 4th and 6th arches	Central part of right subclavian artery	Descending aorta
Caudal to 6th arch	Disappearance of structure	Descending aorta

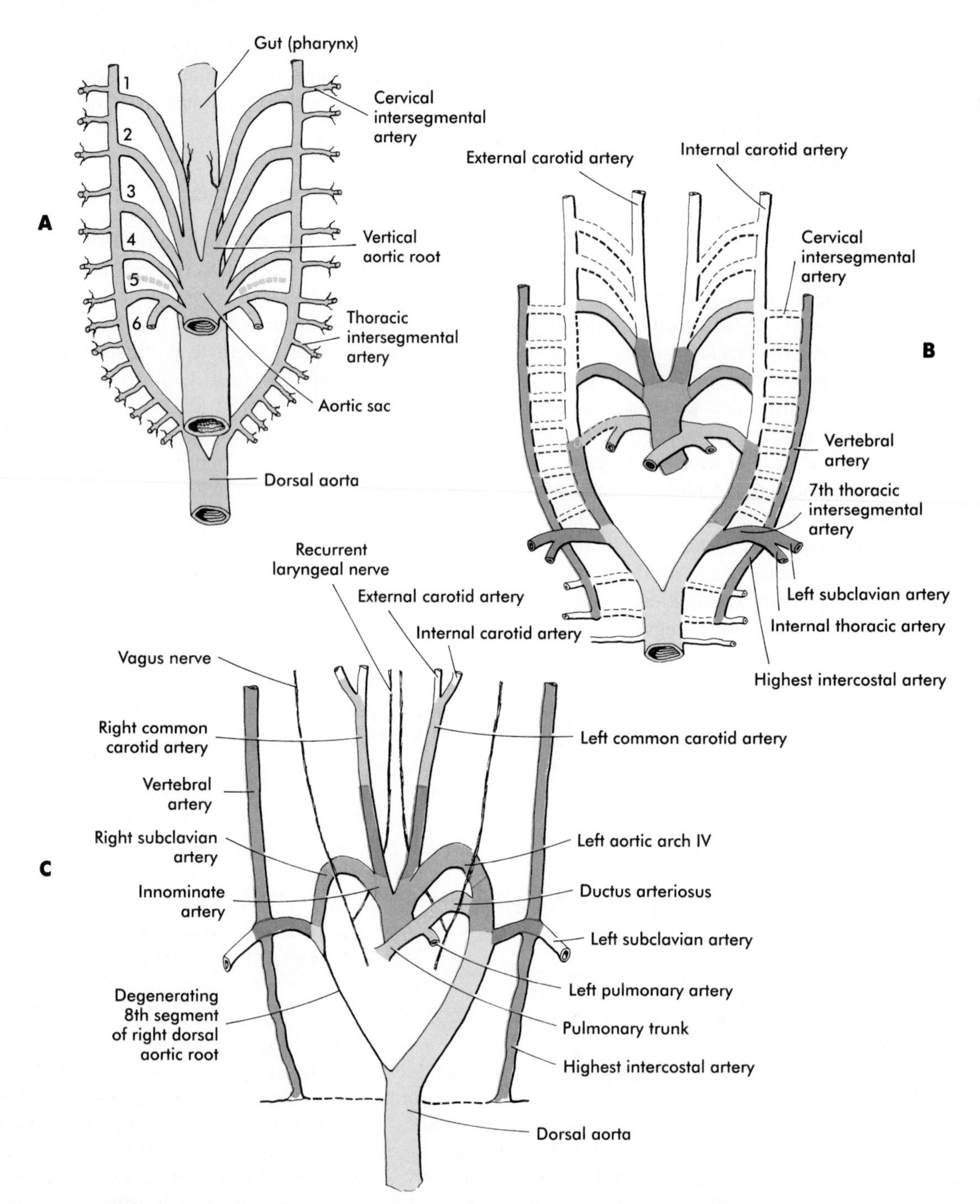

FIG. 18-5 **A,** The embryonic aortic arch system. **B** and **C,** Later steps in the transformation of the aortic arch system in the human. Disposition of the recurrent laryngeal nerve in relation to the right fourth and left sixth arch is also shown in **C.**

vical regions causes components of the first three arches and associated aortic roots to be remodeled into the carotid artery system. (see Fig. 18-5). With the remodeling of the heart tube and the internal division of the outflow tract into aortic and pulmonary components, the fourth arches undergo an asymmetrical adaptation to the early asymmetry of the heart. The left fourth aortic arch is retained as a major channel (arch of the aorta), which carries the entire output from the left ventricle of the heart. The right fourth arch is incorporated into the right subclavian artery.

Embryology textbooks traditionally depict the aortic arch system as consisting of six pairs of vascular arches, but the fifth and sixth arches never appear as discrete vascular channels similar to the first through fourth arches. The fifth aortic arch, if it exists at all, is represented by no more than a few capillary loops. The sixth **(pulmonary arch)** arises as

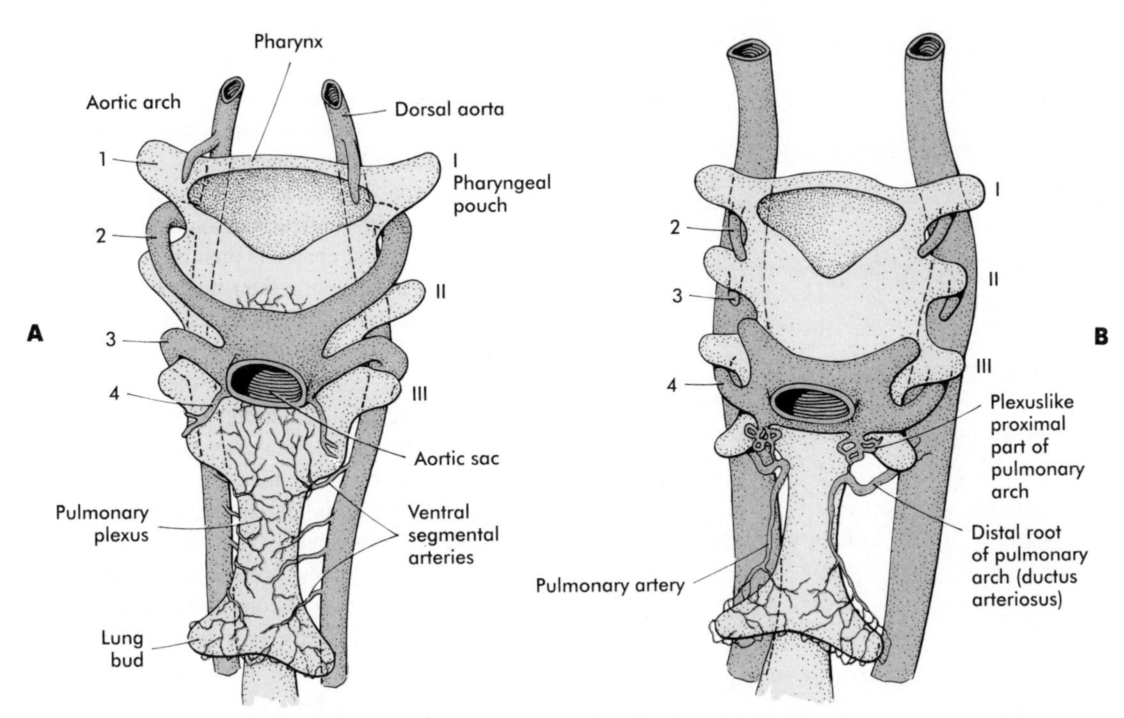

FIG. 18-6 Development of the pulmonary arch, showing the early pulmonary plexus in relation to several ventral segmental arteries associated with the early respiratory diverticulum **(A)** and their consolidation into discrete vessels that establish a connection with the bases of the fourth aortic arches **(B).**
(Based on studies by De Ruiter and others [1990].)

a capillary plexus associated with the early trachea and lung buds. The capillary plexus is supplied by ventral segmental arteries arising from the paired dorsal aortae in that region (Fig. 18-6). The equivalent of the sixth arch is represented by a discrete distal segment (ventral segmental artery) connected to the dorsal aorta and a plexuslike proximal segment that establishes a connection between the aortic sac at the base of the fourth arch and the distal segmental component. As the respiratory diverticulum and early lung buds elongate, parts of the pulmonary capillary network consolidate to form a pair of discrete **pulmonary arteries** that connect to the putative sixth arch. Although the term *sixth aortic arch* is frequently used in anatomical and clinical literature, *pulmonary arch* is a more appropriate term because it does not imply equivalence to the other aortic arches.

Like the fourth aortic arch, the pulmonary arch develops asymmetrically. On the left side it becomes a large channel. Its distal segment, which was derived from a ventral segmental artery, persists as a major channel **(ductus arteriosus)** that shunts blood from the left pulmonary artery to the aorta (see Fig. 18-5). By this shunt the lungs are protected from a flow of blood that is greater than what their vasculature can handle during most of the intrauterine period. On the right side the distal segment of the pulmonary arch regresses, and the proximal segment (the base of the right pulmonary artery) ultimately drains into the pulmonary trunk.

The asymmetry of the derivatives of the pulmonary arch accounts for the difference between the course of the right and left **recurrent laryngeal nerves,** which are branches of the vagus nerve (X). These nerves, which supply the larynx, hook around the pulmonary arches. As the heart descends into the thoracic cavity from the cervical region, the branch point from the vagus of each recurrent laryngeal nerve is correspondingly moved. On the left side the nerve is associated with the ductus arteriosus (see Fig. 18-5), which persists throughout the fetal period, so it is pulled deep into the thoracic cavity. On the right side, however, with the regression of much of the right pulmonary arch, the nerve moves to the level of the fourth arch, which constitutes an anatomical barrier. The positions of the right and left recurrent laryngeal nerves in the adult reflect this asymmetry, with the right nerve curving under the right subclavian artery (fourth arch) and the left nerve hooking around the **ligamentum arteriosum** (the adult derivative of the ductus arteriosus, the distal segment of the left pulmonary arch).

Major Branches of the Aorta. In the early embryo, when the dorsal aortae are still paired vessels, three sets of arterial branches arise from them—**dorsal intersegmental, lateral segmental,** and **ventral segmental** (Fig. 18-7). These branches undergo a variety of modifications in form before assuming their adult configurations (Table 18-3). The ventral segmental arteries arise as paired vessels that

TABLE 18-3 Major Arterial Branches of the Aorta

EMBRYONIC VESSELS	ADULT DERIVATIVES
DORSAL INTERSEGMENTAL BRANCHES (PAIRED)	
Cervical intersegmental (1-16)	Lateral branches joining to become vertebral arteries
7th intersegmentals	Subclavian arteries
Thoracic intersegmentals	Intercostal arteries
Lumbar intersegmentals	Iliac arteries
LATERAL SEGMENTAL BRANCHES	
Up to 20 pairs of vessels supplying the mesonephros	Adrenal arteries, renal arteries, gonadial (ovarian or spermatic)
VENTRAL SEGMENTAL BRANCHES*	
Vitelline vessels	Celiac artery, superior and inferior mesenteric arteries
Allantoic vessels	Umbilical arteries

*Originally paired in areas where aorta is paired.

course over the dorsal and lateral walls of the gut and yolk sac. With the closure of the gut and the narrowing of the dorsal mesentery, certain branches fuse in the midline to form the celiac, superior, and inferior mesenteric arteries.

The **umbilical arteries** begin as pure ventral segmental branches supplying allantoic mesoderm, but their bases later connect with lumbar intersegmental vessels. The most proximal umbilical channels then regress, and the intersegmental branches become their main branches off the aorta. Like their subclavian counterparts in the arms, the initially small arterial branches **(iliac arteries)** supplying the leg buds appear as components of the dorsal intersegmental (lumbar) branches of the aorta. However, after the umbilical arteries incorporate the proximal segments of the intersegmental vessels, the iliac arteries appear to arise as branches off the umbilical arteries (see Fig. 18-20).

Arteries of the Head. The arteries supplying the head arise from two sources. Ventrally, the aortic arch system (first to third arches and corresponding roots) gives rise to the arteries supplying the face **(external carotid arteries)** and the frontal part of the base of the brain **(internal carotid arteries)** (see Fig. 18-5). At the level of the spinal

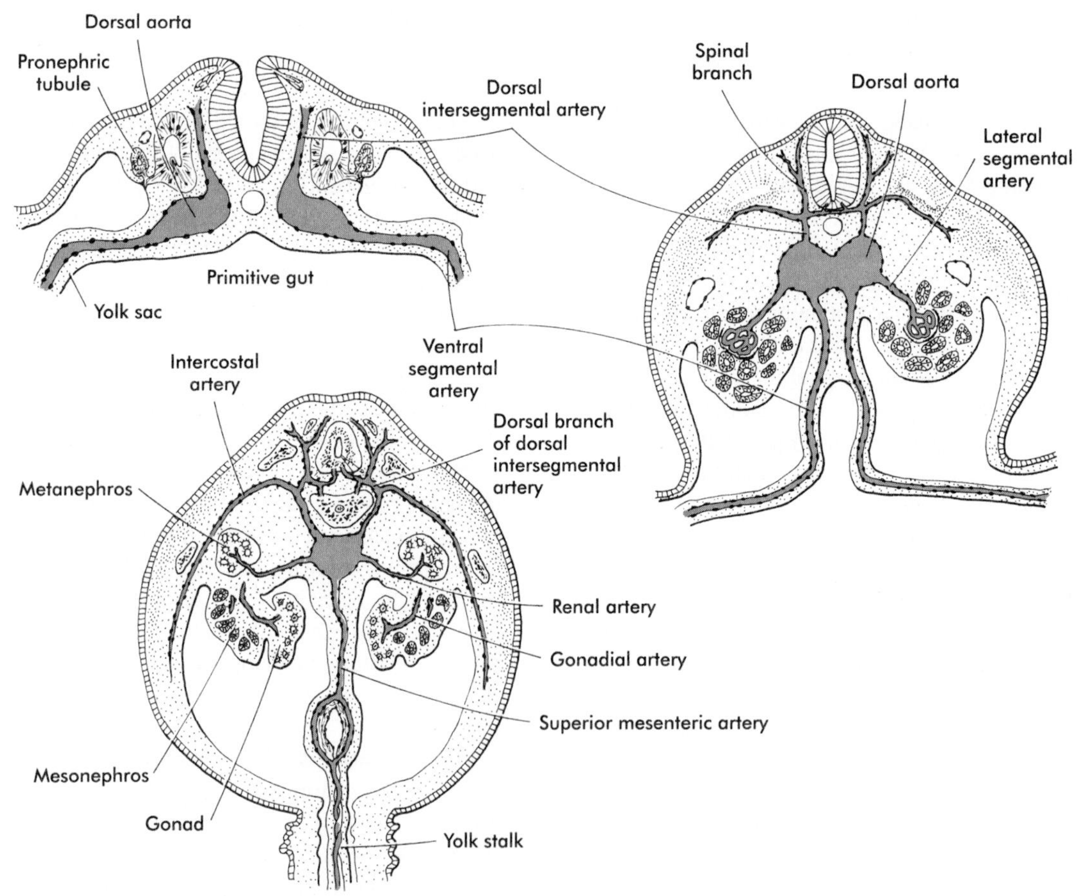

FIG. 18-7 The types of segmental branches coming off the abdominal aorta at different stages of development.

cord, the **vertebral arteries,** which form through connections of lateral branches of the first six dorsal intersegmental arteries, grow toward the brain. Soon they veer toward the midline and fuse, forming the unpaired **basilar artery** (Fig. 18-8). This artery runs along the ventral surface of the brainstem, supplying it with a series of paired arteries. As the basilar artery approaches the level of the diencephalon and the internal carotid arteries, sets of branches from each of these major vessels grow out and fuse, forming **posterior communicating arteries,** which join the circulations of the basilar and internal carotid arteries. Two other small branches off the internal carotid system fuse in the midline to complete a vascular ring **(circle of Willis),** which underlies the base of the diencephalon. The circle of Willis is a structural adaptation that ensures a continuous blood supply in the event of occlusion of some major arteries supplying the brain.

The Coronary Arteries. The origin of the coronary arteries has been controversial, and two main theories have been proposed. According to one, endothelial sprouts grow-

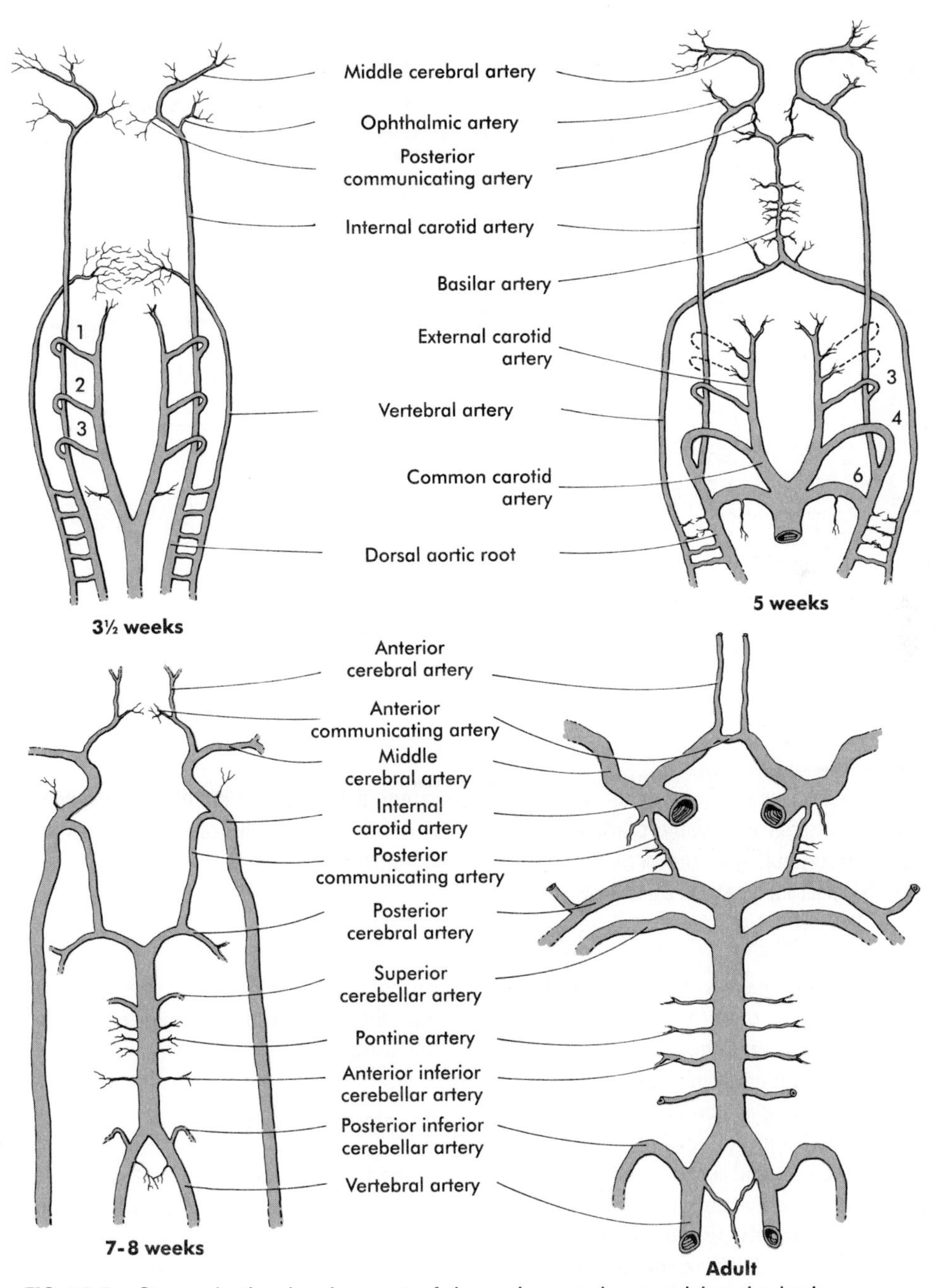

FIG. 18-8 Stages in the development of the major arteries supplying the brain.

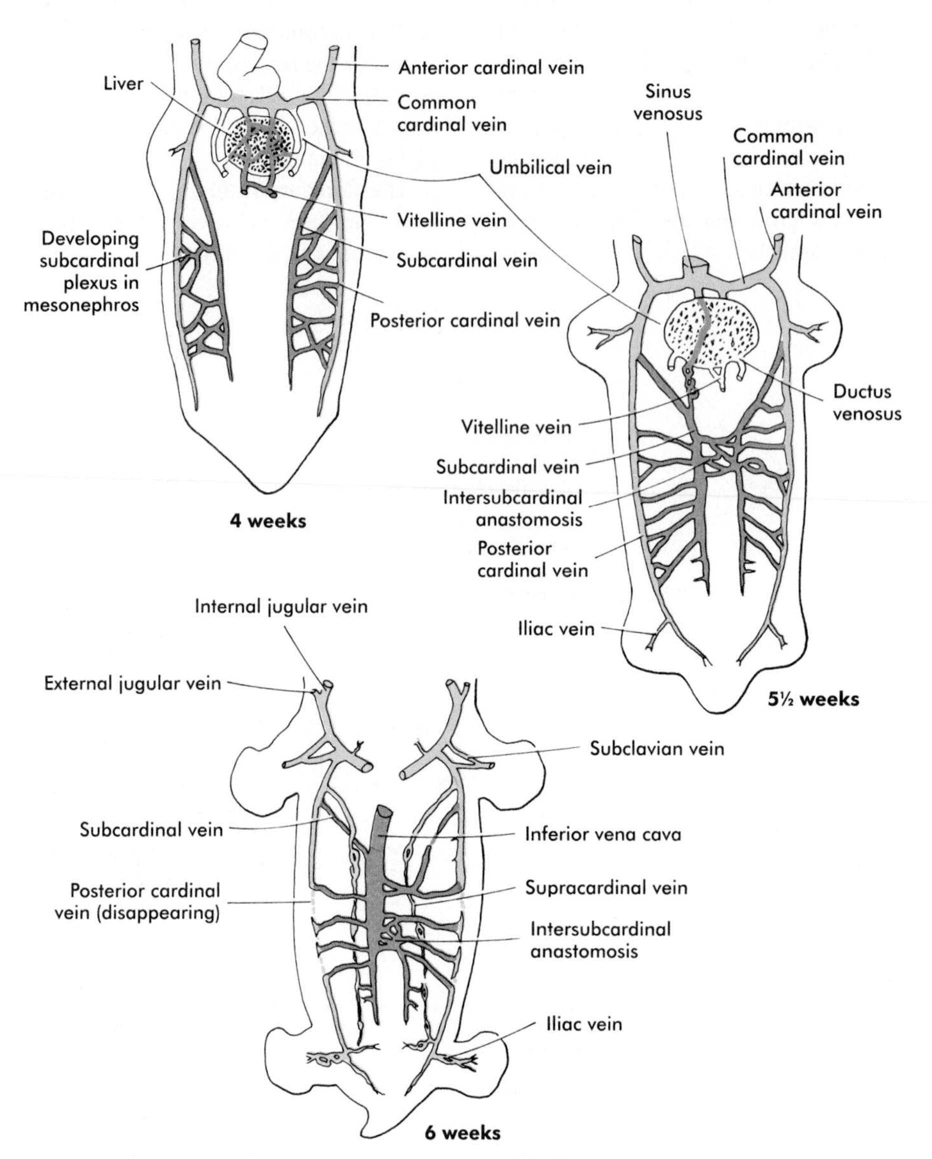

FIG. 18-9 Development of the cardinal vein system in the human embryo. The colors of the original embryonic cardinal veins are carried through in all drawings to facilitate an understanding of the derivations of the adult veins.
(Based on studies by McClure and Butler [1925].)

ing from the base of the aorta make contact with a plexus of vessels in the subepicardial layer of the developing heart. According to the other theory, endothelial sprouts from a capillary plexus associated with the truncus invade the wall of the aorta from without rather than sprouting from the aortic endothelium. The smooth muscle cells of the coronary vessels are of purely mesodermal origin instead of the mixed neural crest and mesodermal origin seen in the aortic arch derivatives. Studies involving the tagging of cells with retroviral markers confirm that progenitor cells of the coronary vessels enter the already beating heart as the epicardial layer envelops the myocardium and that the coronary arteries forming in situ secondarily make contact with the aorta.

Development of Veins

Veins follow a morphologically complex pattern of development characterized by the formation of highly irregular networks of capillaries and the ultimate expansion of certain channels into definitive veins. Because of the multichanneled beginnings and the number of options, the adult

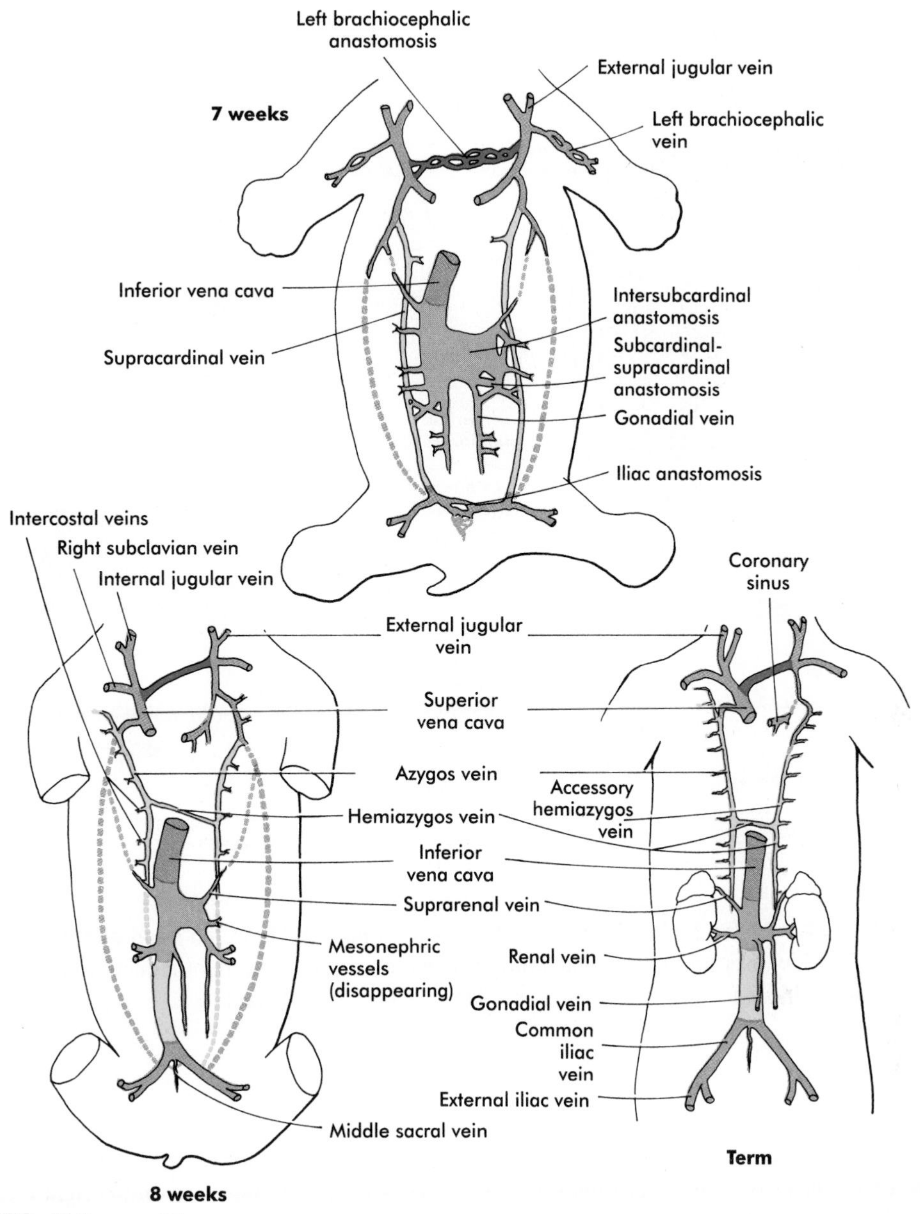

FIG. 18-9—cont'd

venous system is characterized by a higher incidence of anatomical variations than the arterial system (Fig. 18-9). A detailed description of the development of venous channels is beyond the scope of this text.

The Cardinal Veins. The cardinal veins form the basis for the intraembryonic venous circulation. Several sets of cardinal veins appear at different times and in different locations. Within any set of cardinal veins, some segments regress while others persist, either as independent channels or as components of composite veins that also include portions of other cardinal veins.

The earliest pattern of cardinal veins consists of paired **anterior** and **posterior cardinal veins,** which drain blood from the head and body into a pair of short **common cardinal veins** (see Fig. 18-9). The common cardinal veins, in turn, empty their blood into the **sinus venosus** of the primitive heart (see Fig. 18-11).

In the cranial region the originally symmetrical anterior cardinal veins transform into the **internal jugular veins** (Fig. 18-10). As the heart rotates to the right, the base of the left internal jugular vein is attenuated. At the same time a new anastomotic channel, which will ultimately form the

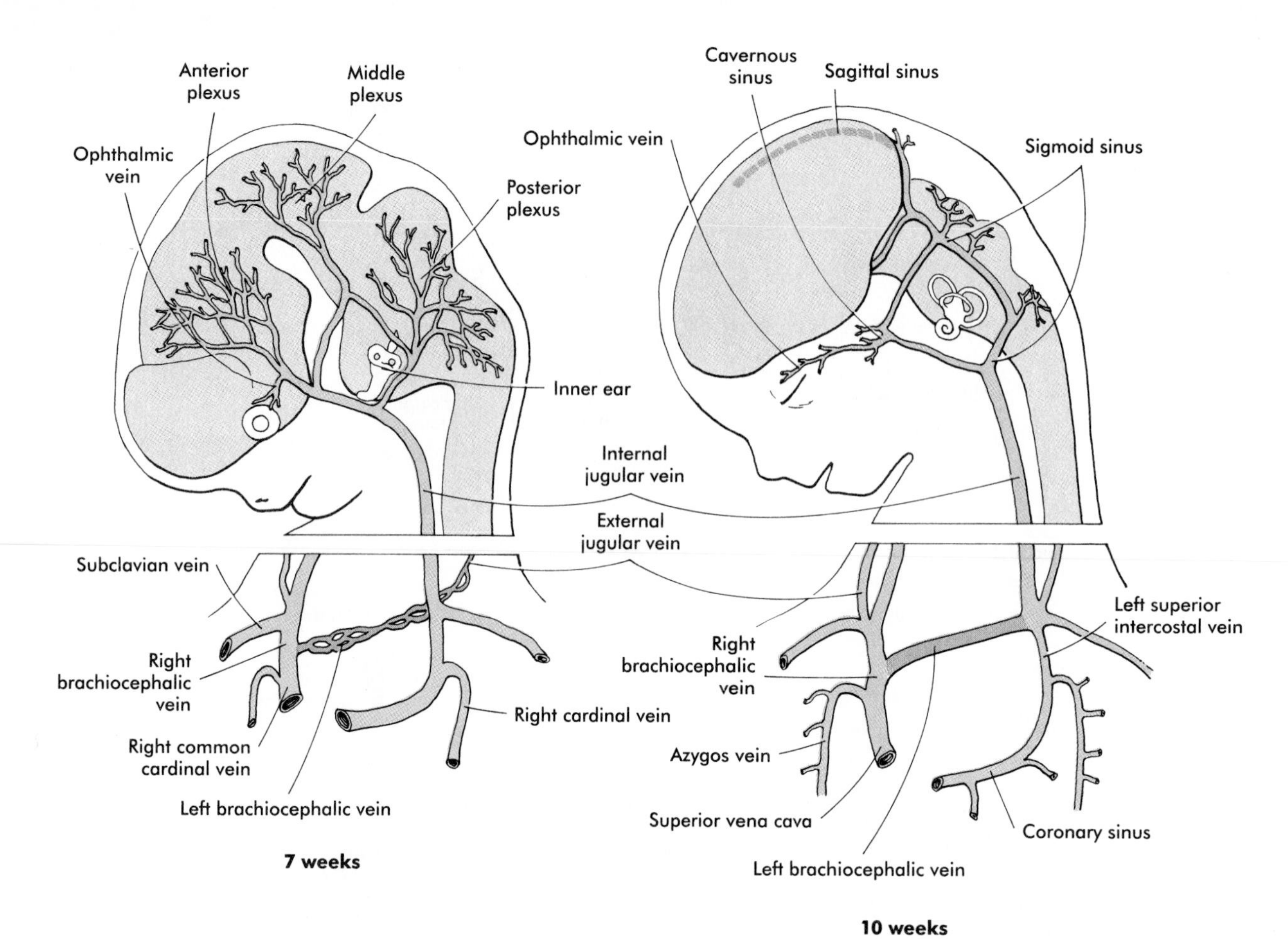

FIG. 18-10 Stages in the formation of the major veins that drain the head and upper trunk. At 7 weeks an anastomosis *(purple)* between the left and right jugular vein forms the basis for establishment of the left brachiocephalic vein.

left brachiocephalic vein, connects the left internal jugular vein to the right one. Through this anastomosis the blood from the left side of the head is drained into the original right anterior cardinal vein, which ultimately becomes the **superior vena cava,** emptying into the right atrium of the heart. Meanwhile, the proximal part of the left anterior cardinal vein persists as a small channel, the **coronary sinus,** which is the final drainage pathway of many of the coronary veins, also into the right atrium of the heart.

In the trunk a pair of **subcardinal veins** arises in association with the developing mesonephros. The subcardinal veins are connected with the posterior cardinal veins and to each other through numerous anastomoses. Both the postcardinal and subcardinal veins drain the mesonephric kidneys through numerous small side branches. As the mesonephric kidneys begin to regress, the veins draining them also begin to break up. At this point a third pair of **supracardinal veins** appear in the body wall dorsal to the subcardinal veins. Over time, all three sets of cardinal veins in the body break up to varying degrees, with surviving remnants incorporated into the **inferior vena cava.** The inferior vena cava forms as a single asymmetrical vessel that runs parallel to the aorta on the right (see Fig. 18-9). Most of the named veins of the thoracic and abdominal cavities are derived from persisting segments of the cardinal vein system.

The Vitelline and Umbilical Veins. The extraembryonic **vitelline** and **umbilical veins** begin as pairs of symmetrical vessels that drain separately into the sinus venosus of the heart (Fig. 18-11). Over time these vessels become intimately associated with the rapidly growing liver. The vitelline veins, which drain the yolk sac, develop sets of anastomosing channels both within and outside the liver. Outside the liver the two vitelline veins and their side-to-side anastomotic channels become closely associated with the duodenum. Through the persistence of some channels and the disappearance of others the **hepatic portal vein,** which drains the intestines, takes shape. Within the liver the vitelline plexus transforms into a capillary bed, allowing the broad distribution of food materials absorbed from

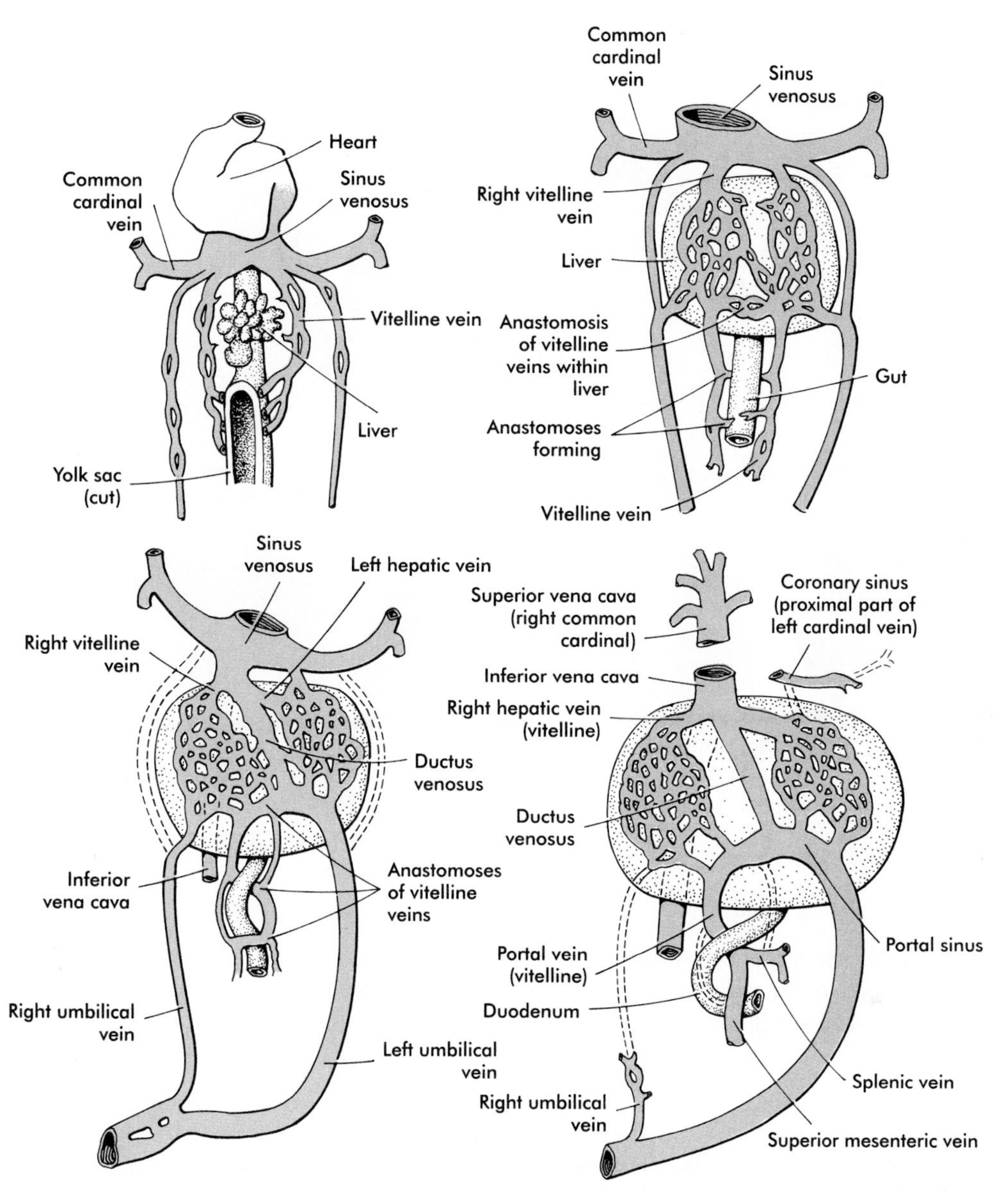

FIG. 18-11 Stages in the development of the umbilical and hepatic portal veins and intrahepatic circulation.

the gut through the functional parts of the liver. From the hepatic capillary bed the blood that arrives from the hepatic portal vein passes into a set of **hepatic veins,** which empty the blood into the sinus venosus.

The originally symmetrical umbilical veins soon lose their own hepatic segments and drain directly into the liver by combining with the intrahepatic vascular plexus of the vitelline veins. Soon a major channel, the **ductus venosus,** forms and shunts much of the blood entering from the left umbilical vein directly through the liver and into the inferior vena cava. The ductus venosus is an important adaptation for maintaining a functional embryonic pattern of blood circulation. Soon the right umbilical vein degenerates, leaving the left umbilical vein the sole channel for bringing blood that has been reoxygenated and purified in the placenta back to the embryonic body. The ductus venosus permits the incoming oxygenated placental blood to bypass the capillary networks of the liver and to distribute it to the organs (e.g., brain, heart) that need it most.

The Pulmonary Veins. The pulmonary veins are phylogenetically recent structures that form independently rather than taking over portions of the older cardinal vein systems. From each lung, venous drainage channels converge until they ultimately form a single large **common pulmonary vein,** which empties into the left atrium of the heart. As the atrium expands, the common pulmonary vein

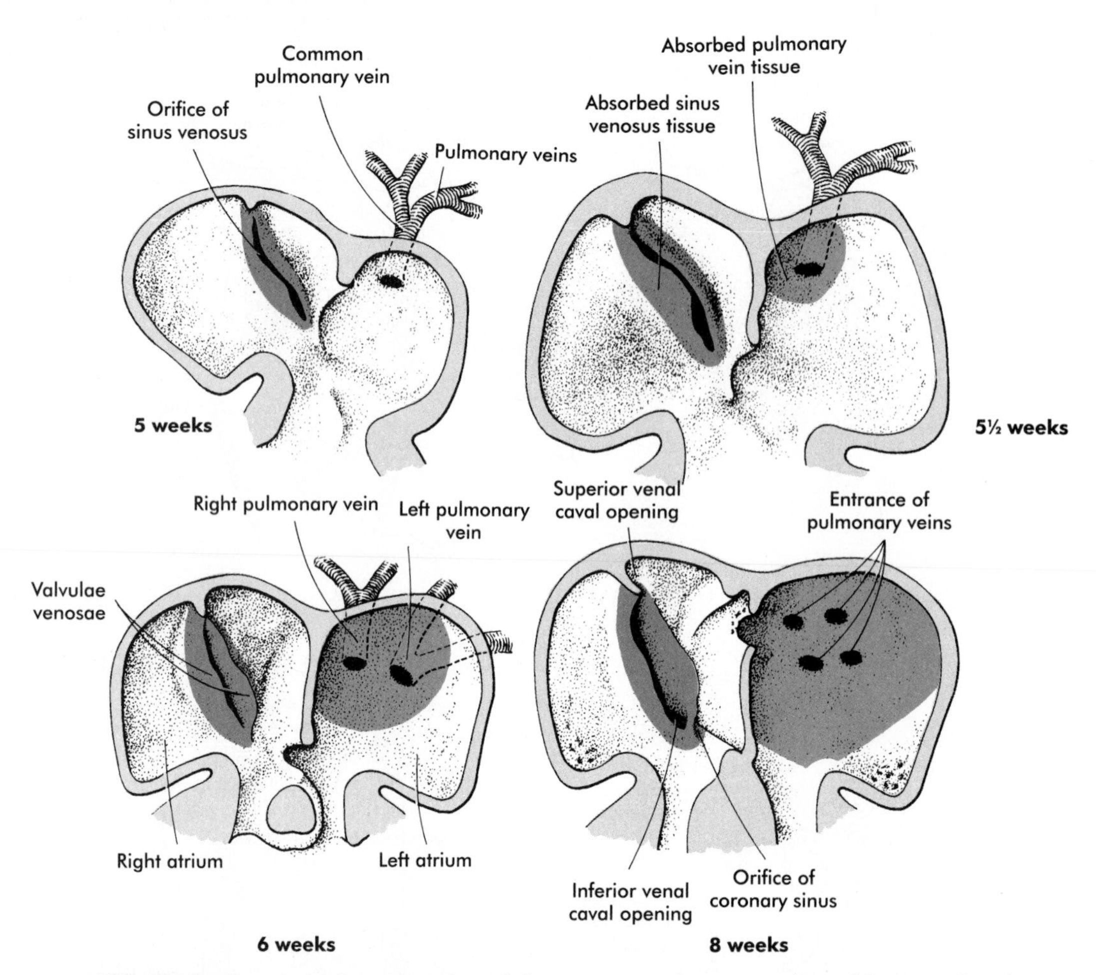

FIG. 18-12 Stages in the absorption of the common pulmonary vein and its branches into the wall of the left atrium and changes over time in the opening of the sinus venosus into the right atrium.

becomes incorporated into its wall (Fig. 18-12). Ultimately the absorption passes the first and second branch points of the original pulmonary veins, resulting in the entrance of four independent pulmonary veins into the left atrium.

Development of Lymphatic Channels

The origin of the lymphatic channels has been poorly understood, with one viewpoint supporting their origin from venous endothelium and another attributing their origin to local mesoderm. Regardless of origin, the lymphatic system first appears as six **primary lymph sacs** starting late in the sixth week of pregnancy (Fig. 18-13). Two **jugular lymph sacs** appear at the angle between the precardinal (future internal jugular) veins and the subclavian veins. In the abdomen a **retroperitoneal lymph sac** forms on the posterior body wall at the root of the mesentery during the eighth week. Somewhat later a **cysterna chyli** forms at the same level but dorsal to the aorta. At about the same time a pair of **posterior lymph sacs** arises at the bifurcation of the femoral and sciatic veins. By the end of the ninth week, lymphatic vessels connect these lymph sacs.

Two major lymphatic vessels connect the cysterna chyli with the jugular lymph sacs. An anastomosis forms between these two channels. A single lymphatic vessel consisting of the caudal part of the right channel, the anastomotic segment, and the cranial part of the left channel ultimately becomes the definitive **thoracic duct** of the adult. The thoracic duct drains lymph from most of the body and the left side of the head into the venous system at the junction of the left internal jugular and subclavian veins. The right lymphatic duct, which drains a smaller area of the cranial part of the body, also empties into the venous system at the original location of the right jugular sac.

LATER DEVELOPMENT AND PARTITIONING OF THE HEART

Partitioning of the Heart

Separation of the Atria from the Ventricles. The endocardial cushions (see Fig. 18-4), which ultimately become transformed into dense connective tissue, form on the dorsal and ventral walls of the atrioventricular canal. As they grow into the canal, the two cushions meet and separate

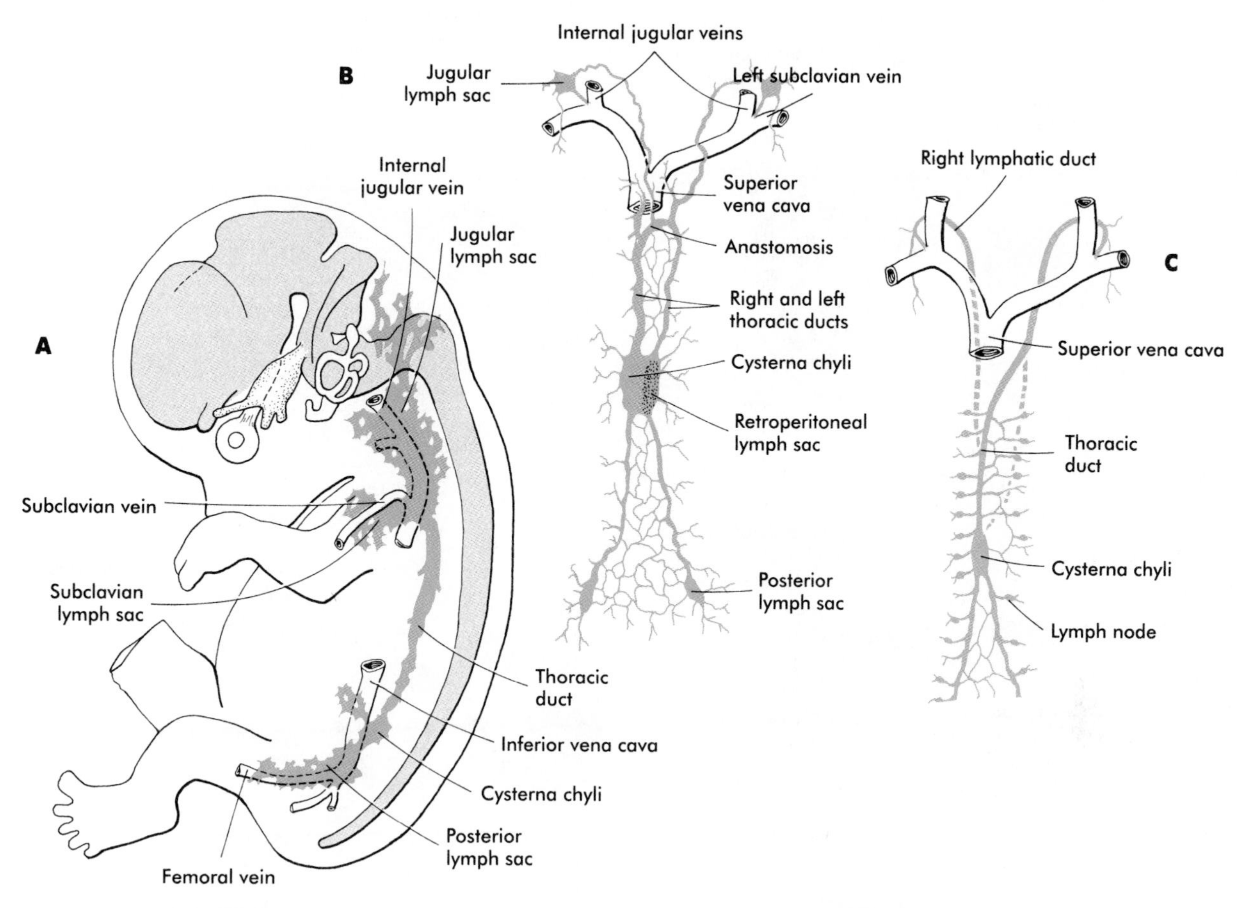

FIG. 18-13 Stages in the development of the major lymphatic channels. **A** and **B** show 9-week-old embryos. **C,** Fetal period. Between **B** and **C** the transformation between the reasonably symmetrical disposition of the main lymphatic channels and the asymmetrical condition characteristic of the adult can be seen.

the atrioventricular canal into right and left channels (Figs. 18-14 and 18-15). The early endocardial cushions serve as primitive valves that assist in the forward propulsion of blood through the heart. Later in development, thin leaflets of anatomical valves take shape in the atrioventricular canal. The definitive valve leaflets do not appear to come from endocardial cushion tissue as much as from invagination of superficial tissues of the atrioventricular groove. The valve that protects the right atrioventricular canal develops three leaflets **(tricuspid valve),** but the valve in the left canal **(mitral,** or **bicuspid valve)** develops only two.

Partitioning of the Atria. While the atrioventricular canals are taking shape, a series of structural changes divides the common atrium into separate left and right chambers. Partitioning begins in the fifth week with the downgrowth of a crescentic **interatrial septum primum** from the cephalic wall of the common atrium (see Fig. 18-15). The apices of the crescent of the septum primum extend toward the atrioventricular canal and merge with the endocardial cushions. The space between the leading edge of the septum primum and the endocardial cushions is called the **interatrial foramen primum.** This space serves as a shunt permitting blood to pass directly from the right to the left atrium.

Circulatory shunts in the developing heart satisfy a very practical need. All incoming blood enters to the right side of the interatrial septum primum. However, because of the late development of the lungs and the poor carrying capacity of the pulmonary vessels during most of the fetal period, the pulmonary circulation cannot handle a full load of blood. If the heart were to form four totally separate chambers from the beginning, the pulmonary circulation would be overstressed and the left side of the heart would not be pumping enough blood to foster normal development, especially in the early weeks.

The problem of maintaining a balanced circulatory load on all chambers of the heart is met by the existence of two shunts that allow most of the circulating blood to bypass the lungs. One shunt is a direct connection between the right and left atria, allowing blood entering the right atrium to bypass the pulmonary circulation completely by passing directly into the left atrium. This shunt permits the normal functional development of the left atrium. If all the blood entering the right atrium, however, passed directly into the

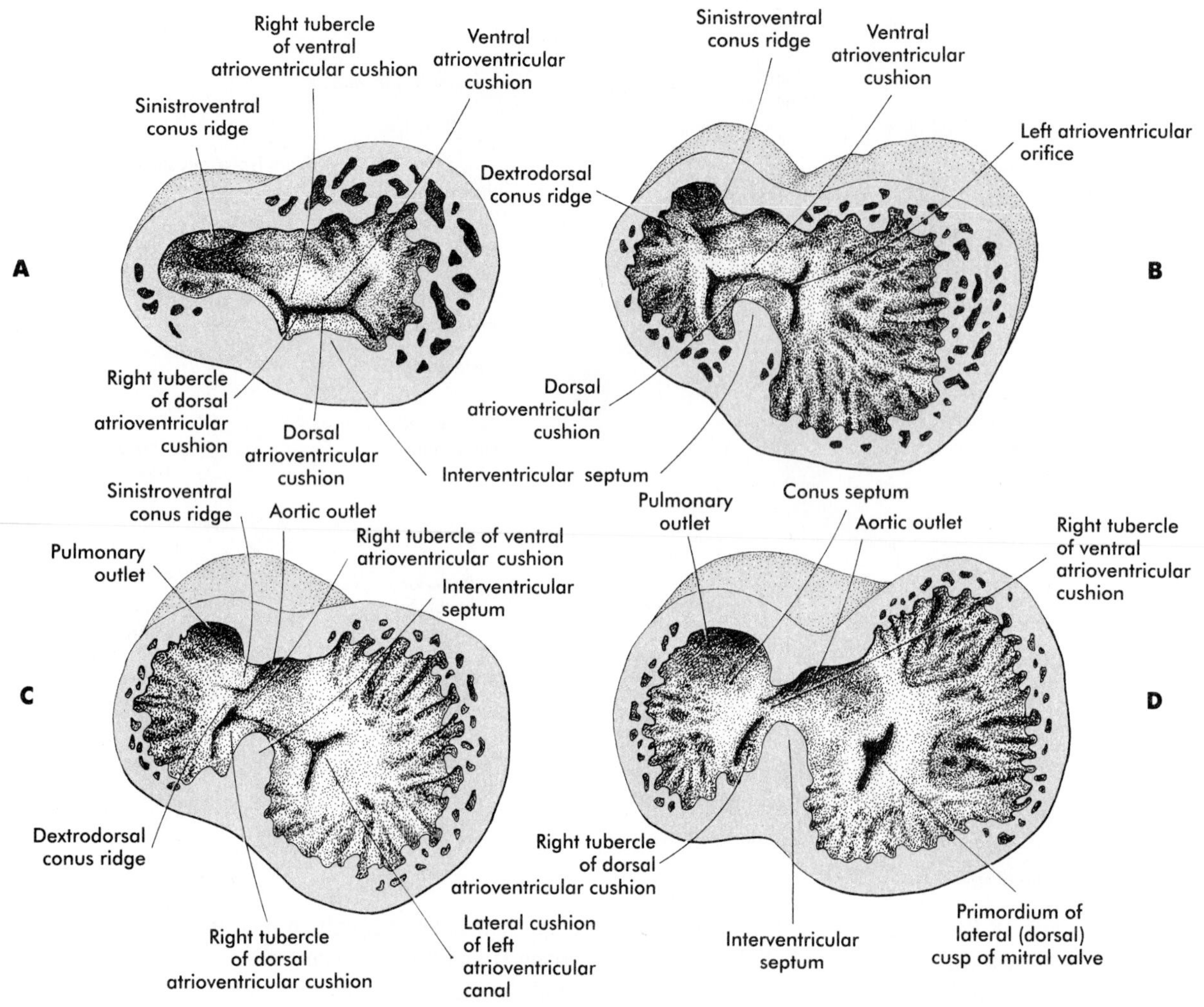

FIG. 18-14 Stages in partitioning the atrioventricular canal from the ventricular aspect.

left atrium, the right ventricle would have nothing to pump against and would become hypoplastic. With the arrangement of the openings of vascular channels into the right atrium, a significant amount of blood also enters the right ventricle and leaves that chamber through the pulmonary outflow tract. The blood leaving the right ventricle, which is far too much to be accommodated by the vasculature of the lungs, bypasses the lungs via the ductus arteriosus and empties directly into the descending aorta. By these two mechanisms, the heart is evenly exercised and the pulmonary circulation is protected.

When the interatrial septum primum is almost ready to fuse with the endocardial cushions, an area of genetically programmed cell death causes the appearance of multiple perforations near its cephalic end (see Fig. 18-15). As the leading edge of the septum primum fuses with the endocardial cushions, obliterating the foramen primum, the cephalic perforations in the septum primum coalesce and give rise to the **interatrial foramen secundum.** This new foramen preserves the direct connection between the right and left atria.

Shortly after the appearance of the foramen secundum, a crescentic **septum secundum** begins to form just to the right of the septum primum. This structure, which grows out from the dorsal to the ventral part of the atrium, forms a **foramen ovale.** The position of the foramen ovale allows most of the blood that enters the right atrium through the inferior vena cava to pass directly through it and the foramen secundum into the left atrium. However, the arrangement of the two interatrial septa allows them to act like a one-way valve, permitting blood to flow from the right to the left atrium but not in the reverse direction.

Repositioning of the Sinus Venosus and the Venous Inflow into the Right Atrium. During the stage of the straight tubular heart, the sinus venosus is a bilaterally symmetrical chamber into which the major veins of the body empty (see Fig. 18-11). As the heart undergoes looping and the interatrial septa form, the entrance of the sinus venosus

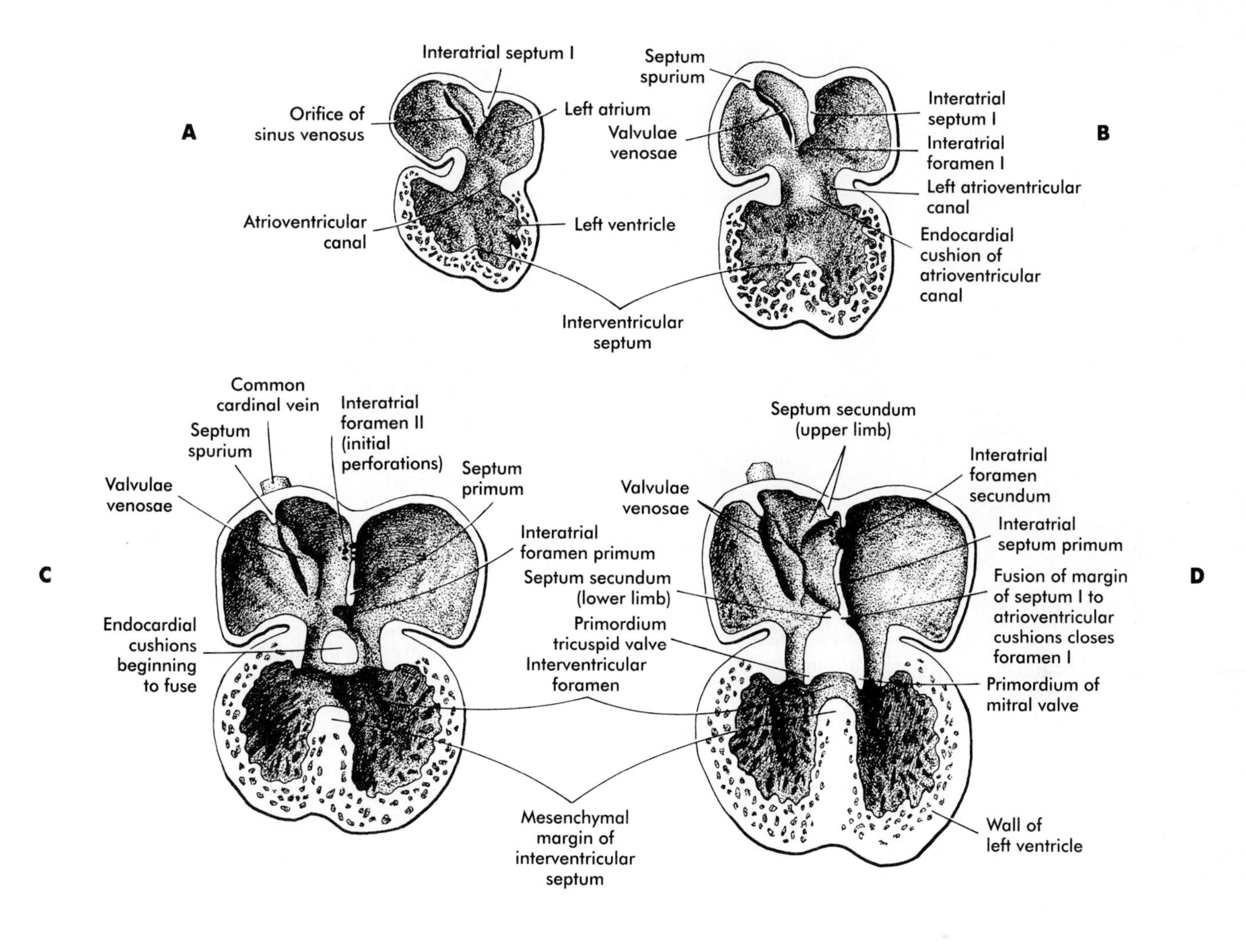

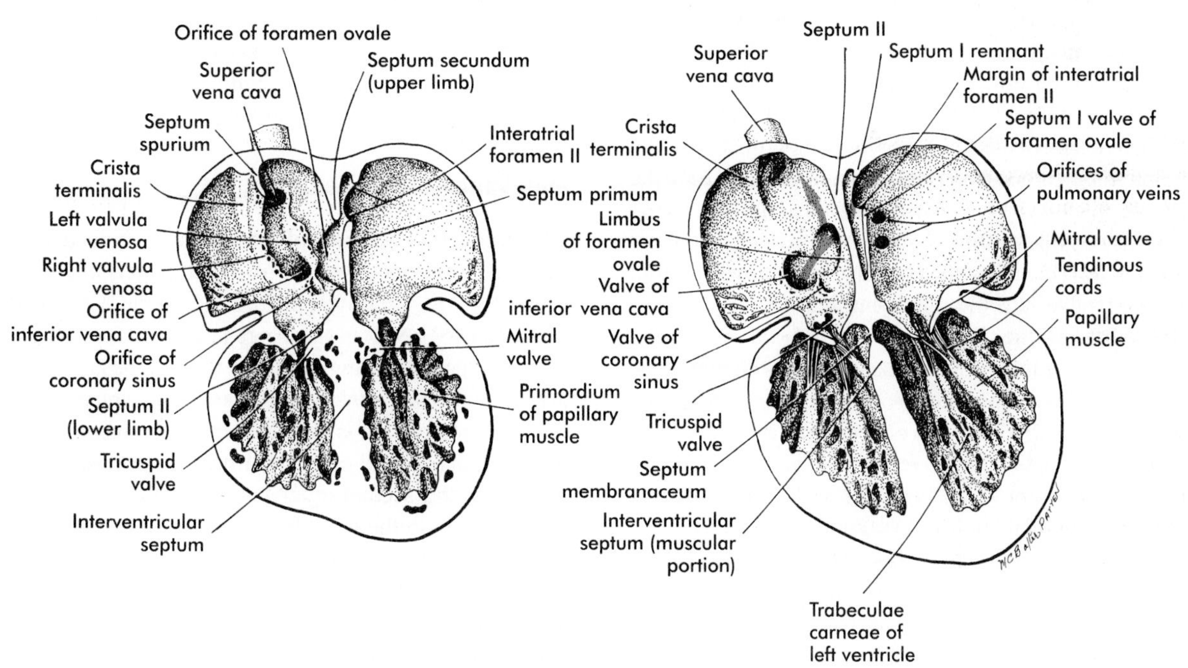

FIG. 18-15 Stages in the internal partitioning of the heart. (After Patten.)

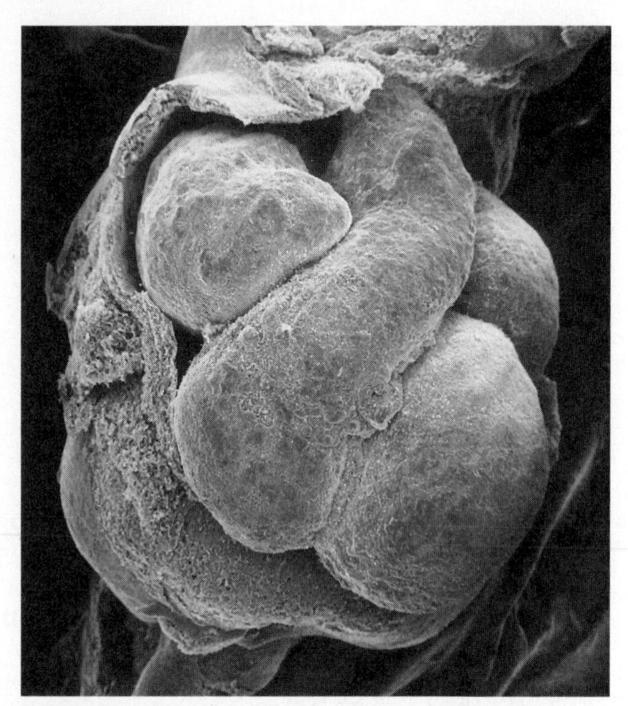

FIG. 18-16 Scanning electron micrograph showing a right oblique view of the heart of a human embryo early in the sixth week. The pericardium has been dissected free from the heart.

(From Jirasek J: *Atlas of human prenatal morphogenesis,* Boston, 1983, Martinus Nijhoff Publishers.)

shifts completely to the right atrium (see Figs. 18-3 and 18-15). As this occurs, the right horn of the sinus venosus becomes increasingly incorporated into the wall of the right atrium, so the much reduced left horn, the **coronary sinus** (which is the common drainage channel for the coronary veins), opens directly into the right atrium (see Fig. 18-12). Also in the right atrium, valvelike flaps of tissue **(valvulae venosae)** form around the entrances of the superior and inferior vena cavae. Because of the orientation of the orifice and its pressure, blood entering the right atrium from the inferior vena cava passes mostly through the interatrial shunt and into the left atrium, whereas blood entering from the superior vena cava and the coronary sinus flows through the tricuspid valve into the right ventricle.

Partitioning of the Ventricles. When the interatrial septa are first forming, a muscular **interventricular septum** begins to grow from the apex of the common ventricle toward the atrioventricular endocardial cushions. The early division of the common ventricle is also reflected by the presence of a groove on the outer surface of the heart (Fig. 18-16). Although an **interventricular foramen** is initially present, it is ultimately obliterated. This is accomplished by (1) further growth of the muscular interventricular septum, (2) the addition to its leading edge of ridge tissue that divides the outflow tract of the heart, and (3) a membranous component derived from endocardial cushion connective tissue.

Partitioning of the Outflow Tract of the Heart. In the very early tubular heart, the outflow tract is a single tube—the bulbus cordis. By the time the interventricular septum begins to form, the bulbus elongates and can be divided into a proximal conus arteriosus and a distal truncus arteriosus (see Fig. 18-2). Although initially a single channel, the outflow tract is partitioned into separate aortic and pulmonary channels through the appearance of two spiral **truncoconal ridges,** which are derived largely from neural crest mesenchyme. These ridges bulge into the lumen and finally meet, separating it into two channels. Partitioning of the outflow tract begins near the ventral aortic root between the fourth and sixth arches and extends toward the ventricles, spiraling as it goes (Fig. 18-17). This accounts for the partial spiraling of the aorta and the pulmonary artery in the adult heart.

Before and during the partitioning process, the neural crest–derived cells of the wall of the outflow tract begin to produce elastic fibers, which provide the resiliency required of the aorta and other great vessels. Elastogenesis follows a gradient, first through the outflow tract, then into the aorta itself, and ultimately into the smaller arterial branches off the aorta.

At the base of the conus, where endocardial cushion tissue is formed in the same manner as in the atrioventricular canal, two new sets of **semilunar valves** form (Fig. 18-18). These valves, each of which has three leaflets, prevent ejected blood from washing back into the ventricles. Cranial neural crest cells and cardiac mesoderm are said to contribute to the formation of the semilunar valves, although not all investigators agree on the neural crest contribution to the valves themselves. As previously stated, the most proximal extensions of the truncoconal ridges contribute to the formation of the interventricular septum. Just past the aortic side of the aortic semilunar valve, the two coronary arteries join the aorta to supply the heart with blood.

Innervation of the Heart

Although initial heart development occurs independently of nerves, three sets of nerve fibers ultimately innervate the heart (Fig. 18-19). Sympathetic (adrenergic) nerve fibers, which act to speed up the heart beat, arrive as outgrowths from sympathetic ganglia of the trunk. These nerve fibers are derived from trunk neural crest. Parasympathetic (cholinergic) innervation is derived from the cardiac component of the cranial neural crest. Neurons of the cardiac ganglia, which are the second-order parasympathetic neurons, migrate directly to the heart from the neural crest. These synapse with axons of first-order parasympathetic neurons that gain access to the heart via the vagus nerve. Sensory innervation of the heart is also supplied via the vagus nerve, but the sensory neurons originate from placodal ectoderm (nodose placode) (see Fig. 5-9). Thus the innervation of the heart has three separate origins.

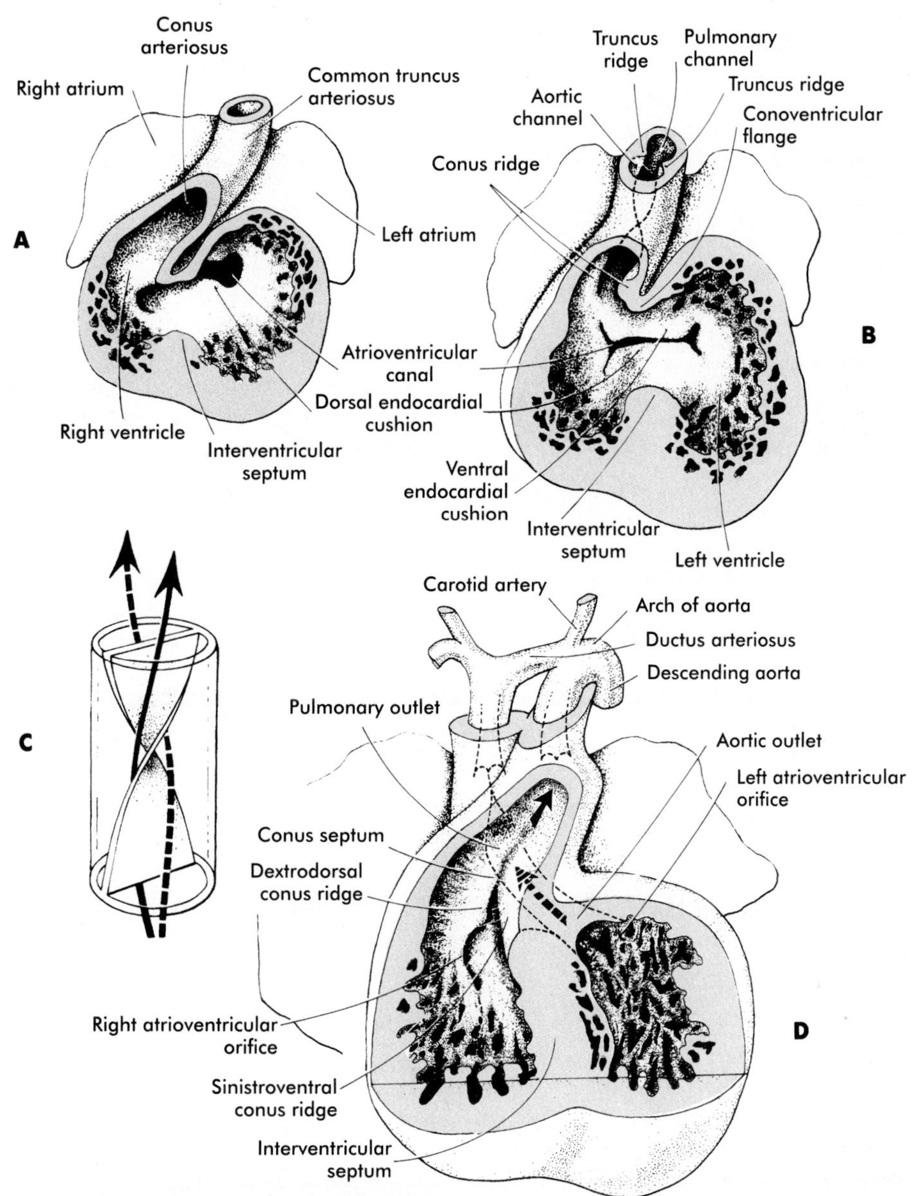

FIG. 18-17 Partitioning of the outflow tract of the developing heart. The truncoconal ridges undergo an 180-degree spiraling.
(After Kramer.)

If the cardiac neural crest is removed in the early chick embryo, cholinergic cardiac ganglia still form. Experiments have determined that the nodose placodes compensate for the loss of neural crest by supplying neurons that take the place of the normal parasympathetic ones.

The Conducting System of the Heart

The normal heart beat is a reflection of a complex of internal pacemakers and a conducting system that rapidly distributes the contractile stimulus throughout the heart. The input of the autonomic nerves, which modulate the heart beat to a faster (sympathetic) or slower (parasympathetic) rate, is also involved.

In very early heart development the location of the pacemaker shifts from the caudal-most end of the left tube of the unfused heart to the sinus venosus. As the sinus venosus is incorporated into the right atrium, the pacemaker, now called the **sinoatrial node,** becomes situated high in the right atrium, close to the entrance of the superior vena cava (see Fig. 18-15). Somewhat later, an **atrioventricular node** forms in the interatrial septal area just above the endocardial cushion tissue. These two nodes are connected by several strands of highly modified cardiac muscle cells that conduct the contractile stimulus from the sinoatrial node to the atrioventricular node. From the atrioventricular node a well-defined **atrioventricular bundle** passes from

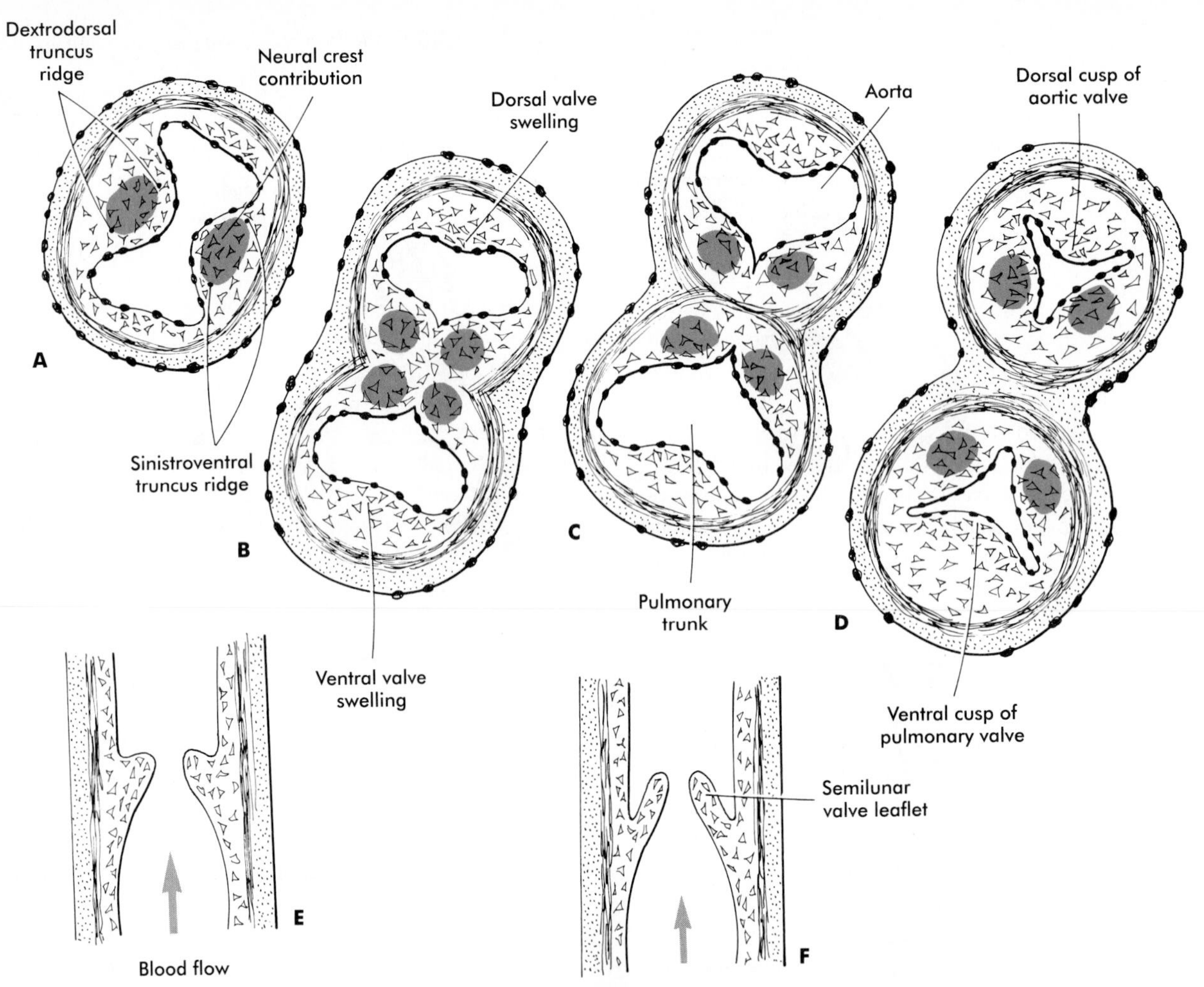

FIG. 18-18 Formation of the semilunar valves in the outflow tract of the heart. Neural crest cells may contribute in part to the formation of the valvular leaflets.

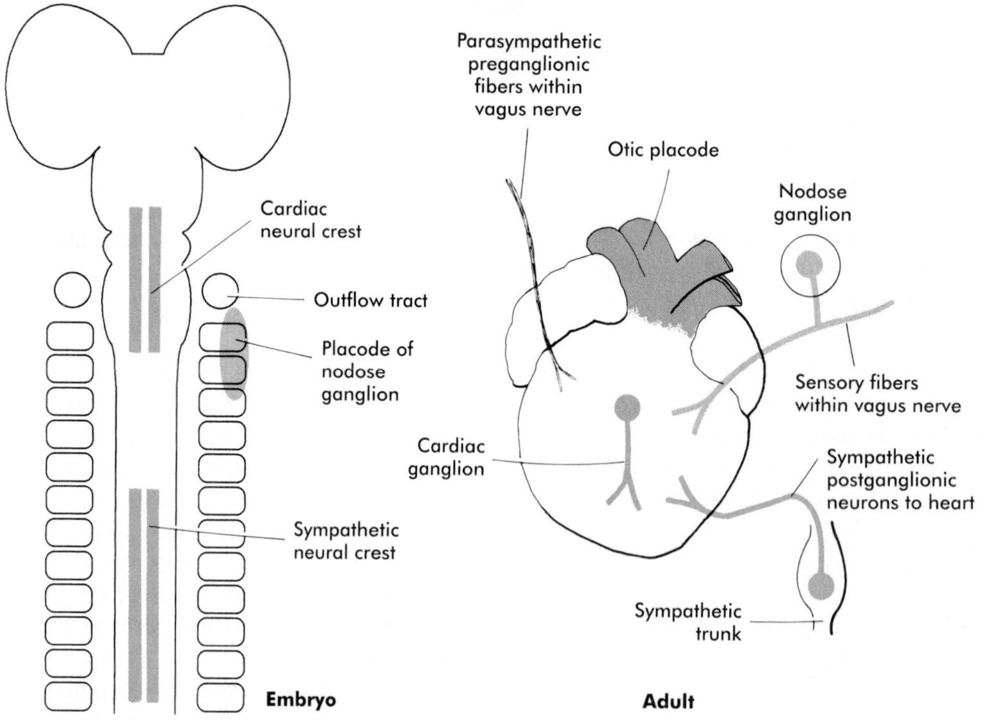

FIG. 18-19 The contributions of cranial and trunk neural crest and nodose placode to the innervation of the avian heart.
(Modified from Kirby [1988].)

the atrium into the ventricle and splits into right and left branches. These branches then distribute conducting tissue **(Purkinje fibers)** throughout the ventricular myocardium.

Many aspects of the embryology of the conduction system remain highly controversial. The conducting bundles consist of highly modified cardiac myocytes that contain large amounts of glycogen. In some mammals, evidence exists that cells of the atrioventricular bundle differentiate from separate precursor cells, possibly from the rings of tissue ultimately forming the cardiac skeleton, and that Purkinje cells are derived from modified ventricular myocytes. Although the conducting system is richly innervated, it forms before nerves enter the heart. Thus the innervation is eliminated as a causative factor in their development. With the recent availability of antibodies specific for cells of the conducting system, better information on the development of the conducting system is likely.

THE INITIATION OF CARDIAC FUNCTION

Because of its accessibility, the chick embryo has taught scientists most of what is known about the earliest functions of the embryonic heart. In recent years the application of ultrasound techniques has allowed the investigation of some aspects of cardiac function in human embryos as young as 4½ to 5 weeks of age.

Bilateral regions of precardiac mesoderm in the chick embryo differentiate into right and left cardiac tubes with functional ventricular, atrial, and sinoatrial regions. Contraction of the embryonic heart begins as the right and left heart tubes begin to fuse, and circulation of the blood begins soon thereafter. In the human embryo, this occurs between 21 and 23 days gestation.

Different parts of the tubular heart have different intrinsic beats, and experimental transplantation studies have shown that the functional characteristics of the regions of the heart depend on extracellular cues. For example, if preatrial tissue is grafted into a preventricular area, it acquires the functional characteristics of ventricular tissue.

As the heart tubes of the chick embryo first fuse in the ventricular region, the fused ventricles have an intrinsic beat of 25 per minute, and the unfused atria do not beat. A few hours later, when the atrial parts of the cardiac tubes have fused, the atrial region beats at a rate of 62 per minute. The atrial beat acts as a pacemaker, causing the ventricle to beat at the same rate. However, if the atrium is separated from the ventricle, the beat of the ventricle slows to an intrinsic rate of only 24 per minute. Finally, when the sinus venosus is consolidated, its intrinsic beat of 140 per minute drives the overall heart beat.

In the early fusing heart a distinct pacemaker region is located in the left sinus venosus region. The pacemaker consists of an aggregation of approximately 60 to 150 cells rather than arising from a single cell. The pacemaker initiates an excitation wave that is propagated through the early heart in about 0.5 seconds. As the heart matures and gets larger, the rate of propagation increases proportionally to the size of the heart. Thus the conduction velocity of the excitation stimulus increases by 100-fold, but the actual conduction time does not greatly change, despite a 1000-fold increase in the mass of the heart in the chick embryo. Increasing numbers of gap junctions between developing cardiac myocytes may partly account for the increasing conduction velocity in the developing heart.

The mechanism by which the tubular heart is able to pump blood in the absence of valves is still not well understood. Once the atrium fills with blood from the sinus venosus, it contracts and sends the blood into the ventricle. A peristalsis-like contraction then moves the blood to the truncus region, from which it leaves the heart and enters the aortic sac. Active contraction of the truncus region prevents the reflux of blood into the heart from the aortic root. Even in the early heart the endocardial cushions of the atrioventricular canal and the outflow tract have a valvelike function. At times the patterns of blood flow in the early heart have been considered to be responsible for the patterns of internal septation of the heart. However, careful studies have shown that many aspects of cardiac morphogenesis are separate from its function.

The blood pressure of an early embryo is very low (0.61/0.43 mm Hg in the 3-day-old chick embryo). As the embryo grows, the blood pressure rises exponentially and then tapers off before birth.

FETAL CIRCULATION

In many respects the overall plan of the embryonic circulation seems to be inefficient and more complex than needed to maintain the growth and development of the fetus. However, the embryo must prepare for the moment when it suddenly shifts to a totally different pattern of oxygenation of blood through the lungs rather than the placenta, making the modifications of the fetal plan of circulation essential.

Highly oxygenated blood from the placenta enters the umbilical vein in a large stream that is sometimes under increased pressure because of uterine contractions. Within the substance of the liver, blood from the umbilical vein under higher pressure passes directly into the ductus venosus, which allows it to bypass the small circulatory channels of the liver and flow directly into the inferior vena cava (Fig. 18-20). Once in the vena cava, it has immediate access to the heart. Poorly oxygenated blood flowing in the inferior vena cava can be somewhat backed up because of the strength of the umbilical blood flow.

Functional evidence exists for a physiological sphincter in the ductus venosus, which forces much of the umbilical blood to pass through hepatic capillary channels and enter the inferior vena cava through hepatic veins when it tightens. This considerably reduces the pressure of the umbilical blood and allows poorly oxygenated systemic blood

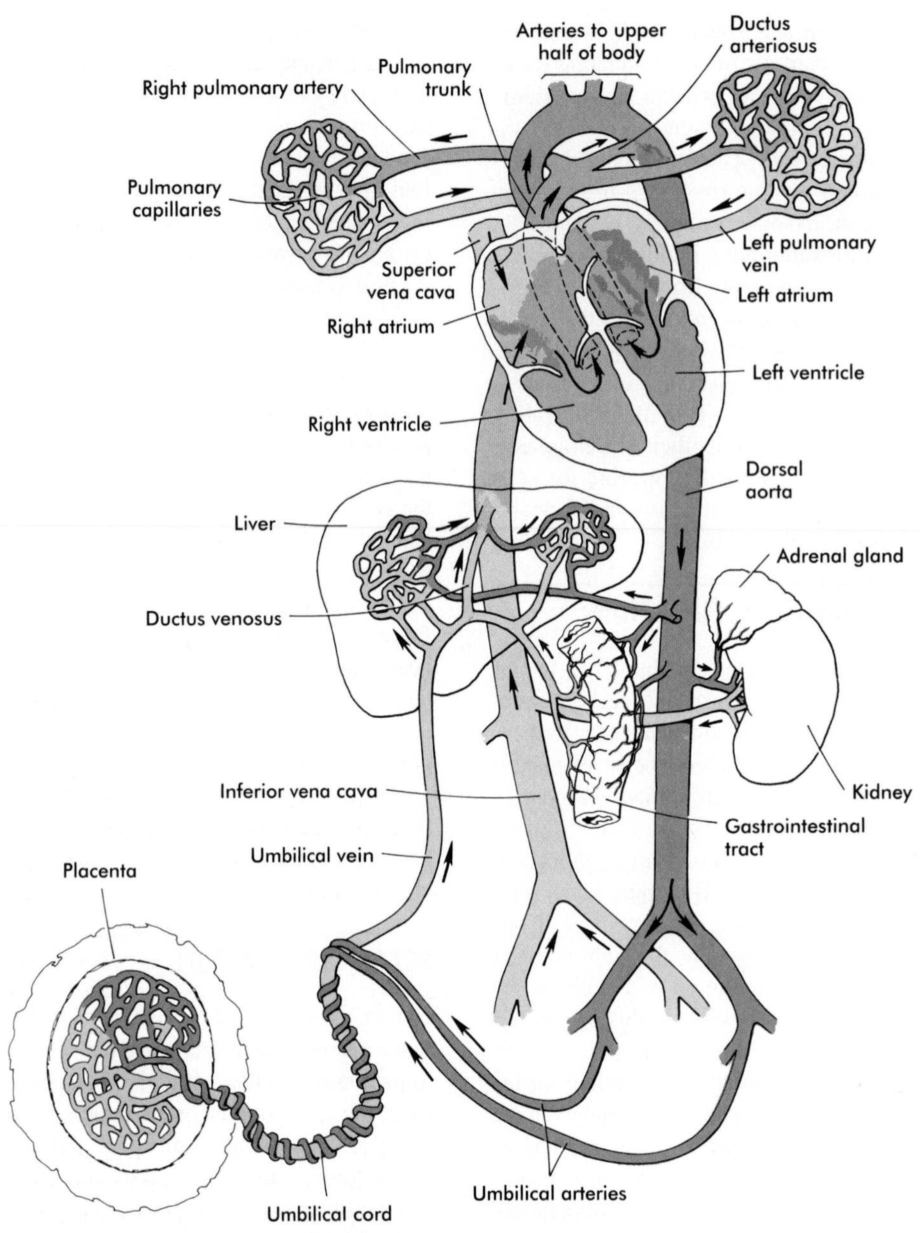

FIG. 18-20 The fetal circulation at term.

from the inferior vena cava to enter the right atrium at a lower pressure. Higher pressure blood entering the umbilical vein from the placenta also tends to prevent blood from the hepatic portal vein from entering the ductus venosus. When the uterus is relaxed and the umbilical venous blood is under low pressure, poorly oxygenated portal blood mixes with the umbilical blood in the ductus venosus. More mixing of umbilical and systemic blood occurs in the inferior vena cava as well.

In the right atrium the orientation of the entrance of the inferior vena cava allows a stream of blood under slightly increased pressure to pass directly through the foramen ovale and foramen secundum into the left atrium (see Fig. 18-20). This is the route normally taken by highly oxygenated umbilical blood entering the body under increased pressure. Because the interatrial shunt of the fetus is smaller than the opening of the inferior vena cava, some of the highly oxygenated caval blood eddies in the right atrium and enters the right ventricle. When low pressure blood (typically poorly oxygenated systemic blood) enters the right atrium, it joins with the venous blood draining the head through the superior vena cava and the heart through the coronary sinus and is mostly directed through the tricuspid valve into the right ventricle.

All blood entering the fetal right ventricle leaves through the pulmonary artery and passes to the lungs. Even in the relatively late fetus, the pulmonary vasculature is not ready to handle the full volume of blood that enters the pulmonary artery. The blood that cannot be accommodated by the pulmonary arteries is shunted to the aorta via the ductus arteriosus. This structure protects the lungs from circulatory overloading yet allows the right ventricle to exercise in preparation for its functioning at full capacity at birth. The control of patency of the ductus arteriosus has been subject to considerable speculation. Patency of both the ductus arteriosus and ductus venosus in the fetus is maintained actively through the action of prostaglandins (prostaglandin E_2 and prostaglandin I_2, respectively).

The left atrium receives a stream of highly oxygenated umbilical blood through the interatrial shunt and a small amount of poorly oxygenated blood from the pulmonary veins. This blood, which in aggregate is relatively highly oxygenated, passes into the left ventricle and leaves the heart through the aorta. Some of the first arterial branches leaving the aorta supply the heart and brain, organs that require a high concentration of oxygen for normal development.

At the point where the aortic arch begins to descend, the ductus arteriosus empties poorly oxygenated blood into it. This mixture of well- and poorly oxygenated blood is then distributed to the tissues and organs that are supplied by the thoracic and abdominal branches of the aorta. Near its caudal end the aorta gives off two large umbilical arteries, which carry blood to the placenta for renewal.

MALFORMATIONS OF THE HEART

Many malformations of the heart occur. Because of the close physiological balance of the circulation, most malformations produce symptoms. Clinically, heart malformations are typically classified as those that are associated with cyanosis **(cyanotic defects)** in postnatal life and those that are not **(acyanotic defects).**

Cyanosis results when the blood contains more than 5 g of reduced hemoglobin/100 ml. Cyanosis is readily recognizable by a purplish to bluish tinge to the skin in areas with a dense superficial capillary circulation. It is associated with **polycythemia,** an increased concentration of erythrocytes in the blood resulting from the overall decreased oxygen saturation of the blood. Long-term cyanosis is associated with a prominent clubbing of the ends of the fingers and decreased growth. In severe cases of cyanosis, children assume a characteristic squatting posture.

Postnatally, cyanosis is associated with the presence of a right-to-left shunt in which venous blood mixes with systemic blood. Some heart defects are acyanotic for many years but then become cyanotic. These defects are initially characterized by a left-to-right shunt in which oxygenated systemic blood refluxes into the right atrium or ventricle. The net result is an increased pumping load on the right ventricle, ultimately leading to right ventricular hypertrophy. Over a long period the increased blood flow through the lungs provokes a hypertensive reaction in the pulmonary vasculature, which effectively increases the pressure in the right ventricle and atrium. When the blood pressure on the right side of the heart exceeds that in the corresponding left chamber, the shunt reverses, and poorly oxygenated blood passes to the systemic circulation, leading to cyanosis. At this point the condition of the patient who has the cardiac lesion often rapidly worsens.

Chamber-to-Chamber Shunts

Interatrial Septal Defects. Several types of anatomical defects in the interatrial septum can result in a persisting shunt between the two atria. The most common varieties are caused by excessive resorption of tissue around the foramen secundum or hypoplastic growth of the septum secundum (Fig. 18-21, *A*). A less common variety is a low septal defect, which is usually caused by the lack of union between the leading edge of the septum primum and the endocardial cushions (Fig. 18-21, *B*). If the defect is the result of a deficiency of endocardial cushion tissue, associated defects of the atrioventricular valves can considerably complicate the lesion. Lack of septation of the atrium results in a **common atrium,** a serious defect that is usually associated with other heart defects. Atrial septal defects are among the most common heart malformations.

Uncomplicated atrial septal defects are usually compatible with many years of symptom-free life. Even during the symptom-free period, blood from the left atrium, which is under slightly higher pressure than that in the right atrium, passes into the right atrium. This additional blood causes right atrial hypertrophy and results in increased blood flow into the lungs. Over many years, pulmonary hypertension can develop. This increases the blood pressure of the right ventricle and ultimately the right atrium. Only a few millimeters of increased right atrial pressure reverses the blood flow in the interatrial shunt and causes cyanosis.

A much more serious condition is **premature closure of the foramen ovale.** In this situation the entire input of blood into the right atrium passes into the right ventricle, causing massive hypertrophy of the right side of the heart. The left side is severely hypoplastic due to the reduced blood that the left chambers carry. Although this defect is usually compatible with intrauterine life, infants typically die shortly after birth because the hypoplastic left heart cannot handle a normal circulatory load.

Persistent Atrioventricular Canal. The usual basis for this defect is underdevelopment of the endocardial cushions that results in a lack of division of the early atrioventricular canal into right and left channels. A number of specific causes of atrioventricular canal defects may exist, ranging from inappropriate expression of types of *Hox-7* genes to defects in the production or reception of inductively active extracellular matrix components (e.g., subunits of adherons) (see Fig. 18-14).

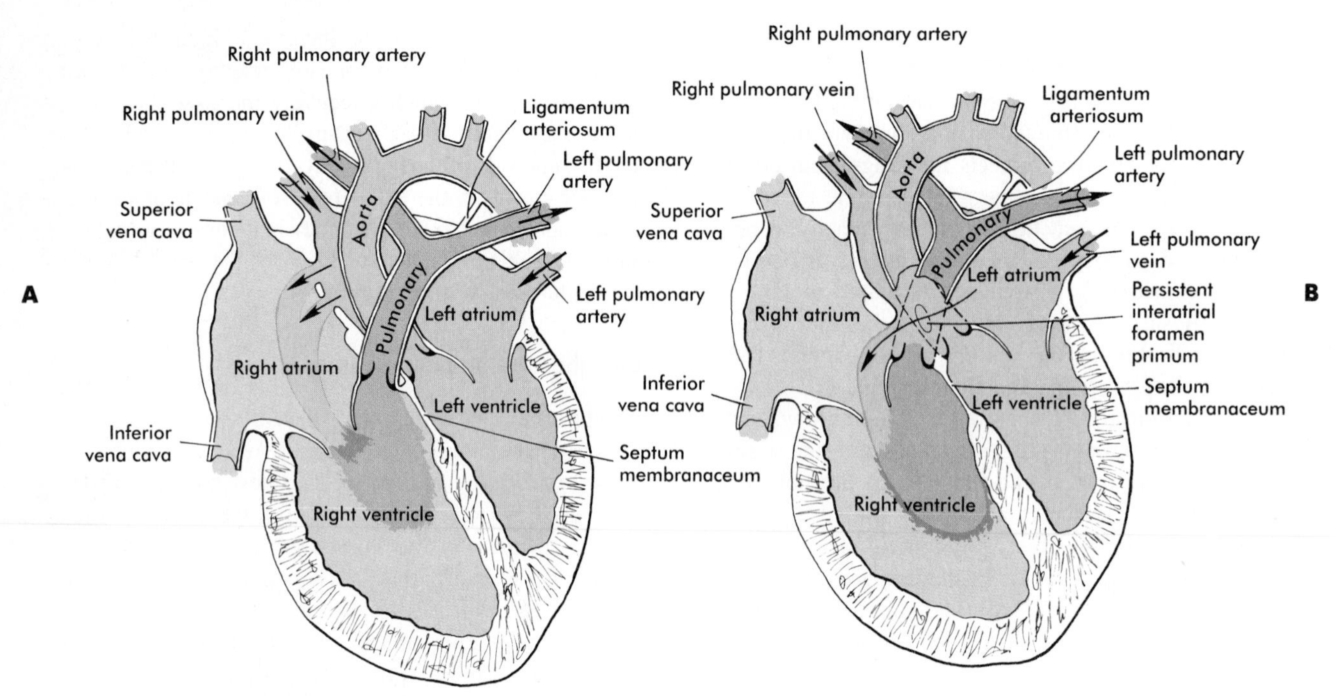

FIG. 18-21 High **(A)** and low **(B)** atrial septal defects in the heart. Red denotes well-oxygenated arterial blood, blue denotes poorly oxygenated venous blood, and purple denotes a mixture of arterial and venous blood.

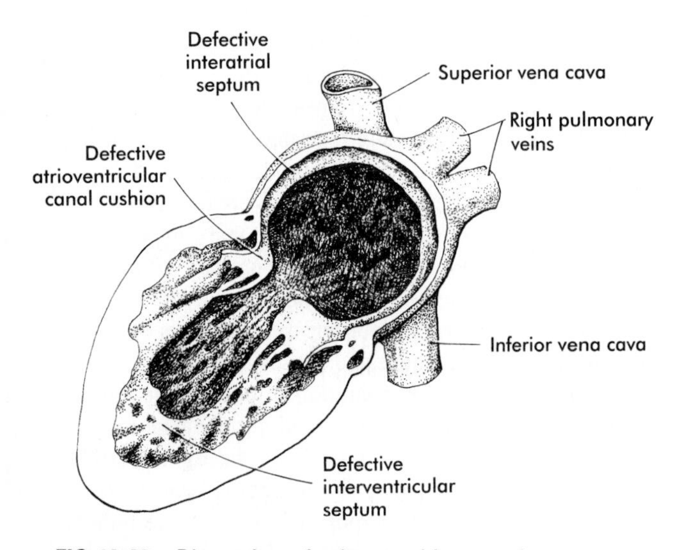

FIG 18-22 Dissection of a heart with a persistent atrioventricular canal in a 12-day-old male.
(Based on studies by Patten [1956].)

A persistent atrioventricular canal is often associated with low interatrial and high interventricular septal defects (Fig. 18-22). This severe defect leads to poor growth and a considerably shortened life. Interestingly, despite the potential for mixing of blood, the predominant shunt direction is left to right, and some patients have little cyanosis.

Tricuspid Atresia. In this condition, the etiology of which is poorly understood, the normal valvular opening between the right atrium and right ventricle is completely occluded (Fig. 18-23). Such a defect alone is incompatible with life because the blood cannot gain access to the lungs for oxygenation. However, children can survive with this malformation, illustrating an important point in cardiac embryology: Often a primary lesion is accompanied by one or more secondary lesions that permit survival, although often at a poor functional level.

In the case of tricuspid atresia, secondary defects have to be shunts that permit blood to enter the pulmonary circulation. One option is a persisting atrial septal defect, which bypasses the hypoplastic right ventricle and shunts the blood into the left atrium. The blood passes to the left ventricle and into the systemic circulation, where it gains access to the lungs through a patent ductus arteriosus. From there the oxygenated blood enters the left atrium, perhaps to be recycled through the lungs again before entering the systemic circulation.

Mitral atresia can also occur, but it is much more rare than tricuspid atresia. Secondary compensating defects again have to be present for survival. Infants with these lesions typically survive only a few months or years.

Interventricular Septal Defect. Defects in the interventricular septum, which are not as common as atrial septal defects, typically occur in the membraneous portion of the septum where several embryonic tissues converge. Because the pressure of the blood in the left ventricle is higher than that in the right, this lesion is initially associated with a left-to-right acyanotic shunt (Fig. 18-24). However, the in-

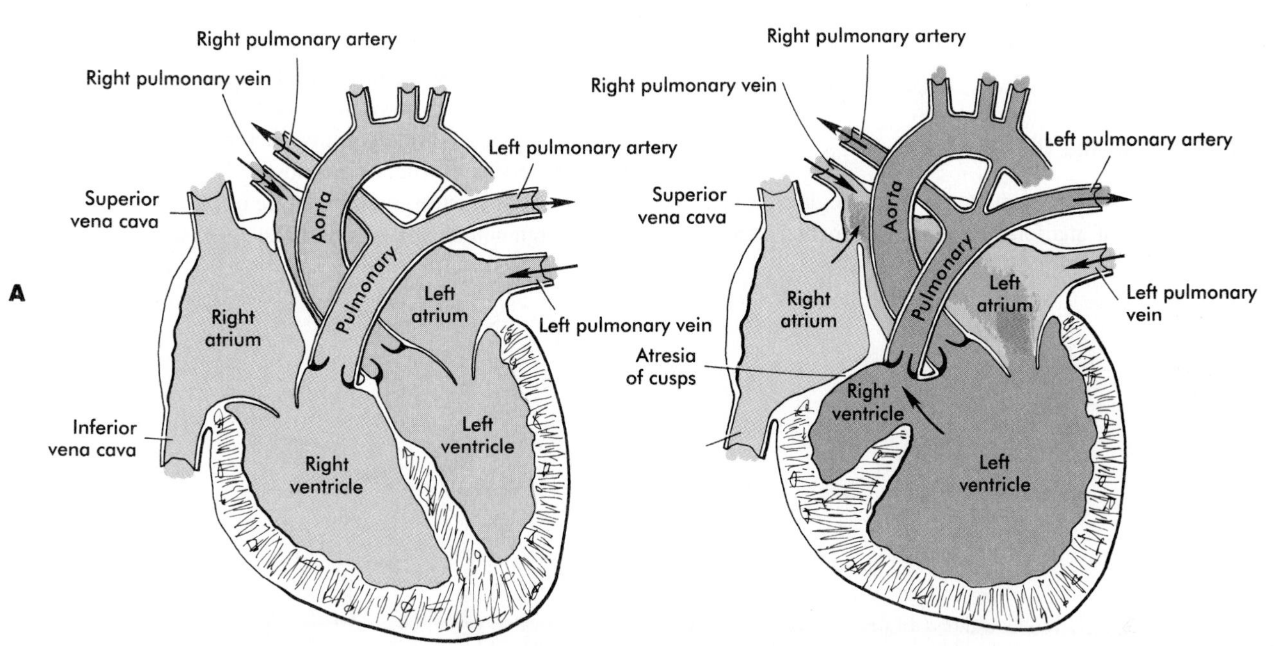

FIG. 18-23 **A,** A normal postnatal heart. **B,** Tricuspid atresia, with compensating defects in the interatrial septum and interventricular septum *(arrows),* which allow this patient to survive.

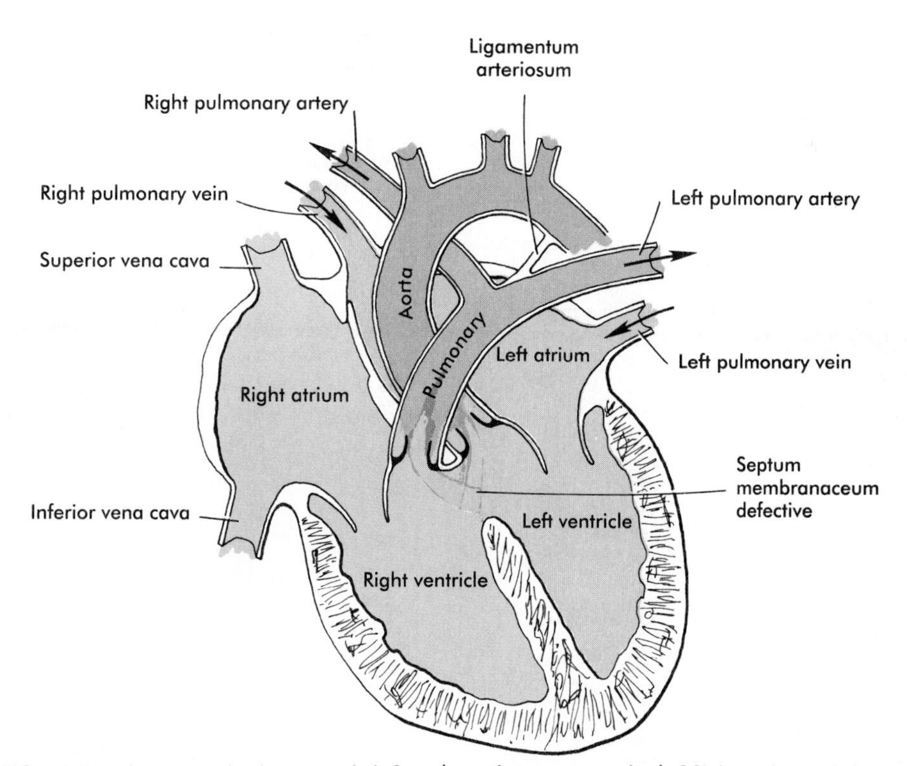

FIG. 18-24 Interventricular septal defect (membranous portion). Mixing of arterial and venous blood occurs in both outflow tracts but especially in the pulmonary artery.

creased blood flow into the right ventricle causes right ventricular hypertrophy and can lead to pulmonary hypertension, ultimately causing reversal of the shunt. The basic pathological dynamics are similar to those for atrial septal defects.

Malformations of the Outflow Tract

The outflow tract of the heart (truncoconal region) is subject to a variety of malformations. Experimental studies have shown a prominent contribution of neural crest cells to this region. Extirpation and transplantation experiments have shown specific requirements for cardiac neural crest cells in the normal development of the cardiac outflow tract (Fig. 18-25). If the cardiac neural crest is removed, ectodermal cells from the nodose placode populate the outflow tract but septation of the outflow tract does not occur, leading to a persistent truncus arteriosus. If mesencephalic or truncal neural crest is grafted in place of cardiac neural crest or if foreign neural crest is grafted lateral to the cardiac neu-

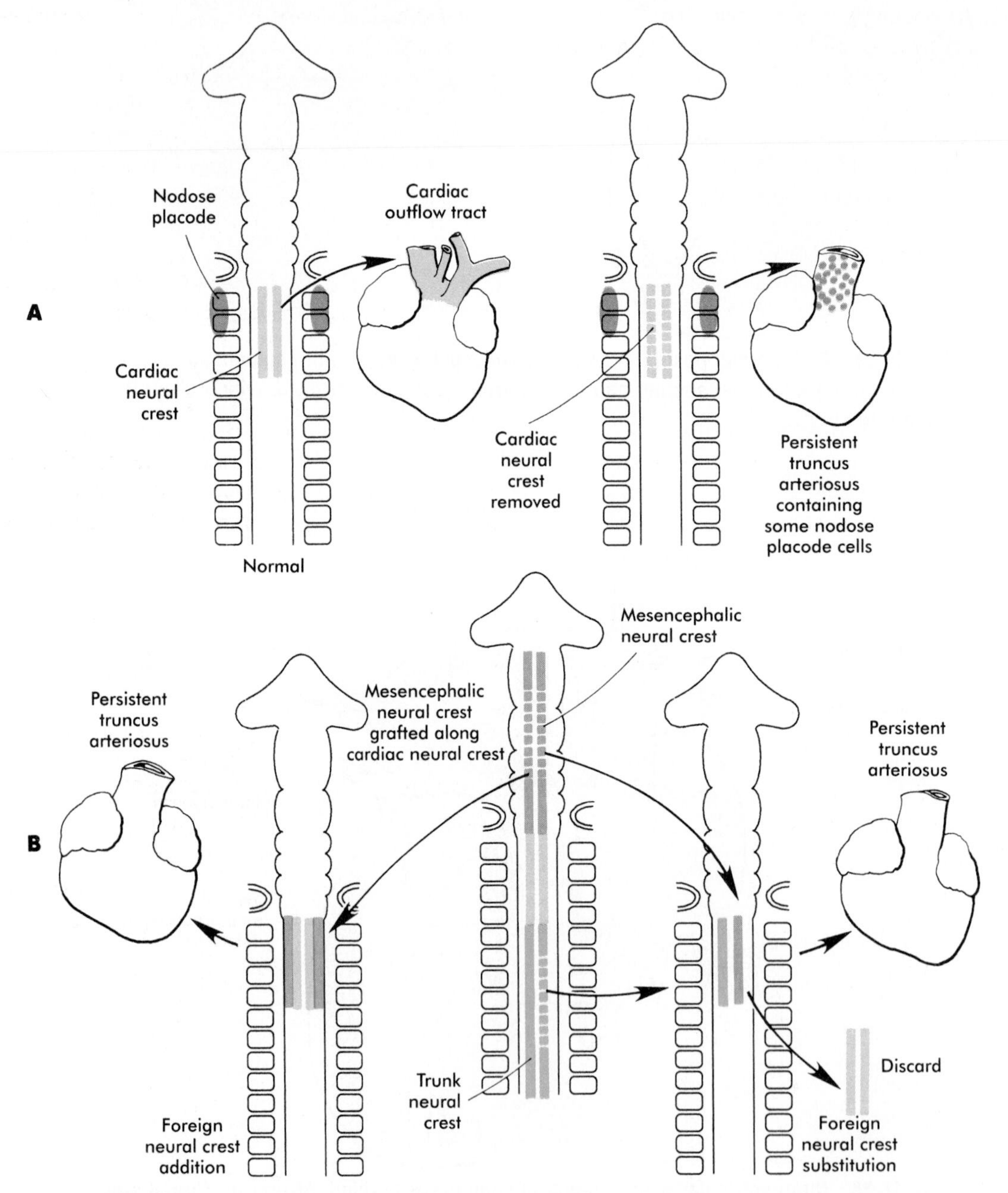

FIG. 18-25 Neural crest and morphogenesis of the outflow tract of the heart. **A,** Normal, showing cardiac neural crest contributing to the formation of the outflow tract in the avian heart. **B,** Removal of the cardiac neural crest leads to the formation of a persistent truncus arteriosus containing cells derived from the nodose placode. (Based on studies by Kirby and Waldo [1988].)

ral crest, persistent truncus arteriosus consistently results. The last experiment shows that foreign neural crest interferes with the normal migration or function of the cardiac neural crest. Although all malformations of this area cannot be attributed to defective neural crest development, circumstantial evidence suggests that this may be a significant factor.

Persistent Truncus Arteriosus. This condition is caused by the lack of partitioning of the outflow tract by the truncoconal ridges (Fig. 18-26). Because of the contribution of the truncoconal ridges to the membranous part of the interventricular septum, this malformation is almost always accompanied by a ventricular septal defect. A large arterial outflow vessel overrides the ventricular septum and receives blood that exits from each ventricle. As may be predicted, individuals with a persistent truncus arteriosus are highly cyanotic. Without treatment, 60% to 70% of infants born with this defect die within 6 months.

Transposition of the Great Vessels. On rare occasions the truncoconal ridges fail to spiral as they divide the outflow tract into two channels. This results in two totally independent circulatory arcs, with the right ventricle emptying into the aorta and the left ventricle emptying into the pulmonary artery. If the condition is uncorrected, the left circulatory arc continues pumping highly oxygenated blood through the left side of the heart and the lungs, whereas the right side of the heart pumps venous blood through the aorta into the systemic circulatory channels and back into the right atrium. This lesion, which is the most common cause of cyanosis in newborns, is compatible with life only if an atrial and a ventricular septal defect and an associated patent ductus arteriosus accompany it. Even with these anatomical compensations, the quality of blood reaching the body is poor.

Aortic and Pulmonary Stenosis. If the septation of the outflow tract by the truncoconal ridges is asymmetrical, either the aorta or pulmonary artery can be abnormally narrowed (Figs. 18-27 and 18-28). The severity of symptoms is related to the degree of stenosis. One of the best-known lesions of this type is the **tetralogy of Fallot,** which is characterized by (1) pulmonary stenosis, (2) a membranous interventricular septal defect, (3) a large aorta (overriding aorta, the opening of which extends into the right ventricle), and (4) right ventricular hypertrophy. Because of the pulmonary stenosis, some poorly oxygenated right ventricular blood leaves via the enlarged aorta, causing cyanosis.

MALFORMATIONS OF THE BLOOD VESSELS

Because of their mode of formation in which one vascular channel is favored within a dense network, blood vessels (especially veins) are subject to numerous variations from normal. Most variations seen in the dissecting laboratory are of little functional significance. Animal experiments suggest that disturbances in the neural crest may be involved in the genesis of certain anomalies of the major arteries. When the cardiac neural crest is removed from early avian embryos, malformations involving the carotid arter-

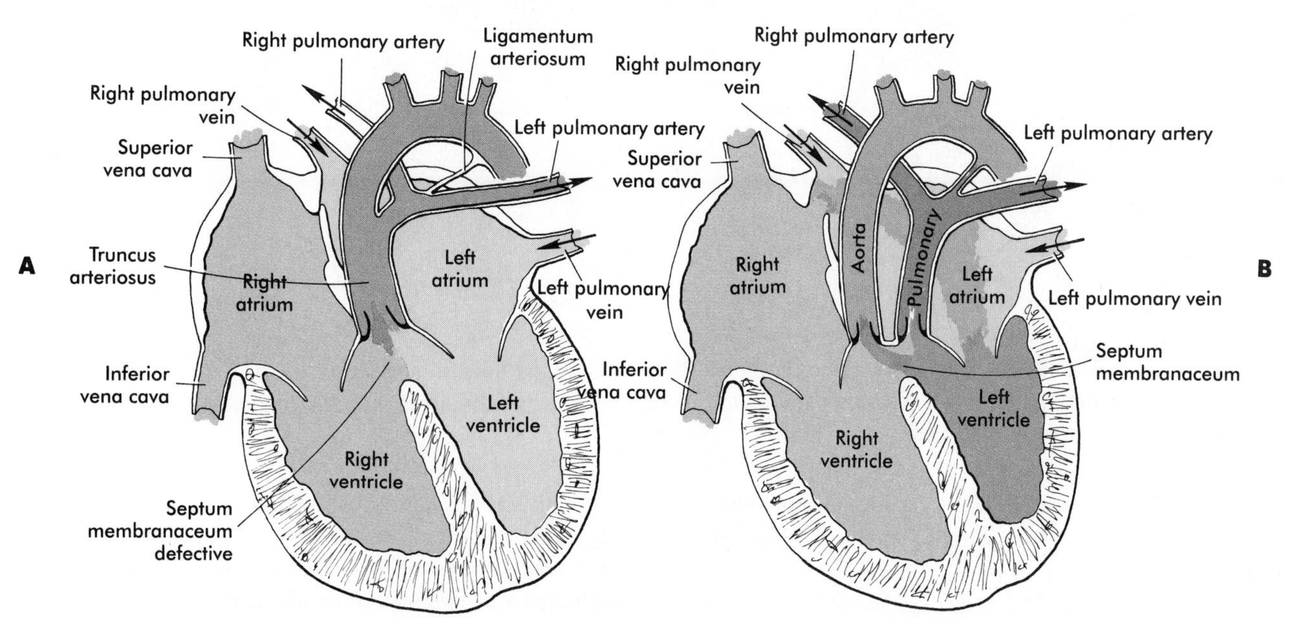

FIG. 18-26 **A,** Persistent truncus arteriosus. A single outflow tract is fed by blood entering from both the right and left ventricles. The membranous part of the interventricular septum is commonly defective. **B,** Transposition of the great vessels due to lack of spiraling of the truncoconal ridges in the early embryo. The aorta arises from the right ventricle and the pulmonary artery from the left ventricle.

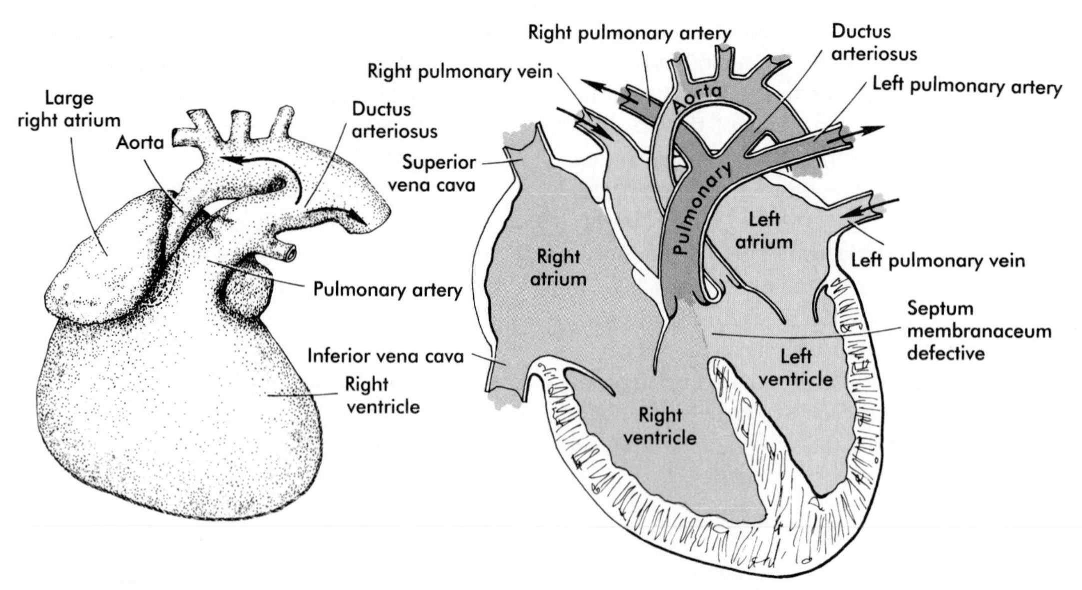

FIG. 18-27 Aortic stenosis. In severe cases, the ductus arteriosus commonly remains patent. *Right:* Mixed arterial and venous blood in the pulmonary artery is shown in purple. Initially, blood from the pulmonary trunk *(purple)* goes through the ductus arteriosus into the aorta, often leading to cyanosis.

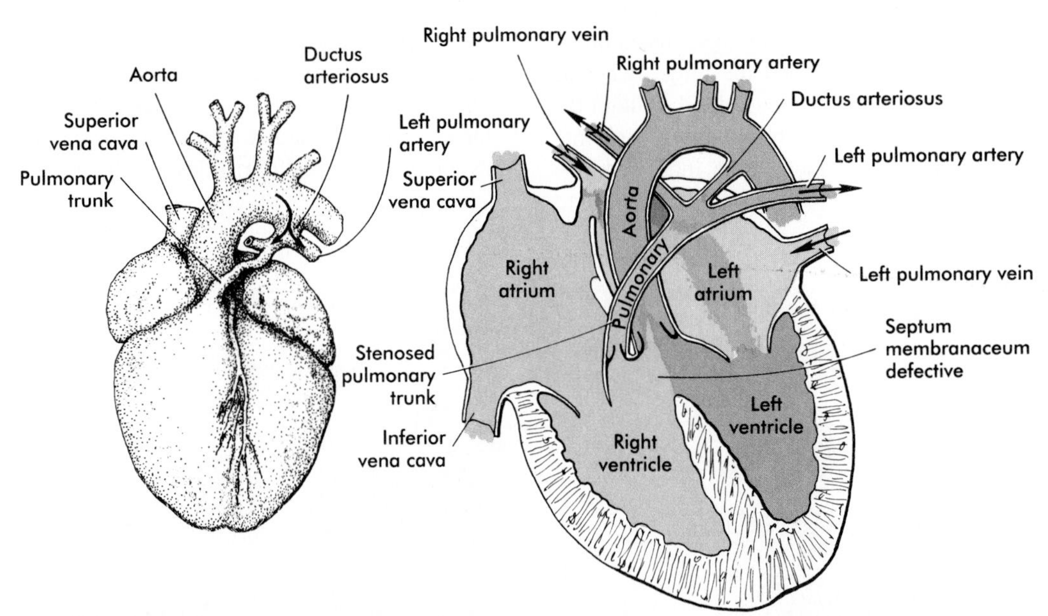

FIG. 18-28 Pulmonary stenosis. *Right:* Patterns of blood flow. In severe cases the ductus arteriosus remains patent, with blood flowing from the aorta into the pulmonary circulation *(arrows).*

ies and arch of the aorta result. Certain malformations of the larger vessels can cause serious symptoms or be significant during surgery.

Double Aortic Arch

Rarely, the segment of the right dorsal aortic arch between the exit of the right subclavian artery and its point of joining with the left aortic arch persists instead of degenerating. This results in a complete vascular ring surrounding the trachea and esophagus (Fig. 18-29, *A*). A double aortic arch can cause **dyspnea** (difficult breathing) in infants while they feed. Even if the condition is asymptomatic early in life, later growth typically narrows the diameter of the ring in relation to the size of the trachea and esophagus, causing symptoms in later years.

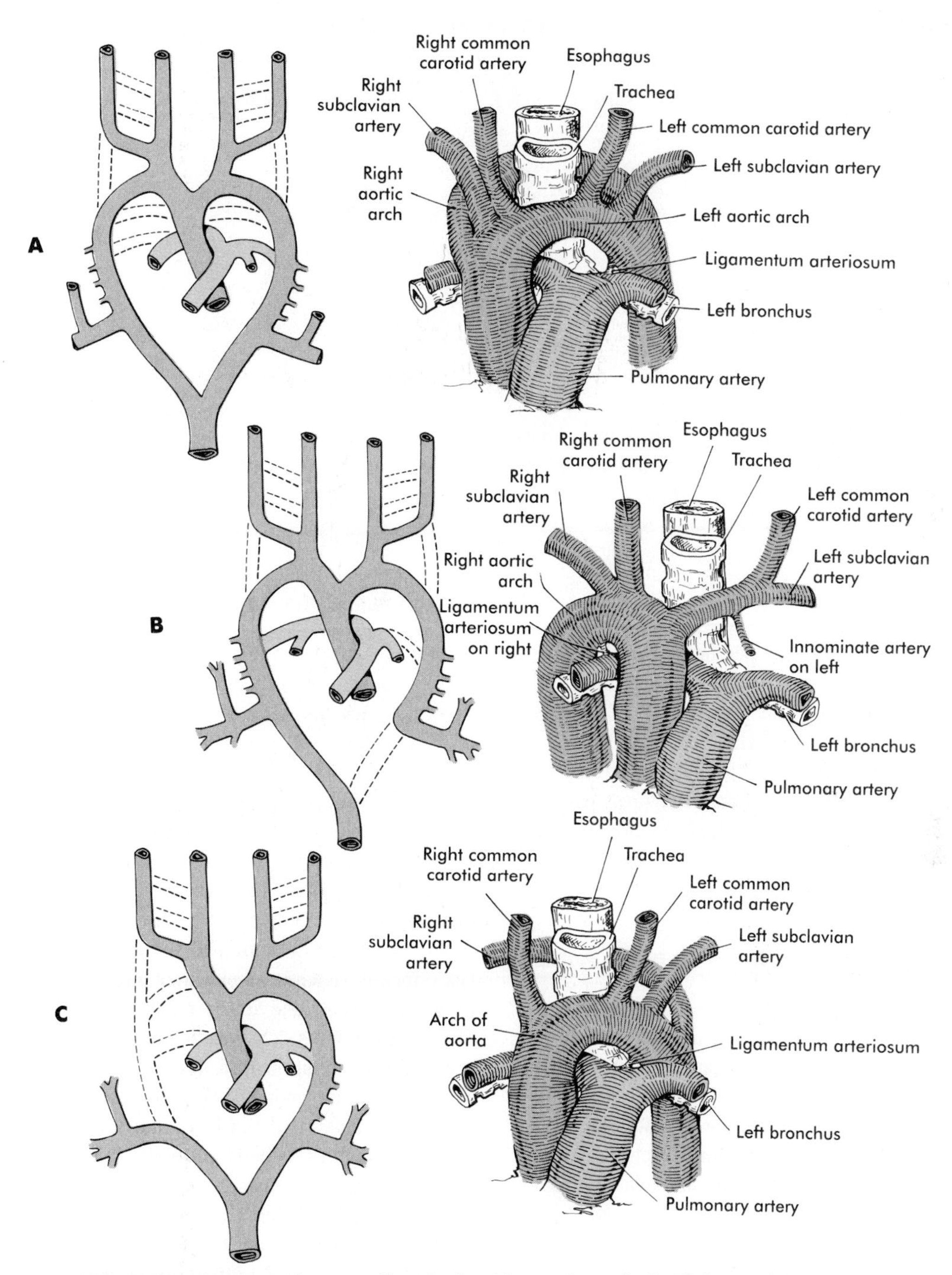

FIG. 18-29 Aortic arch anomalies. **A,** Double aortic arch. **B,** Right aortic arch. **C,** Right subclavian artery from arch of aorta.

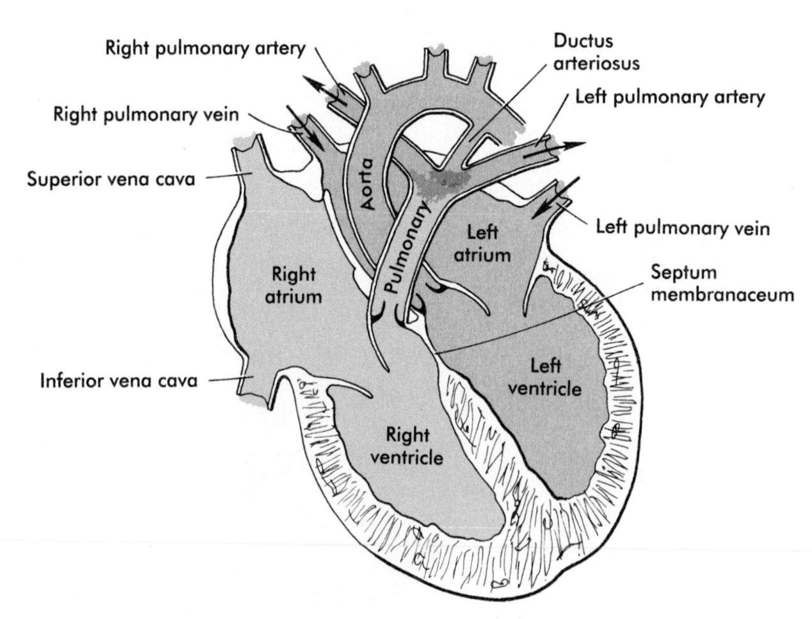

FIG. 18-30 Patent ductus arteriosus showing the flow of blood from the aorta into the pulmonary circulation. Later in life, pulmonary hypertension may result, causing reversal of blood flow through the shunt and cyanosis.

Right Aortic Arch

A right aortic arch arises from the persistence of the complete embryonic right aortic arch and the disappearance in the left arch of the segment caudal to the exit of the left subclavian artery (Fig. 18-29, *B*). This condition is essentially a mirror image of normal development of the aortic arch, and it can occur as an isolated anomaly or as part of complete situs inversus of the individual. Symptoms are typically mild or absent unless an aberrant left subclavian artery presses against the esophagus or trachea.

Right Subclavian Arising from the Arch of the Aorta

If the right fourth aortic arch degenerates between its origin and if the exit of the subclavian artery and the more distal segment (which normally disappears) persists, the right subclavian artery arises from the left aortic arch and passes behind the esophagus and trachea to reach the right arm (Fig. 18-29, *C*). As with a double aortic arch, this condition can cause difficulties in breathing and swallowing.

Patent Ductus Arteriosus

One common vascular anomaly is failure of the ductus arteriosus to close after birth (Fig. 18-30). At least half of the infants with this condition have no symptoms, but over many years the strong flow of blood from the higher pressure systemic (aortic) circulation into the pulmonary circulation overloads the vasculature of the lungs, resulting in pulmonary hypertension and ultimately heart failure.

Coarctation of the Aorta

Another relatively common, nonlethal malformation of the vascular system is coarctation of the aorta, which occurs in two main variants. One consists of an abrupt narrowing of the descending aorta caudal to the entrance of the ductus arteriosus (Fig. 18-31, *B*). The other variant, called **preductal coarctation,** occurs upstream from the ductus (Fig. 18-31, *A*). The former variety **(postductal coarctation)** is by far the most common, accounting for over 95% of all cases. The embryogenesis of coarctation is still not clear. Several underlying causes may lead to the same condition.

In preductal coarctation of the aorta the ductus arteriosus typically remains patent after birth, and the blood supplying the trunk and limbs reaches the descending aorta through the ductus. The vasculature must compensate for a postductal coarctation in a different manner, since the location of the narrowing in this case effectively cuts off the arterial circulation of the head and arms from that of the trunk and legs. The body responds by opening up collateral circulatory channels and connections through normally relatively small arteries that lead from the upper to the lower body. Such channels are (1) the internal thoracic arteries, (2) arteries associated with the scapula, and (3) the anterior spinal artery. The unusually large flow of blood through these arteries passes through segmental branches (e.g., intercostal arteries) into the descending aorta caudal to the coarctation. The increased blood flow in the intercostal arteries causes a distinct notching in the ribs that can be readily seen in radiological images. Despite these compensatory circulatory adaptations the blood pressure in patients

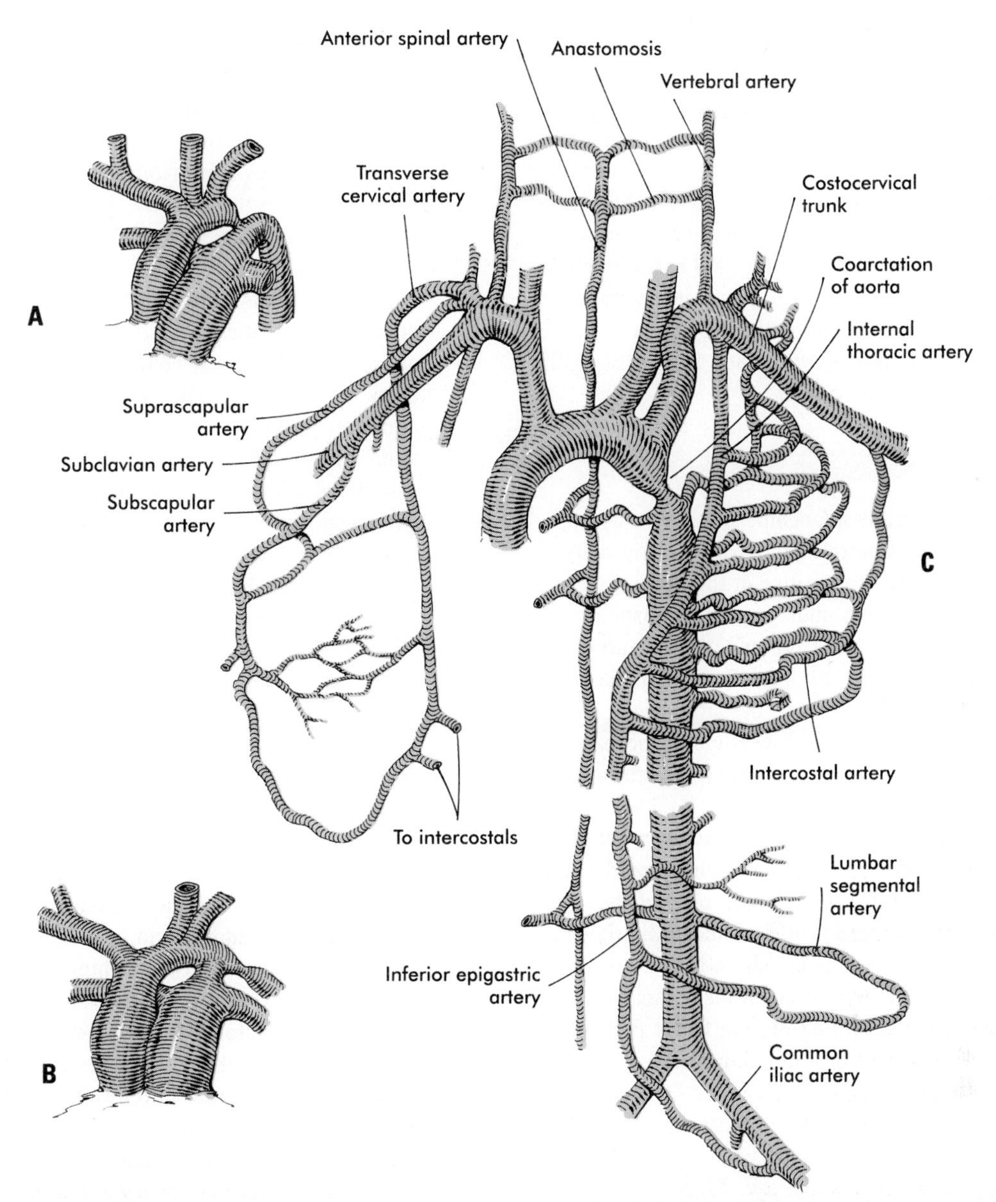

FIG. 18-31 Coarctation of the aorta. **A,** Preductal coarctation. **B,** Postductal coarctation with an accompanying patent ductus arteriosus. **C,** Collateral circulation in postductal coarctation, with enlarged peripheral vessels carrying blood to the lower part of the body.

with a postductal coarctation is much higher in the arms than in the legs.

Malformations of the Vena Cavae

As might be expected from their complex mode of formation (see Fig. 18-9), the superior and inferior vena cavae are subject to a wide range of malformations. Common variants are duplications of the superior and inferior vena cavae or persistence of the left instead of the right segments of these vessels along with the absence of the normal vessel. In most cases these malformations are asymptomatic.

Anomalous Pulmonary Return

Because of the mode of joining of the individual pulmonary veins and the later absorption of the distal part of the pulmonary venous system into the left atrial wall, inappropriate connections of pulmonary veins to the heart can occur. One of the more common is for one or more branches of the pulmonary vein to enter the right instead of the left atrium. In other cases **(total anomalous pulmonary return),** all pulmonary veins empty into the right atrium or superior vena cava. Such a case must be accompanied by an associated shunt (e.g., interatrial shunt) to bring oxygenated blood into the systemic circulation.

Malformations of the Lymphatic System

Although minor anatomical variations of lymphatic channels are common, anomalies that cause symptoms are rare. These typically present as swelling caused by dilatation of major lymphatic vessels. By far the most common major lymphatic anomaly seen in fetuses is **cystic hygroma,** which manifests as large swellings, sometimes even collar-like, in the region of the neck (see Fig. 8-1). Although the embryological basis for cystic hygroma is not certain, excessive local production and growth of lymphatic tissue, possibly originating as pinched off buds from the jugular lymph sacs, is probably the cause.

SUMMARY

1. The heart arises from splanchnic mesoderm as a horseshoe-shaped bilateral primordium. Originally, bilateral endocardial tubes fuse in the midline. The fused cardiac tube then undergoes an S-shaped looping, and soon specific regions of the heart can be identified. Starting with the inflow tract, these regions are the sinus venosus, the atria, the ventricles, and the outflow tract (bulbus cordis). The outflow tract later divides into the conus arteriosus and truncus arteriosus.
2. Atrial endocardial cushions are thickenings between the atria and ventricles. The underlying myocardium induces cells from the endothelial lining of the endocardial cushion to leave the endocardial layer and transform into mesenchymal cells that invade the cardiac jelly. These events are accompanied by the activity of transforming growth factor-β 3. They serve as the basis for the formation of the atrioventricular valves.
3. The earliest blood and associated extraembryonic blood vessels arise from blood islands in the mesodermal wall of the yolk sac. Much of the vasculature of the embryonic body is derived from intraembryonic sources. Endothelial cell precursors (angioblasts) arise from most mesodermal tissues of the body except for the prechordal mesoderm. Embryonic blood vessels form by three main mechanisms: (1) coalescence in situ, (2) migration of angioblasts into organs, and (3) sprouting from existing vessels. Ingrowth of blood vessels into some organ primordia is stimulated by angiogenesis factors.
4. Within the aortic arches the first three pairs of arches form arteries that supply the head. The fourth arch develops asymmetrically, with the left arch forming part of the aortic arch of the adult. The fifth arch never forms. A sixth pair of arches arises as a capillary plexus that connects with the fourth arch. The distal part of the left sixth arch forms the ductus arteriosus, a shunt that allows blood to bypass the immature lungs and directly enter the aorta. Many of the larger arteries of the adult arise from three sets of aortic branches—the dorsal intersegmental, the lateral segmental, and the ventral segmental. The coronary arteries arise from capillary plexuses associated with the truncus. These plexuses are secondarily connected with the aorta.
5. The venous system arises from very complex capillary networks that initially develop into components of the cardinal vein system. Anterior and posterior cardinal veins drain the head and trunk, respectively. They then empty into the paired common cardinal veins and ultimately into the sinus venosus of the heart. Paired subcardinal veins are associated with the developing mesonephros. Paired extraembryonic umbilical and vitelline veins pass through the developing liver and directly into the sinus venosus. The pulmonary veins arise as separate structures and empty into the left atrium. The lymphatic system first appears as six primary lymph sacs. These become connected by lymphatic channels. Lymphatics from most of the body collect into the thoracic duct, which empties into the venous system at the base of the left internal jugular vein.
6. Internal partitioning of the heart begins with the separation of the atria from the ventricles and formation of the mitral and tricuspid valves. The left and right atria become separated by growth of the septum primum and septum secundum, but throughout embryonic life a shunt remains from the right to the left atrium via the foramen secundum and foramen ovale. The sinus venosus and the vena cavae empty into the right atrium, and the pulmonary veins drain into the left atrium. The ventricles are divided by the interventricular septum. Spiral truncoconal ridges partition the common outflow tract into pulmonary and aortic trunks. Semilunar valves prevent the reflux of blood in these vessels into the heart.
7. In addition to sensory innervation, the heart receives sympathetic and parasympathetic innervation. The conduction system distributes the contractile stimulus throughout the heart. The conduction system is derived from modified cardiac muscle cells. The heart begins to beat early in the fourth week of gestation. Physiological maturation of the heart beat follows maturation of the pacemaker system and the innervation of the heart.
8. The fetal circulation brings oxygenated blood from the placenta through the umbilical vein and into the right atrium, where much of it is shunted into the left atrium. Other blood entering the right atrium passes into the right ventricle. Blood leaving the right ventricle enters the pulmonary trunk, which supplies some blood to the lungs and the majority to

the aorta via the ductus arteriosus. Blood in the left atrium empties into the left ventricle and aorta, where it supplies the body. Poorly oxygenated blood enters the umbilical arteries and is carried to the placenta for renewal.

9. Common malformations of the heart consist of atrial septal defects, which in postnatal life allow blood to pass from the left to the right atrium. Ventricular septal defects, which also result in a left-right shunting of blood, are more serious. Defects that block a channel for blood flow (e.g., tricuspid atresia) must be accompanied by secondary shunt defects to be compatible with life. A persistent atrioventricular canal can be attributed to a defect in the formation or further development of the atrioventricular endocardial cushions. Most malformations of the outflow tract of the heart appear to be related to inappropriate partitioning by the truncoconal ridges. The basis for this is frequently found in neural crest abnormalities.
10. Malformations of the major arteries are often the result of the inappropriate appearance or disappearance of specific components of the aortic arch system. Some, such as double aortic arch or right aortic arch, can interfere with swallowing or breathing because of pressure. Patent ductus arteriosus is caused by the failure of the ductus arteriosus to close properly after birth. Coarctation of the aorta, the genesis of which is poorly understood, must be compensated by either a patent ductus arteriosus or the opening of collateral vascular channels that allows blood to bypass the site of coarctation.
11. Because of their complex mode of origin, veins are commonly subject to considerable variation, but these malformations are frequently asymptomatic. Anomalous pulmonary return, which beings oxygenated blood into the right atrium, must be accompanied by a right-to-left shunt to be compatible with life. Malformations of the lymphatic system can cause local swellings such as cystic hygroma, which results in a collarlike swelling in the neck.

REVIEW QUESTIONS

1. A 12-year-old boy tells his doctor that over the past few months he has noticed some difficulty in swallowing when eating meat. The doctor does a physical examination and then orders an upper gastrointestinal x-ray series. After examining the films, the doctor refers the boy for some vascular studies. What is the reasoning behind this decision?
2. An individual with atresia of the mitral valve could not survive after birth without other defects of the cardiovascular system that could compensate for the primary defect—in this case a complete blockage between the left atrium and ventricle. Construct at least one set of associated defects that could physiologically compensate for the disruption caused by the mitral atresia.
3. A man who has not been to a physician since childhood comes into a clinic complaining of fatigue. The examining physician notices cyanosis of the lower part of the body and asks the patient how long his feet have looked that color. He says that his feet started to look purplish at about the time when he began to feel fatigued. On auscultation of the heart (listening with a stethoscope), the physician made a tentative diagnosis. What is it and what was the basis for the decision?
4. What is the embryological basis for a duplication of the inferior vena cava caudal to the kidneys?

REFERENCES

Anderson RH: Simplifying the understanding of congenital malformations of the heart, *Int J Cardiol* 32:131-142, 1991.

Bartelings MM, Gittenberger-de Groot AC: Morphogenetic considerations on congenital malformations of the outflow tract. I. Common arterial trunk and tetralogy of Fallot, *Int J Cardiol* 32:213-230, 1991.

Bartelings MM, Gittenberger-de Groot AC: Morphogenetic considerations on congenital malformations of the outflow tract. II. Complete transposition of the great arteries and double outlet right ventricle, *Int J Cardiol* 33:5-26, 1991.

Bockman DE, Kirby ML, eds: Embryonic origins of defective heart development, *Ann NY Acad Sci* 588:1-466, 1990.

Bolender DL, Markwald RR: *Endothelial formation and transformation in early avian heart development: induction by proteins organized into adherons.* In Feinberg RN, Sherer GK, Auerbach R, eds: *The development of the vascular system,* Basel, Switzerland, 1991 Karger, pp 109-124.

Clark EB, Takao A, eds: *Developmental cardiology: morphogenesis and function,* Mount Kisco, NY, 1990, Futura Publishing.

Coceani F, Olley PM: The control of cardiovascular shunts in the fetal and perinatal period, *Can J Physiol Pharmacol* 66:1129-1134, 1988.

Congdon ED: Transformation of the aortic-arch system during the development of the human embryo, *Carnegie Contr Embryol* 14:47-110, 1922.

Conte G, Pellegrini A: On the development of the coronary arteries in human embryos, stages 14-19, *Anat Embryol* 169:209-218, 1984.

Crossin KL, Hoffman S: Expression of adhesion molecules during the formation and differentiation of the avian endocardial cushion tissue, *Dev Biol* 145:277-286, 1991.

DeHaan RL: Cardia bifida and the development of pacemaker function of the early chick heart, *Dev Biol* 1:586-602, 1959.

DeHaan RL, Fujii S, Satin J: Cell interactions in cardiac development, *Dev Growth Differen* 32:233-241, 1990.

DeRuiter MC and others: The special status of the pulmonary arch artery in the branchial arch system of the rat, *Anat Embryol* 179:309-325, 1989.

Easton H, Veini M, Bellairs R: Cardiac looping in the chick embryo: the role of the posterior precardiac mesoderm, *Anat Embryol* 185:249-258, 1992.

Edwards JE and others: *Congenital heart disease: correlation of pathologic anatomy and angiocardiography,* vols I and II, Philadelphia, 1965, Saunders.

Feinberg RN, Sherer GK, Auerbach R: *The development of the vascular system,* Basel, Switzerland, 1991, Karger.

Fukiishi Y, Morris-Kay GM: Migration of cranial neural crest cells to the pharyngeal arches and heart in rat embryos, *Cell Tissue Res* 268:1-8, 1992.

Goor DA, Lillehei CW: *Congenital malformations of the heart: embryology, anatomy and operative considerations,* New York, 1975, Grune & Stratton.

Heuser CH: The branchial vessels and their derivatives in the pig, *Carnegie Contr Embryol* 15:121-139, 1923.

Ho E, Shimada Y: Formation of the epicardium studied with the scanning electron microscope, *Dev Biol* 66:579-585, 1978.

Hood LC, Rosenquist TH: Coronary artery development in the chick: origin and deployment of smooth muscle cells, and the effects of neural crest ablation, *Anat Rec* 234:291-300, 1992.

Icardo JM: Cardiac morphogenesis and development, *Experientia* 44:909-919, 1988.

Kanjuh VI, Edwards JE: A review of congenital anomalies of the heart and great vessels according to functional categories, *Ped Clin North Am* 11:55-105, 1964.

Kirby ML: Nodose placode provides ectomesenchyme to the developing chick heart in the absence of cardiac neural crest, *Cell Tissue Res* 252:17-22, 1988.

Kirby ML: Role of extracardiac factors in heart development, *Experientia* 44:944-951, 1988.

Kirby ML, Waldo KL: Role of neural crest in congenital heart disease, *Circulation* 82:332-340, 1990.

Kramer TC: The partitioning of the truncus and conus and the formation of the membranous portion of the interventricular septum of the human heart, *Am J Anat* 71:343-370, 1942.

Magovern JH, Moore GW, Hutchins GM: Development of the atrioventricular valve region in the human embryo, *Anat Rec* 215:167-181, 1986.

Manasek FJ, Burnside MB, Waterman RE: Myocardial cell shape as a mechanism of embryonic heart looping, *Dev Biol* 29:349-371, 1972.

Markwald RR, Manasek FJ: Structural development of endocardial cushions, *Am J Anat* 148:85-120, 1977.

Markwald RR and others: Inductive interactions in heart development, *Ann NY Acad Sci* 588:13-25, 1990.

McClure CFW, Butler EG: The development of the vena cava inferior in man, *Am J Anat* 35:331-383, 1925.

Mikawa T, Fischman DA: Retroviral analysis of cardiac morphogenesis, *Proc Natl Acad Sci* 89:9504-9508, 1992.

Noden DM: Origins and assembly of avian embryonic blood vessels, *Ann NY Acad Sci* 588:236-249, 1990.

Noden DM: *Development of craniofacial blood vessels.* In Feinberg RN, Sherer GK, Auerbach R, eds: *The development of the vascular system,* Basel, Switzerland, 1991, Karger, pp 1-24.

Noden DM: Origins and patterning of avian outflow tract endocardium, *Development* 111:867-876, 1991.

Patten BM: The development of the sinoatrial conduction system, *Univ Mich Med Bull* 22:1-21, 1956.

Patten BM: *The development of the heart.* In Gould SE, ed: *The pathology of the heart,* Springfield, Ill, 1960, Charles C Thomas, pp 24-92.

Pexieder T: The tissue dynamics of heart morphogenesis. I. The phenomena of cell death. *Z Anat Entwickl-Gesch* 138:241-253, 1972.

Rosenquist TH and others: Origin and propagation of elastogenesis in the developing cardiovascular system, *Anat Rec* 221:860-871, 1988.

Runyan RB, Potts JD, Weeks DL: TGF-β 3–mediated tissue interaction during embryonic heart development, *Molec Reprod Dev* 32:152-159, 1992.

Sabin FR: The origin and development of the lymphatic system, *Johns Hopkins Hosp Report* 17:347-440, 1916.

Sherer GK: *Vasculogenic mechanisms and epitheliomesenchymal specificity in endodermal organs.* In Feinberg RN, Sherer GK, Auerbach R, eds: *The development of the vascular system,* Basel, Switzerland, 1991, Karger, pp 37-57.

Sinning AR, Krug EL, Markwald RR: Multiple glycoproteins localize to a particulate form of extracellular matrix in regions of the embryonic heart where endothelial cells transform into mesenchyme, *Anat Rec* 232:285-292, 1992.

Sissman NJ: Developmental landmarks in cardiac morphogenesis: comparative chronology, *Am J Cardiol* 25:141-148, 1970.

Steding G and others: Developmental aspects of the sinus valves and the sinus venosus septum of the right atrium in human embryos, *Anat Embryol* 181:469-475, 1990.

Takamura K and others: Association of cephalic neural crest cells with cardiovascular development, particularly that of the semilunar valves, *Anat Embryol* 182:263-272, 1990.

Waldo KL, Willner W, Kirby ML: Origin of the proximal coronary artery stems and a review of ventricular vascularization in the chick embryo, *Am J Anat* 188:109-120, 1990.

CHAPTER 19

The Fetal Period and Birth

After the eighth week of pregnancy the period of organogenesis (embryonic period) is largely completed and the fetal period begins. By the end of the embryonic period, almost all the organs of the body are present in a grossly recognizable form. The external contours of the embryo show a very large head in proportion to the rest of the body and greater development of the cranial than of the caudal part of the body (Fig. 19-1).

The fetal period has often been considered a time of growth and physiological maturation of organ systems, and it has not received much attention in traditional embryology courses. However, recent advances in imaging and other diagnostic techniques have provided considerable access to the fetus. Determining the fetus' pattern of growth and state of well being with remarkable accuracy is now possible. Improved surgical techniques and the realization that surgical wounds in the fetus heal without scarring have led to a new field of fetal surgery.

This chapter emphasizes the functional development of the fetus and the adaptations that ensure a smooth transition to independent living once the fetus has passed through the birth canal and the umbilical cord is cut. Techniques that are used to monitor the functional state of the fetus are also described.

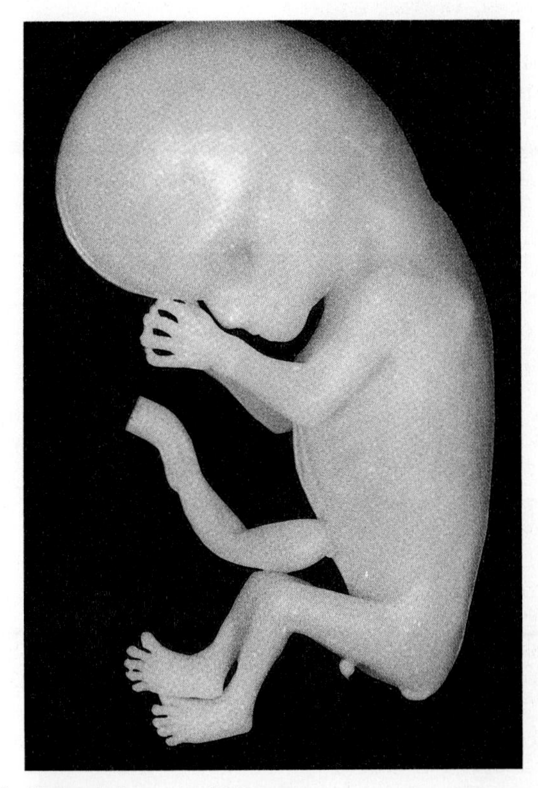

FIG. 19-1 A 9-week-old fetus.
(From England M: *Color atlas of life before birth,* London, 1983, Mosby.)

GROWTH AND FORM OF THE FETUS

Despite the intense developmental activity that occurs during the embryonic period, the absolute growth of the embryo in both length and mass is not great (Fig. 19-2). The fetal period, however, is characterized by intense growth. The change in proportions of the various regions of the body during the prenatal and postnatal growth periods is as striking as the absolute growth of the embryo. The early dominance of the head is reduced as development of the trunk becomes a major factor in growth of the early fetus. Even later, a relatively greater growth of the limbs changes the relative proportions of various regions of the body. During the early fetal period the entire body is hairless and very thin because of the absence of subcutaneous fat (Fig. 19-3). By midpregnancy the contours of the head and face approach those of the neonate, and the abdomen begins to fill out. Beginning at around week 27, the deposition of subcutaneous fat causes the body to round out (Fig. 19-4). (Some of the major developmental landmarks during the fetal period are summarized in the tables on pp. xiii to xv.)

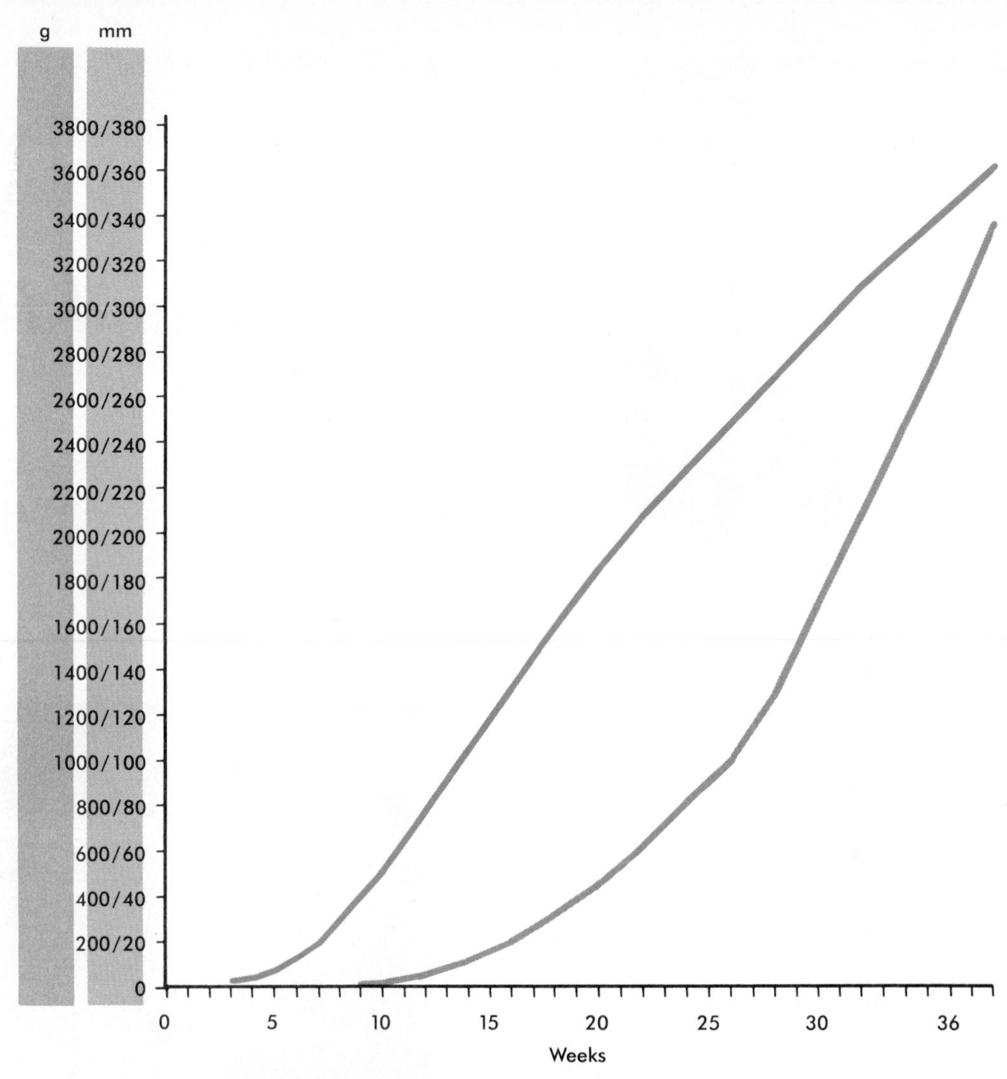

FIG. 19-2 Growth in crown-rump length and weight of the human fetus.
(Data from Patten: *Human embryology,* New York, 1968, McGraw-Hill.)

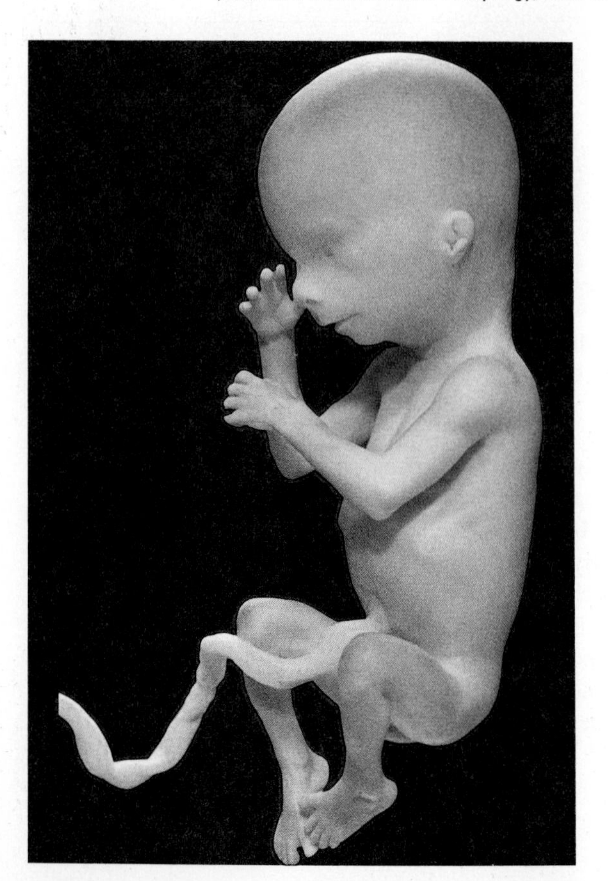

FIG. 19-3 A 14-week-old fetus.
(From England M: *Color atlas of life before birth,* London, 1983, Mosby.)

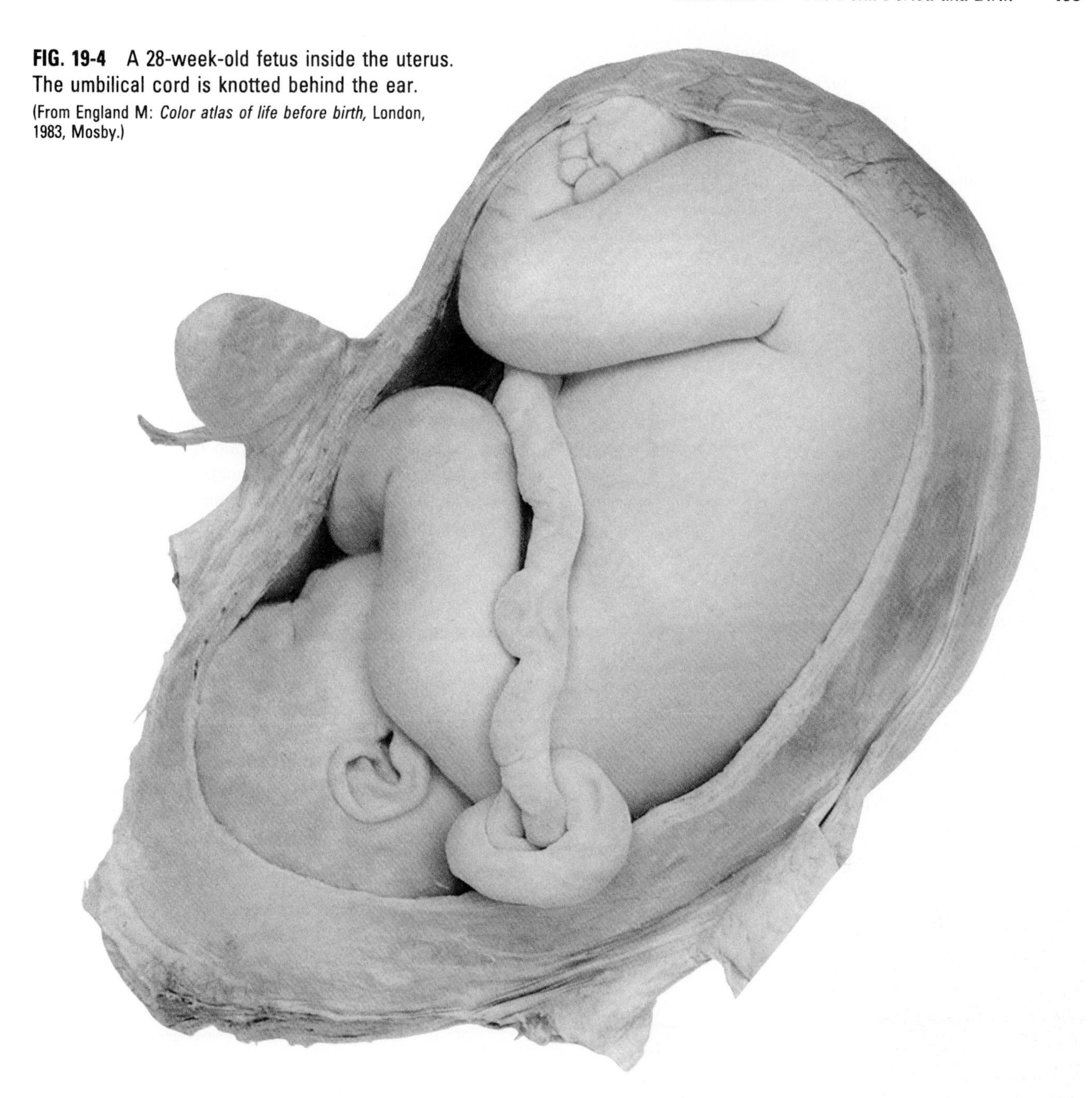

FIG. 19-4 A 28-week-old fetus inside the uterus. The umbilical cord is knotted behind the ear. (From England M: *Color atlas of life before birth,* London, 1983, Mosby.)

FETAL PHYSIOLOGY

The Circulation

The circulation of the human embryo can be first studied at about 5 weeks by means of ultrasound. At that time the heart beats at a rate of approximately 100/min. This probably represents an inherent atrial rhythm. The pulse rate rises to about 160/min by 8 weeks and then drops to 150/min in the 15-week-old fetus, with a further slight decline near term. The pulse rate in utero is remarkably constant, and embryos exhibiting **bradycardia** (slow pulse rate) often die before term. Near term the pulse rate varies to some extent if conditions in the uterus change or if the embryo is stressed. This is probably the result of the functional establishment of the autonomic innervation of the heart (Fig. 19-5).

The heart of the fetus has gross physiological properties quite different from those of the postnatal heart. For example, the myocardial force, the velocity of shortening, and the extent of shortening are all less in the fetal heart. Some gross functional characteristics of the fetal heart are related to the presence of fetal isoforms of contractile proteins in the cardiac myocytes. For example, in fetal heart cells the β-myosin heavy-chain isoform predominates. This is advantageous because a lower oxygen requirement and less adenosine triphosphate (ATP) are needed to develop the same amount of force as the α-myosin isoform in the adult heart.

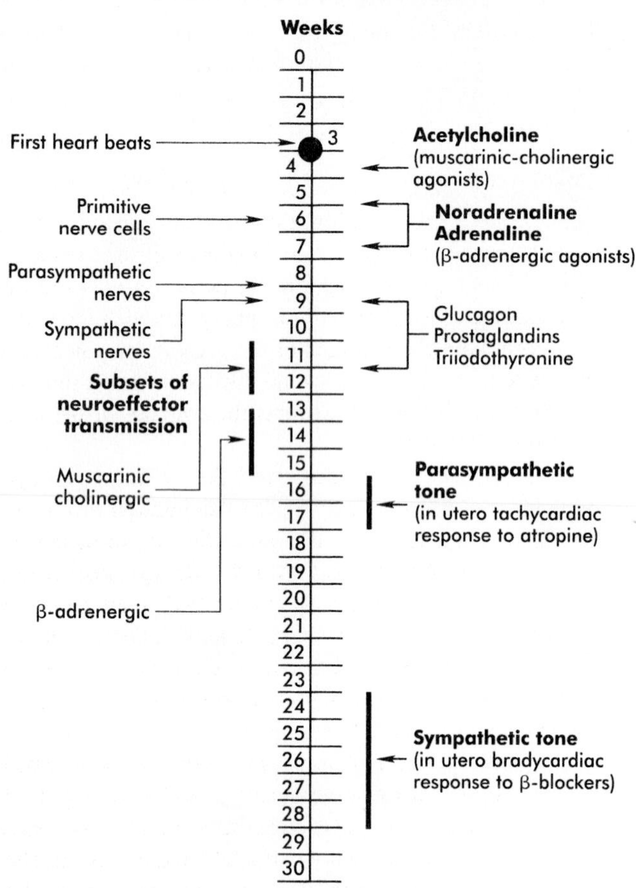

FIG. 19-5 Sequence of events in the autonomic innervation of the heart.
(Based on studies by Papp [1988].)

The **stroke volume** (blood expelled with one heart beat) of the early (18 to 19 week gestation) fetus is very small (less than 1 ml), but it increases rapidly with continued growth of the fetus. In the human fetus the combined ventricular output is about 450 ml/kg/min. The right ventricle of the human fetus has a somewhat greater stroke volume than the left. This is correlated with an 8% greater diameter of the pulmonary artery than that of the fetal aorta.

Quantitative studies have shown a good correlation between blood flow and functional needs of various regions of the embryo. Approximately 40% of the combined cardiac output goes to the head and upper body, thus supplying the relatively great needs of the developing brain. Another 30% of the combined cardiac output goes to the placenta via the umbilical arteries for replenishment. Fig. 19-6 shows the relative amounts of blood that enter and leave the heart via various vascular channels. (The general qualitative pattern of blood flow in the human fetus is presented in Chapter 18.)

Differential streaming of blood within the heart results in different concentrations of oxygen in the chambers of the fetal heart. For example, blood in the left ventricle is 15% to 20% more saturated with oxygen than blood in the right ventricle (see Fig. 18-20). This and the high volume of blood supplying the head via branches of the ascending aorta ensure that the developing heart and brain receive an adequate supply of oxygen.

A key factor in the maintenance of the fetal pattern of circulation is the patency of the ductus arteriosus and the ductus venosus. The classic view was that their patency was maintained by passive means, but active mechanisms are

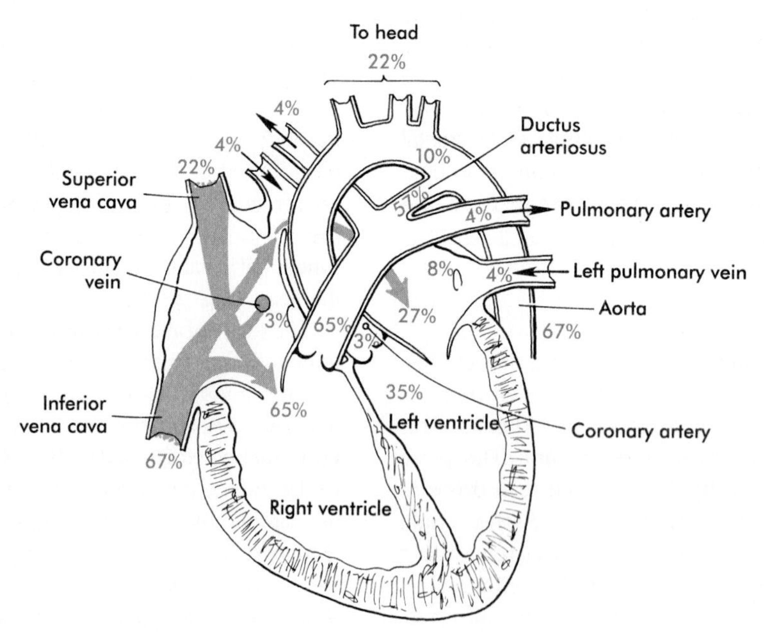

FIG. 19-6 Percentages of blood entering and leaving the heart via various channels.
(Data from Teitel and others [1987].)

clearly involved. Patency of the fetal ductus venosus is maintained through the actions of prostaglandins E_2 and I_2, whereas only prostaglandin E_2 is involved in maintaining patency of the ductus arteriosus.

The Fetal Lungs and Respiratory System

The lungs develop late in the embryo and are not involved in respiratory gas exchange during fetal life. However, they must be prepared to assume the full burden of gas exchange instantly once the umbilical cord is cut.

The fetal lungs are filled with fluid, and the blood circulation to them is highly reduced. To perform normal postnatal breathing, the lungs must grow to an appropriate size, respiratory movements must be carried out continuously, and the air sacs **(alveolae)** must become appropriately configured for air exchange.

Ultrasound analysis has shown that the fetus begins to make gross breathing movements as early as 11 weeks. These movements are periodic rather than continuous, and they take on two forms. One type of movement is rapid and irregular, with varying rate and amplitude. The other form is represented by isolated, slow movements, almost like gasps. The former type is by far the most prominent and is associated with conditions of rapid eye movement (REM) sleep. The fetus alternates periods of rapid breathing (often for about 10 minutes) with periods of **apnea** (cessation of breathing).

Much remains to be learned about the control of fetal breathing, but the breathing movements are known to be responsive to maternal factors, many of which remain to be identified. The amount of breathing (minutes of breathing per hour) is highest in the evening and lowest in the early morning. Fetal breathing increases after the mother has eaten. This may be related to the concentration of glucose in the maternal blood. Maternal smoking causes a rapid decrease in fetal breathing for up to an hour.

Fetal breathing movements are essential for postnatal survival. One obvious function of fetal breathing is to condition the respiratory muscles so that they can perform regular postnatal contractions. Another important function is to stimulate the growth of the embryonic lungs. If intrauterine breathing movements are suppressed in fetal lambs, lung growth is retarded. In vitro studies have shown that proliferation of lung epithelial cells is stimulated by mechanical stretching.

Another important aspect of development of the fetal respiratory system is growth of the upper airway. Although a newborn is about 4% the weight of an adult, the diameter of its trachea is one third that of the adult trachea. Other components of the airway are similarly proportioned. If this were not the case, the physical resistance to air flow would be so great that movement of air would be almost impossible. Even with these adaptations, the resistance of the neonate's airway is 5 to 6 times greater than that of the adult.

During the last trimester of pregnancy, fluid constitutes 90% to 95% of the total weight of the lung. Some fluid is inhaled amniotic fluid, and the rest is secreted by the lungs. The role of the pulmonary fluid in lung development is still not well understood, but a practical problem is its rapid removal after birth (see p. 422).

A functionally important aspect of fetal lung development is the secretion of **pulmonary surfactant** by the newly differentiating type II alveolar cells of the lung starting around 24 weeks gestation. Surfactant is a mixture of phospholipids (about two-thirds phosphatidylcholine) and protein that lines the surface of the alveoli and lowers the surface tension. This reduces the inspiratory force required to inflate the alveoli and prevents the collapse of the alveoli during expiration.

Despite the relatively early initiation of surfactant synthesis, large amounts are not synthesized until a few weeks before birth. At this time the production of surfactant by the type II alveolar cells is higher than at any other period in an individual's life, an adaptation that is an important preparation for the newborn's first breath. A number of hormones and growth factors are involved in the synthesis of surfactant, and the effects of thyroid hormone and glucocorticoids are particularly strong.

Prematurely born infants are often afflicted with **respiratory distress syndrome,** which is manifest by rapid, labored breathing shortly after birth. This condition is related to a deficiency in pulmonary surfactant and can be ameliorated by the administration of glucocorticoids, which stimulate the production of surfactant by the alveolar epithelium.

Fetal Movements and Sensations

Ultrasound analysis has revolutionized the analysis of fetal movements and behavior because the fetus can be examined virtually undisturbed (except for an increase in vascular activity induced by the ultrasound) for extended periods. Earlier studies of fetal movements were principally concerned with the development of reflex responses, and the information was obtained largely by the analysis of newly aborted fetuses (see Chapter 12). Although valuable information on maturation of reflex arcs was obtained in this manner, many of the movements elicited were not those normally made by the fetus in utero.

The undisturbed embryo does not show any indication of movement until about 7½ weeks. The first spontaneous movements consist of slow flexion and extension of the vertebral column, with the limbs being passively displaced. Within a short time, a relatively large repertoire of fetal movements evolves. After study by a number of investigators, a classification of fetal movements has been suggested (box). The first fetal movements are followed in a few days by startle and general movements. Shortly thereafter, isolated limb movements are added (Fig. 19-7). Late in ap-

Major Types of Fetal Movements

General movements. These slow gross movements involve the whole body. Their duration is from several seconds to a minute.
Startle movements. These quick (less than 1 second), generalized movements always start in the limbs and may spread to the trunk and neck.
Hiccups. These are repetitive phasic contractions of the diaphragm. A bout may last several minutes.
Fetal breathing movements. These are paradoxical movements in which the thorax moves inward and the abdomen outward with each contraction of the diaphragm.
Isolated arm or leg movements. These movements of extremities occur without movement of the trunk.
Hand-face contact. This occurs any time the moving hand makes contact with the face or mouth.
Retroflexion of the head. This is a slow to jerky backward bending of the head.
Lateral rotation of the head. This involves isolated turning of the head from side to side.
Anteflexion of the head. This is a normally slow forward bending of the head.
Opening of mouth. This isolated movement may be accompanied by protrusion of the tongue.
Yawn. The mouth is slowly opened and rapidly closed after a few seconds.
Sucking. This burst of rhythmical jaw movements is sometimes followed by swallowing. With this movement the fetus may be drinking amniotic fluid.
Stretch. This complex movement involves overextension of the spine, retroflexion of the head, and elevation of the arms.
Based on studies by Prechtl (1989).

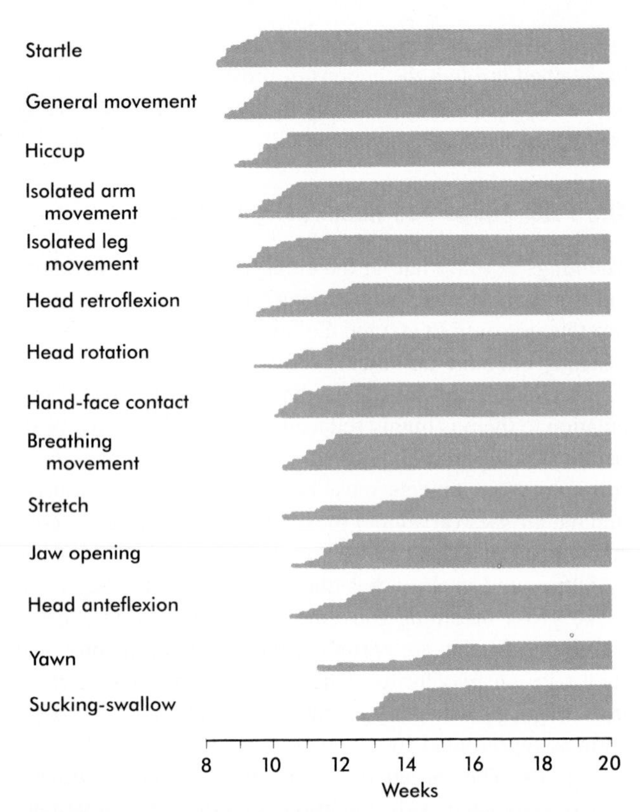

FIG. 19-7 Time of appearance of specific patterns of fetal motor movements. Each corner on the jagged edge at the left of the broad lines represents one fetus. (Data from De Vries and others [1982].)

pearance are movements associated with the head and jaw (Fig. 19-7).

Continuous ultrasound monitoring for extended periods reveals patterns involving many types of movements (Fig. 19-8). At different weeks of pregnancy, some movements are predominant, whereas others are in decline or are just beginning to take shape. Analysis of anencephalic fetuses has shown that although many movements take place, they are poorly regulated. They start abruptly, are maintained at the same force, and then stop abruptly. These abnormal patterns of movements are considered evidence for strong supraspinal modulation of movement in the fetus.

Human fetal activities, as reflected in breathing or general activity level, show distinct diurnal rhythms beginning at about 20 to 22 weeks of gestation. There is a strong negative correlation between maternal plasma glucocorticoid levels and fetal activity. Fetal activity is highest in the early evening, when maternal blood glucocorticoid levels are lowest, and lowest in the early morning, when the concentration of maternal hormone peaks. Studies of women who have been placed on added glucocorticoids or who have been given inhibitors have shown increased fetal activity when maternal corticoid levels are low. Usually when the overall activity is low, the fetus is in a state of REM sleep, but definitions of sleep and wakefulness in the fetus need further clarification.

A number of sensory systems also begin to function during the fetal period. (Reflex responses to tactile stimulation are outlined in Chapter 12.) Near-term fetuses are responsive to loud 2000-Hz stimuli when in a state of wakefulness, but they are unresponsive during periods of sleep. Loud vibroacoustic stimuli applied to the maternal abdomen produce a fetal response consisting of an eye blink, a startle reaction, and an increase in heart rate. Although the fetus is constantly in the dark, the **pupillary light reflex** can usually be elicited by 30 weeks.

The Fetal Digestive Tract

The fetal digestive tract is not functional in the standard sense because the fetus obtains its nutrition from the maternal blood via the placenta. However, the digestive tract must be prepared to assume the full responsibility for nutritional intake after birth. Once the basic digestive tube and

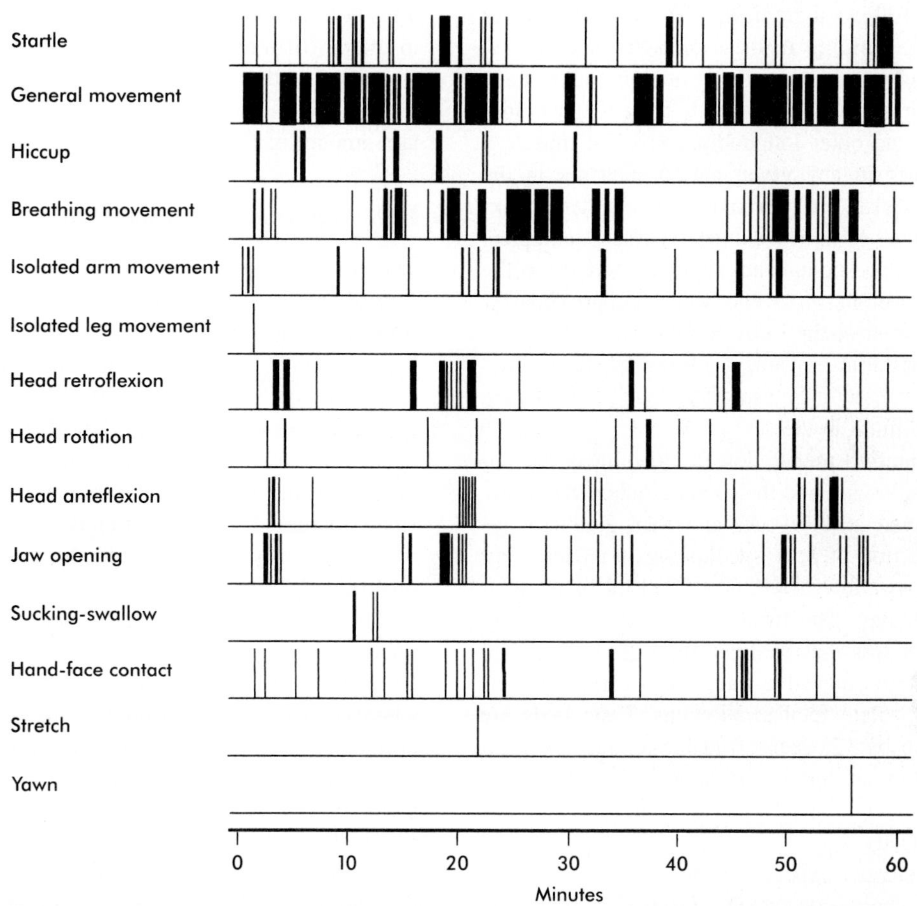

FIG. 19-8 Actogram record of types of movements of a 14-week-old fetus taken over 60 minutes.
(Modified from Prechtl HFR: *Fetal neurology,* New York, 1989, Raven Press.)

glands have formed in the early embryo, the remainder of the intrauterine period is devoted to cellular differentiation of the epithelia of the gut and preparation of the numerous cells involved for their specific roles in the digestive process. Below the epithelium the walls of the digestive tube must become capable of propelling ingested food and liquid. Analysis of development of the fetal digestive tract has concentrated on (1) the biochemical adaptations of the epithelium of the various regions for digestive function and (2) the development of motility of the digestive tube.

The development and differentiation of epithelia or specific regional characteristics of the gut lining typically follow gradients along the length of the segment of the gut specifically involved. In both the esophagus and stomach, differentiation of the mucosal epithelium is well underway starting around 4 months. Although **parietal cells** (HCl producing) and **chief cells** (pepsinogen producing) are first seen at 11 and 12 weeks, respectively, there is little evidence of their secretions during fetal life. In fact, the contents of the stomach are near neutral pH until after birth, but then gastric acid production increases greatly within a few hours.

In the small intestine, villi begin to form in the upper duodenum at the end of the second month, and crypts appear a week or two later. The formation of villi and crypts spreads along the length of the intestine in a spatiotemporal gradient. By approximately 16 weeks, villi have formed along the entire length of the intestine, and crypts appear in the lower ileum by 19 weeks. Villi even form in the colon during the third and fourth months, but they then regress and are gone by the seventh or eighth month.

Individual epithelial cell types, including *Brunner's glands,* which protect the duodenal lining from gastric acid, appear in the small intestine early in the second trimester. Although the presence of most enzymes or proenzymes characteristic of the intestinal lining can be demonstrated histochemically during the midfetal period, their amounts are generally quite small. Activity of a number of the enzymes secreted by the exocrine pancreatic tissue can also be demonstrated between 16 and 22 weeks gestation.

Meconium, a greenish mixture of desquamated intestinal cells, swallowed lanugo hair, and various secretions, begins to fill the lower ileum and colon late in the fourth month.

Differentiation of the neuromuscular complex of the digestive tract also follows a gradient, with the circular layer of smooth muscle forming in the esophagus at 6 weeks. **Myenteric plexuses** (parasympathetic neurons) take shape after the inner circular muscle layer is present but before the formation of the outer longitudinal layer of muscle a couple weeks later in any given region. Starting in the esophagus at 6 weeks, final formation of myenteric plexuses throughout the length of the digestive tract is complete at 12 weeks. The first spontaneous rhythmic activity in the small intestine is seen in the seventh week at approximately the time of formation of the inner circular muscular layer. Recognizable peristaltic movements, however, do not begin until the fourth month. Fetuses older than 34 weeks are able to pass meconium in utero.

Another intrauterine preparation for feeding is the development of swallowing and the sucking reflex. Swallowing is first detected at 11 weeks and then gradually increases. The function of fetal swallowing is unclear, but by term, fetuses swallow from 200 to 750 ml or more of amniotic fluid per day. The swallowed amniotic fluid may contain growth factors that facilitate the differentiation of epithelial cells in the digestive tract. To a certain extent, taste seems to regulate fetal swallowing. Taste buds are seemingly mature by 12 weeks, and the amount of swallowing increases if saccharin is introduced into the amniotic fluid. Conversely, swallowing is reduced if noxious chemicals are added.

Sucking movements appear late in fetal development. Before 32 weeks, there is no sucking. From 32 to 36 weeks, the fetus undertakes short bursts of sucking, but these are not associated with effective swallowing movements. This is the main reason premature infants of this age must be fed through a nasogastric tube. Mature sucking capability appears after 36 weeks.

Fetal Kidney Function

Although the placenta carries out most excretory functions characteristic of the kidney during prenatal life, the developing kidneys also function by producing urine. As early as the fifth week of gestation, the mesonephric kidneys produce small amounts of very dilute urine, but the mesonephros degenerates late in the third month, after the metanephric kidneys have taken shape. Tubules of the metanephric kidneys begin to function between 9 and 12 weeks, and resorptive functions involving the loop of Henle occur by 14 weeks even though new nephrons continue to form until birth. The urine produced by the fetal kidney is hypotonic to plasma throughout most of pregnancy. This is a reflection of immature resorptive mechanisms, which are manifest morphologically by short loops of Henle. As the neural lobe of the hypophysis produces antidiuretic hormone beginning at the eleventh week, another mechanism for the concentration of urine begins to be established.

Intrauterine renal function is not necessary for life of the fetus because embryos with bilateral renal agenesis survive in utero. Bilateral renal agenesis, however, is commonly associated with oligohydramnios (see Chapter 6), indicating that the overall balance of amniotic fluid requires a certain amount of fetal renal function.

Endocrine Function in the Fetus

The development of prenatal endocrine function occurs in several phases. Most peripheral endocrine glands (e.g., thyroid, pancreatic islets, adrenals, gonads) form early in the second month as the result of epithelial-mesenchymal interactions. As these glands differentiate late in the second month or early in the third, they develop the intrinsic capacity to synthesize their specific hormonal products. In most cases the amount of hormone secreted is initially very small; increased secretion often depends on the stimulation of the gland by a higher order hormone produced in another gland.

The anterior pituitary gland develops like many other endocrine glands. Its hormonal products generally act to stimulate more peripheral endocrine glands such as the thyroid, adrenals, and gonads to produce or release their specific hormonal products. Pituitary hormones can be demonstrated immunocytochemically within individual pituitary epithelial cells as early as 8 (adrenocorticotrophic hormone [ACTH]) or 10 (luteinizing hormone [LH] and follicle-stimulating hormone [FSH]) weeks. However, most pituitary hormones are typically not present in the blood in detectable quantities until a couple of months after they can be demonstrated in the cells that produce them. An exception is growth hormone, which can be detected in plasma as early as 10 weeks.

While the anterior pituitary is developing its intrinsic synthetic capacities, the hypothalamus also takes shape and develops its capacity to produce the various releasing and inhibitory factors that modulate the function of the pituitary gland. Regardless of its intrinsic capacities, the hypothalamus is limited in its influence on the embryonic pituitary gland until about 12 weeks, when the neurovascular links between the hypothalamus and pituitary become established.

At each level a generally low intrinsic level of hormone production can be stimulated by the actions of hormones produced by the next higher order gland in the control hierarchy. For example, the amount of thyroid hormone released is considerably increased when thyroid-stimulating hormone (TSH), released by the anterior pituitary, acts on the thyroid gland. The release of TSH by the pituitary is regulated by thyrotropin-releasing hormone (TRH), which is produced in the hypothalamus. Studies on anencephalic fetuses have shown that the anterior pituitary can produce and release most of its hormones in the absence of hypothalamic input, although plasma concentrations of some are reduced.

Among the fetal endocrine glands the adrenal remains the most enigmatic. By 6 to 8 weeks of development the inner cortex enlarges greatly to form a distinct fetal zone, which later in pregnancy occupies about 80% of the gland. By the end of pregnancy, the adrenal glands weigh 4 g each—the same mass as that of the adult glands (Fig. 19-9). The fetal adrenal cortex produces 100 to 200 mg of steroids per day, an amount several times higher than that of the adult adrenal glands. The main hormonal products of the fetal adrenal are Δ^5-3β-hydroxysteroids such as dihydroepiandosterone, which are inactive alone but are converted to biologically active steroids (e.g., estrogens, especially estrone) by the placenta and liver. The fetal adrenal cortex depends on the presence of pituitary ACTH; in its absence the fetal adrenal cortex is small. Conversely, if exogenous ACTH is administered, the fetal adrenal cortex persists after birth.

Despite the prominence of the fetal adrenal cortex, its functions during pregnancy are still not clear. Fetal adrenal hormones influence maturation of the lungs (as prolactin has also been postulated to do), liver, and epithelium of the digestive tract. In sheep, products of the adrenal cortex influence the initiation of parturition, but the situation in primates is considerably less clear. Shortly after birth the fetal adrenal cortex rapidly involutes (see Fig. 19-9). Within a month after birth the weight of each gland is reduced by 50%, and by a year each gland weighs only 1 g. Not until adulthood does the mass of the adrenal glands return to that of the late fetus.

Endocrinology of the fetus is complicated by the presence of the placenta, which can synthesize and release many hormones, convert hormones released from other glands to active forms, and potentially exchange other hormones with the maternal circulation. By 6 to 7 weeks, hormone production (e.g., progesterone) by the placenta is enough to maintain pregnancy even if the ovaries are removed.

One of the earliest placental hormones produced is **chorionic gonadotropin (HCG)** (see Chapter 6). One later function of HCG is to stimulate steroidogenesis by the placenta. The synthesis of HCG by the syncytiotrophoblast of the placenta is regulated by the production of **gonadotropin-releasing hormone (GnRH)** by cells of the cytotrophoblast. Synthesis of GnRH by the placenta supplants its normal production by the hypothalamus and is probably an adaptation that allows earlier and more local control of HCG than could be accomplished by the hypothalamus.

CLINICAL STUDY AND MANIPULATION OF THE FETUS

New imaging and diagnostic techniques have revolutionized the study of the living fetus. Many congenital malformations can be diagnosed in utero with considerable accuracy.

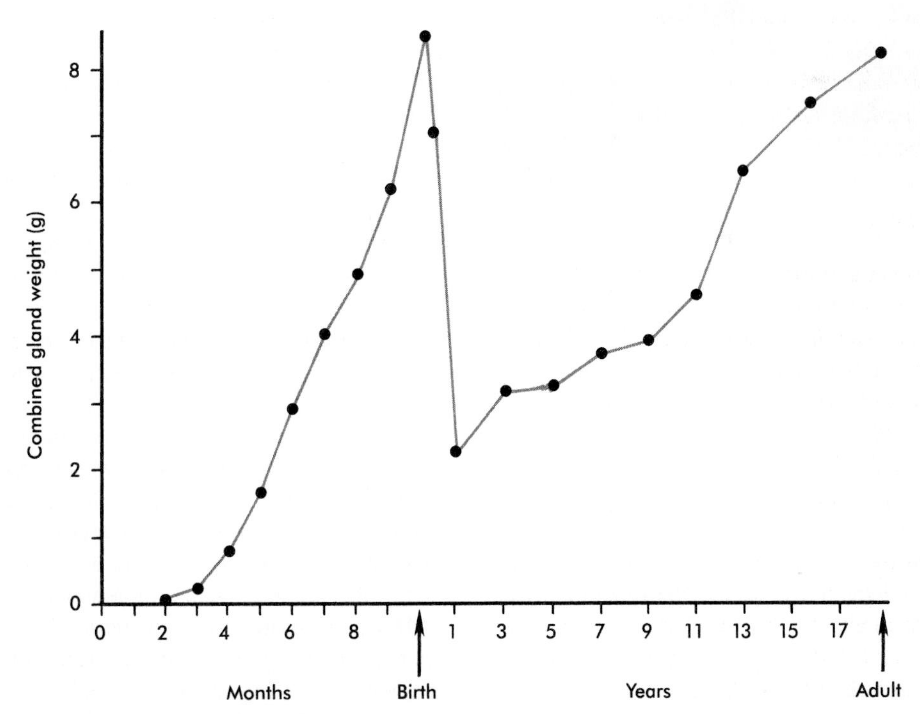

FIG. 19-9 Graph of the weights of the human adrenal glands during prenatal and postnatal development. After birth the weight of the gland decreases dramatically with the reorganization of the cortex of the gland.

(Based on data from Neville AM, O'Hare MJ: The human adrenal cortex, Berlin, 1982 Springer-Verlag.)

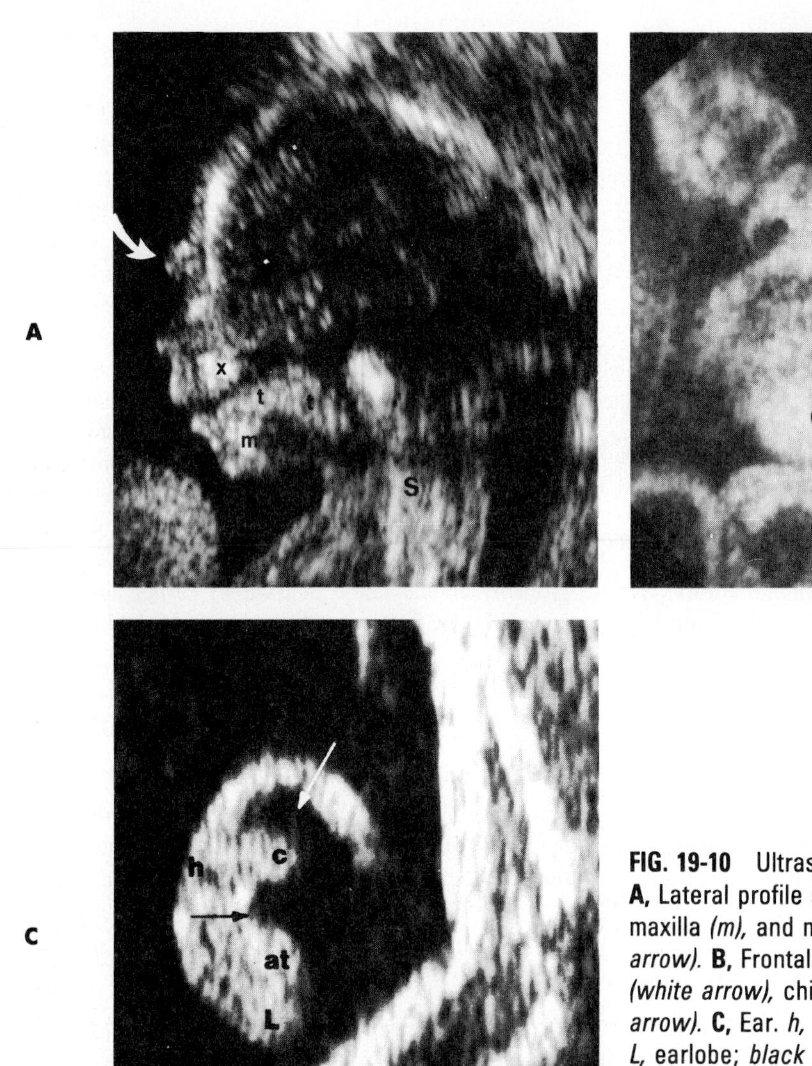

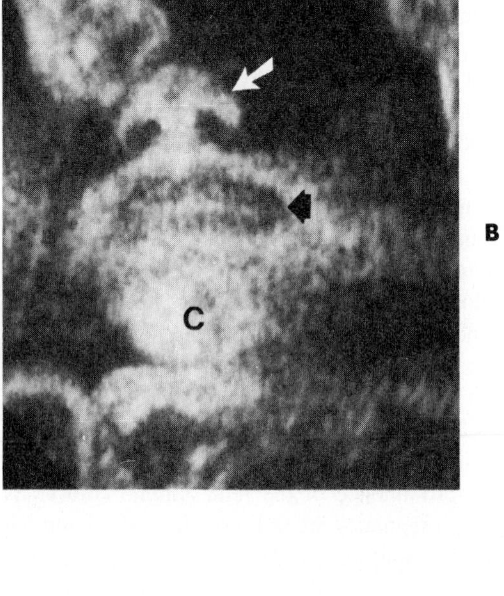

FIG. 19-10 Ultrasound images of the normal fetal head. **A,** Lateral profile showing the nose *(arrowhead),* chin *(C),* maxilla *(m),* and medial aspect of the bony orbit *(curved arrow).* **B,** Frontal view of the face showing the nose *(white arrow),* chin *(c),* and corner of the mouth *(black arrow).* **C,** Ear. *h,* Helix; *c,* antihelical crus; *at,* antitragus; *L,* earlobe; *black arrow,* antihelix; *white arrow,* scaphoid fossa.

(**A** and **B** from Bowerman RA: *Atlas of normal fetal ultrasonographic anatomy,* ed 2, St Louis, 1992, Mosby. **C** from Nyberg DA and others: *Diagnostic ultrasound of fetal anomalies,* St Louis, 1990, Mosby.)

On the basis of this information the surgeon can treat some congenital malformations through fetal surgery much more efficiently than by traditional surgery on infants or older children.

Fetal Diagnostic Procedures

Imaging Techniques. Because of its safety, cost, and ability to look at the fetus in real time, **ultrasonography** is currently the most widely used obstetrical imaging technique (Figs. 19-10 and 19-11). It is useful for the simple diagnosis of structural anomalies and can be used in real time to guide fetal invasive procedures such as chorionic villus sampling and intrauterine transfusions. The major uses of ultrasonography are summarized in the box.

Conventional **x-ray** testing continues to be used in certain circumstances, but because of the potential for radiation damage to the fetal and maternal gonads, its use is less common than in previous years. The use of x-ray testing is relatively limited by its inability to discriminate the details of soft tissues, including cartilaginous components of the skeleton. By injecting radioopaque substances into the amniotic cavity **(amniography, fetography),** clinicians can obtain outlines of the fetus and amniotic cavity. Other imaging techniques, such as **magnetic resonance imaging (MRI), computer-assisted tomographic (CAT) scans,** and **xeroradiography,** produce useful images of the fetus, but their use is limited because of factors such as cost and availability (Figs. 19-12 and 19-13).

Fetoscopy is the direct visualization of the fetus through a tube inserted into the amniotic cavity. This is accomplished principally through the use of fiber optic technol-

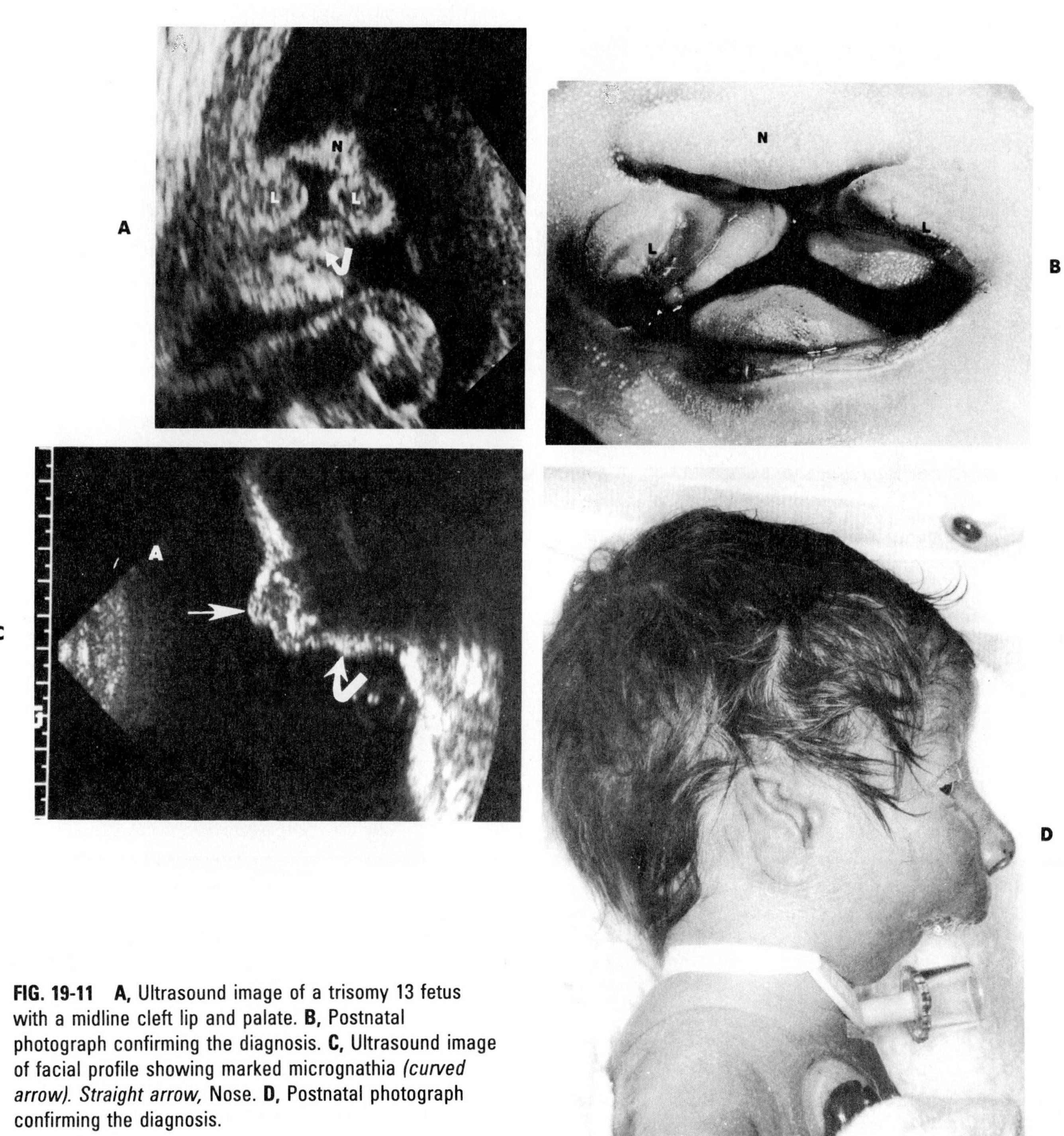

FIG. 19-11 A, Ultrasound image of a trisomy 13 fetus with a midline cleft lip and palate. **B,** Postnatal photograph confirming the diagnosis. **C,** Ultrasound image of facial profile showing marked micrognathia *(curved arrow). Straight arrow,* Nose. **D,** Postnatal photograph confirming the diagnosis.

(**A** and **B,** From Nyberg D, Mahony B, Pretorius D. *Diagnostic ultrasound of fetal anomalies,* St Louis, 1990, Mosby, **C** and **D** from Benson CB and others: *Ultrasound Med* 7:163-167, 1988.)

Uses for Ultrasonography during Pregnancy

- Estimation of age of fetus
- Confirmation of multiple fetuses
- Diagnosis of placental abnormalities, including hydatiform mole
- Localization of placenta in suspected placenta previa
- Diagnosis of ectopic pregnancy
- Documentation of follicular development
- Adjunct to clinical procedures (e.g., amniocentesis, chorionic villus sampling, intrauterine transfusion, intrauterine surgical procedures)
- Detection of congenital malformations of fetus
- Detection of maternal uterine anomalies
- Detection of oligohydramnias or polyhydramnios
- Confirmation of fetal death
- Confirmation of fetal position

A

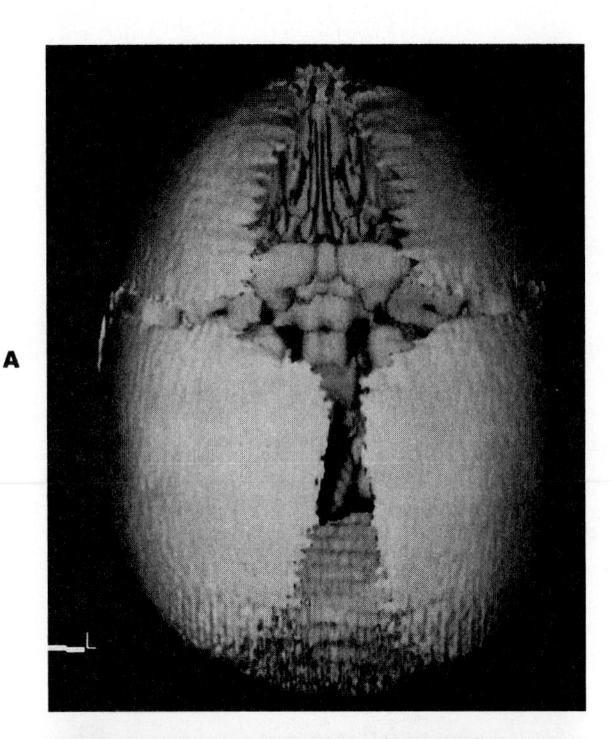

B

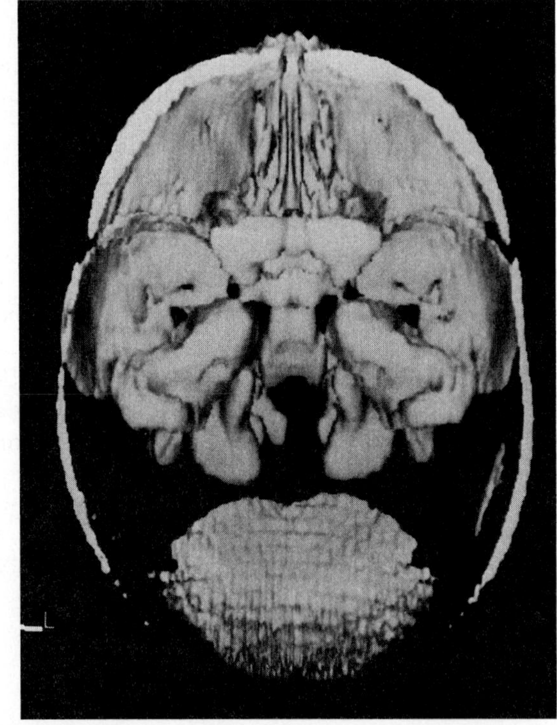

FIG. 19-12 High resolution computer tomography reconstructions of the skull of an 18-week-old fetus. **A,** Focus on superficial bones of the skull. **B,** Deeper bones from the same skull.

(Courtesy R.A. Levy, H. Maher, and A.R. Burdi, Ann Arbor, Mich.)

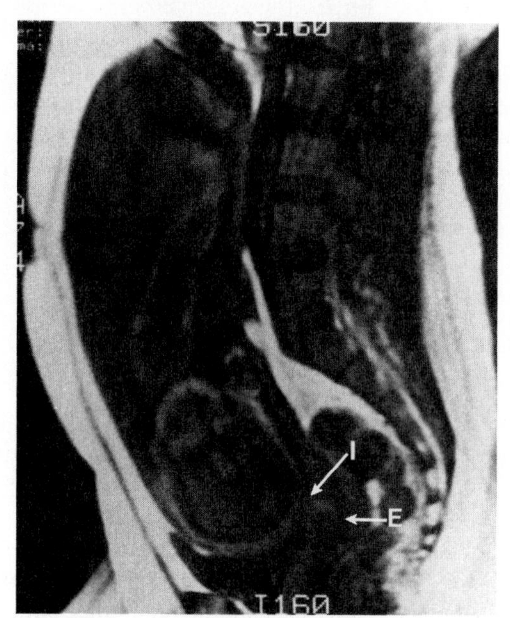

FIG. 19-13 Magnetic resonance imaging image of a normal third trimester fetus inside the uterus. The head of the fetus is near the point of the arrow from *I* (internal cervical os). *E,* External cervical os.

(From Friedman AC and others: *Clinical pelvic imaging,* St Louis, 1990, Mosby.)

ogy. Because of the risk of spontaneous abortion and infection, this technique is not normally used for purely diagnostic purposes but as an aid to intrauterine sampling procedures. Its use has been largely supplanted by other techniques that rely on ultrasonic guidance.

Sampling Techniques. The classical sampling technique is **amniocentesis,** which involves the insertion of a needle into the amniotic sac and removal of a small amount of amniotic fluid for analysis. Amniocentesis is normally not performed before the thirteenth week because of the relatively small amount of amniotic fluid.

Amniocentesis was originally employed for detecting chromosomal anomalies (e.g., Down syndrome) in fetal cells found in the amniotic fluid and for the determination of levels of **α-fetoprotein,** a marker for closure defects of the neural tube. Analysis of the fetal cells in amniotic fluid is also the basis for determining the gender of embryos. This is typically accomplished by the use of a fluorescent dye that intensely stains the Y chromosome. At present, a variety of analytic procedures on amniotic fluid and cells cultured from the fluid are used to detect many enzymatic and biochemical defects in embryos and to monitor the condition of the fetus.

Another widely used diagnostic technique is **chorionic villus sampling.** In this technique, ultrasonography is used as a guide to insert a biopsy needle into the placenta, where a small sample of the villi is removed for diagnostic purposes. This technique is typically used at earlier periods of pregnancy (6 to 9 weeks) than amniocentesis.

With increasing sophistication of fetal imaging techniques, especially ultrasonography, sampling fetal tissues directly is possible. Ultrasonic-guided sampling of fetal blood, mainly from umbilical vessels, is now relatively common for the diagnosis of hereditary and pathological conditions such as immunodeficiencies, coagulation defects, hemoglobin abnormalities, and fetal infections. It is also possible to biopsy fetal skin and even the fetal liver for organ-specific abnormalities.

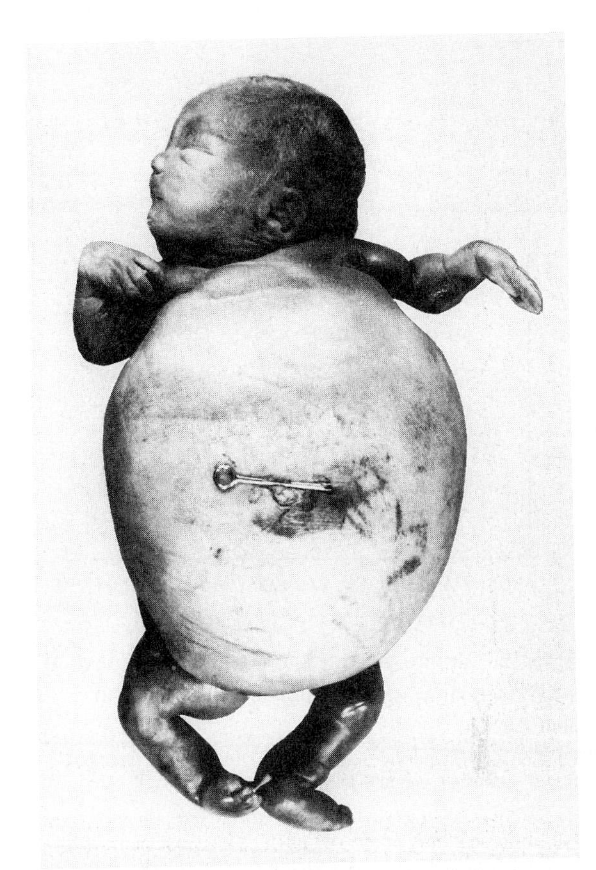

FIG. 19-14 A fetus with great abdominal distention due to megacystis (large bladder) caused by a cloacal plate. (Courtesy Mason Barr, Ann Arbor, Mich.)

Therapeutic Manipulations on the Fetus

Some conditions are better treated in the fetal period than after birth. In some cases involving blockage, severe structural damage to the fetus can be prevented. In other cases the buildup of toxic waste products can be reduced. The recognition that fetal surgery produces essentially scarless results has stimulated some surgeons to consider corrective surgery in utero rather than waiting until after birth.

Fetal shunts can be applied to correct specific conditions in which major permanent damage would result before the time of birth. One such situation is a shunt into the urinary bladder to relieve the pressure and subsequent kidney damage caused by anatomical obstructions of the lower urinary tract. Fig. 19-14 shows the consequence of nontreatment of a persistent cloacal plate, which results in **megacystitis** (enlarged bladder). Fetal shunts have also been used to attempt to relieve the cerebrospinal pressures that result in hydrocephaly (see Fig. 12-34), but the results of these procedures have been equivocal.

Fetal blood transfusions are used for the treatment of fetal anemia and severe erythroblastosis fetalis (see Chapter 6). Earlier, the blood was introduced intraperitoneally. With the increasing sophistication of umbilical cord blood sampling techniques, performing direct intravascular transfusions is now possible.

Open fetal surgery can now be performed because of the diagnostic procedures that allow an accurate assessment of the condition of the fetus. This is still a very new and highly experimental procedure, and its application has been confined to cases of fetal anomalies that would cause grave damage to the fetus if left uncorrected before birth. Currently, the principal indications for open fetal surgery are blockage of the urinary tract, severe diaphragmatic hernia, and some cases of hydrocephalus. Open fetal surgery entails a risk to the mother as well, and the advisability of such a procedure must be carefully considered. With future improvements in procedures, correcting other malformations such as cleft lip and palate or limb deformaties in utero may be possible.

PARTURITION

Parturition, the process of childbirth, occurs approximately 38 weeks after fertilization (Fig. 19-15). The process of childbirth consists of three distinct stages of labor. The first, the **stage of dilatation,** begins with the onset of regular hard contractions of the uterus and ends with complete dilatation of the cervix. Although the contractions of the uterine smooth muscle may appear to be the dominant

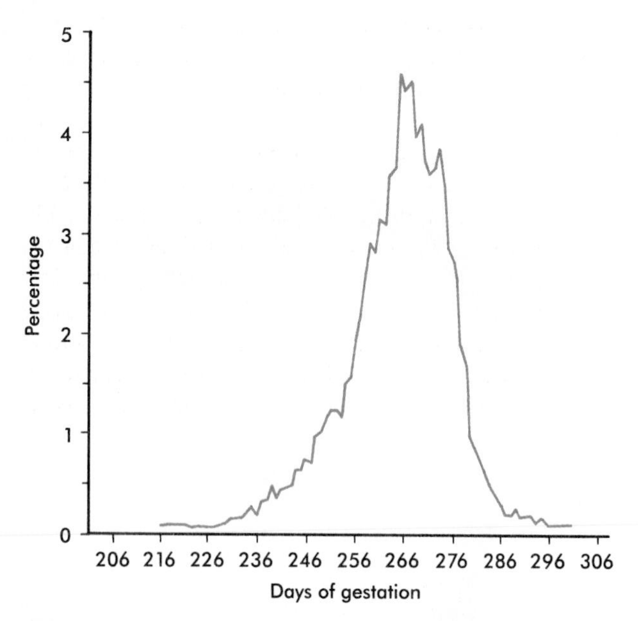

FIG. 19-15 Graph showing the distribution in days of normal pregnancy in 1336 spontaneous, full-term deliveries.

(Modified from Wigglesworth J, Singer D: *Textbook of fetal and perinatal pathology,* London, 1991, Blackwell Scientific.)

process in the first stage of labor, the most important component is the effacement and dilatation of the cervix. During the entire pregnancy the cervix functions to retain the fetus in the uterus. For the childbirth process to proceed, the cervix must change consistency from a firm, almost tubular structure to one that is soft and distensible and not canal-like. This change involves a reconfiguration and possibly removal of much of the cervical collagen. Although many of the factors underlying the reconfiguration of the cervix during the first stage of labor remain undefined, considerable evidence exists for an important role of prostaglandin $F_2\alpha$ in the process. Although variation is great, the average length of the first stage of labor is approximately 12 hours.

The second stage of labor **(stage of expulsion)** begins with complete dilatation of the cervix and ends with the passage of the baby from the birth canal. During this stage, which typically lasts 30 to 60 minutes depending on the number of previous deliveries of the mother, the baby still depends on a functioning umbilical circulation for survival.

The third stage of labor **(placental stage)** represents the period between delivery of the baby and expulsion of the placenta. Typically, the umbilical cord is cut within minutes of delivery, and the baby must then quickly adapt to independent living. During the next 15 to 30 minutes, continued contractions of the uterus separate the placenta from the maternal decidua and the intact placenta is delivered. After delivery of the placenta, major hemorrhage from the spiral uterine arteries is prevented naturally by continued contraction of the myometrium. In actual clinical practice the third stage is commonly abbreviated by intramuscular injection of synthetic oxytocin and external manipulation of the uterus to reduce the amount of uterine blood loss.

The mechanisms underlying the initiation and progression of parturition in humans remain remarkably poorly understood, even though considerable progress had been made in uncovering the stimuli for parturition in certain domestic animals. In sheep, parturition is initiated by a sharp increase in the cortisol concentration in the fetal blood. As a result, placental enzyme activity changes, resulting in the conversion of placental progesterone to estrogen synthesis. This increase in estrogen stimulates the formation and release of prostaglandin $F_2\alpha$.

In humans, there is less dependence on activity of the pituitary-adrenal cortical axis for the initiation of parturition. In fact, spontaneous labor occurs in cases of pituitary or adrenal hypoplasia of the fetus or even in anencephaly, but the timing of parturition typically has a considerably wider range than normal. As in sheep, the local release of prostaglandin $F_2\alpha$ may be important in the initiation of labor in humans. A significant question is whether the initiation of parturition depends on fetal or maternal factors. The evidence favors a dominant fetal stimulus in human parturition, although maternal factors exert a modulating influence. In rare cases of human twins implanted in different horns of a double uterus, one member of the pair may not be born until several days or even weeks after the first delivery.

ADAPTATIONS TO POSTNATAL LIFE

When the umbilical cord is clamped after birth, the neonate is suddenly thrust into a totally independent existence. The respiratory and cardiovascular systems must almost instantaneously assume a type and level of function quite different from that during the fetal period. Within hours or days of birth, the digestive system, immune system, and sense organs must also adapt to a much more complex environment.

Circulatory Changes at Birth

Two major events drive the functional adaptations of the circulatory system immediately at birth. The first is the cutting of the umbilical cord, and the second is the changes in the lungs after the first breaths of the newborn. These events stimulate a series of sweeping changes that not only alter the circulatory balance but also result in major structural changes in the circulatory system of the infant.

Cutting the umbilical cord results in an immediate cessation of blood entering the body via the umbilical vein. This eliminates the major blood flow through the ductus venosus and greatly reduces the amount of blood that enters the right atrium via the inferior vena cava. A consequence of this is a reduction of the stream of blood that was directly shunted from the right to the left atrium via the foramen ovale during fetal life.

After just a few breaths, the pulmonary circulatory bed expands and can accommodate much greater blood flow than during the fetal period. Consequences of this change are a reduced flow of blood through the ductus arteriosus and a correspondingly greater return of blood into the left atrium via the pulmonary veins. Within minutes after birth, the ductus arteriosus undergoes a reflex closure. This shunt, which in prenatal life is actively kept open in great part through the actions of prostaglandin E_2, rapidly constricts after the oxygen concentration in the blood increases. The mechanism for constriction appears to involve the action of cytochrome P-450, but the way it is translated into contraction of the smooth musculature of the ductus is not entirely clear. The principal tissue involved in closure of the ductus is smooth muscle, and the shunt also experiences a breakdown of elastic fibers and a thickening of the inner intimal layer.

Because of closure of the ductus arteriosus, increased pulmonary venous flow, and loss of 25% to 50% of the peripheral vasculature (placental circulation) when the umbilical cord is cut, the blood pressure in the left atrium becomes slightly increased over that in the right atrium. This results in a physiological closure of the interatrial shunt, with the result that all the blood entering the right atrium empties into the right ventricle (Fig. 19-16). Structural closure of the valve at the foramen ovale is prolonged, occur-

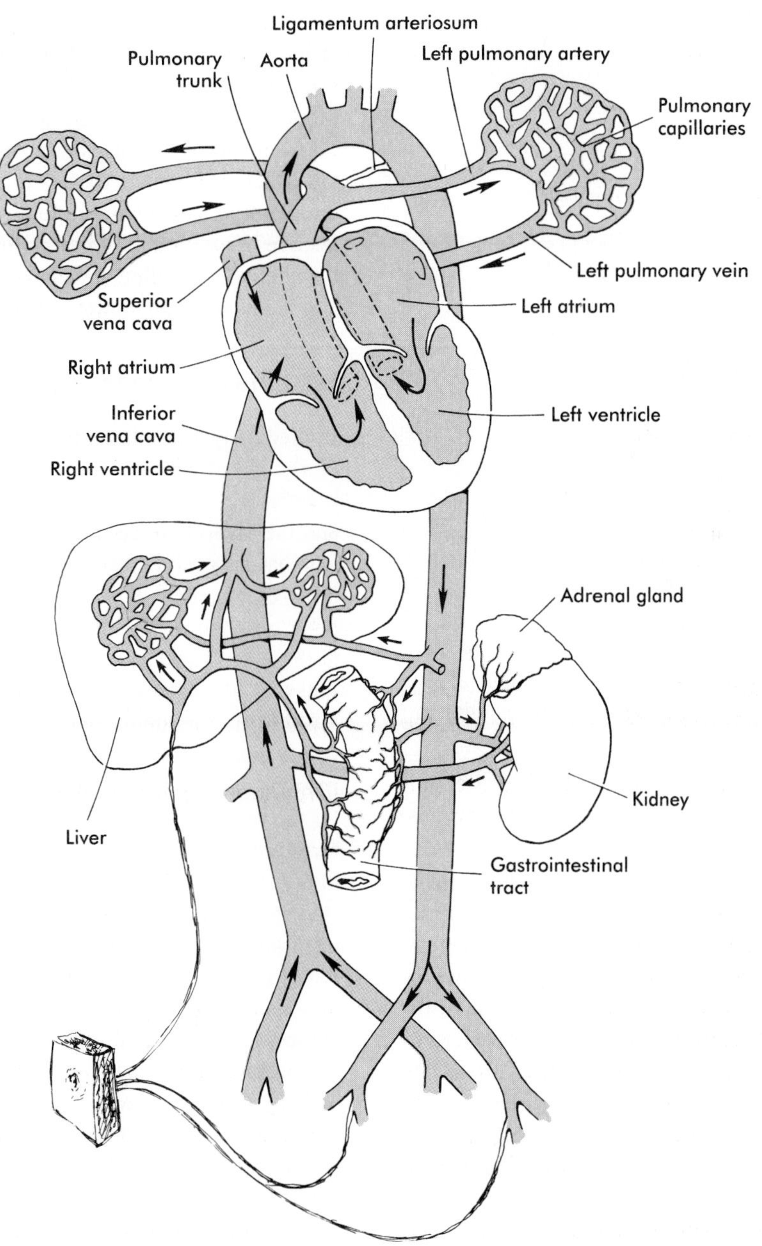

FIG. 19-16 The postnatal circulation showing the location of remnants of embryonic vessels.

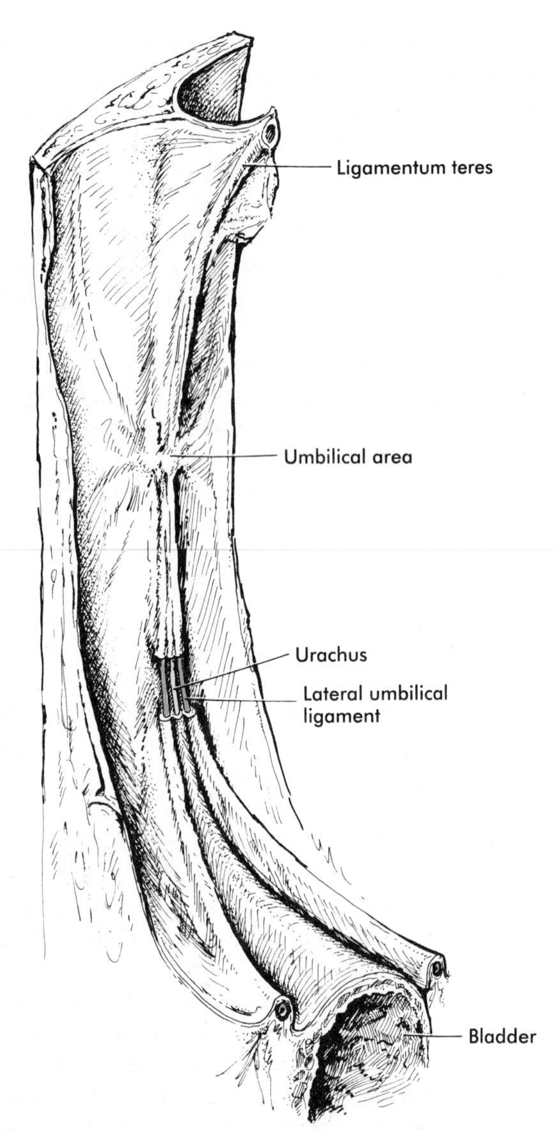

FIG. 19-17 Posterior view of the umbilical region of the abdominal wall showing the two obliterated umbilical arteries flanking the urachus leading from the bladder to the umbilicus. The single strand at the top is the ligamentum teres (remnant of the umbilical vein) leading from the umbilicus to the liver.

ring over several months following birth. Before complete structural obliteration of the interatrial valve, it possesses the property of "probe patency," which allows a catheter inserted into the right atrium to pass freely through the foramen ovale into the left atrium. As structural fusion of the valve to the interatrial septum progresses, the property of probe patency is gradually reduced and ultimately disappears. In approximately 20% of individuals, structural closure of the interatrial valve is not completed, leading to the normally asymptomatic condition of **probe patent foramen ovale.**

Although the ductus venosus also loses its patency after birth, its closure is more prolonged than that of the ductus arteriosus. The tissue of the wall of the ductus venosus is not as responsive to increased oxygen saturation of the blood as that of the ductus arteriosus.

After the postnatal pattern of the circulation is fully established, obliterated vessels or shunts that were important circulatory channels in the fetus are either replaced by connective tissue strands, forming ligaments, or are represented by relatively smaller vessels (see Figs. 19-16 and 19-17). These changes are summarized in Table 19-1.

TABLE 19-1 Postnatal Derivatives of Prenatal Circulatory Shunts or Vessels

PRENATAL STRUCTURE	POSTNATAL DERIVATIVE
Ductus arteriosus	Ligamentum arteriosum
Ductus venosus	Ligamentum venosum
Interatrial shunt	Interatrial septum
Umbilical vein	Ligamentum teres
Umbilical arteries	Distal segments, lateral umbilical ligaments; proximal segments, superior vesical arteries

Lung Breathing in the Perinatal Period

Immediately after birth, the baby must begin to breathe regularly and effectively with the lungs to survive. The initial breaths are difficult because the lungs are filled with fluid and the alveoli are collapsed at birth. On a purely mechanical basis, air breathing is facilitated by a proportionally large diameter of the trachea and major airways. This reduces resistance to airflow, which would be insurmountable if these passageways were proportionally as small as the lungs.

At birth, the lungs contain about 50 ml of alveolar fluid, which must be removed for adequate air breathing. Approximately half of that volume enters the lymphatic system. Of the remainder, perhaps half may be expelled during the birth process. The remainder enters the bloodstream by osmosis.

The alveolar sacs in the lungs begin to inflate on the first inspiration. The pulmonary surfactant, which was secreted in increasing amounts during the last few weeks of a term pregnancy, reduces the surface tension that would otherwise be present at the air-fluid interface on the alveolar surfaces and facilitates inflation of the lungs. With the rush of air into the lungs, the pulmonary vasculature opens, allowing a greatly increased flow of blood through the lungs. This results in an increased oxygen saturation of the blood; the color of the newborn changes from a dusky purple to pink.

Breathing movements in the fetus are intermittent and are irregular even after birth. Many factors can affect the frequency of breathing, but the identity of those responsible for the transition from intermittent to regular breathing remain poorly understood. Factors such as cold, touch, chemical stimuli, sleep patterns, and signals emanating from the carotid and aortic bodies have been implicated. During periods of wakefulness, breathing of the neonate soon stabilizes, but for several weeks after birth, short periods of apnea (5 to 10 seconds) are common during REM sleep.

OVERVIEW

The story of prenatal development is a complex but fascinating one. Many generalizations can be extracted from the study of embryology, but one dominant theme is that of an overall coordination of a large number of very complex integrative processes that range from the translation of information encoded in structural genes, such as the homeobox-containing genes, to the influence of physical factors, such as pressure and tension, on the form and function of the developing embryo.

Sometimes things go wrong. Studies on spontaneous abortions have shown that nature has provided a screening mechanism that eliminates many of the embryos least capable of normal development or independent survival. A simple base substitution in the DNA of an embryo can produce a defect that may be highly localized or have far-reaching consequences on the development of a variety of systems.

With ever greater insight into the molecular and cellular mechanisms underlying normal and abnormal development and with increasingly sophisticated technology, biomedical scientists and physicians can manipulate the embryo in ways that were unimaginable not long ago. It is an exciting era that is intimidating in technological complexity and uncertain in many social and ethical aspects and has an economic impact that is difficult to predict.

SUMMARY

1. The fetal period is characterized by intense growth in length and mass of the embryo. With time, the trunk grows relatively faster than the head, and later the limbs show the greatest growth. The early fetus is thin because of the absence of subcutaneous fat. By midpregnancy, subcutaneous fat is deposited.
2. At 5 weeks, the heart beats at 100/min; the heart rate rises to 160/min by 8 weeks and then declines slightly during the remainder of pregnancy. Some different physiological properties of the fetal heart can be explained by the presence of fetal isozymes in the cardiac muscle. The patency of the ductus arteriosus in the fetus is actively maintained through the actions of prostaglandins E_2 and I_2.
3. The fetal lungs are filled with fluid, but they must be prepared for full respiratory function within moments after birth. The fetus begins to make anticipatory breathing movements as early as 11 weeks. Fetal breathing is affected by maternal physiological conditions such as eating and smoking. Disproportionate growth in diameter of the upper airway is important in allowing the newborn to take the first breath. The secretion of pulmonary surfactant begins at about 24 weeks, but large amounts are not synthesized until just a few weeks before birth. Premature infants with a deficiency of pulmonary surfactant often suffer from respiratory distress syndrome.
4. Fetal movements begin at about 7½ weeks and increase in complexity thereafter. The maturation of fetal movements mirrors the structural and functional maturation of the nervous system. Diurnal rhythms in fetal activity appear at 20 to 22 weeks. The fetus has alternating periods of sleep and wakefulness. Near term the fetus responds to vibroacoustic stimuli, and by 30 weeks the pupillary light reflex can be elicited.
5. The fetal digestive tract is nonfunctional in the usual sense, but maturation of enzyme systems for digestion and absorption occurs. Spontaneous rhythmic movements of the small intestine begin as early as 7 weeks. Meconium begins to fill the lower intestinal tract by midpregnancy. By term the fetus typically swallows over half a liter of amniotic fluid per day.
6. Fetal kidneys produce small amounts of dilute urine. Fetal endocrine glands produce small amounts of hormones that can be histochemically demonstrated in glandular tissue early in the fetal period, but several months often pass before the same hormones can be measured in the blood. The fetal adrenal cortex is very large and produces 100 to 200 mg of steroids per day, but the exact functions of the fetal adrenal remain poorly understood. The placenta continues to produce a variety of hormones throughout most of pregnancy.
7. Many new diagnostic techniques have considerably improved access to the fetus. Among the imaging techniques, ultrasonography has emerged as the most widely used in obstetrics. Through sampling techniques such as amniocentesis and chorionic villus sampling, fluids or cells of the embryo and fetus can be removed for analysis. These techniques allow certain manipulations on the fetus (e.g., fetal blood transfusions, fetal surgery for certain anomalies).
8. Parturition occurs in three stages of labor. The first is the stage of dilatation, which culminates with effacement of the cervix. The second stage culminates with expulsion of the baby. The third stage represents the period between delivery of the

baby and expulsion of the placenta. The mechanisms underlying the initiation of parturition in the human remain poorly understood.

9. After birth and cutting of the umbilical cord, the newborn must quickly adapt to an independent existence in terms of breathing and cardiac function. After the first breaths and severing of the umbilical cord, the pulmonary circulation opens. In response to increased flow into the left atrium, the interatrial shunt undergoes physiological closure, and the ductus arteriosus undergoes a reflex closure. Closure of the ductus venosus in the liver is more prolonged.

REVIEW QUESTIONS

1. A premature infant develops labored breathing and dies within a few days. What is the likely cause?
2. A rare condition that can persist even into adulthood is "caput medusae" (Medusa's head), in which a dark vascular ring with irregular radiations appears around the umbilicus with straining of the abdomen. What is an embryological basis for this condition?
3. A pregnant woman typically first feels fetal movements at about 15 weeks into pregnancy. The movements become more noticeable during succeeding weeks but are commonly reduced during the last couple of weeks before parturition. What is the explanation for this?
4. List some recent medical advances that have allowed fetal surgery to become a reality.

REFERENCES

Avery ME, Wang N-S, Taeusch HW: The lung of the newborn infant, *Sci Am* Apr 1973, pp 75-85.

Barclay AE, Franklin KJ, Prichard MML: *The foetal circulation and cardiovascular system, and the changes that they undergo at birth,* Oxford, 1944, Blackwell Scientific Publications.

Barcroft J: *Researches on prenatal life,* vol 1, Oxford, 1946, Blackwell Scientific Publications.

Barron DH: The changes in the fetal circulation at birth, *Physiol Rev* 24:277-295, 1944.

Bowerman RA: *Atlas of normal fetal ultrasonographic anatomy,* ed 2, Chicago, 1992, Mosby.

Busnel MC, Granier-Deferre C, LeCanuet JP: Fetal audition, *Ann NY Acad Sci* 662:118-134, 1992.

Christ JE: Fetal surgery: a frontier for plastic surgery, *Plast Reconstr Surg* 77:645-647, 1986.

Coceani F, Olley PM: The control of cardiovascular shunts in the fetal and perinatal period, *Can J Physiol Pharmacol* 66:1129-1134, 1988.

D'Alton ME, DeCherney AH: Prenatal diagnosis, *N Engl J Med* 328:114-120, 1993.

Dawes GS: The development of fetal behavioural patterns, *Can J Physiol Pharmacol* 66:541-548, 1988.

Duenhoelter JH, Pritchard JA: Fetal respiration, *Am J Obstet Gynecol* 129:326-338, 1977.

Evans MI and others: Fetal surgery in the 1990s, *Am J Dis Child* 143:1431-1436, 1989.

Fuchs A-R, Fuchs F: Endocrinology of human parturition: a review, *Br J Obstet Gynecol* 91:948-967, 1984.

Grannum PA, Copel JA: Invasive fetal procedures, *Radiol Clin North Am* 28:217-226, 1990.

Johnson P: *The development of breathing*. In Jones C, Nathanielsz P, eds: *The physiological development of the fetus and newborn,* London, 1985, Academic Press, pp 201-210.

Jones CT, Nathanielsz PW, eds: *The physiological development of the fetus and newborn,* London, 1985, Academic Press.

Kitterman JA: Physiological factors in lung growth, *Can J Physiol Pharmacol* 66:1122-1128, 1988.

Lagercrantz H, Slotkin TA: The "stress" of being born, *Sci Am* Apr 1986, pp 100-107.

Larsen T and others: Normal fetal growth evaluated by longitudinal ultrasound examinations, *Early Hum Dev* 24:37-45, 1990.

Lavery JP: *The human placenta: clinical perspectives,* Rockville, Md, 1987, Aspen Publishers.

Liggins GC: Initiation of parturition, *Br Med Bull* 35:145-150, 1979.

Liu M and others: Stimulation of fetal rat lung cell proliferation in vitro by mechanical stretch, *Am J Physiol* 263:L376-L383, 1992.

Manning FA: Fetal breathing movements, *Postgrad Med* 61:116-122, 1977.

Mastroiacovo P and others: Limb anomalies following chorionic villus sampling: a registry based case-control study, *Am J Med Genet* 44:856-864, 1992.

Naeye RL: *Disorders of the placenta, fetus, and neonate: diagnosis and clinical significance,* St Louis, 1992, Mosby.

Nyberg DA, Mahony BS, Pretorius DH: *Diagnostic ultrasound of fetal anomalies: text and atlas,* Chicago, 1990, Mosby.

Polin RA, Fox WW, eds: *Fetal and neonatal physiology,* vols 1 and 2, Philadelphia, 1992, Saunders.

Prechtl HFR: Fetal behavior. In Hill A, Volpe J, eds: *Fetal neurology,* New York, 1989, Raven Press, pp 1-16.

Rigatto H: *Control of breathing in fetal life and onset and control of breathing in the neonate*. In Polin, R, Fox W, eds: *Fetal and neonatal physiology,* vol 1, Philadelphia, 1992, Saunders, pp 790-801.

Siler-Khodr TM: *Endocrine and paracrine function of the placenta*. In Polin R, Fox W, eds: *Fetal and neonatal physiology,* vol I, Philadelphia, 1992, Saunders, pp 74-86.

St. John Sutton M, Gill T, Plappert T: *Functional anatomic development in the fetal heart*. In Polin R, Fox W, eds: *Fetal and neonatal physiology,* vol 1, Philadelphia, 1992, Saunders, pp 598-609.

Stahlman MT, Gray ME, Whitsett JA: The ontogeny and distribution of surfactant protein B in human fetuses and newborns, *J Histochem Cytochem* 40:1471-1480, 1992.

Teitel DF: *Physiologic development of the cardiovascular system in the fetus*. In Polin R, Fox W, eds: *Fetal and neonatal physiology,* vol 1, Philadelphia, 1992, Saunders, pp 609-619.

Van Golde LMG and others: *Synthesis of surfactant lipids in the developing lung*. In Jones C, Nathanielsz P, eds: *The physiological development of the fetus and newborn,* London, 1985, Academic Press, pp 191-200.

Winter JSD: *Fetal and neonatal adrenocortical physiology*. In Polin R, Fox W, eds: *Fetal and neonatal physiology,* vol 2, Philadelphia, 1992, Saunders, pp 1829-1841.

Answers to Review Questions

CHAPTER 1

1. A mediastinal teratoma, which is likely to have arisen from an aberrant primordial germ cell that became lodged in the connective tissue near the heart.
2. In the female, meiosis begins during embryonic life; in the male, meiosis begins at puberty.
3. At prophase of the first meiotic division and at metaphase of the second meiotic division.
4. Chromosomal abnormalities, such as polyploidy or trisomies of individual chromosomes.
5. Spermatogenesis is the entire process of sperm formation from a spermatogonium. It includes the two meiotic divisions and the period of spermiogenesis. Spermiogenesis, or sperm metamorphosis, is the process of transformation of a postmeiotic spermatid, which looks like an ordinary cell, to a highly specialized spermatozoon.
6. Estrogens, secreted by the ovary, support the preovulatory proliferative phase. From the time of ovulation, progesterone is secreted in large amounts by the corpus luteum and is responsible for the secretory phase, which prepares the endometrium for implantation of an embryo.
7. FSH produced by the anterior lobe of the pituitary gland and testosterone produced by the Leydig cells of the testis.

CHAPTER 2

1. The sharp surge of luteinizing hormone produced by the anterior lobe of the pituitary gland.
2. Capacitation is a poorly understood interaction between a spermatozoon and female reproductive tissues that increases the ability of the sperm to fertilize an egg. In some mammals, capacitation is obligatory, but in humans the importance of capacitation is less well established.
3. Fertilization usually occurs in the upper third of the uterine tube.
4. The ZP3 protein acts as a specific sperm receptor through its O-linked oligosaccharides, and much of its polypeptide backbone must be exposed to stimulate the acrosomal reaction.
5. Polyspermy is the fertilization of an egg by more than one spermatozoon. It is prevented through the fast electrical block on the plasma membrane of the egg and by the later zona reaction, by which products released from the cortical granules act to inactivate the sperm receptors in the zona pellucida.

CHAPTER 3

1. The embryonic body proper arises from the inner cell mass.
2. The presence of active maturation promoting factor, which is a complex of cdc2 protein and cyclin, stimulates mitosis.
3. Trophoblastic tissues.
4. Regulation.
5. Cells derived from the cytotrophoblast fuse to form the syncytiotrophoblast.
6. In addition to the standard causes of lower abdominal pain such as appendicitis, the doctor should consider ectopic pregnancy (tubal variety) due to stretching and possible rupture of the uterine tube containing the implanted embryo.

CHAPTER 4

1. The epiblast.
2. Hensen's node acts as the organizer of the embryo.

Through it pass the cells that will become the notochord. The notochord induces the formation of the nervous system. Hensen's node is also the site of synthesis of morphogenetically active molecules such as retinoic acid. If a Hensen's node is transplanted to another embryo, it stimulates the formation of another embryonic axis.

3. Hyaluronic acid and fibronectin.
4. TGF-β and activin.
5. Cell adhesion molecules are lost in a migratory phase. When the migratory cells settle down, they may reexpress cell adhesion molecules.

CHAPTER 5

1. A homeobox is a highly conserved region consisting of 183 nucleotides that is found in many morphogenetically active genes. Homeobox gene products act as transcription factors.
2. A change in cell shape at the median hinge point and pressures of the lateral ectoderm acting to push up the lateral walls of the neural plate.
3. Neuromeres provide the fundamental organization of parts of the brain in which they are present. Certain homeobox genes are expressed in a definite sequence along the neuromeres.
4. The somites. Axial muscles form from cells derived from the medial halves of the somites, and limb muscles arise from cellular precursors located in the lateral halves of the somites.
5. In blood islands that arise from mesoderm of the wall of the yolk sac.
6. The only place where the gut opens directly to the outside of the embryo is in the midgut region, where the ventral part of the gut tube opens into the yolk sac. The region of the mouth is closed off by the stomodeal plate and the region of the future anus is closed off by the cloacal plate.

CHAPTER 6

1. Bilateral renal agenesis is commonly associated with oligohydramnios because urine constitutes a significant component of amniotic fluid in the fetus. Hydramnios can be a sign of a multiple pregnancy and also a sign of anencephaly, since anencephalic fetuses cannot swallow amniotic fluid as a normal fetus does.
2. Because the placental villi (specializations of the chorion) are directly bathed in maternal blood.
3. This depends on the age of the embryo. In an early fetus the molecule may have to pass through as many layers as the following: syncytiotrophoblast, cytotrophoblast, basal lamina underlying cytotrophoblast, villous mesenchyme, basal lamina of a fetal capillary, and the endothelium of the fetal capillary. In a mature placenta the same molecule may pass from the maternal to the fetal circulation by traversing as few layers as syncytiotrophoblast, a fused basal lamina of trophoblast and capillary endothelium, and the endothelium of a fetal capillary.
4. Human chorionic gonadotropin. This is the first distinctive embryonic hormone to be produced by the trophoblastic tissues. Early pregnancy tests involved injecting small amounts of urine of a woman into female African clawed toads *(Xenopus laevis)*. If the woman was pregnant, the chorionic gonadotropin contained in the urine stimulated the frogs to lay eggs the next day. Contemporary pregnancy tests, which can be done using kits bought over the counter, give almost instantaneous results.
5. Many substances that enter a woman's blood are now known to cross the placental barrier. These include alcohol, many drugs (both prescribed and illicit), steroid hormones, and other low molecular weight substances. In general, molecules with molecular weights below 5000 daltons should be assumed to cross the placental barrier with little difficulty.

CHAPTER 7

1. IUDs, RU 486, and the morning-after pill.
2. She had probably taken clomiphene for the stimulation of ovulation. Natural septuplets are almost never seen.
3. Introduction of more than one embryo into the tube of the woman is commonly done because the chance of any single implanted embryo surviving to the time of birth is quite small. The reasons for the poor survival of implanted embryos are still poorly understood. Extra embryos are frozen because if a pregnancy does not result from the first implantation, the frozen embryos can be implanted without the inconvenience and expense of obtaining new eggs from the mother and fertilizing them in vitro.
4. In cases of incompatibility between sperm and egg, poor sperm motility, or deficient sperm receptors in the zona, introducing the sperm directly into or near the egg can bypass a weak point in the reproductive sequence of events.

CHAPTER 8

1. The conditions that result in cleft palate occur during the second month of pregnancy. By the fourth month the palate is normally completely established. It is almost certain that this malformation had already been established by the time of the accident.
2. Although there may be a connection between the drug and the birth defect, proving a connection between an

individual case and any drug, especially a new one, is very difficult. The woman's genetic background, other drugs that she may have taken during the same period, her history of illnesses during early pregnancy, her nutritional status, and so on should be investigated. Even in the best of circumstances, the probability of a specific malformation being due to a particular cause can only be estimated in many cases.
3. A common cause of such malformations is an insufficiency of amniotic fluid (oligohydramnios), which can place exposed parts of fetuses under excessive mechanical pressure from the uterine wall and lead to deformations of this type.
4. Dysplasia of ectodermal derivatives is a likely cause.

CHAPTER 9

1. Cells contain and are capable of expressing more genes than they normally express.
2. Primary and secondary myotubes; yolk sac-derived and liver-derived erythrocytes.
3. Symptoms of a number of conditions characterized by abnormalities of hemoglobin polypeptide chains may be cured if the synthesis of earlier developmental isoforms of hemoglobin polypeptide chains could be synthesized instead of the abnormal adult chains.

CHAPTER 10

1. The dermis. Recombination experiments have clearly shown that the dermis confers regional morphogenetic information on the epidermis, instructing it to form, for example, cranial hair or abdominal hair.
2. They may be supernumerary nipples located along the caudal ends of the embryonic milk lines.
3. In the early embryo, brain tissue induces the formation of the surrounding membranous skeletal elements. If a significant region of the brain is missing, the inductive interaction does not occur.
4. In experiments involving the use of the quail nuclear marker, quail somites were grafted in place of the original somites in chick embryos. The muscles in the developing limbs all contained quail and not chick nuclei.

CHAPTER 11

1. A tear of the amnion during the chorionic villus sampling procedure could have resulted in an amniotic band wrapping around the digits and strangulating their blood supply, causing the tips to degenerate and fall off.
2. This defect is unlikely to be related to the amniocentesis procedure because the morphology of the digits is well established by the time such a procedure is undertaken (usually around 15 to 16 weeks). The most likely cause is a genetic mutation.
3. Muscle-forming cells arise from the somites.
4. The immediate cause is likely the absence of programmed cell death in the interdigital mesoderm. The cause of the disturbance in cell death is currently not understood.

CHAPTER 12

1. The infant has hydrocephalus secondary to the rachischisis. In rachischisis the spinal roots are often fixed to bony structures. With the growth of the vertebral column the brain gets pulled down toward the outlet of the skull, producing a mechanical blockage of cerebrospinal fluid (Arnold-Chiari syndrome). This soon results in hydrocephalus.
2. Congenital megacolon (Hirschsprung's disease), in which a segment of large intestine develops without parasympathetic ganglia. Intestinal contents cannot actively move through such an aganglionic segment.
3. The nerves would be hypoplastic (much smaller than normal), and the spinal cord would be thinner than normal in the area from which the nerves supplying the affected limb arise. The likely cause is excessive neuronal cell death because of the absence of an end organ for many of the axons that normally supply the limb.

CHAPTER 13

1. Along the length of the spinal cord, migrating neural crest cells are funneled into the anterior sclerotomal region of the somites and are excluded from the posterior half. This results in the formation of a pair of ganglia for each vertebral segment and space between ganglia in the craniocaudal direction.
2. A defect in the cranial (specifically cardiac) neural crest that affects the heart, thymus, and parathyroid glands. This infant has a variety of the DiGeorge syndrome.
3. Cranial crest cells can form skeletal elements; trunk crest cells cannot. Migrating cranial neural crest cells have more morphogenetic information encoded in them than trunk crest cells do. (For example, craniocaudal levels are specified in cranial crest, whereas they are not fixed in trunk crest cells.) Cranial crest cells form large amounts of dermis and other connective tissues, whereas trunk crest cells do not.

CHAPTER 14

1. Coloboma of the iris is caused by failure of the choroid fissure to close during the sixth week of

pregnancy. Because the area of the defect remains open when the rest of the iris constricts in bright light, excessive unwanted light can enter the eye through the defect.

2. Some of the secretions of the lacrimal glands enter the nasolacrimal ducts, which carry the lacrimal fluid into the nasal cavity.
3. Hyaluronic acid. Migration of neural crest cells into the developing cornea occurs during a period when large amounts of hyaluronic acid have been secreted into the primary corneal stroma.
4. During the fetal period the middle ear cavity is filled with a loose connective tissue that damps the action of the middle ear ossicles. After birth the connective tissue is resorbed.
5. Like the lower jaw, much of the external ear arises from tissue of the first arch bordering the first pharyngeal cleft.

CHAPTER 15

1. One option is simply acne. Another more significant possibility is a branchial cyst. Branchial cysts are typically located along the anterior border of the sternocleidomastoid muscle. One possible reason for its late manifestation is that the same conditions that resulted in the boy's acne caused a simultaneous reaction in the epidermis lining the cyst.
2. First, all epithelium lining the cyst must be removed or the remnants could re-form a new cyst and the symptoms could recur. The surgeon must also determine that the cyst is isolated and not connected to the pharynx via a sinus, which would result from an accompanying persistence of the corresponding pharyngeal pouch.
3. Ectopic thyroid tissue along the pathway of migration of the thyroid gland from its site of origin is very common. The abnormal thyroid tissue may be located in ectopic sites.
4. By 10 weeks, all the processes of fusion of facial primordia have already been completed. The cause of the defects could almost certainly be attributed to something that influenced the embryo long before the time when the anticonvulsant therapy was initiated—probably before the seventh week of pregnancy.
5. These defects could be a manifestation of fetal alcohol syndrome. They would represent a relatively mild form of holoprosencephaly, which in this case would relate to defective formation of the forebrain (prosencephalon). The defects in olfaction and in the structure of the upper lip could be secondary effects of a primary defect in early formation of the prosencephalon.

CHAPTER 16

1. She was found to have a Meckel's diverticulum containing ectopic gastric mucosa. Her pain was the result of the hydrochloric acid secreted by the ectopic epithelial lining of the diverticulum. After surgery to remove the Meckel's diverticulum, her symptoms permanently disappeared.
2. Esophageal atresia or a tracheoesophageal fistula. In the former, the milk fills the blind esophageal pouch and then spills into the trachea via the laryngeal opening. In the latter, milk may pass directly from the esophagus into the trachea depending on the type of fistula.
3. Congenital pyloric stenosis. Projectile vomiting is a common symptom of this condition, and palpation of the knotted pyloric opening of the stomach confirmed the diagnosis.
4. The most likely diagnosis is a vitelline duct fistula connecting the midgut with the umbilicus. This allows some contents of the small intestine to escape through the umbilicus. Another possibility is a urachal fistula (see Chapter 17), which connects the urinary bladder to the umbilicus through a persistent allantoic duct. In this case, however, the escaping fluid would be urine and would likely not be accompanied by mucus.
5. Imperforate anus. When examining a newborn, clinicians must ensure that there is an anal opening.

CHAPTER 17

1. She probably has testicular feminization syndrome and is a genetic male even though her phenotype is very feminine. She would have internal testes but would lack both male and female genital ducts because the testes produce müllerian inhibitory substance (causing regression of the paramesonephric ducts) and the absence of testesterone receptors results in the inability of the abundant testosterone to maintain the mesonephric ducts. This individual would also have amenorrhea, but because amenorrhea is common in female athletes who train intensively, this condition would not indicate anomalous sexual differentiation.
2. The most likely cause is a urachal fistula connecting the urinary bladder to the umbilicus and allowing the leakage of urine. This is caused by the persistence of the lumen in the distal part of the allantois.
3. Approximately one out of every thousand people has unilateral renal agenesis. Other individuals may have an ectopic kidney. Because ectopic kidneys are more susceptible to infections than normal kidneys, it would be risky to the donor if only an ectopic kidney remains.

4. Bilateral renal agenesis. The first clue was the mother's low weight gain, which could have been the result of oligohydramnios (although this is not the only cause for low weight gain during pregnancy). The appearance of the infant showed many of the characteristics of Potter syndrome, which is caused by intrauterine pressures on the fetus when the amount of amniotic fluid is very low.
5. In normal development the caudal ends of the paramesonephric ducts swing toward the midline and fuse. In this patient's case the point of fusion probably occurred more caudally than normal. This condition is not incompatible with a normal pregnancy and delivery, although in some cases, pain or problems with delivery can occur.

CHAPTER 18

1. The doctor suspected that the boy had a double aortic arch or a right aortic arch, either of which can cause difficulty in swallowing (dysphagia) during childhood, especially when growth spurts are occurring. Another possibility for an embryologically based dysphagia is esophageal stenosis.
2. To survive, an individual with this defect would have to have a means for draining blood entering the left atrium and a means of getting blood into the left ventricle or the systemic circulation. One combination that would compensate is an atrial septal defect, which allows incoming blood to escape from the left atrium, and an associated ventricular septal defect, which allows blood from the right ventricle to pass into the left ventricle. A patent ductus arteriosus added to the first two compensatory defects could also help to balance the circulation by adding blood to the systemic side of the circulation. This would not be very helpful in the physiological sense, however, because the added blood would be unoxygenated blood entering the aorta from the pulmonary artery.
3. The diagnosis is a patent ductus arteriosus with a reversed blood flow. Earlier in life the blood flow through the patient's ductus went from the aorta into the pulmonary circulation because of the pressure differential between the aortic and pulmonary circulation. After years of increased blood flow in the pulmonary circulation, changes in pulmonary vasculature caused pulmonary hypertension. When the blood pressure in the pulmonary arterial circulation exceeded that in the aorta, the left-right shunt reversed to a right-left shunt, causing cyanosis due to the inflow of unoxygenated pulmonary blood into the aorta. However, arterial branches upstream from the entrance of the ductus did not receive venous blood. Therefore the head and upper limbs were not cyanotic.
4. Persistence of the caudal segment of the left supracardinal vein. Normally the caudal segment of the right supracardinal vein persists and becomes incorporated into the inferior vena cava, and the corresponding segment of the left supracardinal vein disappears (see Fig. 18-9). A wide variety of anomalous patterns of the abdominal veins exist.

CHAPTER 19

1. Hyaline membrane disease of the newborn. If the baby was born prematurely, the fetal lungs had not produced sufficient pulmonary surfactant to support normal breathing.
2. This condition is the result of persistence of a patent ductus venosus and umbilical vein. When the individual strains, venous blood fills these vessels and small venous branches radiating from the umbilical area. This is reminiscent of the head of Medusa, with snakes taking the place of hairs.
3. One simple explanation is that with continued growth of the fetus the extremities become so tightly packed in the uterus that there is little room for movement.
4. More refined imaging techniques, such as ultrasonography, that allow more accurate diagnosis in utero of congenital malformations; recognition of the exceptional healing powers of the fetus; improved intrauterine and extrauterine surgical techniques, some assisted by ultrasonographic visualization, that allow direct surgery on the fetus; and increased ability to forestall or prevent premature labor after surgery on the uterus.

Index

B

F

T

U

V